VIRUSES AND INVERTEBRATES

NORTH-HOLLAND RESEARCH MONOGRAPHS

FRONTIERS OF BIOLOGY

VOLUME 31

Under the General Editorship of

A. NEUBERGER

London

and

E. L. TATUM

New York

NORTH-HOLLAND PUBLISHING COMPANY
AMSTERDAM · LONDON

VIRUSES
AND INVERTEBRATES

Edited by

A. J. GIBBS

Research School of Biological Sciences,
The Australian National University, Canberra City,
A.C.T. Australia

1973

NORTH-HOLLAND PUBLISHING COMPANY, AMSTERDAM · LONDON
AMERICAN ELSEVIER PUBLISHING COMPANY, INC. – NEW YORK

Library of Congress Catalog Card Number: 73–75532

North-Holland ISBN: 0 7204 7132 X
American Elsevier ISBN: 0 444 10529 8

134 illustrations and graphs, 19 tables

PUBLISHERS:
NORTH-HOLLAND PUBLISHING COMPANY – AMSTERDAM
NORTH-HOLLAND PUBLISHING COMPANY, LTD. – LONDON

SOLE DISTRIBUTORS FOR THE U.S.A. AND CANADA:
AMERICAN ELSEVIER PUBLISHING COMPANY, INC.
52 VANDERBILT AVENUE, NEW YORK, N.Y. 10017

PRINTED IN THE NETHERLANDS

General preface

The aim of the publication of this series of monographs, known under the collective title of '*Frontiers of Biology*', is to present coherent and up-to-date views of the fundamental concepts which dominate modern biology.

Biology in its widest sense has made very great advances during the past decade, and the rate of progress has been steadily accelerating. Undoubtedly important factors in this acceleration have been the effective use by biologists of new techniques, including electron microscopy, isotopic labels, and a great variety of physical and chemical techniques, especially those with varying degrees of automation. In addition, scientists with partly physical or chemical backgrounds have become interested in the great variety of problems presented by living organisms. Most significant, however, increasing interest in and understanding of the biology of the cell, especially in regard to the molecular events involved in genetic phenomena and in metabolism and its control, have led to the recognition of patterns common to all forms of life from bacteria to man. These factors and unifying concepts have led to a situation in which the sharp boundaries between the various classical biological disciplines are rapidly disappearing.

Thus, while scientists are becoming increasingly specialized in their techniques, to an increasing extent they need an intellectual and conceptual approach on a wide and non-specialized basis. It is with these considerations and needs in mind that this series of monographs, '*Frontiers of Biology*' has been conceived.

The advances in various areas of biology, including microbiology, biochemistry, genetics, cytology, and cell structure and function in general will be presented by authors who have themselves contributed significantly to these developments. They will have, in this series, the opportunity of bringing together, from diverse sources, theories and experimental data, and of integrating these into a more general conceptual framework. It is unavoidable, and probably even desirable, that the special bias of the individual authors will become evident in their contributions. Scope will also be given for presentation of new and challenging ideas and hypotheses for which complete evidence is at present lacking. However, the main emphasis will be on fairly complete and objective presentation of the more important and more rapidly advancing aspects of biology. The level will be advanced, directed primarily to the needs of the graduate students and research worker.

Most monographs in this series will be in the range of 200–300 pages, but on occasion a collective work of major importance may be included somewhat exceeding this figure. The intent of the publishers is to bring out these books promptly and in fairly quick succession.

It is on the basis of all these various considerations that we welcome the opportunity of supporting the publication of the series '*Frontiers of Biology*' by North-Holland Publishing Company.

E. L. TATUM
A. NEUBERGER, *Editors*

Preface

Recently there has been a growing realization by virologists that the boundaries between the traditional branches of virology are counterproductive, and much of value has resulted from pooling of ideas and information particularly in studies of the biochemistry of virus replication. Another point of contact between the different 'virologies' is in studies of those viruses that interact with invertebrates. At least three groups of virologists work with viruses and invertebrates: those who study viruses of invertebrates, and plant and mammal virologists studying vector-borne viruses. These three groups rarely meet, they work in different laboratories, publish their research and review papers in different journals and attend different meetings. This book attempts to cut across these artificial divisions.

The first section is of three chapters which are personal accounts of the history of each of the three traditional branches of virology associated with invertebrates. The remaining 28 chapters are grouped into four more sections. In the first of these sections, one chapter describes the various viruses involved with invertebrates and nine describe the classification and biology of the most important invertebrates involved with viruses. Eight of the latter nine chapters are on arthropods. This is not a biased choice. Very few of the major phyla of organisms seem to be associated with viruses, most known viruses are from only four of the phyla (table), and one of these, the arthropods, provides most of the vectors of viruses of chordates and spermatophytes.

Is this uneven distribution of viruses among different phyla real or false? If it is real and viruses are unevenly distributed then one reason could be that, as viruses must spread from one host individual to another in order to survive, only certain organisms lead lives that favour virus spread. For example most plants are sedentary, and their cells are surrounded by resistant and dead cell walls, and also many primitive plants have no vascular systems, thus viruses may not have prospered in plants until vascular plants, especially spermatophytes, evolved together with phytophagous arthropods. However, it seems unlikely that a similar reason could account for the dearth of viruses in lower animals as there are many viruses of higher animals that infect respiratory and gut surfaces via a contaminated environment, and it is not obvious why similar viruses are not widespread in lower animals.

It is also possible they have not been found because they have not been looked for. The viruses first studied by man were those that caused diseases of man, his animals and his plants. This anthropocentricity of interest has continued so that our knowledge of any particular virus is usually directly correlated to its economic importance to man. This is reflected in this book,

Viruses and phyla

			infect	Viruses transmitted by
Prokaryotes		Bacteria	+++*	—
		Cyanophyta	+	—
Eukaryotes	Animals	Protozoa	+	—
		Porifera	—	—
		Coelenterata	—	—
		Platyhelminthes	—	—
		Nematoda	?	++
		Rotifera	—	—
		Polyzoa	—	—
		Brachiopoda	—	—
		Annelida	?	—
		Arthropoda	+++	+++
		Mollusca	+	?
		Echinodermata	—	—
		Chordata	+++	+
	Plants	Chlorophyta	+	—
		Xanthophyta	—	—
		Bacillariophyta	—	—
		Phaeophyta	—	—
		Myxomycophyta	—	—
		Eumycophyta	++	+
		Bryophyta	—	—
		Pteridophyta	+	—
		Spermatophyta	+++	+

* Numbers of species involved: — none reported, + 1–10 species, ++ 10–100 species, +++ more than 100 species.

and most of the invertebrates described are those that have been studied as vectors of virus diseases of man or his livestock, though also described are some invertebrates such as honey bees and silkworms that are beneficial to man and are known to have their own virus diseases. Similarly the viruses described are mostly those causing diseases of man or his livestock, though there is nowadays a fashionable interest in viruses that might be used to 'biologically control' pest organisms.

The third section of the book is of four chapters on some general topics of interest in virus/invertebrate studies. Missing from this section is a discussion of studies of invertebrate populations and their movements. Recent books and reviews of this subject include those by Clark et al. (1967) and Johnson (1969), and the methods used to study invertebrate populations have been reviewed by Lewis and Taylor (1967).

The fourth and largest section is of eleven chapters and is on the ecology of viruses and particular invertebrates. These chapters illustrate the great differences between the problems confronting a virus ecologist working with vector-borne viruses of plants and those found when studying vector-borne viruses of animals; plants are sedentary and are often studied in crops where all the individuals are of one age and genotype, whereas animals are usually a shifting population of different ages and varying degrees of immunity induced by previous encounters with viruses. One clear point made by these chapters is that although we have some idea of the effect of viruses on their major hosts we usually have very little idea of their direct or indirect effect on vectors. Ch. 25 will be of particular interest to plant virologists because it questions ideas on the mode of transmission of certain aphid-borne viruses, ideas which have been accepted by most plant virologists for many years.

Readers conversant with plant virology will notice that there is no discussion of the ecology of many diseases usually thought to be caused by viruses, for example, all those transmitted by whitefly. This is because plant pathologists have become more cautious about stating that a certain plant disease is caused by a virus since Doi et al. (1968) showed that half of the plant 'viruses' transmitted by leaf- and planthoppers were perhaps mycoplasma-like organisms (see ch. 26). Thus in this book virus-like pathogens are omitted (except in ch. 26), and only those viruses whose particles have been seen are discussed.

The last section of the book is of three chapters on the control of invertebrates and viruses, not only the control of viruses spread by invertebrates but also the control of certain invertebrates by viruses.

It is hoped that work-provoking comparisons will be made by people reading the reviews in this book. For example it is clear that arbovirologists working with mosquito- or tick-borne viruses have much information on the species transmitting particular viruses, yet have little idea of the effect of viruses on the ecology of these vectors, whereas the converse is perhaps true for those studying many aphid- and leafhopper-borne plant viruses. Similarly the review of viruses of honey bees (ch. 22) should be of interest to virologists studying the mysteries of 'latent' virus infections of the Lepidoptera.

Reviews in this book illustrate clearly our lack of knowledge of the fate of viruses inside invertebrates. Viruses that infect invertebrates seem to show host and tissue specificity like that shown by viruses infecting other organisms; a phenomenon about which almost nothing is known. Experiments show that many invertebrates have a non-humoral defence mechanism against foreign material, but the relevance of these experiments to studies on viruses is uncertain as most have been done using parenterally injected foreign materials, whereas most viruses infect invertebrates via the gut. It is also clear that although the gut is the principal way into many invertebrates (especially terrestrial ones) it is well defended (see for example work on bee viruses; ch. 22), so it is difficult to understand how some aphid- and beetle-borne viruses seem to pass through their vectors (from food to saliva) unharmed and without multiplying.

Canberra, June 1972   A. J. GIBBS

List of contributors

Sir Christopher Andrewes – Overchalke, Coombe Bissets., Salisbury, England, U.K. *(ch. 1)*

L. Bailey – Rothamsted Experimental Station, Harpenden, Herts., England, U.K. *(chs. 22, 29)*

A. J. D. Bellett – John Curtin School of Medical Research, Australian National University, Canberra, Australia, 2601. *(ch. 4)*

R. Crichton – John Curtin School of Medical Research, Australian National University, Canberra, Australia, 2601. *(ch. 16)*

L. Dalgarno – Department of Biochemistry, School of General Studies, Australian National University, Canberra, Australia, 2601. *(ch. 14)*

Mary W. Davey – Department of Biochemistry, School of Genetic Studies, Australian National University, Canberra, Australia, 2601. *(ch. 14)*

V. F. Eastop – Department of Entomology, British Museum (Natural History), Cromwell Road, London, England, SW7 5BD, U.K. *(ch. 7)*

P. F. Entwistle – Unit of Invertebrate Virology, Commonwealth Forestry Institute, Oxford, England, OX1 3RB, U.K. *(ch. 10)*

F. Fenner – John Curtin School of Medical Research, Australian National University, Canberra, Australia, 2601. *(ch. 4)*

R. G. Garrett – Victorian Plant Research Institute, Burnley, Victoria, Australia, 3121. *(ch. 25)*

A. J. Gibbs – Research School of Biological Sciences, Australian National University, Canberra, Australia, 2601.   *(chs. 4, 28)*

C. E. Gordon Smith – London School of Hygiene and Tropical Medicine, Keppel Street, London, England, WC1E 7HT, U.K.   *(ch. 30)*

T. D. C. Grace – Division of Entomology, Commonwealth Scientific and Industrial Research Organisation, Canberra, Australia, 2601.   *(ch. 17)*

K. A. Harrap – Unit of Invertebrate Virology, Commonwealth Forestry Institute, Oxford, England, OX1 3RB, U.K.   *(ch. 15)*

B. D. Harrison – Scottish Horticultural Research Institute, Invergowrie, Dundee, Scotland, DD2 5DA, U.K.   *(ch. 27)*

G. D. Heathcote – Broom's Barn Experimental Station, Higham, Bury St. Edmunds, Suffolk, England, U.K.   *(ch. 31)*

H. Hoogstraal – United States Naval Medical Research Unit No. 3, Spanish Embassy, Cairo, Egypt, U.A.R.   *(chs. 6, 18)*

D. J. Hooper – Rothamsted Experimental Station, Harpenden, Herts., England, U.K.   *(ch. 12)*

R. Kisimoto – Central Agricultural Experiment Station, MAF, Konosu, Saitama, Japan.   *(ch. 8)*

K. J. Lafferty – John Curtin School of Medical Research, Australian National University, Canberra, Australia, 2601.   *(ch. 16)*

D. C. Lee – South Australian Museum, North Terrace, Adelaide, South Australia, 5000.   *(ch. 5)*

J. F. Longworth – Unit of Invertebrate Virology, Commonwealth Forestry Institute, Oxford, England, OX1 3RB, U.K.   *(ch. 21)*

I. D. Marshall – John Curtin School of Medical Research, Australian National University, Canberra, Australia, 2601.   *(ch. 20)*

P. F. Mattingly – Department of Entomology, British Museum (Natural History), Cromwell Road, London, England, SW7 5BD, U.K. *(ch. 11)*

L. A. Mound – Department of Entomology, British Museum (Natural History), Cromwell Road, London, England, SW7 5BD, U.K. *(ch. 13)*

D. Peters – Laboratorium Virologie, Binnenhaven 11, Wageningen, The Netherlands. *(ch. 24)*

O. Roivainen – Suotie 11, Mattilan alue, Kerava, Finland. *(ch. 23)*

B. J. Selman – Department of Agricultural Zoology, The University, Newcastle upon Tyne, England, U.K. *(ch. 9)*

R. C. Sinha – Chemistry and Biology Research Institute, Canada Department of Agriculture, Ottawa, Ontario, KIA OC6, Canada. *(ch. 26)*

J. T. Slykhuis – Ottawa Research Station, Canada Department of Agriculture, Ottawa, Ontario, KIA OC6, Canada. *(ch. 19)*

K. M. Smith – 3 Sedley-Taylor Road, Cambridge, England, CB2 2PW, U.K. *(ch. 2)*

G. Surtees – Microbiological Research Establishment, Porton Down, Salisbury, England, U.K. *(ch. 30)*

Marion A. Watson – Bardolphs, Redbourn Lane, Harpenden, Herts., England, U.K. *(ch. 3)*

Contents

For a detailed list of contents the reader is referred to the
first page of each chapter.

Section I

Dramatis personae

CHAPTER 1

Animal viruses

SIR CHRISTOPHER ANDREWES

Contents

1.1 Historical background

Virus diseases infecting man and animals were the objects of studies in the
field and in the laboratory in the days before we knew that there were such
things as viruses. There were even effective measures available against some
of them then: the work of Jenner against smallpox and of Pasteur against
rabies is familiar to all. But it was not until the end of the 19th century that
it was recognised that some pathogens could pass through a bacteria-
retaining filter. In 1898 Beijerinck showed this with a plant virus, tobacco
mosaic virus, and in the same year Löffler and Frosch (1898) showed that
the agent causing foot-and-mouth disease would also pass through such
filters. Not long afterwards an American commission headed by Walter
Reed (Reed et al. 1900) showed that yellow fever was caused by a filterable
agent. In the same year M'Fadyean (1900) proved this for another arthropod-
borne virus, African horse sickness. Dengue was shown to have a viral cause
in 1907 (Ashburn and Craig 1907) and Phlebotomus or sand-fly fever by
Doerr in 1908.

The role of mosquitoes and ticks in transmitting diseases due to protozoa
and helminths had been known since the work of Theobald Smith, Manson
and Ronald Ross in the last century. Discovery of the role of mosquitoes in
yellow fever was the starting point of our knowledge of arthropod-borne
viruses infecting vertebrates. At first, further information was slow to accrue.
A text book with a volume on *Viruses and virus diseases* was published under
the aegis of the Medical Research Council in 1930; this could only add to
those already mentioned four other arthropod-borne viruses – all of them
infecting livestock: blue-tongue, Nairobi sheep disease, Borna disease and
vesicular stomatitis. Today we know of three hundred or more and of these
more than 50 can infect man.

The term arthropod-borne virus is a clumsy one and it duly became
replaced by a telescoped form 'arbovirus'. For a time the word used was
'arborvirus', but it was objected that this might imply some connection with
trees, and the second 'r' was dropped. The term covers those viruses which
have part of their life-cycle in an arthropod and part in a vertebrate. Viruses
such as those causing fowl-pox and myxomatosis may be carried from host
to host only mechanically, by an insect's proboscis, called by Fenner et al.
(1953) a 'flying pin', and are therefore not arboviruses.

For some time it was supposed that the arboviruses formed a natural
taxonomic group. It soon appeared, however, that this was not so: the known
arboviruses could be distributed among a number of genera, now to be

known as alphaviruses, flaviviruses, rhabdoviruses and reoviruses, with many other viruses still awaiting classification. The characters of these genera will be described in ch. 4. They differ in morphology, presence or absence of an envelope or in symmetry: all but one contain RNA – the only known DNA-containing arbovirus is that causing African swine fever. The term 'arbovirus' will continue to be useful but should now be used in a purely biological, not a taxonomic sense.

The study of arboviruses has been a major interest of the International Health Division (I.H.D.) of the Rockefeller Foundation since 1950. Their headquarters were for many years at the Rockefeller Institute in New York, but were transferred to the Department of Epidemiology and Public Health of Yale University in 1965. Work was carried on in associated laboratories in many tropical countries and a rich harvest of knowledge about arboviruses has been steadily gathered.

The Rockefeller I.H.D. centre now acts as a reference centre under the World Health Organization, and endeavours to ensure that different names are not given to the same virus. A difficulty arises in that among the arboviruses there are all degrees of relatedness. Some of them show minor serological differences according to the part of the world from which they come. Some workers are 'lumpers'; others are 'splitters', too eager to give new names to what others would prefer to call serological races. Traditionally, most arboviruses have come to be given names derived from the name of the place where they were first isolated; other bases for naming were, it was felt, hard to discover. One consequence has been that some viruses, such as the members of the Tacaribe group, now included in the arenaviruses, have been named from places, though they are probably not ordinarily transmitted by arthropods. In fact place-names have been used to designate any virus discovered by a worker with arboviruses or in an arbovirus laboratory.

1.2 Yellow fever

Before considering the development of ideas on virus ecology, it would be helpful to look at the history of a few important virus infections. Pride of place must be given to yellow fever: it was the first arbovirus to receive attention; it has been the cause of devastating epidemics affecting human history; further, it illustrates important points in viral ecology.

Yellow fever and malignant malaria between them gave West Africa the title of the white man's grave. Thence epidemics spread at times to Europe; there were serious outbreaks in Spain and Portugal and even a small one in South Wales. Yellow fever took an even more serious toll in South America and the Caribbean with some outbreaks occurring as far north as New York and Boston. The matter cannot now be settled, but most people believe that the virus was introduced into the American continent at the time of the slave trade. One reason for this belief is that South American monkeys, unlike African ones, are by no means immune from fatal infection with the virus: they have not had centuries of contact in which to become naturally adapted.

It was in the Caribbean that light was first thrown on the ecology of the disease. A team from the U.S.A. considered that the infectious agent must be in the air, and Carlos Finlay, a physician practising in Cuba, deduced that the agent concerned was a mosquito, possibly *Aëdes aegypti*. For 20 years he was disbelieved but then a commission under Dr. Walter Reed proved the truth of his deduction (Reed et al. 1900). When this was established, measures for controlling the *Aëdes* could be enforced, and Havana, where yellow fever had been an ever-present menace, was quickly freed from the infection. A first attempt, under De Lesseps, to build a Panama canal had failed, largely because of the illness caused in his labour gangs by yellow fever and malaria. The canal was successfully completed when Major Gorgas, as a first step in the operation, controlled the death-bringing mosquitoes. Later, the Rockefeller Foundation instituted a campaign to free the whole American continent from the disease by exterminating the yellow fever mosquitoes. This was apparently successful for a time: later there was much disappointment when there began to appear cases of jungle fever carried to

Fig. 1.1. Dr. Carlos Finlay who first suggested that mosquitoes transmitted yellow fever. (From Strode 1951.)

foresters from an endemic cycle in the tree-top monkeys and mosquitoes.

It became evident that there are two biological cycles involving yellow fever. There is a primary cycle involving monkeys. Among these infection is transmitted by arboreal mosquitoes: in Africa these are *Aëdes* species, especially *A. africanus*; in South America, the vectors are *Haemagogus* species and others. Infection is carried sporadically to workers in or near the jungle and this is jungle yellow fever. Superimposed on this cycle is urban yellow fever, passing from man to man through bites by *Aëdes aegypti*, a species which breeds in casual water standing in discarded vessels and elsewhere. An attack on urban *A. aegypti* readily controls these infections, but jungle yellow fever cannot be directly attacked. The development by Theiler and Smith (1937) of the 17D strain of yellow fever has, however, fortunately provided us with one of the most effective immunizing agents we know.

In Africa, natives of areas where yellow fever is endemic are relatively resistant. Children may acquire immunity when quite young and there may be no evidence that the virus is present. At times, however, the disease appears outside its usual range and it has caused extensive epidemics in the Sudan and Ethiopia, where the populations have none of the resistance enjoyed by the natives of the virus's headquarters further south. The Ethiopian outbreak of 1960–62 is estimated to have affected 200,000 people causing 30,000 deaths.

1.3 Encephalitis in North America and Japan

A number of arboviruses have caused cases of encephalitis in North America. Two of them are known as Western and Eastern equine encephalomyelitis viruses (W.E.E. and E.E.E.). They have caused serious outbreaks among horses over the last eighty years: in 1938 184,000 horses were affected in the U.S.A. with mortality as high as 90%. When outbreaks in horses occur, some human beings are also afflicted, in 1941 more than 3000. With W.E.E. there are many inapparent infections in man, but infection with the E.E.E. virus is more apt to be fatal and to be followed by serious sequelae. Neither in horses nor man is there viraemia sufficient to allow transmission of infection among those species: spread seems to involve mosquitoes and birds of various kinds: in these infection is inapparent, though house-sparrows and pheasants, neither of which are native American species, may be killed.

A recurring problem in the study of arboviruses is to find how the viruses manage to survive when vectors are not active. The difficulty in North

America concerns the over-wintering of the viruses: in the tropics it may be a matter of survival during a dry season. Several hypotheses were considered, especially by Reeves and his colleagues (1958), to account for the over-winter survival of the W.E.E. and E.E.E. viruses. Could they be carried south by migrating birds and return in the same way in spring? Could hibernating mosquitoes or unusually persistent infection of a bird achieve the same end? Recently, the E.E.E. virus has been recovered from small rodents during the winter (Altman et al. 1967) and it has been shown that there may be low grade infections of small mammals, frogs and even snakes (Thomas and Eklund 1960): infection may be carried by some ectoparasite, but is in any case likely to be stirred into activity when warmer weather comes along. We should then have to visualize a basic cycle in small animals, a superimposed more active mosquito–bird cycle in summer months and on top of that a number of blind-alley infections of men and horses.

The Japanese encephalitis virus poses a similar problem. This mosquito-borne flavovirus infects very large numbers of children in Japan, and though only quite a small proportion of these develop encephalitis, the total incidence of the disease is nevertheless considerable. The ecology in Japan was studied by Scherer and colleagues (1959). The 'reservoir hosts' are probably night herons. During their breeding season, the nestling birds are very vulnerable and virus increases greatly in them: they are 'amplifying hosts'. Man does not come into sufficiently close contact with the herons to run much risk of infection: but heronries exist near piggeries. Pigs are very susceptible and act as 'link hosts' permitting spill-over so that man gets infected. In warmer Asiatic countries the ecology may be rather different.

Fig. 1.2. Dr. Walter Reed who proved that yellow fever was transmitted by mosquitoes. (From Strode 1951.)

In Sarawak pigs are certainly important: the role of birds is rather more doubtful.

1.4 Transmission of disease

1.4.1 Changes in patterns of disease

Viruses can multiply so rapidly that their evolution may well have proceeded much more rapidly than has that of animals and plants. Indeed, changes in arthropod-borne disease may be seen going on within a short space of time. Changes in methods of transmission will be mentioned later. Apart from these, there are many instances of the appearance, waxing and waning of epidemics or changes in the character of symptoms. Often 'new' diseases have swept a country, but how far they were really new one does not know. The *Aëdes*-borne Chikungunya fever which appeared in Tanganyika in 1952 and swept through much of tropical Africa was thought to be new. Even more suggestive is the history of O'nyong-nyong which went through Uganda in 1959–60, affecting five million people with fever and severe joint pains. O'nyong-nyong, like Chikungunya, is due to an alphavirus, but unlike other alphaviruses and indeed unlike most other arboviruses, it is carried by an *Anopheles*, not a *Culicine* mosquito. It is possible that adaptation to a new kind of vector carried with it new potentialities for disease-production in man.

Another apparently new virus infection was Kyasanur forest disease, which appeared in 1957 in the Mysore province of Southern India (Work 1958). The agent is a flavovirus, related to other tick-borne viruses of Europe and Asia. This one turned out to be carried by *Haemaphysalis* ticks and to cause a severe haemorrhagic disease not only in man but also wild langur and macaque monkeys: many of these were found dead in the forests. This genus of ticks is not known to carry other virus infections and it is again possible that the 'new' disease appeared when a novel vector became involved or when an existing one changed its habits.

Rather different is the story of the haemorrhagic fevers of South East Asia associated with infection by the dengue viruses – flavoviruses with at least 4 different serotypes. Dengue has long been known as an unpleasant but not dangerous fever prevalent in many parts of the tropics. As with yellow fever, there may be an urban dengue transmitted by *Aëdes aegypti*, superimposed on a cycle of jungle dengue in the tree-tops, affecting monkeys and probably carried by other mosquitoes. The haemorrhagic fever which appeared in 1954, first in Thailand and the Philippines, caused many deaths especially among children: later it spread to India and elsewhere. The commonest agents isolated were dengue viruses, not demonstrably different from those causing the ordinary dengue fever. The question arose of why the viruses suddenly began to cause fatal illnesses, and it is still unanswered. There is much discussion of the possibility that infection by one dengue serotype can in certain circumstances cause sensitization rather than immunity, so that there is a violent reaction when another serotype is introduced.

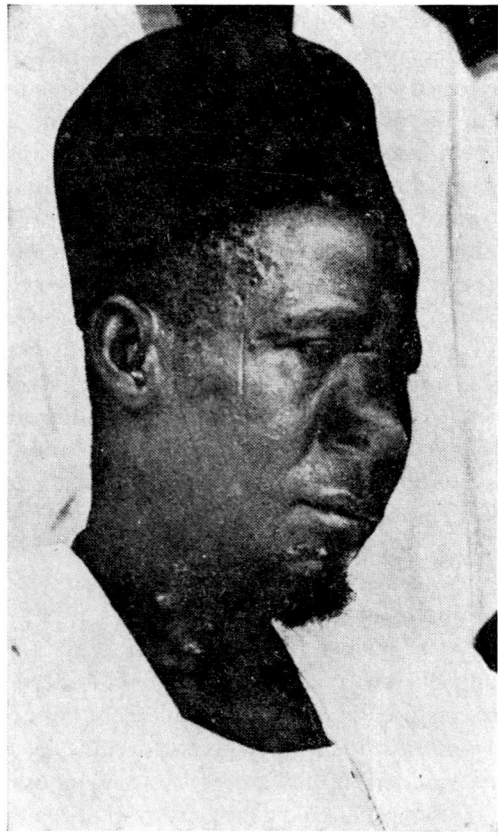

Fig. 1.3. Asibi from whom was recovered the classical Asibi virus much used in yellow fever research. (From Strode 1951.)

1.4.2 Changes in methods of transmission

In many virus genera there are members which are arthropod-borne and others transmitted by other means. Among the animal viruses at any rate, the method of transmission is not a useful taxonomic character. Even in a predominantly arthropod-borne genus such as *Flavovirus*, there are some, particularly among those recovered from bats, which have no known association with arthropods and have even failed to multiply in experimentally injected insects. One can easily imagine that, among social animals, direct methods of transmission could have arisen. A possible analogy is afforded by rabies, a member of a predominantly arthropod-borne genus, *Rhabdovirus*, yet transmitted among social bats by biting or even through infectious aerosols (Frederickson and Thomas 1965).

The tick-borne encephalitis virus of Central Europe can be transmitted by way of milk of infected goats (Van Tongeren 1955), and the mosquito-borne Eastern equine encephalomyelitis virus can pass from pheasant to pheasant by means of playful pecking (Holden 1955). When we see such changes in methods of transmission occurring under our very eyes, who can doubt that over millions of years the thing has happened again and again?

1.4.3 Ecological changes and disease

Of the three hundred or so known arboviruses, relatively few are of evident importance in human or veterinary medicine. Most exist in a state of mutual tolerance in both their arthropod and vertebrate hosts, and it is only when they are accidentally transferred to a strange host that they cause trouble. This often happens as a result of man-made changes in environment, either through migration, introduction of new domestic animals, drainage schemes, scrub-clearance, or even abnormal weather conditions. Infection of man and his livestock may occur as a result of infrequent spill-over of infection, and disease in man will then be sporadic and relatively unimportant: such are the relatively uncommon cases of jungle fever in South America. Often, however, there is a serious upset in equilibrium and a widespread epidemic. A virus reaching a strange host may cause serious disease but usually fails to spread from one of its new hosts to another: the viraemia is often too little or too brief. At times, however, adaptation to the new host is effective and a new cycle of infection is established: an example is urban yellow fever involving *Aëdes aegypti* and man. Outbreaks of epidemic virus disease are often brief: susceptible hosts may all become immunized or the conditions optimal for virus spread may pass away: the disease then becomes sporadic, as has happened to Chikungunya and O'nyong-nyong fever in Africa, their worst epidemic phase being, for the time, over.

1.5 Origin and evolution of arboviruses

One normally thinks of arthropods as vectors of virus infections from one vertebrate host to another. This is a natural anthropocentric view. From the arthropod's point of view, however, the vertebrate is the intermediary, transmitting a virus from one arthropod host to another. As far as origin and evolution are concerned it seems almost certain that one is dealing either with insect–parasites which have become secondarily adapted to living in vertebrates or the other way round. The first alternative seems preferable. Arboviruses seem well adapted to growth in insects in a state of mutual toleration: instances of their pathogenic activity in insects are few, and not conspicuous even then (Mims et al. 1966). On the other hand many of them cause serious or fatal disease in vertebrate hosts, though much less so when natural hosts are concerned. Bawden (1950) suggested that plant viruses might have originated from the insect parasites of plants and the insect parasites of animals have been suggested as possible ancestors of arboviruses infecting vertebrates (Andrewes 1957). It is remarkable that arthropods feature as recurring characters throughout the virological scene. There are the insect-pathogens proper, the many arthropod-transmitted plant-viruses and the arboviruses of vertebrates which form, in fact, a large proportion of known viruses. It is worth recalling that the rickettsiae are believed by many to have been derived from insect symbionts, some of which can well be grouped with them taxonomically.

Insects are much older inhabitants of the earth than are mammals, being their seniors by perhaps six million years. It would be surprising if they had not acquired their own viruses in such a period. One has also to consider, however, that viruses infecting vertebrates may have had quite other origins and that they may have evolved pari passu with their hosts, ever since the latter first deviated from their invertebrate ancestry. It is a striking fact that among the known groups or genera of viruses we have so many instances of morphologically similar viruses infecting vertebrates, insects and plants. Are we to believe in common origins and taxonomic affinities or in convergent evolution? A study of the base-sequences in the nucleic acids of viruses, insects and other hosts may well give suggestive clues.

1.6 A possible role of other invertebrates

Lungworms may play a part in perpetating swine influenza in mid-western America. Shope (1941) presented evidence that virus can be carried in ova of the lungworms, and passed in the pigs' faeces: the lungworm ova are ingested by earthworms in which they pass several stages in development. When the earthworms are eaten by pigs, the lungworms once more reach their lungs, but the virus within them remains latent until activated by some climatic or other stimulus. There is no evidence that the virus can multiply in worms. One group of workers has confirmed Shope's observations but others have failed, and there is no consensus as to the validity of his hypothesis. Later, Shope (1958) suggested that a similar mechanism might be concerned in the survival of swine fever virus.

1.7 Laboratory infections with arboviruses

Many viruses present a hazard to those studying them in the laboratory. An American committee (Hammon 1968) has collected records of over 2700 laboratory infections: 428 of these were with arboviruses, 16 of them fatal. Eight of the deaths were due to yellow fever, but Venezuelan encephalitis virus caused most illnesses – 118 with 1 death. Kyasanur forest virus was the next most dangerous: 65 laboratory infections are reported. Needless to say arthropods were not concerned in the laboratory infections. Greatest danger apparently came from manipulations involving blenders and centrifuges, or from dust from animal cages.

CHAPTER 2

Insect viruses

KENNETH M. SMITH

Contents

2.1 History of the subject

The earliest reference to a disease of insects, now known to be of virus aetiology, is in a poem by Vida in the year 1527 called *The silkworm* in which the author rhetorically enquires the cause of the malady whether it is 'the tainted air's corrupting streams or noxious food the latent poison hold'. In his latter suggestion the poet comes near to the truth, since we know now that the disease of the silkworm was a nuclear polyhedrosis and the chief means of spread is by ingestion of the polyhedra contaminating the food plant.

It was these polyhedra which stimulated the interest of the early workers, such as Cornalia (1856), Maestri (1856) and Bolle (1894), because they were easily visible in the optical microscope. It was Bolle who first established the protein nature of the polyhedra. The exact relationship between the polyhedra and the cause of the disease was the subject of extensive studies and controversy for many years. Von Prowazek (1907) demonstrated that material from diseased silkworms was still infectious after the polyhedra had all been removed by filtration through many layers of filter paper. This was a significant step towards the realization of a 'filterable virus'. Von Prowazek examined his clear filtrate under the microscope and saw that all the polyhedra had been removed. From this he drew the natural, but erroneous, conclusion that the polyhedra were not the carriers of the disease agent. What he did not realize, of course, was that there were many free virus particles in his filtrate in addition to those contained within the polyhedra, and these caused the disease when introduced into healthy silkworms.

In 1924 two workers, Komárek and Breindl, suspected that the causal agent might be contained within the polyhedra. They dissolved the polyhedra in weak alkali and observed them with the optical microscope using dark-field illumination. By this means they could see numerous minute objects showing Brownian movement apparently liberated from the polyhedra. In 1947 Bergold repeated these experiments but using the electron microscope, and showed that the virus particles, consisting of rather short thick rods, were in fact embedded in the matrix of the polyhedral crystals.

In 1926, a French worker, Paillot, discovered an apparently new virus disease in caterpillars of the cabbage white butterfly, *Pieris brassicae*. The disease was characterized by the presence of immense numbers of 'granules' just visible in the oil-immersion lens of the optical microscope. Some time after this, Paillot (1934, 1937) discovered two more such diseases in other insect larvae and named them 'pseudograsserie 1, 2 and 3'. In 1947 this type of virus disease was rediscovered by Steinhaus in a species of 'cutworm' (*Peridroma margaritosa* Haworth). A year later Bergold (1948) confirmed the virus nature of the disease by isolating the virus rods and characterizing them in the electron microscope. The name 'granulosis' was adopted and the single small 'granules' were shown to be crystalline and were called 'capsules' (fig. 2.1).

The existence of a separate and distinct type of polyhedrosis virus the particles of which are near spherical (isometric) in shape instead of the usual rod-shape was first demonstrated by Smith and Wyckoff (1950) in the larvae of *Arctia caja* L. and *A. villica* L. Previously other workers had observed polyhedra in the midgut cells (Ishimori 1934; Lotmar 1941), but the fact that they were of an entirely different nature from the nuclear polyhedra was not realized. It was later discovered that these polyhedra are formed in the cytoplasm of the cells of the midgut and not in the cell nuclei; it is from this fact that the name 'cytoplasmic polyhedrosis' was derived (Xeros 1952; Smith and Xeros 1953).

The insect viruses and virus diseases, then, fall into the two following broad groups: first, the 'inclusion-body' diseases comprising the nuclear and cytoplasmic polyhedroses, the granuloses, and the 'insect pox' diseases; secondly the 'free' viruses which, as their name implies, are not embedded in a crystalline matrix. The insect viruses seem to be the only viruses which are associated with these peculiar crystalline inclusion bodies.

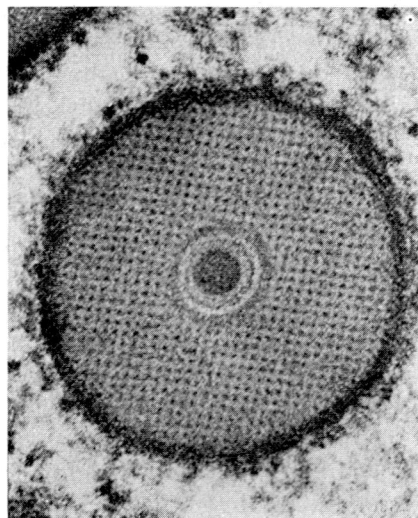

Fig. 2.1. Transverse section through a virus rod in its capsule from the granulosis of *Plodia interpunctella*; note the crystalline lattice of the capsule. × 193,000. (H. J. Arnott and K. M. Smith 1968.)

In the nuclear polyhedroses and granuloses the virus particles are rod-shaped and contain DNA. They are replicated, as their name implies, in the cell nucleus, the tissues attacked being the epidermis, fat body, blood cells and tracheae, rarely the silk glands.

In the cytoplasmic polyhedroses, replication of the virus takes place in the cytoplasm of the midgut and not in the nucleus. The particles contain RNA and are isometric, polyhedral in shape, usually bearing a number of projections.

In both types of polyhedra the numbers of virus particles occluded in the crystals are large, hundreds in the case of the nuclear polyhedra and probably thousands in the case of the cytoplasmic polyhedra. On the other hand, the small crystalline capsules of the granuloses contain, on the average, one or possibly two rod-shaped virus particles.

An unusual and interesting virus called *Tipula* iridescent virus was discovered in 1954 by members of the Virus Research Unit at Cambridge, England, during routine examination of larvae of the fly *Tipula paludosa* Meig. for the presence of another virus, a nuclear polyhedrosis which was first recorded by Rennie (1923).

The *Tipula* iridescent virus (TIV) is unusual in a number of ways. It

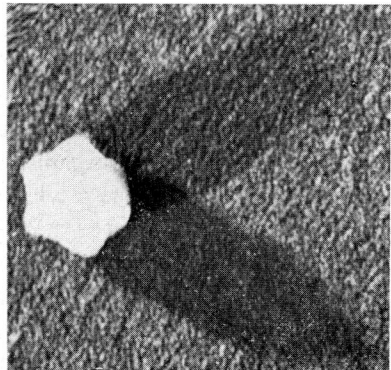

Fig. 2.2. A model of an icosahedron illuminated by two light sources and orientated so that an apex of the hexagonal outline points directly to each light source. This throws two shadows; one is four-sided and pointed, and the other is five-sided with a blunt end. (R. C. Williams and K. M. Smith 1958.)

Fig. 2.3. A particle of the *Tipula* iridescent virus, frozen-dried and shadowed in the same way as fig. 2.2; the similarity between the shadows thrown is evident. × 105,000. (R. C. Williams and K. M. Smith 1958.)

occurs in the infected insect in an astonishingly large quantity, as much as 25% of the dry weight of the larva. It also crystallizes spontaneously within the living insect giving the characteristic iridescence from which the virus got its name.

The virus is a large one measuring 130 nm in diameter, and it was one of the first to be proved unequivocally to be an icosahedron. This was done by double-shadowing. A model of an icosahedron was made and illuminated by two light sources separated 60 °C in azimuth and orientated so that an apex of the hexagonal outline points directly to each light source. Two shadows are thrown, one is four-sided and pointed, and the other is five-sided with a blunt end (fig. 2.2). A particle of TIV, freeze-dried and shadowed in the same way is shown in fig. 2.3. The similarity between the shadows thrown is evident. This indicates with fair certainty that the TIV particle is an icosahedron (Williams and Smith 1958).

Since the first discovery of an iridescent virus, a number of other closely similar iridescent viruses have been reported infecting various types of insect larvae including mosquito and beetle larvae. These viruses, though differing somewhat serologically, are all of the same type, being morphologically similar and showing the same fascinating optical properties.

2.2 Scope and some applications of the study of insect viruses

It might be thought that it would be difficult to isolate sufficient virus for fundamental study from organisms as small as insects. This, however, is not so because insects can be grown in unlimited numbers and the availability of virus is at least as great as that of many plant viruses. The *Tipula* iridescent virus, for example, can be obtained in quantities of 1 g or more without much difficulty, by the systematic inoculation of large numbers of caterpillars of *Pieris brassicae*, the large white butterfly, which by means of artificial light can be raised all the year round.

The necessity for growing large numbers of plants for feeding the insects may soon be avoided by the substitution of synthetic feeding media. This would be a great advantage in cities or where it is not easy to propagate plants in winter time. Furthermore the medium can be sterilized, thus avoiding possible contamination of the insects or their food plants by other pathogens. A suitable diet for the larvae of *Pieris brassicae* has been developed by David and Gardiner (1952). This is a useful insect for the study, as already mentioned, of the *Tipula* iridescent virus, and of its granulosis; as it is a comparatively large insect and yields much virus. Similarly Ignoffo (1964) has developed a successful method for the mass-rearing of the cabbage looper, *Trichoplusia ni* Hübner on a semi-synthetic diet which is basically the pink bollworm medium devised by Vanderzant and Reiser (1956).

A subject in which at present insect virology, together with plant virology, lags behind the study of viruses in the higher animals is the technique of tissue culture. Since this is dealt with elsewhere in this volume it will not be discussed here. Suffice it to say that some insect virologists envisage the day when the growing of plants and the propagation of large numbers of insects will be replaced by synthetic diets and tissue culture methods.

The use of viruses to control pests is not, of course, new, the classic case being the introduction of the myxoma virus into Australia against the rabbit pest. What is more recent is the use of viruses to control insect pests; that this has not been developed earlier is not surprising in view of the long neglect of the study of insect viruses. However, a start has now been made and with some there has been progress beyond the research stage, for example, in a work conference held to discuss and review the progress of research on, registration, and standardization of, the nucleopolyhedrosis virus of *Heliothis zea* (Boddie) a number of resolutions were made (Falcon 1965). It was first pointed out that the virus provides a potential apparently safe substitute for several hazardous pesticides. Among other resolutions it was emphasized that the virus materials destined for commercial use be standardized in content and insecticidal activity, in order that a highly effective, safe and economical insecticidal product may be obtained. In starting what may eventually be a new industry (microbiological control) it is obviously important that, at the beginning, there should not be a number of virus preparations, whether for *Heliothis* or other pests, which have been hurriedly prepared without adequate testing or standardization.

Thus, there is no reason why the insect viruses cannot be used for fundamental study as much as any other group of viruses. With the improved methods of technique now developed in electron microscopy, they are particularly useful for investigating the replication of the virus itself, with the added interest of the unique intracellular inclusions.

2.3 *Some important aspects of the study of insect viruses*

The fact that insect virology has been for so long neglected presents a whole new field of exploration. Some of the important aspects which need investigation are the role played by the intracellular inclusions, the relationship between the virus and the cell organelles, the assembly of the virus particles and the question of transovarial transmission.

Apart from the actual raison d'etre of the intracellular inclusions, there are several other questions of interest and these are exemplified by the cytoplasmic polyhedrosis of the caterpillar of the Monarch butterfly (*Danaus plexippus*). Perhaps the first question that springs to mind is why do only virus particles become incorporated in the crystalline matrix? In the large number of polyhedra examined there was not a single case in which any other cellular component except the virus particle had become incorporated and this seems to be the case with other polyhedroses. How does the crystal 'distinguish' between the virus particles which are readily incorporated and other cellular components which are restricted from inclusion? A possible explanation might be based on the structure of the virus particle and the nature of the protein making up the polyhedron. It may be that the conformation of the protein forming the viral capsid is such that it, and it alone, can become associated with the polyhedral protein.

It is interesting to note the different shapes of the polyhedra that are found in cytoplasmic polyhedroses. The shape ranges from round to diamond-shaped or polygonal to cubic.

The nature of the mechanism that controls crystal shape is a fertile field for investigation. It is already established that with the insect inclusion viruses the crystal shape 'breeds true' (Gershenson 1960), and this also happens with the cubic 'mutant' of the Monarch polyhedrosis. It would be interesting to know at just what point the viral genetics are interposed on the situation to bring about these alternate crystal forms. Is it a change in the sequence or the type of amino acids? Is it an alternate crystallization pattern due to a change in the cellular (or crystal) environment which can be attributed to the mutant virus? Such questions are not only important in understanding insect virology, but they are fundamental questions of molecular biology (Arnott et al. 1968).

Knowledge concerning the assembly of the insect virus particle itself is somewhat meagre. It seems a fact that the particles are assembled, so far as the nuclear polyhedrosis viruses are concerned, in a 'nuclear net' or 'virogenic stroma' which is a mesh or network of strands from which naked virus particles are produced (Smith and Xeros 1953; Harrap and Robertson 1968).

The exact method of replication of the isometric viruses, as typified by the cytoplasmic polyhedrosis virus of the Monarch butterfly, is still somewhat of a mystery. In this instance there are two areas, one of which is called the 'virogenic stroma' and the other, the 'crystallogenic matrix'. Both contain

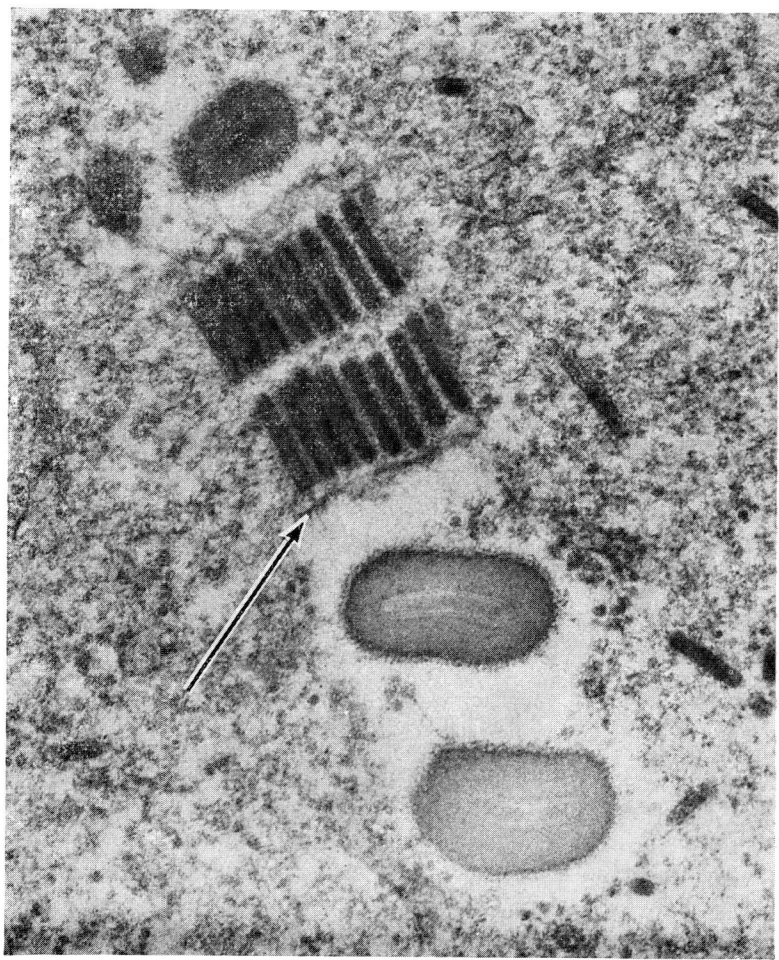

Fig. 2.4. Section through part of a fatbody cell of the larva of *Plodia interpunctella* infected with a granulosis: note the virus rods in regular array and the endoplasmic reticulum which is thought to provide the outer membrane of the virus rod (arrow). × 50,000.

(H. J. Arnott and K. M. Smith 1968.)

virus particles but no normal cellular components. The crystallogenic matrix has many complete virus particles in it while the virogenic stroma contains few complete, but many incomplete particles. The presence of so many empty virus particles suggests that the capsid is produced first and subsequently filled with nucleic acid. This is certainly what appears to happen with the *Tipula* and *Sericesthis* iridescent viruses (Smith 1958; Xeros 1964; Bellett 1965).

What part in the assembly process is played by the normal cell organelles? It seems fairly certain that the smooth endoplasmic reticulum supplies the outer membrane of the virus rods in the granulosis of *Plodia interpunctella* (Hbn.) the Indian meal moth. In this instance the naked virus particles occur in ordered array closely associated with the endoplasmic reticulum. These arrays are a common feature of this granulosis and appear to be important in the mechanism by which the naked rods receive their outer membrane (fig. 2.4) (Arnott and Smith 1968).

Although the fact of transovarial transmission of insect viruses is fairly well established, some controversy may still exist as to whether the virus is carried externally on the chorion or inside the egg itself. Probably both are true and the term 'transovarial' is used for transmission inside the egg and 'transovum' for transmission by adsorption of the virus to the exterior of the egg. By using 'mutant' hexagonal cytoplasmic polyhedra as a marker Hukuhara (1962) found that 75% of the infected progeny of silkworms inoculated with these hexagonal polyhedra contained this type of polyhedra. This is fairly convincing evidence of true virus transmission to the progeny, chiefly because it circumvents the ever present pitfall of latent viruses.

2.4 *Some notable recent advances in insect virology*

During the last few years some unusual new insect viruses have been discovered. Iridescent viruses have now been isolated from dipterous larvae, *Tipula paludosa, Aëdes detritus*; beetle larvae *Sericesthis pruinosa*; and lepidopterous larvae *Chilo suppressalis*, the rice stem borer. These viruses form a related group which has been discussed by Bellett (1968). A new type of insect virus disease named 'densonucleosis' has been described by Vago et al. (1964). The dense nuclei contain large numbers of isometric virus particles. The disease was first noticed in the lepidopterous larva *Galleria mellonella*, the greater wax moth. Efforts to use this virus to control the wax moth in beehives are being carried out in Czechoslovakia (Tchubianishvili 1970).

The artificial division, by virologists, of viruses into those of animals, plants, and insects is being gradually broken down. For example, an entirely new group of insect viruses, called 'insectpox' by Granados and Roberts (1970) has been discovered. The first of these was found infecting the larvae of cockchafer beetles (Vago 1963; Vago et al. 1969). Another has been found in lepidopterous larvae and it is probable that others of this group will be discovered. The virus multiplies in the cytoplasm and the particles are occluded in intracellular inclusions which differ from the intracellular inclusions of the polyhedroses, although they appear to be crystalline. Not only do these insect viruses closely resemble vaccinia and other animal pox viruses but they undergo a similar maturation process (Granados and Roberts 1970).

In a review of vesicular stomatitis and other viruses with large bacilliform particles, Howatson (1970) shows the remarkable morphological resemblance between the particles of vesicular stomatitis virus, sigma virus of the fruit fly *Drosophila* and the viruses of several plant diseases, notably wheat striate mosaic, potato yellow dwarf, sowthistle yellow vein and others. Thus we see already a tendency to collect together viruses infecting all three types of organisms, higher animals, insects, and plants as a 'vesicular stomatitis group' or 'rhabdoviruses' and a 'pox virus group'. Admittedly this grouping is based, at the moment, on morphological resemblance only, but further research may reveal other properties in common.

Certain factors link together the plant and insect viruses; apart from morphological resemblances, as with the sigma virus of *Drosophila* and that of wheat striate mosaic for example, there is the multiplication of what is presumably a plant virus within the body of the insect vector. This certainly implies that the boundaries between an insect and a plant virus are indistinct and doubtful.

The discovery that the RNA of tobacco mosaic virus was infectious incited similar investigation into the infectivity of the nucleic acid of animal viruses. The RNA extracted from the cytoplasmic polyhedrosis virus of the silkworm, *Bombyx mori*, shows a comparatively high infectivity. The infectivity is lost by incubation with RNase but is not affected by DNase (Kawase and Miyajima 1968). The same RNA has been shown by Miura et al. (1968) to be double-stranded.

Also of great interest would be evidence on the nature of the factors which control the host ranges of viruses. There have been claims that mammalian viruses have been propagated in Protista by use of their nucleic acid alone but these reports need confirmation. Kovacs and Bucz (1967) report the

successful propagation of encephalomyocarditis virus (EMC) in Protista and
the same workers have reported the adsorption and uptake of the infectious
particles by yeast and by the protozoan *Tetrahymena pyriformis*.

Other experiments using *T. pyriformis* have been done by Thompson et al.
(unpublished). Three viruses were used, tobacco mosaic virus, *Tipula*
iridescent virus, and an insect granulosis virus inside its capsules. All these
viruses were rapidly taken up into the food vacuoles of the organism and
could easily be seen in the electron microscope by means of thin sections
(fig. 2.5) but there was no evidence of multiplication of the viruses.

The discovery that leaf hopper-borne plant diseases of the 'yellows' type
are not due to virus infection as has always been thought but are associated

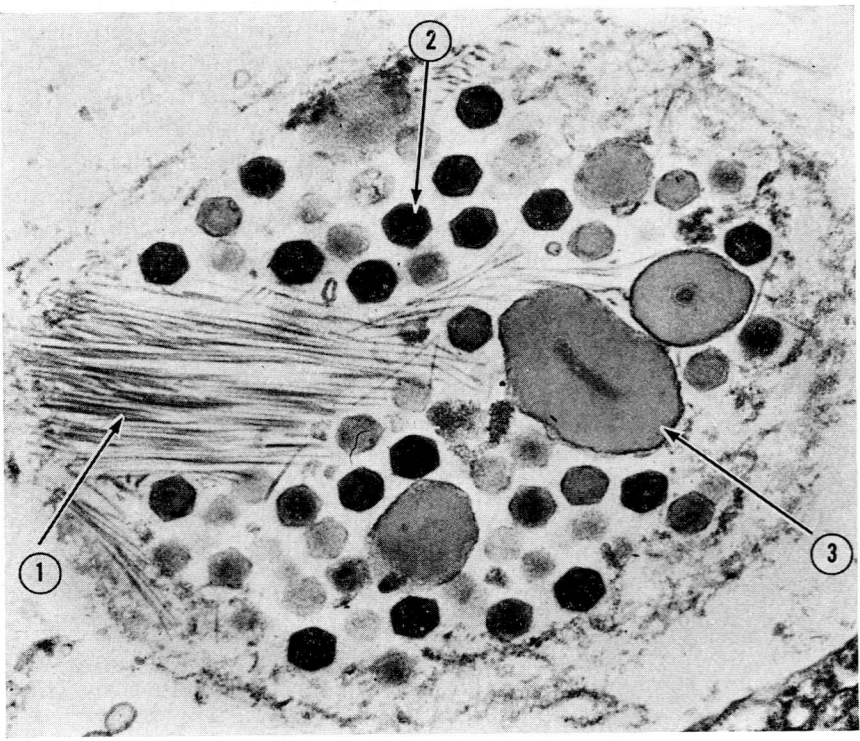

Fig. 2.5. Section through part of the protozoan *Tetrahymena pyriformis* which had been
fed with three viruses, tobacco mosaic virus (arrow 1), *Tipula* iridescent virus (arrow 2),
and the capsules of *Plodia* granulosis (arrow 3); as yet there is no evidence that these
viruses are multiplying. × 51,500. (G. Thompson, R. M. Brown and K. M. Smith.)

with mycoplasma-like organisms, has had a profound effect on some aspects of plant and insect virology. It removes at a stroke much of the evidence of the multiplication of plant viruses in insect vectors as typified by aster yellows, although evidence of multiplication with some 'true' viruses still remains. It also casts doubt on the apparent harmful effect on an insect vector of a plant virus. Peach western X disease has been shown to cause a fatal infection of its leafhopper vector (Jensen 1959) and this disease too, is thought to be due to *Mycoplasma* infection. Similar infections of the insects themselves have also been recorded; *Mycoplasma*-like bodies have been observed in the third instar larvae of the cockchafer beetle, *Melolontha melolontha* L. (Devauchelle et al. 1969).

The whole question of *Mycoplasma* diseases of plants and insects has been reviewed by Maramorosch et al. (1970) (see ch. 26).

Plant viruses

MARION A. WATSON

Contents

> 'Now understand me well – It is provided in the essence of things, that from any fruition of success, no matter what, shall come forth something to make a greater struggle necessary.'
>
> Walt Whitman

3.1 Introduction

Associations between insects and plant diseases have been recognised for more than 70 years, but transmission of autonomous pathogens by specific vectors has been much more difficult to establish. It is difficult even to decide which plant viruses were first known to be transmitted, when, and by whom. In the 19th and early 20th centuries it was customary for many years' work to be accumulated for a single publication (as with *The origin of species*), and anyone who expects to see in retrospect neat monuments to scientific achievement, with tidy columns of references on the reverse faces, will probably be disappointed.

Probably the first epidemic plant disease to be associated with insect-infestation was rice dwarf disease, which was first studied in 1895 by Takata, at the Shiga Agricultural Experiment Station in Japan. Takami (1901) ascribed its spread to infestation by the leaf-hopper *Nephotettix apicalis* Motsch, var. *cincticeps* Uhl. Ando (1910) described the first experiments in field and glasshouse, whereby the transmission of a pathogen was indicated, but Fukushi (1933) generously decided that curly top disease of sugar beet was first publicly recorded as being transmitted by a leafhopper, *Eutettix (Circulifer) tenellus* by Ball (1909), although, even in 1915, Townsend was still uncertain about the cause of the symptoms. Probably Carsner was the first to write of curly-top as a virus disease in 1919.

3.2 Transmission by aphids

Allard described transmission of tobacco mosaic by several species of aphids in 1915, but in spite of support from Hoggan in 1934, this has not been substantiated. Hoggan's careful, well-documented results could not be repeated although many workers (myself among them) have tried hard to do so. It can be supposed that the possibility of transmitting TMV, which is very contagious, while handling and observing the vectors was not recognised in the then state of ignorance about the virus. However, it is possible that

Hoggan had an unusual strain that was aphid-transmitted and never rediscovered. Teakle and Sylvester (1962) recently showed that aphids can infect
leaves when they feed through a heavily contaminated leaf surface, but
Hoggan's experiments were too carefully planned and executed for this to
have explained her results.

Cucumber mosaic probably has priority as the first recognised virus known
to be transmitted by aphids, through the work of Doolittle (1916) and
Jagger (1916). Doolittle and Walker (1928) published a short abstract in
Phytopathology, containing the first reference to the brief time that aphids
remain infective with such viruses and suggesting that they were carried
'mechanically' as contaminents of the mouthparts. Viruses were then referred to as being either 'mechanically' or 'biologically' transmitted.

3.3 The potato crop

The potato crop is as important in the history of aphid-transmission of
viruses as are the Japanese rice and American sugar-beet crops to leaf-hopper
transmission. Diseases of the potato crop were known during the 18th and
19th centuries under such names as 'running out', 'senility', 'degeneration'
or 'crinkle', but it was well into the 20th century before they became associated with aphids.

In 1906 Appel distinguished a condition that he called 'potato leaf-roll',
for which Oortwijn Botjes (1920) identified the vector as *Myzus persicae*
Sulz. Botjes and Quanjer continued to study the potato crop in relation to its
'contagium' diseases and, as Van der Want (1966) pointed out, their work
showed the essential importance of virus diseases for practical agriculture,
not only in potatoes but in other crops, such as sugar beet which contract
virus yellows or 'jaunisse' (Quanjer 1936; Roland 1936).

Quanjer was a great evangelist of the importance of virus diseases and
gave papers at the Vth and VIth International Botanical Congresses. The
first, in 1930, on methods of identification of potato viruses, would have been
remarkable in any event, as it was probably the only paper at any meeting
before or since, that had to be read by the audience, page by page, from
lantern slides; the speaker had acute laryngitis, but equally acute determination to communicate. Other papers on the theme of 'Plant virus differenti-
ation and classification' were given by Drs. James Johnson and Isme Hoggan
of Wisconsin, and Prof. Paul Murphy of Dublin.

My own lasting enthusiasm for aphids and viruses was fired a little earlier at a meeting of the Association of Applied Biologists in 1929, in the tall, dark lecture theatre of the Botany Department at Imperial College, London. Dr. Kenneth Smith lectured on separating the viruses causing mosaic and crinkle diseases into a 'vein-banding' one that was transmissible by aphids, and one causing ring-spot symptoms in tobacco, that was 'needle-transmitted' but not aphid-transmitted. The second he equated with Johnson's 'healthy potato virus' (1925), and later called the two 'potato viruses Y and X' (Smith 1931), by which names they are still known.

Although much was known about virus diseases by 1930, especially those of the potato crop, no paper on transmission by insects had been given at any meeting of the International Botanical Congress, and the VIth International Congress in Amsterdam was no exception. Discussion was confined to nomenclature and classification, but evidently there was ex cathedra discussion, for the following verse, in three languages, appears in a rather frivolous section of the Proceedings:

(Weise: Bier her ...)

> Virus, Abbau
> Was nun soll es sein?
> Wahl ist schwierig wie zwischen Bier und Wein.
> Ist der Virus fortgeschritten?
> Ist der Abbau abgeschnitten?
> Wein – Bier – Wein – Bier,
> Wahl is schwierig hier.

It was accompanied by a picture of two snails (appropriately) approaching the Truth, one labelled 'Virus', the other 'Abbau' meaning degeneration, which was then attributed to every nutritional or pathological condition under the sun, except the right one. 'As if', said Quanjer in 1949 (quoted by Van der Want 1966), 'Robert Koch had not lived ... as if it had not long been recognised that the infection experiment is conclusive in phytopathology'.

3.4 Other vectors

Common insect vectors of plant viruses include several homopterous families other than aphids and cicadellids: white fly, mealy bugs (Pseudococcideae),

3.1.   3.2.

3.3.   3.4.

3.5.   3.6.

Scientists whose work early in the 20th century has influenced later developments in plant virology.

Fig. 3.1.   J. G. Oortwi̱n Botjes, who showed that potato leaf roll is transmitted by *Myzus persicae,*   1920

Fig. 3.2.   Teikichi Fukushi, who said, 'Dwarf disease of rice plant ... multiplies both in the insect body and in the plant tissues, and the virus lies ... beyond the limits of microscopic visibility'.   1934

Fig. 3.3.   Louis O. Kunkel, 'Recovery of ability to transmit results from multiplication of that portion of virus that remains' (after heat treatment of the vectors).   1941

Fig. 3.4.   Haydon H. Storey, 'The physiological basis of inactivity (of vectors) lies in a property of the intestines that prevents movement of virus through its wall'.   1933

Fig. 3.5.   Kenneth M. Smith, '*Myzus persicae:* the arch enemy of the virus world!'   (circa 1935)

Fig. 3.6.   H. M. Quanjer, 'As if Robert Koch had not lived!'   1943

Psyllidae, and Cercopidae; the last two involve a query because the diseases they transmit, Pear decline or 'Psyllid yellows', could be mycoplasmal (Hiroyuki and Schneider 1970). Among the Heteroptera, Tingidae (lace-bugs) and Piesmidae are vectors. Beetles of the Chrysomelid, Apionid and Curculionid (weevil) families transmit some viruses, including turnip yellow mosaic and brome grass mosaic, which vie with tobacco mosaic as viruses most dear to molecular biologists and electron microscopists. Eryophyid mites (Arachnidae) transmit several viruses of fruit trees, cereals and grasses (Slykhuis 1965). Only one virus, but of world-wide range and much economic importance – tomato spotted wilt – is transmitted by thrips; much of the earlier work was done in Australia (e.g. Samuel and Bald 1931).

There are records of transmission by caterpillars, Dipterae and other groups of insects (Carter 1962), but multiplicity of vectors for one virus is rare. The transmission of sowbane mosaic by various insects (Bennett and Costa 1961), and that of tobacco mosaic by slugs, beetles and grasshoppers in experimental conditions (Walters 1952) are exceptions. Many early claims of transmission by unusual vectors, or involving many different orders of insects, still await substantiation.

Abutilon 'infectious chlorosis' disease was described by Morren in 1869, but it was nearly a century before Orlando and Silberschmidt (1946) recognised it as caused by a virus, transmitted by the white fly, *Bemisia tabaci*. This insect had already been detected as the vector of the economically more important 'leaf-curl' virus of cotton (Kirkpatrick 1931). White fly transmit several other viruses of economic importance in India, South America, the West Indies and the Philippines (Tarr 1951), but apparently none in temperate regions.

Cocoa swollen-shoot virus, that almost ruined the cocoa industry in Nigeria and Ghana, and for which drastic and expensive 'cutting out' methods have been used to attempt to control it, is carried by coccids (Pseudococcideae) (Box 1945). Posnette and Strickland (1948) identified several vector-species and strains of the virus, some of which occur also in Trinidad.

3.5 The relation between plant viruses and their vectors

A review, *Transmission of plant and animal diseases by arthropods*, published by Rand and Pierce in Phytopathology in 1920, today makes strange but rather exciting reading, at least for plant virologists. It seems to anticipate much that has become known about the transmission of plant viruses, and yet much of it seems to have been pure imagination, with little and conflicting scientific evidence to support it.

Much was known at that time about the transmission of animal pathogens such as the malaria parasite and yellow fever, which was thought to be a spirochaete. The term virus was already in use for animal pathogens, but Quanjer preferred to call the mysterious filterable plant pathogens *Contagiums* (after Beijerinck's *Contagium vivum fluidum*) and the section is headed *Filterable contagium diseases*.

Smith and Bonquet (1915a) had already published a substantially correct description of curly top of sugar beet. They found or confirmed that the minimum time taken to acquire curly top from infected plants by the vector *Eutettix tenellus* was about 3 hr and the virus sometimes persisted for several weeks in the vector. There was an incubation period of the virus, as 'At least 24 hr and not more than 48 must elapse before the leafhoppers can infect, after obtaining the pathogenic factor', and therefore probably 'curly-top is not carried mechanically, but some development takes place within the body of the insect during the first few hours after feeding on a diseased plant.' This was a modest and reasonable statement of the possibilities, based on results that have since been verified, even though some of the conclusions have not unreservedly been accepted.

However, curly-top was not the only insect-transmitted plant disease agent described in terms of a persistent virus. Another was the extraordinary 'Spinach blight', the record of which comes from a 60 page paper by McClintock and Smith in 1918. Spinach blight was said to be easily transmitted by rubbing healthy plants with sap from diseased ones; it was probably seed-transmitted; its vectors were several species of aphids and the tarnished plant bug, *Lygus pratensis*; it was not transmitted by various other insects including thrips and white flies; it persisted for long periods in the aphids and passed through the moult; it was transmitted by infective mothers to their progeny removed at birth before feeding, and afterwards to the third or fourth generation of aphids reared on non-susceptible hosts; it could be transmitted by manual inoculation of the crushed juices of infective aphids, and there were vector- and non-vector races of the aphids that transmitted it. Its spread in the field could be decreased by spraying, and the authors had started breeding experiments to find blight-resistant varieties of spinach. They compare its transmission to that of the tick-borne Rickettsia organism of Rocky Mountain spotted fever.

All these properties can scarcely have belonged to a single virus but the ideas, since found to apply to many different ones, were there, like fossils waiting to be excavated. Spinach blight seems to have been a sort of Piltdown skull of virology; the results were accepted because they seemed to be reasonable.

Viruses were thought to be sub-microscopic cellular organisms, probably stages in the development of bacteria (Smith and Bonquet 1915b). There was no reason why they should not have alternate cycles of reproduction in plants and insects, and there seemed no other plausible explanation for their long persistence in the vectors, transmission to their progeny and the incubation period, except that virus multiplied in insect tissues as well as in the plants.

Support for the features claimed by McClintock and Smith for spinach blight were soon forthcoming, but each applying to a different virus, none of them spinach blight.

Professor Fukushi first announced the transmission of a plant virus through the egg of its leafhopper vector. He showed in 1933 that nymphs hatched on virus-free plants from eggs laid by viruliferous *Nephotettix cincticeps* females, could infect rice seedlings with rice dwarf virus. In 1940 he showed that infectivity continued through several generations without the intervention of susceptible plant hosts, and considered this as proof that the virus could multiply in the insect as well as in the plant.

Dr. Haydon Storey (1932) showed that the ability of *Cicadulina mbila* to acquire and transmit streak disease of maize, was determined by a genetic factor, inherited in a simple Mendelian manner. Later (1938) he showed that that non-vector insects after acquiring virus, could transmit the disease when they were pricked through the stomach wall so as to release the contents into the haemocoele. He postulated that the heritable property was permeability of the gut wall to the virus. Both these postulates have been confirmed, the first by Black (1943) and the second by Sinha (1963). Storey (1933) was also the first to show that plant viruses could be transmitted by vector to vector injection of haemolymph.

Dr. Kunkel, although working on aster yellows disease (Kunkel 1926), now considered to be mycoplasmal, inspired and influenced very many of the virus-entomologists who succeeded him, in particular a brilliant school headed by Dr. L. M. Black. One of Kunkel's most important contributions (1937) was to show that when viruliferous *Macrosteles* vectors were kept at temperatures of 36 °C their infectivity gradually declined and after a time they were unable to infect. Held at cooler temperatures, they slowly regained infectivity. Kunkel attributed this to inactivation of the disease agent at warm temperatures (36 °C), as happens also in plants, and then multiplication in insect tissues up to an infective concentration. With longer times at 36 °C the inactivation was irreversible. These results were considered by many people to prove that the viruses multipled in the tissues of the vectors.

Dr. Walter Carter (1927) found that curly top virus (*sic*) could be acquired by *Circulifer teneila* when the insects were fed on a resuspended alcohol precipitate in sugar solution. Dr. C. W. Bennett and his colleagues (Bennett 1934; Fife and Frampton 1936) developed the technique, and also showed that curly top virus is acquired from phloem cells, which the hoppers find by probing along a pH gradient. This was important because it accounted for some of the time needed for insects to acquire virus, apart from its possible multiplication.

These results, however, and the relatively short incubation period of curly top in its vector did nothing to support the hypothesis of virus multiplication, and when the properties and composition of virus particles became known (Stanley 1935, 1936; Bawden and Pirie 1937, 1938) ideas of what plant viruses could or could not do, changed. Plant pathologists had no longer any difficulty in supposing that an insect could suck up enough virus particles from plant cells to last them the rest of their lives and even to hand on to their progeny. There were millions of highly stable particles of nucleoprotein in a droplet of sap from tobacco mosaic infected plants, and it was not yet known that plant viruses could differ widely in their properties.

Freitag in 1936, produced his 'negative evidence on multiplication of curly-top virus in the beet leafhopper', which provided an escape route for dissenting pragmatists. In the preface to his 1939 edition of *Plant viruses and virus diseases* Bawden says that he 'tried to distinguish between facts and the deductions made from them', and he did not consider multiplication of plant viruses in insect vectors to be then proved. Fukushi, more in sorrow than in anger, wrote a protesting postscript to his next paper about transovarial inheritance of rice dwarf virus (1940), but in his 1943 edition, Bawden still considered the evidence equivocal and indicated what would amount to proof. Spurred cn to almost superhuman efforts, Black (1950) reported having carried clover wound tumour virus transovarially through 21 generations of the vector *Agalliopsis novella*, representing, he calculated, a dilution of 2.8×10^{-26}. This showed beyond reasonable doubt that at least one plant virus could multiply in its vector. Maramorosch, also, was stimulated into trying Bawden's suggestion (1943) of serially transferring virus by injection from hopper to hopper, to find whether 'vectors inoculated with the minimum concentration ultimately came to have the same concentration as those inoculated with concentrated extracts'. This was demonstrated (Maramorosch 1955), but with the aster yellows agent, now thought to be mycoplasma-like.

Distinguishing the different kinds of aphid-transmission has also used up quite a lot of print since Watson and Roberts (1939) named them 'persistent' and 'non-persistent', according to the lengths of time that aphids remained infective after acquiring virus, and Sylvester (1962) inserted an intermediate category for those such as beet yellows virus that seemed to be neither one nor the other.

The multiplication of a persistent aphid-transmitted virus in its vector was established, within reasonable doubt, by Stegwee and Ponson (1958) for potato leaf roll virus, and the lively exchanges of views that took place (Bradley and Ganong 1955; Bradley ('Our concepts; on rock or sand?'), 1959; Sylvester 1969) about the mechanism of transmission of non-persistent viruses, uncomplicated though it seems, has maintained interest in these also (see ch. 25). However, much is too recent to be considered historical, and will doubtless be discussed in later chapters of this volume, with the aid of electron microscopy, fine-sectioning, analytical centrifugation, and other complex electronic aids that would have surprised Dr. Quanjer.

3.6 The importance of plant virus diseases

Three billion dollars a year is the title of a leading article by Jessie I. Wood in the U.S. Department of Agriculture Year Book for 1953, which contains accounts of several hundred diseases attacking food and fibre crops, about a quarter of them of viral origin. A notable contribution by C. W. Bennett is *Viruses, a scourge of mankind*, describing how sugar-beet growing almost ceased in the Western U.S. because of curly-top, until the development of resistant varieties in the 1930s. Citrus tristeza virus caused the loss of 7 million orange trees in the state of São Paulo, Brazil, alone, and has attacked millions of others in tropical and subtropical countries.

Tropical agriculture has been threatened by many devastating virus diseases. We owe our knowledge and control of a number of them, including those of maize, groundnuts and cassava, to the work of Haydon Storey in Africa, 'the most eminent tropical plant pathologist of our time' (Nutman, 1969); few who knew him, and appreciated his work, could disagree.

Bennett has suggested that the increase in destructiveness of plant virus diseases is caused by development of agricultural enterprises, with movement of plants and plant products about the world. The development of rapid freight transport by air, must be increasing this danger year by year.

In temperate regions probably the best documented of the virus-complexes attacking major crops are those of potatoes. Bawden (1939) quotes figures showing that farmers in England then annually spent around seven hundred thousand pounds for new potato seed from Scotland and Ireland to replace stocks that became virus diseased, but, he says, 'listing the names of the more widely known virus diseases will probably better indicate their economic importance than attempting to show it in largely fictitious' (and at the present time, rapidly changing) 'money values'.

3.7 Coda

Walt Whitman's wise prediction can be seen, even in the few pages presented here, to have been more than justified. Time and space have excluded many who have contributed to what is now the science of plant virology. Persons who doubt the validity of this definition are referred to the Shorter Oxford Dictionary: 3rd edition with revised addenda, Clarendon Press, Oxford, 1955. In the hurley-burley of modern science the aphorism that 'Il n'y a que le premier pas qui coute' seems to have evolved as 'Il n'y a que le dernier pas ...', but only a little thought convinces us of the debt owed to the pioneers who broke difficult ground for others to cultivate.

Section II

Dramatis impersonae

CHAPTER 4

The viruses

A. J. D. BELLETT, F. FENNER AND A. J. GIBBS

Contents

4.1 Introduction

Since the turn of the 20th century the work started by Loeffler and Frosch (1898), Beijerinck (1898) and many others has shown the differences between viruses and other pathogens.

Viruses differ from all microbes, like bacteria, Rickettsiae and Chlamydiae, in their chemical composition and mode of replication. The particles of most viruses contain only one kind of nucleic acid, either DNA or RNA. This functions as a genome supplying the genetic information required for the synthesis of viral proteins, and is itself exactly replicated during viral multiplication. Viruses depend on the protein synthesizing apparatus of the cells they infect; they neither contain (with the possible exception of arenoviruses (sect. 4.5.2.7)) nor synthesize ribosomes and few, if any, code for transfer RNAs. Virus particles assemble from the separately synthesized viral nucleic acid and protein components, and those of certain viruses may incorporate host cell components.

4.2 Composition and properties of virus particles

Virus particles consist of a nucleic acid genome surrounded by a proteinaceous shell. The latter may contain glycoproteins or lipids (often derived from the host cell) and the particles of some viruses are enclosed within crystals of virus-specified protein (e.g. the polyhedroses). Viral nucleic acid functions as a genome, and sometimes also as messenger RNA. The viral genome consists of either deoxyribose nucleic acid (DNA) or ribose nucleic acid (RNA), which may be single- or double-stranded, and either linear or cyclic. The genomes of some viruses consist of several separate linear molecules, which may be contained in a single type of particle or within separate particles of different size or sedimentation coefficient. In exceptional cases the virion contains a small amount of a second type of nucleic acid; in leukoviruses the genome is RNA but DNA is present, whereas in poxviruses the genome is DNA but RNA may be present also. The isolated nucleic acid of some viruses is itself infectious, whereas the nucleic acid of others is not, probably because enzymes within the virus particles are required for infectivity.

The nucleoprotein complexes (nucleocapsids) of most viruses are either tubular, the nucleic acid being surrounded by helically arranged protein subunits, or isometric, the nucleic acid being surrounded by protein subunits arranged as an icosahedral shell (fig. 4.1). Some viruses have a more complex morphology. Either type of nucleocapsid may occur 'naked' or it may be enclosed within other virus-specified protein membranes, or enclosed within an envelope containing host and virus-specified components.

4.3 *Classification and nomenclature of viruses*

Until recently viruses were usually grouped primarily according to their host-organism, i.e. as plant viruses, animal viruses, etc. However, this was not always convenient, if only because several viruses are known to multiply in (or be associated with) more than one type of host, many other viruses that infect plants and vertebrate animals also multiply in their arthropod vectors.

We shall use a classification that is independent of the host (Wildy 1971), and based mainly on the virus particles, both their composition (e.g. DNA or RNA; single- or double-stranded) and their morphology (e.g. tubular or isometric nucleocapsid, with or without envelope). This is a more generally useful classification, but should not be assumed to have any 'natural' or phylogenetic significance.

The naming of viruses is a separate problem from their classification, although clearly related to it. Viral nomenclature is being actively considered by a committee (the International Committee for Nomenclature of Viruses) of the International Association of Microbiological Societies. In this chapter we shall use those generic (group) names so far approved by that committee (Wildy 1971). Many viral groups, especially those of plants and insects, have not yet been discussed or named by that committee, for these we shall use commonly used names. We shall also use cryptograms (Gibbs et al. 1966; Wildy 1971; appendix 4.1) that give a coded summary of the properties of individual viruses and groups. Meanings of the symbols used in the cryptograms are listed in the appendix. The name 'arboviruses' will be used as a general term for all those viruses, irrespective of group, that multiply in both vertebrates and arthropods.

It is impossible in this book to give more than a brief account of the viruses. For further descriptions read Andrewes and Pereira (1967) and Fenner (1973) for viruses of vertebrates, Matthews (1970a) and the C.M.I./A.A.B. Descriptions of Plant Viruses (1970) for viruses of plants.

4.4 Viruses whose particles contain deoxyribose nucleic acid (DNA)

4.4.1 Viruses whose particles contain double-stranded DNA

Many of the viruses of vertebrates and of bacteria, some of the viruses of in-sects, and at least three related plant viruses have genomes which, like those of organisms, are double-stranded DNA. The bacterial viruses, and the verte-brate viruses belonging to two large and important genera (herpes-viruses and adenoviruses) have no associations with invertebrates.

4.4.1.1 Poxviruses [D/2:160–200/6:X/X:V, I/O, Di, Ac, Si]

The poxviruses are viruses of vertebrates, which do not multiply in inverte-brates, although several of them are transmitted mechanically by biting ar-thropods. However, recent studies (Pogo et al. 1971) reveal the close physico-chemical similarities of the poxviruses to the so-called entomopoxviruses de-scribed below. Apart from several ungrouped viruses, the poxviruses can be

TABLE 4.1
The poxviruses

Subgroup type member	Vaccinia	Myxoma[1]	Orf	Sheeppox	Birdpox[1]
Other members	Vaccinia Cowpox Ectromelia Rabbitpox Monkey- pox Variola Alastrim	Californian myxoma Brazilian myxoma Rabbit fibroma Squirrel fibroma Hare fibroma	Orf Bovine papular stomatitis Pseudo- cowpox (milker's nodes)	Sheeppox Goatpox Lumpy skin disease (Neethling strain)	Fowlpox Canarypox Pigeonpox Turkeypox

Other poxviruses of mammals, not yet allocated to subgroups; Swinepox[1], Molluscum contagiosum, Yaba monkey tumor virus[2], Tana virus[2].

[1] Certainly transmitted mechanically by arthropods.
[2] Possibly transmitted mechanically by arthropods.

conveniently arranged in six subgroups (table 4.1); they share a group antigen that is probably internal.

Poxviruses usually produce one or other of two main types of lesion which are relevant as far as arthropod transmission is concerned. Viruses of some subgroups (e.g. the vaccinia and orf subgroups) produce papulo-pustular skin lesions, either localized or more often generalized via the blood stream; these viruses are transmitted either by contact or by the respiratory route. Others produce localized or generalized tumours in the skin. All viruses of the latter type that have been adequately studied are normally transmitted mechanically by arthropods, as has been clearly demonstrated in viruses of the myxoma subgroup (Fenner and Ratcliffe 1965), the birdpox subgroup (Kligler and Ashner 1929) and swinepox (Shope 1940). Probably other pox-viruses that produce protuberant skin lesions, e.g. the Yaba monkey tumour poxvirus (Niven et al. 1961), are also spread by arthropods. Viraemia occurs in many of the diseases associated with papulo-pustular skin lesions (e.g. variola and ectromelia), and in some associated with multiple skin tumours (e.g. myxoma), but viraemia is irrelevant as far as arthropod transmission is concerned (see ch. 20).

The large and complex virions of viruses of the myxoma and birdpox sub-groups are morphologically indistinguishable from those of vaccinia virus, which has been studied in considerable detail (Joklik 1968). Vaccinia virus particles measure about 300 nm × 230 nm × 100 nm, and contain at least 17 different virus-coded proteins distinguishable in poly-acrylamide gels (Holowezak and Joklik 1967), and small amounts of four virus-specified enzymes, a DNA-dependent RNA polymerase, a nucleotide phosphohydro-lase, and two DNases (Nagayama et al. 1970). The complexity of the struc-ture of these particles is illustrated in fig. 4.1. There is a large 'core' which contains at least three major proteins, the four enzymes and the DNA. The internal component of the core is a broad cylindrical element apparently in a tight S-shaped configuration (Peters and Müller 1963). Applied to either side of this core are two 'lateral bodies', and the core plus lateral bodies is sur-rounded by an outer membrane which contains several different proteins and also a lipid which, unlike the cellular lipids found in enveloped viruses, ap-pears to be virus-specific (Dales and Mosbach 1968). It is covered by an irregular arrangement of hollow cylindrical 'threads' about 9 nm in diameter, or, in the orf subgroup, by an apparently continuous thread.

The genome consists of a single linear molecule of double-stranded DNA, with a molecular weight of 160×10^6 (vaccinia) or over 200×10^6 daltons (fowlpox), and a guanosine-cytosine content of 35–36%. The first class of

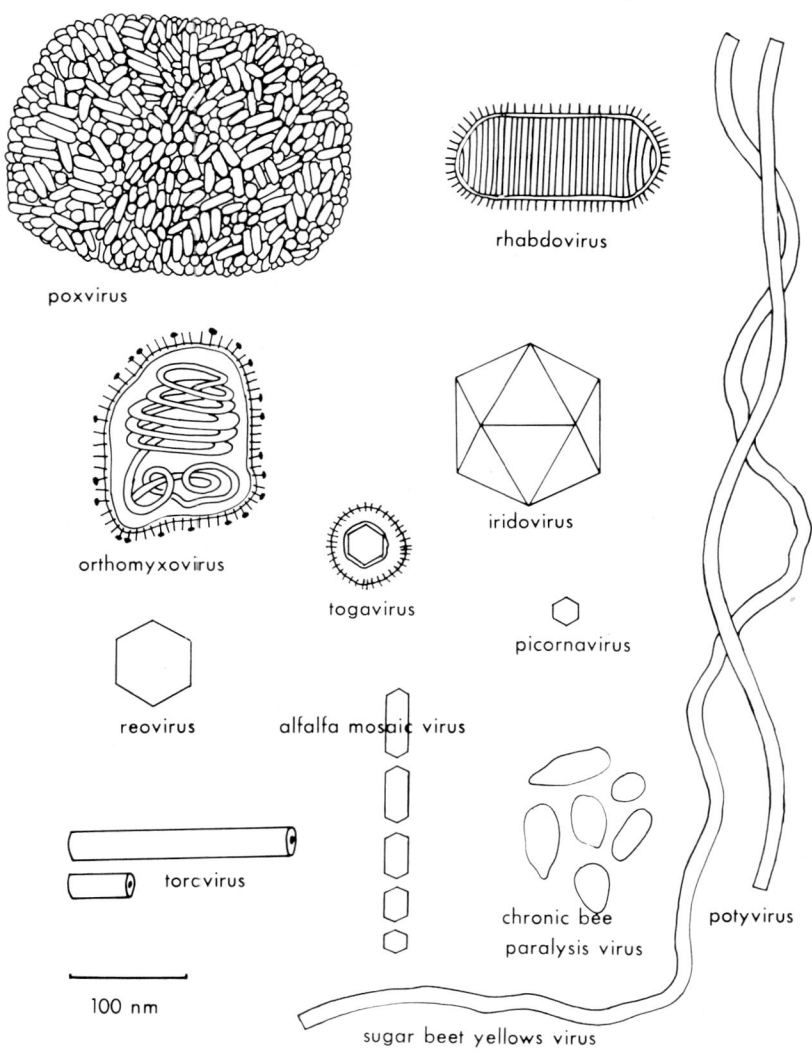

poxvirus

rhabdovirus

orthomyxovirus

iridovirus

togavirus

picornavirus

reovirus

alfalfa mosaic virus

torcvirus

chronic bee
paralysis virus

potyvirus

100 nm

sugar beet yellows virus

Fig. 4.1. A diagram illustrating the different shapes and sizes of the particles of viruses
associated with invertebrates. In this and all other figures in this chapter the bar marker
is 100 nm long.

viral messenger RNA molecules produced in infected cells is transcribed from the viral DNA by an RNA polymerase which is an internal component of the virion. The isolated nucleic acid is not infectious.

Poxviruses multiply in the cytoplasm, and mature virions may occur in large numbers scattered through the cytoplasm, or they may accumulate in large 'inclusion bodies' which are especially prominent in birdpox infections (Eaves and Flewett 1955).

A property of most poxviruses that is important for their transmission by arthropods is their resistance to inactivation.

4.4.1.2 *Entomopoxviruses* [D/*:*/*:X/X:I/O]

The first virus of this type was found in *Melolontha* (Vago 1963). Many others have been found since, but although they are usually grouped together, few studies comparing their nucleic acids, fine structure and serological properties have been reported. They have also been called 'Vagoia-viruses' (Weiser 1965) and 'Spheroidosis viruses' (Goodwin and Filshie 1969).

Their mature particles are oval and about 250×400 nm (Vago 1963; Granados and Roberts 1970; Bergoin et al. 1971), somewhat larger than most poxviruses. Their core resembles that of a poxvirus, although it is concave on one side only compared with the biconcave poxvirus nucleoid (fig. 4.2); it is bounded by two (Grandados and Roberts 1970) or three (Goodwin and Filshie 1969) membranes, and contains a folded cylindrical element (fig. 4.2) like that found in vaccinia by Peters and Müller (1963). The outer part of the particle is covered by a beaded coat of large spherical subunits about 40 nm in diameter, unlike those of vaccinia virus particles.

Entomopoxvirus particles are assembled in DNA-containing areas of the cytoplasm, like poxviruses, and are then occluded in oval protein crystals. These are similar to the type A inclusions of poxviruses, but this similarity may be fortuitous as inclusions are not peculiar to poxviruses and entomopox viruses alone, and other unrelated viruses of insects (e.g. the baculoviruses and cytoplasmic polyhedrosis viruses) also induce the production of polyhedra; polyhedra seem to have been evolved by several unrelated groups of viruses to function both as 'long term survival kits' and also as 'infection timing mechanisms' (see ch. 15).

Entomopoxvirus and poxvirus particles both contain at least four enzymes; a DNA-dependent RNA polymerase, a nucleotide phosphohydrolase, and two DNases (Pogo et al. 1971). These may, however, be common to all DNA-containing viruses that replicate in the cytoplasm. The resemblance of these two groups of viruses is remarkable and further work on their relatedness would clearly be worthwhile, particularly tests for the poxvirus group antigen in entomopoxviruses.

4.4.1.3 *Iridoviruses* [D/2:130/5:S/S:I/*] *and other large, isometric, cytoplasmic viruses*

The iridoviruses have been reviewed by Bellett (1968). Their particles were shown by Williams and Smith (1958) and others to be icosahedral, with apparent diameters of 130 nm to 180 nm depending on the method of preparation. They have sedimentation coefficients of about 2200 S, diffusion coefficients of 1.6×10^{-8} cm^2/sec, intrinsic viscosities of 0.103 dl/g and partial specific volumes of 0.74 ml/g at 20°C indicating particle weights of 1.1–1.3×10^9 dalton.

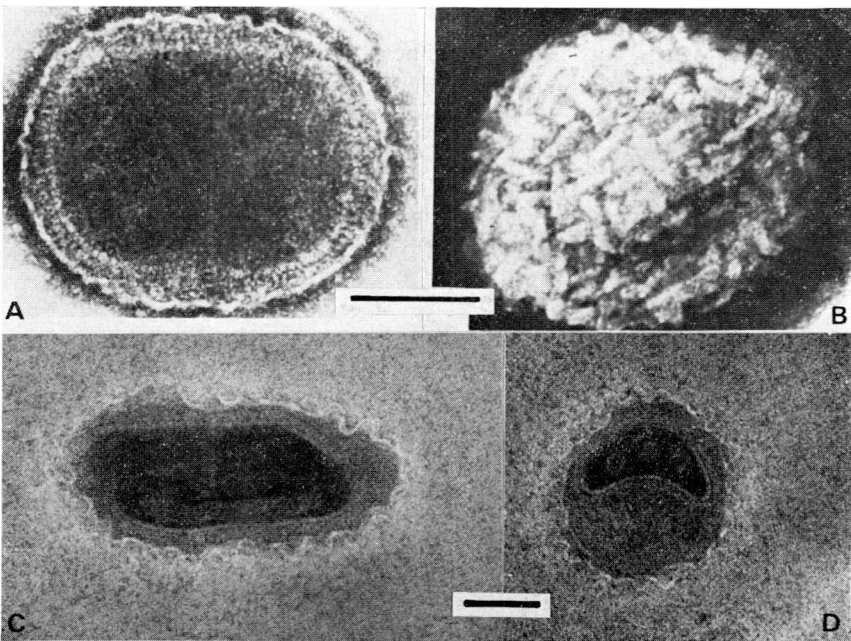

Fig. 4.2.   Poxvirus and entomopoxvirus. Particles of (A) vaccinia virus and (B) fowlpox virus negatively stained to show the structure of the outer membrane, and (C) a longitudinal and (D) transverse section of particles in an inclusion body of the entomopoxvirus of *Othnonius batesi* showing the verrucose surface and coiled inner component of the particles (Goodwin and Filshie 1969). (Courtesy of P. J. Chapple, K. B. Easterbrook and B. K. Filshie.)

The particles have an electron dense, irregular DNA-containing core surrounded by one or more icosahedral shells. The outer shells of *Sericesthis* iridescent virus (SIV) and *Tipula* iridescent virus (TIV) were studied in detail by Wrigley (1969, 1970). He showed that they consisted of morphological subunits close packed about 7 nm apart in a hexagonal array (fig. 4.3). The particles of SIV had an icosahedral edge length of 86.0 ± 2.7 nm, those of TIV 82.4 ± 2.7 nm. By ingenious use of the Goldberg diagram, which categorizes the ways spheres can be packed onto the surface of an icosahedron, Wrigley deduced the most likely structure of the particles, and suggested that if the particles of SIV and TIV are identical they most likely have an outer shell of 1472 subunits, though shells with 1562 or 1292 subunits cannot be excluded.

The structure of the 90 nm diameter core of iridovirus particles is not known, though it consists of DNA and protein (Thomas and Williams 1961; Frist et al. 1965; Glitz et al. 1968).

When iridovirus particles are sedimented they form crystalline pellets, which give Bragg reflection of visible light and appear a beautiful blue–violet or green colour. Sections of these crystals (Williams and Smith 1957) and optical measurements (Klug et al. 1959) show that the particles in these crystals are packed in a face-centered cubic array with an interparticle spacing of 250 nm, which is nearly twice the diameter of the dry particles, and 70 nm greater than their estimated hydrated diameter. Klug et al. (1959) ascribed

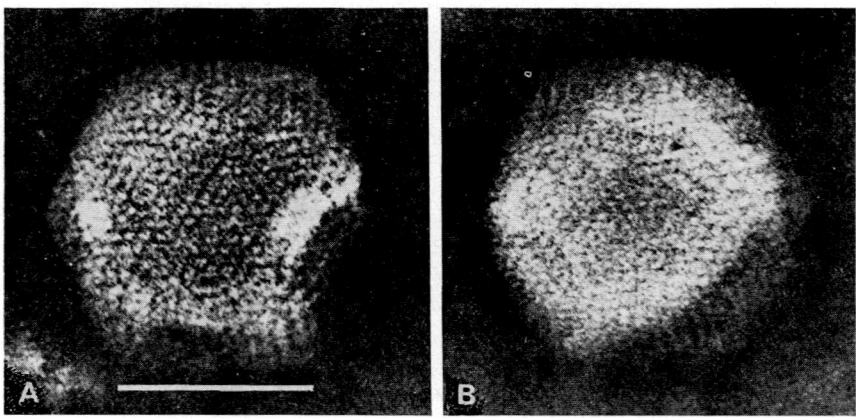

Fig. 4.3.   Iridovirus. Particles of *Tipula* iridescent virus treated with 'Afrin' (A) and negatively stained (B) to show the surface subunit structure. (From Wrigley 1970b.)

this to long-range forces between particles, while Williams and Smith (1957) and Mercer and Day (1965) postulated that the virus particles are surrounded by a tenuous layer which is not visible in electron micrographs.

Iridovirus particles contain 11–16% DNA corresponding to a DNA content for each particle of $135–155 \times 10^6$ dalton; the DNA is probably a single continuous double-stranded base-paired molecule (Bellett and Inman 1967). The base composition of DNAs from CIV, SIV and TIV differ slightly and are close to 31% G + C (Bellett 1968). Bellett and Fenner (1968) using nucleic acid hybridization experiments showed that DNAs from CIV and TIV were 4–17% homologous with SIV, but could not show homology between CIV and TIV DNA; this agrees with other experiments which showed the DNA of SIV to be intermediate in density between those of CIV and TIV.

Iridovirus particles contain little or no neutral or amino polysaccharides, or lipids (Thomas 1961; Glitz et al. 1968), and their infectivity is not affected by ether. They contain about 85% protein, which in polyacrylamide gels gives more than a dozen distinct bands. The unfractionated proteins of TIV and SIV are of similar composition (Kawase and Hukuhara 1967; Glitz et al. 1968; Kalmakoff and Tremaine 1968). SIV is serologically related but not identical to TIV (Cunningham and Tinsley 1968), but CIV and TIV seem unrelated (Fukaya and Nasu 1966); Gibbs, quoted by Bellett (1968), found a serological relationship between CIV, SIV and TIV, but this was later found to be the result of mixed virus stocks, and experiments with uncontaminated stocks confirmed the results of others.

Other iridoviruses have been found (see ch. 21) and some, such as that found in *Wiseana* in New Zealand, are serologically related to TIV (Kalmakoff and Robertson 1970). In addition, viruses with particles of similar shape and size have been found in frogs (Granoff 1969) and fish, but these are ether-sensitive, have DNA with a different base composition and in nucleic acid hybridization experiments no homology has been detected between the DNA of the iridoviruses and one of the frog viruses (Bellett and Fenner 1968).

Similar viruses have also been found in mosquitoes but the relationship of some of these to other iridoviruses is not clear. Some of them when centrifuged form pellets that iridesce blue, and thus probably have particles similar in size to those of TIV (e.g. that from *Aëdes stimulans*; Anderson 1970), but one of these showed no serological relatedness to the iridoviruses (Cunningham and Tinsley 1968).

Other viruses from mosquitoes have larger particles and form pellets that do not iridesce blue. One like this has been studied by Matta (1970) and found to have particles 180–200 nm in diameter, $S_{20,w}$ 4458, 16% DNA, and a total nucleic acid content of about 400×10^6 dalton per particle. Viruses of this type may be widespread for virus-like particles of this size have been found in other organisms including reptiles (Stehbens and Johnston 1966), fungi (Schnepf et al. 1970) and the green alga *Oedogonium* (J. D. Pickett-Heaps personal communication). No relationship of these larger viruses to iridoviruses has been shown.

African swine fever virus is an arbovirus; the natural disease in wart-hogs is transmitted by argasid ticks, in which the virus multiplies and is transmitted transovarially (Plowright et al. 1969, 1970). The virus multiplies in the cytoplasm, and produces large numbers of hexagonal virions about 200 nm in diameter, with an internal core of 80 nm diameter (Breese and De Boer 1966). The genome is double-stranded DNA and it is claimed that infectious DNA can be extracted from the virions (Adldinger et al. 1966); this unexpected result needs confirming.

4.4.1.4 Baculoviruses [D/2:(80–100/3–15):V/(E):I/O] *(The nuclear polyhedrosis and granulosis viruses)*

Particles of these viruses are straight or slightly curved rods $40–70 \times 200–370$ nm, enclosed in a protein sheath. There has been disagreement on the structure of these particles and the membrane (Smith and Hills 1962; Harrap and Juniper 1966). However Beaton and Filshie (1971), using optical diffraction methods to study electron micrographs, found that the isolated particles of several baculoviruses have a stacked disc structure with 12–13 morphological subunits per disc (fig. 4.4). The distance between discs was about 4 nm and the structure repeated every third row (12 nm) so that it can also be considered as a multi-start helix with a pitch of 55 °C like that reported by Smith and Hills (1962). The structure of the core of the particle is still in doubt. It presumably contains the viral DNA either in a filament as proposed by Ponsen (1965), or twisted circular nucleocapsid suggested by Kozlov and Alexeenko (1967).

The virus rods are enclosed singly, or in bundles, within an outer or 'developmental' membrane, the number of rods in a bundle being characteristic of the strain of virus. The bundles of rods are occluded in a polyhedral matrix, the shape and size of which is also characteristic of the strain, and which can be removed by treatment with alkali (Smith 1967). Granuloses normally have a single rod within a polyhedral capsule, although Arnott and Smith (1968) have observed up to nine rods in 'abnormal' capsules. Bergold (1958, 1963a, b) showed the polyhedral matrix to consist of protein subunits

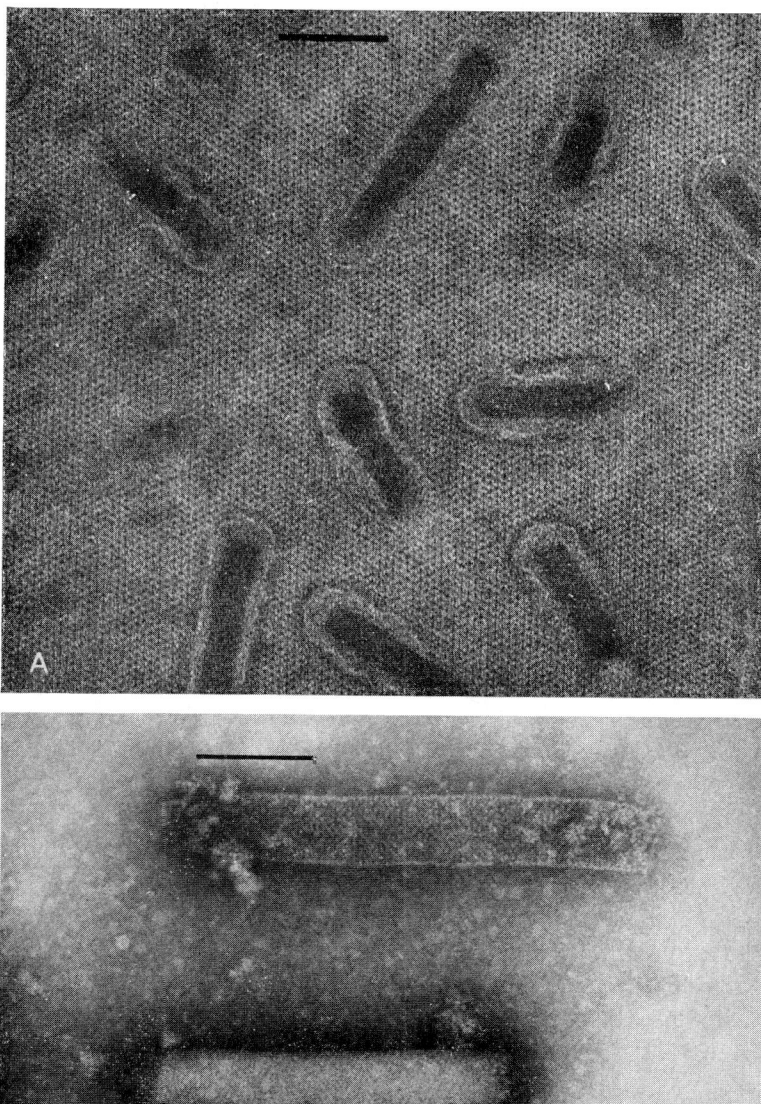

Fig. 4.4. Baculovirus. Particles of *Bombyx mori* nuclear polyhedrosis virus. (A) Section of a polyhedron showing the particles randomly scattered through the crystalline matrix. (B) Particles freed from the polyhedron by alkali and negatively stained; the upper particle seems intact, below it is an envelope from a disrupted particle. (Courtesy of B. K. Filshie and C. Beaton.)

with molecular weights of 300,000–380,000 and diameters of 6.5–9.0 nm arranged in a face-centered cubic lattice. The virus rods neither disturb the crystalline lattice nor act as crystallisation centres.

Baculovirus particles have been shown by many (e.g. Wyatt 1952) to contain DNA, but no RNA. Allison and Burke (1962) calculated the amount of DNA in each particle to be from 56×10^6–76×10^6 dalton. Onodera et al. (1965) and Christyakova (1967) reported finding linear molecules of 2×10^6–5×10^6 dalton in extracted DNA. However, Summers and Anderson (1972) have isolated covalently closed circular duplex DNA with a molecular weight of 100×10^6 dalton from a granulosis virus, and this is consistent with the DNA content of individual particles. Wyatt (1952) found that the DNA contained the usual bases, the proportions of which indicated a double-stranded base-paired structure. This has been confirmed by thermal denaturation and buoyant density determinations of extracted DNA (Onodera et al. 1965; Christyakova 1967). The molar percentage of guanine + cytosine in the DNA varied according to the strain of virus from 35–59%.

Faulkner (1962) and others found RNA associated with the polyhedral protein matrix. Aizawa and Iida (1963) found variable amounts of RNA in the polyhedra of four of eight virus strains tested, and none in the remainder. Whatever the nature and function of this RNA it is apparently not essential, since the polyhedra of some strains do not contain it. In 1963 Gershenson et al. claimed that RNA extracted from silkworms infected with a baculovirus was infective; its infectivity was destroyed by RNase. This unusual observation has not yet been confirmed.

Wellington (1954) reported the compositions of the proteins of a number of the baculoviruses and their polyhedra, and they have also been compared serologically by Tanada (1954) and by Krywienczyk and Bergold (1960a, b, 1961). Bellett (1969) showed that numerical analysis of these results and data on the base composition of the DNA of some of the viruses did not support the traditional separation of the granuloses from the polyhedroses, but indicated that three subgroups were present: 1) the polyhedrosis viruses of *Diprion* and *Neodiprion*; 2) the granuloses of *Choristoneura*, and 3) fifteen polyhedroses of *Lepidoptera* and the granuloses of *Recurvaria* and *Laphygma*, which are more closely related to the polyhedroses in this group than to the granuloses of *Choristoneura*. Cunningham (1968) also found distant serological relationships between a granulosis virus and a number of polyhedrosis viruses. His results suggest at least two further subgroups of the nuclear-polyhedrosis-granulosis group: 4) the nuclear polyhedrosis viruses of four nymphalid species; and 5) the granulosis viruses of *Pieris*, originally included in subgroup 3 by Bellett (1969) on the basis of agglutination tests by Tanada (1954).

Nuclear-polyhedrosis and granulosis viruses are assembled in the nuclei of infected cells, although viral antigens were found in the cytoplasm at early stages of infection by Krywienczyk (1963) and Krywienczyk and Sohl (1967).

4.4.1.5 Papillomaviruses [D/2:5/12:S/S:V/O, Di, Ac, Si]

One of the papilloma viruses, Shope's rabbit papilloma is ordinarily trans-mitted mechanically from one Sylvilagus rabbit to another by mosquitoes and bugs (Dalmat 1958). Although not yet demonstrated experimentally, it is possible that at least some other papillomas that occur on the skin rather than the mucous membranes (Bovine papilloma (Andrewes and Pereira 1967) and deer fibroma (Shope et al. 1958)), may be transmitted mechanically by arthropods as well as by direct contact.

Papilloma viruses cause benign tumors which may become malignant in some species. Human warts never become malignant; 25% of papillomas caused by rabbit papilloma virus eventually undergo malignant changes in cottontail *(Sylvilagus)* rabbits, as do about 75% of papillomas which persist in domestic rabbits for longer than six months.

The morphology of rabbit papilloma virus particles was studied by Finch and Klug (1965). They are icosahedra about 50 nm in diameter, and with 72 capsomers arranged with left-handed skew symmetry (fig. 4.5a). They sedi-ment at 280–300 S, have a density in CsCl of 1.34, and consist of protein and DNA, the latter a single circular double-stranded supercoiled molecule of about 5×10^6 dalton molecular weight, with a guanosine-cytosine content of 49%.

The viral content of human warts, and probably those of other animals, varies with their age, particles being most numerous in young lesions. Within the papilloma itself infective virus particles are most numerous in the kera-tinized layers. The virus multiplies in the nucleus and virions may accumulate there in very large numbers. Papillomas in domestic rabbits rarely contain infective virions, but they do contain infectious viral nucleic acid.

Viral DNA extracted from papillomas or papilloma virions is infectious; because of its supercoiled configuration it renatures in fully infectious form after heat denaturation. Both virions and DNA are resistant to inactivation at ambient temperatures, and are therefore able to survive for prolonged periods on arthropod mouthparts.

4.4.1.6 *Caulimoviruses* [D/2:5/15:S/S:S/Ap]

Viruses of this group, which include cauliflower and dahlia mosaic viruses
(Brunt 1971), are transmitted by aphids in the non-persistent manner (ch. 25).
They are the only plant viruses known to have a DNA genome (Shepherd et
al. 1968; Shepherd 1970), and their particles closely resemble those of the
papillomaviruses. They contain 16% by weight of circular double-stranded
DNA (Russell et al. 1971) with a molecular weight about 5×10^6 dalton, a
G+C content of 41–43%, and a nearest neighbour base sequence frequency
close to that of cauliflower DNA (presumably nuclear). The circular DNA is
not supercoiled and appears to have interruptions in both strands, but as it is
rapidly converted to a linear form during storage it is possible that these
properties are due to a nuclease associated with the DNA. The virus particles
are isometric, about 50 nm in diameter (fig. 4.5B), have a buoyant density in
CsCl of 1.37, and a sedimentation coefficient of 220 S. The structure of the
protein shell is not known.

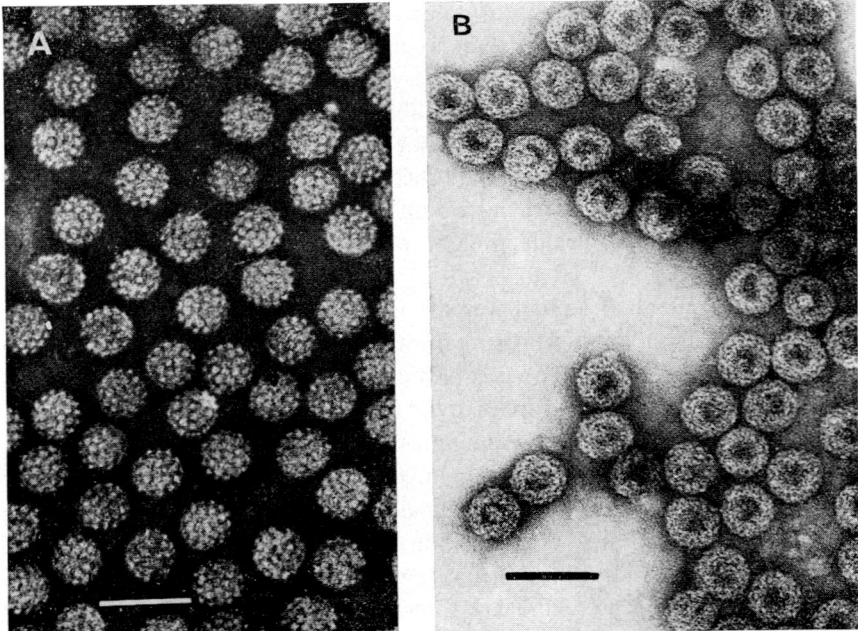

Fig. 4.5. Papillomavirus and caulimovirus. Negatively stained particles of (A) rabbit
papilloma virus. (Courtesy of E. A. C. Follett.) And (B) cauliflower mosaic virus. (Courtesy of B. D. Harrison and I. M. Roberts.)

The caulimoviruses have narrow host ranges, cause mottle and mosaic symptoms in most hosts, and cells of infected plants contain compact rounded inclusions, which consist of particles randomly scattered in a granular matrix.

4.4.2 Viruses whose particles contain single-stranded DNA

The only reported virus of this type associated with invertebrates is *Galleria* densonucleosis virus [D/1:2.2/38:S/S:I/*]. This infectious virus causes a rapidly lethal disease of *Galleria* larvae characterized by swelling and deformation of the nuclei of infected cells. Kurstak et al. (1969) claimed that the virus will transform mouse L cells.

Particles of the virus are isometric and about 20 nm in diameter. Some contain no nucleic acid and sediment at 58 S, the others sediment at 119 S and contain 37% DNA (about 2.2×10^6 dalton per particle; G + C content about 38%); the two particle types are serologically indistinguishable. Kurstak and Côté (1969) claim that the shell of the particle has 42 capsomers. Spectrophotometric studies of the DNA of the virus show that in the particles it is mostly single-stranded but when liberated from the particles it becomes mostly double-stranded to form molecules of molecular weight about 4×10^6 dalton (Barwise and Walker 1970). These results suggest that some of the particles of the virus contain DNA with one particular sequence, and the others contain DNA with the complementary sequence. In this and some other respects this virus resembles the adeno-associated (satellite) viruses [D/1:1.5–1.8/19:S/S:V/*] (Crawford et al. 1969).

4.4.3 Other DNA-containing viruses

Other viruses have been reported whose particles are known, or assumed, to contain DNA. These include a group of viruses with isometric particles about 40 nm in diameter found in the cytoplasm of mid-gut cells of the larvae of various Saturniid moths (Longworth and Tinsley 1968), and a virus found by Stoltz et al. (1968) in Feulgen positive cytoplasmic inclusions in *Chironomus plumosus*. The particles of the latter virus were 160 nm to 180 nm across, similar to those of iridoviruses but surrounded by fine fibrils. Triangular fragments formed by pronase treatment of this virus apparently have 78 subunits (12 per edge), each with a long fibril attached (Stoltz, personal communication; fig. 4.6). It is not clear how closely this virus is related to iridoviruses.

4.5 Viruses whose particles contain ribose nucleic acid (RNA)

4.5.1 Viruses whose particles contain double-stranded RNA

The discovery that some viruses contain double-stranded RNA as their genetic material dates back only a few years, for it was not until 1963 that the reoviruses (Respiratory Enteric Orphan viruses), which infect many species of vertebrates, were shown to contain double-stranded RNA (Langridge and

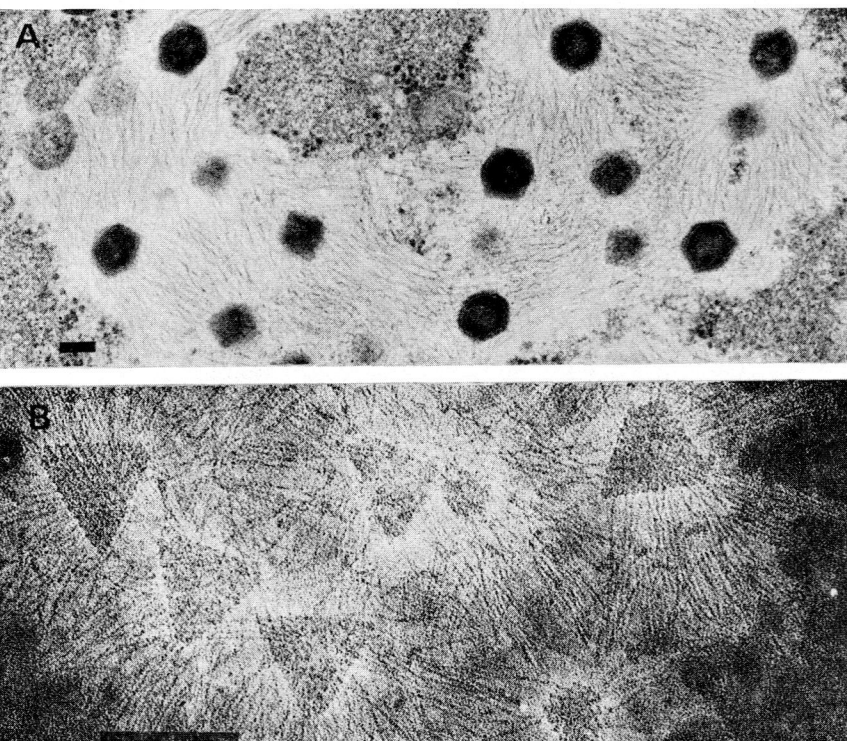

Fig. 4.6.   The icosahedral cytoplasmic deoxyvirus of *Chironomus plumosus*. (A) Section through infected fat body cell. Each virus particle is surrounded by fibrils. (B) Triangular rafts of subunits ('trisymmetrons') released from particles. The trisymmetrons are larger than those of iridovirus particles and along each edge are 12 subunits (rather than 10) to each of which is attached a fibril. (Courtesy of D. B. Stoltz.)

Gomatos 1963). Since then viruses which infect primarily plants (clover wound tumour virus) and the cytoplasmic polyhedroses of insects (see below) have been shown to contain double-stranded RNA. In addition it has been found that some arboviruses, such as blue tongue virus, contain double-stranded RNA; such viruses have been called the orbiviruses.

4.5.1.1 Reoviruses [R/2:15/20:S/S:V/O,Di]

Reovirus type 3 has been recovered in cultured cells inoculated with suspensions prepared from pools of mosquitoes (Miles and Stenhouse 1969) and sentinel suckling mice have been infected in the field (Parker et al. 1965). Mechanical transmission has been demonstrated but no evidence of multiplication was obtained in several species of mosquito (Miles and Stenhouse 1969); the importance of mechanical transmission in the ecology of this very resistant virus is not known. The reoviruses will be described because they are to this extent associated with invertebrates, and because they are the best

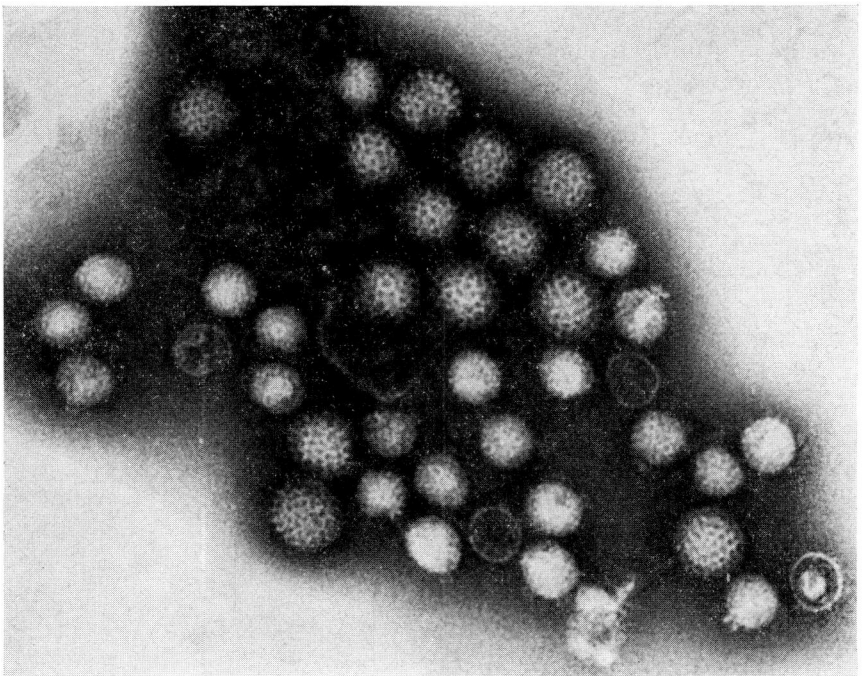

Fig. 4.7. Negatively stained preparation of reovirus showing the double shell of capsomers (compare with fig. 4.8). (Courtesy of D. W. Verwoerd.)

characterized of the isometric viruses containing double-stranded RNA.

Reovirus particles are non-enveloped icosahedra 75–80 nm in diameter. Their capsomers appear to be arranged as trimers or dimers on the triangular facets of the icosahedral surface lattice so that there appear to be symmetrically disposed 'holes' in the capsid when particles are negatively stained (fig. 4.7). The shell of the reovirus particle consists of a double layer of capsomers (Dales et al. 1965). The inner shell has an outer diameter of about 45 nm, and its structure is not known.

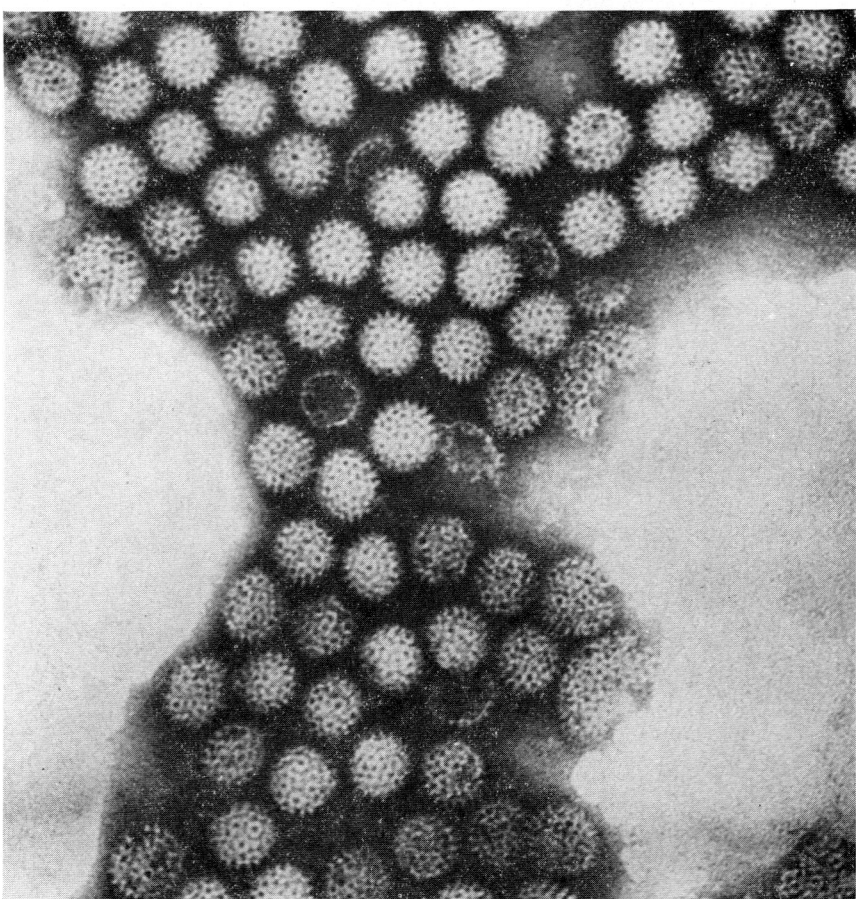

Fig. 4.8. Orbivirus. Negatively stained preparation of sheep blue tongue virus showing the single shell of capsomers (compare with fig. 4.7). (Courtesy of D. W. Verwoerd.)

Reovirus particles consist of RNA and protein. The RNA is double-stranded, and occurs as ten separate pieces which fall into three size classes (Shatkin 1969). In addition, reovirus particles may contain a large amount of adenine-rich single-stranded RNA, probably as a specific 'contaminant' derived from reovirus-infected cells (Krug and Gomatos 1969). The particles contain a virus-coded RNA-dependent RNA polymerase which synthesises viral specific messenger RNA, both in the infected cell and under special conditions in vitro (Skehel and Joklik 1969).

The proteins in the particles have been examined by Smith et al. (1969) who found that the virus particle contains at least seven polypeptides, the sizes of which correspond to certain of the separate pieces of viral RNA. Reoviruses multiply in the cytoplasm and accumulate to form large intracytoplasmic inclusion bodies which may contain crystalline arrays of virus particles.

4.5.1.2 Orbiviruses [R/2:Σ15/*:S/S:V,I/Di]

Recently a large number of arboviruses that infect a variety of vertebrates and multiply in a wide variety of arthropods, have been distinguished from the togaviruses (sect. 4.5.2.5) by their resistance to lipid solvents and by serological tests. Several have been shown to have icosahedral particles, and some of these have been shown to contain double-stranded RNA. There are perhaps 50 arboviruses in this group (Murphy et al. 1971). The best studied representative is blue tongue virus.

Bluetongue virions (fig. 4.8) are non-enveloped icosahedra about 54 nm in diameter, with a single protein shell of 32 capsomers (Els and Verwoerd 1969) that encloses a genome of double-stranded RNA, which occurs in ten pieces (Verwoerd et al. 1970).

Bluetongue virus multiplies and matures in the cytoplasm of both vertebrate and insect cells (Bowne and Jones 1966; Murphy et al. 1971). Diseases caused by these viruses, e.g. bluetongue in sheep, Colorado tick fever, and African horse sickness, are usually severe generalized diseases with viraemia, which is important as a source of virus for arthropods.

The orbivirus group includes several serological subgroups, and they have been isolated from several different types of arthropods (table 4.2).

4.5.1.3 Viruses similar to clover wound tumour virus [R/2:Σ16/22:S/S:S,I/Au]

Clover wound tumour is the best studied (Black 1970) of these viruses, which infect both plants and leafhoppers (see chs. 8 and 26). It has been isolated only once, from a leafhopper collected near Washington, USA. Other viruses of this type are more widespread and include maize rough dwarf (Lovisolo

TABLE 4.2

Serological subgroups of the bluetongue-like viruses and the suspected arthropod vectors

Serological subgroups	Vector or arthropod source of isolates
Bluetongue; 16 serotypes	*Culicoides*
African horse sickness; 9 serotypes	*Culicoides*
Kemerovo and 6 serologically related viruses	*Ixodes, Argas* and other ticks
Changuinola and 4 serologically related viruses	*Phlebotomus*
Colorado tick fever	*Dermacentor*
Epizootic hemorrhagic disease of deer	*Culicoides* suspected
Palyam and serologically related viruses	Mosquitoes and *Culicoides*
Eubenangee and one serologically related virus	Mosquitoes
Corriparta and one serologically related virus	*Culex* and other mosquitoes
Lebombo	*Aëdes*

1971), rice dwarf, rice black-streaked dwarf, and sugar cane Fiji disease viruses, and similar viruses perhaps cause oat sterile dwarf, clover rugose leaf curl and maize wallaby ear diseases.

Plants are systematically infected and characteristically they often develop tumours of the vascular bundles; such tumours are rich in virions. Infected leafhoppers show no signs of disease, though all tissues are infected, and the virus may be transmitted to progeny. Virus particles and antigens are found only in the cytoplasm of infected cells. Virions are about 70 nm in diameter and are all of one type with a sedimentation coefficient of 510 S. They contain 12–15 molecules of double-stranded RNA with a total molecule weight of 16×10^6 dalton (G + C content 38 %) which comprises 22 % of the particle weight (Kalmakoff et al. 1969). Wound tumour virus and the reoviruses have closely similar particles. It has been claimed that they are serologically related, but sensitive and properly conducted serological tests have failed to confirm this (Gamez et al. 1967).

4.5.1.4 Cytoplasmic polyhedrosis viruses [R/2:Σ12/27:S/S:I/O]

These viruses are common in Lepidoptera, and have also been found in Neuroptera. They cause diseases of larvae, inducing the production of polyhedra like those of the baculoviruses (see above; sect. 4.4.1.4), but differ in that the polyhedra are formed in the cytoplasm of infected cells. The shape of the inclusions varies, and depends on the strain of the virus.

Hills and Smith (1959) and Hayasi and Bird (1968a) isolated and purified infectious virus particles from cytoplasmic polyhedra and showed that they had icosahedral symmetry. Arnott et al. (1968), Nishimura and Hosaka (1969) and others found the virus particles to have two concentric icosahedral shells, 35–45 nm and 67–70 nm in diameter, the outer shell having twelve large projections at the vertices (Miura et al. 1969; fig. 4.9). The particles sediment at 370 to 440 S (Hayasi and Bird 1968b; Miura et al. 1968), and have a particle weight of 5–9 × 10^7 dalton. Cunningham and Longworth (1968) described seven cytoplasmic polyhedrosis viruses of very similar structure but apparently of smaller diameter.

Although polyhedra may contain both single- and double-stranded RNA (Hayashi and Kawase 1964), that in the virions is double-stranded (G + C content 43 %). Miura et al. (1968) and Kalmakoff et al. (1969) found 16–23 % RNA corresponding to 12–16 × 10^6 dalton. The RNA shows two components in sedimentation velocity experiments and in length measurements in the electron microscope (Miura et al. 1968; Nishimura and Hosaka 1969; Kalmakoff et al. 1969). Kalmakoff et al. (1969) and Fujii-Kawata et al. (1970) further separated the RNA of silkworm cytoplasmic polyhedrosis virus into 9 or 10 species by electrophoresis. Four of these were found in the major, fast-sedimenting peak isolated from sucrose gradients and the others in the minor peak. The combined molecular weight of the RNA species was 13–15 × 10^6

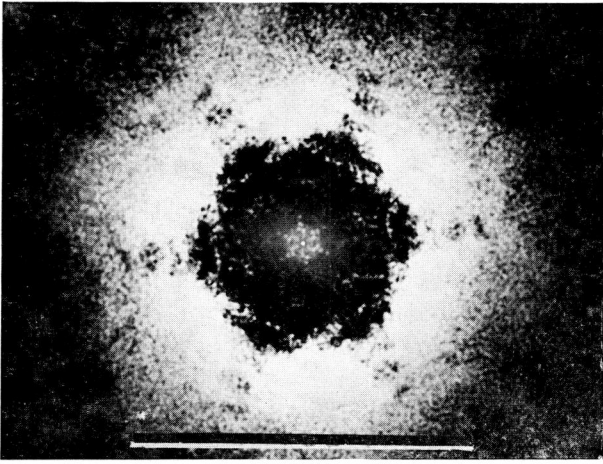

Fig. 4.9.   A particle of cytoplasmic polyhedrosis virus of *Bombyx mori*. The image of the particle has been superposed photographically a total of six times, rotating by 60° around the centre of the image between each exposure. (From Miura et al. 1969.)

dalton, close to the estimated RNA content of each virus particle. There was no adenine-rich RNA like that found in reoviruses, and the number and size of the RNA fragments (apart from the largest, of molecular weight 2.4 × 10⁶) also differed from those of both reovirus and wound tumour virus. It is not certain if the RNA of cytoplasmic polyhedrosis viruses is present in pieces in the virus particles or is broken at specific points during extraction, but Nishimura and Hosaka (1969) observed molecules 6–8 μm long in addition to the two smaller size classes found by others. Kawase and Miyajima (1968) reported that the viral RNA was infectious, however, this needs to be confirmed as Lewandowski et al. (1969) found an RNA polymerase in the virus particles, the product of which was single-stranded RNA which hybridized with the double-stranded viral RNA. It is likely that this enzyme is necessary for the expression of the double-stranded RNA genome, in which case isolated RNA is unlikely to be infectious.

Kawase (1964) reported analyses of the amino acid composition of the viral and polyhedral proteins of two strains of cytoplasmic polyhedrosis virus. The proteins of the two viruses resemble each other closely, but differ from equivalent proteins of the nuclear polyhedroses and granuloses (Wellington 1954; Bellett 1969). Hukuhara and Hashimoto (1966) found that two strains of silkworm cytoplasmic polyhedrosis cross-reacted strongly with each other in serological tests but gave no detectable cross-reactions with silkworm nuclear polyhedrosis. Cunningham and Longworth (1968) and Krywienczyk et al. (1969) have made more extensive serological comparisons of cytoplasmic polyhedrosis viruses. There are at least two serological subgroups.

4.5.2 *Viruses whose particles contain single-stranded RNA*

There are probably more viruses in this class than in any other, and until relatively recently it was believed that all plant viruses contained single-stranded RNA. The virions occur either as naked isometric or tubular particles, or they may be enclosed in a lipoprotein envelope which always contains cellular lipids. Some (the rhabdoviruses) have a more complex morphology.

In this section we will describe the viruses roughly in order of the size of their genome.

4.5.2.1 *Comoviruses* [R/1:2.6/35 + 1.7/28:S/S:S/Cl]
This group of perhaps a dozen plant viruses includes several that have been well studied, such as bean pod mottle, cowpea mosaic (Van Kammen 1971),

and squash mosaic (Campbell 1971). All are transmitted by Halticid or Gale-rucid fleabeetles, and recent work has shown that they may circulate in the flea beetle though not multiply (see ch. 9). All are readily transmitted by sap inoculation, and some are commonly seed transmitted. The virions of these viruses are angular, isometric (about 30 nm in diameter), and of three types (fig. 4.10), which contain about 33, 22 and 0% nucleic acid. Both nucleo-protein particles are needed to cause an infection (Van Kammen 1968), and nucleic acid hybridization experiments show that there is no homology in the sequences of the nucleic acids from the two types of particle; the virus gen-ome is divided between the two particle types.

4.5.2.2 *Orthomyxoviruses* [R/1:(2–4)/1:S/E:V,Ne/O,Ne]
The influenza viruses are now called the orthomyxoviruses. The only reason for including a description of this genus is the fact that one member of it, the type A orthomyxovirus, swine influenza virus, can apparently multiply in the swine lung worm (Shope 1941).

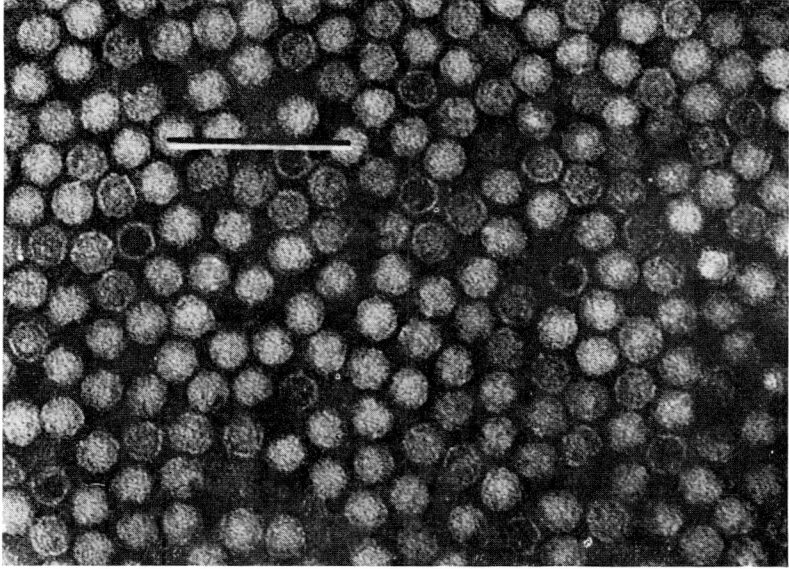

Fig. 4.10. Comovirus. Negatively stained preparation of Echtes Ackerbohnenmosaik-virus. There are three types of particles; nucleoprotein particles that completely or partially exclude stain and protein shells that appear as stain-filled rings; the virus genome is divided between the two types of nucleoprotein particle. (Courtesy of R. D. Woods.)

The virion consists of a tubular nucleocapsid enclosed in a lipoprotein envelope (fig. 4.11). The enveloped particles are 90–120 nm in diameter, and are spherical or elongated ('filaments') with characteristic glycoprotein spikes. Some spikes are viral haemagglutinin, others viral neuraminidase. The viral RNA occurs in several (probably 8) separate pieces, which may be associated with separate pieces of nucleocapsid.

The nucleocapsids are assembled in the nucleus and move to a position beneath the plasma membrane, whence they are released by budding, as enveloped virions.

4.5.2.3 Alfalfa mosaic virus [R/1:1.3/18+1.1/18+0.9/18+0.3/18):U/U:S/ Ap] *and cucumber mosaic virus* [R/1:(1/18):S/S:S/Ap]

These two viruses share many properties (Bos and Jaspers 1971; Gibbs and

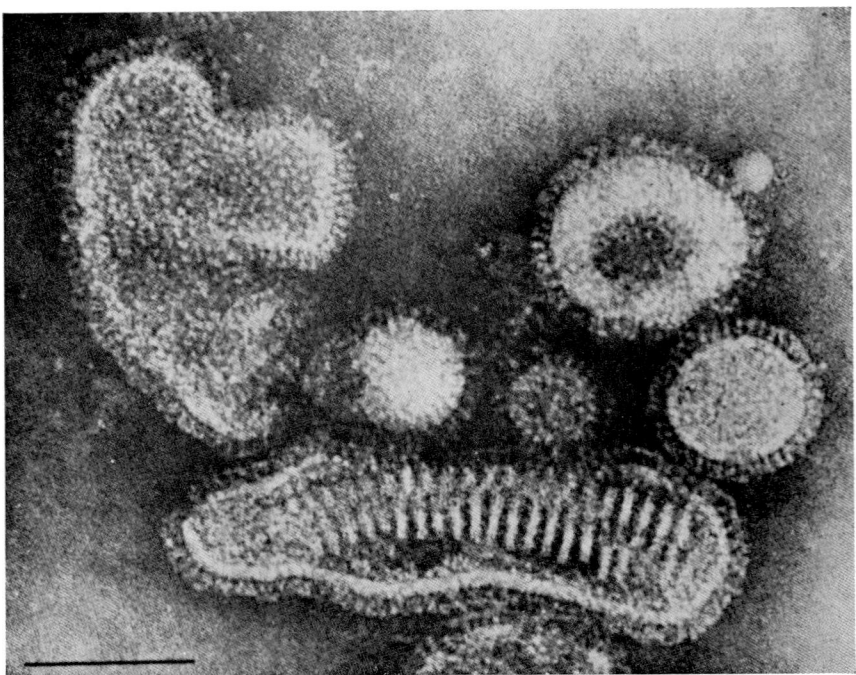

Fig. 4.11.   Orthomyxovirus. Negatively stained particles of influenza virus. The enveloped particles are pleomorphic with a coiled tubular internal component of ribonucleoprotein. Some of the spikes projecting from the envelope are haemagglutinin, others neuraminidase. (Courtesy of J. D. Almeida.)

Harrison 1970) and are perhaps quite closely related even though they have particles of different shape; a difference which may merely reflect small differences in the structure of their protein subunits.

These viruses are widespread and common in most temperate parts of the world causing mosaic and mottle symptoms in a particularly wide range of plants. They are both transmitted by aphids in the non-persistent manner (see ch. 25), are readily transmitted by sap and occasionally through the seed.

The virions of alfalfa mosaic virus are of four main types, one isometric and three bacilliform, 18 nm wide and 18–58 nm long (fig. 4.12). The four types of particles contain RNA molecules with molecular weights of 0.3×10^6, 0.9×10^6, 1.1×10^6 and 1.3×10^6. All or most of these particles are needed to cause infection (Bol et al. 1971). By contrast the virions of cucumber mosaic virus seem to be all of one type. However, there are at least four different species of RNA in these particles and these are of similar sizes to those of alfalfa mosaic virus. Each virus has only one type of protein subunit in its particles and the chemical subunit of each consists of about 290 amino acids.

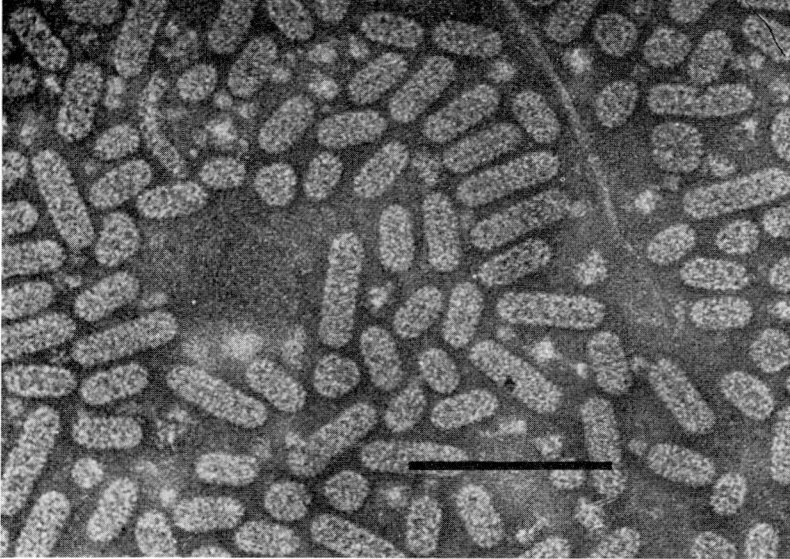

Fig. 4.12. Alfalfa mosaic virus particles, fixed with formaldehyde and negatively stained. The particles are bacilliform and of four different lengths but one diameter. The ends seem to consist of caps with the structure of a half icosahedron bisected at right angles to a three-fold axis, with a variable length of a two/three-fold symmetry net of the same subunits between them. The genome of the virus is divided between the different sizes of particle.

4.5.2.4 Rhabdoviruses [R/1:3–5/2:U/E:V,I,S/O,Di,Ac,Ap,Au]

A number of viruses with complex virions of closely similar morphology are in this group. Their bullet shaped or bacilliform particles are about 200–400

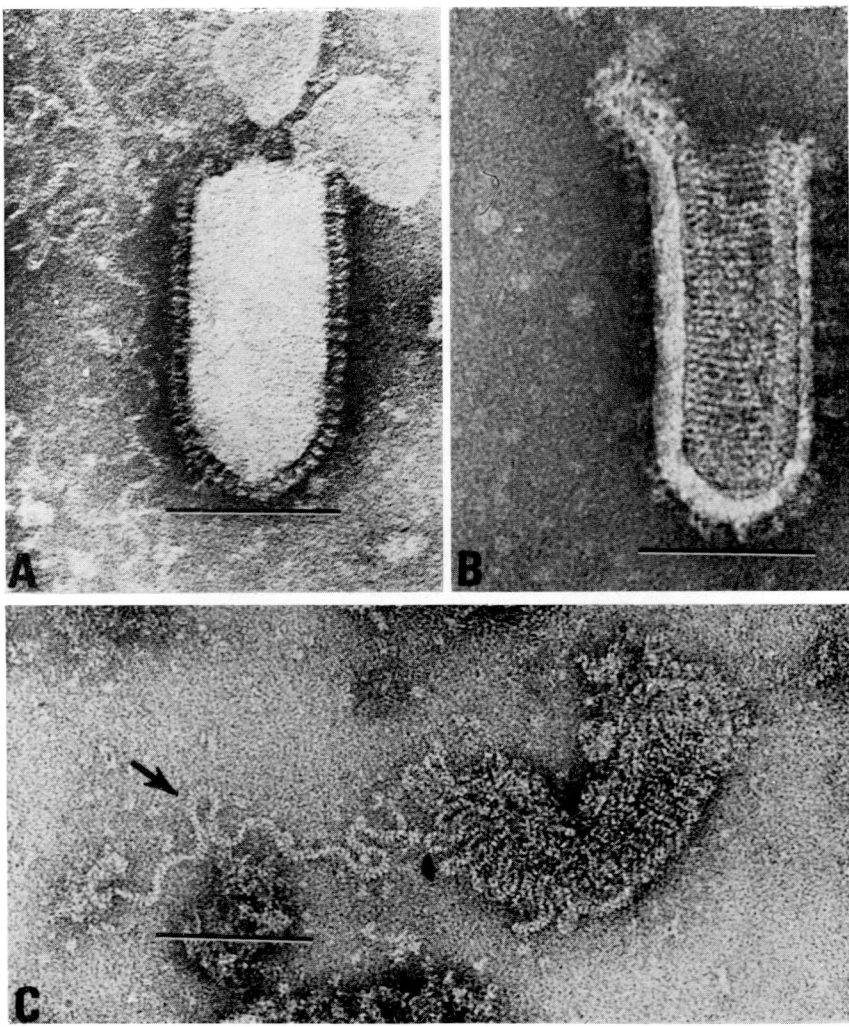

Fig. 4.13.   Rhabdovirus. Particles of vesicular stomatitis virus (A) and (B) penetrated differently by stain and showing the outer membrane and helically constructed core, (C) a liberated core partially unwound. (From Simpson and Hauser 1966.)

nm long and 70–100 nm wide, they have a lipoprotein envelope enclosing a wide short tubular nucleocapsid, which is helically constructed with a basic pitch of 4.5 nm. In the group are viruses of arthropods, of plants and of vertebrates; those found in plants or vertebrates will usually also multiply in invertebrates, and such invertebrates are often vectors of the virus. It is likely that this group will be divided when more is known about the viruses and their relatedness.

The best known animal rhabdovirus is vesicular stomatitis virus (fig. 4.13). Its virion is 175×70 nm and has a spike-covered envelope. There are three major protein species in the virions, and an RNA-dependent RNA polymerase. The virion contains 2% by weight of single-stranded RNA with a total molecular weight of 3.5×10^6 dalton, probably in several pieces. The virus multiplies in the cytoplasm, and leaves the cell by budding from the plasma membrane. Other animal rhabdoviruses include rabies, cocal and Hart Park viruses, also viruses of bats and one of trout.

One of the best known plant rhabdoviruses is potato yellow dwarf virus (Black 1970). It can be transmitted by its Agallian leafhopper vectors (in which it multiplies), or by sap inoculation. Its particles very closely resemble those of vesicular stomatitis virus, not only in appearance but also in the number and composition of their component proteins and nucleic acid (Knudson and Macleod 1972).

Other plant-infecting rhabdoviruses include lettuce necrotic yellows (Francki and Randles 1970) and sowthistle yellow vein virus, both of which seem to multiply in aphids and are transmitted by them in the persistent manner (ch. 24). The structure of their particles and of the related broccoli necrotic yellows virus has been carefully studied (Hills and Campbell 1968; Peters and Kitajima 1970), and all evidence suggests that their particles have a composition close to that of vesicular stomatitis virus; particles of lettuce necrotic yellows virus like those of the latter contain an RNA-dependent RNA polymerase (R. I. B. Francki, personal communication; fig. 4.14).

Another well studied rhabdovirus is the σ virus of *Drosophila*, once thought to be a plasmagene (see ch. 20). Infection of *Drosophila* with σ virus causes sensitivity to CO_2. The virus can be 'stabilized' in female flies and is then transmitted to their offspring. Particles found in the ovaries of stabilized females by Berkaloff et al. (1965) are bullet-shaped, with an internal helical or striated structure and an external membrane covered with radiating spikes. The particles are about 70 nm wide and 200 nm long. Similar particles have been seen in extracts of infected insects. The morphological similarity to vesicular stomatitis virus is obvious. Printz (1967) adapted vesicular stoma-

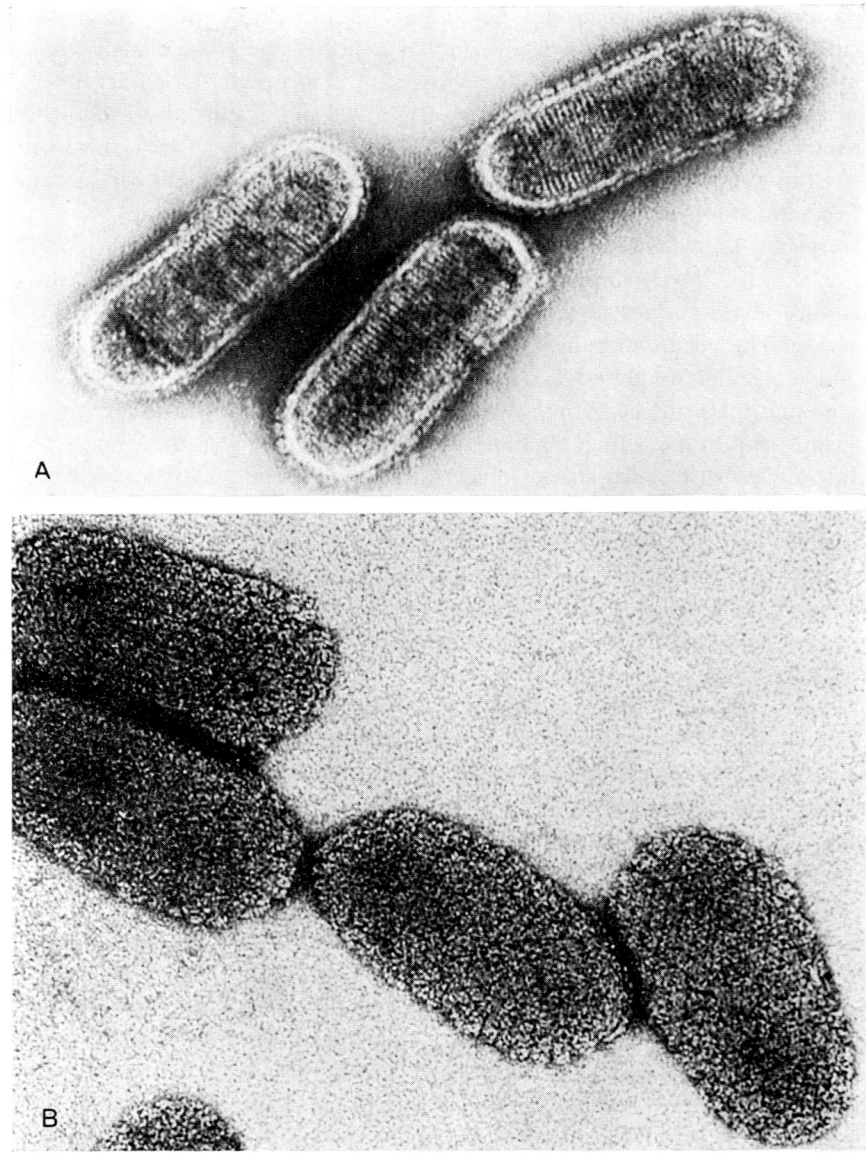

Fig. 4.14. Rhabdovirus. Particles of sowthistle yellow vein virus fixed with gluteral-dehyde and stained with either (A) phosphotungstate to show the helical core and surface projections or (B) uranyl acetate to show the subunits of the envelope. (Courtesy of D. Peters.)

titis virus to grow in *Drosophila* and showed that the adapted strain caused delayed sensitivity to CO_2. σ virus is thus possibly related to vesicular stomatitis virus. However, no serological relationship has been demonstrated and there are some clear differences between the viruses; the immunity against superinfection by σ virus, caused by defective σ in ultra-ρ females (ch. 20) does not protect the flies against vesicular stomatitis virus (Printz, unpublished, quoted by Bernard 1970), and it has not been possible to demonstrate replication of σ virus in vertebrate cells in vitro or in vivo (Ohanessian and Echalier, unpublished).

4.5.2.5 Togaviruses [R/1:3–4/6–7:S/S:V, I/Di, Ac] *and carrot mottle virus* [*/*:*/*:S/S:S/Ap]

The togaviruses include the two best-known genera of arboviruses, now called alphaviruses (= group A arboviruses) and flavoviruses (= group B arboviruses). The viruses belonging to these genera, and their vectors, are listed in table 4.3. Although the flavoviruses are slightly smaller than the alphaviruses

TABLE 4.3A

The vertebrate and invertebrate hosts and geographic distribution of the better-known alphaviruses[1]

| Virus | Host | | Distribution |
	Vertebrate	Invertebrate[2]	
Chikungunya	Mammals	Mosquitoes [argasid ticks]	Africa, Asia
O'Nyong Nyong	Mammals, birds	Mosquitoes	Africa
Sindbis	Mammals, birds	Mosquitoes [argasid ticks]	Africa, Asia, Australasia
Semliki forest	[Mammals]	Mosquitoes [argasid ticks]	Africa
Eastern equine encephalitis	Mammals, birds	Mosquitoes	North and South America
Venezuelan equine encephalitis	Mammals, birds	Mosquitoes	North and South America
Western equine encephalitis	Mammals, birds reptiles	Mosquitoes	North and South America
Ross river	Mammals, birds	Mosquitoes	Australasia

[1] Complete listings can be found in Taylor (1967) and Subcommittee (1970).
[2] Brackets indicate experimental susceptibility, without isolation in nature from that source.

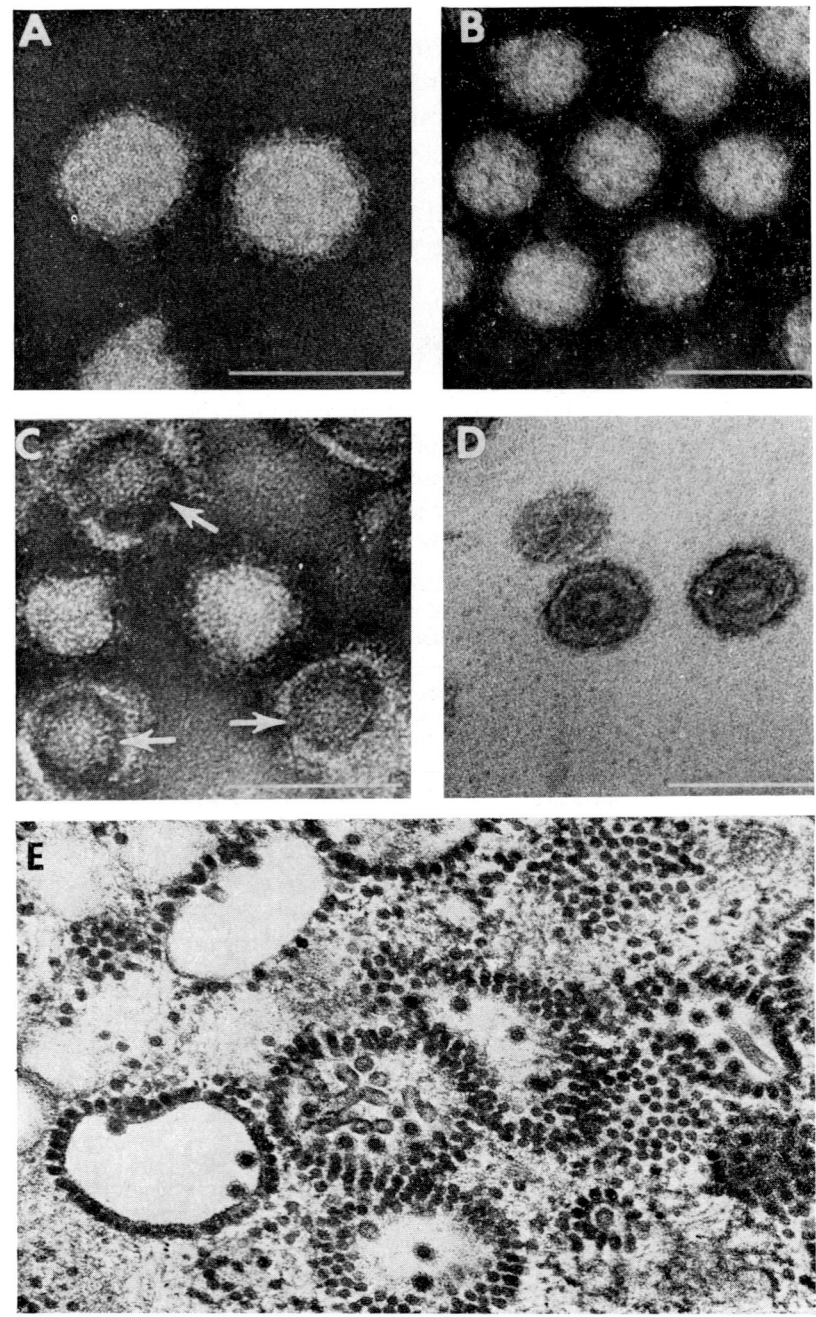

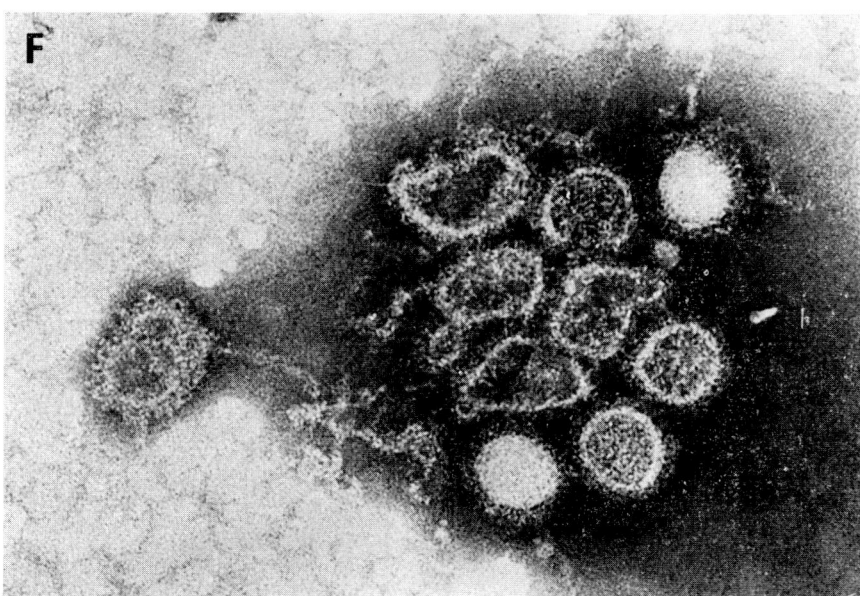

Fig. 4.15.   Togavirus. Alphaviruses (A–E) and one from the Bunyamwera supergroup (F).
(A) Middleburg virions showing poorly defined surface projections from the envelope.
(B) Sindbis virions fixed and stained, showing six-fold symmetry. (C) Damaged Middleburg particles showing the core and envelope. (D) A thin section of Middleburg particles
showing the core cut in two different planes. (E) Section of a chick cell infected with Semliki forest virus showing cores of virions in cytoplasm budding through membranes into
cytoplasmic vacuoles. (F) Negatively stained particles of Inkoo virus. Note envelope with
projections and the thread-like nucleocapsid. ((A)–(C) from Simpson and Hauser 1968;
(E) courtesy of K. B. Tan; (F) courtesy of N. Oker-Blom.)

(WHO Scientific Group, 1967) we shall include them together in this description for they have many properties in common. Each genus is defined by
serological cross-reactivity in hemagglutinin-inhibition tests.

   Most togaviruses cause infections with trivial or no symptoms in their
natural vertebrate hosts, which are usually birds with the alphaviruses and
mammals with the flaviviruses. However, these infections are associated with
high-titre viraemia, which is of major importance in providing the source of
virus for the infection of arthropods.

   The alphaviruses are transmitted by mosquitoes, flaviviruses by mosquitoes and ticks (see chs. 5, 11, 18 and 20). Some tick-borne flaviviruses have
been shown to be transovarially transmitted in the ticks.

TABLE 4.3B

The vertebrate and invertebrate hosts and geographic distribution of the better-known flavoviruses[1]

	Host		
Virus	Vertebrate	Invertebrate[2]	Distribution
Dengue (6 sero-types)	Mammals	Mosquitoes	Asia, Australasia, South America
Japanese encephalitis	Mammals, birds	Mosquitoes	Asia
Murray Valley encephalitis	Mammals, birds	Mosquitoes	Australasia
West Nile	Mammals, birds	Mosquitoes [argasid ticks]	Africa, Europe, Asia
St. Louis encephalitis	Mammals, birds	Mosquitoes [argasid & ixodid ticks]	North America
Yellow fever	Mammals	Mosquitoes	Africa, South America
Wesselsbron	Mammals	Mosquitoes	Africa, Asia, Australasia
Central European tick-borne encephalitis	Mammals, birds	Ixodid ticks	Europe
Louping ill	Mammals, birds	Ixodid ticks	Europe
Russian spring-summer encephalitis	Mammals, birds	Ixodid ticks	Asia
Omsk haemorrhagic fever	Mammals	Ixodid ticks	Asia
Powassan	Mammals	Ixodid ticks	North America
Langat	Mammals	Ixodid ticks	Asia
Kyasanur forest disease	Mammals	Ixodid ticks	Asia

[1] Complete listings can be found in Taylor (1967) and Subcommittee (1970).
[2] Brackets indicate experimental susceptibility, without isolation in nature from that source.

Togaviruses are unlike other enveloped viruses in that they consist of an isometric core 30–40 nm in diameter to which is directly applied a lipid bilayer, from which in turn project glycoprotein spikes (Harrison et al. 1971), so that the enveloped virion has a diameter of 60–80 nm (fig. 4.15). The core looks very like a poliovirus particle (Acheson and Tamm 1970); however, it has been suggested by Horzinek and Mussgay (1969) that it has a T = 3 arrangement of hexamer-pentamer capsomers on its surface. There are two major proteins in alphavirus particles, one in the core and the other in the membrane (Strauss et al. 1968), and possibly a third minor protein associated with the core (Yin and Lockart 1968; Horzinek and Mussgay 1969). By contrast flavovirus particles appear to contain four polypeptides (Westaway and Reedman 1969). In both genera, the viral RNA is a single-stranded molecule with a molecular weight of about 3×10^6 dalton, and which, when isolated from the particles by phenol treatment, is infectious. It probably consists of two single-stranded pieces joined by a short section in which complementary sequences in the pieces form a duplex structure.

Togaviruses multiply in the cytoplasm of vertebrate and insect cells. Cores accumulate around cytoplasmic vacuoles or beneath the plasma membrane and mature by budding through these membranes from which they acquire their lipoprotein envelope. This envelope is readily disrupted by lipid solvents, and infectivity is lost.

Carrot mottle virus has particles like those of the togaviruses, and will be described here. The virus can be transmitted by aphids (in the persistent manner), but only if the plant is also infected with carrot red leaf virus (a leaf roll type of virus, see below). The two together cause a severe disease of carrot

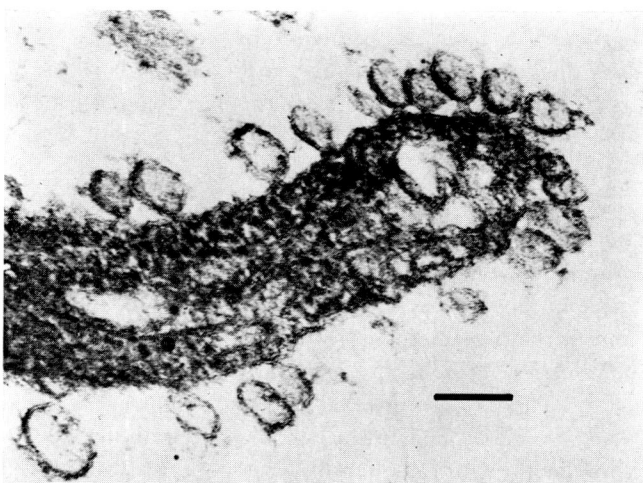

Fig. 4.16. Carrot mottle virus. Section of a palisade cell from a systemically infected leaf of *Nicotiana clevelandii* showing particles attached to and perhaps budding from the tonoplast membrane into the cell vacuole. (From Murant et al. 1969.)

called motley dwarf. Carrot mottle virus may also be transmitted by sap inoculation, the infectivity of sap is destroyed by lipid solvents and the infective agent has a density in sucrose gradients of less than 1.20. Infected plants contain particles that resemble those of the togaviruses, and which are sometimes found budding from, or attached to, all membranes (Murant et al. 1969; fig. 4.16).

4.5.2.6 *The Bunyamwera 'supergroup'*

This 'supergroup' was established by Casals (WHO Scientific Group, 1971) to bring together those minor arbovirus groups linked by distant direct or indirect serological cross-reactions. The groups included in this supergroup, and the numbers of viruses in each group, are: Bunyamwera (14), group C (11), Guama (8), Capim (7), Simbu (16), Bwamba (2), California (10), Patois (4), Tete (3), Koongol (2), and several unassigned viruses. Probably all of these viruses are mosquito-borne. They contain single-stranded RNA (N. Oker-Blom, personal communication). All of those which have been studied have particles approximately 100 nm in diameter, which consist of a dense core and ragged, closely adherent envelope with projections (Murphy et al. 1968a, 1968b). Their infectivity is destroyed by treatment with lipid solvents. The basic structure of the nucleocapsids remains unknown, although a helical internal structure has been resolved in two member viruses (Von Bonsdorff et al. 1969). Several serologically independent minor serogroups and some ungrouped arboviruses have been shown to be morphologically quite similar to Bunyamwera supergroup viruses (e.g., representative viruses of the Phlebotomous fever group, Congo group, Unkuneimi virus, Turlock virus, and Rift Valley Fever virus).

4.5.2.7 *Arenoviruses* [(R)/*:*/*:S/*:V, I/O]

This group was characterized quite recently (Rowe et al. 1970). The best known is lymphocytic choriomeningitis virus, and the group is included here because of several reports that arthropods (Milzer 1942; Reiss-Gutfreund et al. 1962) and perhaps nematodes (Syverton et al. 1947) may be able to transmit the infection, although spread via the respiratory tract is probably more important. The virion is enveloped, 110–130 nm in diameter, and contains characteristic electron dense, ribosome-like granules 20–30 nm in diameter. Lymphocytic choriomeningitis virus contains four different single-stranded RNA molecules (Pedersen 1971). Two of these sediment in sucrose gradients and migrate in polyacrylamide gel electrophoresis at the same rates as host ribosomal RNA species, and their synthesis is sensitive to actinomycin D. It has been suggested that these RNA species, which account for half of the RNA in the virus particles, are derived from the electron dense granules, which are actually host ribosomes. The other two RNA species are presumably virus specified since their synthesis is resistant to actinomycin D. They sediment at 23 S and 31 S, with approximate molecular weights of 1×10^6 and 2×10^6 dalton.

The inner part of the virion develops in the cytoplasm and the envelope is acquired when it buds through the plasma membrane. Infected vertebrates may have a prolonged period of viraemia, and progeny may be infected; congenitally infected mice may have a lifelong viraemia.

4.5.2.8 *Picornaviruses* [R/1:2–3/30:S/S-V, I/O, Di]
These are viruses of animals whose particles are ether resistant and around 30 nm in diameter with a sedimentation coefficient of about 160 S. They include such well known viruses as foot and mouth disease virus and poliovirus. They are usually divided into three major groups; the rhinoviruses, whose particles are acid labile and have a density in CsCl solutions of around 1.4, the enteroviruses, whose particles resist acid and are less dense (1.34), and the caliciviruses, whose particles are of intermediate acid resistance and density, and which have an obvious 32 capsomer structure.

No caliciviruses or vertebrate rhinoviruses are known to be associated with invertebrates. However, sacbrood virus [R/1:2.5/30:S/S:I/O], a picornavirus which kills honeybee larvae (ch. 22), resembles the rhinoviruses in that it is unstable in acid, but in CsCl has the density of an enterovirus.

There is no evidence that enteroviruses of vertebrates can infect invertebrates, but from analogy with faecal bacteria we could expect flies to act as mechanical vectors of enteroviruses accessible to them in faeces, and blood-sucking arthropods could imbibe such viruses during periods of viraemia. Carriage of enteroviruses by flies, mosquitoes and other insects has been demonstrated (Melnick 1949; Gelfand 1961; Maguire and Macnamara 1966). Although this association with invertebrates is probably trivial, both biologically and as far as transmission is concerned, poliovirus has been extensively studied and the description which follows may be useful as a basis for understanding morphologically and chemically similar viruses that are more significantly associated with invertebrates.

Poliovirus particles are non-enveloped icosahedra about 30 nm in diameter. The arrangement of the capsomers has not been clearly demonstrated in electron micrographs but there are probably 60 cleaved subunits arranged on its surface. The virus particles consist of RNA and protein. The RNA occurs as a single linear molecule of about 2.6×10^6 dalton molecular weight. Purified RNA is infectious. There are usually four polypeptides in the virion. By contrast the particles of those viruses of plants whose particles have similar morphology contain only one virion protein.

Enteroviruses multiply in the cytoplasm are released by rupture of the infected cells and are excreted in faeces.

Coxsackievirus A, an enterovirus, has been isolated from mosquitoes (Maguire and Macnamara 1966). Although transmission has been demonstrated up to 13 days after a blood feed, it does not appear to replicate in mosquitoes (Maguire 1970), nor does poliovirus RNA or encephalomyocarditis virus RNA replicate in cultured mosquito cells (Peleg 1969). Most picornaviruses have restricted host ranges. However, Nodamura virus, which is morphologically indistinguishable from the enteroviruses (Murphy et al. 1970), was recovered from a pool of *Culex tritaeniorhynchus* mosquitoes and has been shown to multiply in mosquitoes, ticks and moth larvae (Scherer and Hurlburt 1967).

Several other viruses of invertebrates have been found to closely resemble enteroviruses of vertebrates. These include bee acute paralysis virus [R/1: 2/30:S/S:I/O] (ch. 22), cricket paralysis virus [R/1:*/*:S/S:I/O] (ch. 28) and a possible virus of termites (ch. 28). All resist acid and those tested have a density in CsCl solutions of about 1.34. They seem to be usually confined to the gut, but when they get into the haemocoel (either by injection of small doses, or by feeding large doses) they infect all tissues and cause necrosis of nervous tissues and paralysis.

4.5.2.9 *Tobraviruses* [R/1:2.3/5+(0.6–1.3)/5:E/E:S/Ne]

This group has two members, tobacco rattle (Harrison 1970) and pea early browning viruses, and is one of the two groups of nematode-borne viruses. They are sap-transmissible to a particularly wide range of hosts causing mostly necrotic symptoms. In nature they are transmitted by root feeding nematodes of the genus *Trichodorus* (ch. 27).

These viruses have straight tubular particles of two principal lengths, 180–200 nm, and 50–100 nm long and about 25 nm wide (fig. 4.17), containing about 5% RNA. Although the long particles (or RNA molecules extracted from them) are infective, they produce infective RNA but no virus particles unless short particles (or their RNA) are inoculated into the plant at the same time or later; these and other results show that the gene for the coat protein of the virus is in the RNA of the shorter particles only (Lister and Bracker 1969).

4.5.2.10 *Tobamoviruses* [R/1:2/5:E/E:S/O]

This group of about six closely related viruses includes tobacco mosaic virus, one of the best known viruses. All infect a restricted number of host plants causing mosaics or necrosis. All have tubular particles, 300×18 nm. In nature they seem to be spread mechanically when plants touch, and by seed, and although no efficient vector is known there have been reports of transmission by various plant eating insects.

4.5.2.11 *Nepoviruses* [R/1:2.4/43 + 1.4/30:S/S:S/Ne]

This group of seven or more nematode-borne viruses includes such viruses as tobacco ringspot (Stace-Smith 1970), grapevine fan leaf (Hewitt et al. 1970), arabis mosaic (Murant 1970) and others. They cause ringspot symptoms in a wide range of hosts and are of economic importance in temperate regions. They are frequently seed and pollen transmitted, and are transmitted by root-feeding dorylaimid nematodes of the genera *Longidorus* and *Xiphinema*.

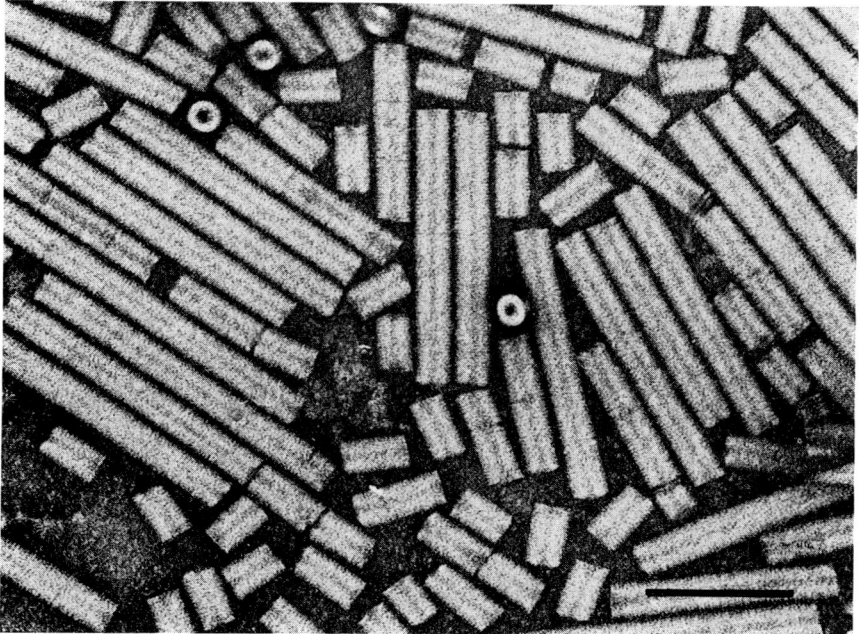

Fig. 4.17.   Tobravirus. Particles of negatively stained tobacco rattle virus. The particles are mostly of two lengths, and the genome of the virus is divided between the two, so that both are needed for full infection. (Courtesy of B. D. Harrison and I. M. Roberts.)

Purified preparations of the viruses consist of isometric particles about 30 nm in diameter with angular outlines, some of which contain about 40% RNA others about 25%, and some none. Like the comoviruses (sect. 4.5.2.1) it seems that both are necessary for infection. The three types of particles are serologically indistinguishable and each particle contains 60 subunits of molecular weight about 54,000 dalton (Mayo et al. 1971).

4.5.2.12 *Tymoviruses* [R/1:2/36:S/S:S/Cl]

This group includes turnip yellow mosaic virus (Matthews 1970b) and at least seven other viruses that cause yellow mosaic diseases in restricted ranges of plants. They are transmitted by Halticid and Galerucid flea beetles (ch. 9), by sap inoculation, and some by seed.

Their particles are isometric and about 30 nm in diameter. Some contain

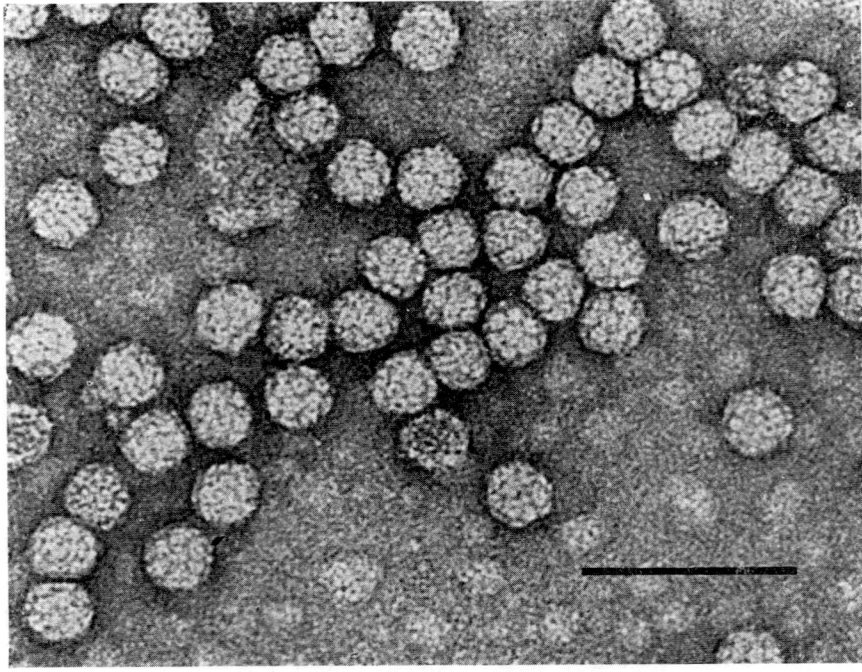

Fig. 4.18. Tymovirus. Particles of turnip yellow mosaic virus showing the grouping of the protein subunits and underlying nucleic acid into 32 major morphological subunits. (Courtesy of J. T. Finch.)

about 35% RNA, these are infective and show 32 distinct capsomers, other non-infective particles contain no RNA (fig. 4.18). This difference gave one of the earliest clues that the nucleic acid of a virus was the infective moiety of the virion (Markham and Smith 1949). The RNA of these viruses has an unusually large cytosine content; 40% of the bases.

In infected cells small vesicles develop in the chloroplasts near, and perhaps attached to, their outer membrane. There is evidence that the viral RNA is transcribed in or on these vesicles.

4.5.2.13 *The potato virus Y group (Potyviruses)* [R/1:*/5:E/E:S/Ap]

This, the largest group of plant viruses is usually called the potato virus Y group, and Harrison et al. (1971) have suggested calling it the potyvirus group. Many of the potyviruses cause economically important diseases of crop plants, and the group includes more than twenty viruses such as bean yellow mosaic (Bos 1970), potato virus Y (Delgado-Sanchez and Grogan 1970), plum pox, sugar cane mosaic and turnip mosaic (Tomlinson 1970), but one of the most completely characterized is tobacco etch virus (Shepherd and Purcifull 1970).

Potyviruses can be transmitted experimentally by inoculation of sap, and in nature their vectors are aphids, by which they are transmitted in the non-persistent manner (ch. 25). They cause mosaic symptoms in restricted ranges of host plants. Their particles are filamentous and mostly 730–900 nm long; the length sometimes depends on the ionic conditions (Govier and Woods 1971). Infected plants contain characteristic pinwheel inclusions (Edwardson 1966).

Two viruses transmitted by eriophyid mites (chs. 6 and 19) and called wheat streak mosaic (Brakke 1971) and ryegrass mosaic [*/*:*/*:E/E:S/Ac] have similar particles but only 700 nm long, and also induce the formation of pinwheel inclusions.

4.5.2.14 *Carlaviruses* [R/1:*/6:E/E:S/Ap]

This is another group of about 10 plant viruses transmitted by sap and by aphids in the non-persistent manner. They too have filamentous particles, but shorter than those of the potyviruses, being around 650 nm long and less flexuous. The group is named after carnation latent virus and includes other well known viruses such as red clover vein mosaic (Varma 1970), potato virus S and the related potato paracrinkle virus (Wetter 1971).

These viruses also usually cause mild mosaics in a limited range of host plants. Infected cells do not contain pinwheel inclusions.

4.5.2.15 Other ungrouped viruses whose particles are known to contain single-stranded RNA

One of these is pea enation mosaic virus (Shepherd 1970b) which causes mosaic and enations in various legumes. It is sap transmissible, and in nature is transmitted by aphids in the persistent manner, though there is no evidence that it replicates in aphids. It has fragile isometric particles about 25 nm in diameter, which sediment as two components, the relationships of which are not known.

Another virus of this type is cocksfoot mottle virus [R/1 :(1/25):S/S:S/Cl] (Catherall 1970). This plant virus is transmitted by larvae and adults of the chrysomelid beetles *Lema melanopa* and *L. lichenis*, adults of which can remain infective for two weeks. This virus is one of the many unclassified plant viruses with isometric particles, which sediment as a single component (110–140S), and which are not degraded by large concentrations of salt (Gibbs 1969).

Yet another virus of this type is bee chronic paralysis virus [R/1 :*/*:U/U : I/0], which is common in colonies of the honey bee (ch. 22). The particles of

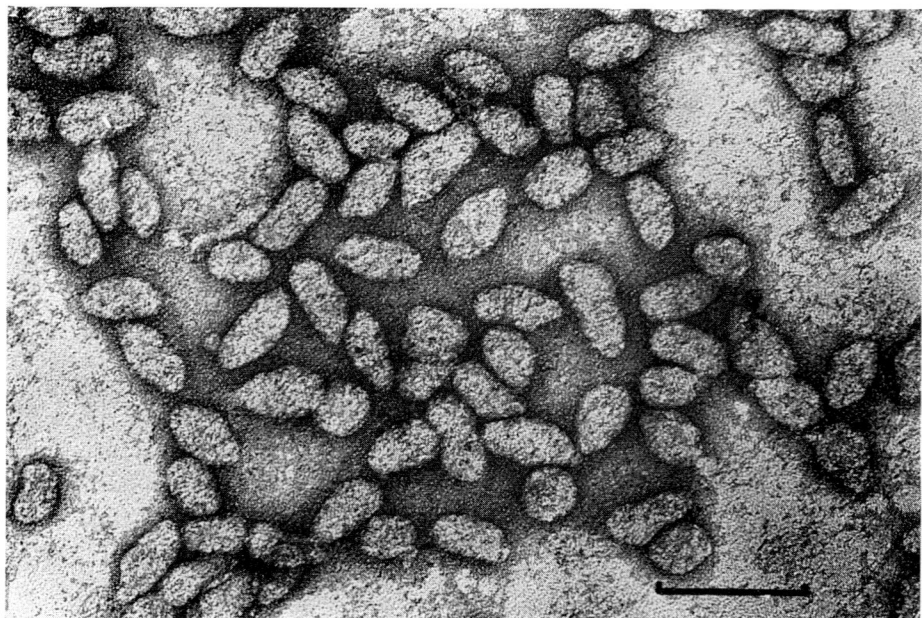

Fig. 4.19.   Chronic bee paralysis virus. Negatively stained particles of the virus, these are pleomorphic and lipid free and of three modal sizes. (Courtesy of R. D. Woods.)

this virus seem to be unlike those of any other virus so far reported, they are of three sizes, all approximately 22 nm wide and ellipsoidal in outline but about 41, 54 and 64 nm long (fig. 4.19) (Bailey et al. 1968). It is not known whether this virus has a divided genome that is partitioned between the different particles.

4.5.3 Viruses whose particles contain RNA of unknown type

One virus in this category seems to be widespread in saturniid moth larvae for it has been found in larvae of *Antheraea eucalyptii* in Australia (Grace and Mercer 1965) and a serologically related virus was reported from *Nudaurelia capensis* in South Africa (Tripconey 1970). The particles of these viruses are isometric, about 34 nm in diameter, and have a sedimentation coefficient of about 215S. Preparations of the *Antheraea* virus also contain many smaller particles (13 nm; 44S; Brzostowski, Gibbs and Grace, unpublished results); although these might be 'cores' of the *Antheraea* virus particles, it is possible that they are another virus or even a satellite virus as they contain nucleic acid.

Another virus in this category is that found in the citrus red mite *(Panonychus citri)* by Smith et al. (1959), and used to try to control the mite. Diseased mites contain many birefringent crystals of unknown composition, and also isometric particles that contain RNA and are 35 nm in diameter (Estes and Faust 1965).

4.6 Viruses whose nucleic acid has not been characterized

4.6.1 Other arboviruses

In addition to the alphaviruses, flaviviruses, rhabdoviruses, the Bunyamwera supergroup and the orbiviruses, all of which contain members which multiply in both vertebrates and arthropods and are transmitted to vertebrates by arthropod bite, many other isolates have been recovered either from vertebrate blood or from haematophagous arthropods. They include some minor serogroups and many that are unrelated to all other arbovirus isolates; their particles have not been studied. Many of these viruses have been considered to be arthropod-borne in nature because A) they are frequently isolated from arthropods; B) antibodies against them are found in wild vertebrates; C) they can be transmitted by arthropods to vertebrates in the laboratory; and D) arthropods can be infected with them by injection.

At present little more has been done than to list the details of their isolation, and their behavior in laboratory animal and cell culture systems (Taylor 1967).

4.6.2 Leaf roll viruses [*/*:*/*:S/S:S/Ap]

This large group of plant viruses includes many of great economic importance such as barley yellow dwarf (Rochow 1970), beet western yellows and potato leaf roll (Peters and Van Loon 1968; Duffus and Gold 1969; Arai et al. 1969), and perhaps also banana bunchy top, bean leaf roll (subterranean clover stunt), beet mild yellowing, turnip mild yellows and at least ten others.

These viruses cause yellowing, reddening and cupping of leaves and phloem

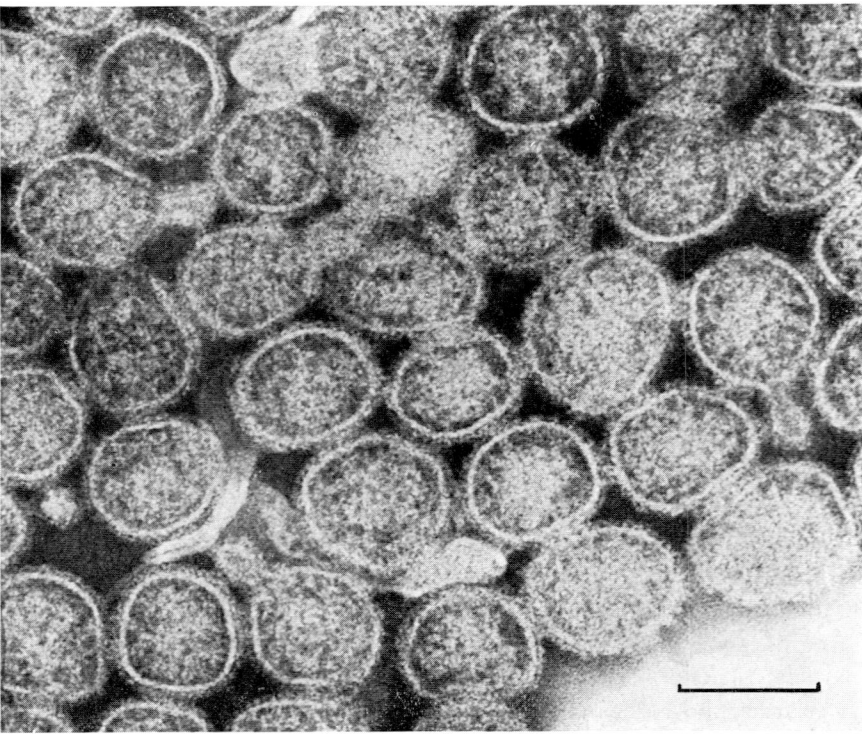

Fig. 4.20.   Tomato spotted wilt virus. Negatively stained particles of the virus, these are pleomorphic and contain lipid, and resemble particles of leukoviruses. (From Van Kammen et al. 1966.)

necrosis. Those that have been purified are found to have isometric particles about 25 nm in diameter (110–120S). None has been transmitted by sap, and, in nature, they are transmitted by aphids in the persistent manner, though there is no convincing evidence that any replicate in aphids.

4.6.3 Ungrouped viruses

Equine infectious anaemia virus is another virus associated with inverte-brates. It causes chronic disease with prolonged viraemia in horses. It appears to be mechanically transmitted by a variety of invertebrates. The enveloped virion 80–120 nm in diameter encloses an electron-dense core 40–60 nm in diameter (Tajima et al. 1969). Its affinities are unknown, but attention has been drawn to possible resemblances to the leukoviruses (Kono et al. 1970).

Another virus which shows some superficial similarity to the leukoviruses is tomato spotted wilt virus [(R)/*:*/*:S/*:S/Th] (Ie 1970). This common plant virus infects a wide range of plants throughout temperate regions. It is transmitted by sap, and, in nature, by thrips (ch. 28). It has complex fragile isometric particles about 80 nm in diameter (530S), these are apparently bounded by a membrane, which sometimes forms rounded protrusions like those seen in early electron micrographs of leukoviruses. Complete particles of the virus are formed in the cytoplasm of infected cells (Milne 1970; fig. 4.20).

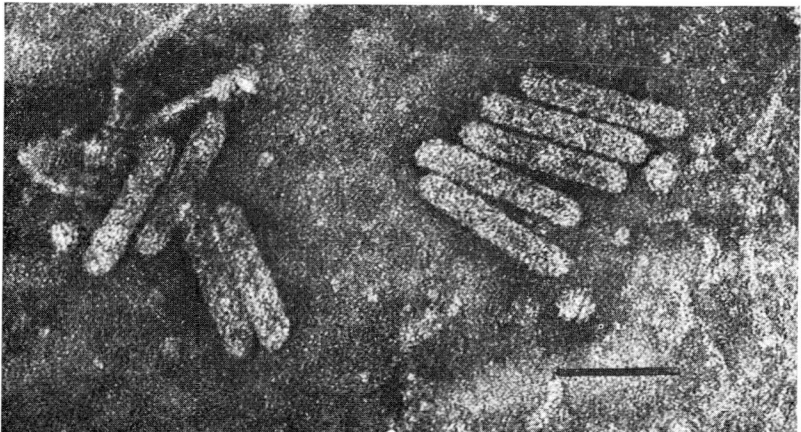

Fig. 4.21.   Cacao swollen shoot virus. Negatively stained bacilliform particles of the virus.
(Courtesy of R. D. Woods.)

Beet yellows (Russell 1970) and citrus tristeza (Price 1970) viruses [*/*:*/*: E/E:S/Ap] are two similar plant viruses. They have long characteristically flexuous filamentous particles with clear helical banding, but whereas those of beet yellows are 1.25 μm long, those of tristeza are over 2 μm long. Both cause yellowing, and are transmitted by aphids in a semi-persistent manner. It is not known what type of nucleic acid they contain, if any.

Similarly the nucleic acid of cacao swollen shoot virus [*/*:*/*:U/V:S/Cc] (Brunt 1970) has not yet been characterized. This virus is of great economic importance and is transmitted in nature by mealy bugs (chs. 10 and 23), and can also be transmitted by sap inoculation. Its particles are bacilliform, about 125 nm long and 28 nm in diameter (fig. 4.21).

4.7 Conclusions

Thus it can be seen that many different types of virus are associated with invertebrates in many different ways. Most generalisations about various types of viruses that have been made in the past have been disproved. Nevertheless, some general trends are obvious from the survey of viruses in this chapter, and some of these trends may have theoretical significance. For instance, the tendency for large RNA-containing viruses to have their genome in several pieces may be because it is impossible to extend the purely post-translational punctuation and control found in the small RNA-containing viruses to larger RNA genomes. This problem does not arise with DNA containing viruses, and the fact that divided viral DNA genomes do not occur suggests that there is selection pressure against split genomes unless they have a compensatory advantage.

The tendency for many viruses of invertebrates to induce the production of inclusion bodies that resist dessication and heat but dissolve slowly in alkaline solutions is probably due to the value of these properties in the survival and transmission of the viruses in nature.

RNA polymerases (and some other enzymes of nucleic acid metabolism) have been found in the particles of many viruses (of both vertebrate and invertebrate animals and plants) that replicate in the cytoplasm and contain either double-stranded DNA or RNA. The polymerase is probably needed in all such viruses to produce at least the earliest viral messenger RNA molecules in the cytoplasm of infected cells. RNA polymerase must also be present in the particles of single-stranded RNA-containing viruses in which the complement of the viral RNA acts as messenger. These conclusions lead us to doubt reports of the infectivity of some viral nucleic acids, for instance that of a cytoplasmic polyhedrosis virus (sect. 4.5.1.4).

Possible explanations for other trends are lacking at present. Fragmented viral RNA genomes in which the pieces are found in separate virus particles instead of being included in a single particle have evolved in plants but not in vertebrate animals although both types of virus can be transmitted by invertebrates. All of the RNA-containing viruses which are confined to vertebrate and invertebrate animals and which have tubular nucleocapsids are enveloped. Similar viruses without an envelope are common in plants, and can be transmitted by invertebrates.

Appendix 4.1 Meaning of the symbols used in the cryptograms

Each cryptogram consists of four pairs of symbols (e.g. tobacco mosaic virus: R/1:2/5:E/E:S/O) with the following meanings:

1st pair

Type of nucleic acid/strandedness of nucleic acid. Symbols for type of nucleic acid: R = RNA; D = DNA. Symbols for strandedness: 1 = single-stranded; 2 = double-stranded.

2nd pair

Molecular weight of nucleic acid (in millions)/percentage of nucleic acid in infective particles. This term gives the composition of infective particles. The genome of some viruses is divided. When different pieces of the genome occur together in one type of particle, the symbol Σ indicates the total molecular weight of the pieces in the particles (e.g. clover wound tumour virus: R/2: Σ15/20:S/S:S,I/Au), but when the pieces occur in different particles the composition of each particle type is listed separately (e.g. tobacco rattle virus: R/1:2.3/5 + (0.6–1.3)/5:E/E:S/Ne).

3rd pair

Outline of particle/outline of 'nucleocapsid' (the nucleic acid plus the protein most closely in contact with it).

Symbols for both properties:

   S = essentially spherical

   E = elongated with parallel sides, ends not rounded

   U = elongated with parallel sides, ends rounded

   X = complex or none of above

4th pair

Kinds of host infected/kinds of vector.

Symbols for kinds of host:

   A = actinomycete   P = pteridophyte

   B = bacterium   S = seed plant

   F = fungus   V = vertebrate

   I   = invertebrate

Symbols for kinds of vector:

   Ac = mite and tick (Acarina, Arachnida)

   Al = white-fly (Aleyrodidae, Hemiptera, Insecta)

   Ap = aphid (Aphididae, Hemiptera, Insecta)

   Au = leaf-, plant-, or tree-hopper (Auchenorrhyncha, Hemiptera)

   Cc = mealy-bug (Coccidae, Hemiptera)

   Cl = beetle (Coleoptera, Insecta)

   Di = fly and mosquito (Diptera, Insecta)

   Fu = fungus (Chytridiales and Plasmodiophorales, Fungi)

   Gy = Mirid, Piesmid, or Tingid bug (Gymnocerata, Hemiptera)

   Ne = Nematode (Nematoda)

   Ps = Psylla (Psyllidae, Hemiptera)

   Si = flea (Siphonaptera, Insecta)

   Th = thrips (Thysanoptera, Insecta)

   Ve = vectors known but none of above

   O   = spreads without a vector via a contaminated environment

In all instances

  * = property of the virus is not known

  ( ) = enclosed information is doubtful or unconfirmed

CHAPTER 5

Acarina (ticks)*

HARRY HOOGSTRAAL

Contents

(continued on page 90)

---

* Throughout this chapter, the names of several authors who have published numerous papers on the subject under discussion are mentioned without specific references. These references may be found in Hoogstraal (1970) and in the Bibliography of ticks and tick-borne diseases (Special Publications, NAMRU-3, 4 vol., Hoogstraal 1970–72), which lists literature to the end of 1969. Where geographical regions are mentioned without literature reference, the following faunal reviews apply: Australia (Roberts 1970), Japan and Korea (Yamaguti et al. 1971), Panama, including Central America and northern South America (Fairchild et al. 1966), Mexico (Hoffmann 1962), Canada (Gregson 1956), Italy (Starkoff 1958), Central Europe (Babos 1964; Feider 1965), Britain (Arthur 1963), Sudan and other African areas (Hoogstraal 1956), Africa (Theiler 1962), Central Africa (Elbl and Anastos 1966), Tanganyika (Yeoman and Walker 1967).

5.1 Introduction

All ticks are obligate ectoparasites. About 800 species are known, most in the tropics and subtropics and fewest in penguin and marine bird nests in circumpolar regions. In temperate forests, prairies and steppes, there may be great population densities of one or two species. Certain tick groups are adapted to desert and semidesert where few other bloodsucking arthropods survive. Two *Amblyomma* species go to sea with marine snakes and lizards, which become infested when on land.

Ticks are reservoirs and vectors for numerous organisms infective for vertebrate animals. Many pathogens are directly transmitted when ticks bite or indirectly by contact with tick coxal fluid, excrement, or crushed bodies. Ticks also cause toxaemia, paralysis, anaemia, and severe irritation to animals.

In this chapter, together with ch. 18, I review tick families of interest to animal virologists. The review of tick biology by Balashov (1968), translated by Hoogstraal and Tatchell (1972), should also be consulted.

The biology of each tick species is unique and although there are relationships between particular tick groups and particular pathogens, closely related species may differ greatly in their ability to maintain and transmit specific viruses (sect. 18.1). Appreciation of virological differences between apparently identical tick populations has led to studies which have revealed species differences and provided much new knowledge of ticks (Hoogstraal et al. 1969; Kaiser and Hoogstraal 1969).

All tickborne diseases of humans and domestic animals, except possibly African relapsing fever, are infections which have spread from wild animals, and therefore must be studied in their natural hosts.

The internal milieu of the body of each tick species has special biochemical and physiological properties in each developmental stage and in the pre-feeding, feeding and postfeeding states. Pathogen transmission ability is influenced by different properties of oögenesis, food digestion, etc., and by histological and histochemical properties, as well as by individual characteristics of the life cycle, ecological preference and tolerance, host attractiveness, survival, water balance, etc. All of these factors influence the vector potential of each species.

The suborder Metastigmata (ticks) contains three families (see ch. 6). The leathery Argasidae (soft ticks) have no scutum (shield) and the fourth palpal segment is apical. In the other two families, a scutum covers part of the larval, nymphal and female dorsum and all of the male dorsum. In the rare family Nuttalliellidae, the scutal and female integumental texture is similar and the fourth palpal segment is apical. In the common, hard-tick family Ixodidae, the rigid scutum is distinct from the integument, and the fourth palpal segment is reduced and usually arises from a ventral pit of the third segment.

5.2 The family Nuttalliellidae

The single species, *Nuttalliella namaqua* Bedford, is known only from several specimens taken under stones and from museum skins of rodents and small burrowing carnivores in South Africa (Little Namaqualand) and South West Africa.

5.3 The family Argasidae

5.3.1 Introduction

Of the 140 species in the family Argasidae (argasid or soft ticks), 131 are highly specialized for parasitizing animals that return periodically to a certain shelter. Argasids avoid dampness and hide in cracks in wood, stone or debris, or burrow into soil, or climb onto walls of caves or stables where their hosts rest (sect. 18.3).

A feast-or-famine annual pattern characterizes argasids associated with birds that nest only for a few weeks, with bats roosting in a cave for only a few days or weeks, and with desert vertebrates that frequently shift resting places. To ensure survival, argasids have great climatic tolerance and are able to drastically reduce their metabolic rate to permit one or more developmental stages to survive sometimes as long as 10 years. They rapidly take in large bloodmeals (ca. 10 times the unfed tick weight).

Soon after hatching, argasid larvae feed to repletion, either within an hour (some *Ornithodoros* species) or for several days (some *Ornithodoros* and all *Argas* species), and then detach from the host. Rapidly feeding larvae usually fall to the ground near where they hatched. Long-feeding larvae may be transported longer distances. After the replete larva detaches, part of the bloodmeal is digested and it moults to a nymph. (Larvae of *Ornithodorus moubata* and *O. savignyi* do not feed). There are two to eight nymphal instars (moults), each feeds once, or sometimes twice. The first nymphal instar of some species moults without feeding. Owing to their short feeding period, nymphal and adult argasids are rarely found on the host. Adults do not mate while on the host. Females feed and oviposit several times before they die. Eggs hatch after two or more weeks.

The life cycle of certain argasid species may be completed in as little as four months though some take many years.

Most argasid species live in arid or semiarid biotopes or places with a long dry season. In heavy rainfall regions, argasids occur in dry caves, tree holes or dry crevices but seldom in burrows. One *Argas* species is known only from snow partridges at 15,000 feet altitude in the Himalayas (Hoogstraal and Kaiser 1972).

Thirty-five different arboviruses have been isolated from argasids, mostly from those of birds, fewer from those of bats and terrestrial mammals (table 18.2).

The 140 species involved are from the genera *Argas* (45), *Ornithodoros* (89), *Otobius* (2) and *Antricola* (4). *Argas* are very thin and flat and most hide in deep cracks. *Ornithodoros* are thicker; some shelter in cracks but most in soil or sand or under debris, or they rest on cave or stable walls. *Otobius* and *Antricola*, which infest wandering hosts, differ from other argasids.

5.3.2 The genus Argas

Eight *Argas* species in Africa, Asia, Europe and North America have thus far yielded 17 different viruses (table 18.2), all from bird-infesting species, except Keterah virus from a bat parasite. Most recent literature on this genus is by Hoogstraal et al. (1968). See also Filippova (1966) and Keirans et al. (1971).

There are seven subgenera of the genus *Argas*. Species in the subgenera *Argas* and *Persicargas* are confined to bird hosts, those in the subgenera *Chiropterargas* and *Carios* to bats, whereas those of the subgenus *Secretargas* infest bats, tortoises, lizards, small insectivores and other animals.

5.3.3 The genus Ornithodoros

Various aspects of this genus have been summarized by Hoogstraal (1956), Walton (1962), Clifford et al. (1964), Kohls et al. (1965), Sonenshine et al. (1966), Van der Merwe (1968), Jones and Clifford (1972).

Fourteen *Ornithodoros* species have yielded 18 different viruses, 12 from seven bird-infesting species and six from seven mammal-infesting species (table 18.2). The 89 known *Ornithodoros* species are placed in seven sub-genera.

5.3.4 The genus Otobius

The two species of the genus *Otobius* live in the external auditory meatus of hosts in deserts and semideserts of western North America. They are active throughout the year and can live for months without feeding. The larva quests for a host from low vegetation and during feeding becomes a pyriform, translucent bladder. It moults to a fiddle-shaped, short legged, spinose, blue gray nymph, which feeds on the same host, then drops to the ground and moults to a spineless adult with non-functional mouthparts. Adults soon mate but do not feed. Females oviposit in several batches, each of about 100 to 300 eggs, over a period of several weeks.

O. megnini parasitizes artiodactyls and also dogs and humans, and has become a serious pest in many parts of the world. It reached South Africa by 1898 and became widely distributed there and in other parts of Africa (Du Toit and Theiler 1964; Uilenberg 1965; Macleod et al. 1970). Various types of secondary infections follow heavy infestations of the tick (Stiles 1944; Jellison et al. 1948; Peacock 1958; Tarshis and Ommert 1961; Peacock 1958; Howell 1970).

O. lagophilus principally infests lagomorphs (Herrin and Beck 1965). This species has been associated with Colorado tick fever virus, Rocky Mountain spotted fever and tularemia.

5.3.5 The genus Antricola

The four members of the poorly known genus *Antricola* parasitize cave-dwelling bats in Central and North America. Larvae have exceptionally large pulvilli (pads) for climbing walls.

5.4 The family Ixodidae

5.4.1 Introduction

There are about 650 species of 'ixodid' or 'hard' ticks divided among 13 genera. This family is biologically more diverse than the family Argasidae. Each postembryonic stage (larva, single nymphal instar and adult) is active and feeds only once, generally for several days. Most adult ixodids mate on the host soon after beginning to feed, but the adults of certain *Ixodes* species do not. The genus *Ixodes* (about 250 species) constitutes the subfamily Ixodinae. Santos Dias (1963) includes four genera (about 307 species) in the subfamily Amblyomminae (sect. 6.4.3–6.4.6) and eight (about 101 species) in the subfamily Rhipicephalinae (sect. 5.4.7–5.4.14). In the small genera *Boophilus* and *Margaropus*, and in a few species of other genera, each active stage feeds on the same animal and the replete, fertilized female drops to the ground to oviposit. This life cycle, involving only one host, seems to be adapted to large mammal hosts that wander for relatively great distances. Most species in other genera have a different host for each of the three stages of the life cycle, though species in the genus *Hyalomma* are more variable in this respect than others. Females drop from the third host after feeding, oviposit once and die, but males remain on the host for weeks or months. Thus the time when feeding females are found indicates the breeding season of the species. The interstage intervals depend on climate.

In some species, the larva and nymph stay on one host then drop to the ground, moult, and later feed as adults on a second, larger host. This type, for example *Hyalomma marginatum*, may feed on migrating birds and may be transported long distances (Hoogstraal et al. 1961, 1963, 1964).

The numerous eggs, usually 1000 to 8000 or more deposited by a female ixodid, reflect the hazards experienced by this group. All the active stages of some ixodid species parasitize the same kind of host. However the size and kind of host used by the immature and adult stages of most species is often quite different: lizard and wild goat, guinea fowl and elephant, mouse and lion, hedgehog and buffalo, etc. Intermediate-size mammals such as hares, cane rats and porcupines may support each stage of species that otherwise feed as immatures and adults on small and large-size hosts. The tick population density frequently decreases dramatically from one stage to the next presumably because of their failure to encounter a suitable host. In the few three-host ticks (such as *Rhipicephalus appendiculatus*) where each active stage can feed on large animals, great numbers survive and the species is usually a virus vector. Most ixodid species have a narrow range of preferred hosts, however they may parasitize other hosts, and this has confused reports and concepts on ixodid–host relationships. Also many experimental studies have used species such as *Ixodes ricinus* and *I. persulcatus*, which have exceptionally wide host ranges and thus are notorious in human and veterinary medicine but are not representative of many other species in this family.

5.4.2 The genus Ixodes

The largest genus in the family Ixodidae, *Ixodes*, contains almost 250 species, 13 of which have yielded 19 different viruses from all continents except Australia (table 18.2). The five viruses from *I. uriae* and Uukuniemi and Sumakh viruses from *I. ricinus* appear to be associated chiefly with birds. The other viruses usually circulate in mammals, or possibly more commonly, in both birds and mammals. Much literature on *I. ricinus*, *I. persulcatus* and on some other *Ixodes* vectors was reviewed by Hoogstraal (1966). The most economically important *Ixodes* have been well studied, however many other *Ixodes* ticks are poorly known and their taxonomy is unsettled. There are over 40 species in North America (Cooley and Kohls 1945), over 60 in Africa (Arthur 1965), 22 in Australia (Roberts 1970) and some in the Palearctic fauna, but fewer in the Orient. Little is known of South American *Ixodes*.

In some *Ixodes* species adults mate on the host (usually on a wandering animal) or mate elsewhere (either on a wandering or fixed-habitat host). In others, only the female feeds, and afterward mates elsewhere (Nuttall 1911). By contrast, the Australian scrub or paralysis tick, *I. holocyclus*, male copulates with feeding females and also pierces the female venter to feed on hemolymph (Moorhouse 1966).

All *Ixodes* species have a three-host life cycle, so far as is known.

5.4.3 The genus Haemaphysalis

Recent information on *Haemaphysalis* is mostly from papers by Hoogstraal and co-authors published in the Journal of Parasitology since 1955 but not referenced here; these papers list many other references to this genus.

Among the 150 species in the genus *Haemaphysalis* found throughout the world, 17 are known to be infected by 9 different viruses (table 18.2). Haemaphysalids also transmit *Rickettsia* spp. causing spotted-fever group typhus diseases in Africa, Asia and the Americas (Hoogstraal 1967) and protozoa (*Babesia*, *Theileria*, etc.) that cause diseases in wild and domestic animals (Neitz 1956). They sometimes cause paralysis and decreased vigour in young and heavily infested animals and more or less severe irritation to humans.

The genus *Haemaphysalis* probably arose in tropical, humid southern Asia with Paleozoic or early Mesozoic reptiles. On the smooth-skinned hosts, adults of these early parasites had clavate palpi and few if any spurs; they were possibly larger (more than 3 mm long) than most contemporary haemaphysalids (1–3 mm long) and resembled the rare *H. vietnamensis*, which is known only from Vietnam highland vegetation.

When mammals and birds evolved, it seems that smaller haemaphysalids evolved. Their palpi became compact and broadened, and various spurs and other hair- or feather-hooking devices developed on the coxae and on various capitular segments.

Relatively few haemaphysalid species spend their entire active life on birds. Immatures and adults of some 25 species sometimes parasitize birds but are common on mammals. Both immatures and adults of well over 100 *Haemaphysalis* species parasitize mammals but seldom, or never, reptiles or birds. The immature stages of species whose adults parasitize wandering artiodactyls, perissodactyls and carnivores usually parasitize smaller hosts such as burrowing rodents and insectivores. Adults of few species are found on domestic cattle, sheep and goats.

About 28 species associated with other mammalian groups (marsupials, insectivores, primates, rodents, hyraxes, hares and rabbits, porcupines) fall into several biological groups. Evidence is available for most of these species to show that all the active stages feed on the same kind of host.

5.4.4 The genus Aponomma

About 26 species of eyeless, sometimes ornamented, *Amblyomma*-like ticks comprise the genus *Aponomma*. All but three species feed on large snakes and varanid lizards. New World iguanid lizards are parasitized by *Amblyomma* rather than by *Aponomma* ticks. Eight Australian *Aponomma* species are grouped separately, two parasitize the marsupial echidna and wombat rather than reptiles. One species has been taken from dogs in Poland.

Little is known about *Aponomma* biology except that adults and immatures are frequently found on a single host. These ticks attach under the scales of snakes or in the axillae or on the underside of the neck of lizards.

5.4.5 The genus Amblyomma

The genus *Amblyomma* contains about 100 species. Most are large, brightly ornamented, tropical ticks with long mouthparts. Of the four species from which seven different viruses have been isolated, the common African *A. variegatum* is notable for being infected by five, including Crimean–Congo hemorrhagic fever virus (table 18.2). Few other *Amblyomma* species have been investigated for viruses.

The taxonomic monograph on *Amblyomma* (Robinson 1926) requires some updating but provides an excellent perspective of the genus. Immature-stage identity in many regions is uncertain. The North American species summaries by Cooley and Kohls (1944) and Bishopp and Trembley (1945) have been brought up-to-date in more recent papers by Kohls. The South American species remain mostly poorly known.

No *Amblyomma* inhabit the Palearctic Region. The few species of southern USA are the only amblyommas from temperate regions. However, nymphs of some species are carried on birds migrating from Africa to Europe and southwestern Asia, where some moult to adults which feed on local wild and domestic mammals and may transmit pathogens acquired in Africa. *Amblyomma*-infested migrating birds also carry several other tick species from which they may receive a variety of infectious agents to share among their parasitic passengers.

All *Amblyomma* probably have a three-host life cycle. A few species may be parthenogenetic (Oliver 1971).

As with *Haemaphysalis*, the hosts of immatures and adults of many *Amblyomma* species differ considerably. Immatures of this group infest birds, small mammals or rodents. Adults of most African species feed chiefly on large herbivores, but rarely on carnivores. Domestic animals are frequently more heavily infested than native animals. Immatures of many species, and often also adults, infest humans.

Host-relationship patterns vary among *Amblyomma* species. However, the ubiquity of records of reptile hosts suggests a concurrent origin in time with *Haemaphysalis* and *Amblyomma*. Equally interesting, though unexplained, is the frequency that amblyommas parasitize relatively hairless wild mammals such as pig, tapir, elephant, rhinoceros, pangolin and armadillo.

Two species of *Amblyomma* found in Central and South America are the only ixodid ticks to infest amphibians. Several amblyommas are known only from certain reptiles, including sea snakes. The huge *A. tuberculatum* parasitizes American land tortoises; larvae also feed on wild and domestic birds. Each *Amblyomma* species of the Galapagos Islands is most numerous on one of three hosts, the giant tortoise, marine iguana, or land iguana, but occasionally infests any of these reptiles.

5.4.6 The genus Dermacentor

This is the last genus of the subfamily Amblyomminae (sect. 5.4.1). It is perhaps phylogenetically the most recent of Amblyomminae genera and contains 31 species. Nine species are known to be infected by nine different viruses (table 18.2), including five members of the Russian spring-summer complex (group B) and by Colorado tick fever virus (table 18.1).

The taxonomic monograph on *Dermacentor* by Arthur (1960c) provides a generic perspective of sorts but should be used with caution (Philip 1960; Kohls 1961). *D. raskemensis* has been redescribed (Dhanda et al. 1971), and I am revising the so-called '*D. auratus*' group.

The *Dermacentor* ticks are mostly moderate to large-size and superficially similar in appearance (sometimes distressingly so!). The dark scutum is often extensively brightened by pale enamelling but this feature, as well as the very variable body and eye size and shape, is unreliable for identification. Immatures are generally not difficult to identify, but few have been properly studied.

Dermacentor ticks are primarily parasites of mammals. Few or none are found on reptiles or birds. Three American species have a single host, a large artiodactyl or perissodactyl. Other species have a three-host life cycle type. Immature stages of the three-host species parasitize insectivores, lagomorphs and rodents, the adults infest large mammals of all types including man.

5.4.7 The genus Rhipicentor*

Only two poorly studied African ticks represent the genus *Rhipicentor* (Theiler 1962), their adults feed chiefly on carnivores, hedgehogs and porcupines. Immature stages are unknown.

----

* *Rhipicentor* and the following genera (sect. 5.4.7–5.4.14) constitute the subfamily Rhipicephalinae (sect. 5.4.11) which is chiefly Palearctic and Ethiopian in distribution, with a few representatives in the Oriental Region. None are indigenous to Australia or the Americas. Adult Rhipicephalinae are associated chiefly with medium- to large-size artiodactyls and perissodactyls, and also carnivores. Adults of a few species parasitize tortoises, hyraxes, rodents, etc.; these species are structurally somewhat atypical and in the case of *Anomalohimalaya* (sect. 5.4.12) remarkably so. In this subfamily, birds are often hosts of immatures and some adults. Birds help maintain some *Hyalomma* (sect. 6.4.9), but no Rhipicephalinae species entirely depends on birds.

5.4.8 The genus Cosmiomma

The only species in this genus, *Cosmiomma hippopotamensis*, is a large, colourful, little known tick found in scattered localities of southern Africa (Arthur 1960c; Theiler 1962; Serrano 1964). Adults are found on black rhinoceros, hippopotamus and domestic goats, but the immature-stage hosts are not known.

5.4.9 The genus Hyalomma

The genus *Hyalomma* consists of 21 recognized species, three of which are each divided into three subspecies. Nine different viruses have been isolated from seven of the species (table 18.2) and five are infected by Crimean–Congo hemorrhagic fever virus (table 18.1).

The chapter on *Hyalomma* by Hoogstraal (1956) reviews the genus and lists most of the earlier literature. More recent reports are in papers by Hoogstraal, by Kaiser and by Balashov (1968). Immature stages remain exceedingly difficult to identify.

Hyalomma probably originated in semidesert or steppe lowlands of Central Asia and are nowadays found in the Palearctic, Oriental and Ethiopian Regions. No species is common at high altitudes or in areas lacking a long dry season.

Hyalomma ticks are generally large, tough and hardy. Adults wait for hosts in the entrance of rodent burrows or under shade plants. When a potential host is sensed, these ticks run to the attack. If they become desiccated while waiting they shelter in a humid place until their water balance is restored.

Most *Hyalomma* have a three-host life cycle though fewer hosts may be parasitized (Balashov 1968). The *H. marginatum* complex and a few other species typically infest two hosts, and other species have a one-host cycle. Parthenogenesis occasionally occurs in several species but is apparently unimportant in maintaining the population.

The three subgenera are *Hyalommasta*, *Hyalommina* and *Hyalomma*. The first is monotypic and a parasite of tortoises. The second consists of five poorly studied species. The third subgenus has 21 members distributed in Africa and Eurasia on a wide range of hosts.

The original ecological and geographical distribution, densities, and hosts of these ticks have been vastly influenced by the presence of numerous domestic mammals and by modified agricultural, rural and urban environments. For instance, optimum conditions for dense populations of *H. marginatum* are created where heavily grazed pastures are bordered by trees supporting large nesting colonies of rooks, a bird that is favoured by agriculture (sect. 18.4.7). Thus, the cattle industry and human tolerance of certain inedible birds foster the contemporary occurrence and incidence of Crimean-Congo haemorrhagic fever.

Most or all hyalommas except the relict species are important economically and are vectors of various pathogens of man and other animals.

5.4.10 *The genus Nosomma*

A single species, *Nosomma monstrosum*, occurs in India and Southeast Asia on man, wild boar, bear, domestic cattle, buffalo, horses and dogs (Arthur and Chaudhuri 1965; Avsatthi and Hiregaudar 1971). Immatures of this large, dark, three-host tick parasitize rodents (Singh 1968).

5.4.11 *The genus Rhipicephalus*

The genus *Rhipicephalus* contains about 63 species, eight of which have been found to be infected by nine or ten different viruses (table 18.2).

Fifty-one rhipicephalid species are Ethiopian, seven Palearctic, and five Oriental. They infest rodents, carnivores, herbivores and other mammals but seldom birds or reptiles. Most African rhipicephalids have a three-host life cycle, utilizing different kinds and sizes of vertebrates as hosts for immatures and adults (see papers by Theiler). About half of the described species are quite localized and poorly known.

5.4.12 *The genus Anomalohimalaya*

A single species, *Anomalohimalaya lama* is found infesting rodents in the alpine desert of Nepal at 12,000 feet altitude (Hoogstraal et al. 1970). This genus appears to be related to *Rhipicephalus*.

5.4.13 *The genus Boophilus*

Boophilus ticks are infected by at least five arboviruses (table 18.2) and are of veterinary importance in most tropical and subtropical areas (Hoogstraal

1956, 1972). These parasites of certain Artiodactyla and Perissodactyla infest a single host, and prospered when man introduced domestic cattle and horses into their habitats. Five *Boophilus* species are now recognized.

The famed 'Texas fever tick', *Boophilus annulatus* has long been considered to be an American species. However, it probably originated in southwestern Asia (where it has been known as *B. calcaratus*) and was transported to the Americas where it is firmly established only in certain Mexican states with low daytime relative humidity (O. H. Graham, personal communication). Population samples from the Americas and those from the southwestern Palearctic and Ethiopian Regions are indistinguishable. All other members of the subfamily Rhipicephalinae (about 100 species and subspecies) are Palearctic, Ethiopian, or Oriental in origin. Thus, it is more likely that *B. annulatus* was introduced into Mexico or Texas from the Palearctic Region.

B. annulatus is known chiefly as a parasite of cattle and deer, and occasional specimens, usually immature, are found on deer in areas where cattle are infested, though deer apparently are not important hosts (O. H. Graham, personal communication). Many papers on the biology of *B. annulatus* in southwestern USA were published in the early years of this century. It is often common in scattered areas of Africa north of the Equator and elsewhere, but has not been well studied in these areas.

Two species are restricted to the Ethiopian Region. *B. decoloratus* occurs in much African savanna and *B. geigyi* in West Africa. The higher abundance rates of these species on domestic cattle and horses than on wild antelopes and buffalo may be due to resistance to boophilids acquired long ago by their natural hosts. Cattle and horses are infrequently exported from Africa, and these ticks are unknown elsewhere.

In the Oriental Region, *B. microplus* infests forest-dwelling antelopes, deer and wild and domestic cattle. It has been transported far and wide, probably chiefly on zebu cattle, and has become a serious economic problem in Australia, New Guinea, Madagascar, Taiwan, South and Central America, and parts of southern Africa. This exceptionally adaptable tick is even found on yaks high in the Himalayas and on the Tibetan high plateau. The important biological studies on *B. microplus* undertaken by CSIRO scientists in Australia were reviewed by Hoogstraal (1970b).

5.4.14 The genus Margaropus

Margaropus ticks are essentially relict boophilids (Hoogstraal 1956). Giraffes are parasitized by two species in Central Sudan and in East Africa. *M. winthemi*, which possibly fed originally on the giraffe or zebra, now infest domestic horses during winter at high altitudes in southern Africa.

CHAPTER 6

Acarina (mites)

D. C. LEE

Contents

6.1 Recognition and classification

The Acarina (or Acari), mites and ticks, is a vast group of mainly minute arachnids. Its members differ from most other arachnids, except spiders, largely in the absence of abdominal segmentation, and they differ from the spiders by having their mouthparts in a discrete gnathosoma and by having that part of the body which carries the legs (podosoma) broadly fused to the posterior abdomen (opisthosoma).

Two major groups are distinguished by the presence or absence of actinochitin in the setae and these are named either the Parasitiformes and Acariformes (Krantz 1970) or the Anactinochaeta and Actinochaeta (Evans et al. 1961). As the former of the two groups includes species called ticks and some species called mites these common names have little taxonomic significance, although some convenience.

Each major group is further divided depending on the position of the external openings of the respiratory system, thus the Anactinochaeta is divided into the Notostigmata (few species), Tetrastigmata (few species), Mesostigmata and Metastigmata (ticks), and the Actinochaeta into the Prostigmata, Astigmata and Cryptostigmata.

6.2 Development and feeding habits

Most mites hatch as a 6-legged larva, followed by one or more 8-legged nymphal stages before becoming an 8-legged adult, but some species are viviparous and some stages may be lost. Although the active stages of a species are usually similar, the parasitic larvae of some Prostigmata differ from the free-living nymphs and adults of the same species, and the non-feeding hypopial nymphs of certain Astigmata differ drastically from other stages. Adults of a few species are polymorphic and the different forms of one sex were often originally recorded as distinct species.

The majority of mites are free-living and use segmented palps and chelate-dentate or stylet-like chelicerae to feed on decaying organic matter or living organisms. The Mesostigmata and Prostigmata ingest only liquid food but probably most Astigmata and Cryptostigmata can ingest solids. Saliva, containing enzymes for pre-oral digestion, is secreted near the mouth which opens into a strongly muscled, suctorial pharynx.

The majority of mites parasitic on invertebrates and plants are species of Prostigmata while mites parasitic on vertebrates are well represented in the

Mesostigmata, Prostigmata and Astigmata. There are no parasitic species in the Cryptostigmata or in the rare Notostigmata and Tetrastigmata. The chelicerae of prostigmatid parasites are stylet-like or scimitar-like while the mesostigmatid and astigmatid parasites have chelate chelicerae, although in the former group they are lengthened and the digits reduced so that they resemble stylets. The saliva of haematophagous species may contain anti-coagulants. Most astigmatid parasites feed on epidermal cells, secretions, fur or feathers. The mesostigmatid and prostigmatid parasites include the majority of species feeding on vertebrate blood or lymph. Trombiculid larvae or chiggers are prostigmatid parasites feeding on vertebrate lymph, while their nymphs and adults are free-living predators. Some chiggers transmit rickettsiae, but are not known to transmit viruses.

6.3 Mites and viruses

Viruses are cited as being associated with mites belonging to three families: Dermanyssidae (sensu Evans and Till 1965), Mesostigmata; Tetrany-chidae (sensu Pritchard and Baker, 1955), Prostigmata; Eriophyidae (sensu Keifer 1964), Prostigmata (fig. 6.1).

Amongst the Dermanyssidae one facultative ectoparasitic species in the Laelapinae is reported to transmit a virus disease of rodents. All other reports refer to three species – *Dermanyssus gallinae* (Degeer), (Dermanys-sinae), *Ornithonyssus bacoti* (Hirst) and *O. sylviarum* (Canestrini and Fan-zago) (Macronyssinae) – in two subfamilies all species of which are obliga-tory, haematophagous ectoparasites of reptiles, birds, or mammals. They are substantially host-specific to one host class, although, for example, *O. syl-viarum* which normally occurs on birds may temporarily infest man. These species are superficially similar to the free-living dermanyssid species, and their morphological adaptations to parasitism are discussed by Evans and Till (1965). Unengorged, the whitish females, 600–800 μm in length, are translucent, but their length increases by up to 30% after a blood meal, when they become red or black showing the colour of ingested blood. Egg laying only follows a blood meal and development from egg to adult takes 5–16 days through 3 stages (larva, protonymph and deutonymph). Some species lay eggs on the host and all the following stages usually stay on the host. Other species lay eggs in the animal's nest or its equivalent, and their non-feeding stages (all larvae and macronyssine deutonymphs) usually also exist away from the host. Females survive between 21 and 150 days without feeding. Females of some species have been shown to be facultatively parthenogenetic; unfertilized eggs produce only males. Further information on the biology of parasitic dermanyssids is given by Bertram et al. (1946); Camin (1953); Furman (1959); Kirkwood (1968); Sikes and Chamberlin (1954); and Skaliy and Hayes (1949).

The Tetranychidae are exclusively plant parasites, and many are important pests of agricultural crops. Two species of *Tetranychus* – *T. urticae*

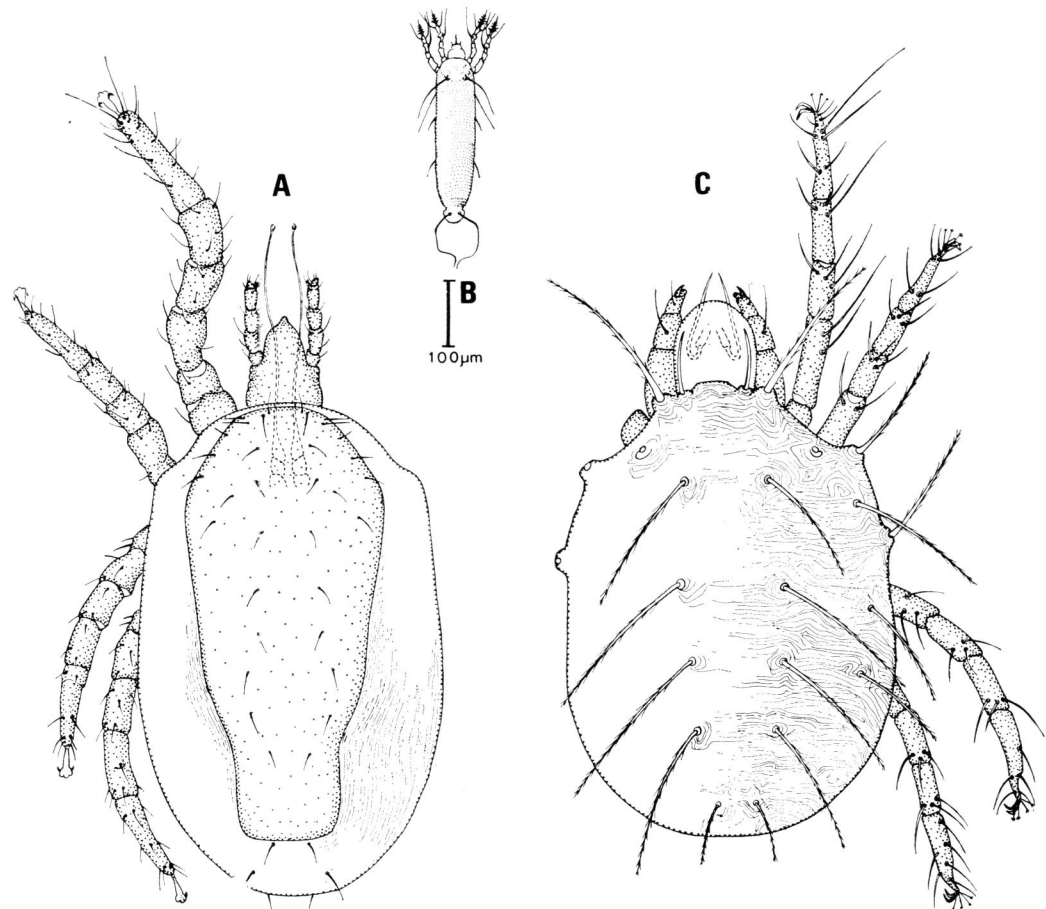

Fig. 6.1.   Acarina associated with viruses. A) *Dermanyssus gallinae* (Degeer); B) *Aceria tulipae* Keifer; C) *Panonychus ulmi* (Koch).

(Koch) and *T. cinnabarinus* (Boisduval), syn. *telarius* (L.) – have been reported to transmit virus diseases, and two species of *Panonychus* – *P. citri*
(McGregor) and *P. ulmi* (Koch) – have been infected with viruses in an
attempt to control their numbers. Their host range is often wide. The above
species are superficially similar to many free-living Prostigmata. The females,
600–800 μm in length, are green, yellow, orange or red in colour. Cheliceral
bases are fused to form an evertible stylophore and the two movable cheliced-
al digits are drawn out into flagelliform stylets. Many species (including all
four listed above) produce silk strands from glands opening on the rostrum
or palps. Development from egg to adult takes 7 (only males) to 27 days
through a larval and one (only males) or two nymphal stages. All stages are
found on the leaves of the host, protected by a silk webbing made by the
mites, except for special polymorphic resistant stages. Thus *Panonychus*
species have a special diapausing winter egg which is laid on the twigs and
branches, while *Tetranychus* species in temperate regions have a special
diapausing winter female which is found in soil, litter or crannies in wood.
Unfertilized females of some species produce only males. Dispersal is aided
by the silk strands in a manner similar to the 'ballooning' of spiders. Further
information on the biology of particular species is given in the following
papers: *T. urticae* by Williams (1954); *Panonychus citri* by Beavers and
Hampton (1971) and Boyce (1948); *P. ulmi* by Blair and Groves (1952) and
Lees (1953).

   The Eriophyidae are exclusively plant parasites, and some are important
pests of agricultural crops. The five species listed by Slykhuis (1962) –
Cecidophyopsis ribis (Westw.), *Aceria tulipae* Keifer, *A. ficus* (Cotte),
Eriophyes insidiosus Keifer and Wilson, *Abacarus hystrix* (Napela) – as
transmitting viruses all belong to this family as defined by Kiefer (1964).
Particular species usually have a limited host range, often one or two similar
perennial species, but *A. tulipae* is exceptional in occurring both underground
in bulbs of onion and tulip as well as on wheat leaves. The eriophyoids (the
superfamily includes two other families) are atypical amongst the Acarina
in that they are maggot-like in appearance and all active stages have only
4 legs. Although possibly more closely allied to the Tetranychoidea than any
other mites, the relationship must be a distant one. The females, 150–250 μm
in length, are white, yellow or orange in colour. The palps form a rostrum,
which is grooved to house the stylet-like chelicerae. Development from egg
to adult takes 6–25 days through a larval and a nymphal stage (which
becomes a resting pseudopupa). Some species have two types of female
(protogyne and deutogyne). The deutogynes are a special diapausing stage

which move from the leaves to the lateral buds, twigs or branches before leaf fall. Hall (1967a) states that some protogynes may produce larvae without any separate egg stage. Unfertilized females of some species produce only males. The males of some species are either rare or not known. All stages are found on the host. Some species induce the host plant to form protective galls, while others are vagrants or rust mites which produce no galls. Gall producers tend to be attenuated and soft, and vagrants tend to be chunkier with heavier dorsal plates. The dispersal of eriophyids is by wind, and on insects and other animals. The mites move off the leaf tips where they stand erect, attached by their anal suckers, and then leap into the air or on to an insect. *A. tulipae* individuals use their anal suckers to group themselves into chains before becoming air-borne. Dispersal by man as a result of budding or grafting, or on fruit, such as wheat kernels, also occurs. An introduction to the biology of eriophyids and their adaptation as parasites of plants is given by Hall (1967b).

CHAPTER 7

Aphids and psyllids

V. F. EASTOP

Contents

7.1 Introduction

Psyllids are usually placed systematically near aphids and coccids, but their closest relatives are probably the Aleyrodidae (ch. 13). Many authors have pointed out similarities between psyllids and leaf hoppers (Auchenorrhyncha). Psyllids, together with aleyrodids, probably represent one evolutionary line, while aphids and coccids constitute another.

7.2 Aphids (Aphidoidea)

7.2.1 The adult

About 3800 species of aphids have been described. They are mostly soft bodied insects from 1–7 mm long and are usually yellow, green, brown or red, when intense the darker colours may appear black. Forbes and Maccarthy (1969) have reviewed the morphology of aphids. Forbes (1964) gave an account of the gut of *Myzus persicae* and Forbes (1969) and Forbes and Mullick (1970) gave accounts of the stylets of aphids. Auclair (1963) summarised the information on aphid nutrition, which may be aided by symbiotic organisms aggregated into organs known as mycetomes or mycetacytes. Malpighian tubes are absent from aphids. Many aphids bear a pair of abdominal organs known as cornicles or siphunculi which secrete triglycerides (Gilbert 1967; Strong 1967). Wax secretion through dermal glands occurs in many aphids and the wax may occur as powder or threads. Aphids secreting wax threads usually have evident abdominal wax glands. The wax may reduce water loss and also prevent contamination by honeydew. Strong (1963) reported analyses of the lipids from 19 species of aphids. By feeding on sap, aphids acquire more water and sugar than they need and excrete it as honeydew. The enormous literature concerning honeydew extends over about 2000 years: two recent papers are by Bacon and Dickerson (1957) and Schäller (1961). Diploid chromosome numbers between 4 and 40 have been recorded for aphids but the commonest diploid numbers are 8, 10, 12 and 14. Recent authors such as Sun and Robinson (1966) have rarely recorded numbers higher than 20. Heie (1967, 1969) gives accounts of fossil aphids. More detailed sources of information about aphids are to be found in the *Handbook of Aphid Technology*, ed. H. F. Van Emden, prepared in conjunction with the International Biological Programme. More than 200 species of aphids have been recorded as vectors of plant/virus diseases

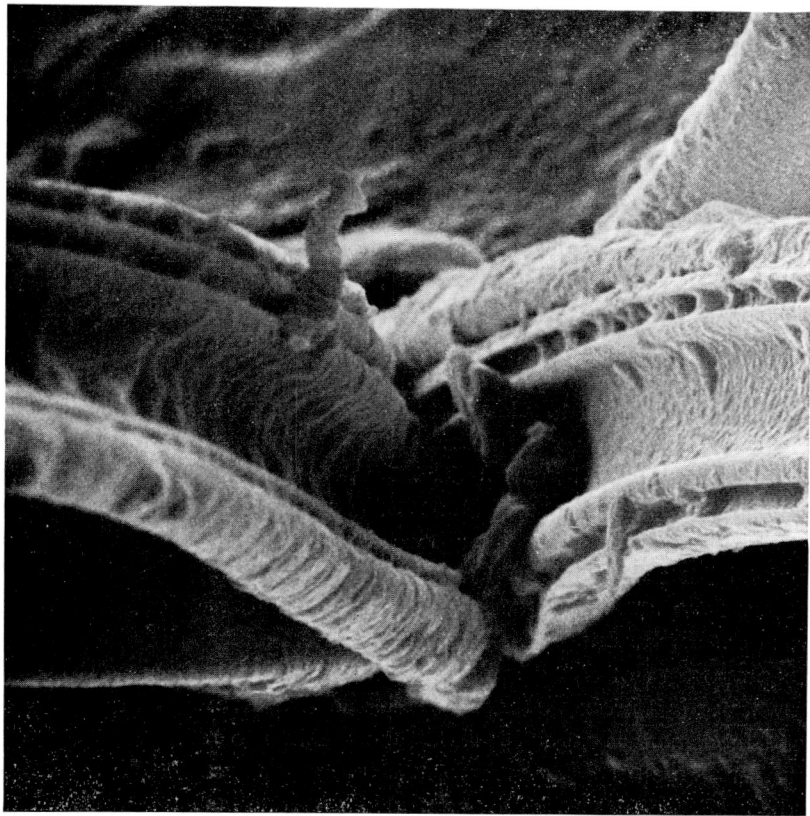

Fig. 7.1. *Tuberolachnus salignus* (Gmelin), maxillary stylets emerging from 'labrum showing the wide food canal and narrow salivary canal associated with locking mechanism. Magnification × 8000. (Courtesy of British Museum.)

(Kennedy et al. 1962) and one of these, *M. persicae*, is the vector of 120 virus diseases. Watson (1967) gives an account of the epidemiology of aphid-transmitted plant virus diseases. Van Emden et al. (1966) have reviewed the ecology of *M. persicae*.

7.2.2 Migration

Johnson (1969) has reviewed the information concerning the flight of aphids and other insects. Aphids take off only when light and temperature are above certain thresholds. Since the rate of development is also affected by temperature, a daily 'double peak' of flight from the host plant is characteristic for many aphids in Western Europe; aphids maturing during the night fly during the morning as light and temperature become more favourable. The midday lull after the accumulated insects have flown is followed during the afternoon by the flight of the individuals whose development has been accelerated by the higher daytime temperatures. In the tropics some species have only one peak of flight activity and this may be in the morning or in the afternoon. *Aphis nerii* flies mostly in the morning in the forest zone of Nigeria while the banana aphid, *Pentalonia nigronervosa*, flies mainly during the afternoon. In Europe there are few aphids in the air at night and this could restrict the range of an afternoon flying aphid. In the tropics two factors may increase the range of afternoon flying aphids. More aphids continue to fly at night and the occurrence of low level jet streams may provide a means of rapid transport for a few hundred miles before nightfall. Aphid populations may travel several hundred miles under suitable conditions and such transport may be a regular occurrence in parts of North America and Australia.

The source of aphids arriving on crops in Britain has seldom been determined but the inverse correlation of virus infection of a crop with the distance from an infested source suggests that under conditions favouring virus spread many of the aphids fly only short distances before alighting. The proportion of flying aphids which travel different distances is not known, nor is it probable that all species behave alike. When the pattern of aphid movement is known it should be possible to reduce the sources of virus. But there is insufficient information on the migration of even the best studied aphids, such as *Myzus persicae*. Enormous numbers of *M. persicae* fly from peaches in Italy, France, Spain and other peach-growing areas in Europe every year and the peaches are reinfested in the autumn, but it is not known where these populations spend the summer nor what proportion of the return migrants in the autumn are the progeny of those individuals which left peach in the spring. Nor is it known what proportion of the British potato crop is infested from populations of *M. persicae* which overwintered parthenogenetically and what proportion originated from *Prunus*, possibly in continental Europe. Similarly large flights of aphids arrive on British cereals in spring or early summer, but the proportion which come from permanent pasture and the proportion which come from other cereals, the proportion which are home grown, and the proportion which have come from overseas, is not known. However, it is probable that most populations of *Aphis fabae* in Eastern Britain at least originate from eggs laid on either *Euonymus*, *Viburnum* or in suburban areas *Philadelphus*.

The tendency of aphids to fly from their larval host soon after maturity aids their colonisation of temporary habitats such as the annual crops used in rotations. As a result of this behaviour the aphid populations on a crop are often not directly descended from the individuals colonising the crop in that area the previous year. For this reason local, geographically limited races and strains are seldom detected in aphids. Races and strains tend to isolate themselves not on geographical grounds but on the basis of behaviour, i.e. host plant selection. The absence of local populations simplifies the picture of virus transmitting ability but the picture is complicated by the presence of biotypes with different host plant preferences and virus transmitting abilities. The absence of local populations also complicates control measures. Fields may not only be reinfected by winged aphids after chemical control but methods such as the release of sterile males would have to cover a very large area to be effective for aphids. Some predators have a vagile biology like that of aphids, while others are much less motile. In southern England *Adalia bipunctata* seems to have populations resident in an area, for instance each year from 1963 to 1971 the dark variety constituted 2–3 % of the population at Kew. *Coccinella 7-punctata* and *C. 11-punctata* appear sporadically and suddenly, show little morphological variation, and seem to behave more like the aphids upon which they feed.

7.2.3 *Life cycle*

Most aphids have at least 7 generations each year, including at least 5 different adult forms or morphs. Lees (1966) and Hille Ris Lambers (1966) have discussed the control of polymorphism. Most aphids overwinter in the egg stage, when the biology is said to be holocyclic. Species or populations which overwinter parthenogenetically without the production of the sexually derived overwintering egg are said to be anholocyclic. Eggs of aphids are 0.35–0.9 mm long and from 0.17–0.6 mm wide. The embryo developing within the egg is armed with an egg-breaker worn on the embryonic cuticle like a skull cap. The form developing from the overwintering egg is known as the fundatrix. Like most aphids it moults four times before maturity. The adult fundatrix is always a parthenogenetic female and in most groups it is wingless. Alate fundatrices almost only occur in groups where apterous viviparous females are unknown. The progeny of the fundatrices may develop into either winged or wingless parthenogenetic viviparous females, depending on the species of aphid, their position in the sequence of young and on the environmental conditions. In most species of aphids the fundatrices are succeeded by one or two apterous parthenogenetic generations after which an alate parthenogenetic viviparous generation occurs. The progeny of this alate generation is again parthenogenetic apterous viviparae and a succession of apterous parthenogenetic viviparous generations may continue until autumn, when males and the usually apterous oviparous females which require fertilisation occur. If the populations of apterous viviparae become overcrowded during the summer, or if the host plant starts to wilt, alate viviparae may be produced. Alate aphids may not be able to fly for all their adult life. In the Aphidinae the wing muscles degenerate after a few days.

About 10% of the known species of aphids alternate between primary host plants on which the overwintering egg is laid and secondary host plants on which the parthenogenetic viviparous generations occur in the summer. This host plant alternation is known as heteroecy in comparison with monoecious aphids which may spend the whole year on a single species of plant. In heteroecious aphids the second, third or sometimes fourth generations of parthenogenetic viviparae on the primary host are winged and fly to the secondary host where a succession of apterous parthenogenetic viviparous generations occurs. If these populations become overcrowded alate viviparae may be produced which fly to other secondary hosts of either the same or a different species, and deposit young which develop into apterous viviparae. In the autumn a return migration to the primary host occurs. In the Aphidinae the return migration is accomplished by gynoparae whose progeny develop into the oviparae on the primary host and mate with the winged males which have flown from the secondary host and were usually the progeny of apterous androparae. In the other sub-families where heteroecy occurs and in the Adelgidae, the return migration is accomplished by sexuparae whose progeny develop into both males and oviparae on the primary host plant. The return migrants occur under conditions of short day length. Cultures maintained for virus transmission purposes need supplementary light during the winter to discourage the production of sexuparae.

The apterous exules of some aphids are dimorphic, a small 'summer dwarf' being produced at high temperatures. These small specimens may have the third and fourth antennal segments fused and the aphids may produce fewer young than the larger specimens which develop at lower temperatures. Small specimens contain smaller embryos than larger specimens and the size of the embryos is correlated with the eventual size of the adults into which they develop. The result is a delay in the effect of environment on a population. Thus populations should be preadapted to experimental conditions for at least two generations if it is intended to compare the results of the experiments with results from specimens reared under different conditions. Some aphids exhibit reproductive diapause at high temperatures while *Periphyllus* and some other genera diapause during the summer as resting first instar larvae known as 'dimorphs'.

Laboratory reared insects tend to be reared under optimum conditions for reproductive rate, not for body size, and aphids received for identification from virologists are usually in the lower half of the body size range of the species. Physiology may be deranged in dwarfs and morphological abnormalities may occur. Some of these abnormalities represent losses but others involve duplication. The number of hairs on a structure may decrease with body size over most of the range but rise suddenly near the lower limit of size. One of the physiological processes that may break down is the barrier to transmission of circulative viruses (Rochow and Eastop 1966). Each vivipara produces from 3–176 young, or even more from some fundatrices. Parthenogenetic viviparae become adult from 5–22 days after birth, largely depending on temperature during development. Some species of aphids live in dense colonies while others live more evenly scattered over the plant. Some aphids live entirely on the aerial parts of plants while others live under the lower leaves, or under earthen shelters constructed by ants at the base of the stem of the host plants. Other species of aphids spend some generations on the roots of their host plants. The subterranean aphids are mostly attended by ants, but often more assiduously in the spring than in the autumn. Way (1963) has summarised the information on the association between aphids and ants. Some aphids shelter under webs constructed by Psocoptera or Lepidoptera, others inhabit the deserted galls of psyllids or other aphids or feed on the outsides of the galls of other insects. *Mimeuria* lives in cysts composed of fungus mycelium on the roots of elm.

The saliva of aphids may be toxic to their host plants, inducing various distortions including closed galls, crumpled leaves, shortened internodes and discolourations, some of which may resemble virus symptoms.

7.2.4 Variation

The common occurrence of parthenogenetic morphs makes it easy to rear clonal material for experimental purposes. Many morphological characters are almost constant in similarly sized individuals belonging to one clone of a species. Some characters may vary within a clone. In *Macrosiphum (Sitobion) avenae (= granarium)* the ultimate rostral segment usually bears 6 accessory hairs, although occasional specimens with 5 or 7 accessory hairs occur as in most species of *Sitobion*. Individuals from the same clone of *M. (S.) avenae* may bear from 2–6 hairs on the anterior half of the sub-genital plate, which is the range of variation found within the species throughout its entire geographical range. *Myzus persicae* bears only 2 hairs on the anterior half of the sub-genital plate rather constantly while individuals of the same clone may bear from 2–7 accessory hairs on the ultimate rostral segment, which is the range of variation for this character found throughout the species.

The acquisition and subsequent loss of insecticide resistance within parthenogenetic populations of aphids is another example of the cytogenetical plasticity within clones of aphids. Both morphological characters such as a small or large number of caudal hairs and biological characters such as the ability to produce winged forms or sexuales have been selected within parthenogenetic populations of aphids. Aphid clones do not correspond to vegetatively propagated plant material. The cytogenetical mechanism of aphid parthenogenesis has been described as endomeiosis. Parthenogenesis in aphids is unisexual, not asexual.

This cytogenetical plasticity is probably the result of the combination of parthenogenesis and polymorphism. In order to respond to the environment by the production of morphs a parthenogenetic organism needs a system of suppressor genes and this system has considerable evolutionary possibility, particularly along neotinic lines.

7.2.5 Ecology

Many aphids are specific to a particular host plant but a few species have a wide range of host plants. Of the 515 British species of aphids, 112 are holocyclic on one genus of tree or shrub; 247 are monoecious on one genus of herb; 54 species are heteroecious between one genus of woody plant and one genus of herb and 50 species are heteroecious between one genus of woody primary host and more than one genus of herbs. Eight British aphids have a wide range of host plants and occur on members of many families of plants. 221 of the British species of aphids lay their eggs on trees and shrubs and 273 species lay their eggs on herbs. Most of the polyphagous aphids have evident host plant preferences. Only *Aulacorthum solani* and *Aphis fabae* show no great preference among their summer hosts. *Myzus persicae* is most commonly received from Cruciferae and Solanaceae; *Aphis craccivora* from Leguminosae; *Aphis gossypii* from Cucurbitaceae and Malvaceae; *Aphis spiraecola* and *Macrosiphum euphorbiae* from Rosaceae and Compositae;

Myzus ascalonicus from Rosaceae, Compositae, Amaryllidaceae and Caro-
phyllaceae; *Myzus ornatus* from Rosaceae, Compositae, Labiatae, Leguminosae and Polygonaceae; *Neomyzus circumflexus* from Liliaceae and
Toxoptera aurantii from Sterculiaceae, Rutaceae, Rubiaceae, Compositae
and Theaceae.

The size of aphid populations may change rapidly (Hughes and Gilbert
1968) because of their high reproductive rate. Shaw (1970) describes quali-
tative changes occurring in populations of *Aphis fabae* during the year. Many
species of aphids are sporadic in appearance, being numerous for a few
weeks or months or sometimes years and are then much rarer for several
years. The abundance of specific parasites and predators follows that of the
aphids.

7.2.6 Natural enemies

Aphids have numerous natural enemies. They are preyed upon by bats,
birds, reptiles, spiders and members of many groups of insects including
Mantodea; Dermaptera; Heteroptera including Miridae, Reduviidae,
Pentatomidae, Nabiidae, Anthocoridae and Microphysidae; Thysanoptera;
Neuroptera; Lepidoptera, Pyralidae; Diptera including the larvae of Cecidomyidae, Syrphidae, Chamaemyidae and Chloropidae and the adults of
Empidae and Dolichopodidae; Hymenoptera, Sphecidae; Coleoptera including Carabidae, Staphylinidae, Elateridae, Cantharidae, Derontidae, Phala-
cridae and Coccinellidae. Mites belonging to the Trombidiidae and Erythra-
oidea are known as exoparasites of aphids. Internal parasites of aphids
include Diptera, Cecidomyidae and Hymenoptera including Ichneumoidea,
Braconidae, Aphidiinae; Cynipoidea, Cynipidae, Charipinae which are
mostly hyperparasites through Aphidiinae; Chalicidoidea, Pteromalidae,
Sphegasterinae and Microgasterinae which are secondary parasites and
Eulophidae, Aphelinae which are primary parasites; Proctotrupoidea,
Ceraphrontidae, Ceraphrontinae and Megaspilinae, mostly secondary parasites through Aphidiinae and Aphelinae primary parasites.

Nematodes are recorded from a few root-feeding aphids and many aphids
are attacked by fungi, particularly by Entomophthoraceae but also by
Cephalosporium of the Mucedinaceae. Bacteria have also been recorded from
aphids. Thompson and Simmonds (1964) provide a catalogue of the predators and parasites of aphid pests. The most troublesome of these natural
enemies in experimental cultures are the predaceous cecidomyids, the parasitic Aphidiinae and the Entomophthoraceae. Coccinellidae and Syrphidae

are more evident in the field. The importance of other groups with active predatory forms such as Dolichopodidae, especially when nocturnal such as Carabidae and Staphylinidae is difficult to evaluate. Nijveldt (1969) has given an account of the zoophagous Cecidomyidae; Stary (1966) of European Aphidiinae; Hodek (1967) of aphid-eating Coccinellidae and Schneider (1969) of aphid-eating Syrphidae. Despite their probable economic importance the taxonomy of many of the natural enemies of aphids is poorly understood. The cecidomyid genera *Aphidoletes* and *Endaphis*; the aphid-eating Chamaemyridae; the braconid genus *Aphidius*; the cynipid sub-family Charipinae; the Proctotrupoidea bred from aphids; the coccinellid genera *Pullus* and *Scymnus* and the trombidiid mites all need revision. The ecological requirements of potential biological control organisms are not likely to be discovered until it is possible to recognise the species accurately. Bad weather may result in the reduction of aphid colonies but how much of the reduction is due to dispersal and how much to mortality, and how much of the mortality is due to ground living predators is not clear.

7.2.7 *Major groups*

The Aphidoidea can be divided into three families, the Aphididae with a number of sub-families, the Adelgidae and the Phylloxeridae. Each of these groups has its own characteristic biology.

7.2.7.1 *Aphididae*
By far the largest group, containing about 3700 of the 3800 species of Aphidoidea which have been described. The Aphididae have viviparous parthenogenetic forms, siphunculi are often present and the cauda is often elongate or knobbed. The vein Rs is present in the fore-wing and the antennae of the alatae are usually 5- or 6-segmented and of the apterae 4- to 6-segmented.

7.2.7.1.1 The sub-family Aphidinae with about 1600 species in the world, contains most of the aphids recognised as 'greenfly' and 'blackfly'. The group has a world-wide distribution but most species are from the temperate regions of the northern hemisphere, roughly the holarctic and northern part of the oriental region. Many virus vectors come in the tribe Macrosiphini ($\pm$ 1100 species) used here to include Myzini and Anuraphidini as their limits are ill-defined.

The primitive biology of the Macrosiphini is alternation between Rosaceae and herbaceous Angiosperms but alternation no longer occurs in some genera and many species. Most of the species live on the 'higher' dicotyledons (i.e. Rosidae (450 species) and Asteridae (630 species)) of Cronquist (1968) and Takhtajan (1969), but 160 species live on Monocotyledons, 120 species on Dilleniidae, 80 species on Caryophyllidae, 45 species on Magnolidae, 32 species on Hamamelidae, 30 species on ferns, 7 species on conifers, 9 on mosses and one species on *Equisetum*.

Typical genera of migrating Macrosiphini are *Anuraphis* heteroecious between *Pyrus* and secondary hosts in the Umbelliferae and Compositae; *Dysaphis*, Pomoidea to Umbelliferae, Campanulaceae and a few other plants; *Brachycaudus*, *Prunus* to Compositae and other plants; *Ovatus*, *Crataegus* and close relatives to Labiatae; *Myzus* and *Phorodon*, *Prunus* to various plants; *Macrosiphum*, *Metopolophium* and *Longicaudus* migrating from *Rosa* to various plants including Gramineae. *Macrosiphum (Sitobion) avenae (= granarium)* known as the English grain aphid in America is well known on cereals. Until recently the somewhat similar species *M. (S.) fragariae (= avenae* sensu Börner nec F.) was known only from Europe and the United States west of the Rockies where it was known as *M. harpagorubus*, when collected on its primary host plant, *Rubus*. *M. (S.) fragariae* has recently been introduced to South Africa and Tasmania. It seems inevitable that aphids which can live anholocyclicly on Gramineae will be transported from airfield to airfield.

Some genera of Macrosiphini have acquired other primary host plants. *Capitophorus* alternates between Elaeagnaceae as primary host and Compositae and *Polygonum* species as secondary hosts; *Cavariella* sp. have *Salix* sp. as their primary hosts; *Ceruraphis* have *Viburnum*; *Hyadaphis* and *Rhopalomyzus* have *Lonicera; Nasonovia* and *Hyperomyzus* have *Ribes* sp. as primary host plants. *Shinjia* alternates between *Viburnum* and *Pteridium* a. *Idiopterus* and some species of *Amphorophora*, *Macromyzus* and *Micromyzus* are known only from ferns. Some *Elatobion* and *Masonaphis* species live on Coniferae. Several species live anholocyclically without producing a sexual generation. Three *Myzus* species, *M. ornatus*, *M. ascalonicus* and *M. cymbalariella (= cymbalariae* Stroyan) are now widespread although they all first appeared during the last forty years. *Neotoxoptera* species are also mostly anholocyclic and their apterae resemble those of *Myzus ascalonicus* and *M. cymbalariella* with which they also share a discontinuous host plant preference for Caryophyllaceae, *Viola* and *Allium*.

Other well-known Macrosiphini include *Aulacorthum solani* and *Macrosiphum euphorbiae* with a wide host plant range; *Amphorophora rubi* ssp.

idaei on *Rubus idaeus* s. str. and *A. agathonica* on *Rubus idaeus* ssp. *strigosus;* *Chaetosiphon* (= *Pentatrichopus)* species on Rosaceae including strawberry; *Brevicoryne brassicae* and *Lipaphis erysimi* (= *pseudobrassicae)* on Cruci-ferae; *Coloradoa, Macrosiphoniella* and *Pleotrichophorus* species on Anthe-midae; *Dactynotus* (= *Uroleucon)* species on Compositae; *Pentalonia nigronervosa,* the banana aphid, on Musaceae, Zingiberaceae and Araceae; and *Rhopalosiphoninus* species on the etiolated parts of various plants.

The 585 species of the tribe Aphidini are readily separable into two sub-tribes. The Rhopalosiphina with 65 species have Rosaceae as the primary hosts, and the secondary hosts are mostly Gramineae. *Rhopalosiphum nymphaeae* alternates between *Prunus* and secondary host plants in many families of Monocotyledons, Dicotyledons and even ferns. The secondary hosts are all aquatic plants and the common factor may be well developed intercellular air spaces. *Rhopalosiphum padi* alternates between *Prunus padus* and Gramineae but is anholocyclic on Gramineae over most of its geo-graphical range which is world wide except for low altitudes in the tropics. *R. maidis* is also anholocyclic over most of its range although it may alternate in Eastern Asia as males seem more common there than elsewhere. *R. rufi-abdominalis* has *Prunus mume* as primary host and goes to the subterranean parts of both Gramineae and dicotyledons, although it too is anholocyclic over most of its geographical range. *Schizaphis graminum,* the 'green bug', is widespread in the warmer parts of the world on small grain cereals and grasses. *Hysteroneura setariae* which originally alternated between *Prunus* and Gramineae in North America is now anholocyclic on Gramineae in South and Central America, appears to have been air-lifted to West Africa during the 1939/45 war and has recently been found in Hawaii, Philippine Is., Hong Kong, Vietnam, India and Eastern Australia. The sorghum and sugar-cane aphid known as *Longiunguis sacchari* may only be a summer form of *Melanaphis pyrarius* which alternates between *Pyrus* and Gramineae in Europe. *Hyalopterus pruni,* the 'mealy plum aphid' alternates between *Prunus* and *Phragmites.* Populations of *Hyalopterus* from peach and apricot may have shorter siphunculi, better developed lateral abdominal tubercles, fewer rhinaria and be less strongly pigmented. As these characters do not always occur in the same combination it is not clear how many species are involved.

The sub-tribe Aphidina contains 520 species but only a few genera, one of them being *Aphis* with about 500 species when used in its broader sense. Like the Macrosiphini the majority of Aphidina live on the 'higher' dico-tyledons. The sub-classes Rosidae and Asteridae each bear more than one-third of the species of Aphidina. No Aphidina are known from Coniferae but the three species of *Ephedraphis* live on Gnetales. Aphidina rarely occur on Gramineae but a few species live on Liliaceae. About 3% of the species of Aphidina live on Caryophyllidae, 2% on Magnolidae and 1% on Hammamelidae. One species is described from *Lycopodium*. The genus *Aphis* has an exceptionally wide range of primary host plants. *Aphis corniella* alternates between *Cornus* species and *Epilobium*, and *Cornus* species are also the primary hosts of the *Aphis helianthi* group. *Aphis fabae* is unusual in having primary host plants belonging to three different families, viz. *Euonynus*, Celastraceae; *Philadephus*, Saxifragaceae or Philadelphaceae; and *Viburnum*, Caprifoliaceae. Most other aphids have primary hosts in only one plant genus. *Aphis gossypii* may be an exception as sexuales have been recorded from both *Rhamnus* and Bignoniaceae, but it is possible that two species of aphids are involved. *Aphis nasturtii* alternates between *Rhamnus* and various herbs. *Aphis grossulariae* belongs to a species complex characterised by an unusually hairy ultimate rostral segment and apterae with pale siphunculi, which are associated with *Ribes* and Onagraceae. *Aphis sambuci* alternates between *Sambucus* and the bases of the stems of Caryophyllaceae, *Rumex* and more rarely Umbelliferae. *Aphis solanella* is a member of the *A. fabae* group and alternates between *Euonymus* and various plants but particularly Compositae and Polygonaceae. Most *Aphis* species however are holocyclic and specific to one species or group of related species of plants. *Aphis idaei* occurs on European raspberry. *Aphis pomi* lives on *Cotoneaster*, *Crataegus*, *Cydonia*, *Malus*, *Pyrus* and *Sorbus*. *Aphis craccivora* belongs to a group of species, sometimes regarded as a distinct genus or sub-genus, *Pergandeida*, mostly associated with Leguminosae. Species resembling *Aphis* but with shorter appendages occur at the centres of continents, often living hidden at the bases of xerophytes. A number of generic names, e.g. *Brachyunguis*, *Protaphis*, *Xerophillaphis* have been proposed and are probably needed in this group as similar looking groups of species may have arisen independently on a number of occasions. Rather similar looking aphids belonging to the tribe Macrosiphini occur in similar situations. *Toxoptera aurantii* lives on many shrubs and is unusual in possessing a functional stridulating mechanism: large colonies producing sounds audible to the human ear. *Toxoptera citricidus* lives mostly on Rutaceae and is well-known as the vector of citrus tristeza virus. *Toxoptera citricidus* has not yet been found in either North America or the Mediterranean area although it is widespread in South America including Trinidad, in Asia and in Africa south of the Sahara.

7.2.7.1.2 The Pterocommatinae is a holarctic group of about 40 rather large hairy aphids associated with Saliceae and attended by ants. Host plant alternation is unknown in the Pterocommatinae.

7.2.7.1.3 The Chaitophorinae contains about 75 species in the holarctic and oriental regions. There are two tribes, the Chaitophorini with 6-segmented antennae which live on shrubs and the Siphini (= Atheroidini) with 5-segmented antennae which live on Gramineae. The 60 species of Chaitophorini are contained in two sub-tribes, the Chaitophorina which live on Saliceae and the Periphyllina which mostly live on Aceraceae but with a few species on members of such related families as Sapindaceae. Host plant alternation is unknown in the Chaitophorinae and most species are holocyclic on and specific to one or to a few closely related species of plants. Some species of *Periphyllus* avoid the nutritional unsuitability of *Acer* leaves during the summer by producing aestivating larvae or 'dimorphs'. These are specially modified first instar larvae bearing unusually long or flattened hairs and which diapause during the summer. Their development continues in the autumn to give rise to sexuparae. Other members of the same species of *Periphyllus* may continue to reproduce parthenogenetically throughout the summer on the same tree, the two cycles running in parallel.

The Siphini contains about 20 species placed in several genera including *Atheroides*. Only *Sipha flava* on sugar cane in the warmer parts of America is of any economic importance.

7.2.7.1.4 About 335 species have been described in the sub-family Lachninae, mostly in the holarctic and oriental regions. The tribe Cinarini contains about 270 species living on Coniferae, with the majority of species described from North America. The sub-tribe Cinarina contains about 225 mostly rather large, brown, hairy aphids feeding on the bark of conifers and particularly of *Abies*, *Larix*, *Picea*, *Pinus* and Cupressaceae. The 48 described members of the sub-tribe Schizolachnina (= Eulachnina) feed on the needles of *Pinus*. The tribe Lachnini contains 44 species living on Fagaceae, Saliceae and more rarely other dicotyledonous trees and shrubs and are perhaps most numerous in the oriental region. The 19 described members of the tribe Tramini live on the roots of Compositae and of a few other herbs, mostly in the palaearctic region. Host plant alternation is unknown in the Lachninae, many of which live associated with ants.

7.2.7.1.5 The sub-family Greenideinae contains about 60 species almost all from an area extending from Japan to New South Wales and from India to the Philippine Isles, but were once apparently more extensive as fossils and recorded from Yugoslavia. The tribe Greenideini with about 50 species, occur on oriental and tropical trees and shrubs. More than half the species of Greenideini occur on members of the sub-class Hamamelidae. Greenideini are described from Anonaceae, *Anona*, *Polyalthia*; Betulaceae, *Alnus*, *Carpinus*; Euphorbiaceae, *Baccaurea*; Fagaceae, *Castanea*, *Castaneopsis*, *Quercus*; Moraceae, *Artocarpus*, *Ficus*, *Streblus*; and Myrtaceae, *Decaspermem*, *Eugenia*, *Melaleuca*, *Psidium* and *Rhodomyrtus*. The Greenideini are unusual in that the oviparae are usually alate. The tribe Cervaphidini (= Setaphidini) is most abundant in South East Asia but *Brazilaphis* occurs in South America and *Meringosiphon* in Western Australia. The most widespread genus *Schoutedenia* (= *Setaphis*) lives on Euphorbiaceae such as *Phyllanthus*, *Emblica* and *Flueggia* in Asia and Africa south of the Sahara. Euphorbiaceae such as Croton and Euphorbia are infested by species of *Eonaphis*. *Anomalaphis* live on Myrtaceae in Australia and *Anomalosiphon* and *Sumatraphis* on Leguminosae and Ulmaceae respectively in south east Asia. Greenideinae have not been tested as virus vectors and apart from the occasional records of *Cervaphis* from the flowers of cacao, rarely occur in economic literature.

7.2.7.1.6 The three species of the genus *Israelaphis* constitute the subfamily Israelaphidinae, live on Gramineae in the Mediterranean region and seem to be intermediate between Cervaphidini and the Drepanosiphinae.

7.2.7.1.7 The sub-family Drepanosiphinae (= Callipterinae, Callaphidinae, Phyllaphidinae) is probably an old group of about 390 species. There are genera native to all the major geographical regions except for the antarctic. The antiquity of the group is suggested by *Neophyllaphis* which occurs on *Podocarpus* and other conifers in South America, Australia, New Zealand, New Guinea, Japan and East Africa and by a group of genera living on *Nothofagus* in South America, Tasmania and the South Island of New Zealand. More than half the species of Drepanosiphinae live on trees of the sub-class Hamamelidae, particularly Fagaceae and Betulaceae.

Members of the tribe Thelaxini avoid the unsuitability of summer leaves by two methods. Species of *Thelaxes* produce alatae in early summer, the progeny of which may be diapausing larvae or which may feed on the outsides of the galls caused by Hymenoptera on oak. The species of *Glyphina* live on Alnus and Betula and may produce their sexuales early in the summer, much of the year being spent in the egg stage.

The tribe Saltusaphidini contains about 50 species characterised by the lack of the evident triommatidion which is a characteristic feature of the eyes of most aphids. These often elongate aphids live on Cyperaceae and some can jump. Apterae viviparae are the usual form found, alatae occurring when the host plant becomes overcrowded. The tribe Neophyllaphidini includes wax covered aphids which live on Podocarpaceae and more rarely other conifers and on Epacridaceae. Apterae viviparae are common and the oviparae may be alate. *Ceriferella* from Epacridaceae are known only from Australia. Members of the tribe Paoliellini live on Combretaceae and Burseraceae in the Ethiopian region and Southern Asia. Somewhat similar genera have been described from Lauraceae in South America.

The Drepanosiphini (= Callaphidini) contains numerous genera mostly living on Fagaceae, Corylaceae, Betulaceae and Ulmaceae, but a few genera are specific to Aceraceae, Leguminosae or bamboos. In some species all the viviparous morphs including the fundatrices are alate but in other species of the same genus apterous viviparae may be produced. Most species live on trees and are not regarded as of economic importance but *Therioaphis* inhabits leguminous herbs and shrubs. Two clones of the old world species *Therioaphis trifolii* appear to have been introduced to North America where they behave as different species. One is the 'Yellow Clover aphid' which probably originated in Western Europe while the form *maculata*, which rarely produces functional sexuales, was probably introduced from the Middle East or India and is known as the Spotted Alfalfa aphid. *Eucallipterus tiliae* on *Tilia* and *Drepanosiphum platanoides* on *Acer* in Europe and *Tinocallis ulmifolii* on *Ulmus* in America cause nuisance in the suburbs by the amount of honeydew they deposit on pavements and parked cars.

7.2.7.1.8 The sub-family Mindarinae contains only the genus *Mindarus* with a few species living on *Abies* and *Picea*. *Mindarus* has a supposedly primitive wing venation and produces only a few generations each year, the overwintering egg being laid early in the summer. The sub-family Phloeomyzinae contains only *Phoeomyzus* with a few species living in wax wool on the bark of poplar trees. All the viviparous morphs are apterous and only the alate oviparae distribute the species.

7.2.7.1.9 The Anoeciinae contains about 30 species. The species of *Anoecia* itself alternate between *Cornus* and the roots of Gramineae or are holocyclic on the latter. As the fundatrices on *Cornus* are similar to the exules on Gramineae it is likely that the primary host, *Cornus*, has been secondarily acquired. The species of *Aiceona* are similar in many respects and live on Lauraceae in Asia. The return migration of *Anoecia* is achieved by sexuparae which deposit small apterous sexuales on the leaves of Cornus. Both their appearance and their biology suggest a position intermediate between the Thelaxini and the Hormaphidinae.

7.2.7.1.10 The sub-family Hormaphidinae contains about 160 species, mostly from South East Asia, which separate into three fairly well defined tribes. The tribe Hormaphidini has about a dozen species in Japan, Europe and North America. In Japan and America galls are formed on *Hamamelis* and in America at least the migrants fly to the leaves of *Betula*. In Europe Hormaphidini are known only from the leaves of *Betula*. The tribe Nipponaphidini contains about 70 mostly Asiatic species alternating between *Distylium* (Hamamelidaceae) as primary host and *Quercus* (Fagaceae), *Ficus* (Moraceae), *Viscum* (Loranthaceae) and several genera of Lauraceae including *Cinnamomum, Litsea, Machilus, Sassafras, Tetradenia* and *Umbellularia*. The generic classification is largely based on the structure of the aleyrodiform exules. Alate exules are rare as in the Hormaphidini and neither tribe is likely to be important in virus transmission. Members of the tribe Cerataphidini (= Oregmini) are also Asian in origin. About 80 species have been described. They have Styrax (Styracaceae) as primary hosts and Gramineae, palms and occasionally orchids as secondary hosts, or are anholocylic. As the effect of environment on morphology is little understood, many of the described species may prove to be synonymous. *Ceratovacuna lanigera* is a pest of sugar cane and *Cerataphis* species occur on palms and orchids. *Astegopteryx nipae, A. rappardi* and *A. rhapidis* sometimes occur on coconut in large numbers.

7.2.7.1.11 The sub-family Pemphiginae (= Eriosomatinae) is characterised by the arostrate sexuales produced by alate sexuparae. Only one egg develops in each ovipara. About 235 species have been described in three well defined tribes. The tribe Eriosomatini contains about 50 species with Ulmaceae as primary host plants, on which galls are formed, and the spring migrants go to the roots of herbaceous dicotyledons, shrubs or Gramineae, or their progeny form woolly masses on the aerial parts of Rosaceae. *Eriosoma ulmi* forms yellow leaf rolls on the *Ulmus carpinifolii/glabra* group and migrates to the roots of *Ribes*. *Eriosoma patchae* crumples the leaves of the same group of elms and spends the summer on the roots of *Senecio*, including *S. cineraria*. *Eriosoma lanuginosum* and *E. pyricola* form closed galls on *Ulmus* and spend the summer on the roots of pear and perhaps other Rosaceae. Another group of *Eriosoma* have the *Ulmus americana/laevis* group of elms as primary host plants and form woolly masses on the aerial parts and the roots of Rosaceae in the summer. *Eriosoma lanigerum*, the 'wooly apple aphid' or 'American Blight' belongs to this group. *E. lanigerum* is probably Eastern Asiatic in origin but may have been introduced to Europe via America. The species of *Gobaisha* form closed galls on *Ulmus* and use the roots of Labiatae as secondary host plants. *Tetraneura* species form closed galls on Ulmus and spend the summer on the roots of Gramineae. *Watabura nishiyae* forms galls on *Ulmus japonica* and the exules which develop on the roots of apple have one of the tarsal claws much smaller than the other and thus resemble coccids.

The tribe Pemphigini contains about 120 species placed in two sub-tribes. The Pemphigina form closed galls on *Populus* and migrate to the roots of herbaceous dicotyledons. *Pemphigus bursarius*, the lettuce root aphid is a well known example. The Prociphilina have a variety of primary hosts but the secondary hosts are usually the roots of Coniferae but occasionally the roots of monocotyledons. The primary host plants of Prociphilina include *Fraxinus*, *Crataegus*, *Tilia* and *Lonicera* on which 'leaf nests' are formed. These primary hosts were probably secondarily acquired. *Mimeuria ulmiphila* is said to migrate from leaf nests on *Acer campestre* to the roots of *Ulmus* or *Rubus*. It lives in cysts composed of fungus mycelium and is said to be able to feed on both the roots of the host plants and on the mycelium of the fungus which ferments the sugars of the honeydew. The fundatrix on *Acer* is said to live from May until November continuously producing alatae.

The tribe Fordini contains 65 species and the primary host plants are Anacardiaceae. Members of the sub-tribe Fordina alternate between galls on *Pistacia* and the roots of Gramineae or occasionally herbaceous dicotyledons. The biology is adapted to the winter rainfall regime of the Mediterranean region and the complete cycle often extends over two years. The gall may not open until the autumn of the first year and the migrants deposit their young under stones to await the plant growth stimulated by the winter rain. A number of generations of exules occur throughout the winter and sexuparae return to *Pistacia* the following summer. Some of the species are now widely distributed as anholocyclic populations on the roots of Gramineae (e.g. *Aploneura* and *Geoica* spp.) or more rarely dicotyledons (*Smynthurodes betae* (= *Trifidaphis phaseoli*)). *Aploneura ampelina* occurs on the roots of vines and is sometimes confused with *Viteus vitifolii (= Phylloxera vastatrix)*. Members of the *Melaphidina* form galls on *Rhus*, their primary host and migrate to mosses.

7.2.7.2 Adelgidae

All the known primary host plants of the 45 described species of Adelgidae belong to the genus *Picea* and the secondary host plants of all species are also Coniferae. A 'pineapple gall' is produced on *Picea* and woolly masses on the secondary hosts. The complete cycle extends over two years during which time fewer generations are produced than in most Aphididae. The return migrant is a sexupara and the timing of the cycle is similar to that of the Fordini. The second winter is spent as an immature form on the secondary host plant. The taxonomy of the group is difficult. One population may give rise to very different looking individuals and some species probably consist of many populations with slightly different host plant preferences. Carter (1971) has given an account of the winged forms and of the biology of the British species and a check list of the world fauna.

7.2.7.3 Phylloxeridae

About 60 species of Phylloxeridae have been described, mostly living on *Quercus* (Fagaceae), *Carya* (Juglandaceae), *Populus* or *Salix* (Saliceae). Host alternation as understood in other Aphidoidea does not occur but *Phylloxera florentina* and *P. quercus* are said to spend the summer on deciduous oaks and the spring and autumn on evergreen oaks. All generations are oviparous as in the Adelgidae. *Viteus vitifolii (= Phylloxera vastatrix)* is the best known phylloxerid and is the subject of an enormous literature. Coombe (1963) gives an account of this aphid in Southern Australia and references to recent literature. *Aphanostigma piri* is a pest of pears in the Mediterranean and *A. iwaksuiense* is said to be very injurious to pears in Korea and is also recorded from Japan.

7.3 Psyllids (Psyllidae)

7.3.1 The adult

Psyllids (fig. 7.2) are small to medium sized insects with a wing span from 1.5–15 mm. They resemble aphids but are of a rather harder consistency. About 1500 species of psyllids have been described and placed in 180 genera.

The anatomy of psyllids is of the general hemipterous type but the genae are produced into cones in many species. The juxtaposition of the fore-gut with the hind-gut suggests a filter chamber mechanism. The outer surfaces of the apices of the mandibular stylets bear up to 20 transverse striae making the tip appear serrate in lateral view (fig. 7.3). Most psyllids have piercing mouthparts but *Atmetocranium myersi* from New Zealand is said to have 'non-piercing' mouthparts. The tarsi of adult psyllids consist of two well developed segments of about equal length terminated by a pair of claws but the claws are reduced in the New Zealand genus *Hemischizocranium*. Most psyllids have the hind legs enlarged and modified for leaping, but in the Indian genus *Apsylla* the hind legs are little longer than the others and the antennae bear sensoria of a type not found elsewhere in the Psyllidae. In most members of the large sub-families Psyllinae and Triozinae, a single rhinarium is borne subapically on antennal segments IV, VI, VIII and IX.

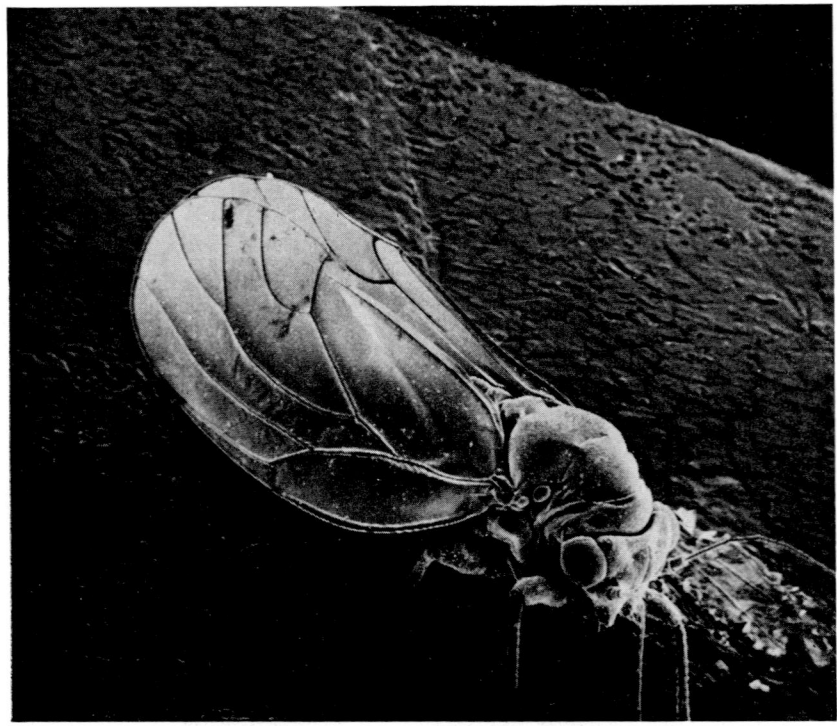

Fig. 7.2. *Psylla peregrina* adult. Magnification × 25. (Courtesy of British Museum.)

More numerous rhinaria may occur on antennal segments III to VII in members of other sub-families. Stridulation has been described for several species (Campbell 1964).

Young adults are usually green or yellow but old adults are often brown or black. The adults of all known psyllids are winged. The wings are commonly transparent but are maculate or opaque in some species. The females are often slightly larger than the males and the wing shape and pigmentation of the body may also exhibit sexual dimorphism. The pigmentation develops slowly in psyllids. The adults of species with only one generation per year (univoltine) such as *Psylla melanoneura* and *P. ulmi* are green

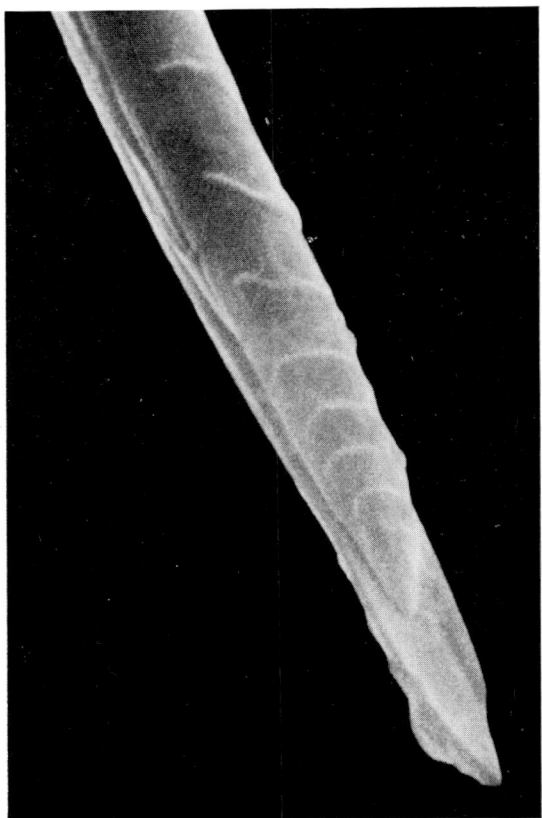

Fig. 7.3.   Tip of the stylet bundle of the psyllid *Trioza urticae*. Magnification × 575. (Courtesy of British Museum.)

when they first appear in early summer and the brown pigmentation develops during the next few weeks. The dark areas increase in size and specimens which overwinter are almost black by spring. Multivoltine species such as *Psylla pyricola* may have pale summer and darker winter generations.

7.3.2 The life cycle

The eggs of psyllids are 200–300 μm long and 100–160 μm wide, may be sculptured or smooth, or sculptured on one side and smooth on the other. The eggs may bear a tail-like projection at one end or a short basal projection at the other by which the egg is fixed to the leaves, and through which it may be able to absorb moisture.

Male and female psyllids usually develop in equal numbers. Differences in their activity and life expectancy results in more females than males being found. Trap catches indicate that the females of some species have only a short period of flight while they are young but males of all ages occur in the traps. *Psylla myrtilli* is probably parthenogenetic since many females but only one male has been found.

The two largest genera of psyllids are *Psylla* and *Trioza*, each of which contains about 250 species. Many of these are univoltine. Some of the species overwinter and spend nine months of their year as a second instar larva, some species overwinter as adults and others in the egg stage. Some species of *Psylla* and *Trioza* have two, three or four generations per year. *Trioza fletcheri* which forms leaf galls on *Gmelina* spp. has eight generations per year in India. A *Paurocephala* that produces pit galls and a *Diaphorina* with free living larvae have 9 generations per year and a *Psylla* with free living larvae has 11 generations a year in India.

Some species of *Caillarda*, *Euphalerus*, *Psylla* and *Trioza* overwinter in the egg stage and in India a species of *Trioza* is recorded with a prolonged egg stage during the summer. The last stage nymph seems to be the most common resting stage in the tropics but in temperate regions the adults, second instar larvae and eggs are common as overwintering stages. There are usually 5 nymphal instars but specimens have become adult after 4 and 6 moults under experimental conditions.

Non-diapausing eggs take 3–14 days to hatch, depending largely on temperature. The nymphal period may last from 11 to 260 days. There is a period of 5–12 days after the final moult before egg laying begins. Adults live for 13 or more days in captivity but probably much longer in the wild, commonly for 1 or 2 months and for 8 months in species which overwinter as adults. In some species both males and females overwinter but in *Paratrioza cockerelli* the males seldom live for more than one month while the females overwinter and may lay eggs for nearly six months. A female psyllid may lay from 39–1475 eggs but 200–500 eggs are usual. The eggs may be laid during 3–179 days at a rate of up to 72 per day, but 10–50 eggs per day for 10–20 days is more usual. Most eggs hatch.

7.3.3 Ecology

As far as is known, all psyllids feed directly on plants but the biology of many genera, including *Atmetocranium*, has not been studied. Many psyllids are free living on the leaves and shoots of dicotyledons and particularly on trees and shrubs. More than 200 species of psyllids have been described from Myrtaceae, mostly in Australia. The Leguminosae bear 84 species of psyllids; the Compositae 76 species; Rosaceae 42 species; Saliceae 40; Anacardiaceae 26; Moraceae 24; Lauraceae 21 and 20 species of psyllids have been described from Rhamnaceae. No other family of plants is known to support as many as 20 species of psyllid. The saliva of some psyllids is toxic to their host plants and characteristic deformations may appear after 7–10 days. Some species form closed galls while others live in pits on the leaf or twig or otherwise distort the leaves, shoots or even roots of their host plant. Other species live hidden under a characteristic scale or test known as a lerp. The nymphs of some free-living species belonging to several subfamilies secrete long filaments of wax while other free-living species have green or brown nymphs barely dusted with wax.

The immature forms of most species of psyllid develop on only one genus of host-plant, but adults are often collected from plants on which they do not breed. No host plant alternation with a succession of generations (such as occurs in aphids) is known from psyllids but some psyllids overwinter as adults on evergreen shrubs and return to their deciduous or herbaceous breeding hosts to lay their eggs in the spring. Some authors have considered the winter host nutritionally essential while others have suggested that the exercise on the migratory flight is essential for the development of the eggs. Others have thought that the evergreens provide only winter shelter and some nutrition. Adults of *Trioza urticae* may overwinter in or around nettle beds, living on whatever plants are in leaf at the time or may migrate to conifers. In Japan the galls of *Trioza ukogi* may open in late October and the adults emerge and lay eggs at the bases of the buds or the adults may remain in the galls until April or May. While most psyllids breed on only

one genus of host plants, *Trioza nigra* is recorded from both *Styrax* (Styracaceae) and *Symplocos* (Symplocaceae).

The nymphs of *Mesohomatoma hibisci* and of *Leptynoptera sulphurea* are
gregarious and of *Optomopsylla formicoformis* mimic ants. The nymphs of
Eucalyptolyma may live under the lerps made by species belonging to other
genera of psyllids or in leaves tied together by caterpillars. Nymphs may
dispose of their honeydew by excreting it into bags of wax but adults flick
away the globules with their forewings. *Paurocephala* and *Diaphorina* may
be attended by ants and otiose ant attendance is common in many groups as
in other Hemiptera.

Individual adult psyllids may not move much in a crop for several days.
Old individuals are capable of flight as is shown by the species that overwinter as adults on evergreen plants. Psyllids occasionally cause domestic
annoyance when entering houses in search of hibernating sites and have
been found in the sea nearly 100 miles from land.

Psyllids have many natural enemies. Their predators include mantids,
chrysopids, pentatomids, nabiids, anthocorids, mirids, Lepidoptera including
gall eating species, coccinellids, ants, empids, syrphids, cecidomyids and
arachnids. Psyllid parasites include ichneumonids, braconids, chalcids
(signiphorids), eulophids, pteromalids, the exo-parasitic first instar larvae
of cyclotomids (Lepidoptera) and the secondarily parasitic ceraphrontids
(callicerids). Dunstan (1927) attempted to control *Psylla mali* with a fungus
disease.

7.3.4 Diseases associated with psyllids

Diaphorina citri and *Trioza erytreae* are recorded as vectors of virus diseases
of citrus. The biology of *Trioza erytreae* is probably better understood than
any other psyllid as the result of recent work in South Africa by J. R.
Blowers, H. D. Catling and V. C. Moran, and their published work also includes references to earlier studies of the biology of psyllids. *Psylla pyricola*
has been sporadically important in America since 1884 as the vector of
sudden death of pear. Swirski (1954) and Marshal (1959) give accounts of
the biology of *Psylla pyricola*, which is also regarded as of economic importance in the mediterranean region but *Psylla pyri* is said to be more
important in central Europe. Only a few species of psyllids have been tested
as virus vectors and these few species have only been tested with a few
viruses. Perhaps because of this no general features of the 'vector species'
have been detected.

The biology of *Psylla mali* has been studied by Brittain (1923) and Minkiewicz (1927) in America and Europe respectively. The feeding damage of *Paratrioza cockerelli*, known as psyllid yellows of potatoes, was once thought to be symptoms of a virus, but is now regarded as caused by a toxaemia from which protection is obtained by the prior feeding of the psyllids on a virus source (Staples 1968). The toxic saliva of *Mesohomatoma tessmanii* is suspected of being the cause of leafless twig of Cacao.

Trioza apicalis is recorded as a pest of carrots, *T. tripunctatus* as a pest of American blackberry, *T. alacris* of *Laurus nobilis* and *T. litseae* as a pest of vanilla, a plant upon which it feeds but does not breed. *Apsylla cistellata* damages mango, *Euphyllura olivina* and perhaps other *Euphyllura* are pests of olives, *Paurocephala gossypii* occasionally occurs in large numbers on cotton, *Phytolyma* species cause galls on the young shoots of *Chlorophora* and make it difficult to rear seedlings in forest nurseries. *Diclidophlebia* species may be important on *Triplochiton* and *Cardiaspis* and *Glycaspis* have caused concern on *Eucalyptus* in Australia.

7.3.5 Systematics and sources of information

The last catalogue of the world fauna of Psyllidae was published nearly sixty years ago (Aulmann 1913). References to the morphology and anatomy of psyllids may be found in Dobreanu and Manolache (1965) and Vondraček (1957). H. J. Müller (1956) gives accounts of the species regarded as pests and references to economic literature.

The data provided by Ossiannilsson (1963) brought the British List (Kloet and Hincks 1964) up to date. Regional taxonomic accounts are available as follows: Central Europe (Haupt 1935); Czechoslovakia (Vondraček 1957 and Lauterer 1965); Rumania (Dobreanu and Manolache 1962); Spain (Ramirez Gomez 1956, 1960); European Russia (Loginova 1964); Finland (Lindberger and Ossiannilsson 1960); Sweden (Ossiannilsson 1970); Poland (Klimaszewskii 1967); Switzerland (Schaefer 1949); Central Asia (Loginova 1960, 1964); Mongolia (Klimaszewskii 1964); Japan (Miyazaki 1963, 1964, 1965, 1969); Congo (Vondraček 1963); South Africa (Capener 1970); Oriental region (Crawford 1919); China (Yu 1956); India (Mani 1959 and Mathur 1935); Philippine Islands (Uichanco 1921); Fiji (Tuthill 1943b); Hawaii (Zimmerman 1948); Micronesia (Tuthill 1964); Borneo (Crawford 1920); Sumatra (Crawford 1928); New Zealand (Tuthill 1952); Australia (Tuthill and Taylor 1955; Taylor 1960; Moore 1970); South America (Crawford 1925); Peru (Tuthill 1959); Mexico (Tuthill 1944b–1950); Puerto

Rico (Caldwell and Martorell 1952); Cuba (Tuthill 1945b); U.S.A. (Tuthill 1943 and Jensen 1951); Ohio (Caldwell 1938); Alberta (Strickland 1938).

Acknowledgements

I am indebted to M. C. Burns, D. Hollis and B. R. Pitkin for obtaining the stereoscan micrographs.

CHAPTER 8

Leafhoppers and planthoppers

RYOITI KISIMOTO

Contents

8.1 Introduction

Leafhoppers (Cicadellidae) and planthoppers (Delphacidae) are second only to aphids in importance as vectors of plant pathogenic viruses, and also transmit diseases associated with mycoplasma-like organisms (MLOs) in phloem tissue. Though they are taxonomically distinct groups, they have been studied together, but most attention has been paid to leafhoppers, and little work on planthoppers has been done outside northern Europe and parts of Asia. Although there are interesting differences between the two groups in their biology, behaviour, etc. as vectors of plant diseases the two groups are closely similar. The number of species of Cicadellidae known to be vectors much exceeds those of the Delphacidae; Nielson (1968) listed 119 Cicadellida species as vectors of diseases of cereals, vegetables, shrubs and trees, but so far only 17 Delphacids have been reported as vectors and they only transmit diseases of Graminaceous plants. Paucity of known Delphacidae vectors may depend on limited interest in this group, but also reflects the fact that Delphacids feed mostly on graminaceous and a few other monocotyledonous plants. The few planthopper vectors are, however, very interesting as they are widely distributed, the viruses they transmit show a comparatively high degree of vector specificity (Oman 1969), and the diseases caused by these viruses are of economic importance in cereal growing areas.

Recent reviews exclusively on leafhoppers have been published by Nielson (1968) and DeLong (1971); they covered most topics and particularly the details of the characteristics of different species.

8.2 Classification

The order Hemiptera is divided into two major suborders, the Homoptera and the Heteroptera. The Homoptera (except for one small group) are divided into the Auchenorrhyncha and Stenorrhyncha, in the former the rostrum, feeding beak, is attached to the neck while in the latter to the chest. The insects described in this chapter come in the Auchenorrhyncha, which has more than 22 families including spittlebugs (Cercopidae), cicadas (Cicadidae), treehoppers (Membracidae), and two super families, leafhoppers (Cicadelloidea) and planthoppers (Fulgoroidea). In the Fulgoroidea are included more than 15 families, among which the Delphacidae (Araeopidae) is one of the largest and includes the planthoppers. The large and robust species of the Cicadellidae, such as those in the subfamilies Cicadellinae, Aphrodinae and Cyponinae are often called sharpshooters.

Large Cicadellida adults are up to 12 mm long, small ones as little as 2.5 mm, and slender and fragile. Delphacidae are usually smaller than the Cicadellidae. The general shape of leafhoppers and planthoppers is differ-

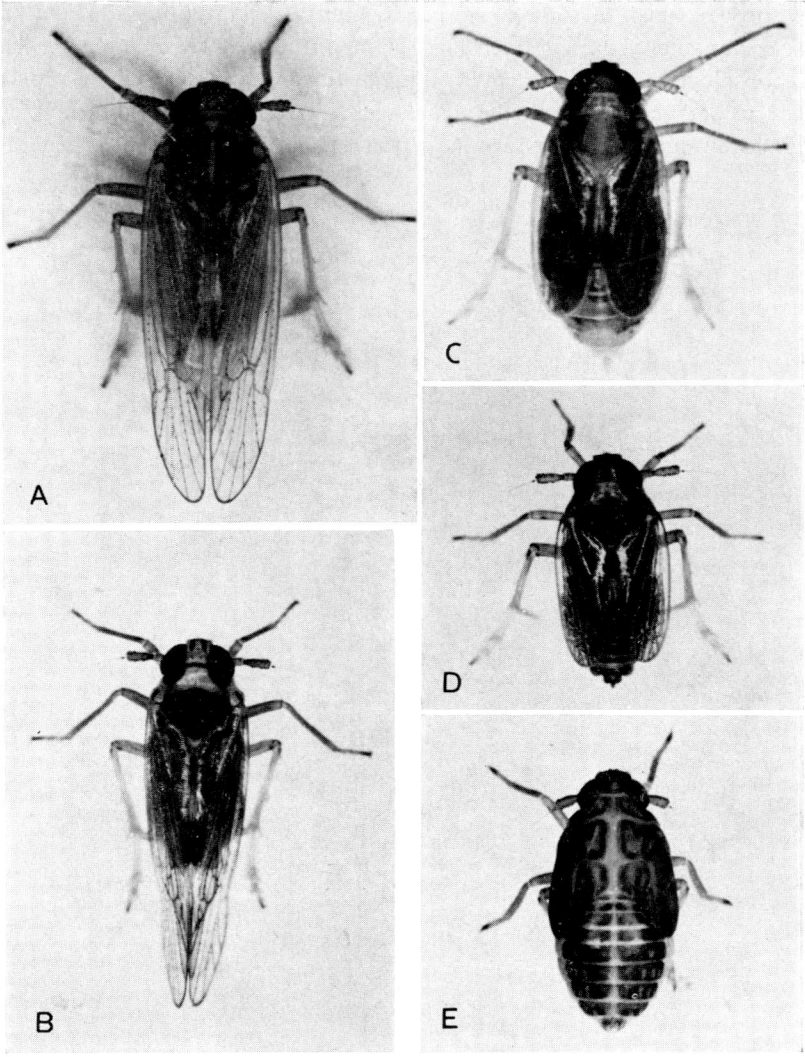

Fig. 8.1.   The small brown planthopper, *Laodelphax striatellus* (Fallén). (A) macropterous female, (B) macropterous male, (C) brachypterous female, (D) brachypterous male, (E) fifth stage nymph.

ent (figs. 8.1 and 8.2). In leafhoppers, the vertex is smooth and convex, sometimes obtuse-angle with compound eyes at the extremes, and the antennae are thin and minute, bristle like with small basal segments. The fore-wings, or tegments, are not transparent and have various coloured patterns of green, grey, brown, black or sometimes red. By contrast planthoppers have a narrow, rectangular frons between large compound eyes, their antennae have two conspicuous basal segments equipped with many round sensorial plates in the second segment, and the distal segments reduced to a thin filament. Their fore- and hind-wings are usually hyaline or membraneous, generally

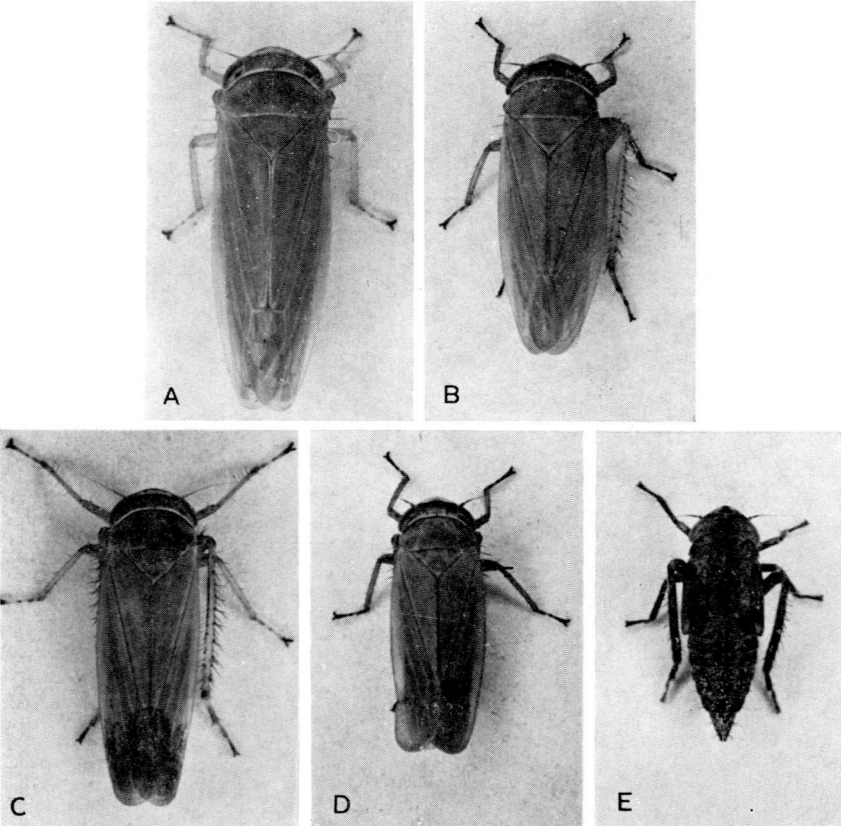

Fig. 8.2. The green rice leafhopper, *Nephotettix cincticeps* Uhler. Female of long day form (A), female of short day form (B), male of long day form (C), male of short day form (D) and fifth stage nymph (E).

much less colourful than those of leafhoppers and are brown, grey, black or whitish.

Ideas on the taxonomic position and hence the names of even well-known and widely distributed species have often been changed and this has resulted in much synonymy so that when reading past papers it is important to realize that, for example, the name *Delphacodes pellucida* has been changed to *Javesella pellucida*, *Delphacodes striatella* to *Laodelphax striatellus*, and *Sogata orizicola* to *Sogatodes orizicola*. Many names of Cicadellida vectors have also been changed (Nielson 1968).

Various morphological characters are used for identification, such as details of the external reproductive organs, and, in particular, details of the well chitinized structures such as the aedegus (penis) and the paramere of male planthoppers and the first gonocoxa of female planthoppers, and the aedegus, styles and pygofer of male leafhoppers and the seventh abdominal sternum of female leafhoppers. Care must be taken when using these characters, as Müller (1957) has demonstrated that the form of the penis depends on the photoperiod during nymphal development so that two described *Euscelis* species were actually long-day and short-day forms of the same species.

8.3 Biology

8.3.1 Egg development

Both leafhoppers and planthoppers have three main developmental stages, egg, nymph and adult; the nymphal instars showing typical exopterygote metamorphosis. The banana-shaped eggs are usually deposited into host plant tissue, singly or in clusters, usually in the parenchyma of the underside of leaf, petiole, mid-rib of leaves in tree, shrub or herbaceous host plants. In graminaceous host plants, the stem and leaf sheath near the mid-rib are mostly preferred. The proximal end of the egg is often glued to the plant tissue with a paste-like material secreted at oviposition. Eggs do not resist drying and soon shrivel when the plant wilts.

The incubation period depends on temperature but is usually several days. In most species the incubation period lasts about half that of the nymphal period, excluding the diapause, and hence the insect may be able to survive in the egg stage for a long period even on host plants that have been treated with insecticides.

The blastoderm is located at the distal end of the egg and elongates with the cephalic lobe oriented towards the distal end until blastokinesis when the embryo turns over so that its head is at the proximal end. After blastokinesis the embryo is the complete nymphal shape and has clear red eyes. The full grown embryo hatches from the proximal end of the egg. The egg increases in length by about 15% during development, presumably by taking in water.

8.3.2 Nymph development

There are five nymphal stages allowing a stepwise increase of body size and changes in morphological characters. The nymphs live freely and feed on the underside of leaves or the basal parts of host plants, which are apparently favourable as they are most moist. Nymphs of planthoppers particularly prefer humid conditions within dense vegetation near water. Each generation without diapause takes about two to four weeks, and the number of generations in the year depends on the total effective temperature of a given locality. The estimated theoretical 'developmental zero' is 10 °C for *Laodelphax striatellus* and *Nilaparvata lugens* and 14 °C *Nephotettix cincticeps*.

8.3.3 Adult

Adults usually emerge during the morning. After their wings have expanded, they become quite active, even though still in the teneral period, or premigratory stage. The reproductive organs are not fully developed at emergence in either male or female. Mochida (1971) observed with *Javesella pellucida* that germ cells developed into spermatozoa in the testicular tubules soon after emergence and transferred to vas deferens, so that the testicular tubules gradually decrease in size, and the vas deferens enlarges to about one and half times its size at emergence (fig. 8.3).

The female reproductive organs are also not fully developed at emergence, and contain several small oocytes in the vitellarium of each ovariole. The spermatheca is empty and transparent or pale yellow (fig. 8.4). Each ovariole is divided histologically into several parts, the terminal filament, the germarium, the vitellarium and the pedicel. In the germarium there are many oocytes and nurse cells, in the vitellarium the oocytes develop as the yolk and chorion form so that there are a series of oocytes at various stages of development (fig. 8.5).

Nasu (1963) estimated the number of oocytes per ovariole of the green rice leafhopper, *Nephotettix cincticeps*, as 56–63 in the overwintering generation and 48–60 in the first generation. As there are 17–23 ovarioles per female, the

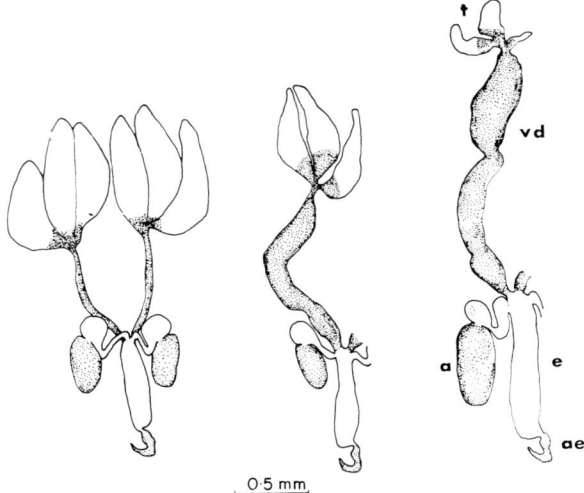

Fig. 8.3.   Male reproductive organ at emergence, 5 days old and 21 days old (from the left) of *Javesella pellucida* (F.)   a, accessory gland; ae, aedegus; e, ejaculatory duct; t, testicular tubules; vd, vas deferens. (From Mochida 1971.)

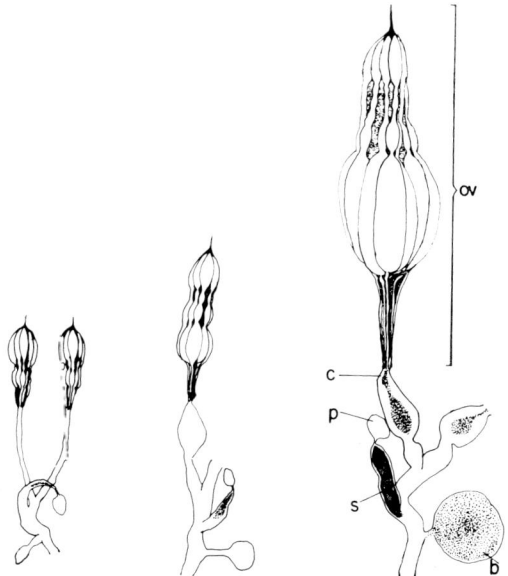

Fig. 8.4.   Female reproductive organ at emergence, 1 day old and 3 days old (from the left) of brachypterous form of *J. pellucida* (F.). b, bursa copulatrix; c, calyx; ov, ovary; p, pouched gland; s, spermatheca. (From Mochida 1971.)

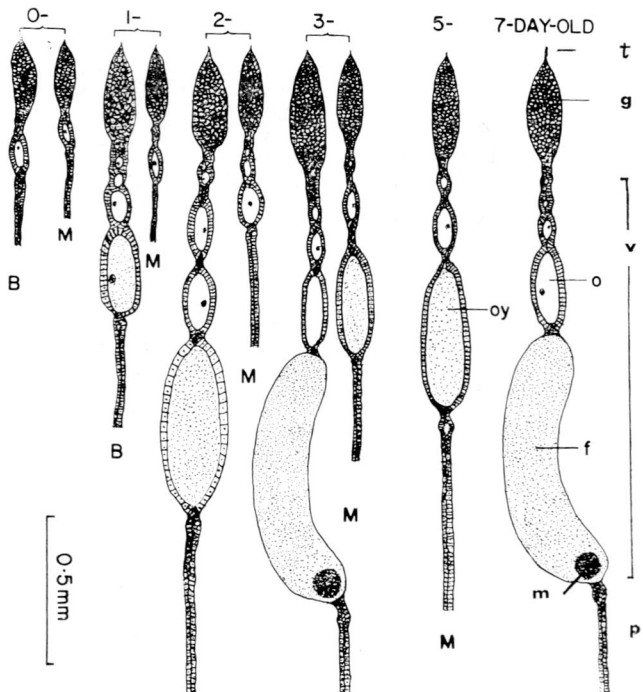

Fig. 8.5. Development of ovariole at various days after emergence of brachypterous form (B) and macropterous form (M) of *J. pellucida* (F.). f, full grown egg; g, germarium; m, mycetome; o, oocyte without yolk; oy, oocyte with yolk formation; p, pedicel; t, terminal filament; v, vitellarium. (From Mochida 1971.)

total number of oocytes per female is between 800 and 1400. However the average numbers of eggs actually laid were only 303.5 and 37.5, in the overwintering and first generations respectively which means that between about 25 % and 3 % of the oocytes became eggs. Fecundity seems, therefore, not to depend on the number of oocytes.

The mycetome, an organ which harbours symbiotes, is easily recognized from the beginning of embryonic development, and is a pale yellow or orange yellow organ at the caudal tip of embryo. It moves into the abdomen after blastokinesis and is kept throughout the life of the insect. Mycetocytes are found in the female reproductive organs at the basal part of the vitellarium where it is connected to the pedicel (the duct of ovariole through which eggs pass into the lateral oviduct). The symbiotes stain well with eosin, and can be seen within the oocytes, having penetrated shortly before the yolk was deposited.

Nasu (1965) found that there were two types of symbiotes, L (low) and H (high) which differ from each other in electron density, and also noted that rice dwarf virus particles attach selectively to the surface membrane of L-symbiotes but not to H-symbiotes. He claimed that this was the main way in which rice dwarf virus was transovarially transmitted in *N. cincticeps*; virus thus carried into oocytes presumably multiplies in the cytoplasm of mycetome during embryonic development. However the possibility of virus being transferred into oocytes in some other way before transference of mycetocytes has not been excluded.

Mating can occur any time after the teneral period, which is usually only a few days and is particularly short in brachypterous females. Mating lasts a very short time, only a few seconds. Males call in a special way, with fluttering of wings, before copulation. Ossiannilsson (1949) found that there is a tympanal apparatus, apparently homologous with that of Cicadidae, in many species of Aucherorrhyncha, and he described several distinct calls of which the common song and the male courtship call were commonest.

Migrating planthoppers caught in light traps are mostly unmated, by contrast some leafhoppers caught in light traps are mated and some contain full grown eggs. The adults live for a long time in captivity sometimes more than two months but in natural conditions they live for a much shorter time owing to natural enemies or other external influences.

8.3.4 Overwintering

Most leafhoppers overwinter as adults or eggs and some as nymphs (Beirne 1956; Nielson 1968; DeLong 1971). By contrast planthoppers mostly overwinter as nymphs or eggs and few as adults. Overwintering stages are in diapause. Nymphal diapause of *N. cincticeps*, *Laodelphax striatellus* and others is induced by short photoperiods at lower temperatures during nymphal development (Kisimoto 1958, 1959a, b). Egg diapause is also induced by short photoperiods during nymphal development or even in *Nilaparvata muiri* after the mother planthopper has emerged (Kisimoto, unpublished). However the obligatory ovarian diapause in *Stenocranus minutus*, a monovoltine species, has been shown to be induced by photoperiods longer than 16 hr (Müller 1958).

The nymphal diapause of leafhoppers and planthoppers is not so mandatory as egg or pupal diapause in other insects. Diapausing nymphs resume development when returned to long photoperiods. Chilling, of course, is effective for completion of diapause even under short photoperiods; so that under natural conditions diapause is completed before the coldest shortest days of winter. Diapausing nymphs feed sporadically but can survive for considerable periods without food, but need water. Humid places among winter weeds are favoured for overwintering. In the ecology of those viruses which overwinter mainly in hopper vectors but are not transmissible transovarially, weather affects the overwintering generation and hence controls the abundance of the virus disease in the following year; after warm winters the vectors appear before susceptible host plants do, and hence vectors but not virus survive the winter.

8.3.5 *Number of generations per year*

In high altitudes with short summers, most species have only a single generation each year, but in the subtropical or tropical areas even the same species may have six or more generations. *Javesella pellucida* has only one generation a year in Scandinavia and England, but two in Germany and France. Similarly, *Laodelphax striatellus* has only a single generation each year in Sweden, but in northern Japan it has two, the number of generations increases with increasing summer length until six generations are found in southern Kyushu, in Japan and in the south of Kiangsu, China (Pu 1963). A subtropical strain of the same species is found in Amani Is, at about 28° N, and this has a less recognizable nymphal diapause. Similarly, in Israel, no diapause was found (Harpaz 1961). The factors which determine the distribution of a given species are complicated, but low cold hardiness, absence of diapause, lack of an appropriate food plant during the winter, often inhibit permanent breeding of tropical species in northern areas, even though the summer is long enough for one or more generations and the species may migrate there every year from the tropics.

8.3.6 *Polymorphism*

Many planthoppers species have brachypterous and macropterous (short and long winged) forms of both sexes. The macropterous form has normally functioning wings, while the brachypterous form has short forewings and rudimentary hind-wings (fig. 8.1). Kisimoto (1965), working with *Nilaparvata lugens* and others, compared the two forms in several characters. The brachypterous form has a slightly larger body and a shorter pre-ovipositional period, which does not increase during unfavourable conditions (such as low temperatures and food shortage) as much as that of the macropterous form.

The brachypterous form usually predominates in populations growing in favourable conditions such as low population density and plentiful fresh food of preferred species and after a normal nymphal period, whereas the macropterous form is common under unfavourable conditions, such as overcrowding, food shortage, etc. In some species, such as *Javesella pellucida* brachypterous males develop in conditions similar to those for brachypterous females, however for other species particular species-specific factors determine whether brachypterous males develop. For example, a particular population density is the stimulus for *Nilaparvata lugens*, and diapause during nymphal development is the stimulus for *Laodelphax striatellus*.

Wing polymorphism seems to be connected with the seasonal life cycle. Dual life, that is, dissemination of the macropterous form by migration and population growth through the brachypterous form is a useful behaviour pattern for insects feeding on annual plants or even perennials with short growing periods. The dominancy of a particular wing-form is also characteristic of a particular species. Brachypterous forms predominate in sedentary species that feed on perennial grasses or those species that are rare in crops. This is perhaps because agricultural practices usually alter flora and fauna of farms abruptly and insect pests and their populations usually start from a few immigrant individuals, thus agricultural pest species are usually those with predominantly macropterous forms, and brachypterous forms are less likely.

In leafhoppers seasonal polymorphism is rather common (fig. 8.2). Müller (1957) demonstrated the effect of photoperiod on seasonal polymorphism in size, pattern and even the shape of the penis. He found that *Euscelis plebejus* and *E. incisus* which had been described as different species were the late summer form and spring form of the same species now called *E. plebejus*, and the West-European species *E. lineoratus*, *E. stictopterus*, *E. bilobatus* and four other related species were seasonal modifications of a single species *E. lineoratus*. He proposed a concept of 'form cycles', the series of modifications caused by seasonal changes. In the species having many generations a year, seasonal polymorphism is useful for distinguishing different generations. For example in *Nephotettix cincticeps*, a vector of the rice yellow dwarf disease, adults of the overwintering generation can be distinguished morphologically, and it has been found that they are more important than other generations as vectors as they migrate into nursery beds or early transplanted rice fields.

8.4 Flight activity, swarming and migration

Leafhoppers and macropterous planthoppers fly quite actively after a certain
teneral period. Take-off is considered to depend on internal drives and the
ready availability of suitable host plants does not prevent it (Raatikainen
1967), though in nature swarms and large migrations are usually associated
with seasonal changes of vegetation.

Leafhoppers and planthoppers are weak flyers, they do not seem to fly in a
definite direction and so are influenced by even a light wind. In northern
areas Delphacids fly during daytime (Raatikainen 1967) and also sometimes
during the night when the temperature is high. In fact, leafhoppers and planthoppers may be collected in light traps in warm areas. Yellow pan water
traps are also effective for collecting planthoppers but less effective for leafhoppers. These differences perhaps reflect differences in flying activity.

Aerial density depends on the interaction of many factors such as the
periodicity of take-off from breeding sites, and the duration of flight, etc.
(Johnson 1969). Analyses of the aerial density of *Nilaparvata lugens* and
Nephotettix cincticeps using a Johnson/Taylor suction trap (Ohkubo et al.
1971) clearly showed peaks in density at dawn and dusk of the former species
and one peak at dusk of the latter. The controlling factor seemed to be light
intensity. In late August, *Nilaparvata lugens* flew at light intensities between
1–200 lux, with a maximum at 100 lux, but later in the season light intensities
of 50–4000 lux were needed before flight was stimulated, perhaps because of
the lower temperatures, as the dawn flight was less than the dusk flight. On
cloudy and cool days sporadic flights occurred throughout the day. Winds
also inhibit take-off so that where the temperature was lower than 17 °C and
winds more than 2.3 m/sec at 8 m above the ground, few were flying. By
contrast *Nephotettix cincticeps* flies after sun-set with a light intensity between 0.1–20 lux (maximum 1–2 lux), which is considerably less than that
needed by *Nilaparvata lugens* in late August.

The height above the ground at which these insects fly has not been investigated enough. A tow net 10–20 m above ground level is effective for
collecting planthoppers swarming near the ground level (Raatikainen 1967),
but Glick (1960) caught leafhoppers and planthoppers at heights of 200–4000
ft using an aeroplane. This may seem rather high but using tethered insects
flying in a wind tunnel (B. Johnson 1958) Ohkubo (1968) found that *N.
lugens* would fly continuously for 4–5 hr, sometimes for more than 10 hr.

Flying insects are often borne aloft in convection currents, and, if weather
conditions are suitable, they can be carried for long distances, depending on
their capacity to keep flying.

Work in the U.S.A. has shown that leafhoppers, such as the potato leaf-hopper, *Empoasca fabae*, the six-spotted leafhopper, *Macrosteles fascifrons* and the beet leafhopper, *Circulifer tenellus*, regularly migrate for several hundreds of miles. *Empoasca fabae* and *Macrosteles fascifrons* overwinter in the permanent breeding areas in the southern United States as adults or eggs. In spring and early summer currents of warm southern air forming on the east side of a north to south cold front that is moving east carry the insects from the southern permanent breeding areas at around 30° N to the northern summer breeding area of 45° N–50° N, where the air flow is usually stopped by the polar/Canadian anticyclone (Wallis 1962; Pienkowski et al. 1964). Migration of *Circulifer tenellus* is a little different from the other two. Ferti-lized female *C. tenellus* overwinter and breed one or two generations on wild desert annuals in the western United States. The important spring migration starts when the host plants wither and die for lack of water and the leaf-hoppers migrate into beet growing areas, up to several hundreds miles away carrying with them curly top disease; the direction and distance the leaf-hoppers move are determined by the winds blowing at the time of migration. Some insects return in the autumn to the overwintering host plants in the desert but there is no concerted migration. Overgrazing, range fires or farm abandonment favour the wild annuals that are the preferred hosts of the leaf-hopper (Cook 1967).

Transoceanic migrations of the white back planthopper, *Sogatella furci-fera*, and the brown planthopper, *Nilaparvata lugens*, which are very im-portant pest insects of rice in Japan and other South-East Asian countries, have recently been shown to be important (Kisimoto 1971). Large swarms of planthoppers, mainly of *Sogatella furcifera*, have been seen from a weather survey ship in the Pacific Ocean, 500 km south of the Japanese mainland (Asahina et al. 1968) and others have been caught by tow nets at various locations in the East China Sea. These observations were correlated with mass flights of the insects into paddy fields from late June to mid July, par-ticularly in the western part of Japan. Subsequently, insects have been trapped in tow nets (10–15 m high) above the sea and on land at Chikugo, Fukuoka and it was found that a mass flight usually occurred in the warm sector near the cold front of a frontal system moving east in the rainy, Bai-u, season. Warm moist south-west winds blowing for considerable time seemed to carry most insects, but the speed and height at which the insects fly, etc., are un-known. In North America Huff (1963) analysed the relation between mete-orological conditions and migration of *Empoasca fabae* and found that most migrated when there was a persistent moist southerly flow of at least 36 hr duration of maritime tropical air from the Gulf States, and this was associated with a cold front.

8.5 Feeding

8.5.1 Feeding behaviour

The feeding behaviour of these insects is one of the factors that makes them important vectors of viruses and MLOs. Leafhoppers and planthoppers pierce and suck plant juices using fine and complicated mouthparts. The needle-like stylet is composed of the outer mandibular stylets and the inner maxillary stylets, and is surrounded with a labium and labrum. It is widely believed that the plant tissue is probed at random and that if the stylets reach a satisfactory position in the tissue the insect sucks, but, if not, it withdraws halfway and pierces again in a different direction so that laterally ramifying stylet sheaths are formed. Day et al. (1952) confirmed this by comparing the feeding habits of several leafhoppers which appeared to have tissue preferences. However Naito et al. (1967a) and Sogawa (1970) found that of the stylet sheaths in rice plant tissue 70.3% of those of *Nephotettix cincticeps* successfully reached the vascular bundles, and 87.3% and 72.5% of those of *Nilaparvata lugens* feeding on normal and nitrogen deficient rice plant, respectively were successful. Other stylet sheaths terminated in the mesophyll or parenchyma. Thus it seems reasonable to suppose that the leafhoppers and planthoppers select the probing site before piercing, but after piercing penetration of stylet is unselective. Leaf and planthoppers pierce much more quickly than aphids and also injure tissues more.

Although the way in which viruses are transmitted by hoppers is not yet understood, injury of the tissue during feeding is presumably important. The injuries depend on size of stylet and number of probes made. Day et al. (1952) pointed out that the amount of injury to the tissues appears to increase with the size of leafhopper, and that vector species tend to be small. Naito et al. (1967b) found that the number of probes per day was 43.7 for females and 40.7 for male adult *Nephotettix cincticeps*. There were few probes at 5 °C and their number increased linearly with temperature, reaching a maximum of 80.1 per day at about 30 °C. In *Nilaparvata lugens*, Sogawa (1970) reported 17 and 25 per day on normal and nitrogen deficient rice plants. There is however no reliable data on the relationship between the number of probes and the quantity of juice taken in, though it seems that frequent probing is related to restlessness of the insects and an insufficient supply of juice.

Leafhoppers and planthoppers usually suck juice and excrete drops of honey dew continuously while the stylets are inserted, and 85% of the time *Nephotettix cincticeps* keeps its stylets inserted (Naito et al. 1967b). *Orosius argentatus*, a vector of witches' broom disease of lucerne, excretes about 65% of the amount it ingested within 30 min (Day et al. 1951). The amount of honey dew excreted was estimated from the total amino acids and sugar contained by female individuals of *Nilaparvata lugens* (Sowaga 1970), and it was found that about 13 μl/day were taken in by those fed on normal rice plants and about 3 μl/day by those fed on nitrogen deficient plants; a large proportion of the body weight of this insect, which is about 2.6 mg at emergence.

8.5.2 Salivary gland and saliva

Feeding of hemipterous insects often has a greater effect on plant tissues, increasing oxidation, outgrowths and malformation, than could be the result merely of removing nutrients from the plant. These effects may result from the injection of saliva while feeding.

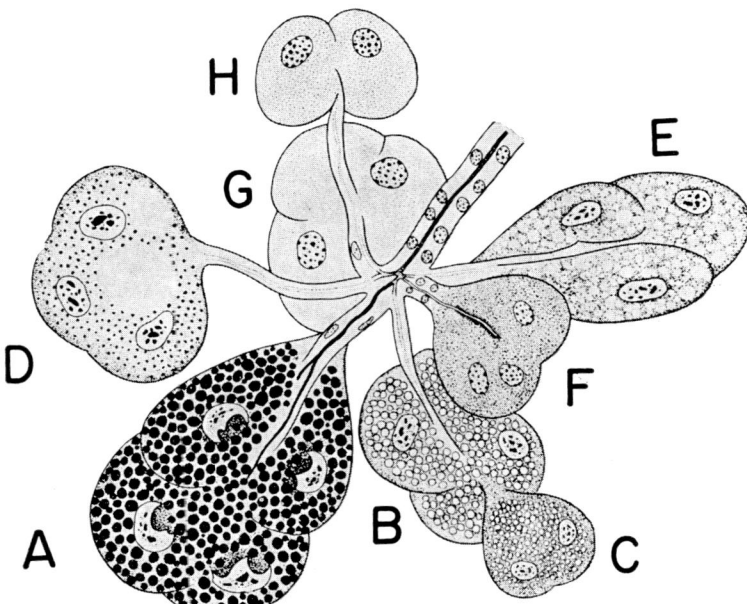

Fig. 8.6.   Diagramatic section of the principal salivary gland of the small brown plant-hopper, *Laodelphax striatellus* (Fallén). (From Sogawa 1965.)

A pair of soft white salivary glands lies on either side of the oesophagus. Sogawa (1967a, b, 1968a, b) demonstrated that hoppers produce two types of saliva, watery and coagulable, like other hemipterous insects. Coagulable saliva makes the salivary sheath and the other carries digestive enzymes. The salivary glands of leafhoppers and planthoppers each consist of a principal gland and an accessory gland. In planthoppers the principal gland consists of eight kinds of follicles (fig. 8.6A–H) each of which has 2–8 secretory cells, whereas the accessory gland has two large main secretory cells and several non secretory 'ejective cells'. In leafhoppers, the principal gland consists of six masses of different kinds of secretory cells (named by latin numerals), the anterior lobe consisting of type I and II cells and the spherical appendage and the posterior lobe consisting of types III–VI cells. In the posterior lobe each cell mass is arranged concentrically like the petals of a flower. The accessory gland is divided into two portions, head and tail, and is connected to the duct. Variation between different species in the structure of the anterior lobe has been found (fig. 8.7).

Sogawa (1967a, b, 1968a, b) suggested that in *Nephotettix cincticeps* the salivary sheath material originated as two distinct secretions, protein from type IV cells and, probably, mucolipids from type V cells. In *Laodelphax*

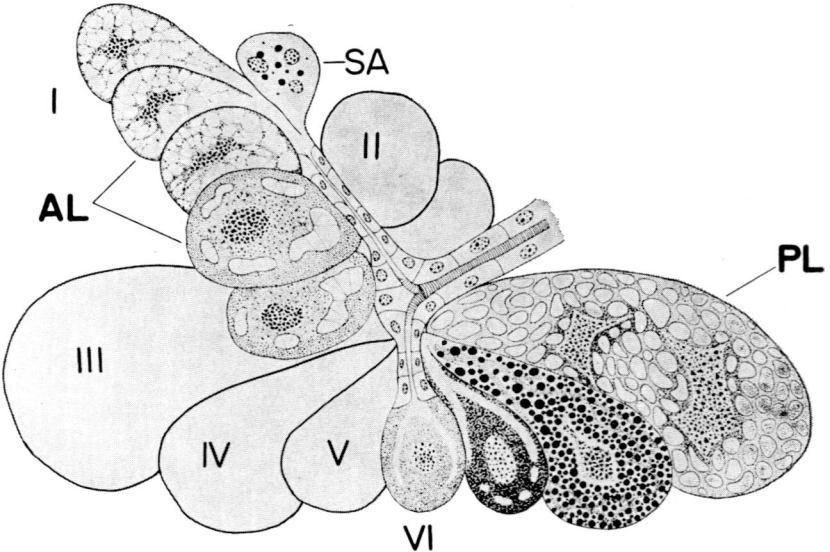

Fig. 8.7. Diagramatic section of the principal salivary gland of the green rice leafhopper, *Nephotettix cincticeps* Uhler. AL, anterior lobe; PL, posterior lope. (From Sogawa 1965.)

striatellus the sheath is perhaps formed by a combination of proteinaceous precursor elaborated in type A follicles, unsaturated lipid from type G and H follicles, and, probably, certain mucosubstances from the accessory gland. Phenolase activity was detected in the type V cells in leafhoppers and in the type E follicles in planthoppers, and may help form the sheath. In the watery saliva of both leafhoppers and planthoppers has been found α-glucosidase, which hydrolyses sucrose, trehalose, etc. and β-glucosidase, which hydrolyses phenolic glucosides. In leafhoppers both α- and β-glucosidase activities were localized in the type III cells, while in planthoppers α-glucosidase was demonstrated in the type G follicles and β-glucosidase in the type B and D follicles. The presence of the different enzymes does not seem to correlate with the experimentally determined feeding preferences of one planthopper for Mitsuhashi et al. (1969) found that *L. striatellus* artificially fed through stretched Parafilm M survived best on sucrose solution at a concentration of 5% or on glucose but not so well, but not at all well on fructose, maltose, raffinose, trehalose and soluble starch.

Virus particles have often been found in the salivary glands by electron microscopy, (Nasu 1965) or by fluorescent antibody staining (Sinha 1965) but no differences in the distribution of the virus in different follicle or cell types have been noted.

8.6 *Relation between vectors, host plants and viruses*

Most leafhoppers and planthoppers have wide host ranges but some of them are strictly monophagous. Under experimental feeding tests very wide host ranges are often shown but in nature the hosts on which they actually breed are fewer. The important virus vector species usually have wide host ranges. Kennedy et al. (1962) postulated that there are selection pressures between vector aphids, host plants and aphid-transmitted viruses, to favour viruses that have the same host plants as their aphid vectors, to favour spread among the less favoured host plants of aphids, and to favour polyphagous aphids as vectors. These postulates seem to be applicable to leafhoppers and planthoppers as well.

Though most of the important vectors are common species, the vector species do not necessarily prefer the economically important host plant of the virus diseases. For example *L. striatellus* transmits the rough dwarf virus of maize in Israel and Italy and the black streaked dwarf virus and the rice stripe virus in maize in Japan yet the vector does not breed on maize in either country; in Israel it breeds on Bermuda grass and migrates to irrigated maize fields when Bermuda grass withers, and in Japan it breeds and acquires the black streaked dwarf virus from overwintering wheat and barley, rice stripe virus overwinters in the vector. In these instances, temporary feeding on the host plants during the seasonal migration results in infection of the valuable crop plant, while nearby wild host plants, such as, *Digitaria adscendens* Henr., *D. violascens* Link., *Lolium multiflorum* Lam., *L. perenne* L., etc. are quite free from the viruses though not immune.

Examples of vector specificity are common in leafhopper- and planthopper-transmitted plant viruses and their vectors. An extreme situation is represented by the transovarial passage in *Laodelphax striatellus* of rice stripe virus and in *Nephotettix cincticeps* of rice dwarf virus; more than 90–95% of the offspring produced by infective females are congenitally infected. However even this high rate of transmission cannot be solely responsible for survival of the virus and even with virus/vector combinations that give 90% of congenitally infected offspring the proportion of infected individuals will decrease to about 53% after six generations, which is in one or two years in tropical or temperate regions, unless there are no benefits to infected individuals, and no such beneficial effects have been reported.

In Japan few trials had been made to find whether rice stripe virus had vectors other than *Laodelphax striatellus* since Kuribayashi's (1931) original report, then Shinkai (1966) found another vector when he obtained transmission of the virus by *Unkanodes sapporona*, which also transmits rice black streaked dwarf and the northern cereal mosaic viruses. Within a few

Fig. 8.8.   *Unkanodes albifascia* (Matsumura), a newly discovered vector of the rice stripe, rice black-streaked dwarf and northern cereal mosaic viruses. Less migratory species feeding on graminaceous weeds.

years, *Unkanodes albifascia* (Ishii 1966; Shinkai 1967), *Muellerianella fairmairei* (Ishii 1967) and *Terthron albovittatus* (Shinkai, personal communication) were shown to be able to transmit all or some of the three viruses (fig. 8.8); some virus/vector combinations have not yet been tested. None of these new vectors breed commonly in either rice, wheat or barley fields, though some of them breed readily on wheat or barley in the laboratory. So far, no infected individuals of these species have been collected in the field. Generally, they are Palearctic species of northern areas and their populations are usually small throughout the year. Furthermore they are mostly brachypterous, which accounts for their small migratory ability (fig. 8.8).

Is it useful to distinguish the common vectors of a virus and newly discovered vectors such as those discussed above, and call the latter accidental or fortuitous vectors as suggested by Oman (1969)? The outbreak of virus diseases of economical importance seems to largely depend on attributes of the virus concerned, such as its severe effect on and rapid spread in crops combined with slight or no effect on wild plant reservoir hosts which are preferred hosts of the vectors. After selection imposed on both virus and vector, it is possible that accidental or fortuitous vectors might become much more important.

Although vector specificity does not now seem as absolute as was once thought (Oman 1969) the specific ability of a particular vector group to transmit a certain virus seems definite. As mentioned above, *Laodelphax striatellus* is distributed widely eastern and western Eurasia but it is minor species in northern Europe and not known to be a vector of European wheat striate mosaic (Lindsten 1961) the species implicated is *Calligypona marginata*. *Laodelphax striatellus* and *Javesella pellucida* are found in Italy and transmit maize rough dwarf virus; *Laodelphax striatellus* more efficiently than *Javesella pellucida*. In Japan, *J. pellucida* is a minor species in cool mountainous regions but its ability to transmit viruses has not been investigated. It is an interesting example of parallelism that though *Laodelphax striatellus* and *Javesella pellucida* are both found throughout Eurasia, the former is commonest in the east and the latter in the west, though each transmits a particular set of virus diseases that are transmissible and non-transmissible through the ovary. It is also noteworthy that the symptoms of rice stripe virus and European wheat striate mosaic are much alike, also that those of black streaked dwarf virus resemble those of oat sterile dwarf virus and maize rough dwarf virus (Harpaz 1961), and likewise pseudorosette (zakuklivanie) in Siberia resembles the northern cereal mosaic, both being transmitted by *Laodelphax striatellus*.

List of Delphacida vectors of plantpathogenic virus diseases

Dicranotropis hamata Boheman	Oat sterile dwarf (Oat dwarf tillering)
Javesella dubia (Kirsch.)	European wheat striate mosaic
Javesella obscurella (Boheman)	European wheat striate mosaic
	Oat sterile dwarf
Javesella pellucida (Fabricius)	European wheat striate mosaic
	Oat sterile dwarf
	Maize rough dwarf
Laodelphax striatellus (Fallén)	Rice stripe, rice black-streaked dwarf,
	Northern cereal mosaic of Japan
	Oat pseudorosette (Zakuklivanie)
	Maize rough dwarf
Muellerianella fairmairei (Perris)	Northern cereal mosaic of Japan
Nilaparvata lugens Stål	Grassy stunt of rice plant
Peregrinus maidis Ashmead	Maize mosaic virus I (Corn stripe, enanismo rayado)
	Maize rough dwarf
Perkinsiella saccharicida Kirkaldy	Fiji disease of sugar cane
Perkinsiella vastatrix Breddin	Fiji disease of sugar cane
Sogatella furcifera (Horváth)	Dying-off of Pangola grass
Sogatella vibix (Haupt)	Maize rough dwarf
Sogatodes cubanus (Crawford)	Hoja blanca of rice plant
Sogatodes orizicola (Muir)	Hoja blanca of rice plant
Terthron albovittatus (Matsumura)	Rice stripe
	Northern cereal mosaic of Japan
Unkanodes albifascia (Matsumura)	Rice stripe
	Rice black-streaked dwarf
	Northern cereal mosaic of Japan
Unkanodes sapporona (Matsumura)	Rice stripe
	Rice black-streaked dwarf
	Northern cereal mosaic of Japan

CHAPTER 9

Beetles – phytophagous Coleoptera

B. J. SELMAN

Contents

9.1 Chrysomelid beetles as plant virus vectors

Most plant viruses are transmitted by insects with sucking mouthparts but a few are transmitted by insects with biting mouthparts. These vectors are almost exclusively from the orders Orthoptera, Dermaptera and Coleoptera. The most important of these orders is the Coleoptera or beetles.

The beetles of importance as virus vectors are the plant feeding species and though phytophagous species are found in most of the superfamilies in the order, the great majority are found in the two great superfamilies of the Phytophaga, the Chrysomeloidea and the Curculionoidea. The most important virus vectors are from the family Chrysomelidae of the superfamily Chrysomeloidea (Table 9.1). These species can transmit viruses that are apparently incapable of being transmitted by other insects. At present few of the Curculionoidea are known to be able to transfer plant viruses and these

TABLE 9.1
The classification of the family Chrysomelidae

Family	Group	Sub-family
Chrysomelidae	Eupoda	Sagrinae
		Donaciinae
		Criocerinae
		Zeugophorinae
		Orsodacninae
		Synetinae
	Camptosoma	Megascelinae
		Megalopodinae
		Clytrinae
		Cryptocephalinae
		Clamisinae
	Cyclica	Lamprosomatinae
		Eumolpinae
		Chrysomelinae
	Trichostoma	Halticinae
		Galerucinae
	Cryptostoma	Hispinae
		Cassidinae

are mainly viruses which can be transmitted mechanically by any species or object damaging the plants.

The Chrysomelidae are phytophagous beetles, each species feeding on a strictly limited range of host plants with a few species confined to a single host species. It is the second largest family of beetles with some 55,000 species in seventeen subfamilies. Of these some 24 species are known to be able to transmit plant viruses and of these all but four of the species belong to the subfamilies Galerucinae and Halticinae (the flea beetles). The four exceptions are the Colorado beetle, *Leptinotarsa decemlineata* Say and *Phaedon cochleariae* F. from the Chrysomelinae and the cereal leaf beetles *Lema melanopa* (L.) and *Lema lichenis* Voet from the Criocerinae. This distribution almost certainly represents the world distribution and interests of the researchers and not the properties of the insects. Indeed it may well be found that the curculionids are at least as important vectors of plant viruses as are the chrysomelids. Recently Cockbain (1971) reported that four species of weevils including two *Apion* species and two *Sitona* species are the main vectors of broad bean stain virus (BBSV). The *Apion* species were by far the most efficient vectors. Other species of beetles fed on the beans including the chrysomelid *Phaedon cochleariae* but did not transmit BBSV. Some of the other subfamilies of Chrysomelidae e.g. Cassidinae, Hispinae and Eumolpinae may eventually be shown to be important virus vectors. So far it is only those few species which are common pests of temperate and mainly annual crops which are known to transmit viruses. Furthermore they are mainly species which attain large populations on their host plants. One exception is *Lema melanopa* L. which transmits cocksfoot mottle virus (CFMV) (Serjeant 1967), despite occurring in only small numbers on the host. It is likely that many more low density feeders e.g. some species of the Chrysomelinae will be found to be efficient virus vectors.

9.2 Hosts

9.2.1 Host range of Chrysomeloidea

The Chrysomeloidea family tree may be aptly likened to climbing a tree (Crowson 1954). The most primitive species are those whose larvae live on decayed wood, more advanced are those whose larvae feed on the sound dead wood of the trunk, thence those that attack the living wood and out into the twigs. These are the cerambycids. Some of the more advanced have larvae that live in galls on young growing shoots and feed exposed on the outside of the leaves. These are the chrysomelids. A side-branch of these, the bruchids, feed on the carpels and the seeds. Most advanced are those that have 'returned' to feed on and in the roots. These are the chrysomelids and the cerambycids. The habitat range of the weevils or Curculionidae, the largest family of insects is similar. The chief difference between the life histories of the curculionids and the chrysomelids is that curculionid larvae are mainly concealed feeders while chrysomelid larvae are mainly exposed feeders. However the larvae of flea beetles, Halticinae, which are probably the most destructive group of chrysomelids attacking farm crops, are root feeders or leaf miners and therefore usually confined to an individual host plant.

Usually it is the adult chrysomelids which are important as vectors of viruses, moving readily from plant to plant, many being strong fliers.

9.2.2 Host selection

Most chrysomelids will only feed on a strictly limited series of host plants, and host selection seems to be determined by the interaction of attractant and inhibitory substances in the plants. Many of these hosts are poisonous and the poisons are often incorporated into the blood of the beetles, which thereby becomes obnoxious to predatory birds. Occasionally chrysomelids will feed on plants other than their normal hosts, for example many *Leptinotarsa* species will feed when transferred to other solanaceous plants including potato species other than the Irish potato (*Solanum tuberosum* L.). Sometimes when transferred the larvae die though the adults thrive or the larvae will thrive but the adults, though feeding, fail to become sexually mature (Busnel and Chevalier 1938). There is no evidence that beetles feed preferentially on virus-infected plants.

The young adult beetles of many chrysomelids prefer to feed on the host on which they fed as larvae. In *Lema melanopa* it is the stage of development of the plant which determines which of the alternative hosts is selected. In the field in Britain *L. melanopa* will on emergence move from grass hosts at the field borders to winter sown cereals which at that time are at the tillering stage and later to the spring sown cereals usually when they are at the 2–3 leaf stage. Here most of the eggs are laid (Benigno, personal communication). All cereals become progressively less attractive to *L. melanopa* as the plants mature.

9.2.3 The biology of the Chrysomelidae

The body-form of the Chrysomelidae is surprisingly uniform in marked contrast to the great diversity of their biology and especially of their larval development. The whole group shows a polyphyletic trend from leaf feeding larvae towards root feeding larvae. The adults of many species are good fliers, some migrating long distances. A few species of Galerucinae and Chrysomelinae are exceptions to this, being secondarily apterous. Balachowsky (1963) has given a very fine account of the individual life histories of all the chrysomelids of importance to European agriculture while Maulik (1919, 1926, 1936) collected together the life histories as then known of the world's chrysomelids.

The larvae of the Donaciinae live in water on the stems of water plants, where they pupate. The larvae of the Sagrinae are found within large stem galls in which they pupate. The Criocerinae are mainly grassland feeders, the larvae encasing their bodies in dry or fluid excreta or even making cases of lengths of grass stems. In *L. melanopa* (L) the eggs are usually laid on the upper surface of a leaf. They hatch in ten days to three weeks and the larvae feed on the leaves. After four to five weeks and three instars they pupate in the ground forming an earth cocoon lined with a viscous secretion. Other species e.g. *Lema lichenis* Voet pupate on the host plant. The pupal stage lasts for approximately one week. Usually there is one generation per year but in favourable conditions there may be more. The larvae of the asparagus beetle *Crioceris asparagi* (L.) consume the berries while other species of Criocerinae feed on flowers, fruits, etc. The adults pass the dormant period in plant detritus.

The larvae of the Zeugophorinae, Orsodacninae and Synetinae are mainly leaf and petiole miners and have reduced legs. The larvae of *Syneta* are root feeders. The adults of many of these species feed on buds and flowers in the spring or beginning of the rainy season, usually on trees and other woody plants.

The Clytrinae and Megalopodinae typically lay their eggs on the ground and on plant stems and leaves from where they are picked up by ants and carried to their nests where the larvae apparently feed on plant debris. Many of the larvae of these two families and of the Cryptocephalinae and Chlamisinae have abdomens curved under their bodies in the manner of a cockchafer larva. The abdomen is enclosed within a shell-like case of excreta which is continually added to, and within which they pupate. The adults feed on shrubs and herbs and can be major pests especially of small trees, e.g. pistachio nut trees.

The Hispinae and Cassidinae are closely related with similar life histories. The Hispinae adults are often very spiny. The larvae are leaf miners hollowing out the leaves until they can look like transparent envelopes. A few primitive species are found almost enclosed between the stem and the leaf bases of Palmaceae. The Hispinae are important pests of palms and cereal crops in the tropics. The adult Cassidinae are never spiny and the larvae are surface feeders encased in a shell-like covering. The cast skins of the previous instars are retained on the forked tail of the larva which is held over the back like an umbrella. This is flanked by the faeces which are also retained like strings of sausages. The eggs are usually laid in groups within an ootheca. The newly hatched larvae feed on the parenchyma on the underside of the leaves, later moving to the upper surface. They pupate on the undersurface of the leaves. There are usually one or two generations per year. The dormant period is in the adult. The Cassidinae are important pests of vegetable and broad-leaved crops, especially in the tropics.

The Halticinae and Galerucinae are very close in their taxonomy. The Halticinae are distinguished by a spring-like structure of chitin within the swollen hind femora. This is the jumping organ. The biologies of the Halticinae and Galerucinae are very different. Most galerucid larvae feed on the outer surfaces of leaves though a few species are root feeders and others leaf miners and seed and berry eaters. Some, e.g. *Ceratoma trifurcata*, feed on nitrogen-fixing root nodules. Most halticid larvae feed either on or within roots, while a minority of species are leaf miners. The adults and surface-feeding larvae of the Galerucinae carefully consume the parenchyma, leaving the veins until the leaves are skeletonised. The adults tend to eat right through the leaves while the larvae tend to feed on the underside leaving the upper epidermis as a transparent membrane. Pupation is usually in the soil. A few species feeding as larvae on aquatic plants pupate on the leaves. Pupal cases are rarely made. They are usually dormant as adults, sometimes grouped together in large numbers. Larval dormancy is known especially in the Mediterranean species. Eggs may be laid anywhere depending upon the species and they may be laid in an ootheca. Some species, including some of the most important agricultural pests, have different adult and larval host plants.

The Halticinae usually lay their eggs on the soil surface near the host plant's stem. The larvae pupate within the leaves or stems or in the soil. A few species are dormant as larvae though most are dormant as adults. The typical adult halticid eats a hole through the leaf between the veins and then enlarges this, leaving the characteristic shot-hole damage.

The Eumolpinae, Chrysomelinae and Lamprosomatinae have adults which are often superficially similar in morphology and often similar in habits. Eumolpid adults like criocerids cut a slot in the leaves when feeding. The eumolpid damage is however highly characteristic especially when feeding on fruit and broad leaves where the directions of the slots are random. Chrysomelinae adults tend to eat an area of a leaf, rarely cutting a slot. Eumolpid larvae are, as far as they are known, root feeders; while Chrysomelinae larvae feed on leaf surfaces. The larvae of the Lamprosomatinae are also surface feeders and often resemble thorns. The larvae of many species of Chrysomelinae have defensive glands on the dorsal surface. In favourable conditions species of these three sub-families may have two or more generations per year. The dormant period is in the adult. Some species of Chrysomelinae are facultatively viviparous; some species merely retain the eggs until hatching, whereas others may retain them until the larvae are first instar and about to moult to the second instar. Viviparity in the Chrysomelinae is associated with high altitudes and latitudes, i.e. habitats where the time for development is short. Facultative parthenogenesis occurs occasionally.

9.2.4 The biology of the Curculionidae

Most adult curculionids feed on the surface of stems, leaves or flowers of plants whilst their larvae are concealed feeders with reduced appendages forming mines or galls within the flower buds, seeds, stems, leaves or roots of their host plants. A few larvae e.g. *Cionus* spp. are surface feeders and are slug-like with thoracic and abdominal pseudopodia, the body covered by a viscous secretion from the gut. Both the adults and larvae of the hop root weevil, *Epipoloeus caliginosus* are root feeders. The larvae of *Otiorrhynchus* are leaf miners in the first instar and live subsequently either in or on roots. These biological changes are accompanied by morphological changes. The eggs are usually laid singly in the soil or in holes cut by the female's rostrum in the flowers, leaves or stems of the host plant. Curculionids are usually bisexual but facultative parthenogenesis is common especially in the genus *Ceuthorrhynchus*. Pupation is usually in the soil or in a mine or gall within the host. Often they construct pupal cocoons from silk or from the secretion of the Malpighian tubules and these cocoons may be open networks. The dormant period when present is usually in the adult. The surface feeding adults of many species feed at strictly defined times of the day or night, hiding at other times in the soil or dense vegetation e.g. *Otiorrhynchus* spp.

adults. Thus the damage caused by weevils is often easier to see than the beetles causing it. The apionids are often placed in a separate family the Apionidae. Most apionids, like chrysomelids but unlike the curculionids, lack a well developed proventriculus. The larvae are usually miners and sometimes form galls. They develop in seeds, stems or roots and are serious pests especially of the Leguminosae and Malvaceae. The adults are usually surface feeders often making very large numbers of small holes in each leaf attacked.

Other weevils, e.g. *Sitophilus* spp. are major pests of stored grain. The bark beetles, the Platypodinae and Scolytinae bore into bark and the outer layers of wood where they excavate extensive mines. Some curculionid species are aquatic, e.g. *Lissorrhoptrus*, and a few have adults living in the littoral zone of the sea. Many species are notorious pests including the cotton boll weevil, *Anthonomus grandis*, rice weevils, *Calandra oryzae*, granary weevil, *C. granaria*, and pea and bean weevils, *Sitona* spp. Balachowsky (1963) gives an account of the life histories of those species of importance to European agriculture.

Many weevils are monophagous or confined to closely related plants. The adults are often apterous or have limited powers of flight and are generally much less mobile over distances than are the chrysomelids though they may move more readily within a confined area.

9.3 *The viruses transmitted by beetles*

The particles of all of the viruses so far known to be transmitted by beetles contain single stranded RNA and most are isometric. The one exception is perhaps tobacco mosaic virus (TMV) which has tubular particles and which can be transmitted by any mechanical means and thus by many of the insects feeding on the tobacco plant e.g. *Epithrix parvula* (Fab.) (Schuster 1963). The beetle-transmitted viruses are listed in table 9.2. The properties of most of the viruses transmitted by beetles, and the numerous papers on beetle transmission have been reviewed by Walters (1969).

TABLE 9.2
Beetles transmitting plant viruses

Beetle	Virus
Apionidae	
Apion vorax Herbst	Broad bean stain
Apion aethops Herbst	Broad bean stain
Curculionidae	
Sitona lineatus (L.)	Broad bean stain
Sitona hispidulus (Fab.)	Broad bean stain
Chrysomelidae	
Criocerinae	
l. *Lema melanopa* (L.)	Cocksfoot mottle
Lema melanopa (L.)	Phleum mottle
Lema lichenis Voet.	Cocksfoot mottle
Lema lichenis Voet.	Phleum mottle
Chrysomelinae	
Phaedon cochleariae (Fab.) [especially larvae]	Turnip yellow mosaic
Leptinotarsa decemlineata Say [especially larvae]	Potato spindle tuber
Halticinae	
s. *Phyllotreta* spp.	Radish mosaic
s. *Phyllotreta* spp.	Turnip crinkle
s. *Phyllotreta undulata* Kutsch	Turnip yellow mosaic
s. *Phyllotreta ata* (Fab)	Turnip yellow mosaic
s. *Phyllotreta ata* v. *cruciferae* (Goeze)	Turnip yellow mosaic
Halticinae continued	
Phyllotreta nemorum (L.)	Turnip yellow mosaic
s. *Psylliodes* spp.	Turnip crinkle
Psylliodes affinis (Payk.)	Dulcamara mottle
Epithrix spp. especially E. *fuscula* Crotch	Egg plant mosaic
Epithrix parvula (Fab.)	Tobacco ring spot
Epithrix cucumeris (Harris)	Tobacco ring spot
Epithrix cucumeris (Harris)	Tobacco mosaic
Systena taenita Say	Tobacco mosaic
Disonycha triangularis Say	Potato spindle tuber

Beetle	Virus
Galerucinae	
l. *Diabrotica undecimpunctata* Jac.	Squash mosaic
s. *Diabrotica undecimpunctata* Jac.	Radish mosaic
Diabrotica undecimpunctata Jac.	Cowpea chlorotic mottle
l. *Diabrotica soror* Lec.	Squash mosaic
l. *Acalymma trivittatum* (Mann)	Wild cucumber mosaic
Acalymma vittatum (Fab.)	Squash mosaic
s. *Ceratoma trifurcata* Forst.	Cowpea mosaic
s. *Ceratoma trifurcata* Forst.	Arkansas cowpea mosaic
l. *Ceratoma trifurcata* Forst.	Trinidad cowpea mosaic
Ceratoma trifurcata Forst.	Bean pod mottle
Ceratoma trifurcata Forst.	Southern bean mosaic
l. *Ceratoma trifurcata* Forst.	Cowpea southern bean mosaic
l. *Ceratoma trifurcata* Forst.	Severe bean mosaic
Ceratoma trifurcata Forst.	Cowpea chlorotic mottle
l. *Ceratoma ruficornis* Ol.	Trinidad cowpea mosaic
	Bean pod mottle
s. *Ootheca mutabilis* (Sahlb)	Nigerian cowpea yellow mosaic
Ootheca mutabilis (Sahlb)	Sann hemp mosaic (syn. cowpea strain tobacco mosaic)

l = long persistent infection of ~ 14 days
s = short infection of 1–2 days

9.4 Feeding

9.4.1 Mode of feeding

Most chrysomelid beetles both adults and larvae differ from many other leaf-biting insects by their mode of feeding. When small numbers of beetles are feeding on leaves they eat the parenchyma between the vascular bundles leaving these more or less intact. This gives the typical slotted (figs. 9.3 and 9.4) and shot-hole damage (figs. 9.1 and 9.2), and often the damage is confined to only one side of the leaf. Thus chrysomelid beetles (*Diabrotica* species) are the sole vectors of the bacterial wilt, *Erwinia tracheiphila* which infects only vascular bundles (Rand and Enlons 1916; Rand and Cash 1920).

When large populations of beetles are feeding the amount of damage is usually greater and *Leptinotarsa decemlineata*, for example, can consume almost the entire potato plant above ground, stems and leaves. It is significant that many experimenters have obtained more virus transmission with

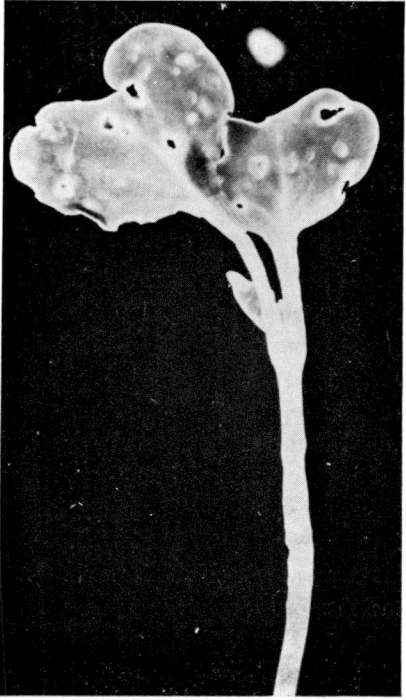

Fig. 9.1. Cauliflower leaf damaged by *Phyllotreta undulata* Kuts. showing typical feeding damage.

Fig. 9.2. Crucifera seedling showing typical 'shot hole' damage of the cotyledons by *Phyllotreta* sp.

small numbers of beetles feeding for short periods than with larger numbers feeding for long periods which seriously damage the plant (Campbell and Colt 1967). However at low feeding densities, percentage infection is correlated with the amount of leaf consumed (Dale 1953). Most other biting insects either feed along the edges of the leaves eating inwards or start by biting out the protruding vascular bundles. Their feeding is often much cleaner than that of many chrysomelids, such as the galerucids, which often crush the surface of the leaf into a juicy mess. Regurgitation seems to be rare except when beetles, e.g. *Leptinotarsa* spp. have been forced to feed on plants other than their usual host. The feeding site is frequently contaminated by faecal matter which may be more or less solid, or fluid as in *Lema melanopa* whose larva is entirely encased in a drop of fluid excrement. Thus faecal contamination of feeding damage is common.

It is frequently claimed (Walters 1969; Smith 1965) that beetles which have no salivary glands regurgitate fluids during feeding and in so doing may transmit viruses acquired in previous meals and 'that this is supported by the fact that the larvae of white butterflies, *Pieris brassicae* and *P. rapae*, which are biting insects possessing salivary glands and do not regurgitate after feeding were unable to transmit' turnip yellow mosaic virus (TYMV). However examination of the evidence suggests that this explanation is unlikely. The 'salivary' glands of lepidopterous caterpillars are mandibular glands and not labial glands as are typical insect salivary glands. The larvae and adults of many Coleoptera have similar gnathal glands usually thought to be associated with the mandibles. Recently Srivaskava (1959) showed that these glands in many Coleoptera are maxillary glands opening to the exterior on the inner surface of each maxilla close to the base of the mandibles. He found well developed maxillary glands in the larvae of *Galeruca tanaceti* L. and in the larvae and adults of *Phaedon cochleariae* F., a species with larvae which are even better transmitters of TYMV than the adults (Markham and Smith 1949). Selman (unpublished) has found well-developed maxillary 'salivary' glands (fig. 9.8) in the larvae of Chrysophtharta spp.: Chrysomelinae, and these appear to have the same structure as the salivary glands of other insects. 1 found that it is often difficult to find 'salivary' glands in the adults of many chrysomelid beetles because they are slender and coiled within the head; this needs further investigation. Adult weevils have also been shown to have paired coiled 'salivary' glands (Dennel 1942) which are so closely applied to the lateral walls of the crop to be thought at one time to discharge directly into the crop. Jeannel (1949) states that the salivary glands of chrysomelid larvae are at first digestive, losing this function before pupation when they produce silk, and resuming their digestive role in the adult. Srivastava (1959) was unable to detect the presence of amylase in gland extracts of either the adults or larvae of *Tenebrio molitor*. Thus it is clear that the presence or absence of 'salivary' glands is not responsible for the ability or otherwise of biting insects to transmit viruses. This is not to say that differences in the secretions of the glands is not the determining factor. It is interesting that TMV can apparently be inactivated by amylase in both the saliva and gut of the bush crickets *Tettigonia viridissima* and *T. cantans* (Schmutterer 1961), yet can pass through the gut and remain active in other biting insects (Walters 1952; Break 1957) including the chrysomelid *Epithrix cucumeris* (Orlob 1963) and lepidopterous caterpillars (Smith 1941). Of these only the lepidopterous caterpillars are not vectors of TMV. Amylase has been found in the midgut of most insects, but several workers have failed to

Fig. 9.3.   *Lema melanopa* (L.) feeding on a young oat plant.

Fig. 9.4.   Close up of 'slot' damage to the leaf of an oat plant by *Lema melanopa* (L.).

demonstrate amylase in several species of Lepidoptera and Coleoptera, though amylase is present in at least some field pest species of lepidopterous caterpillars (Ishaaya et al. 1971).

Similarly Walters (1969) claimed that 'Viruses retained by beetle vectors for prolonged periods of time are transmitted through the process of re-gurgitation of virus during feeding'. However I have never seen fluid re-gurgitated naturally by feeding chrysomelid beetles. Fluid is produced by both chrysomelid and weevil adults when they are gassed or anaesthetised and sometimes when roughly handled; some species exhibit reflex bleeding when threatened. Freitag (1956) etherized his cucumber beetles to obtain drops of regurgitated fluid, and this was infective with squash mosaic virus. That beetles can be made to regurgitate fluid which is infective is no proof that this is the natural mode of transmission. The marked differences between the infectivity of different species of chrysomelids feeding on the same host indicates that the determining factors remain to be discovered.

9.4.2 The gut of chrysomelid beetles

The gut of adult chrysomelid beetles is simple. The foregut has a short oesophagus and pharynx which either opens gradually into the crop or opens abruptly into the floor of the crop, as in *Leptinotarsa* spp. The crop is thin-walled and opens directly into the midgut. The proventriculus is absent and is represented by a skirt of epithelium marking the change from the thin-walled non-secretory foregut to the thick-walled, secretory midgut. The midgut cells are columnar with a distinct brush border. Secretion of digestive juices is reported to be holocrine in *Galerucella* larvae (Poyarkoff 1910), a proportion of the midgut cells discharging part of their contents into the lumen of the gut followed by the breaking away and disintegration in the lumen of the whole cells. These cells are immediately replaced by the development of new ones. If viruses transmitted by beetles infect or pass through the cells of the gut wall, then the repeated shedding of the epithelium would have a considerable influence on the ability of the beetle to transmit the virus. Indeed this might account for some of the highly variable results obtained in virus transfer experiments.

In *Paropsis* spp. which bite lumps off the host leaf I have found no discernible digestion in the foregut and the anterior third of the midgut, but thereafter the plant cell masses in the gut can be seen to be digested, losing their morphological integrity within a few hours of being swallowed. It is unlikely that food in the gut could on its own act as a virus reservoir over the periods of 7–14 days during which many chrysomelids can remain infective, as few beetles retain food in their gut for longer than 24 hr and in many species the time is far shorter.

9.5 Beetles and viruses

9.5.1 Interaction between plant viruses and their beetle vectors

There is evidence which indicates that adult chrysomelid beetles feeding on plants infected with viruses can have a shorter life span than others. Benigno (personal communication), working on *Lema melanopa* carrying cocksfoot mottle virus (CFMV) and phleum mottle virus (PMV), found that both these viruses significantly increased the mortality of the beetles. This was especially marked for CFMV (table 9.3). Benigno caged single beetles for 4 days on infected leaves kept moist by cotton plugs. The beetles were then transferred to healthy leaves for 4 days and then to fresh healthy leaves for two further periods of 5 days each. The total mortality 14 days after removal from the infected or control leaves is shown in table 9.3.

TABLE 9.3

The mortality of *Lema melanopa* 14 days after removal from plants infected with CFMV, PMV or both viruses together

| | Beetles infected with | | | |
	CFMV	PMV	CFMV + PMV together	Uninfected control
Deaths/total no. of beetles	31/50	28/50	11/50	7/50

The lower mortality of beetles offered both infected cocksfoot and infected timothy grass is curious. Unfortunately the results of this type of experiment are notoriously difficult to interpret because of the difficulty of associating cause and effect. However, it seems that beetles cannot transmit more than one virus at once. Thus Benigno found that *L. melanopa* would transmit either CFMV or PMV from a mixture of the two, but never the two together. When a beetle was fed first on a plant infected with one virus and then on a plant infected with the other virus, then only one virus was transmitted. The order of feeding did not seem to affect the results. The percentage of insects transmitting virus was the same as in experiments using only a single virus. Slack and Fulton (1971) found that they could recover either tobacco ring spot virus (TRSV) or bean pod mottle virus (BPMV) from the faeces of *Ceratoma trifurcata* fed on plants infected with both viruses but never the two together though both viruses could be recovered simultaneously from regurgitated fluid. This is interesting because *C. trifurcata* is not a vector of TRSV. The factors limiting the transmission of viruses by beetles are unknown.

9.5.2 *The blood as a reservoir of virus*

It has been suggested that the blood of beetles could act as a reservoir of virus (Slack and Fulton 1971) and viruses have been found in the haemo-

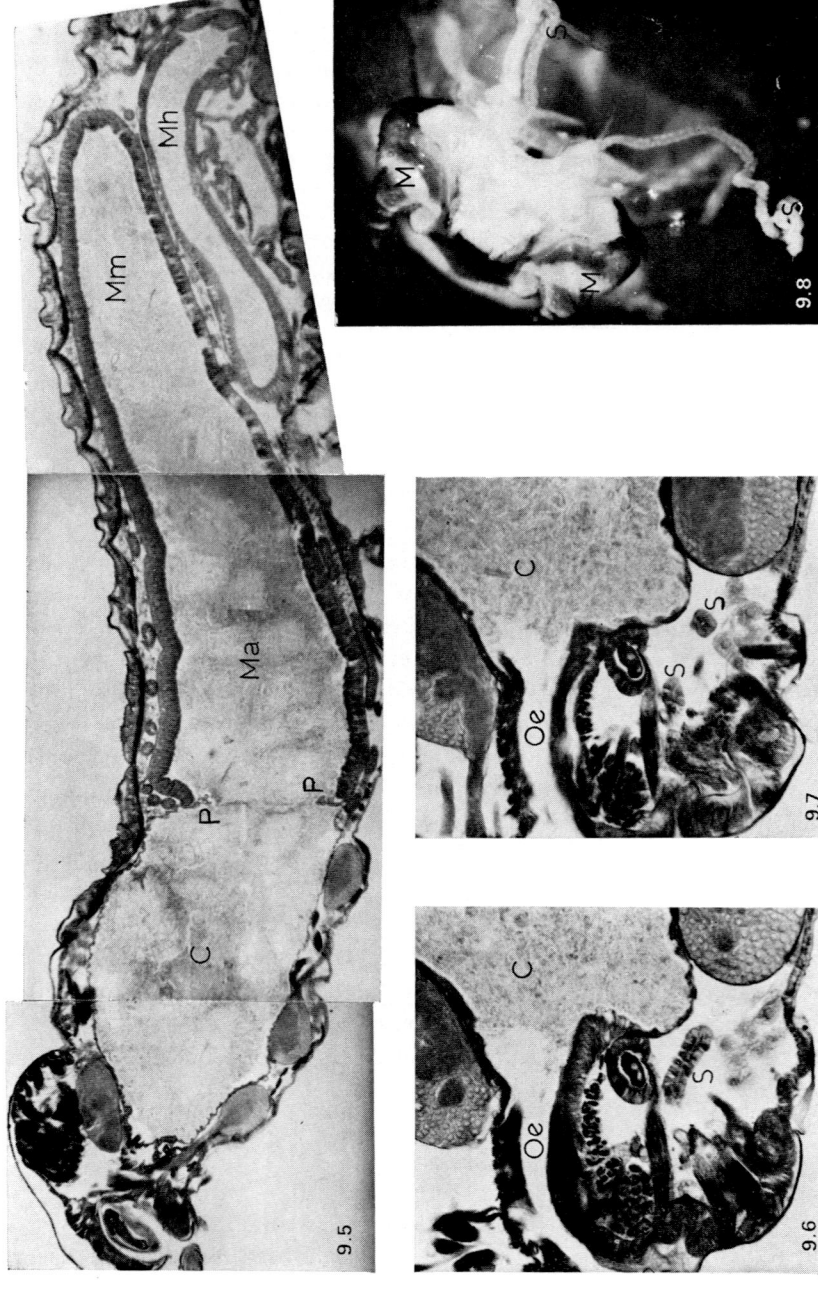

lymph of chrysomelid beetles (Freitag 1956; Slack and Fulton 1971), (table 9.4).

TABLE 9.4
Virus recovery from haemolymph of viruliferous chrysomelids

Vector	Virus
Ceratoma trifurcata (Forst.)	Southern bean mosaic virus (SBMV)
Ceratoma trifurcata (Forst.)	Cowpea strain of SBMV (CP–SBMV)
Ceratoma trifurcata (Forst.)	Cowpea mosaic virus (CPMV)
Diabrotica undecimpunctata	
Diabrotica undecimpunctata Mann	
Diabrotica undecimpunctata	Squash mosaic virus (SqMV)
Diabrotica howardi (Mann)	
Acalymma vittatum (Fab.)	
Acalymma trivittatum (Mann)	

CP-SBMV injected into the haemocoel of *C. trifurcata* made the beetles viruliferous (Slack and Scott 1971). The patterns of transmission by beetles acquiring CP–SBMV by feeding or by injection were similar, and virus recovery from the haemolymph of injected beetles followed the same trend as recovery following natural acquisition. Slack and Scott (1971) recovered active virus from all parts of the gut wall from 24–48 hr after injection into the haemocoel and up to 8 days after acquisition by feeding. In two isolated cases, after 10 and 20 days respectively, active virus was recovered from the posterior mid-gut only. Thus it appears that in at least a few species of chrysomelids the haemolymph acts as a virus reservoir and that this virus can be transmitted at intervals. However Slack and Fulton (1971) could find no virus in the blood of *C. trifurcata* feeding on plants infected with TRSV or BPMV though the beetles readily transmitted BPMV.

Fig. 9.5.   Longitudinal section of the first instar larva of *Chrysophtharta obovata* (Coleoptera: Chrysomelidae). C, crop; P, proventriculus; Ma, anterior midgut; Mm, mid midgut; Mh, hind midgut. Note the progressive breakdown of the plant cells along the length of the midgut, the thin walled crop and the wide open proventriculus.
Figs. 9.6 and 9.7.   Longitudinal sections of the head of *Chrysophtharta obovata* showing the narrow oesophagus and the maxillary salivary glands S. In fig. 9.6. the salivary gland is in longitudinal section showing the chitin duct and the large glandular cells. In fig. 9.7. the salivary gland is in oblique and transverse section.
Fig. 9.8.   Dissection of the head and thorax of the larva of *Chrysophtharta obovata* to show the paired maxillary salivary glands. M, maxilla; S, salivary gland.

9.5.3 Acquisition time

Dale (1953) found that *Ceratoma ruficornis* beetles became efficient vectors of CPMV after less than five minutes feeding. The longer the acquisition feed the more consistent was virus transmission. Regression analysis showed no significant difference in infectivity between beetles given 5 min, 3 hr or 24 hr acquisition feeds. A significant difference ($P = 5\%$) was found between the 5 min and 3 hr-feeding periods as against prolonged acquisition feeds. The average infectivity of the beetles given prolonged acquisition feeds decreased more quickly when taken from the infected plants than in beetles given short acquisition feeds, though a few individuals remained infective longer (14 days against 11 days). After 24 hr the beetles seemed to approach the limit of their virus-retaining capacity as longer feeds did not make them any more infective.

Benigno (personal communication) working with *Lema melanopa* found that the beetles could become infective with CPMV after acquisition feeds as short as 3 min, while 15 min gave 20% infection and 48 hr 30%. With PMV *L. melanopa* gave 20% and 0% infection after acquisition feeds of 48 hr and <15 min respectively. *Ceratoma ruficornis* is an efficient vector of CPMV immediately after feeding while *Lema melanopa* adults and *Phaedon cochleariae* larvae normally only transmit PMV and TYMV respectively after 24 hr have lapsed following feeding (Markham and Smith 1949; Catherall 1970). It has been suggested that initially the virus is transmitted purely mechanically on the mouthparts, but later infections are by virus that has persisted in the insect, and perhaps arrived by some pathway through the insect's body; the former masking any latent period that the latter may have. Indeed the experiments of many workers including Dale (1953) show that the beetles often have a high initial non-persistent infectivity followed by a fall and then a later increase to well above the initial level.

9.5.4 Summary of infectivity experiments

Virus can be recovered from the crop contents and faecal matter of both vector and many non-vector beetles. It seems that the virus enters the blood of most vector beetles but the pathway back to the plant is unknown. This might be through the maxillary glands, the epithelial cells of the gut or the Malpighian tubules. The cyclical regeneration of the gut lining may account for some of the variability in the results obtained in transfer experiments with beetles. The viruses have not been transmitted from a beetle to its progeny in those chrysomelids tested. The beetles remain infective for several days but seldom longer than three weeks. An infection acquired by beetle larvae is invariably lost during metamorphosis; infective larvae never develop into infective adults. There is no evidence for the multiplication of the viruses within the beetle host but infection of the beetle with a plant virus appears to shorten the life of adult beetles.

One difficulty in working with adult chrysomelid beetles is that they can be very long lived; individuals of some species may survive for several years. In many species adults show marked physiological and colour differences in different seasons and years. For experimental work it is obviously desirable to use beetles of the same age which have been kept in constant laboratory conditions for some time before experimentation. Unfortunately much of the past work has used adults captured in the wild and this may be responsible for some of the variability of the results.

9.6 *The pathogenic viruses of beetles*

Beetles not only transmit plant viruses but are infected by insect pathogenic viruses. Most viruses infecting beetles are non-inclusion viruses and most have been found in lamellicorn beetles including both pasture scarabs and rhinoceros beetles. Other viruses which do not fit into any of the virus categories so far defined, have been found infecting the larvae of *Melolontha melolontha*. An excellent account of these viruses can be found in Smith (1967). Recently viruses with spindle-shaped inclusion bodies have also been found in *Melolontha* spp (Vago and Bergoin, 1968). So far the pathogenic viruses of chrysomelid beetles are unknown almost certainly because few workers have looked for them.

Key to the subfamilies of the Chrysomelidae excluding the Megascelinae (after Gressit and Kimoto (1961))

1	a	Head and vertex projecting strongly forward and mouth directed posteriorly below, and often partly hidden by prosternum	2
	b	Head normal with vertex not projecting and with mouth directed forwards and downwards	3

2 a	Pronotum and elytron with broad marginal expansions, the former often covering the head	*Cassidinae*
b	Pronotum and elytron without broad marginal expansions, but often with spines; head never covered	*Hispinae*
3 a	Antennae closely inserted on front of head; elytron not very rigid	4
b	Antennae not closely inserted, separated by frons or vertex; elytron generally somewhat rigid	5
4 a	Posterior femur not greatly enlarged, no jumping organ within	*Galerucinae*
b	Posterior femur strongly swollen, jumping organ within	*Halticinae*
5 a	Prothorax not completely margined laterally; eyes prominent and head more or less strongly constricted behind them	6
b	Prothorax completely margined laterally; eyes not very prominent and head not strongly constricted behind them	12
6 a	Posterior femur large and strongly swollen, often armed with teeth	7
b	Posterior femur usually not greatly enlarged, rarely armed with teeth	8
7 a	Antennae long, with most segments longer than broad	*Sagrinae*
b	Antennae short, barely reaching beyond humerus, distal segments usually broader than long, more or less dentate	*Megalopodinae*
8 a	Antennal insertions separated by width of frons	9
b	Antennal insertions not separated by width of frons, relatively close	*Donaciinae*
9 a	Tarsal claws usually bifid or toothed internally; prothorax frequently toothed laterally	10
b	Tarsal claws simple, not toothed; prothorax never toothed laterally	*Criocerinae*
10 a	Side of prothorax with a prominent swelling or with 2–3 fairly distinct teeth	11

b	Side of prothorax plain, evenly moulded, body rather flat and narrow	*Orsodacninae*
11 a	Side of prothorax with a prominent swelling anterior to constricted base; body not very flat	*Zeugophorinae*
b	Side of prothorax rounded with 2–3 sharp teeth; body rather flattened	*Synetinae*
12 a	Middle 3 visible abdominal sternites constricted in central part; body subcylindrical	13
b	Middle 3 visible abdominal sternites not constricted; form of body usually more or less ovate or rounded, often strongly convex and constricted anteriorly	15
13 a	Antennae relatively short and serrate	14
b	Antennae usually longer and slender, not serrate	*Cryptocephalinae*
14 a	Prothoracic pleuron without antennal groove; body surface smooth	*Clytrinae*
b	Prothoracic pleuron with groove for reception of antennae; body surfaces rough or tuberculate	*Chlamisinae*
15 a	Wing venation not reduced; cubital veins present; clypeus not divided into two parts	16
b	Wing venation greatly reduced; cubital veins lacking; clypeus divided into two parts	*Chrysomelinae*
16 a	Prothorax as broad as elytra basally, its side grooved for reception of antenna; abdomen grooved for reception of hind leg	*Lamprosomatinae*
b	Prothorax generally narrower than elytra basally, its side not grooved for reception of antenna; abdomen not grooved for reception of hind leg	*Eumolpinae*

CHAPTER 10

Coccoids

P. F. ENTWISTLE

Contents

10.1 General biology

Coccoids feed exclusively by sucking the sap of plants with stylet-like mouth-parts in a manner typical of the order Hemiptera, and more especially their particular suborder Hemiptera-Homoptera. Consequently they feed in a similar way to other homopterous plant virus vectors, notably aphids, whiteflies and leaf hoppers (see chs. 7, 8 and 13).

Though they have much general morphological homogeneity coccoids are usually grouped in several families in the superfamily Coccoidea. The classification is based mainly on the adult female principally because of a lack, until recently (Boratynski and Davies 1971), of studies on the male. The superfamily Coccoidea is unusual in two respects. Firstly because male instars three and four are, respectively, prepupa and pupa, four being similar to the true pupa of endopterygote insects, whereas the development of all other Hemiptera is typically exopterygote. The male prepupa and pupa do not feed, neither do the male adults as they lack developed mouthparts and are ephemeral, living, according to species, from a few hours to a few days only. Their mesothoracic wings are functional but the metathoracic pair is reduced to hook-like halteres. The second unusual feature of the coccoids is that the female is neotenic, and has three juvenile instars before becoming reproductively mature in instar four. Often she is concealed beneath secretions of the dorsum which may be either scaly (in scale insects) or of a powdery or flocculent wax (mealybugs and some others). She lacks wings, is generally sessile with, in most families, either no legs or non-functional legs.

Coccoids are dispersed by the much more mobile juvenile instars, which have longer functional legs, notably in the first instar or 'crawler'. They are in general very small insects and early instars are often carried by air currents, and this is the main mode of dispersal, though, where suitable host plants are sufficiently close they may walk from one plant to another.

Coccoidea may be found almost wherever there is terrestrial plant life but the tropics, subtropics and mediterranean climatic areas are richest in species, especially where there are many different plant species. Probably plants in all phanerogam orders are attacked and all parts, from the roots to the fruits, have their adapted species. In addition to great specific diversity some species are very numerous particularly in plant monocultures so that the Coccoidea include some notable pest species, e.g. the San José scale *(Quadraspidiotus perniciosus)* which attacks deciduous fruit-tree species and the red scales of citrus *(Aonidiella surantii* and *Chrysomphalus ficus)*. Accidental dispersal on infested plants has resulted in some species becoming almost cosmopolitan and there are now strict quarantine regulations to try to stop the spread of others. Many species are polyphagous and this must be very advantageous to an animal which disperses randomly, and will be especially useful for those species living in diverse plant communities, such as tropical rain forest, but not so necessary in those living in generally less diversified temperate areas. Some monophagous species are known, e.g. *Cryptococcus fagi* on *Fagus sylvatica* and *Physokermes abies* on *Abies excelsa*.

Like many other Homoptera, honeydew, rich in amino acids, amides, proteins, many minerals and B-vitamins (Way 1963), is excreted by some species and is generally given as the reason why such insects are attended by ants. The ants and coccoids may be very dependent on one another as Way (1954) noted for the ant *Oecophylla longinoda* Lat. and the scale *Saissetia zanzibarensis* on clove *(Jambosa caryophyllus)* in Zanzibar. Here at times of ant population recession excess scales are consumed and when, conversely, ant populations are increasing additional colonies of scale are established by the ants. Thus ants are sometimes involved in the dispersal of coccoids, a type of behaviour which appears to reach its peak in *Acropyga (= Rhizomyrma)* ants the females of which each carry a mealybug *(Neorhizoecus coffeae* (Laing)), presumably a young fertilized female, between their jaws during the mating flight. All shades of relationship exist, however, and some coccoids appear never to be attended by ants.

There are fifteen families of coccoids (Imms 1957 – but other taxonomic accounts may not be strictly in agreement), but although species of all families are similar in feeding behaviour and general ecology only the Pseudococcidea, the mealybugs, have been implicated in the spread of plant viruses. Consequently the following account relates entirely to members of this family and specially emphasises the more important vector species.

10.2 The viruses transmitted by Pseudococcidae (mealybugs)

Planococcus citri (Risso) was first reported to be a vector of tobacco mosaic virus (TMV) by Olitsky (1925) and this was confirmed more recently (Newton 1953) when TMV but not Storey's leaf curl virus was transmitted from tobacco plants jointly infected with the two viruses: Storey's leaf curl is readily transmitted by white fly. *Pseudococcus maritimus* (Ehrh.) has been recorded as a vector of an unspecified mosaic disease (Elmer 1922; 1925).

Pineapple wilt disease was long considered to result from toxins introduced by mealybugs during feeding. In a recent reappraisal Carter (1963) has concluded a transmissible latent factor is also involved which 'is presumed to be a latent virus'. Altogether three main factors appear to be involved in pine-apple wilt. Firstly there is no doubt that some time after virus-free mealybugs feed on plants their growth is depressed and normal colour characteristics are accentuated and fruiting becomes premature. Secondly initial feeding by mealybugs on infected plants appears to activate the latent factor and mealy-bugs feeding later can, thirdly, synthesise a secretion which can cause wilting to a degree dependent on the numbers of mealybugs involved, and which can also affect uninfected plants. Both the latent factor and the wilting secretion can be transmitted by *Dysmicoccus brevipes* (Ckl.) and the closely related *D. neobrevipes* (Beardsley) which are phloem feeders. *Phenacoccus solani* (Ferris), which does not normally feed on pineapple and is purely a mesophyll feeder can transmit only the latent factor. 'It is possible that the latent factor moves and multiplies in the mesophyll' (Carter, op. cit.) whilst presumably phloem feeding is a prerequisite to the insect synthesised wilting secretion.

Mealybugs, however, are best known as the vectors of certain virus dis-eases of cocoa (*Theobroma cacao* L.) (see chs. 4 and 24). In West Africa they also transmit the same viruses from alternative species of host plants (Tinsley 1971). Hence, though of very much wider distribution mealybugs are mainly known as plant virus vectors only within the range of cocoa cultivation, be-tween about 15 °N and 15 °S of the equator.

10.3 Morphology and biology of Pseudococcidae

The general features and the developmental pattern described for the Coccoi-dea are germane to mealybugs (fig. 10.1). Mealybugs are usually small (sel-dom longer than 3.0 mm) oval to elongate-oval soft bodied insects. The name mealybug is especially applicable to the adult females which usually have the dorsum and to a less extent the venter covered with a mealy white waxy secretion. In juvenile stages this is less marked. The wax cover is thin-nest in the intersegmental areas and often is sparse along at least a part of the dorsal midline. Marginal filaments of wax, the numbers of pairs of which varies with the species, are also usual. Long caudal filaments are the only waxy adornment of adult males. The adult females of different species are distinguished by the relative lengths of the lateral wax filaments, which arise from the cercarii, by variations in the nature of the dorsal wax and by the colour of the body beneath the wax. Precise identification depends upon

P. F. Entwistle

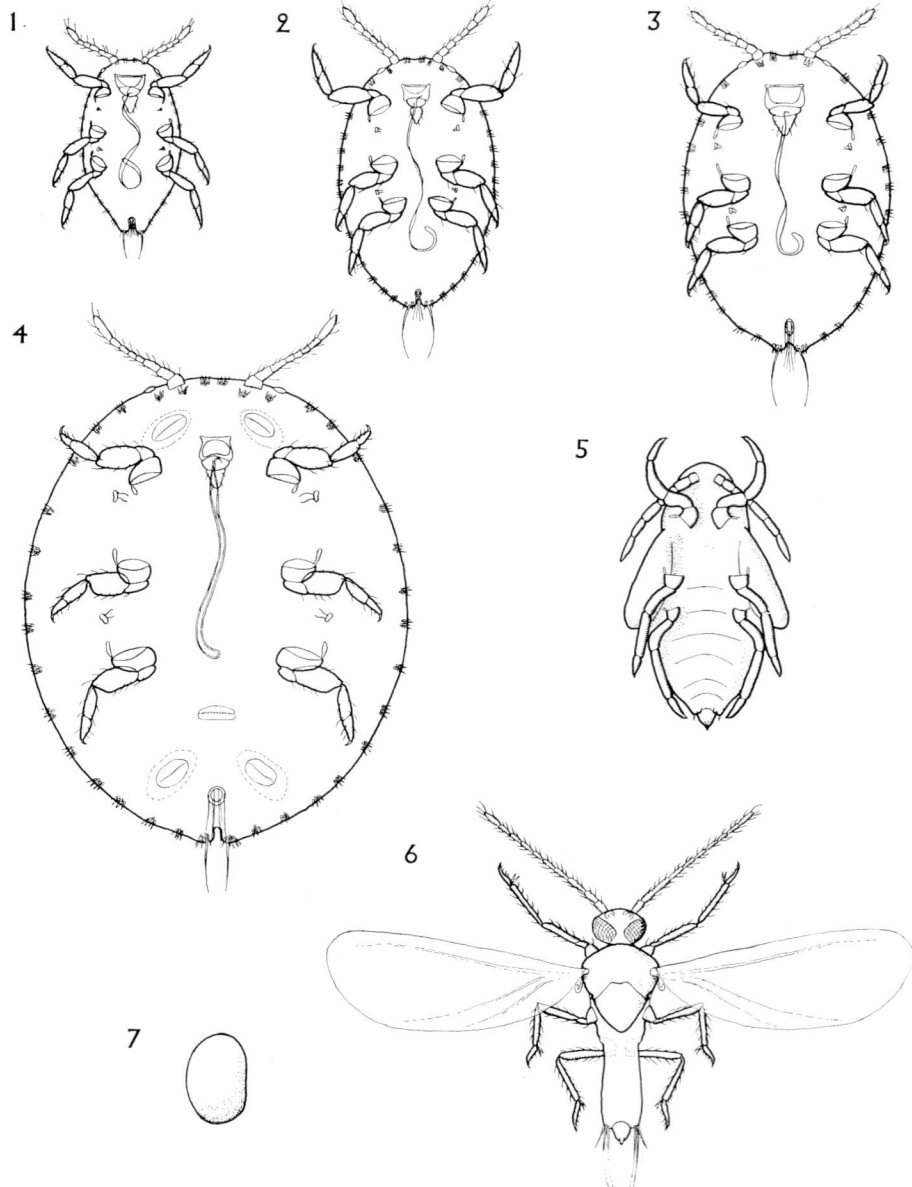

Fig. 10.1. Stylized life cycle of *Planococcoides njalensis*. 1. First instar nymph; 2. Second instar nymph; 3. Third instar female nymph; 4. Adult female; 5. Male pupa; 6. Adult male; 7. Egg.

microscopic characters only visible in cleared, stained preparations, preferably of pregravid final instar females. Unlike many other families of Coccoidea, the adult female mealybugs have functional legs (fig. 10.2), though in some species they do not move unless disturbed. By contrast the first instar nymph is very mobile and between brief feeds moves actively until settling down before moulting. The second and third instars have legs progressively shorter in relation to body size and show progressively less activity. Other structural differences between female instars and the first two male instars are slight. The adult female may be recognised by the presence of a vulva. In *Saccharicoccus sacchari* (Ckll.) the male is dimorphic being either macropterous or apterous, and the latter form is structurally more like the juvenile instars and the adult (neotenic) female, though it has no mouthparts (Rao 1943; Beardsley 1960). The structure of pseudococcid mouthparts is discussed below in relation to feeding.

Many of the vector species of cocoa virus diseases fall into the tribe Planococcini which has been reviewed on the basis of females (Ezzatt and

Fig. 10.2. Adult female of *Planococcoides njalensis*.

McConnel 1956). The morphology and taxonomy of adult males has been studied by Afifi (1968). There is no general account of mealybug biology and ecology, but the work of McKenzie (1967), though on a regional basis, is of great assistance.

Mealybugs have been recorded on cocoa wherever it is grown and even in greenhouses in temperate areas. Altogether 42 species are known from this host (Entwistle 1972) but there are useful data on the biology of only a few of these, especially where it relates to the host plant *Theobroma cacao*. Five species are particularly important and each tends to be dominant, or co-dominant in some particular region of the world. The general status of some species is summarised in table 10.1.

TABLE 10.1
Importance and distribution of the main mealybug vectors of cocoa viruses

	Geographical distribution	Occurrence on cocoa	Geographical areas where dominant on cocoa
Ferrisia virgata	Tropicopolitan climate, also some mediter-ranean areas (Egypt etc.)	Worldwide	Possibly never dominant except on young trees and seedlings
Planococcus citri	Tropicopolitan and mediterranean (45 °N to 40 °S)	Worldwide	Nigeria, Ceylon and E. Archipelago
P. kenyae	E. Africa to Ghana	Ghana to Congo	Nigeria
P. lilacinus	Far East, Aden, Madagascar and Mauritius	Ceylon and E. Archipelago (Madagascar?)	Ceylon and Java
Planococcoides njalensis	Sierra Leone to Congo Republic	West and Central Africa	Sierra Leone to Nigeria

The males are often overlooked in the field, partly because they are so short lived and partly because before pupation they may move away from the principal feeding sites on the host plant to, for example, in *Planococcus lila-cinus* (Ckll.) the backs of leaves. Consequently it has often been assumed that sexual reproduction is of secondary importance. James (1937) found reproduction of five mealybug species he studied to be exclusively sexual. The ratio of males to females fell as low as 1:5 in *Pseudococcus longispinus* (Targ.-Toz.) but as each male was able to fertilize twenty or more females the supply of males was adequate. James also reported that delayed fertilization in *Planococcus citri* resulted in an increased proportion of the progeny being male and more recently the sex ratio was shown to vary during the oviposition period (Nelson–Rees 1960). Female parthenogenesis (thelytoky) is known in some species, e.g. *Phenacoccus solani* (Lloyd 1952). Seasonal parthenogenesis also seems to occur and, for example, Kirkpatrick (1927) studying the life history of *Planococcus kenyae* (Le Pelley) on coffee in Kenya found males only during the colder weather about August and assumed, though he did not prove, that reproduction at other times of the year is parthenogenetic. Other species of coccoids are known to have both sexual and parthenogenetic races (Hughes–Schrader 1948) as is likely also in the mealybug *D. brevipes*, males of which are common in Trinidad (Kirkpatrick 1953) but unknown in Hawaii (Beardsley 1960). However, Beardsley (1959) found in Hawaii a closely similar species, *Dysmicoccus neobrevipes*, in which the normal mode of reproduction appears to be bisexual. The *D. brevipes* studied by Kirkpatrick (op. cit.) resembles *D. brevipes* in some characters and *D. neobrevipes* in others. Clearly taxonomic exactitude is an essential companion to comparative biology. *Planococcoides njalensis* (Laing), a mealybug of major importance as a vector in West Africa, almost certainly uses both modes of reproduction. In Ghana eight of nine females reared in isolation laid eggs (Strickland 1951a) and about Adiopodoume, Ivory Coast, some insects gave parthenogenetic birth to only female offspring, though in the Bingerville area fertilization was consistently necessary to reproduction (Magnin 1953). The cytology of mealybugs has been discussed by McKenzie (1967).

Pheromones responsible for attracting the male to the female have been demonstrated in *Planococcus citri* (Gravits and Wilson 1968).

In some species the young are produced ovoviviparously (the eggs usually hatch within a few minutes of deposition) and in others oviparously (when laid the embryo is less well developed and does not hatch for some days). For instance in *Planococcoides njalensis* egg laying takes 15–20 min and the nymph is bounded by a thin yellow refringent membrane which in about 10 min 'dries or evaporates leaving the young 'crawler' lying free', about 45 min elapsing between egg deposition and the nymph first walking (Strickland 1951a). The egg of *Planococcus lilacinus* also hatches within about an hour (Van der Goot 1917) but, by contrast, that of *P. citri* has an incubation period of 3–5 days. *Ferrisia virgata* (Ckll.) occupies an intermediate position, generally with a brief incubation of 1–2 days, but in Trinidad the eggs always appear to hatch immediately they are laid (Kirkpatrick 1953). Eggs of oviparous species are usually deposited in a fluffy white ovisac composed of tangled waxy threads secreted from terminal abdominal glands. The ovisac is secreted during laying and ultimately it may be longer than the body of the parent. In ovoviviparous species a rudimentary ovisac composed of a few wax threads may be secreted *(Planococcoides njalensis; Dysmicoccus neobrevipes)* while occasionally in some oviparous species the ovisac may be absent, e.g. *Planococcus kenyae* (a feature which helps to distinguish between this and the very closely similar *P. citri* in which the ovisac is as long as the body).

Fecundity varies widely: *Planococcoides njalensis* lays only 30–40 eggs whilst *Ferrisia virgata* produces ten times this number, and the number may be influenced by genetic and environmental factors as *Planococcus citri* is estimated to produce 150–200 eggs in the Ivory Coast but 300 in Trinidad.

There is considerable inter- and intraspecific variation in developmental rates. In West Africa *P. citri* completes the life cycle (from egg to egg) in 40–50 days (7–9 generations per year) and *Planococcoides njalensis* in 66–105 days (3–5 generations per year). Though in West Africa *Planococcus citri* is not dominant over *Planococcoides njalensis*, its reproductive potential is enormously greater (*Planococcus citri*, 8 generations per year and 175 offspring per female; *Planococcoides njalensis*, 4 generations per year and 35 offspring per female – the calculation is left to the interested reader). Perhaps this is an indicator of greater success in *P. njalensis* in the conditions concerned. Further developmental data are summarised in Entwistle (1972).

Mealybugs are difficult to rear in captivity. *Ferrisia virgata* can be reared on cocoa seedlings in the greenhouse (or screenhouse) but most other species are less successfully reared under these conditions. Some mealybug species can be reared successfully on potatoes, especially if these are sprouting. *Planococcus citri* and *Dysmicoccus brevipes* were reared in this way, in the absence of ants, in Trinidad and were reported to have withdrawn their stylets more easily than from cocoa (Kirkpatrick 1950). In the absence of ants cultures of conspicuous honeydew producers may be overgrown by sooty moulds, though these do not necessarily always kill the mealybug (Strickland 1951b). However, fungicidal treatment may be necessary, as some moulds kill insects. Mealybugs will probe through a creased Parafilm membrane, and may be fed on artificial media in this way. *Planococcus citri* has been so cultured and males but not females grew normally; the adult females were abnormally small and laid only 20–30 eggs (Gothilf and Beck 1966). Feeding of coccoids through Parafilm membranes was also the subject of comment by Salama and Salch (1971) and the technique is obviously promising.

10.3.1 The mealybug–ant relationship

Strickland (1951b) has written about the relationship of ants and mealybugs and indicated that a great deal remains to be investigated.

He considered (1951a) that the short legged and more sluggish mealybugs, which are strongly negatively phototropic and ovoviviparous (e.g. *Planococcoides njalensis*, *Paraputo anomalus* (Newst.), *Delococcus tafoensis* (Strick.) and *Cataenococcus loranthi* (Strick.)) are almost invariably associated with ants of the genera *Crematogaster* and *Pheidole*. Other mealybug species (*Planococcus citri*, *Pseudococcus hargreavesi* Laing, *P. concavocerarii* James, *P. longispinus*, *Ferrisia virgata* and *Tylococcus westwoodii* Strick.) which have legs of normal pseudococcid length and are generally far more active and mostly oviparous are 'occasionally but certainly not obligatorily associated with ants'.

The biological effects of ants on the mealybugs they attend are almost completely unknown and even the effect of ants on the developmental and multiplication rate of the mealybugs has not been studied.

10.3.2 Mealybug feeding

Little is known of the feeding habits of Coccoidea, though Smith (1926) studied *Pseudococcus longispinus* on potato and *Aspidiotus hederae* Vallot (family Diaspididae) on a species of *Cordyline* and Glass (1944) *Phenacoccus colemani* (Ehrh.) and *P. comstocki* (Kuw.) on apple. More recently *Planococcoides njalensis*, *Ferrisia virgata* and *Phenacoccus madierensis* Green have been studied on cocoa (Entwistle and Longworth 1963).

Like other Hemiptera, the mandibles and maxillae of coccoids have a piercing and sucking function and are slender flexible stylet-like structures. These are often nearly as long as the body, though in feeding they are not necessarily extruded to their full length. Penetration of the host plant seems always to be intra-cellular in contrast to some aphids in which it may be intercellular. The path of mealybug stylets may occasionally be seen in cell walls. In what tissues feeding takes place seems to depend upon two things. Firstly the predilection of the species is important. *Phenacoccus madierensis* always feeds in the cortex of cocoa, whilst *Planococcoides njalensis* mainly feeds in the phloem of stems. Secondly the site of feeding may influence the choice of tissue; thus *Ferrisia virgata* probably feeds better from phloem in leaves (on which it prefers to settle) than in stems. Stylet length seems not to be a limiting factor for both in *Phenacoccus madierensis* and *Ferrisia virgata* stylets are long enough to reach stem vascular tissue.

When not in use the stylets lie looped in a membrane bounded sac, the crumena, lying between the central nervous system and the ventral body wall and connected by a channel with the labium. The stylets are thrust by contraction of protractor muscles into the plant; while these muscles relax, and the stylet slack is taken up by the contraction of retractor muscles, the stylets are held firmly by a modified muscular region of the labium. The sequence of operations is repeated until the required depth of penetration is attained. It seems probable all four stylets are not pushed in simultaneously but first the mandibular stylets, one after the other, then the maxillae (Imms 1957). The stylets of *Planococcoides njalensis* locate the phloem in stem tissue apparently by a process of trial and error, probing in a fresh direction when they encounter sclerenchymatous bundle caps and primary xylem (Entwistle and Longworth 1963). It has been suggested that enzymatic secretions by the insect may aid penetration of the plant especially, as in the case of *Aspidiotus hederae* on a *Cordyline* species, where host tissues are resistant (Smith 1926). However, it is unlikely that this is important with mealybugs on cocoa. The time taken for extrusion of the stylets of *Planococcoides njalensis* into soft agar to a length which would be sufficient to reach the vascular bundle were it on a plant stem is similar to the minimum time necessary to transmit virus, suggesting a similar rate of penetration of agar and the physically much more resistant plant.

Thus the way in which stylets penetrate the plant is known, but the means by which their direction is controlled is not, though the theories of directional control in Hemiptera advanced by Pollard (1969) should be consulted.

Inside the plant the stylets are surrounded by a stylet sheath which is thought to result from the interaction of salivary secretions of the insect and plant tissues. However, a sheath is also formed when mealybugs probe into agar and it may also, with the stylets, bridge an air space between adjacent blocks of agar. It is thus more likely that the stylet sheath material in mealybugs is of purely insect origin. Its purpose is not known but sheaths are produced by many Hemiptera-Homoptera.

Though mealybugs may be present in large numbers on cocoa, they are not known to cause physical damage though they are on record as being directly injurious to many other economic plants. However, *Rhizoecus falcifer* (Kunck.), a root feeder, is thought to contribute to yield decline in parts of

Trinidad (Strickland 1945). Apart from this, mealybugs are important to cocoa only as vectors of virus diseases.

10.3.3 Virus acquisition in relation to sites of feeding

Transmission experiments with *Planococcoides njalensis* and *Planococcus citri* using West African viruses have shown all feeding stages can acquire and transmit virus and no consistent results have been produced to show clearly if one stage is a more efficient vector than another. New Juaben has occasionally been transmitted to beans (cocoa beans (Amelonado type) are convenient for infection and can then be grown on as indicator plants, symptoms usually appearing in the first leaves) by *Planococcoides njalensis* in between 15 and 20 min, though optimal transmission is not usually attained until 2–4 hr. Mealybugs fed below leaves exposed to $C^{14}O_2$ became radioactive within 24 hr and autoradiography of stem sections in the feeding region showed radioactivity to be restricted to the phloem nearest the cambium (Longworth 1964 reporting experiments by J. R. S. Lawton). Additional evidence that virus is acquired from the phloem lies in the long acquisition feeds necessary to make vectors infective. On average 17.5 min are needed for the stylets to extend to 0.5 mm into agar, a depth sufficient to reach the phloem in a young cocoa stem. Thus, as withdrawal is a rapid process, it seems probable the greater part of the lengthy acquisition feeding period is from the phloem. This may be necessary to acquire an infective quantity of virus particles. (Indicator beans germinated in the dark and kept there for several weeks show a higher numerical level of infection than ones raised normally in sunlight. This is probably the result of 'dark' beans being more sensitive to virus, perhaps itself a result of deranged protein synthesis. This is a line of investigation worth pursuing for it may be found that 'dark' beans can be infected following abnormally brief acquisition feeds). By comparison with the lengthy 'normal' acquisition feeds, periods of only 15–20 min may occasionally suffice to infect new plants, a length of time in which it seems possible stylets could penetrate to the phloem. Virus moves initially in young cocoa seedlings at a rate comparable with that of viruses for which there is evidence that they must be injected directly into the phloem (Bawden 1950; Thresh 1957). It therefore seems that mealybugs acquire virus from the phloem and must deposit it into the phloem to infect a plant. Mealybugs which do not feed in the phloem, such as *Phenacoccus madierensis*, are not vectors. A similar reason why in pineapples *Dysmicoccus brevipes* and *D. neobrevipes* transmit both the wilting secretion and 'latent factor', whilst

Phenacoccus solani transmits only the latter and is unable to synthesize the wilting secretion, has been given by Carter (1963, and see above).

Additionally vector species seem to reach optimum transmission when the acquisition feed has been at the preferred feeding site. Thus, when *Planococcoides njalensis* and *Ferrisia virgata* crawlers were released on viruliferous cocoa seedlings and allowed to settle at will, many more of the former settled on the stem than on the leaves and vice versa. Transmission rates for *Planococcoides njalensis* for stems and leaves were respectively 50.0 and 6.3% and for these tissues for *Ferrisia virgata* were 3.9 and 48.1% (Longworth 1964). On young stems it has been shown many fewer *F. virgata* manage to feed in the phloem than do *Planococcoides njalensis* (Entwistle and Longworth 1963) and though reciprocal studies have not been made on leaves it seems possible that the explanation of differential vector efficiency with feeding site is that mealybugs are best adapted to phloem location at their preferred feeding sites.

10.4 Discussion

Whatever the origin of cocoa viruses, whether initially in the Neotropics or independently in the several regions of cocoa cultivation currently afflicted, the viruses now exist in a number of species of woody plants in West Africa and probably throughout the tropics, especially in the Tiliales (Bombacaceae, Sterculiaciae and Tiliaceae) and Malvales (Malvaceae).

The evidence of many years transmission studies suggests that several of the mealybugs consistently associated with cocoa, *Planococcoides njalensis*, *Planococcus citri*, *P. kenyae*, *Pseudococcus concavocerarii* and *P.* sp. nr. *ghani* can transmit many isolates of both CSSV and CMLV. Others, such as *P.* sp. nr. *masakensis*, not found on cocoa but occurring on wild hosts of cocoa viruses can transmit virus from the wild host to cocoa but either do not feed on cocoa long enough to acquire virus or else feed on non-viruliferous tissues. It must be remembered that the capacity to survive and breed on cocoa has little to do with the ability to act as a vector of virus from another host to cocoa and, indeed, vice versa.

In the face of this mealybugs assume a wider significance as vectors of plant viruses and should not be thought of only as vectors of viruses in cocoa. Though these viruses cause severe symptoms, including death of the seedlings of *Corchorus* spp. and *Bombax costatum*, most hosts other than cocoa are more tolerant of infection, and may show no symptoms with some virus isolates. It is unlikely that plants showing few or no symptoms, especially when they are species of no economic importance, will be tested for virus infection and this may have mitigated against detection of the true extent of the vector relationship of mealybugs to plant viruses. The vector capacity of root-feeding mealybugs has been investigated only for a *Geococcus* species in Ghana which was found unable to transmit virus. Indeed the study of these mealybugs has been greatly neglected and their possible importance almost wholly overlooked.

The clear importance of aphids and planthoppers as vectors of crop plants may well have deterred studies on the possible vector activity not only of mealybugs but of coccoids in general. Only half the coccoid species on cocoa trees in Ghana are pseudococcids and over 70 non-mealybug species are known to attack cocoa in various parts of the world. Many of the more important non-mealybug species have been tested for their ability to transmit viruses, but all have failed to transmit; the possible reasons for this failure have, however, not yet been investigated, and even the tissues in which such important genera as *Stictococcus* (Stictococcidae), *Ceroplastes* and *Gascardia* (Coccidae) feed are unknown.

CHAPTER 11

Diptera

P. F. MATTINGLY

Contents

11.1 Introduction

This is an order of Class Insecta deriving its name from the fact that only the front pair of wings functions as such. Except for two small families in which both pairs of wings are largely or wholly suppressed the hind wings are modified to form specialized organs (halteres) serving to maintain equilibrium and neuromuscular coordination during flight. The enhanced powers of controlled flight conferred by these organs give the Diptera a unique role in the transmission of disease, particularly of blood parasites and pathogens. All other arthropod vectors of these organisms are wingless except for the reduviid bugs in which the powers of flight are relatively little used. Behaviourally speaking the bloodfeeding Diptera are typical predators, alternating long periods of rest with brief raids on the host. Ecologically speaking they conform more closely to parasites since they are numerically much more abundant than their hosts and thus form the base of the so-called pyramid of numbers (Mattingly 1969). Three sub-orders are recognized, the Cyclorrhapha, Brachycera and Nematocera. The first of these is generally considered to be the most highly evolved, the last the most primitive.

11.2 Sub-order Cyclorrhapha

These are stoutly built flies of which the common housefly *Musca domestica* is a familiar example. Many feed on human food and excrement and serve from time to time as mechanical vectors of viruses particularly when these are already present in abundance. Too many different kinds are involved to allow of discussion here. Polioviruses alone have been recovered from more than 30 genera (Greenberg 1971). Bloodfeeding Cyclorrhapha are confined to two small wingless families and the large family Muscidae which includes specialized bloodfeeders with piercing mouthparts such as the tsetse flies, horn flies and stable flies *(Stomoxys)*. The last named, in particular, may serve from time to time as mechanical vectors, e.g. of poxviruses or vesicular stomatitis, but they do not appear to be extensively involved. More typical muscids have highly specialized sponging mouthparts used for mopping up fluid from moist surfaces. In a few of them the mouthparts are equipped with small teeth capable of drawing a little blood by excoriating the skin. It is generally agreed that bloodfeeding in this family has evolved secondarily with woundfeeding perhaps an intermediate stage.

Bloodfeeding does not in itself impose requirements differing greatly in purely mechanical terms from those associated with feeding on plants. More significant is the ability to locate an active, dispersed host and in this respect the Diptera are particularly well equipped. At the same time the females of many bloodfeeding species also feed extensively on plant juices or exudates while some are genetically dimorphic with respect to the need for a blood-meal. The males of most species feed exclusively on plant sources. Some conclude from this that the capacity for bloodfeeding has been secondarily acquired by the few families which possess it. Others argue that, despite the example of the muscids, it is a primitive character in the Diptera. The argument is unresolved (Downes 1958; Mattingly 1965). It seems likely, at any rate, from the very close resemblance between the mouthparts of blood-feeding Brachycera and Nematocera that the habit existed before these sub-orders became fully differentiated.

11.3 Sub-order Brachycera

Clegs and horseflies belonging to the family Tabanidae are the most familiar members of this sub-order. Bloodfeeding is confined to this family and to a few members of the little known family Rhagionidae. Tabanids are mechanical vectors of trypanosomes *(Trypanosoma evansi)* and of Tularaemia and may serve as mechanical vectors of some viruses. The only arbovirus recovered from them to date is Jamestown Canyon in the California group (DeFoliar et al. 1970). Their role in the transmission of these viruses remains to be demonstrated. All Diptera so far shown to be involved in arbovirus transmission belong to the Nematocera.

11.4 Sub-order Nematocera

These are mostly small, rather fragile insects deriving their name from their slender antennae. Among the various non-bloodfeeding families only the Tipulidae (Craneflies, Leatherjackets) are of particular virological interest. Bloodfeeding is confined to the families Simuliidae, Psychodidae, Ceratopogonidae and Culicidae. The Simuliidae include important vectors of protozoan and nematode blood parasites and arboviruses have been isolated from them on three occasions, but there is no evidence at present of their involvement in transmission. The other three families all include known virus vectors but on present evidence the great majority of these are Culicidae. The latter differ from the other families in their greatly elongated, flexible mouthparts which allow penetration of individual bloodvessels and direct imbibition of circulating blood (capillary feeding). In the other families, as in the Tabanidae, the mouthparts are used to excavate a small pit or sump in which the blood collects and from which it is imbibed (pool feeding). This difference affects the uptake or otherwise of parasitic nematodes and the transfer of some Protozoa (avian trypanosomes). It might have some relevance for the mechanical transmission of viruses.

Two broad categories of larval habitat are encountered, on the one hand soil, detritus, animal droppings and decaying organic matter (most Tipulidae, many Ceratopogonidae and Psychodidae) and on the other aquatic habitats (Simuliidae, Culicidae and the remaining members of the other families mentioned). The principal mode of larval feeding in both groups is the seemingly indiscriminate ingestion of microorganisms and non-living organic particles. Larvae of some tipulids feed on the living tissues of higher plants, both terrestrial and aquatic. A few are fungivorous and the larvae of some aquatic species are predatory. Among the bloodfeeding families predation, including cannibalism, and necrophagy are found in some groups with spatially restricted larval habitats. A few Culicidae *(Culex bitaeniorhynchus* group, *Anopheles* Series *Paramyzomyia)* are specialized feeders on filamentous green algae but in general there is little evidence of association with particular classes of food organisms other than that conferred by the factor of availability.

11.4.1 Family Psychodidae

Generally considered the most primitive of the bloodfeeding families of Nematocera, this includes two sub-families, the Psychodinae and Phlebotominae. A few of the former (*Pericoma* spp.) are said occasionally to bite man but all known disease vectors belong to the Phlebotominae. The latter are minute insects commonly known as sandflies, a name also applied at times to simuliids and ceratopogonids with consequent possibilities of confusion. Three main genera are commonly recognized, *Phlebotomus* and *Sergentomyia* in the Old World, *Lutzomyia* in the New. *Phlebotomus* is associated mainly with mammalian hosts and has a mainly temperate and subtropical distribution. *Sergentomyia* is the dominant genus in the tropics and replaces *Phlebotomus* in the Far East and Australia. It is associated largely with reptilian hosts. *Lutzomyia* contains a higher proportion of forest species than the other genera which occur mainly in open country. It includes both mammal and reptile feeding species. Birds are seldom mentioned as hosts although they are regarded as of some importance in the Soviet Union.

Phlebotomine larvae are exclusively terrestrial. Their breeding places reflect the adult habitat. Those of some *Lutzomyia* are found both on the forest floor and in collections of litter in the tree-tops, the adults being canopy feeders (Thatcher 1968; Williams 1970). Those of non-forest species are found in detritus in crevices in soil, rock or walls and floors of buildings or in rodent burrows or similar refuges. (For a review see Hanson 1961).

Although *Phlebotomus papatasii* is the only known vector of sandfly fever virus it is clear from distributional evidence that other vectors must be involved. Six of the twelve viruses currently placed in the sandfly fever and Changuinola groups have been recovered from sandflies in nature. The vectors of the others are unknown though one (Itaporanga) has been recovered from mosquitos. A number of viruses recovered from *Phlebotomus* and *Sergentomyia* spp. have still to be categorized (Barnett 1962). Yellow fever virus has twice been recovered from wild caught sandflies but there is no evidence of their involvement in transmission. Transovarian transmission of sandfly fever virus has been postulated but confirmation is required (Smith 1964). The putative role of sandflies in the transmission of vesicular stomatitis virus is particularly interesting and is discussed separately below. Sandfly vector relationships are unique in several respects. On the negative side they differ from all other bloodfeeding Nematocera in having no known association with parasitic nematodes. On the positive side they include the only known winged vectors of rickettsias or rickettsia-like organisms (*Bartonella bacilliformis* of man in South America). They are best known as vectors of blood flagellates *(Leishmania)* and here again they are unique among Nematocera apart from a recently discovered, perhaps casual, association of ceratopogonids with avian trypanosomes (Bennett 1961). Sandfly-borne trypanosomes are confined, so far as is known, to reptiles and terrestrial amphibians but the leishmanias also include parasites of mammals among them major causative organisms of human disease. It is in this context that they have been mainly studied.

In the cutaneous and muco-cutaneous forms of leishmaniasis (Oriental Sore, Espundia) the organisms are confined to the skin or the mucous membranes of the nose, mouth and throat where they produce ulcerating lesions, an unusual feature which may lend some weight to the hypothesis that leishmanias, unlike the blood sporozoans and nematodes, originated extraneously as parasites of the insect gut, perhaps in combination with the habit of wound feeding. A comparable example may be provided by the arthropod-borne nematode blood parasites. These are believed to have evolved from parasites of the orbit and naso-pharynx following their introduction into dermal lesions by dipteran vectors feeding on eye secretions and the moist surfaces of the mucosa and of wounds and sores (Anderson 1959).

The recovery of vesicular stomatitis virus from sandflies (and in one case from mosquitos, in which it has been shown to multiply) raises interesting problems. The confinement of the virus to dermal and mucosal lesions would suggest either a very primitive virus-vector system or one in which this type of vector is involved only casually. It resembles the rhabdoviruses, rather than the arboviruses, in structure and it was at one time thought that it might be acquired as a contaminant of pasture or fodder (Jonkers 1967). As against this the explosive nature of epizootics in cattle seems to suggest the involvement of arthropod vectors and it has been suggested that woundfeeding sandflies could fulfil this role (Hanson 1968). There is little present evidence of woundfeeding by sandflies but this may merely reflect lack of observation. The pattern of their involvement in leishmaniasis lends some colour to the suggestion and recent recoveries of vesicular stomatitis virus from arboreal and other forest mammals may be felt to support it.

Leishmanias and sandfly-borne trypanosomes lack the capacity to migrate to the salivary glands and are injected by mechanical pressure, following blockage of the foregut, or pass out of the hindgut with subsequent invasion of skin lesions. Infectivity of some leishmanias is apparently dependent on prior uptake of plant substances by the vector (Adler and Theodor 1957). The mode of transmission recalls the transmission of plague by blocked fleas and it is interesting that sandflies resemble these insects, if only superficially, in a number of other features among them the close association with the host environment, short 'hopping' flight and larval habits and breeding places.

Sandfly-borne sporozoans include a primitive *Hepatozoon* of snakes and lizards and *Plasmodium mexicanum* of lizards. The latter, though considered to be one of the most primitive members of the genus, is transmitted in the normal manner.

11.4.2 Family Ceratopogonidae

These are very small midges often constituting a serious pest to man and domestic animals. Bloodfeeding species are confined to the genera *Culicoides* and *Leptoconops* and the subgenus *Lasiohelea* of genus *Forcipomyia*. Other *Forcipomyia* spp. and some *Atrichopogon* feed on the body fluids of larger insects including mosquitos. Other genera are predatory or phytophagous.

Breeding places range from damp soil or rotting organic matter to fully aquatic habitats. Flight is generally limited but wind assisted dispersals up to 4 miles are on record. For bionomics see Kettle (1969). Despite their economic importance and extensive man-assisted distribution the epizootiology of African horse-sickness and blue-tongue viruses is still not well understood. The classic picture of transmission by *Culicoides* still holds the field but recent evidence suggests that mosquitos may also be involved. Until more is known little can usefully be said in the present context. Buttonwillow virus of lagomorphs has been shown to involve *Culicoides* as vectors but knowledge of its epizootiology is still incomplete (Hardy et al. 1970). Six new viruses, involving man, stock and lagomorphs as hosts, have recently been recorded from *Culicoides* (Burge et al. 1971) but the extent of their involvement has still to be ascertained. Three of them were also recovered from other arthropods (ticks, mosquitos). Several human viruses (JBE, EEE and possibly VEE) have been recovered from *Culicoides* but their role as vectors is still unproven.

Among filariids transmitted by *Culicoides* the larvae of *Dipetalonema streptocerca* of man and apes, and *Onchocerca gibsoni* and *reticulata* of domestic stock are largely confined to the skin, rarely entering the bloodstream, and so depend on pool feeding vectors. *Dipetalonema perstans* and *Mansonella ozzardi* of man, on the other hand, are fully evolved blood parasites. Interestingly *Icosiella neglecta* of frogs, transmitted by *Forcipomyia velox*, develops well in the primitive psychodid *Sycorax silacea* although, as noted previously, no member of that family has yet been incriminated as a nematode vector. Sporozoa transmitted by *Culicoides* include *Parahaemoproteus* of birds which closely resembles the related genus *Haemoproteus*, also an avian parasite, transmitted by hippoboscids, but differs in having smaller oocysts, fewer sporozoites and more rapid development in both vertebrate and vector, illustrating the contrasting needs of parasites with winged and those with wingless vectors (Bennett et al. 1965). Other Sporozoa transmitted by *Culicoides* include *Akiba caulleryi* of domestic poultry and *Hepatocystis kochi* of African monkeys. The association with avian trypanosomes has already been noted.

11.4.3 Family Culicidae

This family is here restricted to the mosquitos as in the current world catalogue (Stone et al. 1959); some authors include the non-biting dixid and chaoborid midges. Nearly 3000 species of mosquitos are known and over 150 different viruses have been recovered from some 180 of these. The great majority of these are typical arboviruses. Exceptions include vesicular stomatitis virus, already mentioned, and the closely related Cocal virus. These are listed by Taylor (1967) with the arboviruses because both have been recovered from mosquitos and the former has been shown to multiply in them. It seems, however, that the epizootiology of Cocal virus in forest rodents is inconsistent with mosquito transmission (Jonkers et al. 1965).

Nodamura virus, apparently a quite typical picornavirus, has been recovered from mosquitos in nature and shown to multiply in them and be transmitted by them in the laboratory (Murphy et al. 1970). Taylor consequently includes this also in the arboviruses. Other picornaviruses recovered from mosquitos include Coxsackie viruses and encephalomyocarditis virus, the latter recovered from some 10 different species and in both hemispheres. The possibility of mosquito transmission seems to merit further investigation. Among poxviruses there is good evidence for mosquito transmission of rabbit myxoma and birdpox, sometimes at remarkably long intervals after infection, though it seems that neither can multiply in mosquitos and transmission is purely mechanical. Reoviruses have been recovered from several kinds of mosquitos in nature and transmission has been achieved after some days incubation. Limited multiplication in the vector is not entirely precluded but on present evidence simple persistence and mechanical transmission seem more probable. Trans-stadial maintenance of reovirus, as of a number of arboviruses, has been achieved from larva to adult followed, in the case of the arboviruses, by successful transmission by bite (Peleg 1965). There is at present no evidence that it occurs in nature. Attempts at transovarian transmission with mosquitos have proved negative or inconclusive. Mosquito-borne arboviruses have been found to multiply, after parenteral inoculation, in a remarkable variety of arthropods but attempted transmission to a hemipteran predator by feeding it on infected mosquitos was unsuccessful. Mosquito tissues have been successfully cultured in vitro and such cultures hold much promise for the study of virus replication.

Factors governing multiplication of viruses in mosquitos and their transmission have been reviewed by Day (1955), Chamberlain and Sudia (1961) and Smith (1964). A barrier to uptake of virus occurs in the midgut epithelium and can be bypassed by parenteral injection. A further barrier appears, in some cases, to prevent invasion of the salivary glands. Among extrinsic factors temperature is important for its effect on the rate of multiplication of virus in the poikilothermic vector. This and other factors are also important in governing the reproductive turnover and life expectancy of the vector though their detailed implications have been studied, at present, mainly by the malariologists (Detinova 1968). Genetic studies of susceptibility of mosquitos to infection with viruses have been almost entirely neglected though such studies have proved rewarding in relation to other pathogens (Wright and Pal 1967).

Three subfamilies of mosquitos are currently recognized but one of these, the Toxorhynchitinae, consists entirely of non-bloodfeeding species and need not be considered here. The others both include important virus vectors and will be dealt with individually.

11.4.3.1 Subfamily Anophelinae

The two small genera *Bironella* and *Chagasia* are of no known virological interest and need not be discussed. The remaining genus, *Anopheles*, is the third largest in the family with some 350 recognized species. At the time of writing just over 40 different viruses have been recorded from this genus in contrast to the 140 or so recorded from the subfamily Culicinae. A supposed association of culicine mosquitos with viruses, comparable to the highly specific association of *Anopheles* with human malaria, led at one time to a neglect of anophelines as potential virus vectors. In consequence the number of recorded recoveries of virus from the latter is still comparatively small and it is of particular interest to try to determine how far these imply merely casual recovery and how far they are genuinely indicative of regular transmission. The evidence presented here seems to lend some weight to the first hypothesis though both are no doubt partly true. Six subgenera of *Anopheles* are currently recognized, three of them very small and confined to the Neotropical region. Among the latter subgenus *Lophopodomyia* is of no known virological interest and need not be discussed. In contrast subgenus *Kerteszia* is of special interest, not only as a source of six different arboviruses but also because the included species all breed extensively in epiphytic bromeliads a habit unknown elsewhere in the Anophelinae. In consequence they are largely arboreal though descent to the ground takes place to a sufficient extent in parts of the range for three species to be important vectors of human malaria (*Anopheles bellator* in Trinidad, Venezuela and Brazil, *An. cruzii* in Brazil and *An. neivai* in Colombia). *An. cruzii* is also considered to be a vector of simian malaria and of occasionally transmitting this to man. *An. bellator* was responsible for an extensive malaria epidemic following invasion of shade trees in the cocoa plantations in Trinidad by epiphytic bromeliads. Most of our knowledge of *Kerteszia* biology is based on the investigations which followed (Pittendrigh 1950, etc.). Yellow fever virus has been recovered from *An. neivai* and this species may play some part in the arboreal reservoir. On the other hand repeated recoveries of Guaroa virus from the same species, coupled with its recovery from man, probably reflect its association with human malaria. It may well be significant that this virus is a member of the Bunyamwera group (see below). The other *Kerteszia* viruses (Anopheles A and B, Boracea, Tacaiuma) are ungrouped or belong to very small, oligotypic groups. Their vertebrate hosts are unknown except in the case of Tacaiuma, antibodies to which have been recovered from man, horse and rodents, presumably reflecting the terrestrial component of the behaviour of the vector *(Anopheles cruzii)* though the reservoir may be arboreal since the same virus has been recovered from the culicine genus *Haemagogus.*

The last of the smaller subgenera, *Stethomyia*, has been relatively little studied. Little is known of its biology except that its members breed, like the great majority of anophelines, in ground water of various sorts. The pattern of relationships within the viruses recovered from it seems distinctively anopheline. One (Pixuna) belongs to Group A, two (Maguari, Wyeomyia) belong to the Bunyamwera group, one (Lukuni) is grouped with Anopheles A and the fifth (Tembe) is ungrouped. The very little that is known of its hosts seems to suggest a possible association with birds which is not, in general, characteristic of Anophelinae. When more is known about it its role in the general biology of anopheline viruses may prove to be very interesting.

Setting aside those which are still incompletely categorized, or whose recovery requires confirmation, some 25 arboviruses have been recovered from the three larger subgenera of *Anopheles* (*Cellia, Nyssorhynchus* and *Anopheles* s. str.); 8 of these belong to group A, 4 to group B, 5 to the Bunyamwera group, 2 to the California group, 1 to the Bwamba group and 5 are ungrouped. Bearing in mind the fact that both are much smaller than group B a special association with group A and the Bunyamwera group seems indicated. This is particularly true of the Bunyamwera group. Of the 15 viruses belonging to this group which have been recovered from mosquitos 8 have been recovered from *Anopheles*, 4 of them (Cache Valley, Chittoor, Guaroa, Tensaw) repeatedly. No such regularity of association exists with respect to group A except in the case of O'nyong-nyong virus which has been shown to be transmitted by anophelines as major, perhaps exclusive, vectors to man. It is arguable that the association of other group A viruses with *Anopheles* is a casual one reflecting the large extent to which these feed on domestic ungulates. The latter are important hosts not only of group A viruses but also of *Anopheles* as shown by the many tens of thousands of precipitin tests performed in the course of malaria surveys.

This is not, of course, the whole story. Such tests represent a highly biased sample mainly reflecting host availability in the peridomestic environment.

Fig. 11.1.   Boatline through Moriche Swamp to Camp Bush Bush, Trinidad, the site of a
classic study of mosquito-borne forest viruses. (From Downs et al. 1968.)

Our knowledge of wild hosts of *Anopheles* is much less extensive though their potential epizootiological importance is evident from the existence of para-site-vector systems such as bat and rodent malaria in Africa and simian malaria in Asia. The recovery of rabbit myxoma virus from anophelines in North America, Europe and Australia similarly reflects an association with lagomorph hosts of which we should have been unaware but for the intro-duction of the virus.

11.4.3.2 Subfamily Culicinae
Two tribes are recognized, the Sabethini and Culicini.

11.4.3.2.1 Tribe Sabethini. Sabethine mosquitos, though mainly tropical, are found in all the major zoogeographical regions. Genera *Trichoprosopon* in the New World and *Maorigoeldia* in the Old are believed to be the most primitive genera of Culicinae. The only pathogens known to be transmitted by Sabethini are viruses and these are found only in the New World tropics. Breeding places are confined entirely to small containers such as leaf axils, pitcher-plant pitchers, small or very small tree holes, cut, split or insect-bored bamboos and fallen leaves or husks. Many neotropical species breed in the leaf axils of epiphytic bromeliads. Adult activity is almost entirely diurnal and some species, especially those with bright metallic colours, are markedly heliophilic.

At the time of writing 19 different arboviruses have been recorded from Sabethini, embracing all five New World genera *(Trichoprosopon, Wyeomyia, Phoniomyia, Limatus* and *Sabethes)*. Three of these viruses are ungrouped, one belongs to the Guama group, three belong to group C and the other twelve are equally divided between groups A and B and the Bunyamwera group. The vectors concerned are mainly forest mosquitos and their vertical distribution as between forest floor and canopy consequently plays a key role in their vector relations. Vertical distribution is subject to seasonal and diur-nal variations and is affected, in particular, by thinning of the understorey whether natural or the result of clearing by timber felling or for planting. Despite this individual species can usually be categorized, on the basis of a reasonable number of samples (taken, usually, on platforms with human or animal bait), as either acrodendrophilic (occurring mainly in the canopy) or terrestrial or as relatively indiscriminate. Such categorization can provide valuable insights into the epizootiology of any virus which the species con-cerned transmits. This is shown with (probably deceptive) simplicity in the case of the few sabethines for which the relevant parameters (vector, virus, vertebrate host and vertical distribution) are known. Thus the acrodendrophilic *Sabethes chloropterus* and *S. belisarioi* have yielded St. Louis, Ilheus and yellow fever viruses, the first two essentially bird viruses and the third a virus of arboreal primates. In contrast the terrestrial *Limatus durhamii, L. flavisetosus* and *Trichoprosopon digitatum* have yielded the rodent viruses Guama, Pixuna and Venezuelan equine while *Wyeomyia medioalbipes*, which is relatively indiscriminate, has yielded the rodent viruses Caraparu and Venezuelan equine and the bird virus Ilheus. The picture is complicated by the fact that all these viruses, except Pixuna, have also been recovered from man, often, no doubt, through his habit of tree-felling. Otherwise it is entirely consistent.

11.4.3.2.2 Tribe Culicini. Some 120 different arboviruses have been recovered from culicine mosquitos belonging to at least 130 species and 9 different genera. Surprisingly, all but 5 of these have been recovered from the two large genera *Aedes* and *Culex* (though many of them are, of course, shared with other genera). It is also a surprising fact, calling for explanation, that the number of arboviruses recovered from *Culex* spp. exceeds by about 40% the number recovered from *Aedes* despite the larger size of the latter (about 800 spp. compared to 650). Many arbovirus systems certainly remain to be discovered and the possibility of biased sampling cannot therefore be ruled out. However, present indications seem to suggest that, if anything, *Aedes* has been more favoured in this respect than *Culex*. As an example the 5 viruses of the Dengue complex recovered from *Ae. aegypti* in the human domestic environment clearly distort the picture.

Contributory factors may include a greater degree of ornithophily, or possibly of host plasticity in general, on the part of *Culex* and differences in phenology reflecting in turn differences in the ecology of the egg and adult stages. The distinctive feature of *Aedes* is the possession of drought resistant eggs allowing exploitation of temporary container or ground pool habitats subject to more or less rapid desiccation. This has led them to become the dominant mosquitos in grasslands in all parts of the world and permitted them to extend far into the arctic tundras where they are found within a few degrees of the pole. Species of this kind, belonging mainly to the subgenus *Ochlerotatus*, overwinter in the egg and are thus unsuited to the winter maintenance of virus but they are putative major vectors of viruses of the California group in the northern U.S.A. and southern Canada. *Ochlerotatus* is poorly represented in the tropics but a few New World species are important vectors. They include *Aedes taeniorhynchus*, breeding in enormous numbers in salt-

Fig. 11.2.   Pygmy woman holding a leaf of cocoa-yam (*Xanthosoma sagittifolium*), Bwamba County, Uganda. The axils of this and other broad-leaved food plants provide breeding places for *Aedes simpsoni*, a major vector of rural yellow fever. Their cultivation and distribution as articles of commerce have played an important part in the dissemination of the disease in Africa. (From Haddow 1948.)

marshes from New England down to Brazil, and the tropical swamp forest species *Ae. serratus*. Each of these has yielded some dozen different arboviruses but, as often with numerically very abundant species, their role in the viral ecosystem is not always very clear. In tropical Africa the role of *Ochlerotatus* is taken over by subgenus *Aedimorphus* with no one species, unless possibly *Ae. cumminsii*, emerging as a major vector. In the South African lowveld on the other hand *Ae. (Neomelaniconion) circumluteolus*, with similar breeding places, is a vector of great importance with currently some 12 different arboviruses to its credit. Further east the dominant ground pool species, occurring in vast numbers in swamp forest and similar habitats, are found in the subgenera *Neomacleaya* and *Verrallina*. To date these have yielded only a single virus (Baberu from the *Ae. butleri* group in Malaya). The various crabhole breeding subgenera found in southern Asia and Melanesia have so far produced only a very few recoveries (Japanese B and a member of the Bunyamwera group from *Ae. curtipes* in Sarawak). Interestingly, however, the crabhole breeding *Ae. (Skusea) pembaensis* has yielded, besides a member of the Bunyamwera group, the only known African member of the California group (Lumbo, in Mozambique).

Among container breeding *Aedes* the role of *Ae. (Stegomyia) aegypti* in the transmission of yellow fever, Dengue and Chikungunya to man needs little comment. Wild populations of this species occur in Africa but they are not known to be involved in virus transmission. The putative forest vector of yellow fever is the acrodendrophilic *Ae. (St.) africanus*, breeding in treeholes. *Ae. (St.) simpsoni*, breeding in the leaf axils of broad-leaved food plants such as plantains, yams and pineapples, serves as a transfer vector and even, sometimes, as a major vector in its own right in rural areas. *Ae. (St.) albopictus*, recently shown to be a sibling species complex, plays a still somewhat ill defined role in the transmission of Dengue in Asia and members of the *Ae. (St.) scutellaris* complex are putative vectors of the Dengue viruses in the Pacific area. *Ae. (Finlaya) leucocelaenus* is thought to play a part in the maintenance of jungle yellow fever in Central and South America and there have been recoveries of virus from a number of other species of *Stegomyia*, *Finlaya* and *Howardina* but in general such isolations have been less numerous than might have been expected by those for whom *Ae. aegypti* is the prototype virus vector.

Although species breeding in natural containers are to be found in most of the subgenera of *Culex* isolations from such species are at present confined to the recovery of Ntaya virus from *C. (Culiciomyia) nebulosus* in Cameroun. The more numerous species breeding in ground waters are generally to be

Fig. 11.3.   Mosquito catchers ascending to catching platforms, Tree Station No. 3, Bush Bush Forest, Trinidad. (From Aitken et al. 1968.)

found in more permanent bodies of water than *Aedes*. As a corollary they often show a greater tolerance of organic pollution and this has permitted some of them to encroach on the human peridomestic environment, at times living in domestic containers. The process has reached its climax in *C.p. fatigans* (= *quinquefasciatus* Auctt.) the tropical representative of the *Culex pipiens* complex and the major vector of human filariasis in many parts of the world. This and other members of the complex are also urban vectors of several viruses, among them West Nile, St Louis and Japanese B. *C. pipiens* s.str. is largely ornithophilic and probably plays a part in the maintenance of avian viruses. Other vectors, belonging to a different section of subgenus *Culex*, are less domesticated but play an important part in the transmission of viruses such as Western Equine, Japanese B and Murray Valley, encouraged in this by man through the provision of extensive breeding grounds such as irrigated pastures, ricefields and accumulations of polluted water in the environs of cities. These include such species as *C. tarsalis* in North America, *C. tritaeniorhynchus* and other members of the *C. vishnui* complex in southern and eastern Asia and *C. annulirostris* in Australia. Finally in a quite different category are the wild species, particularly *C. (Melanoconion) portesi*, serving as vectors of Venezuelan equine and other viruses to forest rodents (Jonkers et al. 1968). The unravelling of this story has been a model of sustained enquiry throwing light not only on this but on other forest viruses which man has ill advisedly domesticated.

Among the smaller culicine genera *Aedeomyia*, the smallest, with only 6 species, despite its cosmotropical distribution, has yielded 2 viruses, Alfuy, apparently a bird virus, in Australia and Gamboa, with unknown hosts, in Panama. This is a highly specialized swamp breeding genus. Adults are found at ground level in open swamps but only in the canopy and understorey in adjacent forests. *Deinocerites*, another highly specialized genus breeding in crabholes and adapted to the littoral habitat, has yielded only one virus, St Louis. Its only known hosts are man and horses. The small neotropical genus *Haemagogus* includes major vectors of jungle yellow fever. These are small, brightly coloured, diurnal species combining aedine and sabethine features and utilizing container habitats. The principal vector of enzootic yellow fever, *H. spegazzinii*, is very largely arboreal, almost 100% being taken in the canopy in closed canopy forest and of the order of 80% in forests of more open type. Other species involved as vectors are also acrodendrophilic though less extensively so. The only other viruses recovered from this genus are Ilheus and a virus related to Embu.

The New World genus *Psorophora* has a much wider distribution, ranging

Fig. 11.4. Sylvan Yellow Fever in the New World. a) Virgin rain forest inland from Almirante, Panama. Establishment of the study area here followed a fatal case contracted during clearing for a road survey. (From Trapido and Galindo 1957.) b) Collecting *Haemagogus* mosquitos brought to ground level by selective thinning of canopy for cocoa planting, Wauchope, Costa Rica. (From Galindo and Trapido 1955.)

from southern Canada to the Argentine. Like *Haemagogus* it exhibits both aedine and sabethine features but the breeding places are purely aedine, being confined to temporary ground pools. Survival between successive floodings is dependent on drought resistant eggs which are pushed into minute crevices in the soil. The adults are notoriously bloodthirsty and often constitute a serious pest. Fifteen different viruses have been recovered from this genus, including at least a dozen from the single species *Ps. ferox*, but the nature and extent of its involvement are not always very clear. An exception is Ilheus which has been repeatedly recovered from *Psorophora* spp. Adults of the latter sometimes occur in such numbers in swamp forest as to be the dominant canopy mosquitos. Nevertheless they are basically terrestrial and it may be suspected that in this case swamp birds are the principal hosts. Other viruses recovered from *Psorophora* are associated with forest rodents and marsupials and with domestic ungulates (Western equine and Tensaw in the United States, Venezuelan equine elsewhere).

The remaining small genus combining aedine with sabethine characteristics is *Eretmapodites*, which is confined to the Ethiopian region. Breeding in this genus is confined to container habitats including rotting husks and snail shells to which some species are specially adapted. The larvae are often predatory on those of other mosquitos as are those of some sabethines and *Psorophora* spp. The adults are diurnal. Three of the viruses recovered from *Eretmapodites* (Middelburg, Rift Valley, Spondweni) are familiar chiefly as pathogens of man and stock in South Africa but recent discoveries suggest a possible involvement with wild reservoirs in Central Africa. Little is known regarding the other viruses recovered from *Eretmapodites* (Nkolbisson, Okola, Simbu and strains related to Nyando) and the virus relations of this genus still represent an interesting and largely unexplored field.

Genus *Culiseta*, though with some sabethine features, shows little resemblance to *Aedes*, more to *Culex* which it also resembles more closely in its biology. Breeding takes place mainly in permanent or semi-permanent ground waters. The genus as a whole appears to be largely ornithophilic. *C. melanura* is believed to be the principal maintenance vector of eastern equine virus in swamp birds in the United States and *C. tonnoiri* may play a similar role in relation to Whataroa virus in New Zealand. A few species have invaded the human peridomestic environment and *C. inornata* is a pest of stock in some parts of North America. Recoveries of virus from this species (Cache Valley and Jamestown Canyon and Jerry Slough in the California group) are also consistent with mammal feeding though the recovery of Turlock virus might suggest a significant degree of bird feeding. Other viruses recovered from *C. melanura* (Flanders, Hart Park) seem to be essentially bird viruses and although western equine has been recovered both from this species and from *C. inornata* their role in its complex epizootiology is likely to be quite different. Some species hibernate as adults in colder parts of the range and show a surprising degree of activity at low temperatures. On both counts they would seem to possess considerable potential as winter refuges for virus.

The most striking feature of genus *Mansonia* is the possession by the larvae of piercing respiratory siphons. These are used to obtain oxygen from the air spaces in the roots of aquatic plants and specialized waterside trees. Utilization of floating plants often renders possible the exploitation of the entire surface area of extensive bodies of water and adult populations may be extremely large. Cultivation of aquatic plants for fodder, or simply their accumulation through neglect, often engenders close contact with man. In southern Asia *Mansonia* spp. are major vectors of brugian filariasis to man and to the various wild animals which form the forest reservoir of the disease. Some 30 arboviruses have been recovered from this genus, 14 of them, in addition to encephalomyocarditis virus, from *M. venezuelensis*, the largest number from any one species of mosquito. *Mansonia* spp., in common with many other culicines, are mainly nocturnal. Mosquitos of this kind show well marked patterns of activity characteristic of the species concerned. These have been widely studied from the point of view of relations with the vertebrate hosts (Haddow et al. 1961; Wharton 1962).

CHAPTER 12

Nematodes

D. J. HOOPER

Contents

12.1 Introduction

Nematodes are a large ubiquitous group of invertebrates with representatives in almost every kind of environment. They occur in soil, fresh-water and the sea at all latitudes from Antarctica to Equatorial regions. They require moisture to be active but some stages of some species can withstand dry conditions for long periods. Many that live in soil or water are free-living, some feeding on bacteria, algae or fungi, others preying on other small animals including other nematodes. Some are parasitic on higher plants, including crops, which they severely damage; a few of these also transmit plant viruses.

Some nematodes are parasitic in vertebrates, including man, domesticated and wild animals, fish and reptiles; others parasitize arthropods and molluscs. Some have a life cycle that may include hosts from more than one of these groups and they may transmit viruses.

Nematodes are usually worm-like (vermiform), cylindrical, fusiform to filiform, rarely saccate; the shape of females is often more adapted to parasitism than that of the males, which are usually smaller; fig. 12.1 shows the basic structures of the body. Adults of forms occurring in soil range from 0.2 mm to just over 1 cm long; the average is about 1 mm. Animal parasites are often larger: females of the common pinworm, *Enterobius vermicularis*, of man are 1 cm long; adults of the pig lungworm, *Metastrongylus elongatus*, which transmits swine influenza virus, are 5 cm long; the common intestinal parasite, *Ascaris lumbricoides*, reaches 40 cm and the giant kidney-worm, *Dioctophyme renale*, over 1 m long.

The nematode body is non-segmented, basically bilaterally symmetrical and usually circular in cross section; it is covered by a multilayered, flexible, cuticle which may be smooth or ornamented with spines, plates or transverse annulations – these last sometimes give a segmented appearance. The cuticle is often colourless to white, transparent or translucent so that the body contents are easily seen. Animal parasitic forms often have a thicker creamy white cuticle. Beneath the cuticle the hypodermis forms chords that protrude dorsally, ventrally and laterally into the longitudinal blocks of muscles immediately beneath it, thus separating them into four basic sets in transverse section (fig. 12.1).

The main part of the nervous system is a commissure ('nerve ring') that encircles the oesophagus usually in the isthmus region. Several nerves proliferate anteriorly from the nerve ring connecting with papillae and setae of the cephalic region and with amphids (lateral cephalic sensory organs);

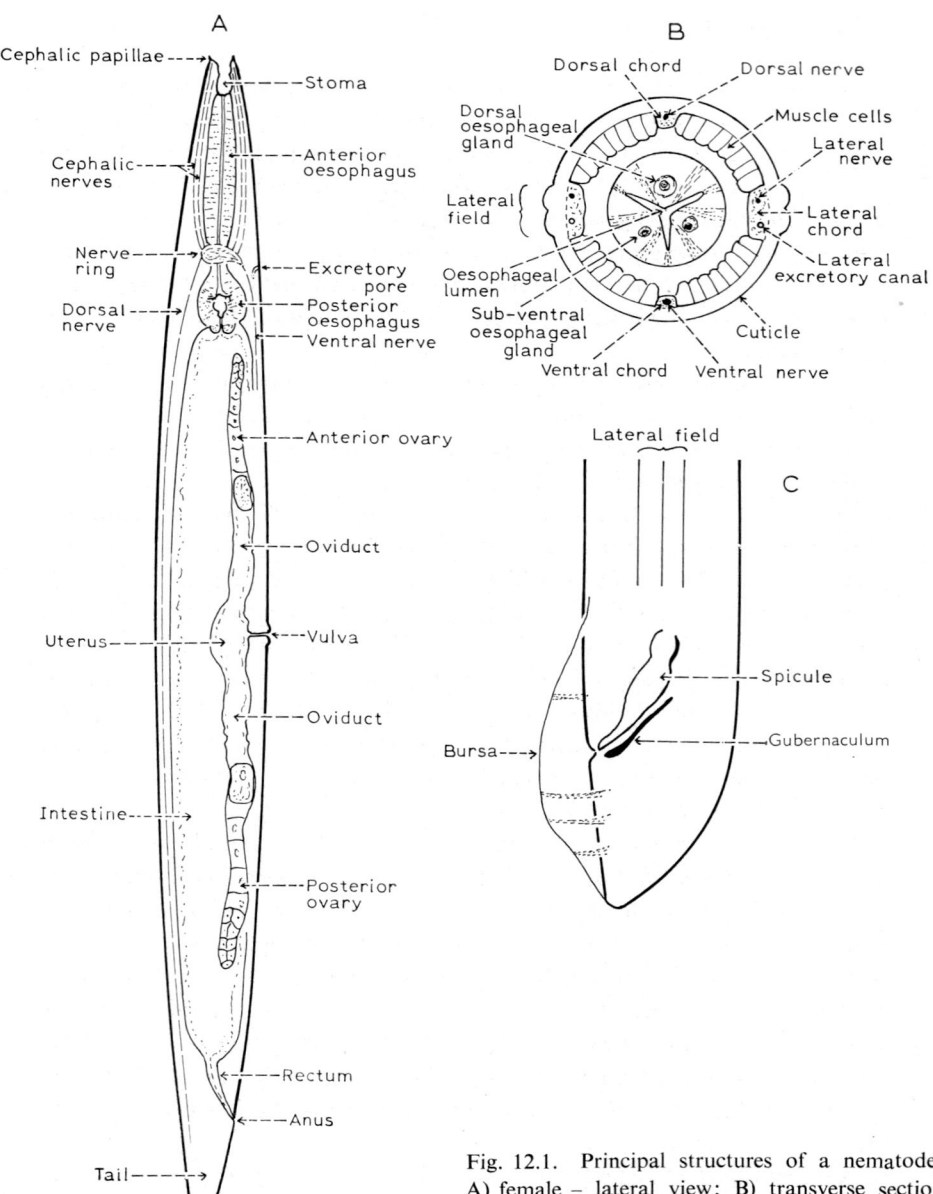

Fig. 12.1. Principal structures of a nematode:
A) female – lateral view; B) transverse section
of oesophageal region; C) male tail – lateral view.

posteriorly the body is served by proliferations from a nerve in each of the four hypodermal chords, the one in the ventral chord being the main body nerve.

The mouth is anteriorly terminal and surrounded basically by six lips that may be modified or elaborated and/or bear pronounced papillae. The stoma (buccal capsule) differs between groups, from a simple unarmed chamber to one with small denticles, large teeth, or both, or it may have a hollow axial stylet as found in most plant parasitic forms. The oesophagus (pharynx) is usually triradiate and is a muscular/glandular pumping organ with three or more glands that secrete digestive enzymes. Many have a valve-like structure, sometimes called the 'cardia', at the junction of the oesophagus and intestine. The intestine is usually a simple straight tube that leads to a posterior, ventral anus via the rectum in the female or the cloaca in the male.

The female reproductive system basically consists of paired opposed gonads as shown in fig. 12.1. The ventrally placed gonopore (vulva) is usually at or behind the mid-body region; sometimes one gonad, usually the posterior, is reduced or absent and the vulva then tends to be more posterior.

The male has one, sometimes two, testes connected via a vas deferens with the cloaca near the posterior end. Within the cloaca there are usually paired evertible copulatory spicules and sometimes on either side of the cloacal opening there are cuticular flaps (caudal alae) forming a characteristic bursa. Forms without a bursa often have ventral or subventral papillae in the posterior region.

Reproduction is usually bisexual but some groups have few or no males and reproduction is hermaphroditic or parthenogenetic.

The typical nematode life cycle has four juvenile stages (often called larvae), and the adult stage is reached after the fourth moult. The first stage develops in the egg and may moult into the second stage before hatching; in endoparasitic species of plant nematodes second stage larvae are often the ones that invade the host but in animal parasites it is usually the third; in some nematodes, one stage, usually a juvenile or larval stage, can survive long periods of adverse conditions, including desiccation.

More detailed information on the morphology and biology of nematodes is given in Chitwood and Chitwood (1950), Hyman (1951), Grassé (1965), Chitwood (1969), Zuckerman et al. (1971). Nematodes as crop pests are reviewed in Jones and Jones (1964) and Southey (1965).

12.2 Classification

Some authorities regard Nematoda as a class within the Aschelminthes, grouping them with other pseudocoelomates such as Rotifera, Gastrotricha and Echinodera, whereas others regard them as a separate phylum. Maggenti (1971) reviews the situation and argues in favour of a separate phylum for Nematoda. An outline classification of the Nematoda to sub-order level is given in table 12.1.

Nematodes that transmit viruses between plants are in two sub-orders of the Dorylaimida, class Adenophorea, whereas most other forms parasitic on plants are in the order Tylenchida, class Secernentea. Maggenti (1971) suggests that plant parasitism developed independently in these two widely divergent groups, that the Adenophorea are the older group and that within

TABLE 12.1
Classification of the phylum Nematoda

Class – Adenophorea

Order	Sub-order	Bionomics
Chromadorida	Chromadorina	
	Cyatholaimina	
Araeolaimida	Araeolaimina	
	Tripyloidina	Aquatic or moist habits,
Monhysterida	Monhysterina	many marine; saprophagous,
Desmodorida	Desmodorina	algal feeders and/or
	Draconematina	predatory
Desmoscolecida		
Enoplida	Enoplina	
	Oncholaimina	
Dorylaimida	Dorylaimina*	Saprophagous, predatory,
	Diptherophorina*	algal feeders; few
	Alaimina†	ectoparasitic on plant roots
Mononchida	Mononchina	
	Bathyodontina	Mainly predatory
Isolaimida		
Trichosyringida	Trichosyringina	Animal parasites, some
Trichinellida	Trichinellina**	groups require transport
Dioctophymatida	Dioctophymatina	or intermediate hosts

Table 12.1 (*continued*)

Class – Secernentea		
Order	Sub-order	Bionomics
Tylenchida	Tylenchina Aphelenchina	Fungivorous, plant and insect parasites; few predatory
Rhabditida	Rhabditina**	Mainly saprophagous; few animal parasites
Strongylida	Strongylina Trichostrongylina Metastrongylina**	Animal parasites, some requiring an intermediate host
Ascarida	Ascaridina Heterakina Oxyurina	
Spirurida	Spirurina Camallanina Filariina	Animal parasites requiring an intermediate host

 * Plant–parasitic virus vectors only recorded in these two sub-orders.
** Animal–parasitic virus vectors only recorded in these three sub-orders.
 † Although the Alaimina is sometimes placed in the Dorylaimida it is omitted by Coo-
   mans and Loof (1970) and would perhaps be better as a separate order – Alaimida.

this class the Dorylaimida contains two divergent lines, one containing
the genera *Longidorus* and *Xiphinema*, which transmit nepoviruses (see ch.
27), is in the family Longidoridae, sub-order Dorylaimina, whereas the
other line, containing *Trichodorus*, which transmits tobraviruses (see ch. 27),
is in the sub-order Diphtherophorina.

Metastrongylus elongatus, a vector of swine influenza virus, is in the order
Strongylida, class Secernentea. This group is close to the Rhabditida from
which it is generally accepted that Tylenchid parasites have evolved.
Strongyloides ratti, which transmits swine influenza virus to laboratory rats
and mice, is in the order Rhabditida, class Secernentea, and has a free-living
form similar to non-parasitic rhabditids. *Trichinella spiralis*, a vector of
lymphocytic choriomeningitis; is a highly specialised nematode in the order
Trichinellida, class Adenophorea.

12.3 Morphology of plant–parasitic virus vectors

Longidorus and *Xiphinema* are large nematodes, 2 mm–1 cm long, with a long hollow axial spear (100–300 µm) the anterior part of which (odontostyle) is formed in the anterior oesophageal wall during the previous moult (see Coomans and De Coninck 1963), and takes up its operational position when moulting to the next stage. During moulting the old cuticle, with the cuticular linings of the amphid pouches, stoma and oesophageal lumen together with the old odontostyle, are shed; the posterior basal part of the spear (odontophore) is reformed during the moult. Fig. 12.2 shows the basic differences between the spear regions of *Xiphinema* and *Longidorus*. *Longidorus* has a simple attachment, without prongs, of the odontostyle to the rather weakly developed odontophore; the spear-guide ring is well developed surrounding the anterior part of the odontostyle. *Xiphinema* has small prongs at the attachment of the odontostyle to the much more strongly developed odontophore, which has a flanged and knobbed base. Surrounding the base of the odontostyle are two guide rings formed by an eversible guiding sheath. The amphids in *Longidorus* are large pouches of variable shape but with minute pore-like openings to the exterior, whereas in *Xiphinema* the amphid pouches are usually goblet shaped with a wide opening almost extending across the head at the base of the lip region. The Longidoridae also contains the genus *Paralongidorus*, which resembles *Longidorus* except that the amphid openings are slit-like, usually extending more than halfway across the head in lateral view. All three genera have a typically Dorylaimoid oesophagus with a narrow anterior part and a well developed, wider, cylindrical, muscular region.

Electron micrographs have been made of the ultrastructure of some species – *Xiphinema index* by Wright (1965), Roggen et al. (1967); *X. americanum* by Lopez-Abella et al. (1967); *Longidorus elongatus* by Taylor et al. (1970) and *L. macrosoma* by Aboul-Eid (1969). These confirm the above mentioned differences between *Longidorus* and *Xiphinema* in spear guide apparatus and odontophore structure. It is interesting that all these authors

---

Fig. 12.2.   A) *Longidorus elongatus* – anterior region; B) *Xiphinema diversicaudatum* – spear structure; C) *Trichodorus christiei* – anterior region (partly after Taylor and Robertson 1969, 1970; Hirumi et al. 1968.)   agr: anterior guide ring; aoes: anterior oesophagus; aos: anterior outer spear; dpw: dorsal pharyngeal wall; gr: guide ring; gs: guiding sheath; int: intestine; is: inner spear; odp: odontophore; ods: odontostyle; phl: pharyngeal lumen; pgr: posterior guide ring; pos: posterior outer spear; poes: posterior oesophagus; spm: stylet protractor muscles; wph: wall of pharynx; ws: wall of stoma.

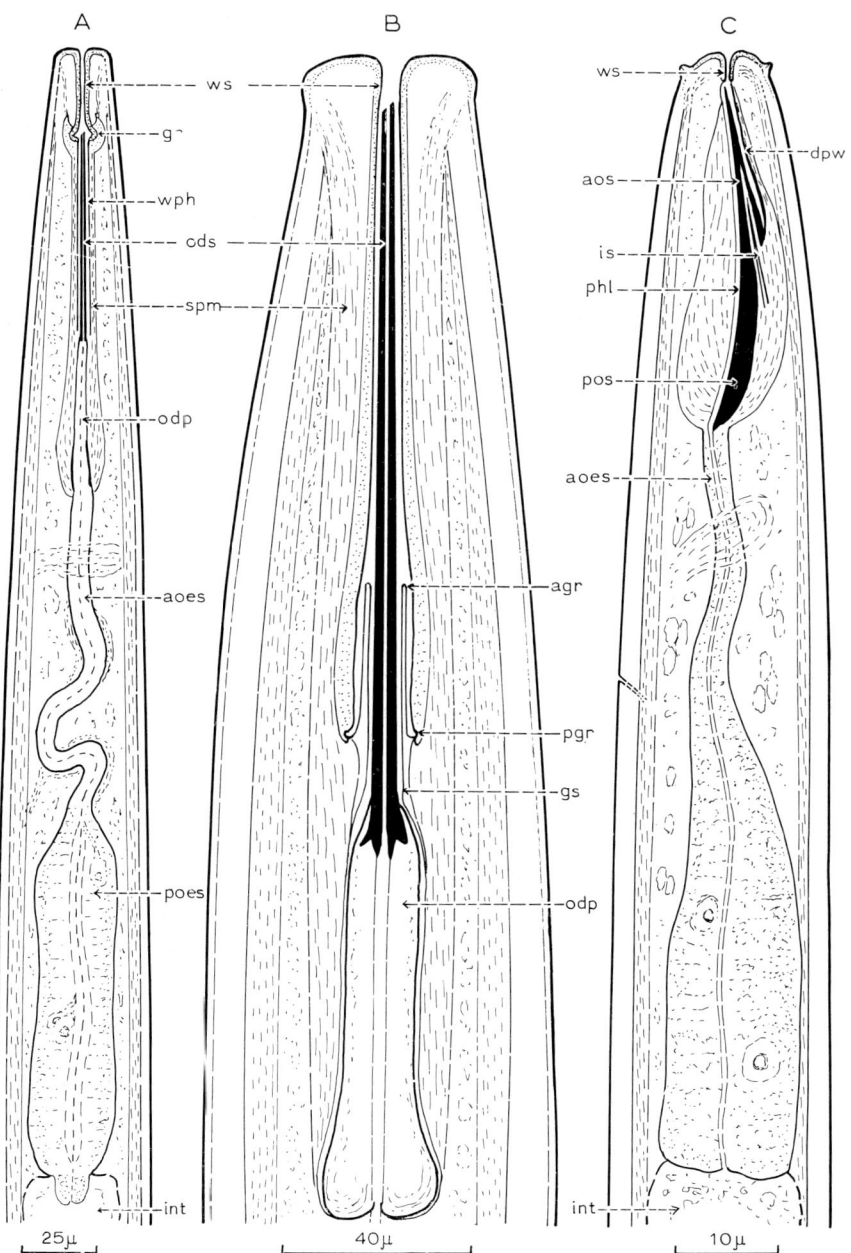

found the odontostyle had an eccentrically placed lumen with a dorsal slit-like aperture throughout its length possibly indicating that it is derived from an enlarged tooth found in related groups. Stereoscan micrographs of the anterior regions of *Longidorus* and *Xiphinema* are shown in figs. 12.3 and 12.4.

Trichodorus are short, 0.5–1.2 mm, stout, cigar-shaped nematodes with a rather thin loose cuticle that has a swollen, inflated, appearance under certain conditions. The stoma is armed with an elongate, non-axial, ventrally curved mural tooth (fig. 12.2) ranging 30–150 µm long between species. Ultra-structure studies by Hirumi et al. (1968) show that in *T. christiei* the tooth, 31–47 µm long, consists of an outer spear with a solid anterior end that projects into the stomatal lumen; the middle region of the dorsal side of this outer spear fuses with the dorsal pharyngeal wall; also in this middle region there is an oblique dorsal opening in the outer spear through which is inserted the anterior half of an inner spear the base of which extends into the surrounding dorsal pharyngeal tissue. Bird (1971) describes a similar spear structure for *T. porosus* and although Raski et al. (1969) did not de-scribe an inner spear for *T. allius*, it is illustrated in their photomicrographs (Bird 1971) and similarly an inner spear is illustrated by Taylor (1971) for *T. pachydermus*. According to Seinhorst (1970), this inner spear is also readily seen in *T. nanus* with light microscope and is similar to the replace-ment stylet in the juveniles of other species, but he could not see a re-placement stylet or vestigium in adults of several other *Trichodorus* species including *T. pachydermus*. How the spear is formed is uncertain; observations by Morton and Perry (1968) indicate that no portion is shed during moulting and Bird (1971) considers the stomatal armature of *T. porosus* should be termed an onchiostyle to distinguish it from the odontostyle because it is 1) formed in situ in first stage juveniles; 2) moulted spears have not been observed in cast cuticles, and 3) a new spear has not been seen developing in the oesophagus.

A narrow pharyngeal lumen, ventrally adjacent to the spear, connects the stoma to the oesophagus, which has a narrow anterior part and an enlarged, glandular, weakly-muscular basal portion that abuts or slightly overlaps the intestine.

The development and structure of the stomatal armature and structure of the oesophagus differs greatly from that in most other dorylaimids and emphasises the disparity between *Trichodorus* and the longidorid virus vectors; this difference is also reflected in the different types of virus trans-mitted by these two groups (see ch. 27) and perhaps justifies Coomans and

Loof (1970) forming the sub-order Diphtherophorina to contain *Trichodorus* and related genera.

12.4 Biology of plant–parasitic virus vectors

Of some 30 described species of *Longidorus* (Needle nematodes), only *L. elongatus*, *L. attenuatus* and *L. macrosoma* are known to transmit nepoviruses; also of some 60 described species of *Xiphinema* (Dagger nematodes), only *X. americanum*, *X. coxi*, *X. diversicaudatum*, *X. index* and *X. italiae* are known to be virus vectors (see chs. 4 and 27). All of these species cause, or are associated with, damage to plant roots. They are root ectoparasites and use their long hollow stylet to penetrate root tissue, sometimes reaching the vascular system; feeding is usually at or just behind the root tips and causes necrosis, distortion and/or galling.

Longidorus elongatus has a wide host-range including many crops and weeds (Whitehead and Hooper 1970; Taylor 1971). It is widely distributed in Western Europe, parts of Asia and North America. In addition to transmitting virus (see ch. 27), it seriously affects the growth of strawberries (Sharma 1965; Wyss 1970a) sugar beet (Whitehead and Hooper 1970) and peppermint (Konicek and Jensen 1961). It probably has one generation per year; egg laying coincides with active root growth of host plant, with a two to fourfold increase on favoured hosts (Thomas 1969) and it usually occurs in moist sandy to medium loams (D'Herde and van den Brande 1964; Wyss 1969). Wyss (1970b) found development was faster with increased temperature, and the life cycle was completed in 9 weeks at 30 °C.

L. attenuatus is common in sandy soils in eastern England, especially below plough depth, and damages sugar beet roots (Whitehead and Hooper 1970); it also occurs in other parts of Western Europe (Wyss 1969).

L. macrosoma usually occurs in heavy clay soils (D'Herde and van den Brande 1964), often associated with rosaceous plants, and sometimes together with *Xiphinema diversicaudatum*.

Longidorus elongatus and *L. attenuatus* seem to reproduce mainly parthenogenetically because males are usually rare, whereas males of *L. macrosoma* are common and seem functional (Dalmasso 1970).

Xiphinema americanum is widespread in the United States and in the Mediterranean region; it also occurs in India and Australia. It may consist of a group of four closely related species, as Lima (1968) has indicated. It is associated with the roots of various plants, especially trees and woody perennials (Taylor 1971) and takes about a year to complete each generation (Griffin and Darling 1964).

X. index occurs in warmer soils of the United States, the Mediterranean region and India; it seems to have a small host range but is frequent on grape vine (*Vitis* spp.) and has presumably been distributed on grape stocks. Although *Xiphinema index* completed its life cycle in 22–27 days at 30 °C (Radewald and Raski 1962), Cohn and Mordechai (1969) found that it took 7–9 months at 20–23 °C. *X. index* is similar in appearance to *X. diversicaudatum* but rather smaller and males are rare. *X. index* does not occur naturally in temperate soils but Cotton et al. (1971) found that it could survive the winter, including freezing temperatures, and subsequently produce dense populations.

X. diversicaudatum is widespread in Europe, parts of the United States, Asia and Australia. It has a wide host range of woody and herbaceous crop plants and weeds, especially rosaceous ones including strawberry, raspberry, hawthorn and rose and often causes severe root galling. It usually occurs in moist soils, especially heavy clays (see Taylor 1971). Flegg (1968) found that its generation time was about 2 years with adults possibly surviving for 5 years; males are common.

X. coxi is a little known species recorded in East Germany and Florida (Tarjan 1964) and East Anglia (Whitehead in litt.).

X. italiae is fairly common in the Mediterranean region: Cohn (1969) found that it occurred mainly in sandy soils in Israel associated with *Prunus* spp., grapevine and citrus. Males are rare for both *Xiphinema coxi* and *X. italiae*.

The genus *Trichodorus* has some 35 described species of which eleven transmit tobraviruses (see chs. 4 and 27).

Trichodorus spp. are root ectoparasites and are commonly known as 'Stubby root nematodes'; they usually feed at or just behind root tips checking their growth and causing necrosis, stunting and/or swelling.

T. christiei (? a junior synonym of *T. minor* that is common in Australia on various crops) has been extensively studied as it is a serious pest of many crops in the United States where *T. allius* and *T. porosus* also occur. Ayala et al. (1970) briefly review the literature on *T. christiei*; they found that the optimum temperature for reproduction of this species was 16–24 °C; 21–24 °C for *T. allius* and 24 °C for *T. porosus*. All these three species seem to be polyphagous as over 90% of the many plants they tested were hosts. Populations of *T. allius* increased to levels nine times greater on Glurk tobacco plants infested with California tobacco virus than healthy plants of the same variety and age. Rohde and Jenkins (1957) found that *T. christiei* completed its life cycle in 21–27 days at 22 °C and 16–17 days at 30 °C; Morton and Perry (1968) found it took 17–18 days at 27 °C. Ayala et al. (1970) indicated that at 20 °C the development of *T. allius* second stage larvae to adults or from adults to fourth stage larvae took 22 days but the time for the complete life cycle was not obtained. Males are very rare for all of these three species, reproduction seems to be parthenogenetic and populations of *T. christiei* recover more quickly after soil fumigation than European species.

The other 8 *Trichodorus* species listed in ch. 27 occur in Western Europe, including Britain (Sturhan 1967; Whitehead and Hooper 1970; Taylor 1971). Little detailed information is known about the reproduction and feeding of some of these species but most of them seem to occur in light sandy soils; *T. primitivus* sometimes also occurs in heavier soils. Several species may occur together; in samples from sandy soils in eastern England *T. pachydermus* and *T. primitivus* are generally most common (Whitehead and Hooper 1970) as they are also in fields in eastern Scotland (Cooper 1971) whereas *T. pachydermus* and *T. similis* are most common in West Germany (Sturhan 1967; Wyss 1969); however other species are more common locally, such as *T. teres* (syn. *T. flevensis*) in sandy polder soil in Holland (Kuiper and Loof 1962) and *T. nanus* in calcareous sandy soils in eastern Scotland (Cooper and Thomas 1970). Cooper (1971) found that many of the Scottish soils infested with *Trichodorus* are deficient in copper and manganese; also, in vitro tests showed that solutions of cupric sulphate (16 ppm Cu) were toxic to *Trichodorus* spp., especially *T. pachydermus*, but not to other soil nematodes. Whitehead and Hooper (1970) found that *T. anemones*, *T. cylindricus*, *T. pachydermus*, *T. primitivus*, *T. similis* and *T. teres* aggregated and fed at the root tips of several crop plants. Pitcher (1968) found that *T. viruliferus* aggregated and fed at the elongating zone of the 'extending' type of apple roots and Pitcher and McNamara (1970) reported this species feeding on such apple roots in the winter at 5 °C. The life cycle of *T. viruliferus* takes about 45 days at 15–20 °C.

Wyss (1971) showed that the feeding process of *T. similis* on tobacco or rape roots in vitro lasts only a few minutes and is in five stages: 1) probing; 2) the cell wall is perforated by rapid thrusts of the stylet; 3) the stylet is thrust more slowly and saliva flows into the cell and coagulates the cell cytoplasm into a mass about the point of perforation, a short feeding tube, probably of coagulated saliva, develops between the nematode's stoma and the cell contents; 4) heavy stylet thrusts pierce the mass of coagulated cytoplasm and it is then immediately pumped by the oesophagus into the intestine; 5) the nematode detaches its head from the cell, the feeding tube remains as a sealing plug.

Males are common in populations of the 'European' species listed in table 27.1 except that those of *T. nanus* are very rare and those of *T. teres* are absent from some populations but present, though apparently non-functional, in others (Loof 1965).

12.5 Animal–parasitic virus vectors

Pig lungworms, *Metastrongylus* spp. (class Secernentea; sub-order Metastrongylina), live in the bronchioles damaging lung tissue and sometimes causing pneumonia, especially in young animals. Male lungworms are about 2.5 cm long and females about 5 cm long. Eggs pass up the bronchi to the pharynx and then down the intestine and are voided with the faeces. First stage larvae develop in the eggs on the ground but do not hatch until ingested by a suitable species of earthworm. These larvae, about 0.3 mm long, then penetrate the wall of the anterior gut and enter the hearts and dorsal blood vessel of the earthworm where they moult twice to give third stage larvae that grow to about 0.5 mm long and retain their second stage cuticle as a protective sheath. No further development takes place in the earthworm, where they can remain alive for up to 4 years, but if the earthworm dies they can survive for only days in the soil. Pigs become infected by eating infected earthworms or soil; the larvae penetrate the pig's intestine and move into the lymphatic system and then into the bloodstream. Eventually they move to the lungs where they develop to adults and the females lay eggs about 4 weeks after infection.

Infective larvae of *Metastrongylus* spp. can act as a reservoir and intermediate host for swine influenza virus or hog cholera virus (Shope 1941, 1958). The viruses are apparently in a 'masked' phase in nematodes outside the mammalian host and become activated in the infected host when the virus–host relationship is changed by some stimulus such as inclement weather or infection by other parasites such as migrating *Ascaris suum* larvae.

Strongyloides ratti (class Secernentea, sub-order Rhabditina) larvae also acted as a reservoir and intermediate host for swine influenza virus when inoculated to laboratory rats and mice (Shotts et al. 1968). *Strongyloides* spp. are intestinal parasites of various vertebrates except fish; they have a free-living (rhabditiform) heterosexual generation which, under certain conditions, produces a parasitic (filariform) generation for which males are unknown. More than one free-living generation may occur; environmental and/or genetic factors seem to influence the incidence of filariform, infective, larvae. The infective larvae usually penetrate the skin of the host and migrate via the circulatory system and the lungs to the mouth and then down to the intestine where they become embedded in the mucosa and develop into parthenogenetic females; eggs are passed out with the faeces.

Trichinella spiralis obtained from muscle tissue of guinea pigs infected with lymphocytic choriomeningitis, transmitted this virus when fed to or injected into other guinea pigs (Syverton et al. 1947). Woodruff (1968) also reports that circumstantial and experimental evidence indicates that *T. spiralis* can transmit lymphocytic choriomeningitis in mice.

T. spiralis (Pork trichina-worm) is parasitic in various omnivorous and carnivorous animals, especially rats, pigs and man, who usually becomes infected by eating infested pork inadequately cooked. The adult worms live in the small intestine, the males are about 1.3 mm long and the females 3 mm long. The females are viviparous and release first stage larvae, about 1500 per female; the larvae are about 0.1 mm long and penetrate the wall of the intestine and pass into the lymph and blood vessels and migrate to various parts of the body. These migrating larvae cause a disease called trichinosis or trichiniasis and heavy infestations may kill the host. Only larvae that enter voluntary musculature such as the diaphragm develop any further and these enter the muscle cells and develop to fourth stage larvae, about 0.4 mm long, which then coil in a quiescent condition and the host forms a cyst around them. The larvae develop to adults only when infected tissue is eaten and they are released into the intestine of the new host; they then burrow into the mucosa and become adults in 3–4 days. The male dies shortly after copulation and the female produces larvae for about six weeks.

12.6 Discussion

As noted in 12.4 only a small proportion of the described species of *Longidorus* and *Xiphinema* are known to transmit plant viruses. These genera have been studied for only a few years and other species can be expected to be shown to be virus vectors. The genus *Paralongidorus* with some 20 described species has very similar morphology to *Longidorus* and some of these could also be expected to be virus vectors; most of the described species are from warmer soils and none had been implicated as vectors. Other long-speared

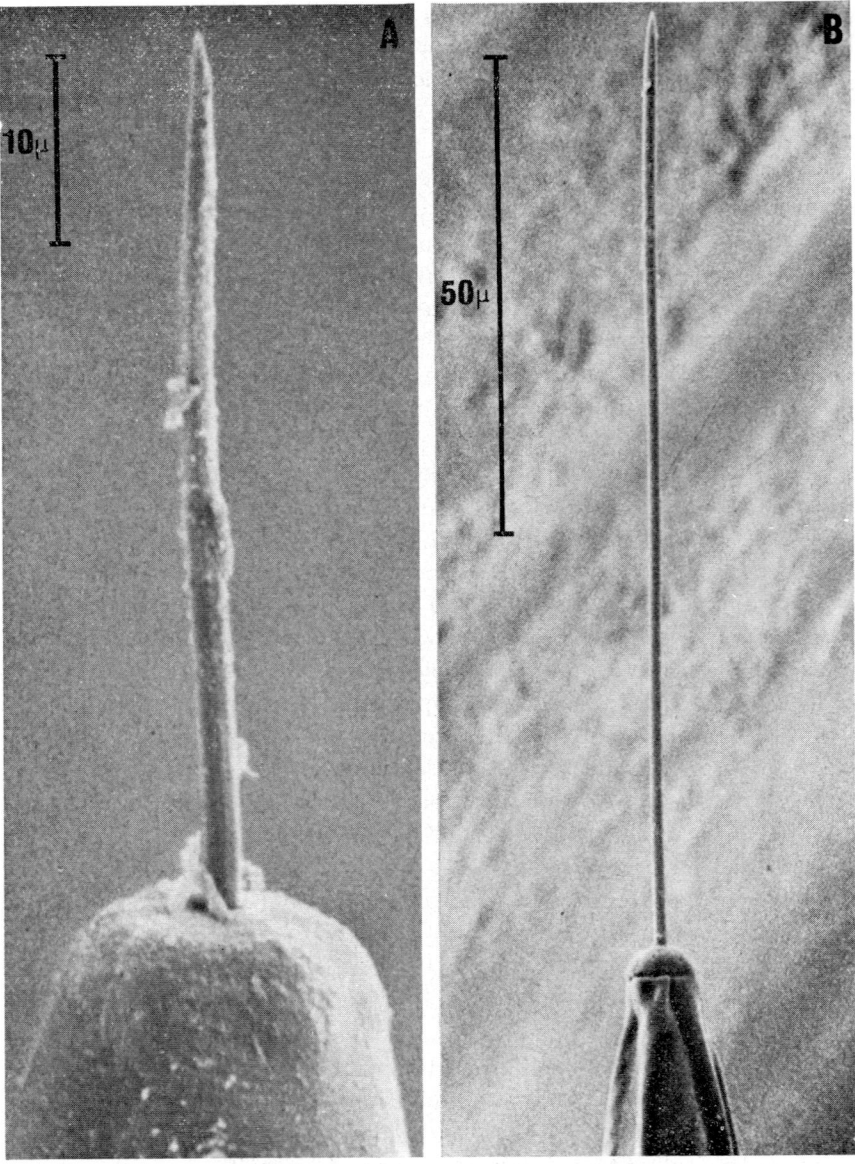

Fig. 12.3.   A) *Longidorus elongatus* ♀, anterior end with odontostyle extruded; the odonto-style has a dorsal groove throughout its length. B) *Xiphinema diversicaudatum* ♀, anterior end in lateral view with odontostyle extruded; slit-like aperture just behind the lip region is the opening to the amphid pouch. (Photographs by Mrs. S. Clark, Rothamsted Experimental Station, using a scanning electron microscope.)

Dorylaimid nematodes that might be suspected of being vectors of nepoviruses are species of *Longidorella* and *Xiphinemella*; they have similar
oesophageal structure to the longidorids but much shorter bodies, usually
under 2.5 mm long.

Fritsche and Kegler (1968) reported that a *Eudorylaimus* sp. could transmit
chlorotic leaf spot virus of apple but this report has not been confirmed.
Eudorylaimus spp. and related genera are very common in some soils. They
are short-speared dorylaimids but they are not generally regarded as plant
parasites.

Only about one-third of the known species of the genus *Trichodorus* have
so far been implicated as virus vectors; specificity for the tobraviruses they
are known to transmit is much less than in *Xiphinema* and *Longidorus* (see
ch. 27). Other *Trichodorus* spp. can therefore be expected to be implicated as
vectors. The related genus *Diphtherophora* has so far received little attention;
species of this genus are usually rather few in soils and have not been associ-

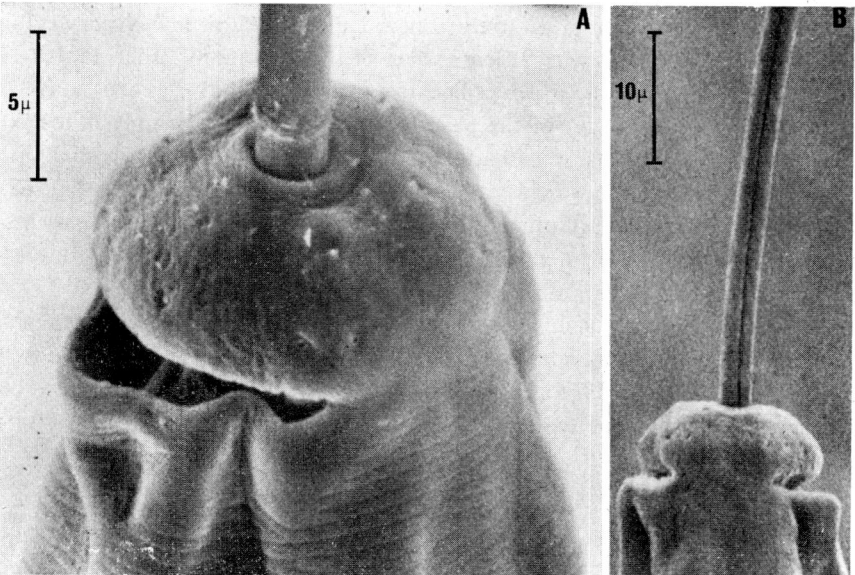

Fig. 12.4. A and B, *Xiphinema diversicaudatum* ♀, A) sublateral view of head region
showing extruded odontostyle, papillae on lip region and large opening to amphid pouch.
B) dorsal view of anterior end showing extruded odontostyle with its dorsal groove and
the lip region offset by the large amphid openings on each side. (Photographs by Mrs. S.
Clark, Rothamsted Experimental Station, using a scanning electron microscope.)

ated with plant damage but they are presumably root ectoparasites and might be virus vectors.

As indicated in sect. 12.2 plant-parasitic virus vectors are only known in two distinct suborders of the Dorylaimida (class Adenophorea), whereas most of the plant parasites are in Tylenchida (class Secernentea). Maggenti (1971) thinks that the Dorylaimida are the older group; therefore the virus–vector relationship has presumably evolved over a very long period of time. Such a relationship does not seem to have been established in the Tylenchida but presumably the passive transfer of virus from infected to healthy plants by migrating parasites, especially those with long spears such as *Belonolaimus*, *Dolichodorus* and *Hemicycliophora* spp., cannot be excluded. Also, free-living nematodes might act as carriers and disseminators of plant viruses as some are already known to act in this role for fungi (Jensen 1967), bacteria and bacteriophages (Chantanao and Jensen 1969); and they might even carry mycoplasma as these can survive ingestion by nematodes (Jensen and Stevens 1969).

Of the three animal-parasitic virus vectors mentioned in sect. 12.5 two are in the class Secernentea and the third in Adenophorea. Similar virus–vector relationships may exist with at least some of the very many other parasites, but this may be difficult to detect because of the fact that the virus may be 'masked' in certain stages of the parasite and become active only in a host that has other infections or is in adverse conditions. Synergistic relationships between animal parasites and virus infections are known (Woodruff 1968) and the incidence of poliomyelitis in people is correlated with *Toxocara* (sub-order Ascaridina) infections, suggesting that it is involved in the dissemination of the virus (Woodruff 1969).

Although a close relationship between a pathogenic bacterium and nematode *Neoaplectana* (sub-order Rhabditina) which parasitises insects is known and exploited (Welch 1965), there are many other nematodes that parasitize insects and that may be worth studying as possible vectors of insect pathogenic viruses.

CHAPTER 13

Thrips and whitefly

L. A. MOUND

Contents

13.1 Thrips

Accounts of the morphology of Thysanoptera are readily available in entomological textbooks such as Imms Textbook of Entomology or the Grassé Traité de Zoologie, and more general accounts dealing with biology have been published by Priesner (1964), Stannard (1968), and Ananthakrishnan (1970). Faunistic studies which facilitate the identification of species are available for eastern North America (Stannard 1968); California (Cott 1956; Bailey 1957); Europe and Egypt (Priesner 1964a, b); India (Ananthakrishnan 1970). Literature dealing with thrips for other parts of the world is widely scattered, but Sakimura (1962) has given a review of the information on the relationship between thrips and viruses.

Adult Thysanoptera can be recognised by the presence on the feet of a large eversible bladder instead of the normal insectan tarsal claws, and both the larvae and the adults can be distinguished from all other insects by the absence of the right mandible. The Thysanoptera can be divided into two main groups, the Terebrantia whose females have a saw-like ovipositor with which eggs are laid into plant tissue, and the Tubulifera whose females lack an ovipositor and merely attach the eggs to the surface of their food. The Tubulifera contain a single family, the Phlaeothripidae. Females of species in this family have the last (tenth) abdominal segment tubular and usually much longer than the preceding segment (eggs are not laid through this tube but emerge from a pore between the ninth and tenth segments), and the forewings when present have no longitudinal veins even along the anterior margin. The Terebrantia contain four families, the Aeolothripidae, Heterothripidae, Merothripidae, and Thripidae. Females in these families have a short non-tubular tenth abdominal segment, and the forewings have one or more longitudinal veins. The Aeolothripidae are often facultatively predatory on other small insects and mites, have relatively broad wings which are usually banded with pigment and most species are found in the Holarctic and the Australian regions. The Heterothripidae are a group of flower infesting species from the Americas, particularly their tropical areas, and the Merothripidae are minute usually wingless insects with a weak ovipositor not uncommon in leaf litter from the tropics and sub-tropics. The majority of Thysanoptera are included in the families Thripidae and Phlaeothripidae, and the only thrips which have been proven to be virus vectors belong in the Thripidae. The two records quoted by Heinze (1959) from the Phlaeothripidae are unproven.

As stated above the right mandible of the Thysanoptera is absent. The left mandible is an elongate solid and acute stylet, and the laciniae of the maxillae are also developed into elongate stylets. These three stylets are protruded through the mouth cone during feeding although the method of ingestion is not really understood. Recent studies with the stereo electron microscope suggest that the textbook accounts of ingestion are inadequate (Mound 1971). The maxillary stylets are C-shaped in section and the lateral margins are tongued and grooved (fig. 13.1, 13.2, 13.3, 13.4). When the stylets are protracted they fit together to form a tube with a sub-apical aperture in much the same way as do the stylets of aphids or other Hemiptera. It seems likely that thrips suck their food through this tube, and that they are more specific in their feeding sites than textbook references to rasping and scraping would suggest. The structure of the maxillary stylets is essentially the same in all groups of Thysanoptera with interlocking complex apices (figs. 13.4 and 13.6), but the stylets are much broader in species which feed on large fungal spores (figs. 13.5 and 13.6).

Thysanoptera feed on a wide range of substrates. As mentioned above some if not all Aeolothripidae are facultatively predatory, but the only predatory Thripidae are species of the genus *Scolothrips* and *Parascolothrips*. In the Tubulifera, *Leptothrips* and *Karnyothrips* species are predatory on mites, *Podothrips* species are predatory on scale insects on grass stems, and *Aleurodothrips fasciapennis* is widespread in the tropics attacking whitefly and scale insects on trees. Most Terebrantia feed on the tissues of green plants, some only in flowers and others mainly on leaves, although the Merothripidae probably feed on fungal hyphae. Of the family Phlaeothripidae, the species in the sub-family Megathripinae apparently all feed on fungal spores on dead wood or in leaf litter, although the species in the Phlaeothripinae are more diverse. A few groups in this sub-family are found in flowers but the majority of species are either leaf feeding or fungal hyphae feeders. Many leaf feeding Tubulifera cause the leaves they attack to curl or roll up or even to form galls. The hyphal feeders are found on freshly dead wood but in the tropics and sub-tropics are more common in leaf litter. The excreta of most species is unknown but the spore-feeding Megathripinae commonly contain a black bolus of indigested spore walls in their hind-gut, and leaves attacked by *Heliothrips* and *Caliothrips* species often bear black spots of faecal material. The larvae of these Heliothripini, when disturbed, raise the hind end of the body and exude a drop of faecal matter.

Many plant feeding thrips and possibly some fungus feeding species, are monophagous, but each instance of suspected monophagy should be critically examined. Faulty taxonomy has undoubtedly increased the number of

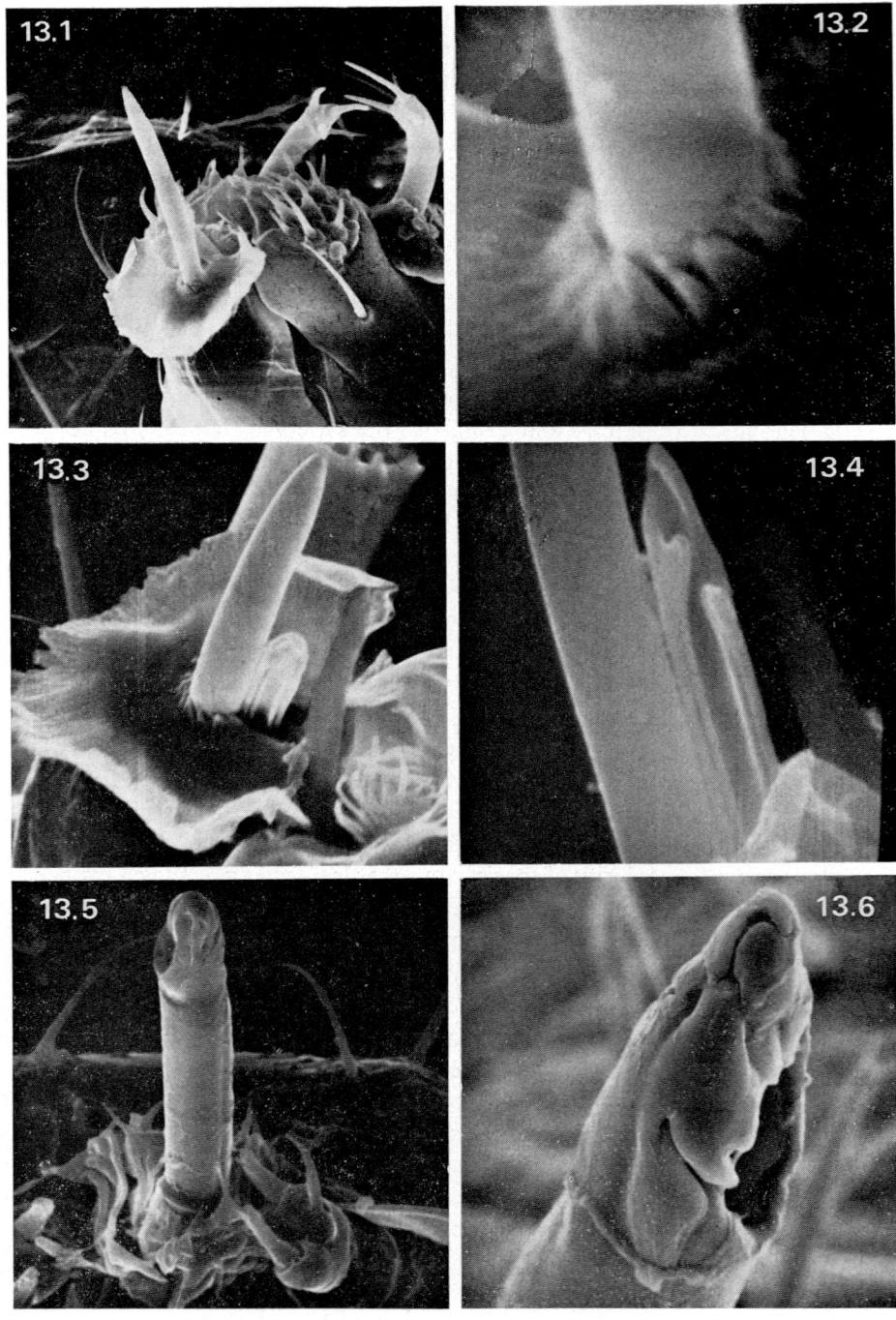

reported instances of monophagous thrips because of the variation in structure and appearance in many species. Individuals growing on a particularly nutritious host, or at low population levels, are likely to be larger than individuals from dense populations or from weakly growing host plants. Such differences have been interpreted by taxonomists as representing differences between species. Particularly confusing is the fact that large individuals may have some structures hypertrophied which in small individuals are only weakly developed or absent, e.g. stout spines instead of fine setae, large tubercles instead of minute papillae. However, even a polyphagous species is likely to have distinct host preferences, and it is because of such a preference that *Thrips tabaci* is well known as the onion thrips. It is possible that such host preferences could be manipulated by plant breeders to develop crop varieties unattractive to thrips. For example, hairy leaved cotton varieties which have been tested in both Sudan and India for their resistance to leaf hopper attack also tend to have fewer thrips on their leaves than glabrous leaved varieties.

Adults of many Thysanoptera are wingless and such specimens may not have ocelli, the three simple eyes between the compound eyes. Even when present, wings are not always fully developed, and many species have micropterous or hemimacropterous forms. The development of functional wings is apparently related to deterioration in the food supply in some fungus-feeding Phlaeothripidae, in these the macropterae and micropterae correspond to alate and apterous aphids. In other species wing length may be genetically rather than environmentally controlled. *Thrips nigropilosus*, the chrysanthemum thrips which is a pest of pyrethrum leaves in East Africa, is unusual because the wing length is continuously variable and

Feeding stylets of Thysanoptera. Figs 13.1–13.4, *Thrips tabaci* adult female:

Fig. 13.1.   Mandible projecting from mouth cone, sensoria of labium on right, and labial palps on upper right ($\times$ 1250).

Fig. 13.2.   Detail of 1 to show valves around the mandible protruding through the mouth aperture in the labrum ($\times$ 14,000).

Fig. 13.3.   Foreshortened view of one maxillary stylet partially concealed behind mandible ($\times$ 3100).

Fig. 13.4.   Inner face of maxillary stylet to show tongue and grooves on lateral margins and complex apex ($\times$ 7000).

Figs. 13.5–13.6, *Idolothrips spectrum*, a spore-feeding Megathripine:

Fig. 13.5.   Paired maxillary stylets united to form feeding tube, labial palps visible below ($\times$ 1000).

Fig. 13.6.   Apex of united stylets to show feeding aperture ($\times$ 3000).

[Stereo electron microscope pictures by B. R. Pitkin, British Museum (Nat. Hist.).]

strictly controlled morphs do not occur. When wings are fully developed they have a narrow membrane bearing around the margin a fringe of long hairs. These hairs greatly extend the effective surface area of the wing and may enable the insect to remain airborne for long periods with little muscular effort. In large species in which take-off is most readily observed the wings are held vertically above the body and the insect then jumps and flies vertically upwards, apparently in a spiral. Small species are more difficult to study, but many leaf-feeding Thripidae have highly developed thoracic musculature which is associated with their ability to jump. Once airborne, a thrips will be distributed by air currents as it is too small for directional flight, and the author has studied specimens netted from aeroplanes at about 1000 ft over both Sudan and South Australia. In view of this, and because most species are active and fly readily in warm sunny weather, it is not surprising that many have rather wide distributions. It is not known if the ability to fly is lost during an individual's life as in some aphids, but leaf-feeding species on trees tend to fly less readily than species on the leaves of herbs or in flowers. This activity and readiness to fly probably accounts for failures to control the spread of tomato spotted wilt virus with insecticides under field conditions. The insect vectors are not difficult to kill but reinfestation from elsewhere may be continuous under warm conditions. The ecology of most Thysanoptera is poorly known, but in Europe a number of species will remain on their hosts until the atmospheric conditions are suitable for flight and as a result a very large number of individuals may take to the air in a short space of time. These mass flights have been studied in *Limothrips cerealium*, the grain thrips (Lewis 1964), which is popularly known as the thunder fly. Similar behaviour has not been studied in tropical species, but the author has occasionally seen collections of several individuals of a single species from completely spurious 'finding places' in the tropics, which suggests that these individuals had taken part in a mass flight from their true host.

Apart from some of the Megathripinae, in which the larvae develop inside the adult and emerge from the egg as soon as this is laid, most thrips lay eggs which take several days to hatch. The eggs are rather large, about one-third as long as the abdomen, and only one is laid at a time. The first two larval instars are active with antennae held out in front of them, and they feed freely. These produce a relatively inactive prepupa which has short horn-shaped antennae, and this is followed by the pupa which has the antennae fused down the sides of the head and also bears long wing buds. Tubulifera have two pupal instars instead of the single one found in the Terebrantia.

Neither the prepupa nor the pupa is known to feed. In most species of Tubulifera the immature instars can be found living with the adults, but in many Terebrantia, particularly species with only one generation in the year, the second instars drop to the ground to complete their development. However, all the known virus vectors apparently pupate on their host plants, and these can be expected to have a life cycle of less than four weeks at 25 °C. In Britain fungus-feeding species are usually more abundant in winter, but in Central Europe where snow is more common this is not so. Flower-feeding species which are monophagous tend to be univoltine as most plants have a restricted flowering period, but in Britain leaf-feeding species may also be univoltine. Pest species, however, are more adaptable organisms and usually have more than one generation each year. Thrips populations are often remarkably free from parasites and predators, although considerably more is known about parasites than the review by Best (1968) suggests. The size of a population appears to depend largely on climatic factors and the nutritional status of the host. Emigration may be important, possibly stimulated by the host's condition, but other organisms apparently have little effect.

The recognition of individual species of thrips is frequently difficult and sometimes not possible. About 5000 species have been described, but recent work on tropical faunas suggests that the group probably comprises more than 10,000 species. In Europe, where the fauna is relatively well studied, readily distinguished but undescribed species can still be found in Spain and southern France. Even in Northern Europe some of the polyphagous species which are variable in structure may yet prove to be composed of more than one biological entity. For example, *Thrips inopinatus* is specific to *Solanum dulcamara* and has only been distinguished from the polyphagous *Thrips fuscipennis* in the last ten years, and discussion still continues as to whether *T. hukkineni* is a valid species or merely a large form of *T. physapus*, the common species in Compositae flowers. This is particularly relevant to virologists as all four species of thrips which have been proven to be virus vectors are variable in size and colour. The dark and light forms of *Frankliniella occidentalis* are genetically determined, but they have a similar ability to transmit tomato spotted wilt virus (Sakimura 1970). *Thrips tabaci* females vary from small pale to large dark insects possibly in response to the temperature during development, and *Frankliniella fusca* varies from orange brown to dark brown and also has long and short winged forms. No experiments are reported on the vector ability of the forms of these two species, but the pale form of the vector species *Frankliniella schultzei*, often called *sulphurea* or *dampfi*, is apparently unable to transmit spotted wilt (Sakimura 1970). For these reasons the identity of a suspected vector may not be easy to establish without further studies on the local ecology of the insect and its variability. More than one species may occur in a single flower or on one leaf, and this can lead to difficulties in correctly relating the males, females and larvae of each species. Aggregations of one sex, either females or more rarely males, are sometimes due to differences in the period of development of the sexes. Such differences can also result in the adults of one species being present in a flower at the same time as the larvae of a second species. It is likely that many species of thrips can reproduce in the absence of males, and in a few species males are rare or quite unknown. The absence of males for any given species in a particular area may indicate that the species is introduced. This is by no means always true but it is a useful indication when attempting to work out the natural distribution and host range of a species.

Heinze (1959) lists 17 species of Thripidae and 2 species of Phlaeothripidae which have been associated with suspected viruses, but most of these records have little factual basis. Sakimura (1962) in a review of the literature on thrips-borne viruses, points out that apart from tomato spotted wilt virus, which is known to have four thrips vectors, there are only three other cases, and each of these needs further clarification. The first case is a report by Bondar (1924) of a 'Mosaico' of cassava in Brazil. However, Bondar specifically stated that this was mechanical damage by *Scirtothrips manihotis* feeding on young leaves, not a virus disease. This type of leaf curling is not uncommon with large populations of *Scirtothrips* species on other plants. More confusing is the record of a pistacia virus carried by *Liothrips pistaciae* (an undescribed species) in southern Russia (Kreutzberg 1940). Similar damage to pistacia trees is described by Kreutzberg (1955) when publishing a description of a new species of thrips, *L. jakhontovi*. The earlier report is not referred to in the second paper, although the observations were all carried out in Turkmenia in 1935–37. The insects referred to are almost certainly the same species and the 'pistacia rosette virus' is probably due to mechanical damage by the thrips. Finally, the record of a thrips transmitting a sunflower mosaic virus in Argentina with a list of almost 100 alternative host plants (Traversi 1949) is extremely dubious, particularly as a whitefly and an aphid are also recorded as vectors. This virus is probably tobacco mosaic virus, and the reported transmissions fortuitous.

Tomato spotted wilt is the only virus which has been shown clearly by experiment to be transmitted by thrips. Four species are known to be vectors, *Thrips tabaci*, *Frankliniella fusca*, *F. occidentalis* and *F. schultzei*, and Best

(1968) has given a review of the relationships of the virus to its vectors and host plants. The virus was discovered in Australia ten years before it was found elsewhere, but this is unlikely to indicate that it originated there. The genus *Frankliniella* contains numerous species which are largely restricted to the New World, particularly to Central America and the West Indies. *F. schultzei* is now widespread throughout the tropics but may have originated in South America, *F. fusca* is common in eastern North America and *F. occidentalis* is common in western North America. *Thrips tabaci* is almost cosmopolitan but the writer has only seen males of this species in reasonable numbers from Iraq and Saudi Arabia. As the favoured host plant of the insect is *Allium cepa* which has its centre of distribution in Western Asia, it is possible that *T. tabaci* is native to that area. Unfortunately, the origin of the vectors appears to be unlikely to throw much light on the original locality of the virus, particularly as both the virus and all the vectors are physiologically and morphologically plastic organisms.

13.2 Whitefly

Homoptera of the family Aleyrodidae are usually called whitefly, although rather perversely one of the most widespread pest species is called the citrus blackfly. The ecology and taxonomy of this group is probably less understood than that of any other easily recognised family of insects. The majority of species cannot be recognised as adults due to their uniformity of structure, and species are usually described from the fourth instar larvae or pupal cases. Unfortunately the appearance of these pupal cases is frequently affected by the structure of the leaf of the host plant, with the result that taxonomists have on several occasions described several closely related monophagous species when in practice there exists in nature only one variable and polyphagous species. This variation is discussed more fully below but it must be borne in mind when consulting literature on the Aleyrodidae. Most species of whitefly are found in the tropics and are as yet undescribed, but the fauna of Central Europe (Zahradnik 1963) and Britain (Mound 1966) is well known and as a result of recent studies about 30% of the whitefly of Western Africa have probably been described (Mound 1965b; Cohic 1970). Other areas from which many species have been described are Brazil (Bondar 1923), Malaya (Corbett 1935), and Madagascar and Taiwan (see Takahashi bibliography 1963). Some of the common species from India have been treated by Singh (1931), but apart from some generic studies by Russell including the im-

portant genus *Trialeurodes* (1948) the American whitefly have been little
studied since Quaintance and Baker initiated the modern classification of
the group (1913–14 and 1917).

Morphologically the aleyrodids seem to be degenerate psyllids, although
ecologically they can be regarded as the tropical equivalent of the aphids,
opportunist insects with transient populations. These three groups, together
with the coccids are placed in the Sternorrhyncha. The aleyrodids differ from
these insects in the position of the anus and in the structures associated with
it. The anus opens on the dorsal surface of the abdomen just ventral to a
tongue-like projection called the lingula. These lie in a shallow depression,
the vasiform orifice, which is partially closed dorsally by the operculum.
These structures are important in preventing the body from being soiled by
the large quantities of honeydew, the sugary excreta, which whitefly produce
from their anus. Whitefly also differ from psyllids in having fewer antennal
segments and fewer veins on the forewing, but the feeding apparatus is
similar to that of the other Sternorrhyncha. The two mandibles have ten to
twenty transverse serrations externally near the apex, and the two maxillae
are closely approximated to enclose a feeding channel and probably a salivary canal. When the adult is resting the stylet bundle is retracted into the
head, but there is no clear evidence that the immature instars can withdraw
their stylets from plant tissue.

The family Aleyrodidae is divided into two sub-families the Aleyrodinae
and the Aleurodicinae. The pupal cases of the Aleurodicinae have one or
more pairs of compound pores which produce wax rods on the dorsal surface
of the pupal case, and these include the largest whitefly. Most of them are
neotropical but a few species are found in Australia and the Oriental region.
The Aleyrodinae includes the majority of whitefly which are found throughout the tropics as well as the species which are found in temperate regions.
These do not have compound wax pores on the pupal case, although they do
have numerous simple pores which often secrete wax, and the adults have
one less vein on the forewing than the Aleurodicinae. Several species of the
Aleyrodinae have been seen to lay their eggs in a circle, or parts of a circle,
as they rotate around their mouthparts, and these often leave a thin dusting
of wax on the leaf around the eggs. Most Aleurodicine adults produce large
quantities of wax and this sometimes hangs in a thick curving mass from
either side of the abdomen. The eggs of these species are surrounded by
lines of wax which are often arranged in a loose spiral like a finger print. The
eggs have a short subterminal stalk which is inserted into the leaf tissue, but
in some aleurodicine species the stalk is longer than the egg.

Eggs of whitefly are usually laid on the lower surface of leaves and the first instar larvae are minute but with relatively long legs and antennae. These can crawl actively although they probably do not leave the leaf. The second, third and fourth instar larvae have atrophied legs and antennae and do not move at all. Sometimes with very dense populations whitefly larvae may be found on the upper surface of a leaf or even on the petioles but they are usually restricted to the lower surface. The fourth instar larva ceases feeding after a time and the adult develops inside this instar which is then referred to as the pupal case. The fourth instars of many species produce large quantities of wax from the margins and dorsal surface, and in these species the cast skins of earlier instars may be found on the dorsal surface of the pupal case. In most species the adult emerges through an inverted T-shaped split in the dorsal surface of the pupal case, but in a few species the apices of the T are joined by additional sutures and the adult emerges through a 'trap door'. Pupal cases from which parasites have emerged can be recognised by an irregular circular hole which is chewed by the emerging parasite.

Newly emerged adults are not waxy and the wings are usually translucent. The white waxy powder which covers the wings and bodies of most adult whitefly is secreted from abdominal glands after the adult has emerged, and this wax is removed from the glands and spread over the insect mainly by the hind legs. Many species have dark spots on their wings but these spots do not develop until several hours after emergence, and a few species are not white at all. The citrus blackfly, *Aleurocanthus woglumi*, has black wings and little wax, *Bemisia giffardi* from citrus in the Far East and Australia has pale yellow wings, an undescribed *Dialeurodes* species from Nigeria has red wings, and the author has seen an undescribed species in South Australia with blue wings.

One of the most confusing and interesting aspects of whitefly biology is the variation in the structure of the pupal case in relation to the structure of the host plant leaf. This has been demonstrated in *Bemisia tabaci* and *Trialeurodes vaporariorum* and probably occurs in several other species. Individuals of *Bemisia tabaci* raised on hairy leaved plants have several pairs of long setae on the dorsal surface of their pupal cases, but individuals raised on glabrous leaves have few or no long setae. The progeny of a single virgin female can produce either of these forms, and the progeny of one form can be transferred to a different host to produce the opposite form in the succeeding generation (Mound 1963). As a result the whitefly on glabrous leaved cotton in southern Nigeria look very different from the whitefly on hairy leaved tobacco in an adjoining field, and it was assumed that the insects on the two crops were as distinct from each other as the two viruses with which they are associated. This view was reinforced by the fact that under experimental conditions it can be very difficult to transfer whitefly from one species of host plant to another. However other plant feeding insects, even when they are known to feed on a range of plants, frequently show strong preferences for one or more species of plant, and they may also show similar conservatism when attempts are made to transfer them from one plant to another. This conservatism has been interpreted as indicating 'ecological biotypes' or 'races' (Costa 1969: 103), but these terms probably underestimate the plasticity of whitefly behaviour under field rather than experimental conditions. The most effective method of transferring whitefly from one host plant to another is to put the two plants together in a small cage so that the insects can move across under their own volition.

In bright sunlight, or with a high daytime temperature, adult whitefly fly actively, particularly if disturbed. Hourly catches with sticky yellow traps in Nigeria indicated that *Bemisia tabaci* had two periods of activity, an extensive one during the morning and a short one in the late afternoon. However these catches might have been a measure of settling activity rather than flying activity as this whitefly is strongly attracted to yellow (Mound 1962). Young cassava plants in pots which were placed in fields of grass about two feet high and about three hundred yards from a cassava field in the morning were usually found to bear adult whitefly by late afternoon. From this it seems that this species is a typical member of the aerial plankton, with a flight behaviour pattern which enables it to be widely distributed by air currents and a feeding behaviour pattern which enables it to accept a wide range of host plants. However not all aleyrodids have this type of biology. For example, populations of the British heather whitefly, *Tetralicia ericae*, and also the ivy whitefly, *Siphoninus immaculata*, are very local and sedentary although their host plants are common (Mound 1966: 400). Unfortunately there is very little evidence on how whitefly find their hosts, or on what factors make different hosts acceptable, but species apparently differ both in their colour sensitivity and in their olfactory sense (Mound 1962). The host plants of some species are related botanically, e.g. *Pealius quercus* lives on several genera of Corylaceae and Fagaceae, although in other species the host plants are not related botanically but their leaves are physically similar e.g. *Aleurotrachelus jelinekii* on *Viburnum tinus* (Caprifoliaceae) and *Arbutus unedo* (Ericaceae).

Many whitefly, particularly pest species, show a strong preference for feeding and laying eggs on young leaves. This is probably an adaptation to the high level of soluble nitrogen in immature leaves, although the behaviour is modified on certain plants, e.g. tobacco which has sticky hairs on the young leaves which trap small insects. In warm conditions the life cycle from egg to adult may be completed in three weeks, although in temperate climates most species take much longer to develop and only have one generation each year. These species pass the winter as larvae either on the plant or on dead leaves on the ground, and the summer temperatures or the intensity of winter frost is likely to be the most important factor in determining population size. In the tropics however, where whitefly can breed almost continuously, the size of a population is probably determined mainly by the nutrient status of the host plants and to a lesser extent by the activity of predators and parasites. A young vigorously growing plant, or a crop after an application of a nitrogenous fertiliser, supports more whitefly than an old or slowly growing plant. The number of whitefly may also be affected by the form of the host plant, in dry or windy places low growing bushy plants have more whitefly than exposed plants, and hairy leaved cotton varieties in Sudan bear higher whitefly populations than similar glabrous leaved varieties (Mound 1965a).

Most parasites of whitefly are small wasps belonging in the family Aphelinidae. These lay their eggs in the second or third instar larvae and there is usually only one parasite in each whitefly pupal case. Parasitised pupal cases of *Trialeurodes vaporariorum* turn black, but in most species the appearance of the whitefly pupal case is little affected by a parasite. The adult parasite chews its way out of the host through a circular hole, whereas the adult whitefly emerges through a T-shaped split. The most common predators of whitefly are coccinelid beetle larvae and typhlodromid mites, although under humid conditions pupal cases may be attacked by entomophagous fungi. A few whitefly species are attended by ants, and in Nigeria colonies of the common grass living species *Neomaskellia bergii* may be concealed under shelters made from soil and plant fragments by ants of the genus *Crematogaster*. If the ants are prevented from tending the colonies and removing the honeydew the whitefly may die due to the accumulation of a sooty mould fungus. This type of sooty mould is one of the harmful effects of whitefly pest species. Citrus fruit and tomatoes can be made unsaleable by the accumulation of moulds which grow on the sugary honeydew excreted by the larvae, but an actual reduction in yield may also occur with very dense populations although this can be difficult to measure (Mound 1965c).

Heinze (1959) gives a long list of proven and suspected aleyrodid vectors of plant viruses, but both Costa (1969) and Varma (1963) have pointed out that only a few of these records bear close examination. Probably all the records of *Bemisia* species in the virus literature refer to *Bemisia tabaci*, due to the variability in structure of the pupal case discussed above. Certainly *B. goldingi*, *B. inconspicua*, *B. nigeriaensis*, and *B. rhodesiaensis* are other names for *tabaci*, and *manihotis* and *vayssierei* very probably are. Similarly *Trialeurodes natalensis* is the same insect as *T. vaporariorum*, although the record of this species as a vector of tobacco leaf curl virus may itself be a mistake (Costa 1969). *T. abutilonea* is a distinct species which is known to carry sweet potato yellow dwarf virus in Maryland, but the records of *Aleurothrixus flocossus* and *Aleurotrachelus socialis* as virus vectors are very doubtful. Thus only three species of whitefly, *Bemisia tabaci*, *Trialeurodes abutilonea* and *T. vaporariorum*, can be accepted unequivocally as vectors of plant viruses. Costa (1969) has given an excellent review of the relationships between whitefly and plant viruses, and only one aspect deserves further discussion here. Several authors have commented that female whitefly are more efficient vectors than males, but these observations have not been correlated with longevity or food uptake studies. In some whitefly species the adult males have a much shorter life than the females and it seems likely that they feed less actively. Thus they will not have the same opportunities to transmit a virus as females which must take in sufficient food to mature their eggs.

The identification and host relationships of whitefly borne viruses appear to be as little understood as the biology of the vectors. Varma (1963) lists and describes 28 viruses associated with whitefly, and Costa (1969) attempts to place the known viruses into two major groupings which he calls the mosaic type and the leaf curling type. However Costa also indicates that a disease can cause different symptoms in different host plants and the host relationships are by no means rigid. The African cotton leaf curl is particularly damaging to the cultivars of *Gossypium barbadense* but it can be transmitted to some cultivars of the American cotton *G. hirsutum*. Similarly the Californian cotton leaf crumple virus attacks mainly *hirsutum* cultivars but can be transmitted to *barbadense* varieties. These complications make it difficult to consider the origin of the whitefly borne viruses, but there is some evidence that cotton leaf curl spread from West Africa to Sudan between 1920 and 1930, and cassava leaf curl is known to have spread across Nigeria from east to west between 1930 and 1940. As there is no evidence that the native African whitefly, *Bemisia hancocki*, is a virus vector (Mound 1965b: 142), this suggests that *B. tabaci* is an introduced species, possibly from India, which has enabled several viruses to become much more widespread.

Section III

Diorama

CHAPTER 14

Virus replication

L. DALGARNO AND MARY W. DAVEY

Contents

In this chapter we discuss the replication of certain arthropod-borne and insect-pathogenic viruses. The main emphasis will be placed on togaviruses (Wildy 1971); this term includes alphaviruses (group A), flaviviruses (group B) and the ungrouped arboviruses (see Casals and Clark 1965); only passing reference will be made to arboviruses of the reovirus- and rhabdovirus-type. Of the insect-pathogenic viruses only the baculoviruses (i.e. nuclear poly-hedrosis and granulosis viruses), cytoplasmic polyhedrosis and the iridovir-uses (Wildy 1971) will be discussed.

14.1 Togaviruses

14.1.1 Growth of togaviruses in the insect

Togaviruses, probably the smallest and simplest of the viruses whose parti-cles contain RNA and have an outer lipid-containing membrane, are trans-mitted in nature between susceptible vertebrate hosts by haemophagous ar-thropods (see ch. 20); they multiply in both their vertebrate and invertebrate hosts. The natural invertebrate hosts are mosquitoes and ticks, though a particular togavirus can multiply in insects from a variety of different orders without preadaptation (Hurlbut and Thomas 1960); titres reach a high level without apparently damaging the tissue and virus persists for long periods in a viable state. The experimental host range in invertebrates is largest in alpha-viruses and less in other togaviruses (Hurlbut and Thomas 1960). Hurlbut and Thomas proposed (1960) that togaviruses originated in arthropods and during evolution have become more intimately adapted to certain hosts losing their earlier broad host spectrum (see review by Schlesinger 1971).

Under normal conditions, the susceptible mosquito or tick vector is in-fected when it takes a blood meal from a viraemic vertebrate. The extent to which the mosquito gut cells become productively infected is one of the factors determining whether the mosquito will transmit the virus (see ch. 20). Initially the epithelial cells of the posterior midgut are infected; subsequently replication occurs in the fat body and virus is then released into the haemo-lymph to infect other tissues. All organs, particularly the salivary glands and nerve tissues become infected and support secondary virus growth (La Motte 1960; Chamberlain and Sudia 1961; Doi et al. 1967).

After a certain time, termed the extrinsic incubation period, virus is found in the saliva and can be transmitted to the vertebrate host by insect bite. The length of the extrinsic incubation period decreases at higher temperatures.

Thus yellow fever virus growing in *Aëdes aegypti* has an extrinsic incubation period of four days at 37 °C and eight days at 25 °C; at 18 °C no transmission occurs after 30 days; virus is still viable however since transmission occurs after a further six days incubation at room temperature (Davis 1932). The extrinsic incubation period can be reduced by 2–3 days if virus is injected directly into the haemolymph since the initial site of multiplication in the gut is by-passed (McLean 1955).

The growth pattern of a variety of togaviruses in both ticks and mosquitoes shows the following characteristics (see review by Mussgay 1964): after ingestion of the virus there is an initial decrease in titre followed by an increase until a peak is reached; virus titre then remains relatively constant for a long time. The highest titre is usually 4–6 days after infection. McLean (1953) studied the growth of Murray Valley encephalitis (MVE) in the culicine mosquitoes, *Culex annulirostris*, *C. fatigans*, *Aëdes vigilax* and *A. vittiger*. Growth curves showed an initial drop of 2 dex below the input titres of approximately 3 dex infective doses ('dex' converts the number preceding it into 10-based antilogarithm; Haldane 1960). After 2–4 days, titres increased reaching a maximum 1–2 dex higher than the initial dose. Virus was first transmissible at 6–14 days after infection; mosquitoes then remained infective for several weeks.

The multiplication of togaviruses to high titres causes no reduction in the lifespan of mosquitoes infected with western equine encephalitis (WEE) (La Motte 1960) or Japanese B encephalitis (JBE) (Thomas 1963) and no pathological changes are seen in electron micrographs of infected salivary gland tissues (Whitfield et al. 1971; cf. Mims et al. 1966). These observations can be contrasted with the known cytocidal effects of togavirus growth in cultured vertebrate cells.

14.1.2 Susceptibility of insect cell lines to togavirus infections

The capacity of togaviruses to multiply in cells of both insect and vertebrate origin, and to induce quite different responses in each, gives considerable importance to comparative studies of virus replication in these cells. Such studies are now possible with the recent development of continuous insect cell lines (see ch. 17). The use of such lines allows virus multiplication to be assayed using a single cycle of growth and permits quantitative estimates of latent period, the percentage of cells productively infected and virus yield per cell.

Continuous lines from the mosquitoes *Aëdes aegypti* (Grace 1966; Singh

1967; Peleg 1968b), *A. albopictus* (Singh 1967), *Anopheles stephensi* (Schneider 1969) and *Culex quinquefasciatus* (Hsu 1971) as well as from the moth *Antheraea eucalypti* (Grace 1962a) have been used in virus studies. No continuous lines have been reported for tick tissues although primary lines are used (Rehacek 1965). The mosquito lines show marked differences in susceptibility to togavirus infection. Singh's (1967) lines support the growth of a broad range of togaviruses; four alphaviruses (Chikungunya, eastern equine encephalitis, Semliki Forest and Venezuelan equine encephalitis) and three flavoviruses (St. Louis encephalitis, WEE and yellow fever) grow in both lines; the 13 togaviruses which fail to grow are mainly tick-borne (Buckley 1969). The *Aëdes albopictus* line can be infected with a greater variety of togaviruses and to higher titre than can the *A. aegypti* line (Paul and Singh 1969).

The continuous *A. aegypti* line (Peleg 1968b) supports the growth of alphaviruses (Semliki Forest, eastern equine encephalitis and West Nile) (Peleg 1968b, 1969a, b) and at least one flavovirus (Kunjin) (Davey and Dalgarno in preparation). West Nile, EEE and SFV also grow in a primary *A. aegypti* line (Peleg 1968a). O'nyong-nyong virus isolated from anopheline mosquitoes did not grow in a cell line from the culicine mosquito *A. aegypti*; it did grow in a line established from *Anopheles stephensi* (Varma and Pudney 1971). Buckley (1971) reports however that the same virus grows readily in *Aëdes albopictus* (Singh 1967) and in *Anopheles stephensi* (Schneider 1969) cell lines. Thus lines from Culicine mosquitoes are not necessarily refractory to infection with viruses isolated from Anopheline mosquitoes.

Using Grace's *Aëdes aegypti* and *Antheraea eucalypti* lines, Converse and Nagle (1967) report 2 dex increases in infectivity 24 hr after infection with yellow fever virus. The *Antheraea* line also supports growth of JBE (Suitor 1966) and some other mosquito-borne flavoviruses (Yunker and Cory 1968). The flavoviruses, Kunjin, MVE, JBE and West Nile but not the alphaviruses, are reported to establish persistent infections in Grace's *Aëdes aegypti* line (Rehacek 1968a). In these experiments the titre of the flavoviruses tested was at relatively low constant levels for several months and rarely increased above the input level. We have failed to demonstrate significant growth of Kunjin, MVE or SFV in this *A. aegypti* line or in several clones derived from it (Davey and Dalgarno unpublished results).

No continuous line of tick cells has been reported but in a primary line from *Hyalomma dromedarii* (Koch) tick-borne flavoviruses and all alphaviruses tested (SFV, Sindbis, WEE and Venezuelan equine encephalitis) replicate rapidly; mosquito-borne flavoviruses grow poorly in these cells (Rehacek 1965).

Thus the susceptibility of mosquitoes and ticks to togavirus infection broadly parallels the susceptibility of the corresponding cultured cells. There are at least two exceptions however. Colorado tick fever virus, a reovirus which fails to grow in any arthropod other than ticks, grows in cultured *Aëdes albopictus* cells (Yunker and Cory 1969). EEE can replicate in several insects but will not grow in any ticks examined; however, it grows to a high titre in primary cultures of *Dermocentor pictus* and *Ixodes ricinus* fat body and hypodermis cells (Rehacek 1965).

14.1.3 Quantitative studies of togavirus growth in vitro

No adequate one-step growth experiments have been reported for a toga-virus in an insect cell except for a recent preliminary report by Whitney and Diebel (1971). Peleg (1969a) demonstrated that cultured *Aëdes aegypti* cells released some SFV at 8 hr after infection, and most at 2 days. Titres then gradually fell over a long period apparently stabilizing (at a level 3 dex less than the maximum titre) at 20 days after infection and maintained this level until the end of the experiment at 150 days. Singh and Paul (1968) reported finding maximum titres of Sindbis virus 3–4 days after infection of *A. albopictus* cells. In both these sets of experiments no samples were taken to esti-mate the latent period. It also seems likely that only a small percentage of cells were initially infected and that multiple cycles of infection caused the late appearance of maximum titres.

We have compared the single cycle growth of Semliki Forest virus in cul-tured *A. albopictus* cells and vertebrate cells (fig. 14.1). The latent period, du-ration of the rise period and the maximum titre reached in the insect cell line at 28 °C are similar to those of the same virus growing in vertebrate cells at 37 °C. Stevens (1970) showed that in *A. albopictus* cells, dengue–2 and Sindbis viruses respectively reached titres of 3×10^4 pfu/ml at 48 hr and 2×10^8 pfu/ml 30 hr after infection. Although the latent period cannot be assessed exactly from these experiments, the result is comparable to that obtained by us with an alphavirus (SFV, fig. 14.1) and a flavivirus (Kunjin, unpublished results) grown in both *A. albopictus* cells and a vertebrate cell line.

Comparing growth of SFV in *A. albopictus* cells at 28 °C and 37 °C, the latent period (1–2 hr) and the final titre reached (10^8 pfu/10^6 cells) are similar at the two temperatures (fig. 14.1A; Davey and Dalgarno, in preparation). In vertebrate cells at 28 °C, SFV grows more slowly (latent period of 7 hr) than in vertebrate cells at 37 °C (fig. 14.1B). Lowering the temperature has the same effect on both cell types; in *A. albopictus* cells at 20 °C, SFV growth is slowed (latent period of 7 hr, fig. 1A) by about the same extent as vertebrate cells at 28 °C (fig. 14.1B; Dennett and Dalgarno, in preparation); however, the maximum titre at 20 °C and at 28 °C is the same (fig. 14.1A).

In a previous study of togavirus replication in cultured mosquito cells it was reported that most of the infectious virus produced is retained in association with the cell (Rehacek 1968a). The results of Peleg (1969a) and Davey and Dalgarno (unpublished results) with togaviruses, and of Mirchamsy et al. (1970) with African horse sickness virus (a reovirus), show that similar amounts of extracellular and cell-associated virus are obtained from mosquito cell lines.

SFV at a multiplicity of 1–5 productively infects 2.8–8 % of *A. aegypti* cells (Peleg 1969a). The percentage of virus-producing cells declines and 6–13 days after infection only 0.2–0.9 % of cells produced virus; 20–140 days after infection less than 0.1 % produced virus. The estimated mean virus yield was 1–6 pfu per infected cell throughout the experiment and there was no obvious

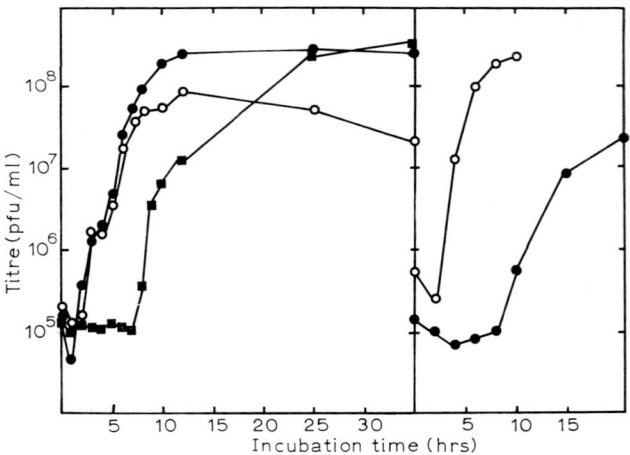

Fig. 14.1. Left: *Aëdes albopictus* cells grown in Buckley's (1969) medium were concentrated to 5 × 10^6 cells/ml and infected with SFV (m.o.i. ∼1) for 1 hr at 28 °C. Cells were washed (× 3) in growth medium, resuspended in the same medium at 5 × 10^5 cells/ml and incubated at 20 °C (■—■), 28 °C (●—●) and 37 °C (○—○). Right: Vero cell monolayers were adsorbed with SFV (m.o.i. ∼15) for 1.5 hr at 37 °C (○—○) and 28° (●—●). The monolayers were then washed (× 3) with PBS, 5 ml of growth medium (M199–LAH) added, and then incubated at 28 °C or 37 °C. In both fig. 14.1 left and right, virus was assayed on Vero monolayers; total virus (CAV + EV) is plotted.

cell damage, nor any indication that interferon-like substances were produced to maintain the low level of infection.

In experiments with VSV, a rhabdovirus, Yang et al. (1969) found that at different multiplicities similar proportions of *Antheraea eucalypti* and mammalian cells were infected.

14.1.4 Biochemical studies

The major events in the replication of alphaviruses have been studied in a number of vertebrate cell lines. However, for invertebrate cells little information exists other than that obtained from a few ultrastructural studies.

Alphaviruses grow rapidly in cultured vertebrate and insect cells; flavoviruses usually grow less rapidly (Stollar et al. 1967; Trent et al. 1969). For SFV the latent period in CEF (chick embryo fibroblast) cells is 2–3 hr and maximum yields are obtained at 6–8 hr (Taylor 1965; Mecs et al. 1967). In CEF cells, SFV and Sindbis infections markedly inhibit host macromolecule synthesis late in infection. SFV substantially inhibits host RNA synthesis 6–7 hr after infection, when most viral RNA is produced (Taylor 1965). The rate of host protein synthesis in Sindbis-infected cells declines 5 hr after infection (Pfefferkorn and Clifford 1964). Despite the extensive involvement of membranes in togavirus replication, Sindbis causes a decrease in phospholipid synthesis in CEF cells (Pfefferkorn and Hunter 1963b; Waite and Pfefferkorn 1970b); this may result from virus-directed inhibition of host RNA or protein synthesis (Waite and Pfefferkorn 1970b). In BHK cells, phospholipid synthesis is much less affected by virus infection and inhibition is only apparent 8–10 hr after infection by which time virus replication is substantially finished. Inhibition of phospholipid synthesis is not therefore a necessary result of virus replication.

The genetic material of two alphaviruses (SFV and Sindbis) consists of an RNA molecule of M.W. $3.5–4.5 \times 10^6$ daltons and sedimentation coefficient 40–45S (Cartwright and Burke 1970; Dobos and Faulkner 1970; Levin and Friedman 1971). An RNA molecule of similar molecular weight is obtained from SFV-infected *Aëdes albopictus* cells (Davey and Dalgarno, in preparation). The flavovirus St. Louis encephalitis has RNA of molecular weight 3.3×10^6 as determined by sedimentation measurements (Trent et al. 1969).

When CEF cells are infected with the alphavirus WEE, infectious viral RNA is first detected 2 hr after infection and mature virus 1 hr later (Wecker and Schonne 1961). In cells infected with the flavovirus dengue-2, virus specific RNA synthesis precedes virus release by 6 hr (Stollar et al. 1966). Apart from viral RNA the other virus-specific RNA species found in togavirus-infected mammalian cells may include: a) 26S RNA of little or no infectivity with the same composition as 45S RNA. This RNA is found in SFV infected *A. albopictus)* cells (Davey and Dalgarno, in preparation); it is barely detectable in KB cells infected with dengue virus (Stollar et al. 1967). 26S RNA can be preferentially inhibited when Sindbis-infected cells are treated with puromycin early in infection; there is no concomitant inhibition of infectious viral RNA production (Scheele and Pfefferkorn 1969b);   b) polydisperse, partly double-stranded 'replicative intermediate' of approximately 22S which is thought to be template for RNA synthesis;   c) a ribonuclease-resistant RNA (approximately 20S) which resembles the resistant core produced by treating replicative intermediate with ribonuclease (Sonnabend et al. 1964; Sreevalsan and Lockart 1966; Friedman and Berezesky 1967; Mecs et al. 1967; Sonnabend et al. 1967; Friedman 1968b).

As yet the function, if any, of 26S RNA is not known. Radioactive labelling experiments show that 20–22S RNA is labelled first, then 26S RNA and finally 45S RNA (Friedman and Berezesky 1967; Mecs et al. 1967; Cartwright and Burke 1970). When labelling is at different times after infection, 20S RNA is the major product labelled early, but more 26S than 20S RNA is labelled later (Cartwright and Burke 1970). 26S RNA is formed early after infection at a time when no 45S RNA is apparently labelled; at later times the radioactivity of the 45S species is a relatively constant and smaller proportion of that in the 26S species, suggesting a rate-limiting conversion of 26S to 45S.

Sreevalsan et al. (1968) propose that 26S RNA is either a conformer (i.e. the same molecule but in a different conformation) of the viral RNA molecule, or is produced when a particular labile region near the centre of the molecule cleaves. Mecs et al. (1967) suggest that 26S RNA may be an extended messenger form of 45S viral RNA. RNA species with similar properties have been isolated from bacteria infected with MS2 RNA bacteriophage (Kelly et al. 1965) where it was concluded that 20S RNA (analogous to 26S RNA) was a conformer of 27S bacteriophage RNA. The results of Dobos and Faulkner (1970) and Cartwright and Burke (1970), however, favour the view that 45S RNA is two molecules of roughly equal molecular weight which separate when heated.

26S RNA is found in a ribonuclease-sensitive 65S cytoplasmic particle which may be messenger RNA bound to a 40S ribosome sub-unit (Friedman and Berezesky 1967). Ben-Ishai et al. (1968) report that 26S RNA is associated with the cell membrane, also consistent with a role as messenger RNA. After labelling for 60 min with RNA precursor, the most prominently labelled viral structure in the cytoplasm sediments at 140S and contains only 42S RNA (Friedman and Berezesky 1967; Ben-Ishai et al. 1968). This structure is the nucleocapsid seen in electron micrographs of infected cells.

Assuming a molecular weight of $3-4 \times 10^6$ daltons, the RNA of togaviruses can code for 10–12 proteins of molecular weight approximately 30,000. Sindbis-infected cells contain 12–16 new polypeptides (Strauss et al. 1969) and between five and eight polypeptides have been isolated from SFV, JBE and SLE-infected cells (Friedman 1968c; Hay et al. 1968; Trent et al. 1971; Shapiro et al. 1971). Some of the Sindbis-induced polypeptides found by Strauss et al. (1969) may be large precursors or host-proteins as their total molecular weight is about twice what could be coded by viral RNA. Evidence that post-translational cleavage of large precursor proteins like that found in polio-infected cells (Jacobson and Baltimore 1968; Summers and Maizel 1968) may also occur in SFV-infected CEF cells is provided by Burrell et al. (1970).

Two proteins of molecular weight approximately 50,000 and 30,000 are the principal components of the virions of Sindbis and SFV (Strauss et al. 1968; Acheson and Tamm 1970). The larger protein, a glyco-polypeptide (Strauss et al. 1970; Burge and Strauss 1970), is found in the envelope and is rich in hydrophobic amino acids; the smaller protein is rich in basic amino acids and is associated with the nucleocapsid. At least three and possibly four proteins are found in the virions of flavoviruses. Trent and Qureshi (1971) report three proteins in the SLE virion; two of these are in the envelope and have molecular weights of 63,000 and 8500 and the third (M.W. 18,000) is associated with the nucleo-capsid. Dengue-2 virus is reported to have a glycoprotein (M.W. 59,000) as the major envelope protein; in addition there is a capsid protein (M.W. 8000) perhaps tightly bound to the envelope and a lysine-rich, histidine-free core protein (M.W. 13,000) (Stollar 1969). Three proteins of similar molecular weight and distribution within the virion to those of dengue have been found in JBE (Shapiro et al. 1971); Kunjin virus is reported to have four proteins and differs from other flavoviruses in that one of the proteins is very large (M.W. 120,000) and is associated with the envelope (Westaway and Reedman 1969).

Sindbis-specific proteins synthesised 1–12 hr after infection constitute 7–8% of the total protein of the CEF cell; the amount of membrane protein alone is equivalent to 3% of the total chick cell protein (Strauss et al. 1969). In Sindbis-infected cells not all the synthesized envelope and nucleocapsid protein is incorporated into virions (Scheele and Pfefferkorn 1969a); the envelope protein, unlike the nucleocapsid protein, does not exist free and is rapidly incorporated into the membrane fraction of infected cells (Burge, quoted by Scheele and Pfefferkorn 1969a). During virus maturation, protein synthesis is rapidly inhibited and infectious particles stop forming even though excess proteins are present and virus-specific RNA continues to be made (Scheele and Pfefferkorn 1969a).

Virus replication appears to be intimately associated with membranes. The bulk of SFV-specific protein is synthesised on membrane-bound polyribosomes, unlike proteins in uninfected CEF cells (Friedman 1968a). The replicative intermediate, replicative form (Friedman 1968b) and an SFV-induced RNA polymerase (Martin and Sonnabend 1967) are found associated with lipid-containing structures. Actinomycin-resistant, SFV-induced RNA synthesis has been shown by high resolution autoradiography to be close to the cell membrane and to the surfaces of the vesicle-lined 'cytopathic vacuoles' found in infected cells (Grimley et al. 1968; Grimley and Friedman 1969). These structures (CPV-1) are first seen during the log phase of virus growth 2–3 hr after infection. CPV-1 differ in size and time of appearance from other vacuoles (CPV-2) which appear 3–4 hr after infection. Whether such structures are necessary for virus replication or are the result of cellular disorganization is not known.

The release of WEE virus from CEF cells is rapid requiring only 0.4–2 min (Rubin et al. 1955). Sindbis virus release is prevented by incubating infected cells in a medium of low ionic strength and is speeded by exposing such cells to medium of normal ionic strength. Release is not dependent on osmotic pressure, protein synthesis or cellular energy metabolism (Waite and Pfefferkorn 1970a).

14.1.5 Ultrastructural studies

Electron microscopic studies of togavirus development confirm and extend the conclusions obtained from biochemical studies. There is no evidence that the host nucleus is involved in either the early synthetic, or later morphogenetic stages of virus formation although nucleocapsids can sometimes be seen in the nucleus of infected cells (Murphy et al. 1968; Janzen et al. 1970).

When grown in vertebrate cells, the major groups of togaviruses differ distinctly in certain aspects of their morphogenesis. Alphavirus nucleocapsids appear in the cytoplasm and are enveloped when they bud through the plasma membrane of the cell (Morgan et al. 1961; Mussgay and Weibel 1962; Saturno 1963; Chain et al. 1966; Acheson and Tamm 1967; Erlandson et al. 1967; Higashi et al. 1967; McGee-Russell and Gosztonyi 1967; Lascano et al. 1969; Murphy et al. 1970; Murphy and Whitfield 1970). Usually far more nucleocapsids are formed than are enveloped and late in infection large numbers may be found aligned on membranes, or in crystalline arrays in the cytoplasm. A series of characteristic cytopathic effects follow and later the cell lyses. In cells infected with flavoviruses there is a massive proliferation of intracytoplasmic membrane. Few free nucleocapsids are seen in the cytoplasm. Virus particles mature when the nucleocapsids bud through the cytoplasmic reticulum. Enveloped virus particles accumulate within distended cisternae of the endoplasmic reticulum and may be released from the cell when the vesicle migrates to the cell surface or when the cell ruptures (Ota 1965; Yasuzumi and Tsubo 1965; Murphy et al. 1968a).

There have been few studies on the morphogenesis of other types of togavirus (Casals and Clarke 1965); however there have been studies of Guaroa virus (Southam et al. 1964) and representatives of the California and Bunyamwera groups (Murphy et al. 1968b, c). No free nucleocapsids accumulate and budding is through cytoplasmic membranes particularly those of the Golgi complex and the endoplasmic reticulum which proliferate during infection. Virus particles accumulate individually in vacuoles or in groups within membranes. The development of Powassan virus, a tick-borne flavovirus is similar to that of the alphaviruses VEE and EEE (Abdelwahab et al. 1964).

Electron microscopic studies of salivary glands from EEE-infected *Aëdes triseriatus* (Whitfield et al. 1971; see also Bergold and Weibel 1962) indicate that virus morphogenesis in vertebrate and invertebrate cells is similar (Murphy and Whitfield 1970); nucleocapsids are first found in the cytoplasm; they are enveloped when they pass through intracytoplasmic or plasma membranes. One difference, however, is that no free nucleocapsids accumulate in the insect cells; they appear to be enveloped immediately they are formed. As distinct from flavovirus infections of vertebrate cells there is no development of a massive membranous system in mosquito cells late in infection. Vesicle-lined vacuoles like those found in alphavirus infected-CEF cells and termed CPV-1 by Grimley et al. (1968) are present, but are rare. No other obvious changes are found in *A. triseriatus* salivary gland cells through 31 days of infection and cytopathic effects are absent.

Figs. 14.2–14.4 show electron micrographs of Singh's (1967) *A. albopictus* cell line sampled during a single cycle infection with the alphavirus Ross River (RRV; Doherty et al. 1964) (Raghow, Bartley and Dalgarno, in prepa-

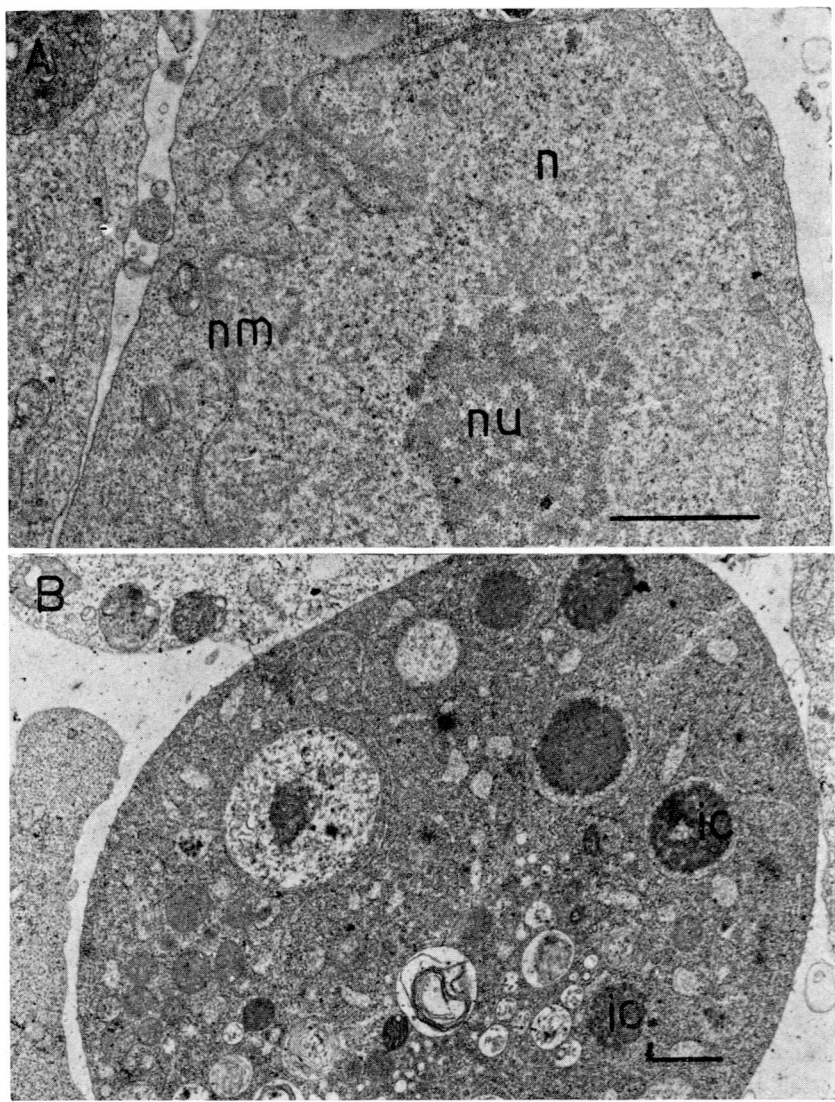

Fig. 14.2.   Cultured *Aëdes albopictus* cells 6 hr (A) and 14 hr (B) after infection with Ross River virus. n, nucleus; nu, nucleolus; nm, nuclear membrane; ic, cytoplasmic inclusion body. Bar represents 1 μm.

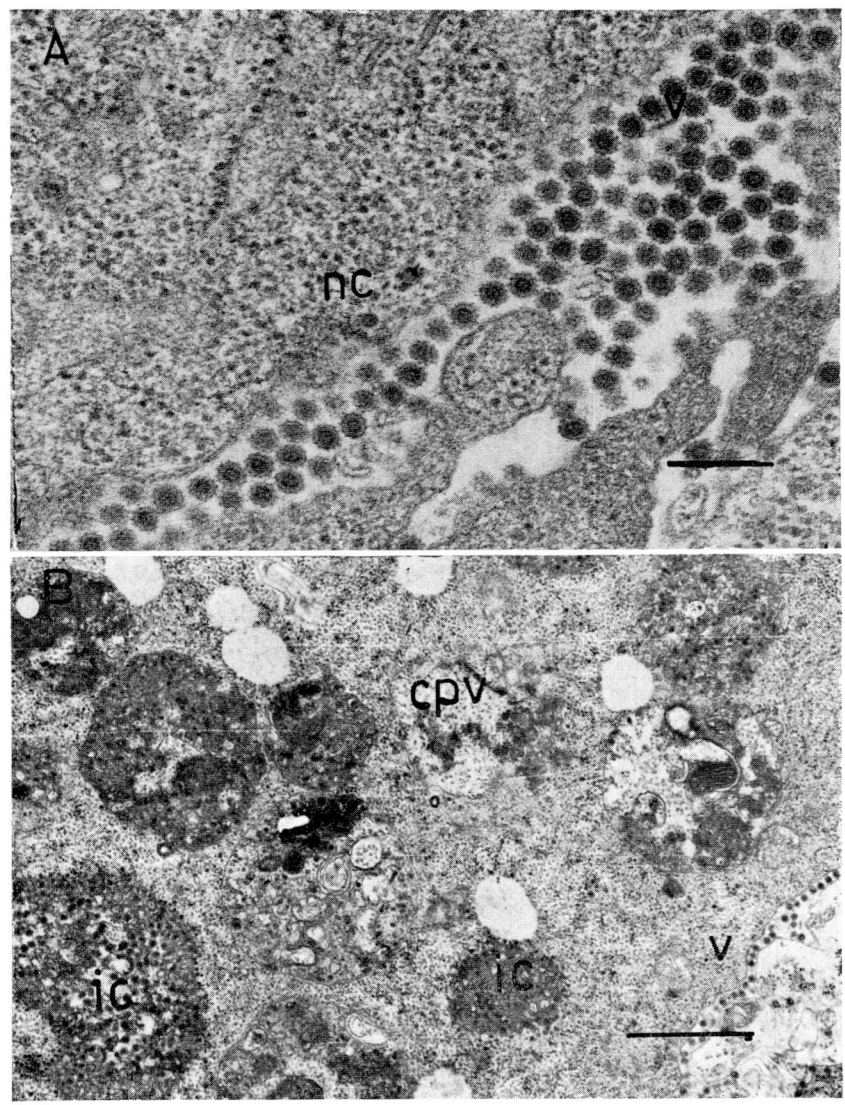

Fig. 14.3.   Cultured *Aëdes albopictus* cells 22 hr (A) and 26 hr (B) after infection with Ross River virus. nc, nucleocapsid; v, enveloped virus; cpv, cytoplasmic vacuole; ic, cytoplasmic inclusion. Bar represents 0.2 μm (A) and 1 μm (B).

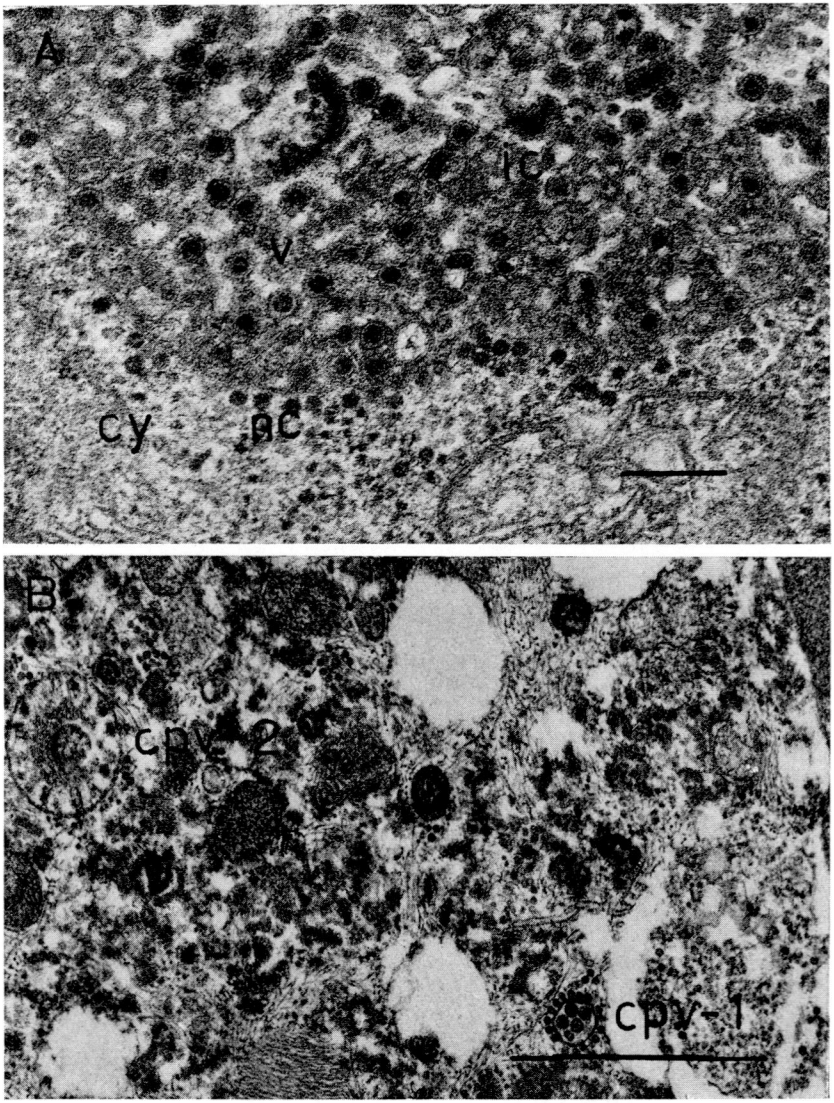

Fig. 14.4. Cultured *Aëdes albopictus* cells 26 hr after infection (A); cy, cytoplasm; v, enveloped virus; nc, nudeocapsid; ic, cytoplasmic inclusion. BHK cells 12 hr after infection with Ross River virus (B); represents structures similar to the 'cytopathic' vacuoles described by Grimley et al. (1968) (see text). Bar represents 0.2 μm (A) and 0.5 μm (B).

ration). As with EEE (Whitfield et al. 1971) nucleocapsids do not accumulate in RRV-infected cells, nor do membranes proliferate much. However, unlike EEE, maturing RRV is observed in large cytoplasmic inclusions; these structures are not seen 6 hr after infection (fig. 14.2A) but are seen at 14 hr (fig. 14.2B). At 26 hr after infection; close to the time of maximum virus titre, the inclusions appear to lose electron-dense material and to develop vacuoles (fig. 14.3B). Virus particles are enveloped both within the cytoplasmic inclusions (fig. 14.4A) and at the cell membrane. Late in infection much enveloped extracellular virus accumulates (fig. 14.3A). Fig. 14.4B is an electron micrograph of RRV-infected BHK cells 12 hr after infection (Raghow, Dennett and Dalgarno, unpublished results); RRV induces morphological changes and cytopathic effects which are similar to those induced by SFV in CEF cells (Grimley et al. 1968).

Although animal cell cultures can carry togaviruses for long periods (Chambers 1957; Lockart 1960; Beasley et al. 1961; Weibenga 1961; see review by Walker 1964), infection of cultured vertebrate cells is generally cytopathic and the cells are destroyed. Most light microscopic (Lam and Marshall 1966), fluorescent antibody (Doi et al. 1967) and electron microscopic studies (Yang et al. 1969; Janzen et al. 1970; Whitfield et al. 1971) support the view that arthropod-borne viruses are less damaging to the invertebrate than to vertebrate cells. As well as the studies mentioned above (Whitfield et al. 1971; Raghow et al. in preparation), it has been shown that Chikungunya virus does not cause cytopathic changes in *A. aegypti* salivary glands (Janzen et al. 1970). Similarly VSV, a virus which is cytolytic to vertebrate cells causes no cytopathic effect when growing in cultured *Antheraea eucalypti* cells (Yang et al. 1969).

However, cytopathic effects may be observed in infected insect organs and cultured cells. Thus SFV causes some cytopathic changes in the salivary glands of *Aëdes aegypti* (Mims et al. 1966). However, *A. aegypti* is not the natural vector of SFV and the effect of SFV on its vector's salivary glands is not known (see Chamberlain 1968). Cultured *A. albopictus* cells became more granular and rounded 4–6 days after infection with JBE, West Nile and dengue viruses (Banerjee and Singh 1968; Singh and Paul 1968; Buckley 1969); but these viruses had no cytopathic effect on Singh's *A. aegypti* cells (Singh 1968; Buckley 1969). Cells infected with dengue and West Nile viruses characteristically form syncytia (Paul et al. 1969; Suitor and Paul 1969); this effect can be prevented by inactivating the virus with antisera and is only observed in cells grown in plastic containers. Suitor (1969) reports that under a 1% agarose overlay, a clone from Singh's *A. albopictus* line will plaque

JBE virus giving titres approximately 1 dex less than the TCD_{50} of the same stock assayed in hamster kidney cells. These results suggest that togavirus-infected insect cells only show cytopathic effects when stressed in a particular way; a similar effect is commonly noted in studies of togavirus-infected vertebrate cells (Porterfield 1959).

14.2 Baculoviruses

Most studies reported to date on the multiplication of the two major groups of insect-pathogenic viruses – the baculoviruses and the cytoplasmic poly-hedrosis viruses – have been confined to examining the process of infection in the whole insect. A description of the biochemical steps in virus replication awaits the development of reliable methods for growing insect-pathogenic viruses in invertebrate cell lines in 'minimal' medium. There have been successes in this direction and following the first report by Trager (1935), a number of observations of polyhedral virus growth in cultured insect cells have been made (Grace 1958, 1962a; Gaw et al. 1959; Martignoni and Scallion 1961; Vago and Bergoin 1963; Vaughn and Faulkner 1963; Mitsuhashi 1967; see review by Grace in this volume, ch. 17). A continuous *Antheraea eucalypti* line (Grace 1962a) has been successfully exploited in detailed studies of iridovirus replication (see section 14.4).

14.2.1 Host range

Over one hundred nuclear polyhedrosis viruses (NPV) from different insect species have been reported; most in Lepidoptera and the rest in Hymenoptera, Diptera and Orthoptera (see reviews by Aizawa 1963; Ignoffo and Hink 1971). Granulosis virus (GV) infections (see review by Huger 1963) are apparently confined to Lepidoptera (Smith 1967).

14.2.2 Sites of replication

Insects are infected with NPV or GV when they ingest contaminated food. The initial site of baculovirus replication is probably the gut (see ch. 15; also Laudeho and Amargier 1963; Heimpel and Adams 1966; Harrap and Robertson 1968; Summers 1969; Tanada and Leutenegger 1970; Cunningham 1971). Subsequent to this primary infection, NPV multiplies in the hypodermis, fat body, blood, tracheal epithelium, muscle, nerve tissues and pericardial cells. GV multiplies in the fat body, hypodermis and tracheal matrix; tissue tropisms may vary with different strains of virus and different hosts. Initially both NPV and GV replicate in the nucleus of the infected cell.

In the primary gut infection, occlusion of virions in a polyhedron or capsule occurs only rarely (Summers 1969; Harrap 1970; Tanada and Leutenegger 1970; Cunningham 1971). Since free virions are likely to initiate infection more readily than the occluded form, this modification of the developmental pathway in the gut probably helps establish infection (Harrap 1970).

14.2.3 Growth kinetics

The growth kinetics of NPV or GV in single cells are not known; this information depends on developing methods for the growth and assay of baculoviruses in continuous tissue culture systems (see 14.2). In the experiments reported below, the durations of the latent periods are maximum values since multiple cycles of infection occur when whole insects are infected. Aizawa (1959) has examined the time-course of infectious virus particle synthesis in silkworm larvae at 25 °C. After inoculation of an NPV suspension into the body cavity, the virus titre of the haemolymph supernatant decreases for several hr; 10 hr later it starts to increase logarithmically reaching a maximum titre at about 50 hr. The titre then remains almost constant at 6–7 dex (LD_{50}) until just before death; similar growth curves are obtained with infected silkworm pupae. The kinetics of appearance of total infective virus (free virus + virus released from polyhedra by alkali treatment) in the larval and pupal body are almost identical (Aizawa and Kawarabata 1963). Summers (1971) has used a micro-injection technique to infect gut cells synchronously; in these experiments GV 'progeny' was observed 24 hr after infection. Other results also suggest that a single cycle of virus growth occurs within 24 hr of infection (Tanada and Leutenegger 1970; Wager and Benz 1971).

The time for NPV infection to kill cabbage-looper larvae is inversely related to temperature and represents a constant proportion of the duration of the larval stage as larval development is also temperature dependent (Canerday and Arant 1968). It is estimated that the LD_{50} of *Heliothis* NPV in *Heliothis* larvae is approximately 400–700 NPV virions per mg body weight (Allen and Ignoffo 1969). There is a linear relationship between temperature and the development of NPV disease in *Prodenia litura* larvae (Okada 1969). By extrapolating from virus growth rates at moderate temperatures, Okada concluded that the minimum temperature permitting NPV growth is 9 °C. NPV replication is decreased when the development of larvae is arrested by ligaturing the head, even though the environmental temperature is maintained at 30 °C (Stairs 1970), indicating that virus development may be influenced more by the metabolic rate of the host cell than by the environmental temperature.

14.2.4 Ultrastructural studies

The initial stages of GV infection in gut cells have been examined by Summers (1969, 1971). Release of the virion from the protein matrix in which it is embedded precedes infection (Martignoni 1967, quoted in Summers 1971; see also Faust and Adams 1966). This appears to occur as the result of a dislocation of the crystalline capsid protein in a position adjacent to the embedded virus rod (Summers 1969; 1971). The released virion, still within its outer membrane, then attaches to membranes of gut columnar cell microvilli. The outer membrane is lost during entry of the virion since only non-enveloped virions are seen inside the cell shortly after infection. These are randomly distributed in the cytoplasm and subsequently may be seen close to the outer nuclear membrane. Summers found no evidence that enveloped virions were phagocytosed. NPV seems to have a similar entry mechanism: Harrap (1970) observed empty balloon-like membranous projections attached to the outer membrane of NPV-infected cells.

It seems that the next stage is the attachment of the GV particle, free of its outer membrane, perpendicular to the nuclear membrane, apparently at a nuclear pore (Summers 1971); this can occur within 2 hr of infection. This process presumably enables viral DNA to enter the nucleus, since empty inner-membranes can be seen attached to the outer nuclear membrane. Since the ends of the NPV rod are morphologically different it is likely that one end specifically attaches to a nuclear pore. Teakle (1969) has observed a structure resembling a nipple at only one end of the NPV rod; a 'claw setting' is present at both ends. In summary, the entry of both polyhedrosis and granulosis viruses to the organelle in which they replicate, the nucleus, appears to be in two stages involving the sequential removal of the outer and inner membranes at the cell- and nuclear membrane respectively.

Several people have studied the ultrastructure of NPV and GV infected cells (Hughes 1952; Smith and Xeros 1953, 1954; Day et al. 1958; Bird 1964; Tanada and Leutenegger 1968, 1970; Kislev et al. 1969; Krieg and Huger 1969; Summers and Arnott 1969; Wager and Benz 1971). These observations are summarised below.

One of the earliest and most characteristic changes in the NPV-infected cell is the enlargement of the nucleus at the expense of other parts of the cell so that it may eventually fill almost the entire cell. Within the nucleus appears an electron-dense, granular mass (the virogenic stroma) which may occupy a large part of the nucleus. In GV infections this material is also observed but apparently in a more dispersed form. The virogenic stroma may be the site of viral DNA synthesis since virus rods appear within it, seemingly at random. Rods longer than the normal virion and also filamentous forms, have been reported; the inner membrane may be present or absent on these structures (Krieg and Huger 1969). In NPV infections a number of rods are often found associated with a membranous structure in the nucleus and aligned in a parallel, ordered array; these arrays are not seen in GV infections. An outer ('developmental') membrane then progressively encloses a single or variable number of virus particles; the mean number differs with the virus strain and host and may range from 1–19 (Smith 1967). Heimpel and Adams (1966) and Tomkins et al. (1969) have shown that when two particular different strains of NPV mixedly infect susceptible larvae only one strain is found in each nucleus (see also Longworth and Cunningham 1968; Mathad et al. 1968). One strain encloses only single rods within the outer membrane; thus a number of embedded, singly-enclosed rods make up the polyhedron. In the other strain 'bundles' of rods (each bundle enclosed by an outer membrane) are embedded in the polyhedron. In GV infections the number of particles enclosed in the outer membrane and subsequently encapsidated by protein is one or occasionally two. During particle maturation double-layered membranes form in the nucleus and cytoplasm; these may be involved in the generation of developmental membranes. Dense fibrillar material also appears in the nucleus; this may be derived from such membranes (Krieg and Huger 1969). Tanada et al. (1969) have described a strain of NPV which causes cellular hypertrophy, an increase in the number and an alteration in the appearance of mitochondria, folding of the nuclear membrane and the formation of electron-transparent areas in the cytoplasm. Cellular proliferation may also be induced by NPV (Petre and Ploaire 1969).

While the virus rods mature, protein gradually accumulates in the nucleus. Enveloped virions are occluded by this material which assumes an ordered crystalline form around the virions and a specific shape characteristic of the virus strain rather than the host; this shape is commonly polyhedral. Virions are occluded in polyhedra in an apparently random fashion and no disturbance to the regularity of the crystalline protein matrix is seen in complete polyhedra (Engstrom and Kilkson 1968). Not all virus particles are occluded even in tissues other than the midgut. The yield of NPV in a number of hosts ranges from $2-430 \times 10^8$ polyhedra per gram of tissue; for GV growing in *Cydia pomonella*, 200×10^8 capsules per gram of host tissue are reported (see Ignoffo and Hink 1971).

14.2.5 Biochemical studies

The intracellular site of synthesis of both virion and polyhedral protein has been examined in NPV-infected *Bombyx mori* larvae using fluorescent antibody prepared against both proteins (Krywienczyk 1963; Krywienczyk and Sohi 1967). Fluorescence to both proteins is visible in the cytoplasm of infected cells 4 days after infection, i.e. before the appearance of nuclear polyhedra. As the polyhedra develop, cytoplasmic fluorescence decreases and that in the nucleus intensifies. Fluorescence disappears from the cytoplasm at a late stage in infection when the nuclei are packed with polyhedra.

Total protein synthesis in the fat body of *B. mori* larvae increases after infection with NPV (Shigematsu and Noguchi 1969). The synthesis of silk proteins in the silk gland of *B. mori* larvae is greatly decreased by infection with NPV; this decrease appears to coincide with the increasing synthesis of polyhedral proteins (Watanabe and Kobayashi 1969).

Carnegie and Beaudreau (1969) have noted a difference in buoyant density between the DNA of an NPV from *Hemerocampa* larvae and that of the host cell (1.710 and 1.695 g/cc respectively). Using this difference to separate viral and host DNA's they followed the change in rate of viral DNA synthesis during the course of infection using 2 hr pulses with P^{32} over a 5-day period. At 30 °C the mean lethal period is 134 hr. Most viral DNA is synthesized between 72 and 96 hr after infection and host DNA synthesis is unaffected. However, after the fifth day of infection host DNA synthesis decreases. Using autoradiography, Morris (1968) showed that the rate of DNA synthesis in NPV-infected *Lambdina* larvae increases up to a time just before polyhedra form; it then decreases and is barely detectable in nuclei containing mature polyhedra. There is an increase in both the rate of DNA synthesis and the activity of DNA polymerase in NPV-infected silkworm pupae at about 20 hr after infection (Onodera et al. 1968); a maximum in both enzyme activity and rate of synthesis of DNA is reached 60 hr after infection. No increase in polymerase activity is found when either actinomycin or chloramphenicol is injected together with virus. It is not clear whether the DNA polymerase is a pre-existing host enzyme(s) because, as the authors point out, uninfected cells contain such small amounts of the enzyme that its properties are difficult to compare with those of the enzyme from infected cells.

14.3 Cytoplasmic polyhedrosis viruses

14.3.1 Host range

CPV infections are mainly restricted to lepidopterous larvae of which a large number of species are susceptible. Approximately 100 species infected with CPV's have been reported, but it is unlikely that all are different since many are readily transmissible to other species. CPV is reported in Diptera, Hymenoptera and Neuroptera (Kellen et al. 1966; Stoltz and Hilsenhoff 1969; Longworth and Spilling 1970; Ignoffo and Hink 1971; see Review by Smith 1963). Virus usually only grows in larval gut cells although Kellen et al. (1966) report replication in the hypodermis and other tissues of *Culex tarsalis*. CPV will grow in cultured cells: Vago and Bergoin (1963) grew CPV in fibroblasts from *Lymantria*. A latent CPV can be activated in a continuous *Antheraea eucalypti* cell line by infecting cultures with *Bombyx mori* NPV (Grace 1962b).

14.3.2 Kinetics of virus growth and polyhedron production

As with baculovirus infections, there is little quantitative information on the kinetics of production of viral components. CPV-specific proteins can be detected in silkworm larvae as early as 6 hr after infection (Kawase and Miyajima 1969; see below). Cytoplasmic polyhedra are first detected in the midgut of silkworm larvae about 15 hr after infection at 25 °C; subsequently polyhedra become larger and more numerous, reaching a maximum about 48 hr after infection (Miyajima and Kawase 1968).

14.3.3 Ultrastructural studies

Electron-microscope studies show that CPV only matures in the cytoplasm of infected cells; the nucleus remains morphologically unchanged in the early stages of infection (Bird 1965, 1966; Arnott and Smith 1968; Stoltz and Hilsenhoff 1969; Longworth and Spilling 1970). Stoltz and Hilsenhoff (1969) have shown that in *Chironomus plumosus* larvae, virus matures in the cytoplasm close to dense, protein-rich areas they termed the 'virogenic stromata' (Xeros 1956). These regions enlarge during the course of polyhedron development and appear to be free of normal cytoplasmic components such as mitochondria, endoplasmic reticulum and ribosomes (Stoltz and Hilsenhoff 1969). Although its exact role is unknown, this may suggest that the 'virogenic stroma' (VS) is the site of accumulation of virus-specific proteins (Stoltz and Hilsenhoff 1969). Virions are distributed throughout this region; they become occluded in a thick layer of crystalline protein giving rise to coated virions. The protein surrounding several adjacent virions fuses to give particles embedded in a homogeneous crystalline matrix; the outlines are then 'rounded off' to give smooth surfaced polyhedra which are small relative to baculovirus polyhedra.

The free CPV virion has 12 projections or spikes; these may protrude through a membrane surrounding the virion and can sometimes be seen in electron micrographs. The projections are probably formed before the virions are occluded in polyhedra; once occluded neither the projections nor the membrane are easily distinguished (Stairs et al. 1968).

In a study of CPV infection of larvae of the monarch butterfly, *Danaus plexippus*, Arnott et al. (1968) distinguish between the virogenic stroma and a complex, reticulated, fibrillar region termed the 'crystallogenic matrix' (CM). The VS is found early and the CM later in infection; the VS is free of maturing virions whereas the CM contains approximately the same density of particles as are found in polyhedra. Thus polyhedra and the CM are the only structures in which virions are found; unlike Stoltz and Hilsenhoff (1969) the authors define the VS by its lack of virus particles. There is no evidence, however, that these two types of structure are not part of a 'developmental continuum' (Arnott et al. 1968). The ultrastructural studies on CPV and NPV development obtained to date emphasize the need for an electron microscope study correlated with a single-step growth curve for these viruses.

Cytoplasmic polyhedra vary in shape and size. Octahedral, tetrahedral and spherical 'polyhedra' have been reported (see, for example, Hukuhara and Hashimoto 1966). This symmetry has been shown in some instances to be a genetic property of the virus strain. The diameter of polyhedra varies greatly and may reach 10 μm (Arnott and Smith 1968; Stairs et al. 1968). Size differences may in some instances be related to different effective periods of infection when different parts of the midgut are examined (Aruga et al. 1963b). Differences in polyhedral symmetry have been used to study interference between CPV's in dual infections (Aruga et al. 1961, 1963a). Not all virions are occluded in polyhedra; in *Malacosoma* larvae 10 days after infection, up to three times as many virions may be free as are occluded (Hayashi 1970). Similar observations have been made in NPV infections (see 14.2.3). The yield of CPV is reported to range from $20-46 \times 10^8$ polyhedra per gram of host tissue (Ignoffo and Hink 1971) with as many as 10,000 virions per polyhedron (Arnott et al. 1968).

14.3.4 Biochemical studies

Each CPV virion contains nine double-stranded RNA segments with a total molecular weight of near 13×10^6 daltons (Kalmakoff et al. 1969; see ch. 4); it does not contain the A-rich RNA found in reovirus. The virion has an associated RNA polymerase activity which can be readily assayed in virions derived from polyhedra by alkali treatment. This enzyme has a temperature optimum of 27 °C and catalyses the actinomycin-resistant synthesis of single-stranded RNA using double-stranded CPV RNA as template (Lewandowski et al. 1969). It is suggested that after entry into the host-cell, the virus-associated polymerase transcribes messenger RNA from the RNA's in the virion. The messenger is then translated to provide the proteins required for the replication of CPV RNA (Lewandowski et al. 1969). Whether the two strands of the viral RNA replicate symmetrically or asymmetrically is not known.

The origin of the polymerase is of interest. Kawase and Miyajima (1968) claim that isolated CPV RNA is infectious; this implies that the host cell contains a polymerase capable of transcribing single-stranded RNA from double-stranded CPV RNA. However, Lewandowski et al. (1969) were unsuccessful in repeated attempts to extract infectious RNA from purified CPV or CPV-infected midguts. This suggests that the host cell does not contain an RNA polymerase able to transcribe the CPV genome, and that the virion-associated polymerase is necessary for the infectivity of CPV RNA.

To determine where CPV-specific proteins were synthesized in the cell, Kawase and Miyajima (1969) prepared fluorescent antisera to both virion and polyhedral proteins; the cytoplasm fluoresced with both antisera in infected *Bombyx mori* larvae; there was no nuclear fluorescence. Viral and polyhedral proteins were first detected 6 and 9 hr after infection respectively.

Kawase and Kawamori (1968) have examined the time course of viral RNA synthesis in the midgut of *B. mori* infected with CPV. Twenty-four hr after infection, the time at which disease symptoms are first noted, isotope is incorporated into viral RNA. Labelling 72 hr after infection yields viral RNA with three times the specific radioactivity of ribosomal RNA. The results of studies on the subcellular localisation of viral RNA synthesis are conflicting. Hayashi and Retnakaran (1970) imply that the RNA is synthesised in both the nucleus and cytoplasm; those of Watanabe (1967) suggest that it is synthesised in the nucleus.

14.4 Iridoviruses

14.4.1 Host range and infectivity

Although in nature *Tipula* iridescent virus (TIV) rarely infects insects other than *Tipula* larvae, Smith et al. (1961) have shown that it can be transmitted to 21 insects from three orders: Coleoptera, Lepidoptera and Diptera. *Sericesthis* iridescent virus (SIV) can be transmitted to several members of the Coleoptera and Lepidoptera although it is normally found in *Sericesthis pruinosa* (Day and Dudzinski 1966).

The lethal doses of SIV and TIV for six different insect species differ greatly; they range from 6 particles to 5×10^7 particles per insect; *Tipula* spp. are more susceptible to TIV than to SIV (Glitz et al. 1968). Day and Mercer (1964) report a particle to infectivity ratio of close to 1 for SIV in *Galleria*, and Bellett (1965a) a ratio of 80:1 for SIV in cultured *Antheraea eucalypti* cells.

14.4.2 Effect of temperature

The growth of iridoviruses shows sharp temperature optima in both whole insects and cultured cells. Optima range from 20–25 °C depending on the host and the experimental conditions (Day and Mercer 1964; Tanada and Tanabe 1965; Day and Dudzinski 1966). *Galleria* larvae inoculated intrahaemocoelically with TIV and reared at 23–25 °C die from virus infections; at 30°C and higher most larvae survive to adulthood. When reared for more than one day at 25 °C and then transferred to 37 °C larvae die from virus infection (Tanada and Tanabe 1965). Optimal growth of SIV in cultured *Antheraea* cells occurs at 20 ± 1 °C; at 25 °C non-infectious virus is formed (Bellett and Mercer 1964; Bellett 1965a). This relatively low temperature optimum is not found for NPV, GV or CPV replication in whole insects.

14.4.3 Virus replication and maturation

The first obvious site of TIV replication in a variety of insects is the fat body; as the disease progresses, other tissues are invaded (Anderson et al. 1959; Smith et al. 1961). The first step in infection is the phagocytic engulfment of the virus particle. Leutenegger (1967) found SIV particles in phagocytic vesicles of *Galleria* haemocytes 2 hr after infection at 22 °C; most virus had apparently been engulfed after 3 hr. TIV particles are seen in phagocytic vesicles 1 hr after infection; after 8 hr virus is present in lysosomes (Young-husband and Lee 1969, 1970). The formation of a stable SIV–*Antheraea eucalypti* cell complex is complete in 10 hr (Bellett 1965b).

Autoradiography of TIV- and SIV-infected insect cells shows that viral DNA is synthesised in cytoplasmic foci or 'viroplasmic centres' (Leutenegger 1964; Bellett 1965b; Leutenegger 1967; Younghusband and Lee 1969, 1970). Clark et al. (1965) conclude that MIV replicates in the cytoplasm of *Aëdes* cells. Fluorescent antibody studies show that coat protein is also synthesized in the cytoplasm (Bellett 1965b). *Antheraea* DNA synthesis is slightly in-hibited by SIV infection (Leutenegger 1967; Bellett 1968).

Each infective unit of SIV can initiate a separate focus of virus synthesis which enlarges and may eventually fill the cytoplasm (Bellett 1965b). Virions at various stages of development are observed in granular masses in the cyto-plasm (Leutenegger 1967; Younghusband and Lee 1969) which correspond to these foci (Leutenegger 1967). In cultured *Antheraea eucalypti* cells in-fected at multiplicities of less than one, more foci are formed than expected, perhaps because of virus reactivation (Bellett 1968); at multiplicities of two or three, extra foci are not found (Bellett 1965b). All foci within a multiply infected cell begin to synthesise virus at about the same time although differ-ent cells start virus replication at different times (Bellett 1965b). This pheno-menon is similar to that found in pox virus-infected cells (Cairns 1960) and may indicate that there is an initial critical event which simultaneously trig-gers synthesis at all foci within a single cell.

Virus replication occurs more rapidly in *Galleria* haemocytes in situ than in cultured *Antheraea* cells; in haemocytes, SIV DNA synthesis is detected 4.5 hr after infection and progeny virus is seen after 12 hr (Leutenegger 1967). SIV-specific antigen and virus-specific DNA both appear in the cytoplasm of cultured *Antheraea* cells 2 days after infection at 20 °C; maximum levels are reached in 4 days (Bellett 1965b). Infectious virus is also detected after 4 days and is continuously produced and released from cells until the eighth day after infection. The yield is about 500 infectious units per cell; about two-thirds of this virus is released. The large paracrystalline iridescent masses of virus characteristic of insect infections are not observed in cultured *Anthe-raea* cells (Bellett 1965b).

Cytopathic effects are reported 5–6 days after infection of *Antheraea* cul-tures with SIV at 21 °C; by this time 90% of the cells are rounded, the cyto-plasm becomes more granular and vacuolated and both the cell and nucleus enlarge (Bellett and Mercer 1964). Between 4 and 7 days after infection of *Pieris* haemocytes with TIV, the haemocytes are greatly enlarged; at 7 days some burst and release their contents (Oliveira and Ponsen 1966). Whether this is the usual mechanism of virus release is not known (see below).

There have been three alternative suggestions of the way iridovirus particles are formed, all based on structures seen in electron micrographs of infected cells. Smith and Hills (1959) proposed that first empty capsids are synthesised and subsequently their cores filled with DNA. Bird (1962) suggests that the core is formed first and then enclosed by the capsid. Xeros (1964) proposed that an incomplete capsid is filled with core before final completion of the capsid occurs. The electron micrographs of Bellett and Mercer (1964) show virus particles apparently in several stages of assembly from almost empty shells through partly filled, to complete virions in dense cytoplasmic areas; no free cores were observed. Younghusband and Lee (1969) show what appeared to be both free TIV cores and empty capsids in areas of virus maturation. Thus the morphogenetic pathway for the iridovirus virion is not yet known.

Iridovirus particles are released when they bud through the cell membrane of both cultured *Antheraea* cells (Bellett and Mercer 1964; Hukuhara and Hashimoto 1967) and *Galleria* haemocytes (Younghusband and Lee 1969). Virus released in this fashion has both capsid and membrane (Bellett and Mercer 1964). Whether this is the normal release mechanism is not clear; budding has not been observed from the fat body, the major site of iridovirus accumulation (Bellett, personal communication).

Acknowledgement

The experiments reported by the authors were assisted with a grant from the Australian Research Grants Committee.

CHAPTER 15

Virus infection in invertebrates

K. A. HARRAP

Contents

15.1 Introduction

An exhaustive account of the modes of infection of invertebrates by viruses
and their resulting pathological effects would constitute a work of immense
size. This contribution therefore does not claim to be a comprehensive
survey, but aims to outline current knowledge of virus infection in invertebrates using examples when appropriate. Later chapters in this book (chs.
18–28) will describe in more detail many of the virus invertebrate interactions
discussed in this chapter.

Viruses associated with invertebrates fall broadly into two categories,
namely those which are pathogenic only to the invertebrate ('true invertebrate viruses') and those which are disseminated by invertebrate vectors and
are pathogenic to other animals and plants ('invertebrate vector viruses'). In
this latter category the virus may not even multiply in the invertebrate vector
and simply be transmitted mechanically from one plant or animal host to
another. In virus–vector relationships of this type however it is not always
easy to discount completely the possibility of some virus replication in the
invertebrate tissues especially as it may be very restricted in extent.

15.2 Invertebrate virus diseases

15.2.1 Types of virus infection

Most virus infections in invertebrates have been described from the class
Insecta though there are accounts of virus infections in a Platyhelminth, an
Annelid and in Arachnids. The most familiar group of virus infections are
the so-called inclusion-body diseases in Lepidoptera, Hymenoptera and
Diptera. The characteristics of these viruses have been reviewed previously
(Smith 1967; Vago and Bergoin 1968). Nuclear polyhedrosis viruses were
probably first observed in the mid-nineteenth century. As the name implies,
the inclusion-bodies are polyhedral in shape, ranging in diameter from 0.5
μm to 15 μm, and are very refractive and easily visible under the light
microscope. The polyhedra are formed in a greatly enlarged cell nucleus and
contain many virus particles. The virus particles contain DNA and are rod-
shaped, being 50–60 nm by 250–300 nm. An envelope surrounds either a
single virus particle or groups of up to twenty or so, depending on the particular virus. The enveloped virus particles are embedded randomly in a
crystalline proteinaceous matrix, organized as a face-centred cubic lattice,

which accounts for the bulk of the polyhedron (fig. 15.1). The polyhedra are extremely effective in preserving the infectivity of the enclosed virus particles in adverse environmental conditions. The polyhedra may be dissolved in weakly alkaline solutions liberating both the virus particles, which

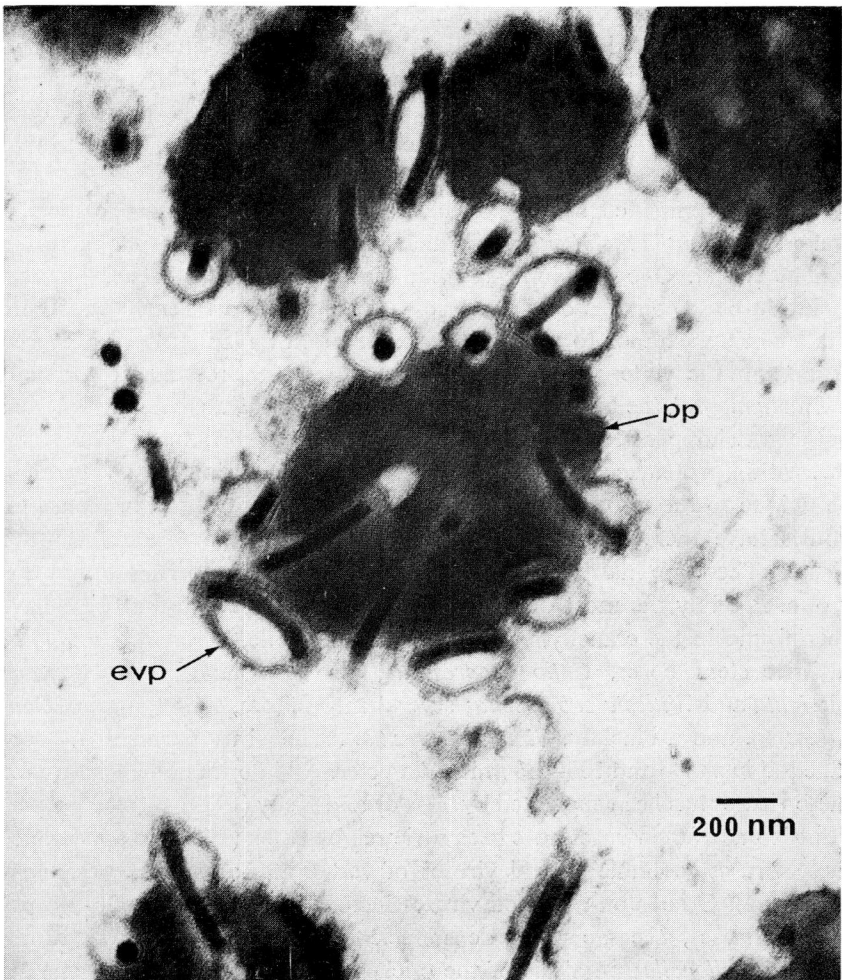

Fig. 15.1. Nuclear polyhedrosis virus developing in the nucleus of a fat body cell of the small tortoiseshell butterfly larvae, *Aglais urticae*. evp: enveloped virus particle; pp: polyhedron protein.

can be purified by conventional methods such as density gradient centrifugation, and a proteinaceous solution from which protein can be precipitated at its isoelectric point, typically about pH 5.8. In another group of inclusion-body viruses, granulosis viruses, the inclusion-body is smaller and capsule-shaped and contains only a single virus particle. Otherwise their properties are similar to those of nuclear polyhedrosis viruses. A third group of inclusion-body viruses is the cytoplasmic polyhedrosis viruses. In this case the polyhedra are formed in the cytoplasm of infected cells. The enclosed virus particles are spherical, 50–70 nm in diameter and contain RNA. Here again the infectivity of the virus particles is well protected by the proteinaceous matrix of the polyhedron which has a cubic lattice structure similar to that of the nuclear polyhedra. A further type of inclusion-body virus, pox-like viruses, is exemplified by the so-called *Vagoiavirus* of the European cockchafer, *Melolontha melolontha*, in which two types of cytoplasmic inclusion can be found. One is spindle-shaped, the other spherical. The spherical inclusion-body contains ovoid or brick-shaped virus particles 400 × 250 nm similar in appearance to vertebrate pox viruses. Similar viruses have been reported in the winter moth, *Operophthera brumata*, the red hairy caterpillar, *Amsacta moorei*, the blacksoil scarab, *Othnonius batesi*, a *Culicoides* sp., and in the grasshopper *Melanoplus sanguinipes*. The incorporation of virus particles within proteinaceous polyhedral or capsular inclusion-bodies seems to be unique to these four groups of viruses. Viruses not found in inclusion-bodies have therefore been given the unfortunate title of non-occluded viruses. There are many viruses with widely differing properties within this group. Possibly the most distinctive are the iridescent viruses. These have been found in the cranefly, *Tipula paludosa*; a beetle, *Sericesthis pruinosa*; the rice stem borer, *Chilo suppressalis*; two grassland pests, *Wiseana cervinata* and *Witlesia sabulosella*; several mosquitoes, *Aëdes* spp.; a *Culicoides* sp. and a blackfly, *Simulium ornatum*. The virus particles are icosahedral in shape and 130–180 nm in diameter. An interim nomenclature for these viruses has been suggested by Tinsley and Kelly (1970) and the properties of some of the viruses have been reviewed by Bellet (1968). In appearance these viruses resemble several viruses of vertebrates and other organisms (see ch. 4). Other non-occluded viruses include acute and chronic bee paralysis viruses, bee sacbrood, densonucleosis of the greater wax moth, *Galleria mellonella*; paralysis of the cricket, *Gryllus bimaculatus*; Malaya disease of the rhinoceros beetle, *Oryctes rhinoceros*; virus diseases of the larvae of several Saturniid moths and of a Lasiocampid from Uganda, *Gonometa podocarpi*; a virus of a grasshopper, *Melanopus bivattatus*; and a virus of the fruit fly, *Drosophila melanogaster*, to mention only a few.

15.2.2 Pathological effects of virus infection

Many of the virus diseases affecting invertebrates kill the host though there is no reason to believe that this is typical. Indeed it is likely that interest was only aroused when the host died and this led to a description of the virus disease. Large numbers of virus diseases which are not fatal must exist in invertebrates.

Insect larvae can be infected experimentally with virus by injection into the haemocoel but it is reasonable to assume that in nature the most usual route of virus infection is through the gut. Symptoms of disease are not apparent for a number of days after ingestion of the virus and behavioural changes are often the first sign of metabolic malfunction. In nuclear poly- hedrosis and granulosis virus infections the larvae may cease feeding and leave the food plant. Smirnoff (1965) describes how such infected larvae of the nun moth, *Lymantria monacha*, characteristically gather at the top of trees, hence the original name 'Wipfelkrankheit virus' (tree top disease virus). Larvae of the Swaine jack-pine sawfly, *Neodiprion swainei*, live in colonies feeding on the previous year's needles and migrate in an orderly fashion to new food. Abnormal movements can be seen in these larvae only 15–18 hr after infection with nuclear polyhedrosis virus. All members of an infected colony simultaneously raise the front part of the body and lower it again. These movements differ from typical defense reflexes. Poplar and willow sawfly larvae normally feed in single rows on the edge of the upper surface of leaves but after nuclear polyhedrosis infection they begin to feed solitarily on either surface. Such larvae often become sluggish, the integument changes in appearance and becomes fragile, and the larvae may be found hanging characteristically in an inverted position. Death occurs soon afterwards and the skin often ruptures liberating masses of polyhedra. In Lepidoptera, fat body, hypodermis, haemocytes, tracheal matrix, muscle, nerve ganglia and pericardial cells can all be susceptible to nuclear polyhedrosis virus infection. In Hymenoptera virus replication is usually confined to the gut tissue. The virus replicates in the cell nucleus which becomes greatly enlarged. Rod- shaped naked virus particles are found within a dense mass of 'viroplasm' in the centre of the nucleus whereas virus particles elsewhere in the nucleus are enveloped. Enveloped virus particles are 'occluded' by crystalline poly- hedron protein, mainly in the region between the nuclear membrane and the 'viroplasm' (fig. 15.2), and later the entire cell contains mature poly-hedra. In granulosis virus infections the fat body cells are always infected with virus and the epidermis and tracheal matrix frequently so; the nuclear membrane disintegrates during the infection and virus is observed through-out the entire cell.

In cytoplasmic polyhedrosis virus infections the larvae usually cease feeding and are greatly reduced in size. Occasionally they change colour during the infection. Virus infects the midgut epithelial cells and the gut becomes fragile. Polyhedra are frequently regurgitated and voided with the faeces. Cytoplasmic polyhedrosis virus infections do not kill as quickly as nuclear polyhedroses and granuloses and some larvae survive, pupate and become adults. The spherical virus particles develop in areas of 'viroplasm' in the cell cytoplasm and protein is deposited around them to form the polyhedra.

Larvae of *Melolontha melolontha* infected with an occluded 'pox-like' virus are white in appearance, the haemocytes and the fat body, which becomes distended, are the susceptible tissues. Henry et al. (1969) described the symp-toms of a similar virus infection in the grasshopper, *Melanoplus sanguinipes*, which resulted in general torpor and prolonged development. Infected nymphs were frequently swollen with inclusion bodies in the enlarged fat body tissues. Adults which developed from less heavily infected nymphs were pallid, and sexually immature.

Iridescent virus infections are easily recognized by the characteristic iri-descent indigo to green colour of the larvae due to the presence of masses of virus particles, in quasi-crystalline arrays, in the cytoplasm of fat body cells. The skin, muscles and wing buds may also be infected. Matta and Lowe (1970) found that only the fat body and imaginal discs of the mosquito, *Aëdes taeniorrhyncus* contained iridescent virus particles. The larvae could be infected by contaminating the water with virus and noticeable symptoms were first seen in third instar larvae infected in the first instar. The pink-white appearance which appeared first in the lateral portion of one of the first four abdominal segments gradually spread to the entire larva. Behaviour was normal until just before death.

The symptoms of 'bee paralysis' in the honey bee *(Apis mellifera)* are a trembling of the wings and legs and an inability to fly. Hairlessness, an oily appearance, dysentery, and an unpleasant smell are also recorded symptoms though Bailey (1965) considers that these may be due to secondary factors. Bailey et al. (1963) isolated two viruses from this condition. Acute bee paralysis virus (ABPV) kills bees quickly after severe symptoms but is found

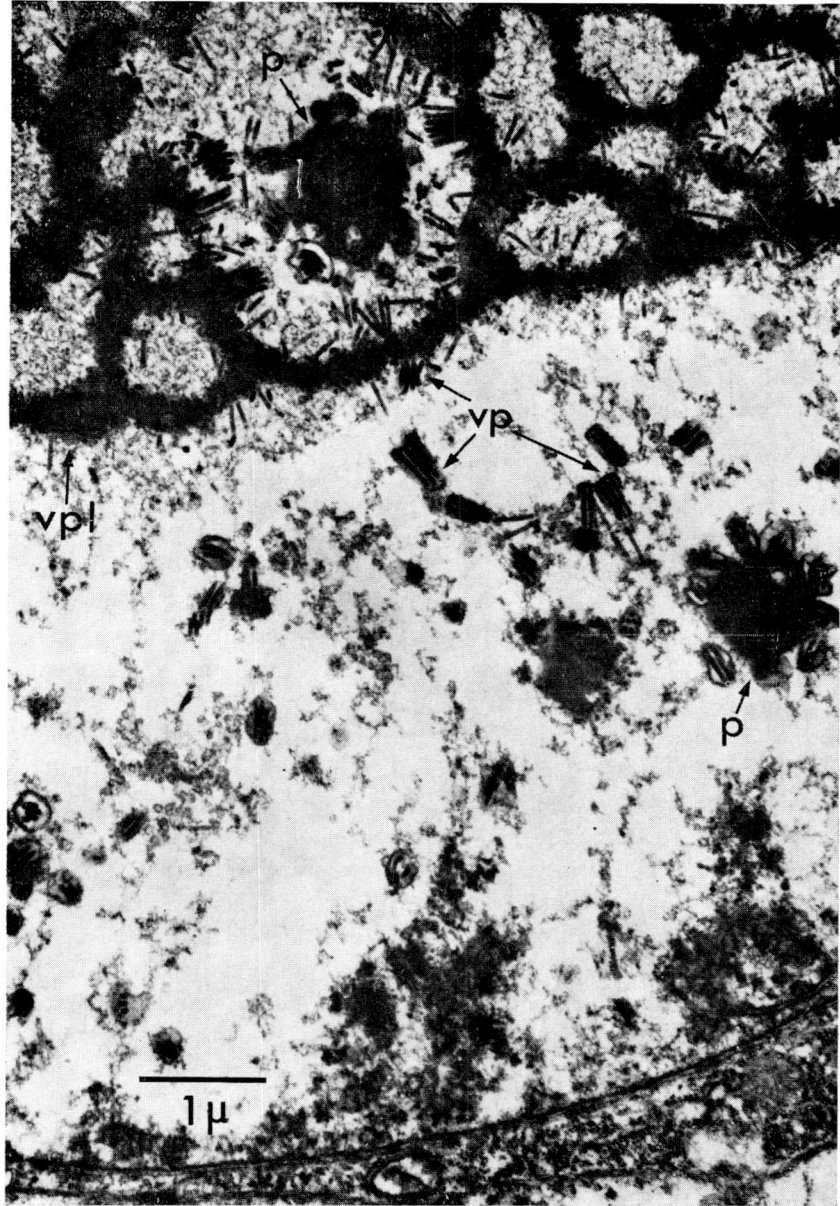

Fig. 15.2.   Nuclear polyhedrosis virus in the nucleus of a fat body cell of the gipsy moth larvae *Porthetria dispar*. vpl: 'viroplasm'; vp: naked and enveloped virus particles; p: developing polyhedron.

in both paralysed and apparently healthy bees, whereas chronic bee paralysis virus (CBPV) seems to be the cause of outbreaks of 'bee paralysis'. Another virus of bees, sacbrood virus, attacks only the larvae which become smooth, glistening, pallid and distended. The virus infects fat body, muscle and tracheal end cells (Lee and Furgala 1967). Another virus which causes spastic paralysis affects European red mites; affected individuals tremble and are unable to right themselves when overturned. The mites are a darker red than normal and have blackened areas dorsally.

Other non-occluded virus diseases are associated with a wide range of different pathological conditions. For example, a spherical virus, 50–60 nm in diameter, infecting muscle, heart and glandular tissue causes paralysis in the crab, *Macropipus depurator*. Paralysis is also caused in a virus infection of the cricket, *Gryllus bimaculatus*, in the field crickets, *Teleogryllus oceanicus* and *T. commodus*, and by densonucleosis virus infection in *Galleria mellonella*. Frequent diarrhoea and swelling of the abdomen are found in the 'Malaya disease' of *Oryctes rhinoceros* (Huger 1966). The fat body disintegrates and the larvae become translucent with the exception of chalky-white bodies under the integument. Spherical particles about 164 nm in diameter and rod-shaped particles of 200×70 nm were found in infected fat body tissue. Krieg and Huger (1960) have studied a disease producing similar symptoms in *Melolontha melolontha* previously known as 'Wassersucht', a name derived from the watery transparent appearance of the diseased larvae due to fat body degeneration. The larvae show irregularity in locomotion and extreme shrinkage due to dehydration. This virus has isometric RNA-containing particles between 60 and 75 nm in diameter, and replicates in the cytoplasm of the fat body and certain haemocytes. Diarrhoea was also found associated with a virus infection of four saturniid species by Longworth and Harrap (1968). Moribund larvae were normal in colour, though flaccid, and there was a considerable loss in body weight before death. The isometric DNA-containing virus particles were detected in the cytoplasm of the mid-gut epithelium cells. The sigma virus of *Drosophila melanogaster* apparently causes only one symptom, inability to recover from a brief exposure to carbon dioxide, whereas uninfected flies can be kept for long periods in a pure carbon dioxide atmosphere.

Several studies have been made on metabolic changes associated with virus infection. Morris (1968a, 1968b) found that DNA synthesis increased in nuclear polyhedrosis infection until polyhedra formed and then it decreased and eventually ceased in those nuclei containing mature polyhedra. In nuclear polyhedrosis virus infections of Lepidoptera the amount of DNA and both nuclear and cytoplasmic RNA increased initially and then decreased, whereas in cytoplasmic polyhedrosis virus infections the amount of DNA was scarcely affected while cytoplasmic RNA decreased and nuclear RNA remained constant. In Hymenoptera, nuclear polyhedrosis virus infections in the mid-gut resulted in increased amounts of DNA but the amount of RNA decreased initially and increased again later. In virus infections of this type large amounts of lysine, arginine and histidine-rich material accumulated in the nuclei in the early stages of infection. Martignoni and Milstead (1964) have reported a decrease in total solids and total protein in the cutworm *Peridroma saucia* infected with both nuclear polyhedrosis and granulosis viruses. In cytoplasmic polyhedrosis virus infection, however, Watanabe (1970) found that the daily change of soluble protein content in infected mid-gut was similar to that in healthy larvae up to a late stage in infection. The pattern and concentration of proteins (excluding polyhedron protein) on agarose gel electrophoresis was similar in both healthy and diseased mid-guts, though the concentration of mid-gut proteins tended to decrease in diseased larvae. Incorporation of ^{14}C-glycine into all protein fractions was greater in diseased mid-guts throughout most of the virus infection period. It is suggested that an active synthesis of the mid-gut proteins as well as polyhedron protein is induced by infection with cytoplasmic polyhedrosis virus and continues to a late stage in infection. Other metabolic changes occurring during virus infection of insects have been summarized by Aizawa (1963).

Some virus infections of insects cause abnormal cell proliferations. Diseased epidermis of the fall webworm, *Hyphantria cunea*, infected with nuclear polyhedrosis virus was observed by Watanabe (1968). A 'multilayered structure' resulted from a cell proliferation which, after polyhedral formation in the cells, detached from the hypodermis and disintegrated in the body cavity. Tumours have also been recognized in nuclear polyhedrosis virus infections of the European spruce sawfly, *Gilpinia hercyniae*, in cytoplasmic polyhedrosis virus infections and in *Tipula* iridescent virus infections of the silkworm *Bombyx mori*. The extent or significance of such insect 'tumours' is not yet clear.

It is obvious that the pathological effects of virus infection in invertebrates are very diverse and descriptions are incomplete. Few diagnostically valuable symptoms can be recognised in invertebrate virus infections in contrast to the situation in virus diseases of higher animals. At present the symptoms of a virus disease in invertebrates give few clues of either the nature of the virus or its location within the host animal.

15.2.3 Infection of the host

In nature, virus is likely to be acquired either per os or trans-ovarially. Jaques (1962) considered that the most usual source of infection of the occluded insect viruses was the contamination of foliage with polyhedra from insect cadavers. Non-occluded viruses are probably transmitted in a similar way. However, other factors undoubtedly are involved in the dissemination of virus and the ecology of the different viruses will be discussed in chs. 18–28.

Transovarial transmission has been reported for several insect viruses. It is said that such transmission occurs both as a result of contamination of the surface of the egg and by the virus actually being contained within the egg (Smith 1967). Most examples of transovarial transmission which have been described involve the occluded insect viruses. Small concentrations of nuclear polyhedrosis virus do not always kill larvae and larvae infected late can develop into normal fertile adults. However, larvae from eggs laid by such adults succumb to virus infection after emerging. Nuclear polyhedrosis virus has been said to be transmitted as a surface contaminant of the egg though this view is not universally accepted. Transovarial transmission of cytoplasmic polyhedrosis virus infections in an occult state in the egg has also been claimed on the basis of experiments employing two virus strains which produce polyhedra of different shape. There is no evidence that the iridescent viruses from *Tipula paludosa* and *Sericesthis pruinosa* are transmitted through the egg. Many other instances of transovarial transmission have been described but at the present time there is little evidence that the phenomenon is anything more than that the larvae emerging from the egg are infected by virus acquired from the surface of the egg.

No one knows what determines the susceptibility of insects to virus infections. It is clear however that larvae of earlier instars are more susceptible to virus infection than those of later ones. This has been demonstrated by Morris (1962) in nuclear polyhedrosis virus infections of the Oak looper and by Stairs (1965b) in nuclear polyhedrosis virus infections of *Malacosoma disstria*, where some fourth instar larvae survived 2×10^9 times more virus than would kill first instar larvae. Allen and Ignoffo (1969) found that the LD_{50} for 8-day larvae was three times that for 3-day larvae. However, this represented less than twice the difference when the dose was calculated in terms of the number of polyhedra per milligram of body weight. They estimated that 370–690 virions were required per milligram of body weight for an LD_{50}.

Environmental factors can also affect the susceptibility of insects to virus infection. For example, conditions of stress such as overcrowding can result in increased expression of virus disease. Usually the larvae of Lepidoptera and Hymenoptera are more susceptible to infection by nuclear polyhedrosis virus than other stages. However Vail and Hall (1969) have shown that pupae of the cabbage looper, *Trichoplusia ni*, can be infected by injection, though adults, often deformed, emerge unless large infective doses are used. The amount of larval mortality expected from a specific virus dosage was predictable whereas the combined mortality of larvae, pupae and adults was not.

There have been several reports of simultaneous infection of an insect host by nuclear polyhedrosis and granulosis viruses and by nuclear polyhedrosis and cytoplasmic polyhedrosis viruses. The effect of mixed infections of granulosis and nuclear polyhedrosis virus infections has been claimed to be synergistic by some workers and antagonistic by others. However when Tanada and Hukuhara (1968) examined granulosis virus infection in the armyworm they found one strain which acted synergistically with nuclear polyhedrosis virus infection and one which did not. In polyhedrosis and granulosis infections in the spruce budworm primary infection with one virus interfered with secondary infection with the other virus and Bird (1959) suggested that the susceptible cell would accept either virus separately but not both together. Adjacent cells could be infected with different viruses but usually there were blocks of cells with the same virus. The confusing situation concerning double infections in insect hosts underlines the difficulties experienced when working with whole animals rather than cultured cells. Convincing evidence of virus interference or synergism must therefore await experiments with virus susceptible insect cell lines.

Insect viruses were initially considered to be very host-specific but it is now known that cross infection is common. Indeed, most insect viruses have not been properly characterized using the data from biophysical criteria normally used for other viruses. The practice of providing names according to the insect host from which the virus was first isolated is clearly a recipe for confusion; the Invertebrate Virus Subcommittee of the International Committee on Nomenclature of Viruses has the task of suggesting an alternative system. It is possible therefore that many named insect viruses are not simply cross-infective but are related strains or even identical. Nevertheless cross-infection by occluded insect viruses has been claimed by several workers. Aizawa (1962b) showed that the nuclear polyhedrosis virus of *Bombyx mori* could lethally infect the larvae of *Galleria mellonella*. Perhaps more significantly the nuclear polyhedrosis virus of *Porthetria dispar*, a lepidopteran, infects *Hemerobius* sp., a neuropteran (Sidor 1960). In cytoplasmic polyhedroses, cross-infection has been described by Tanada and Chang (1962) though there is increasing evidence of the similarity of several cytoplasmic polyhedrosis viruses (Cunningham and Longworth 1968). Many other examples of cross-infection of insect viruses were reviewed by Smith (1967).

Studies of cross-infection are also hampered by the problem of 'latency' of viruses in insect populations. Insects experimentally exposed to infection with one virus may die of a different virus infection (Longworth and Cunningham 1968). It is tempting to believe however that all cases of 'latency' of insect viruses can be explained by inapparent infections of virus becoming overt infections. In such a situation data from cross-infection and virus specificity studies must remain equivocal until the virus has been properly identified. This point has been emphasized previously by Ignoffo (1968). The problems associated with 'latency' of insect virus infections have also been reviewed by Smith (1967).

It is particularly unfortunate that a majority of insect virus workers are still bound to the idea that virus infection in invertebrates can only be measured by mortality. Virus diseases are frequently both recognized and quantified on the basis of the death of the host. Little, if any, regard has been paid to sub-lethal virus infections. Resistance to virus infection has been reported in insects, however. For example in granulosis virus infections of *Pieris brassicae*, in cytoplasmic polyhedrosis virus infection of *Bombyx mori* and in American foulbrood in honey bees. These reports raise the question as to what factors limiting virus replication may be at work in an invertebrate virus infection. Could a form of immune response or inhibition of virus infection occur in insects? This subject is discussed in ch. 16.

The gut constitutes the major route for virus invasion in the insect larva and one might expect some protective system at this portal of entry. Indeed there have been reports that gut juice has virus-inhibitory properties (Aizawa 1962a). Many viruses in the arthropods are known to multiply in the midgut epithelial cells, those which apparently do not however must traverse the gut wall in some way if the gut is the common portal of entry. This fact was noted by Matta and Lowe (1970) during histological examination of iridescent virus infection in mosquitoes: 'It is interesting to note that virus must pass through the gut wall but does not infect it or cause any abnormality. This 'eclipse phase' is commonly found in the majority of insect viruses and has not yet been satisfactorily explained'. It is unfortunate that virus-cell interactions have not been closely investigated at the portal of entry, a fact noted by Heimpel and Harshbarger (1965): 'Researchers have initiated most virus studies at the nuclear level, ignoring the presence of resistance encountered by the virus from the time it is ingested until it arrives inside the cell'. Evidence of a viral invasion sequence from the portal of entry is scanty but recent work has given some indication of what may occur at the cellular level.

15.2.4 *Infection of cells and tissues*

It has already been noted that the most usual route of entry for a virus in an insect host is per os. If cells of the alimentary tract of the insect are susceptible to a given virus infection the virus must attach to gut epithelial cells or be actively taken in by cells, presumably by pinocytosis. There have however been virtually no attempts to determine how viruses infect midgut epithelial cells. Stoltz and Hilsenhoff (1969) observed cytoplasmic polyhedrosis virus particles in the microvilli of the midgut cells of the dipteran *Chironomus plumosus* (fig. 15.3) but they considered that this might represent virus liberation. More surprising however, is the acceptance by many workers of the idea that in the Lepidoptera for example, nuclear polyhedrosis viruses infect per os but do not replicate in the gut tissue. This concept undoubtedly stems from that era of histological examination when the use of electron microscopic techniques was the exception rather than the rule. Even so, it is tempting to agree with Bellet (1968) when he considered that 'insect virology is an attitude of mind' if the polyhedron has so influenced thinking in this field that its absence from a cell is considered to signify the absence of virus replication. Some workers have undoubtedly been puzzled by the apparent 'immunity' of the midgut epithelium to infection by certain groups of viruses. Consequently they devised elaborate theories of infection mechanisms based on a maximum of fantasy and a minimum of experimentation. Frequent mention has been made to infective virus 'units' derived from the virus particle which pass through or penetrate the gut wall and are absorbed by susceptible cells. Infectious components found in supernatants of virus preparations after centrifugation at high *g* forces have also been considered significant in this respect.

It is clear that inclusion-bodies break down in the gut. Estes and Faust (1966) consider the polyhedra to have a framework of silicates. Alkaline chloride solutions can solubilize silicates and such solutions are used very effectively for releasing virus particles from polyhedra in vitro. Faust and Adams (1966) suggest that gut juice solubilizes polyhedra primarily by a pH effect together with enzymic degradation. Virus particles are thereby released in the gut lumen and are exposed to its adverse effects. This means that the cell infection must occur rapidly to be successful. The area of infection is likely to be restricted since the infectivity of the virus particles may last only for that limited period of time between removal of the polyhedron protein and inactivation of the virus particles by exposure to adverse environmental conditions in the gut lumen.

Early studies suggested that nuclear polyhedrosis virus did not replicate in the gut tissue of Lepidoptera. Benz (1963), who worked with the nuclear polyhedrosis virus of *Malacosoma disstria*, found that the virus first replicated in the fat body followed in order by hypodermis, tracheal matrix, muscular sheath, nerve, muscles, ganglia and pericardial cells. By contrast Harrap and Robertson (1968) found that nuclear polyhedrosis virus first infected

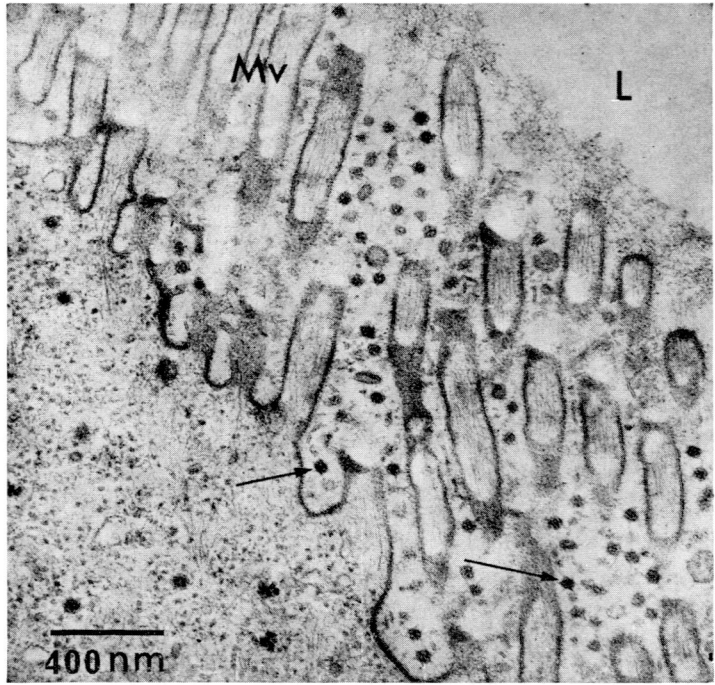

Fig. 15.3. Virus particles of the cytoplasmic polyhedrosis of *Chironomus plumosus* associated with microvilli in the gut lumen. Mv: microvilli; L: lumen of gut. Arrows indicate virus particles with six projections of spikes. Bar = 400 nm. (Stoltz and Hilsenhoff 1969.)

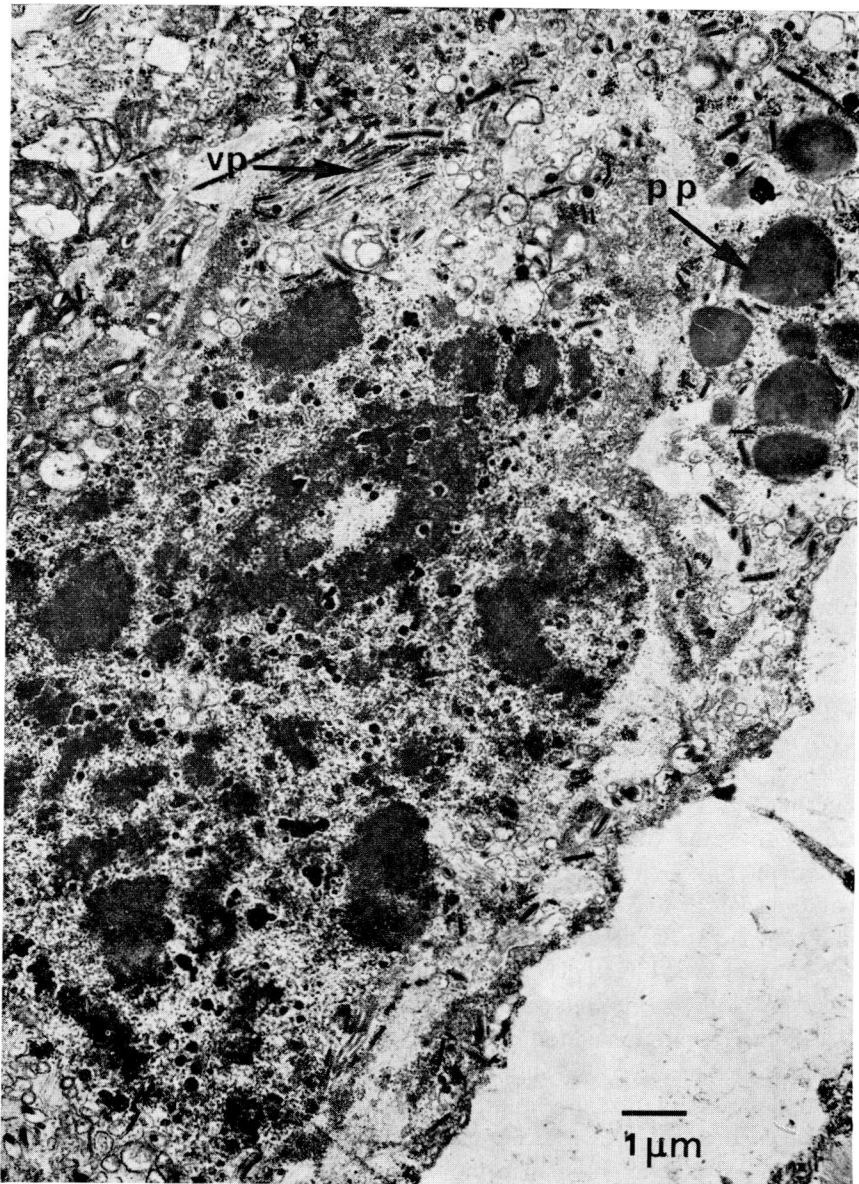

Fig. 15.4.   Nuclear polyhedrosis virus in a midgut columnar cell of *Aglais urticae*. pp:
polyhedron protein; vp: virus particles.

midgut tissue. Larvae of the small tortoiseshell butterfly, *Aglais urticae*, were killed and dissected at intervals after infection with nuclear polyhedrosis virus. Virus infection was detected in the nuclei of columnar cells of the midgut before any evidence of virus infection was observed in other tissues by either light or electron microscopy. Only at the later stages of infection was virus replication detected in cells of the tracheal epithelium and then in the fat body. Virus replication in the gut columnar cells is of an unusual type however; it would not be seen by light microscopy and as the area of infected cells is small it is not always easy to find. Both naked and enveloped virus particles were observed in infected nuclei together with small crystalline masses of what was probably polyhedron protein as it possessed the characteristic lattice dimensions. However, the virus particles were not occluded by the polyhedron protein in these nuclei and typical polyhedra did not form (fig. 15.4). The appearance of the cell was completely different therefore from that of cells of, for example, the tracheal matrix, fat body and hypodermis. The columnar cell was not distended and an obvious network of 'viroplasm' was absent. The area of cytoplasm between the nucleus and the basal lamina contained many enveloped virus particles which were possibly moving towards the basal lamina as groups of virus particles were found adjacent to the basal cell membrane. Virus particles were also found outside the columnar cell within the basal lamina and close to tracheal epithelium cells which were in an early stage of infection (fig. 15.5) (Harrap 1970). Further evidence that the gut columnar cells were the initial replication site for nuclear polyhedrosis viruses was that virus particles were observed in the microvilli at the apex of the columnar cells (Harrap 1969, 1970). Virus particles within the microvilli were unenveloped whereas those not in the microvilli possessed an envelope (fig. 15.6). The virus envelope possibly could attach to and fuse with the plasma membrane of a microvillus thus allowing the nucleocapsid to enter the microvillus. This stage of the process was not observed directly however though structures which resemble residual virus envelopes were seen attached to microvilli. It seems likely therefore that some essential function can be assigned to the virus envelope in the initial stages of infection. Summers (1969) using similar methods observed an analogous situation in granulosis virus infections of the cabbage looper, *Trichoplusia ni*, in which he observed 'a disruption or fusion' of the virus envelope with the microvillar membrane. He considered that the virus envelope was lost at the cell surface because virus particles within the microvilli were unenveloped. Summers also observed naked virus particles attached to the nuclear membrane with their long axis perpendicular to nuclear pores.

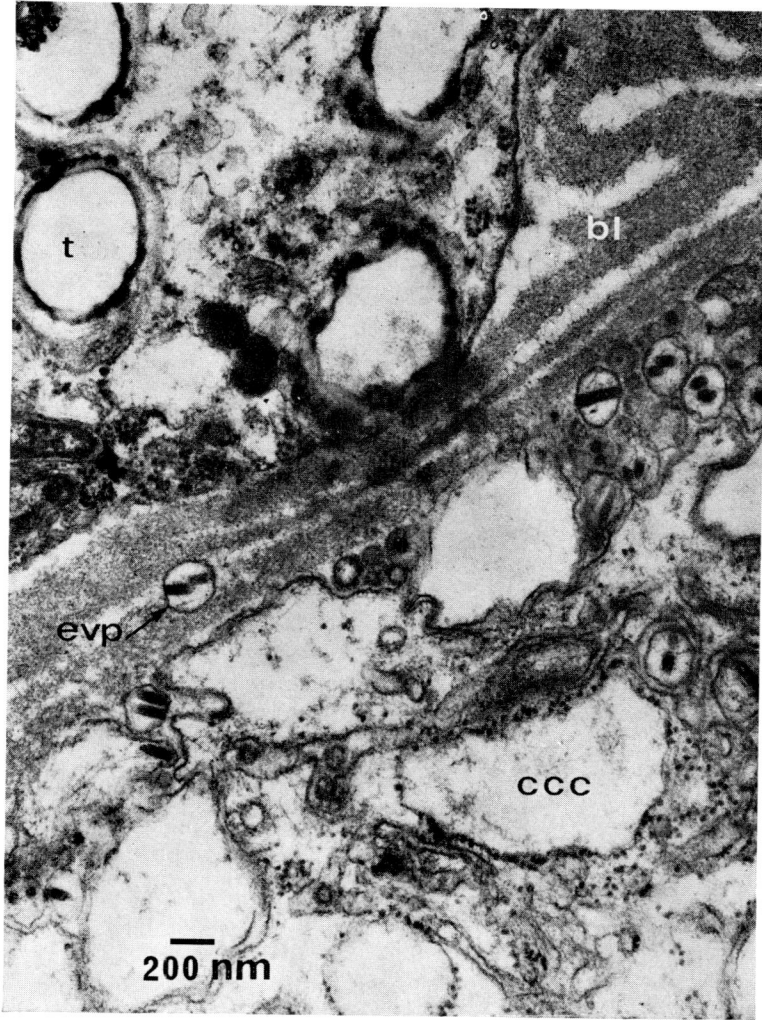

Fig. 15.5.   Nuclear polyhedrosis virus particles in the cytoplasm at the base of an infected columnar cell of *Aglais urticae* and in the basal lamina close to a nearby tracheal epithelium. cell. evp: enveloped virus particle; ccc: columnar cell cytoplasm; bl: basal lamina; t: tracheole.

288 K. A. Harrap

Structures similar to empty inner membranes, probably the virus capsid, were similarly attached to nuclear pores (fig. 15.7). This suggests that the virus nucleic acid is released into the nucleus without entry of the complete virus particle.

Clearly it is reasonable to regard these events as representing the infection

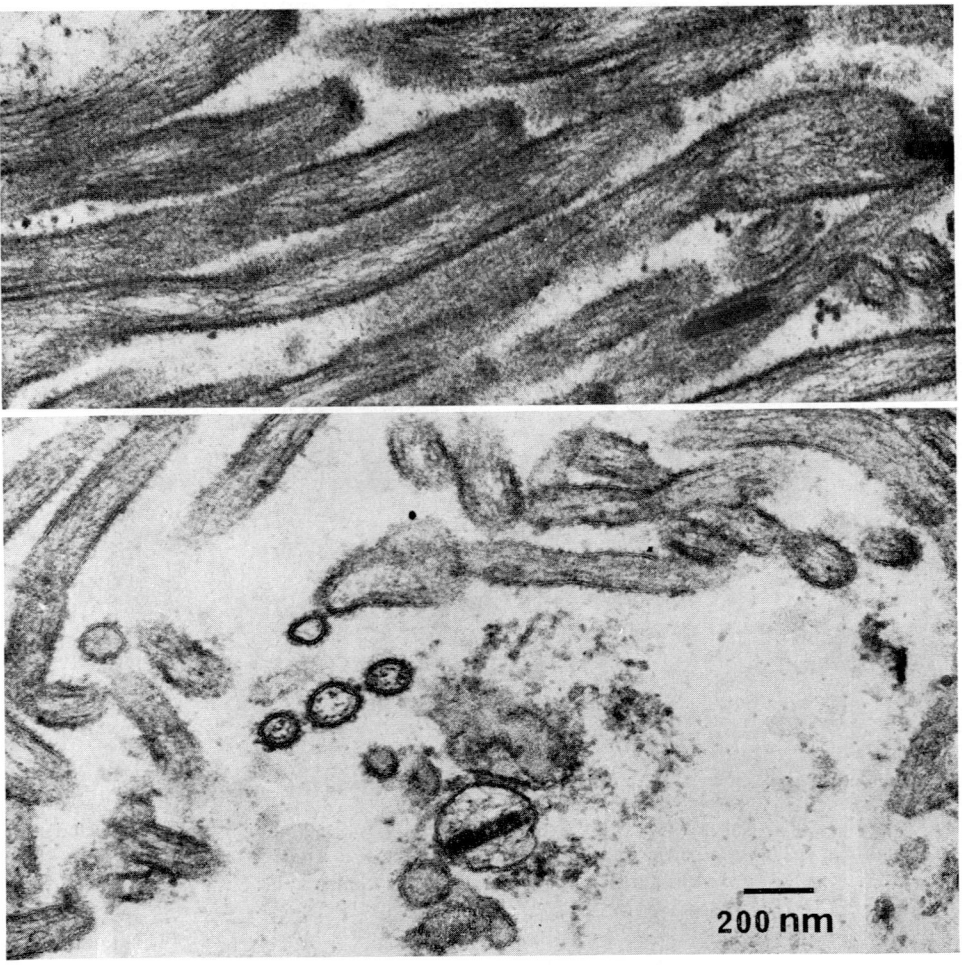

200 nm

Fig. 15.6. Nuclear polyhedrosis virus particles associated with the microvilli of *Aglais urticae* gut columnar cells. Virus particles outside the microvilli are enveloped whereas those within are not.

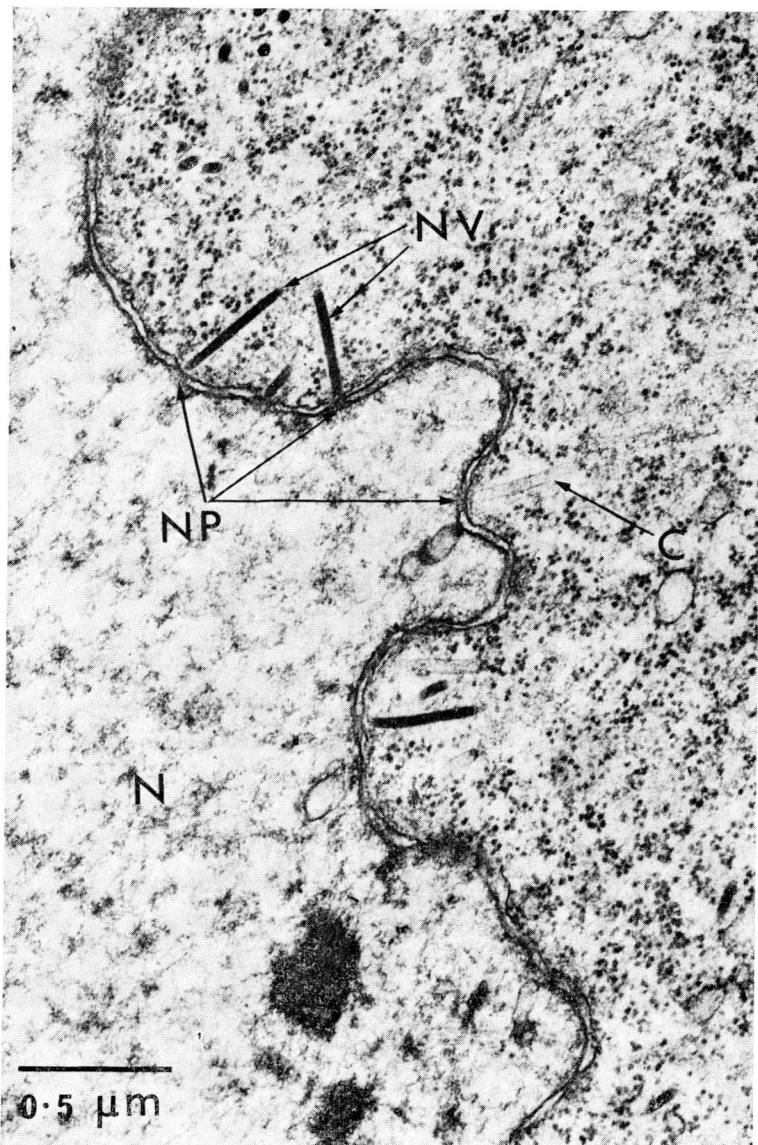

Fig. 15.7. Intact particles of *Trichoplusia ni* granulosis virus at the nuclear pores of a midgut epithelial cell. N: nucleus; NV: naked virus particles; NP: nuclear pores; C: virus capsid. (Summers, 1971.)

pathway of the virus as virus particles were not found in other tissues of the insect larva. Virus particles which do not infect through columnar cell microvilli are probably destroyed in the gut lumen. It appears that only a few gut cells are infected perhaps because of some host defense mechanism. The most unusual feature of nuclear polyhedrosis virus infection in the gut columnar cells however is the absence of occlusion of virus particles into polyhedra. This means that every virus particle can infect another cell whereas in cells in which polyhedra form the majority of the virus particles are removed from further involvement in the spread of virus in the host and are preserved for future infection of other individuals. The absence of such a process in susceptible cells at the site of entry of the virus is clearly advantageous for the initial spread of the virus in the host. Infected columnar cells can therefore be regarded as primary centres responsible for furthering the early stages of infection, and their number, and the efficiency of virus replication within them, may be vital to the progress of the disease. Whether or not infected columnar cells fulfil such a role during the entire course of the disease is not known. Progress of the disease within the host may thus depend upon those virus particles not occluded within polyhedra. The way in which virus is released from infected cells is unknown; it may simply result from cell breakdown. The function of the envelope around the virus particle is not known, but it is obviously essential as enveloped virus particles are found in both the cytoplasm of columnar cells and the basal lamina and enveloped virus particles can also be found in the cytoplasm of fat body

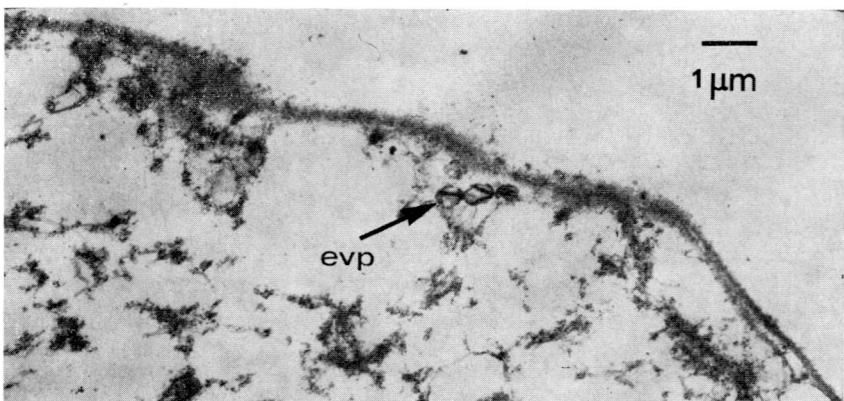

Fig. 15.8.   Nuclear polyhedrosis enveloped virus particles (evp) in the cytoplasm of a *Porthetria dispar* fat body cell.

cells (fig. 15.8). Perhaps the enveloped virus particle should be thought of as the complete infective particle, the virion.

There is little information about infection of cells and tissues by other invertebrate viruses.

15.3 *Invertebrates as vectors of virus diseases*

Association between viruses and invertebrates is not limited to pathogenic disorders produced only in the invertebrate. It is probably true that, as far as man is concerned, those viruses of plants and higher animals carried by invertebrate vectors could justifiably claim more attention than those which are pathogenic for invertebrates alone. Plants are relatively static and are largely dependent on vectors for virus spread, the undamaged plant is virtually immune to virus infection and some form of cell injury such as is caused by an insect vector is a prerequisite of virus infection. It is tempting to suggest that some plant viruses have evolved from viruses of the invertebrate vector.

Many invertebrates have a similarly important role in the spread of virus diseases of higher animals. The health risk from invertebrate-borne viruses in tropical countries is very considerable and the control of blood-sucking insects has been a major concern in several parts of the world for many years. The relationship between the animal virus and the invertebrate vector, like that of a plant virus, can range from one purely of contamination to a subtle biological relationship. In many cases the relationship is little or only imperfectly understood though such knowledge could profoundly affect our attempts to control virus diseases disseminated in this way.

15.3.1 *Virus diseases of plants*

Plant viruses can be transmitted by aphids, leafhoppers, mites, whiteflies, mealybugs, thrips, treehoppers, grasshoppers, beetles and nematodes (see later chapters). In theory of course any type of organism feeding on, or parasitizing, an infected plant could act as a vector and transmit a virus to another healthy plant but not all do. The reason for this may often lie in the inoculation process. For example, the aphid, *Myzus persicae*, can acquire tobacco mosaic virus by feeding through a membrane into a purified suspension of the virus and can, by probing, transmit infective virus to a buffer solution. However it cannot transmit the virus to plants (Pirone 1967). Many plant viruses, such as those transmitted by mites and some of those transmitted by aphids, do not persist in the vector and there is no evidence that they multiply in the vector. However the length of time that the vector remains infective is no guide as to whether or not the virus is replicating in the vector and the viruses transmitted by, for example, beetles, mealy bugs and nematodes, may persist in the vector for long periods apparently without multiplying. These viruses are outside the province of this chapter and are described in other chapters. More difficult to decide is the position of those viruses which persist for some time in aphid vectors. Kennedy et al. (1962) suggested that the former classification of persistent and non-persistent viruses in aphids was no longer tenable as increasing numbers of intermediates between the two extremes were being reported. Instead a classification system based on the route of virus transport was proposed and so far three routes are claimed, two of these have been called 'stylet-borne' and 'circulative', and the circulative viruses which multiply in the vector tissues, are termed 'propagative'.

The relationship between the aphid and those viruses which persist in the aphid, 'circulative' viruses, is incompletely understood. In most the aphid is not infective immediately after feeding on the infected plant, presumably because the virus must pass into the insect gut, through the gut into the blood, and thence to the salivary glands before it can be transmitted to another plant. However in some, such as pea enation mosaic virus, the aphid can transmit immediately. In the plant circulative viruses are found mainly in phloem tissue and are not usually mechanically transmissible. A good test of whether a virus is in fact circulative appears to be its retention after moulting. Nault et al. (1964) showed that pea enation mosaic virus was retained by the aphid vector after moulting and as the interior linings of the fore-gut, and hind-gut, the stylets and the skin of the insect were lost on moulting it was inferred that the virus does not persist in these particular tissues. Evidence for the multiplication of pea enation mosaic virus in the aphid vector is conflicting, however and a similar situation exists with potato leafroll virus. It is difficult to prove whether or not a circulative virus multiplies in the vector tissues that is, whether it should be called a 'propagative' virus. Nevertheless in the case of barley yellow dwarf virus, Paliwal and Sinha (1970) conclude that although virus can be recovered from the gut, haemolymph and salivary glands of the aphid, *Macrosiphum avenae*, no virus multiplication occurs.

Another virus which persists in its vector is tomato spotted wilt virus which is transmitted by *Thrips tabaci*. The virus is acquired by the larvae. There is a distinct latent period before the virus can be transmitted by the vector and an increase in percentage infection is recorded as the feeding period is increased. The virus is often retained throughout the life of the vector but transmission can be erratic and as yet there is no evidence of multiplication of the virus in its vector.

Viruses transmitted by whiteflies are usually retained in the vector for long periods. For example *Euphorbia* mosaic virus is transmitted by *Bemisia tabaci* after an incubation period of from 4–48 hr. In some whitefly-virus associations, for example bhendi yellow vein mosaic, it has been found that if feeding time is prolonged the transmission efficiency of the vector is increased. The viruses are therefore likely to be circulative.

With a circulative virus there is a general opinion that the vector acquires virus from an infected plant, that the virus spreads through the gut epithelium into the haemolymph and thence to the salivary glands and so the virus can be transmitted to another plant. The major barrier in such a cycle is the gut epithelium. How does the virus traverse it if no multiplication occurs in the vector? In view of the difficulty in detecting a low level of virus multiplication in a vector and bearing in mind the evidence for limited replication in gut cells of insect viruses previously considered not to multiply in that tissue it would seem wiser to reserve judgement on this issue in the present state of knowledge.

Plant viruses which have definitely been shown to multiply in the vector are said to be propagative, but there are few unequivocal examples. Duffus (1963) observed an extremely long and temperature dependent latent period for sowthistle yellow vein virus in the aphid vector *Hyperomyzus lactucae*. In serial infection experiments plants infected initially had a long virus incubation period whereas plants infected later in a series showed symptoms sooner than those infected earlier in the series, suggesting that the amount of virus being introduced by the aphid was increasing with time. Aphids could still transmit the virus 52 days after they had been removed from an infected plant. The bacilliform virus particles of sowthistle yellow vein virus were observed in the nuclei of salivary gland cells of *Hyperomyzus lactucae*, and in the nuclei of infected plant cells (Richardson and Sylvester 1968). The injection of infected haemolymph will cause infection in healthy aphids and infected aphids are said to have increased mortality. Limited transovarial transmission of this virus has also been claimed. Recent work has demonstrated virus particles in many tissues of the aphid. (Richardson and Sylvester 1970). Another virus of similar morphology associated with the same aphid vector is lettuce necrotic yellows virus. Virus particles have been observed in several tissues of the vector including muscle, brain, fat body, tracheae, salivary gland and alimentary canal (O'Loughlin and Chambers 1967). Particles resembling pea enation mosaic virus have been seen in the fat body cells and gut lumen of the aphid *Acyrthosiphon pisum* (Shikata et al. 1966). No such particles were observed in nonviruliferous aphids. Although this and other information suggests that pea enation mosaic virus may be propagative further evidence is required. It is similarly likely that potato leafroll virus multiplies in *Myzus persicae* as Stegwee and Ponsen (1958) found that aphids could be infected by injection of haemolymph and once infected retained their infectivity. When the virus was serially transmitted in aphids to give a theoretical dilution end-point of 10^{-21}, the aphids still contained virus with an end-point of 10^{-4}. It seems therefore that the virus multiplies in the vector.

More examples of propagative viruses are known with leafhopper-borne viruses. Indications that plant viruses multiply in their leafhopper vectors were first obtained in the 1930's though some of the work was undertaken using diseases which, as recent work has shown, may be due to mycoplasmas. Extracts of viruliferous leafhoppers such as *Agallia constricta* infected with wound tumour virus were serially transferred in virus-free leafhoppers. This gave evidence that the virus must have multiplied in the leafhopper as leaf-hoppers containing theoretical dilutions of 1 in 10^{18} or more contained as much virus as the original leafhoppers (Black and Brakke 1952). The fact that the virus is transovarially transmitted also suggests that the virus multi-plies in the vector. It is also possible to demonstrate association of virus particles and leafhopper tissues directly by electron microscopy and fluo-rescent antibody techniques. Fukushi and his co-workers may well have been the first to demonstrate a plant virus in the tissues of its insect vector by electron microscopy (Fukushi et al. 1962; Fukushi et al. 1963). Particles of rice dwarf virus were observed in the fat body, blood, gut epithelium, salivary glands and Malphigian tubules of *Nephotettix cincticeps*. The virus is transovarially transmitted and virus particles were also found in the ovariole. Wound tumour virus has also been shown to be widely distributed in tissues of viruliferous *Agallia constricta* (Shikata and Maramorosch 1967). The virus was often seen in large concentration, even as microcrystals, in the cytoplasm of fat body cells. In the salivary gland only small amounts of virus were observed. Using fluorescent antibody staining methods Sinha (1965) detected virus first in the filter chamber in the intestine of *Agallia constricta*. Later, other tissues contained virus antigen. The viral antigen was detected within the cells and was not just adsorbed to their surfaces which again suggests that the virus was replicating. Hirumi et al. (1967) observed wound tumour virus particles in the cytoplasm of cells of the nervous system, especially in ganglion cells and Granados et al. (1968) found wound tumor virus particles in the cytoplasm of plasmatocytes and spherical cells of the haemolymph. Thus it seems that wound tumour virus is widely distributed in the tissues of the vector, but infects the intestinal filter chamber first. Other instances of plant viruses which multiply in their leafhopper vector are potato yellow dwarf virus in *Agallia constricta* and American wheat striate mosaic virus in *Endria inimica*. Leafhopper nymphs are generally more efficient vectors than adults and with increasing age adults become less efficient vectors. Sinha (1963) found that puncturing the abdomen of adults before they acquired virus increased vector efficiency, and he considered it possible that the gut wall increasingly resisted infection with increasing age. In much earlier work Storey (1932, 1933) punctured the gut walls of individuals of a strain of *Cicadulina mbila* which is not a vector and demonstrated that they could then transmit maize streak virus. High temperature also prevents the spread of wound tumour virus from the intestine to other tissues of the leafhopper. Races of *Agallia constricta* can be produced by selective breeding which are either efficient or inefficient in transovarial transmission of wound tumour virus; efficient transovarial transmitters are also efficient transmitters of virus to plants. Conversely virus mutations may occur. Black (1953) demonstrated that potato yellow dwarf virus cultured in plants for many years by grafting was no longer transmissible by the leafhopper vector.

Those plant viruses which replicate in their insect vectors infect most tissues of the vector. These viruses are of fundamental interest to a virologist as they are able to replicate in such widely different taxonomic hosts. It is not known what if any of the host cell components are incorporated into the virus particle and what governs its ability to infect such different types of cell. Answers to the important question of virus specificity, (Oman 1969), may well be found in the study of plant viruses which replicate in their vector (see ch. 14). Studies of these viruses in cell culture systems derived from vector species may make a significant contribution to our understanding of this problem at the molecular level (fig. 15.9) (see ch. 17). Using fluorescent antibody techniques Chiu et al. (1970) demonstrated viral antigens of potato yellow dwarf virus in the nucleus of *Agallia constricta* cell cultures and wound tumor virus antigen in the cell cytoplasm. Cells in culture are very susceptible to infection, maximal virus adsorption of wound tumor virus occurring in 2 hr. Specific fluorescence can be detected after 12 hr. Cultured cells of non-vector leafhoppers may also be infected. Similarly sow-

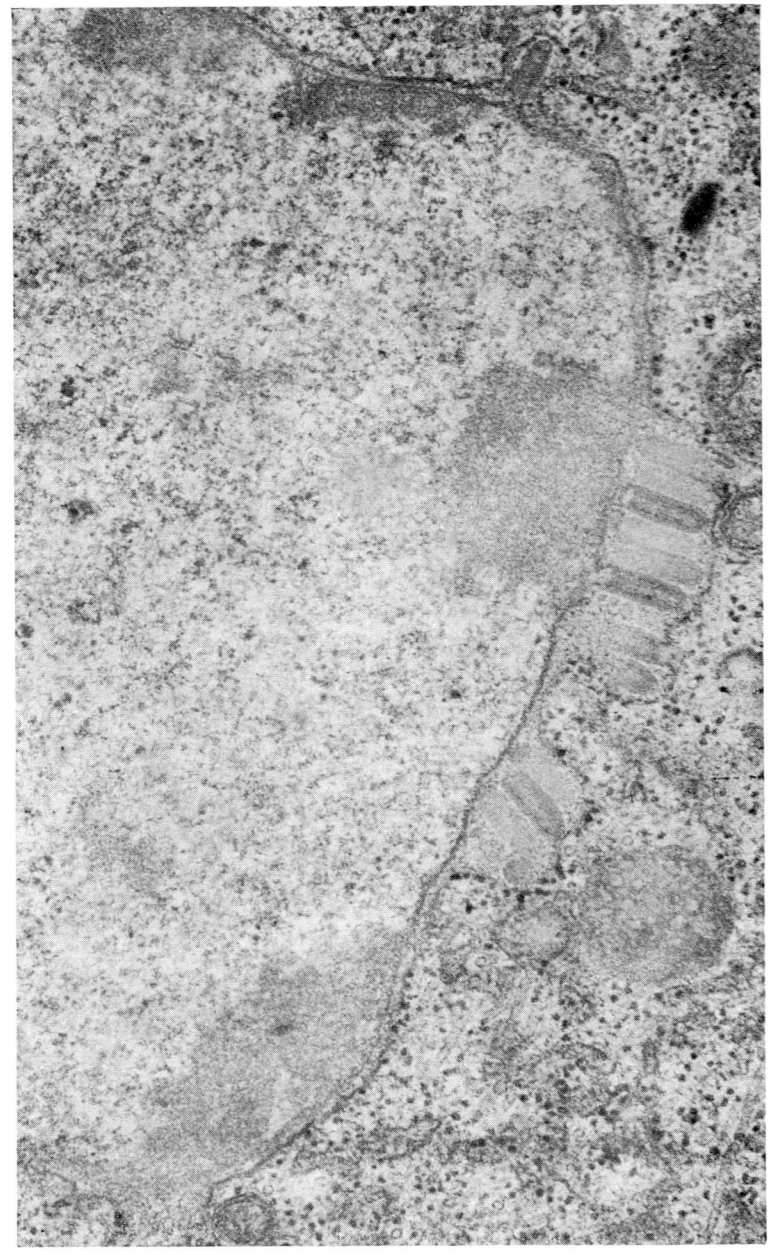

Fig. 15.9. A cell of an *Aceratogallia sanguinolenta* monolayer infected with potato yellow dwarf virus. (Chiu et al. 1970.)

thistle yellow vein virus has been grown in primary cultures of ovarian and embryonic tissue from the aphid vector *Hyperomyzus lactucae* (Peters and Black 1970). The future application of such techniques could lead not only to a very sensitive assay method for the infectivity of plant viruses but also to the characterization of the specific proteins produced during virus replication.

15.3.2 Virus diseases of higher animals

Transmission of viruses from arthropods to vertebrates may be either mechanical or involve a biological association. When transmitted mechanically the viruses are acquired when the insect probes in infected epidermal or subcutaneous cells or from blood capillaries. The virus is transmitted to a healthy individual during further biting or probing. Mechanical transmission is of great importance where high virus titres are reached in the skin, such as with pox viruses and papillomas. Myxomatosis is a good example of a virus disease which is mechanically transmitted principally by mosquitoes but also by other arthropods such as the rabbit flea. Flies might be expected to mechanically transmit enteric viruses but so far there is no convincing evidence of this.

Most of the vector-borne viruses affecting higher animals infect the vector and there is no doubt that this is essential for transmission. Such viruses are called arboviruses (arthropod – borne viruses) and, by definition, multiply in tissues of the vector. They usually cause no obvious pathological effect. Despite the vast amount of information concerning the properties of arboviruses and their disease conditions in man the amount of information concerning the association of these viruses with their invertebrate vectors is quite small. The majority of arboviruses are transmitted by mosquitoes and ticks. Virus multiplies in the tissues of the arthropod after being acquired in a blood meal and is transmitted to the next host in saliva. This interval between biting and transmission is known as the 'extrinsic incubation period' and represents a period of virus multiplication in the arthropod. Arboviruses can be passaged through generations of mosquitoes, by injection, without loss of vertebrate virulence.

Virus is acquired in both ticks and mosquitoes by sucking up infected blood. Ticks feed for longer periods than mosquitoes and cause more laceration and haemorrhage in the tissues. Digestion proceeds rapidly in the mosquito and is usually complete in 48 hr whereas in the tick digestion is much slower. Maximum titres of Japanese encephalitis virus in the midgut wall of *Culex pipiens* are found 7–21 days after infection. The virus appears in the haemolymph in the second week and its concentration decreases after the third week. Many tissues are invaded during this viraemic period and the virus concentration is at a maximum in the salivary glands in the third week. High titres of virus persist here throughout the life of the insect (La Motte 1960). The highest titres of tick-borne encephalitis are found in the gut of *Ixodes persulcatus* 25 days after infection feeding and large concentrations are also recorded in the salivary glands and ovaries. Transmission of tick-borne arboviruses from larva to nymph and nymph to adult is also found. Virus may also be egg transmitted in ticks but this does not seem to occur in mosquitoes.

Virus is transmitted in the saliva of both mosquitoes and ticks during the insect bite, though reliable estimates of the amount of virus present in given quantities of saliva are not available. How the specificity of arboviruses for their vectors is determined is not understood but some arboviruses will multiply in a wide range of species when injected into the haemocoel, spreading to the salivary glands and being transmitted to other hosts. Invasion of the gut cells may therefore be the key event in the determination of vector specificity. Nodamura virus is an exception to the vector specificity usually associated with arboviruses as it can multiply in mosquitoes, ticks and lepidopterous larvae. It is found in swine in Japan and is transmitted by mosquitoes (Scherer et al. 1968). The virus is antigenically unrelated to other arboviruses.

The temperature of the host can be expected to have a considerable effect on an arbovirus. The 'extrinsic incubation period' depends on temperature and because of this a potential vector may die before the virus can be transmitted. The variation in temperature to which the virus is exposed in the vector may have a selective effect on the population of virus particles. Such selection could cause attenuation of the virulence of the virus in another host.

Most of the published work concerning arbovirus infection of cells has made use of vertebrate cell culture systems and the amount of information available about the virus-arthropod association at a cellular level is scanty (see ch. 14). Bergold and Weibel (1962) observed virus-like particles in salivary glands of *Aëdes aegypti* infected with yellow fever virus and these were similar in appearance to the virus-like particles observed in infected brain and liver of mice and in HeLa and KB cells. Blue-tongue virus, which causes a disease in sheep and cattle has also been observed in the salivary glands of the vector *Culicoides variipennis* after intrathoracic inoculations (Bowne and Jones 1966); virus appeared to be synthesized in dark staining masses of unknown composition found in the salivary gland syncytium. More recently Janzen et al. (1970) observed Chikungunya virus in the salivary glands of *A. aegypti*. The particles were found in the nucleus, in the cytoplasm, and on membranes around cytoplasmic vesicles, from which they appeared to acquire an outer envelope. The particles were frequently found with secreted material in the distal portions of salivary gland lobes. No other pathological change was found. By contrast Mims, Day and Marshall (1966) found pathological changes in the salivary glands of *A. aegypti* infected (either naturally or artificially) with Semliki Forest virus; there was a decrease in the secretory regions of glandular cells, and these cells were unable to exclude trypan blue. The changes were maximal three weeks after infection.

There have been few studies of the effect of viruses on the cells of arthropod vectors despite the amount of time devoted to other aspects. Attempts should also be made to trace the infection pathway of these viruses in their vectors in more detail. Such studies in the arthropod could perhaps provide clues about the specificity of the virus–vector association and useful information about the properties of the viruses themselves. Infection of arthropods by arboviruses has been reviewed by Hurlbut (1965).

15.4 Conclusions

Probably the most important common denominator to emerge from a survey of virus infection in invertebrates is the importance of the gut as the route of entry and site of replication of virus in an arthropod. It may be that the ability to infect the gut is a prerequisite for virus invasion in any arthropod whether for an arthropod–pathogenic virus or for a propagative vector-borne virus. The gut cells represent a natural 'barrier to infection' and if there is a 'defensive mechanism' in arthropods perhaps it is most likely to be associated with the gut cells.

One enigma of virology is undoubtedly that of virus specificity. It is not clear why viruses replicate in some host organisms and not others, or only in some cells or tissues. A study of viruses which can multiply in cells of both plants and animals may provide information on this subject.

Knowledge of virus infection in invertebrates has developed from many different and often unconnected fields of interest. The resulting fusion of ideas may yet provide a great impetus to the study of virus-invertebrate relationshisp and, in broader terms, of the nature of virus specificity.

CHAPTER 16

Immune responses of invertebrates

K. J. LAFFERTY AND R. CRICHTON

Contents

16.1 *The concept of immunity in relation to invertebrate systems*

In discussing the immune responses of invertebrates we must be very clear from the outset regarding our use of the term immune. Some immunologists would argue that the way in which invertebrates react to foreign materials introduced into their tissues bears little relation to what would be defined as classical immunity amongst the vertebrates. This view stems from a tendency to define the immune response in terms of mechanisms that are well developed amongst the vertebrates; the ability to produce circulating antibody, and the ability of animals to show an accelerated and functionally more effective response to a second challenge with the same antigen. On the basis of these criteria there is little evidence that invertebrates possess a 'classical' immune system. However, if we set our definition of immunity in terms of its functional significance to the biological system, we are led to a wider concept of immunity that is relevant to the biology of both vertebrate and invertebrate animals.

Biological systems are highly ordered and inherently unstable structures, that under normal conditions can only exist in a metastable state; to achieve this degree of stability they must continually utilize energy in an attempt to maintain their ordered structure. The primary requirement for the maintenance of life is therefore, an inbuilt ability on the part of the living system to regulate its metabolism and thus maintain its order within tolerable limits.

The biological system does not only possess a vulnerable instability because it is prone to spontaneous disorganization. The same final effect, the death of the organism, will result if its order is severely disrupted as a result of being invaded by foreign environmental elements, or if damaged components are not removed from the tissues. Thus, the second essential requirement for the maintenance of life is the possession of a system whose function is the maintenance of the body's integrity. For this second system to operate within the tissues of an animal, it must possess some mechanism for distinguishing between normal healthy components and unwanted 'self' or foreign material.

The following discussion of invertebrate immune responses is confined to those mechanisms involving the differential recognition of 'self' and 'foreign' components, and the manner in which this recognition might be achieved. The subject matter is divided into two major sections, one dealing with cellular and the other with humoral aspects of immunity. In the final section we have attempted to bring the two together and indicate how they may be related. Other factors, such as the chemical and physical nature of secretions or protective mechanisms related to anatomical structure are not considered, although they undoubtedly contribute to nonspecific resistance. For information on this aspect of invertebrate resistance, the reader is referred to other reviews (Feng 1967; Brooks 1969; Tripp 1969).

16.2 Cellular responses

In general, the cellular responses initiated by the introduction of foreign materials into the tissues of invertebrates follow much the same pattern in the members of the various invertebrate phyla. The usual response following the recognition of foreign material, is an attempt on the part of the animal's fixed or free-floating phagocytic cells to ingest the foreign substance. The type of response seen is largely dependent on the size of the foreign particle. Colloidal materials and small particles such as carbon or bacteria are rapidly ingested by phagocytic cells. Larger particles, such as foreign tissues or parasites too large to be ingested, may be encapsulated by phagocytic cells which surround the foreign material, thus isolating it from the host tissues. Both these responses involve some form of specific recognition of foreign material in the sense that neither appears to be activated by normal 'self' components.

16.2.1 Phagocytosis by free-floating haemocytes

The blood or haemolymph of most invertebrate animals contains free-floating cells, the haemocytes. These cells often form a heterogeneous population consisting of morphologically distinct cell types. In the majority of invertebrates studied, one or more of these cell types are actively phagocytic.

Cameron (1932) studied the response of earthworm haemocytes to the injection of India ink, carmine, colloidal iron and a number of bacteria. All of these materials were rapidly phagocytosed. Within 5 min, material could be seen in the cytoplasm of non-granular haemocytes. By 2 hr there was marked phagocytosis by all haemocyte types. Similar results have been obtained following the injection of foreign material into the larvae of insects (Cameron 1934) or a number of molluscan species (Tripp 1969). In insects the predominant phagocytic cell is the plasmatocyte, a large pleomorphic non-granular haemocyte. However other haemocytes may also show some phagocytic activity (Jones 1962). In those species lacking a circulatory system, such as the coelenterates and sponges, foreign materials are taken up by phagocytic cells present in the mesogloea (Phillips 1963; Cheng et al. 1968).

The phagocytic response is not simply a reaction to any 'non-self' material introduced into the haemolymph. The degree of foreignness appears to play an important part in determining the outcome of the response. Cameron (1932) introduced spermatozoa obtained from other earthworms into the coelom of *Lumbricus terrestris* recipients. In these experiments he compared the response of recipient haemocytes to contact with allogeneic sperm (obtained from other worms of the same species), with the response observed when donor and recipient worms belonged to different genera. In the intergeneric transfers, germ cells were phagocytosed by free haemocytes. However, the process was somewhat slower than the phagocytosis of small particles such as bacteria. By 24 hr, donor spermatozoa could be seen enclosed in well defined phagocytic vacuoles within the cytoplasm of free-floating haemocytes. Allogeneic cells were not attacked by phagocytes up to 4 days after transfer to the recipient animal.

Hilgard and Phillips (1968), working with the sea urchin *(Strongylocentrotus)*, demonstrated that haemocytes eliminated both bovine and human serum albumens from the coelomic fluid much more rapidly than the animal's own proteins. Clearly, foreign materials that are similar to the animal's own constituents elicit considerably less vigorous responses than more foreign materials. More will be said on this subject when we discuss the response of invertebrates to the transplantation of tissues.

Phagocytosis does not always occur when materials that one might expect to be truly foreign are presented to the phagocytic system. In his study of the in vitro phagocytosis of marine bacteria by oyster haemocytes, Bang (1961) showed that most, although not all strains of bacteria were phagocytosed. Although we must be careful in extrapolating from in vitro studies to the situation in vivo, there are situations where foreign organisms can lead an unmolested existence in their invertebrate host. The symbiotic algae of *Tridacna*, for example, are seldom seen within the cytoplasm of host cells (Feng 1967). Schwartz and Townshend (1968) described an infection of the Japanese beetle *(Popillia japonica)* by *Bacillus popillae*. This infection (type A milky disease) is characterized by the failure of the animal's haemocytes to phagocytose the invading bacteria. The reason for the lack of reactivity on the part of the host haemocytes is not known.

Results reported by Vago and Vasiljevic (1956), and McKay and Jenkin (1970a) suggest that the uptake of foreign material by phagocytic cells is a two step process. Following injection of *Bacillus* spores into the silkworm *(Bombyx mori)*, Vago and Vasiljevic (1956) found little evidence of phagocytosis in animals maintained at 4 °C. Under these conditions approximately 80% of the phagocytes had spores attached to their surface, but only 11% contained spores in the cytoplasm. When animals were held at 24 °C no phagocytes were found with attached spores, while 80% contained intracellular spores. Similar results were reported by McKay and Jenkin (1970a) using an in vitro system to study the adherence and subsequent phagocytosis of sheep erythrocytes by crayfish haemocytes.

It would thus appear that the initial recognition step involves binding of foreign material to the cell surface. As this step is relatively temperature independent it might be controlled by ionic bonds or van der Waals adsorptive forces. The subsequent movement of material from the surface into the interior is apparently a function of the metabolically active cell.

16.2.2 Phagocytosis by fixed cells

In addition to the circulating haemocytes, many invertebrate species also possess a system of fixed phagocytic cells with activity complementary to the circulating phagocytes. The organization of fixed phagocytic cells exhibits varying degrees of complexity in different species. They may be found widely distributed throughout connective tissue and lining blood sinuses, grouped to form loose aggregates as in the case of insect pericardial cells, or more definitely arranged into discrete phagocytic organs which function as true blood filters.

In a comprehensive study of the distribution of phagocytic cells in molluscs, Cuenot (1914) reported that most species possessed a system of widely distributed fixed cells, generally found in intimate association with the circulatory system. In the chiton, a marine mollusc, these cells are found lining blood sinuses and associated with connective tissue throughout the animal. They are especially abundant in both the foot and gills of this species. When foreign materials such as carbon, colloidal gold or the protein ferritin are injected into the haemocoel of the chiton, they are rapidly removed from the haemolymph by fixed phagocytes (R. Crichton, V. A. A. Killby and K. J. Lafferty, unpublished data). Fig. 16.1 shows the distribution of labelled cells in a foot sinus, 24 hr after the injection of colloidal gold. Gold was located within membrane bound cytoplasmic vacuoles that are morphologically similar to the phagocytic vesicles and secondary lysosomes of mammalian macrophages (see Cohn and Fedorko 1969). Although the colloidal gold was very prominent in the cytoplasmic vacuoles of the fixed phagocytes, very few of the free-floating haemocytes ingested the colloid. Injected carbon and

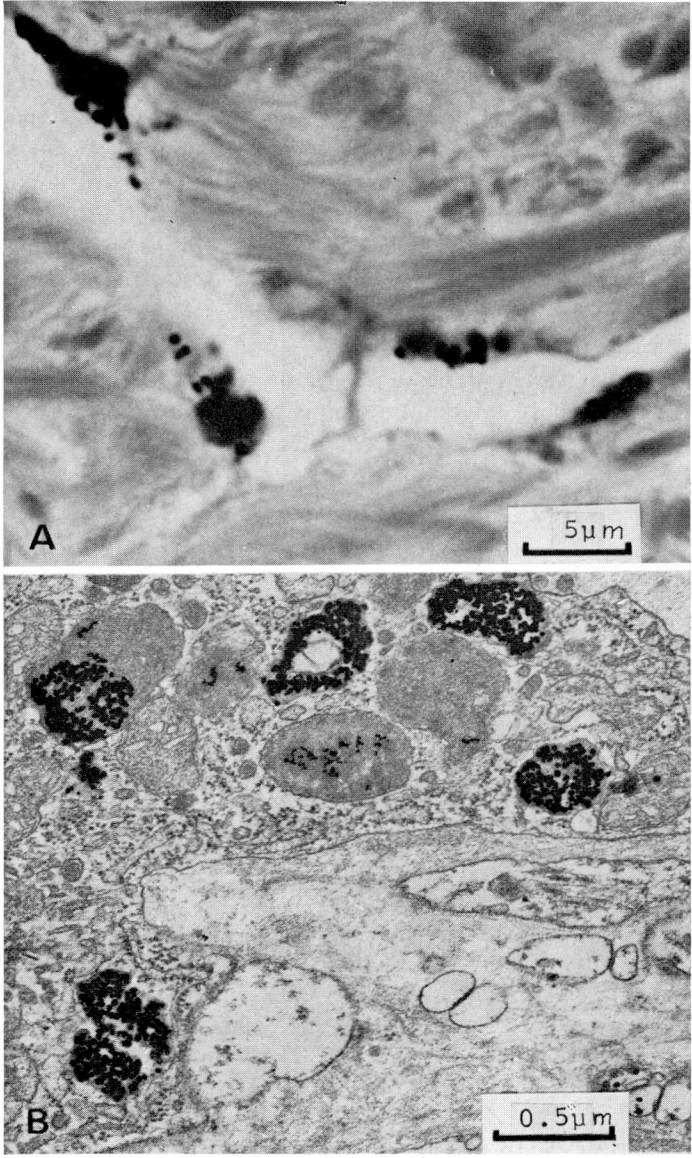

Fig. 16.1. Fixed phagocytic cells in the chiton *Liolophura gaimardi*, 24 hr after the injection of colloidal gold.   A) General view showing distribution of cells in a foot sinus. B) Electronmicrograph showing location of gold in membrane-bound cytoplasmic vacuoles.

ferritin were handled in much the same way; the foreign material being rapidly phagocytosed by fixed cells whilst relatively little was seen associated with free-floating haemocytes within the first 24 hr.

The pericardial cells of insects (fig. 16.2) have been considered by some authors to be analogous to the vertebrate reticuloendothelial system (Lesperon 1937; Wigglesworth 1970). These cells were originally distinguished by their morphological similarity and their ability to concentrate colloidal dyes. Such cells may be distributed throughout the body of the insect, but are most frequently found grouped in the vicinity of the dorsal vessel. They usually form clusters along the ventral surface of the dorsal diaphragm and around the heart itself (fig. 16.2).

In his study of wax moth larvae, Cameron (1934) found that injected colloidal iron and colloidal dyes were rapidly taken up by the pericardial cells.

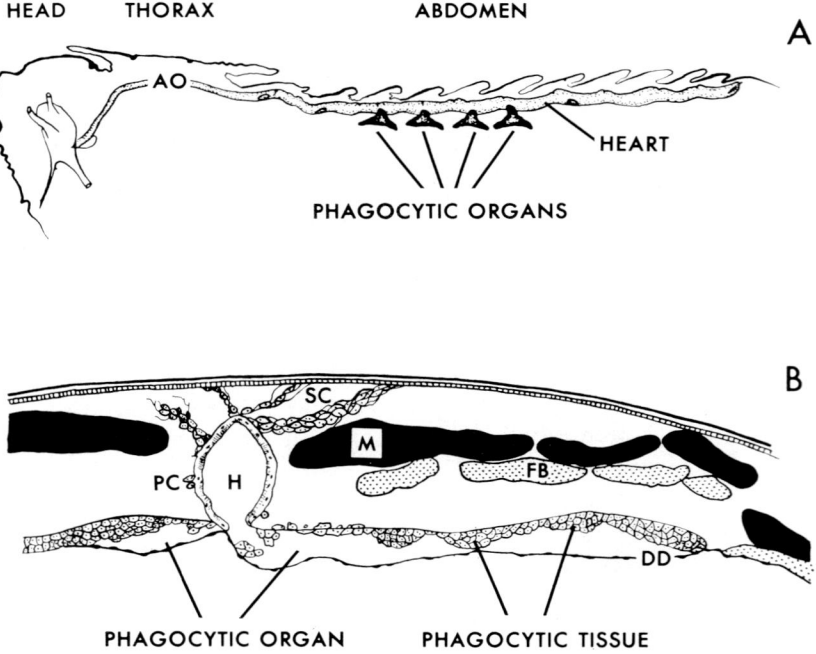

Fig. 16.2. Reconstruction showing the location and structure of phagocytic organs and pericardial cells in the cricket *Gryllotalpa*. A) Median longitudinal section. B) Transverse section through the phagocytic organs. DD: dorsal diaphragm; FB: fat body; H: heart; M: muscle; PC: pericardial cells; SC: suspensory cells. (After Nutting 1951.)

He found no evidence of carbon uptake, but suggested that some bacteria may localize in this region. Later workers were unable to confirm Cameron's suggestion that bacteria could be phagocytosed by pericardial cells. The investigations of Lesperon (1937) and Lison (1942) indicated that the pericardial cells can ingest only a restricted range of foreign materials, being unable to take up particles larger than 16–20 Å. In view of their apparent restricted phagocytic activity it is misleading to suggest that this cell system might be analogous to the vertebrate reticuloendothelial system.

The phagocytic organs found in some gastropod molluscs (Cuenot 1914) and in orthopteroid insects (Nutting 1951), function as true blood filters and can effectively remove foreign material from haemolymph circulating through them. In insects, these organs consist of lateral diverticula emanating from the ventral side of the heart in the anterior segment of the abdomen (fig. 16.2). The flow of haemolymph through these organs is controlled by primitive valves situated at the junction of the phagocytic organ and the heart. Blood flowing into the phagocytic organ percolates through the dorsal wall which consists of a spongy cellular matrix. Recent studies of Hoffman et al. (1968), indicate that this cell mass consists of phagocytic reticular cells associated with a network of supporting fibres and developing haemocytes, which would also be expected to carry out their phagocytic function in this area.

16.2.3 The fate of phagocytosed material

The interaction of the phagocytic system with foreign material is dependent on the recognition of foreign material. Recognition is at least in part mediated by binding or adhesion of foreign material with the surface of the phagocytic cell. Following binding, the cell responds by ingesting the foreign particles or colloid. Evidence suggests that this is an active process which may in itself be triggered by the binding of material to the cell membrane (Bennett 1956). Subsequent events show considerable variation in different invertebrate animals and may also vary according to the particular combination of phagocytic cell and foreign particle.

Cameron (1934) described the degeneration of some bacteria and foreign cells in cytoplasmic vacuoles of wax moth amoebocytes. He also mentioned the formation of a melanin-like pigment in close association with the ingested particles. Tripp has described the intracellular degradation of vertebrate erythrocytes, bacteria and some yeast cells by snail and oyster amoebocytes (see Tripp 1969). Although the process of digestion has not been investigated in detail, it is probably mediated by hydrolytic enzymes associated with 'lyso-somal' bodies contained in the cytoplasm of phagocytic cells (fig. 16.1).Lyso-somal enzymes have been demonstrated in both clam and sea star haemo-cytes (Janoff and Hawrylko 1964). In oyster haemocytes, discrete particles exhibiting hydrolytic enzyme activity were thought to be lysosomes (Eble 1966). Proteins such as haemoglobin are rapidly degraded following their ingestion by insect pericardial cells (Wigglesworth 1970).

Not all ingested organisms are destroyed by phagocytes. In the wax moth, certain bacteria such as *Proteus vulgaris* although rapidly phagocytosed, con-tinue to grow intracellularly and cause the disruption of the phagocytic cell within 12–24 hr (Cameron 1934). Some fungi and bacteria may also survive and multiply within molluscan haemocytes (Mackin 1951; Michelson 1961).

In some species foreign material is removed by cells, which having taken up the material subsequently migrate through epithelial surfaces to the ex-terior. This has been observed in oysters (Tripp 1960; Feng 1965), and a gastropod snail (Tripp 1961) following the introduction of a variety of soluble and particulate materials. Foreign materials introduced into the me-sogloea of sponges are eliminated by the migration of particle laden cells called archaeocytes into ex-current canals of the sponge (Cheng et al. 1968).

16.2.4 Encapsulation

Phagocytosis is usually observed when relatively small foreign particles are introduced into the tissues of invertebrates. However, Cameron (1934) de-scribed a related phenomenon in which clumps of small particles were sur-rounded by haemocytes, which collectively form a nodule. Fig. 16.3 shows the structure of a nodule formed in the haemocoele of the American cock-roach following the injection of yeast. The process of encapsulation differs from nodule formation only in degree and the term is usually applied to the formation of a thick coating of haemocytes around foreign metazoan para-sites, tissue transplants, or inert objects such as cotton thread, introduced into the tissues of invertebrates. Encapsulation has been shown to occur in insects, molluscs, annelids, crustaceans, acarina, sponges and larval echino-derms. However the process has been most extensively studied in the re-sponse of insects and molluscs, to invasion by metazoan parasites (Salt 1963; Brooks 1969; Poinar 1969).

The general features of the encapsulation process are similar in most ani-mals that have been studied. Misko (1968) observed capsule formation fol-lowing introduction of the parasitic worm *Heterotylenchus* into the body

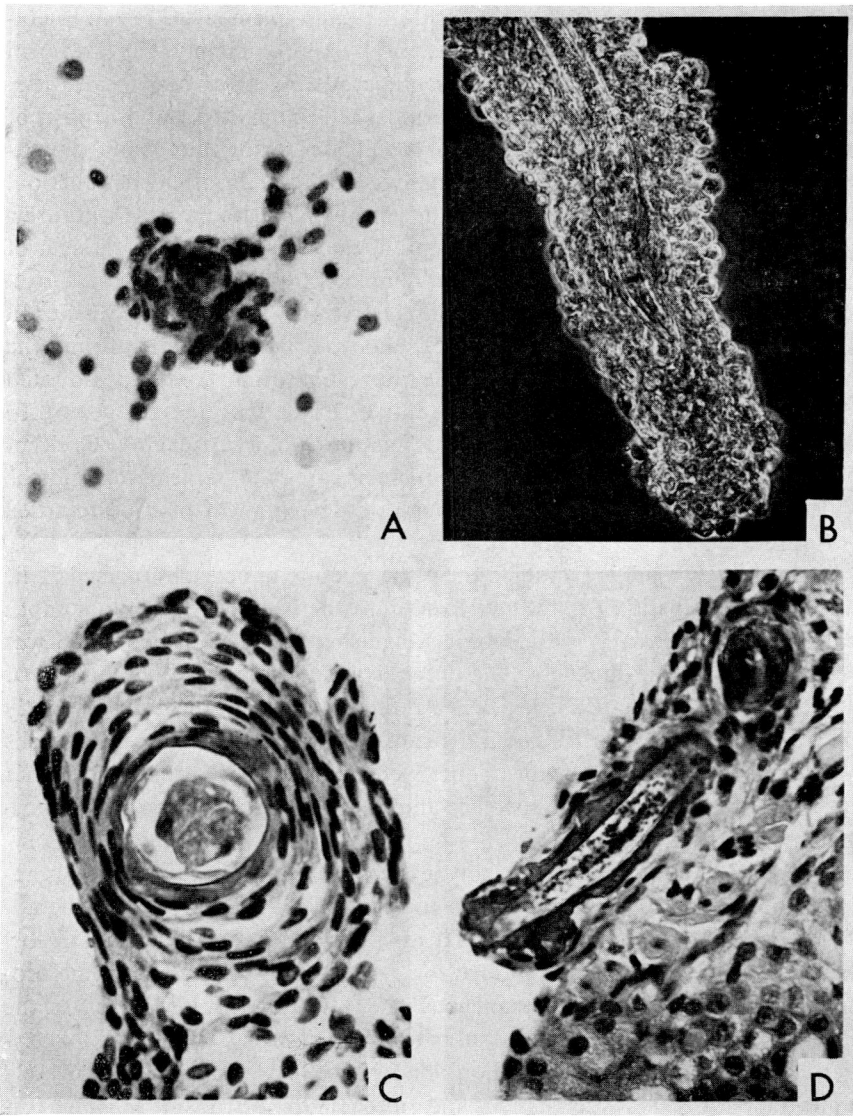

Fig. 16.3.   A) A nodule formed about an aggregate of yeast cells in the haemocoele of
the cockroach *Periplaneta*.   B) Capsule formed in the body cavity of *Periplaneta*, 9 hr
after the introduction of the parasitic worm *Heterotylenchus*.   C) Transverse section
through B.   D) Late stage capsule enclosing parasite remnants. (Fig. 16.3A courtesy of
Madeleine Ryan; B, C, and D from Misko 1968.)

cavity of the American cockroach. Within 15 min haemocytes began to collect around the surface of the parasite, and by approximately 9 hr the worm was almost completely enveloped by numerous layers of haemocytes (fig. 16.3). As the capsule formed, the innermost cells elongated and a deposit of dark pigment (thought to be melanin) accumulated at the surface of the parasite. The 'melanized' parasite was usually dead 24–48 hr after introduction. After prolonged periods, the size of the capsule decreased as cells withdrew from its outer portion. In the final stages of the reaction only a thin layer of capsular cells surrounded the parasite remnants (fig. 16.3).

Encapsulation does not always completely wall off a foreign parasite. In some cases discontinuous clusters of haemocytes may aggregate around the parasite; in other instances a haemocytic response may only be mounted against part of the parasite's surface (Salt 1955; Bartlett and Ball 1966). In some host–parasite combinations the development of a capsule may be either a slow or rather incomplete process. In such cases the parasite may not be killed by the cellular response, although its development may be retarded (Muldrew 1953).

Grimstone et al. (1967) studied the ultrastructure of capsules formed 72 hr after the introduction of araldite slivers into the haemocoele of flour moth caterpillars. Three layers could be distinguished in the capsule, which was some 50–60 cells in thickness. The outer layer was composed of rounded or somewhat spindle-shaped cells. Beneath these, was a thicker layer of tightly packed and flattened cells. The inner-most layer consisted of flattened cells showing signs of degeneration; some were necrotic. The cells in the inner layer contained many cytolysosomes and gave a strongly positive acid phosphatase reaction.

Encapsulation does not invariably lead to the death of the trapped organism. Some trematode metacercariae for example, do not appear to be adversely affected by encapsulation. This is probably in part due to their low level of metabolic activity, although the cyst wall may also afford some protection to the encapsulated organism. Fisher (1961) has pointed out that the larvae of young hymenopterous parasites are very sensitive to a lack of oxygen. In this situation, the formation of a tight cellular capsule could severely limit the parasite's oxygen supply and suffocate the organism. Melanin formation can also play a part in the destruction of an encapsulated parasite. Melanization does not always follow capsule formation in insects, but when it does occur, organisms that can survive the physical effect of encapsulation are destroyed (Schell 1952). The deposition of melanin, when it occurs, is always associated with a haemocytic response to the foreign organism and may be triggered by the degeneration of the innermost layer of capsular cells. It is not known whether melanin production is directly responsible for the death of the organism, or whether this is a result of some other toxic process.

16.2.5 Transplantation reactions in invertebrates

Tissue transplantation has been a common technique used by insect physiologists to study the function of various organs and tissues, and the general experience has been that insect tissues are not rejected by allogeneic recipients (animals of the same species as the tissue donor). However this is not the case for all tissues. Shatoury (1956) attempted intraspecific transplantation of *Drosophila* lymph glands. In all cases the transplanted tissues were encapsulated and melanized by the recipient. The allotransplantation of insect tumours has also been shown to excite a haemocytic response in the recipient, leading to the encapsulation of the transplanted tissue (Harker 1958). There may be some connection between Harker's finding and the common experience that lymph glands cannot be transplanted within the species, since Shatoury and Waddington (1957) have suggested that the cells involved in both benign and malignant *Drosophila* tumours originate from the lymph glands.

Tissue transplants are rejected more frequently when the donor and recipient animal belong to different genera. Kopec (1911) transplanted gonads among Lepidopteran species. These did not survive, and were usually encapsulated by the recipients. Salt (1960) demonstrated that testes of the flour moth *(Ephestia)* were rapidly enveloped in a haemocytic capsule following their transfer into the body cavity of the tomato moth *(Diataraxia)*. In some cases the intergeneric transplantation of tissues may be successful, and Salt (1961) when reviewing the situation among the insects suggested that tissues of related species and genera excite the haemocytes of the host to a lesser extent than those of widely separated species. Scott (1971) found no gross haemocytic response to the allotransplantation of nerve cord in cockroaches. However, the implantation of nerve cord obtained from the American cockroach *(Periplaneta)* into the body cavity of another cockroach *(Nauphoeta)* excited a rapid host response that resulted in the complete encapsulation of the transplanted tissues.

Recent studies of transplantation reactions among the annelids have excited considerable interest because they provide the first unequivocal evidence of adaptive cellular immunity outside the vertebrates. Using the earthworm *(Eisenia foetida)*, Duprat (1967) and Cooper and Rubilotta (1969) independently demonstrated that body wall transplants excited a haemocytic response in the recipient that brought about a chronic rejection process. The striking discovery made by Duprat was that allografts transferred between animals originating in different geographical localities were much more vigorously attacked. The rejection process in these 'inter-racial' combinations was similar to that seen when transplants were made between earthworms belonging to different generic groups (Cooper 1968).

Second transplants between different 'races' of *Eisenia* were rejected in an accelerated manner and the prospective recipient could be sensitized either by a prior body wall transplant or by the injection of coelomocytes from the graft donor. The heightened reactivity following sensitization was associated with the circulating haemocytes. Transfer of these cells from a sensitized donor would confer on unsensitized recipients of the same 'race', the ability to reject primary 'inter-racial' grafts in an accelerated manner.

Duprat did not examine the specificity of these reactions in detail. It could be argued that the heightened state of reactivity exhibited by a sensitized animal results from non-specific haemocyte stimulation, in a manner similar to that described by McKay and Jenkin (1970c) for crayfish haemocytes (see following section). However more recent studies reported by Cooper (1971) have extended Duprat's initial findings to show that second-set rejection is a specific process.

Cooper transplanted tissues from earthworms of two different genera *(Allolobophora* and *Eisenia)* to the body wall of *Lumbricus* recipients, previously sensitized to *Eisenia* tissues by a body wall transplant. These recipients rejected the *Allolobophora* grafts in the normal first set manner (median survival time, 43 days). However, the *Eisenia* transplants were rejected after 15–18 days; first-set *Eisenia* grafts have a survival time of 34 days on normal *Lumbricus* recipients. When *Lumbricus* worms were injected with coelomic fluid containing coelomocytes from *Lumbricus* donors previously sensitized to *Eisenia* grafts, these recipients rejected *Eisenia* grafts in an accelerated fashion (Bailey et al. 1971). Thus prior sensitization of *Lumbricus* recipients with *Eisenia* tissues or *Eisenia*-sensitized coelomocytes confers on the recipient the ability to mount a specific secondary response to the tissues of the sensitizing donor.

Apart from work on the annelids and insects little has been done on transplantation reactions in other invertebrates. However the available information indicates that among both the molluscs (Tripp 1961) and echinoderms (Ghiradella 1965), allogeneic tissues incite little or no reaction in their host. Intergeneric transplants on the other hand, incite a host response that destroys the graft.

16.2.6 Factors affecting the cellular response

Numerous environmental, physiological, ecological and genetic factors can play an important role in modulating the response of invertebrates to foreign materials. Hu (1939) reported that the microfilaria *Wuchereria bancrofti* was not encapsulated in its mosquito host *(Culex pipiens)* during the winter season, although the host responded in the summer; this difference in reactivity was attributed to temperature. Walker (1959) found the encapsulation response of *Drosophila* to be most efficient between 18 and 20 °C. Likewise, the physiological state of an animal can have a pronounced effect on its ability to mount a defence reaction. Muldrew (1953) found that only 37% of parasite eggs were encapsulated by what he termed 'unhealthy' saw flies, whereas 80–100% were encapsulated by normal animals. Van den Bosch (1964) has pointed out that the nutritional status of the host can influence its capacity to encapsulate foreign parasites; animals that have been starved being considerably less efficient than well-fed specimens. This effect may be related to the number of circulating haemocytes since Shapiro (1967) found fewer circulating haemocytes in starving wax moth larvae than in well-fed animals.

In one invertebrate species antibacterial immunity may be non-specifically enhanced by agents which have been shown to increase the activity of phagocytic cells. Following immunization of crayfish *(Parachaeraps bicarinatus)* with alcohol-killed *Pseudomonas* vaccine, McKay and Jenkin (1969) were able to demonstrate an increased resistance of vaccinated animals to subsequent challenge with the viable pathogen. The response was non-specific in that it could also be induced by vaccines prepared from a number of unrelated gram negative bacteria, and by an endotoxin preparation of *Salmonella typhimurium*. Subsequent studies revealed that vaccinated animals were more efficient than normal animals in clearing injected bacteria from their circulation. More striking was the finding that living bacteria were rapidly killed following their introduction into vaccinated animals, whereas they continued to multiply in normal crayfish (McKay and Jenkin 1970b). This bactericidal effect could not be attributed to the production of a humoral factor (see following section), but resulted from a change in the phagocytic activity of the crayfish haemocytes. In vitro studies demonstrated that vaccination caused a four-fold increase in the phagocytic activity of haemocytes as measured by their ability to ingest foreign erythrocytes (McKay and Jenkin 1970a). This change in phagocytic activity was found to parallel the change in resistance of vaccinated crayfish to challenge with viable organisms (McKay and Jenkin 1970c). 'Activated' phagocytes from immune crayfish were also found to degrade ingested red cells more rapidly than did haemocytes from normal animals. This effect may be similar to that described for mammalian cells, where it has been shown that the administration of endotoxin can increase the metabolic activity of phagocytes, raise their concentration of lyso-somal enzymes, and stimulate their phagocytic activity (Whitby et al. 1961; Auzins and Rowley 1962).

The physiological state of phagocytic cells can also influence the susceptibility of invertebrate animals to virus infection. Ponsen and De Jong (1964) reported that wax moth larvae, *(Galleria mellonella)* following the injection of India ink, became much more susceptible to infection by nuclear poly-hedrosis virus. This effect they attributed to blockade of the phagocytic system, which in normal larvae destroys a proportion of the injected virus. An attempt to demonstrate an acquired state of resistance to viral infection, by prior injection of non-infectious virus, proved unsuccessful (Bailey and Gibbs 1964).

One of the most important factors influencing the interaction between a parasite and its invertebrate host is the ecological relationship between the two organisms. As early as 1912 Timberlake suggested that in insects, a haemocytic response occurs only in hosts to which the parasite is unaccustomed and unadapted. This view needs to be modified slightly in that some parasites can elicit a response in their usual host, but often only at certain stages of their life cycle (Salt 1963).

Newton (1952) has shown a strain specificity in the reaction of a gastropod mollusc to a metazoan parasite. When miracidia of a Puerto Rican strain of *Schistosoma mansoni* penetrate a Brazilian strain of *Australorbis glabratus* the parasites are rapidly encapsulated and destroyed. In contrast, a Puerto Rican strain of this snail failed to mount a haemocytic response against the Puerto Rican parasite.

The mechanisms used by 'usual' parasites to circumvent the host response throw some light on the way in which host haemocytes recognize foreign material. Salt (1966, 1970) has studied the relationship between the flour moth caterpillar *(Ephestia kuehniella)* and a parasitoid wasp *(Nemeritis canescens)* that is normally associated with the flour moth. The egg of *Nemeritis* excites no cellular response when introduced into the haemocoele of the flour moth. This situation is quite specific since the flour moth rapidly encapsulates other foreign materials introduced into its haemocoele, and the egg of *Nemeritis* is readily encapsulated if introduced into the tissues of an unusual host. This specific protection from the attack of *Ephestia* haemocytes is associated with the surface structure of the egg. If the surface is abraded or altered by extraction with lipid solvents, the damaged egg is rapidly encapsulated. Salt has shown that the egg of *Nemeritis* acquires a 'protective' coating as it passes through the calyx of the female wasp. This coating is not an integral part of the egg and it would appear that the production of this material represents an adaptation of *Nemeritis* to its usual host.

Another way in which a parasite may avoid the immune response of its host is by generating some material which is either introduced into the host along with the egg, or is produced by the egg itself. This material appears to paralyse the host amoebocytes so that they no longer respond (Streams and Greenberg 1969).

Since it is not uncommon for an invertebrate animal to be successfully parasitized by several different organisms, it is possible that the invertebrates may have a recognition mechanism that is functionally similar to that of the vertebrates. That is, certain patterns can be recognized as foreign and stimulate a haemocytic response whilst other quite discrete patterns are not recognized in this way. Amongst the latter group would be those patterns presented by normal 'self' components. Thus the adaptation of a parasite to an invertebrate host (where this adaptation involves the selection of a compatible surface) might be similar to the manner in which some parasites are thought to adapt to their vertebrate host (Sprent 1969). The adapting parasite may select an exterior pattern that is so similar to a normal host component that it will not stimulate an effective haemocytic response in its specific host.

16.3 *Humoral factors in immunity*

Recent interest in comparative immunology has led to a considerable expansion in the literature dealing with humoral aspects of invertebrate immunity. Information has accumulated largely as a result of attempts to induce in invertebrate animals, a state of immunity comparable with that elicited in vertebrates. Both naturally occurring and induced substances with reactivity similar to vertebrate antibody have been demonstrated in a large number of species. However, these substances have been shown to be either non-protein, or to be structurally dissimilar to vertebrate immunoglobulins. There is little evidence to suggest that they form part of a generalized immune mechanism contributing to natural resistance or defence. More detailed discussion of humoral aspects of invertebrate immunity may be found in reviews by Stephens (1964), Briggs (1964), Brooks (1969) and Tripp (1969).

16.3.1 Naturally occurring humoral factors

Normal haemolymph from a number of invertebrates has been shown to exhibit antimicrobial activity against some bacteria (Briggs 1958; Stephens 1962; Johnson and Chapman 1970). In most cases the nature of the active material is unknown, although lysozyme activity has been demonstrated in both oyster (McDade and Tripp 1967) and wax moth haemolymph (Chadwick 1970), and may account for the observed antimicrobial activity in some instances. Hink and Briggs (1968) partially purified a bactericidal principle present in the haemolymph of normal wax moth larvae. Although not completely characterized, this material did not appear to be protein.

After investigating the virulence of some 32 species of bacteria in wax moth larvae, Chadwick (1967) found all but 7 to be non-pathogenic. Normal haemolymph was found to exhibit bactericidal activity against 20% of the non-pathogenic species. Although not proving that bactericidal activity is responsible for the non-pathogenicity of the 5 bacterial species concerned, these results indicate that bactericidal factors may play an important role in natural resistance to infection with some bacteria. Clearly, other mechanisms must also contribute to resistance, since the wax moth can effectively deal with bacteria that are not inactivated by normal haemolymph.

16.3.2 Induced humoral factors

Induced humoral responses following the introduction of bacterial products or other agents have been reported in a variety of invertebrate species. Haemolymph from the sipunculid worm *Sipunculus nudus* experimentally infected with the marine ciliate *Anophrys magii* (Bang 1966), and from male cockroaches infected with *Tetrahymena pyriformis* (Seaman and Robert 1968), was found to contain a factor which would rapidly immobilize the respective ciliates in vitro. Similar immobilizing activity directed against the miracidia of *Schistosoma mansoni* was obtained in tissue extracts of snails infected with the mature schistosomes, or injected with *S. mansoni* eggs (Michelson 1963, 1964).

Although none of these responses has been investigated in great detail, the available evidence indicates that any similarity with a vertebrate immune response is probably superficial. In most cases these responses have proved to be relatively non-specific. Michelson (1963) reported miracidia-immobilizing activity in extracts from snails infected with a nematode unrelated to the test organism, while Bang (1966) demonstrated that haemolymph from worms injected with a marine bacterium, would also immobilize *Anophrys*. In addition, both authors provide evidence which indicates that the factor or factors elaborated are not entirely specific in their activity. The specificity of the cockroach response to *Tetrahymena pyriformis* was not investigated (Seaman and Robert 1968).

Induced bactericidal activity following the injection of bacterial vaccines, has been found in a number of invertebrate species (Briggs 1958; Stephens 1962; Gingrich 1964; Acton et al. 1969) and despite reports to the contrary (Cooper et al. 1969; Weinheimer et al. 1969; McKay and Jenkin 1969) there is sufficient evidence to suggest that this response may occur widely although not universally in invertebrates. In most instances, both the inducing stimulus and the factors elaborated have been shown to be non-specific. However, Stephens (1959, 1962) was able to demonstrate a state of protective immunity in wax moth larvae, which resulted from the production of a bactericidal factor specific for the immunizing organism.

Wax moth larvae vaccinated with heat-killed or formalized *Pseudomonas aeruginosa*, were able to withstand challenge doses of from 10–100 LD_{50} of the virulent organism. Resistance developed rapidly to reach a maximum 16–24 hr after vaccination, but was relatively short lived, having disappeared by 72 hr. Haemolymph from resistant larvae was found to possess bactericidal activity against the immunizing organism, and changes in this activity were closely correlated with changes in the resistance of larvae to challenge. Protection could be passively transferred to normal larvae by means of haemolymph obtained from immune animals. The bactericidal factor was shown to be heat stable, dialysable, unaffected by trypsin, and of relatively small molecular weight (Stephens and Marshall 1962). The activity of this induced bactericidal factor is specifically directed against *P. aeruginosa*, as haemolymph bactericidal for one strain (P11–1) was almost inactive against a related strain (L.S.H.T.M.). In addition, the titre of bactericidal activity against *Salmonella dysenteriae* (for which normal haemolymph has a natural activity), was not increased following vaccination with *Pseudomonas aeruginosa* (Stephens 1962).

Although wax moth larvae can respond specifically following vaccination with *P. aeruginosa*, further studies by Chadwick (1967) have shown that similar responses could not be induced with other bacterial species. Thus the specific response to *P. aeruginosa* is perhaps atypical and not an example of a general immune mechanism triggered by the recognition of foreign organisms within the tissues.

16.3.3 Opsonic factors in invertebrate haemolymph

One important property of vertebrate antibody is its opsonic activity, that is, its ability to combine with foreign particles and facilitate their uptake by phagocytic cells. Recent experiments have now shown that invertebrate phagocytic activity may also be modulated by humoral opsonins. Tripp (1966) reported enhanced phagocytosis of rabbit erythrocytes when erythrocytes suspended in dilute haemolymph were added to oyster amoebocyte cultures in vitro. Similar results were obtained by Stuart (1968), who studied the phagocytosis of human red cells by octopus haemocytes; red cells were phagocytosed when added to haemocytes suspended in haemolymph or when cells previously incubated with haemolymph, were washed and added to haemocyte cultures in haemolymph-free medium. No attempt was made to investigate the specificity of the opsonic factor in either oyster, or octopus haemolymph.

The phagocytosis of formalized yeast or formalized sheep erythrocytes by cultures of haemocytes from the snail *(Helix aspersa)*, was also shown to depend on a factor present in normal haemolymph (Prowse and Tait 1969). This factor could be absorbed from haemolymph by the homologous but not by the heterologous cell, indicating that the opsonic activity was specific for at least the two particles tested. The chemical structure of the material responsible has not been investigated.

In studying the phagocytic activity of crayfish haemocytes, McKay and Jenkin (1970a) reported that prior sensitization of sheep or human erythrocytes with haemolymph, greatly enhanced their in vitro uptake by haemocytes. Crayfish haemolymph contains natural agglutinins for erythrocytes of both species, and these show some specificity. Adsorption of haemolymph with either type of red cell removes the agglutinin for the homologous cell, while causing a reduction in titre for the heterologous erythrocyte. When absorbed haemolymph was used to sensitize erythrocytes the specificity of the opsonin was found to parallel that of the agglutinin, and McKay and Jenkin (1970a) suggest that the same molecule may be responsible for both activities.

Natural agglutinins showing varying degrees of specificity for bacteria, vertebrate erythrocytes and other heterologous cells, have been demonstrated in the body fluids of a large number of invertebrate species (Tripp 1969). Recent reports on the isolation and characteristics of haemagglutinins from the horseshoe crab (Marchelonis and Edelman 1968) and from the Murray mussel (Jenkin and Rowley 1970), indicate that the material responsible is protein. If these molecules also function as opsonins, it is possible that a recognition mechanism based on specific opsonins may be widespread amongst the invertebrates.

16.4 Summary and conclusions

The immune system of invertebrates depends primarily on the activity of phagocytic cells. In those species with a circulatory system, some or all of the blood cells or haemocytes are phagocytic. The phagocytic activities of these cells are often complemented by a system of fixed phagocytic cells, whose organization may show considerable variation in different invertebrate groups. Fixed phagocytic cells may be diffusely scattered throughout the connective tissue of the animal, or collected into well defined phagocytic organs designed to filter the animal's haemolymph.

Phagocytic cells recognise and ingest foreign materials. Ingested material may be digested intracellularly, or in some instances transported within cells to the exterior. In situations where the foreign material is too large to be ingested, haemocytes collect around the particle, forming a tight capsule. This capsule effectively isolates the foreign body from the host tissues, and in the case of metazoan parasites, usually results in the death of the trapped organism.

As a group, the invertebrates seem not to possess an adaptive immune system, like that of the vertebrates. However, at least one invertebrate, the earthworm, has now been shown to be capable of adaptive cell mediated immunity.

There is no evidence that invertebrate animals produce humoral factors similar to vertebrate antibodies, and humoral responses induced in a number of invertebrate species following vaccination with bacteria, or other parasitic organisms, usually lack specificity. One known example of a specific and adaptive humoral response following vaccination with a particular bacterium appears to be an isolated phenomenon.

Humoral opsonic factors may be important in the immune recognition process of invertebrates. However, regardless of whether recognition is by soluble opsonins or by receptors in the surface of phagocytic cells, it is not yet known how this recognition operates.

Salt (1970) suggested that invertebrate haemocytes recognize only the membrane lining the haemocoele, and are unresponsive to contact with this or closely related surfaces. According to this hypothesis all foreign materials are recognized because they are not similar to 'self' components. However, a recognition system of this type, because of its inherent 'on/off' nature, would not allow *graded responses*, and foreignness, when recognized, would initiate a response of the same intensity regardless of whether the material was slightly foreign or very foreign. There could be no degree of discrimination. The results of transplantation studies indicate that graded responses are observed in invertebrate systems. Among the insects it appears that tissues of related species and genera excite the haemocytes of the host less than those of unrelated species.

For the system to show graded responses, it is more likely that the recognition of foreign patterns is a positive process, and the unresponsiveness to self components results from discrete gaps in this recognition spectrum. A parasite could then adapt to its invertebrate host, by assuming a surface pattern similar to one of the 'self' gaps in the recognition system. In this way several parasites with different surfaces could lead an unmolested existence in the one host. A decision between these two hypotheses must await the results of further experimental study.

CHAPTER 17

Cultured cells in virus research

T. D. C. GRACE

Contents

17.1 Introduction

The major advances in animal virology over the past 25 years have, to a great extent, been made possible by the ability to grow animal cells in vitro, in large numbers. Before about 1948 most virus studies were carried out in mice, rabbits, chick embryos, etc. Following the research of Enders, Weller and Robbins (1949) on polio virus replication in tissue culture, virologists were quick to realize the potential of using cultured cells in their studies. Rapid progress was soon made both in devising simple media in which cells would grow for long periods and in techniques for growing cells in large numbers, such as suspension cultures, roller tubes, etc. The use of trypsin to dissociate tissues (Dulbecco 1952) was a big advance as then large numbers of various types of cells could be grown. The use of cell cultures has made it possible for virologists to isolate and identify viruses and study their replication quantitatively.

Unfortunately, insect virology has lagged far behind both in the ability to grow insect tissues in vitro for any length of time, and also in knowledge of insect viruses. Although the first attempt to grow insect tissue was made in 1915 (Goldschmidt 1915), it was not until 1962 that the first insect cell line was established (Grace 1963). Since 1962, 40 lines have been established (Hink 1971; see table 17.1). Of these, 11 lines are from the tissues of moths, 20 from mosquitoes, 3 from leafhoppers and 6 from the vinegar fly *(Drosophila melanogaster)*. It is true to say, certainly in the case of insect and tick

TABLE 17.1

Species	Tissues of origin	Reference
Lepidoptera		
Antheraea eucalypti	Pupal ovaries	Grace, T. D. C. 1962
Bombyx mori	Larval ovaries	Grace, T. D. C. 1967
Bombyx mori	Ovary	T. D. C. Grace, unpublished
Carpocapsa pomonella	Minced embryos	Hink, W. and Ellis, B. 1971
Carpocapsa pomonella	Minced embryos	Hink, W. and Ellis, B. 1971
Choristoneura fumiferana	Minced larvae	Sohi, S. S. 1968
Heliothis zea	Minced adult ovaries	Hink, W. and Ignoffo, C. 1970
Malacosoma disstria	Larval hemocytes	Sohi, S. S. 1971
Samia cynthia	Pupal hemocytes	Chao, J. and Ball, G. 1970
Spodoptera frugiperda	Pupal ovaries	J. L. Vaughn, unpublished
Trichoplusia ni	Minced adult ovaries	Hink, W. F. 1970

Species	Tissues of origin	Reference
Diptera		
Aëdes aegypti	Last instar larvae	Grace, T. D. C. 1966
Aëdes aegypti	Minced larvae	Singh, K. R. P. 1967
Aëdes aegypti	Homogenized embryos	Peleg, J. 1969
Aëdes aegypti	Minced larvae	Varma, M. G. and Pudney, M. 1969
Aëdes aegypti	Minced larvae	Varma, M. G. and Pudney, M. 1969
Aëdes aegypti	Minced larvae	Varma, M. G. and Pudney, M. 1969
Aëdes albopictus	Minced larvae	Singh, K. R. P. 1967
Aëdes vexans	Minced pupae	Sweet, B. H. and McHale, J. S. 1970
Aëdes vittatus	Newly hatched larvae	Bhat, U. K. M. and Singh, K. R. P. 1970
Aëdes w-albus		K. R. P. Singh and U. K. M. Bhat, unpublished
Anopheles gambiae	First stage larvae	M. G. R. Varma and M. Pudney, unpublished
Anopheles stephensi	Minced larvae	Schneider, I. 1969
Anopheles stephensi	First instar larvae	Pudney, M. and Varma, M. G. R. 1971
Anopheles stephensi	First instar larvae	M. G. R. Varma, unpublished
Culex molestus	Adult ovaries	Kitamura 1970
Culex quinquefasciatus	Adult ovaries	Hsu, S. H., Mao, W. H. and Cross, J. H. 1970
Culex salinarius	Neonate larvae	Schneider, I. 1971
Culex tritaeniorhynchus	Ovarian	Hsu, S. H. 1971
Culex tritaeniorhynchus	Neonate larvae	Schneider, I. 1971
Culiseta inornata	Minced adult	Sweet, B. H. and McHale, J. S. 1970
Drosophila melanogaster	6–12 hr embryos	Echalier, G. and Ohanessian, A. 1970
Drosophila melanogaster	Embryo	Karpakov, V. T., Goosdev, V. A., Platova, T. P. and Polukarova, L. 1969
Drosophila melanogaster		C. Richard-Molard, unpublished
Drosophila melanogaster	Embryo	Schneider, I. 1972
Drosophila melanogaster	Embryo	Schneider, I. 1972
Drosophila melanogaster	Embryo	Schneider, I. 1972
Homoptera		
Aceratogallia sanguinolenta	Embryo	Chiu, R. and Black, L. M. 1969
Agallia constricta	Minced embryo	Chiu, R. and Black, L. M. 1967
Agallia quadripunctata	Minced embryo	Chiu, R. and Black, L. M. 1967
Agalliopsis novella	Minced embryo	Chiu, R. and Black, L. M. 1967

tissue culture, that the primary aim, and in most instances the only aim, of culturing the cells was to use them as a means of studying the replication of viruses which are either pathogenic to insects or grow in insects and are pathogenic to man, animals or plants. Yunker (1971) has given emphasis to this point, in relation to the arboviruses when he stated 'that approximately a third of all publications dealing with the application of arthropod tissue culture to studies of arboviruses appeared last year (1969), another third appeared in 1968; the remainder were published at intervals during the preceding three decades'.

In other sections of this book the ways in which viruses replicate in whole insects or insect cell cultures will be described. The emphasis in this section will be to describe the types of insect cells which grow in culture and to compare them with vertebrate cells used in the study of viruses.

A recent review of insect cells and tissue culture by Brooks and Kurtti (1971) gives the most up-to-date information and a fairly complete list of publications in this field. Two other publications which discuss the uses of insect cell cultures in insect, animal and plant virus research are the reviews by Grace (1968b, 1969), and the review by Black (1969) which gives a detailed account of the use of cell cultures in plant virus research.

Before it is possible to discuss the types of cells which have been grown in vitro, it is necessary to say something about the sources of the tissues and the conditions under which the cells are cultured.

For virus studies it is of course preferable to culture those tissues in which the virus grows in the insect. Unfortunately, the only tissues that have been cultured at all successfully and from which cell lines have developed are the ovaries and haemocytes. The majority of cell lines now existing were initiated from embryos or larvae which were minced or treated with trypsin to dissociate the cells, so that it is impossible to tell the cell types that eventually transformed into cell lines. Even for those lines whose primary explants were ovaries or haemocytes, it is not possible to identify categorically the type(s) of cell which grew. However, the cell types that are found in primary cultures appear to be very similar, irrespective of the species, instar (embryo, larva, pupa or adult) or tissue from which they arose.

One of the major difficulties in culturing insect tissues, and one which has discouraged many from trying, is the small size of most insects. It is most difficult, if not impossible to get enough tissue to culture from many species. If specific organs or tissues are not required, large numbers of insects can be bred or caught and simply minced, and this is usually done when starting cultures of embryonic tissues. Another major difficulty often encountered is that it is not possible to maintain some insects all the year round, either because they will not breed in the laboratory or the food they eat is only available in certain seasons, e.g. the silkworm. This problem is being overcome as advances are made in producing artificial diets.

The stage in the life cycle when the tissues are taken can also be important. In general, tissues from the immature stages, (embryos or larvae) produce better growth and survive longer in culture than do adult tissues. Of the 21 lines of cells of mosquito tissues most were initiated from either newly hatched larvae or embryos. However, tissues from much later stages have also been used successfully. I used tissues from fourth instar larvae just about to pupate (Grace 1966). Sweet and Dupree (1968) developed a line of cells from pupal tissues of *Aëdes vexans* and another line from adult tissues of *Culiseta inornata*. The best instar from which to culture cells of *Drosophila*, which is the only other member of the Diptera from which cell lines have been established also varies. Horikawa and Fox (1964) maintained that eggs 8 hr old were the best. During the first 2 hr after fertilization no cells were obtained which would multiply in vitro. Echalier and Ohanessian (1968) used eggs between 8 and 14 hr old while Lesseps (1965) found eggs 11 hr after fertilization most suitable. Schneider (personal communication) successfully developed a line of cells from embryos 20–24 hr old, i.e. very near to hatching.

Hirumi and Maramorosch (1964 a, b, c) showed that in the leafhopper *(Macrosteles fascifrons)*, in which the embryo takes 11 days to develop, the only satisfactory age to culture the tissues was between the 7th and 8th days, when the embryo was undergoing blastokinesis. Similarly, Chiu and Black (1967) found the best stage to culture embryos of another leafhopper *(Agallia constricta)* was at the 7th day of development. Other species in which the best time to culture the tissues has been determined are the housefly *(Musca domestica)* – 6 hr after fertilization (Eide and Chang 1969), and the cockroach *(Blabera fusca)* – at the time when the dorsal vessel has begun to function, which is at a relatively late stage in embryogenesis.

In Lepidoptera, the ovaries from diapausing pupae or from spinning larvae have consistently given much better growth in vitro than ovaries from either younger larvae or developing adults. Embryos of moths have only rarely been used in cultures. This is not because they would not be suitable but because it is so much easier and practical to use tissues from pupae.

By contrast with insects, it appears that cells from all stages of ticks, of either sex, grow in culture (Rehacek 1971). Cells grow best when taken from metamorphosing nymphs several days after engorgement, when the developing adult appendages can be clearly seen. If the tissues are taken from either pre-imaginal stages or at a late stage of development, cell growth is usually much poorer. The tissues are either cut into small pieces or separated into cells by gentle pipetting or by the action of trypsin. If only gentle agitation is used it is possible to separate the haemocytes from the other cells.

17.2 Factors affecting cell growth and morphology

Many factors may affect the morphology of the cells in culture and of these the most important are the medium in which the cells are grown, and the temperature.

17.2.1 The medium

A change has occurred over the past 4–5 years in our ideas of what constitutes an adequate insect cell culture medium. Previously it was emphasized that it was essential that the medium should be based on an analysis of insect haemolymph in both its chemical and physical characteristics, and that haemolymph must be added for prolonged growth (Grace 1954; Day and Grace 1959; Jones 1966). These ideas have been shown to be wrong, as several media, notably those formulated by Mitsuhashi (1967) for leafhopper tissues and by Kitamura (1970) for mosquito tissues are based on vertebrate culture media. They are usually supplemented with hydrolysates of yeast or lactalbumin to increase the amino acid concentrations to a level approaching that of the haemolymph. It has also been found unnecessary to add haemolymph, if from 10–20% vertebrate serum is added. All the dipteran cell lines are now maintained in media containing vertebrate sera. Even my original *Aëdes aegypti* line (Grace 1966) is now grown in medium containing foetal calf serum (Nagle, Crothers and Hall 1967; Converse and Nagle 1967; Hsu, Liu and Suitor 1969).

Most of the moth lines are grown at present in either Grace's medium (Grace 1962) or in media which were formulated from analyses of insect haemolymph. In most instances the haemolymph has been replaced by bovine serum (Yunker, Vaughan and Gory 1967; Sohi 1969; Sohi and Smith 1970).

The amount of serum added, whether it be from an insect or vertebrate, may affect the behaviour and the shape of the cells. The serum appears to be necessary for the cells to attach to the surface of the culture flask, as with vertebrate cells (Willmer 1965). This is important, especially in primary cultures, as only attached cells divide. In many of the mosquito lines the cells tend to form monolayers attached to the surface but in many of the moth lines, the cells, in stationary cultures, tend to be only loosely attached and some remain floating. Unattached cells are usually round or spindle-haped. Vago and Chastang (1962) studied the effect of different concentrations of haemolymph and calf serum on the growth of cells from ovarian tissue of the silkworm *Bombyx mori*. In media containing either 15% haemolymph, 10% haemolymph and 5% calf serum, 5% haemolymph and 10% calf serum, or 2% haemolymph and 13% calf serum, the cells multiplied, migrated and lived equally well, but in 15% calf serum, the initial migration, multiplication and survival of cells were all decreased. I have tested different concentrations of *Antheraea eucalypti* heat-treated haemolymph (0–50%) on the growth of *A. eucalypti* cells (Grace 1968b). With less than 2% or more than 10% haemolymph (without added bovine plasma albumin) the cells became round, ceased dividing and degenerated, in 50% haemolymph the cells died in 2–3 days. Nowadays 1% haemolymph is used but it is also necessary, at this concentration, to add 1% bovine plasma albumin for continuous growth.

17.2.2 Temperature

The shapes of insect cells in culture are affected by temperature. Within the normal temperature range (24–30 °C) cells do not vary much in shape, but at higher or lower temperatures their shape and behaviour varies. As the temperature is lowered the cells become more round and produce fewer cytoplasmic processes (which may indicate less activity of the cells). At 4 °C all the cells are round and although they remain attached to the substrate they detach more easily than at higher temperatures. The cells continue to multiply at a slow rate at low temperatures but because they move very little, small colonies form. When the cells are returned to their normal growth temperature they become more firmly attached and once again produce processes and move actively. Cells can be left at temperatures around 4 °C for about 2–3 weeks without harmful effects. When the temperature is increased above normal the cells react differently. They detach from the surface of the culture vessel at temperatures above 32 °C, but do not become round, as at low temperatures, but remain flattened with ruffled edges. At 37 °C the cells start to degenerate after about 48 hr and will not survive longer than 7 days. Mitsuhashi (1968) studied the effect of temperature on the growth of cells from the rice stem borer *Chilo suppressalis* and found that 24 °C was the optimal temperature for growth. At either 20 °C or 30 °C the cells did not multiply as rapidly and more cells died.

Insect cells, like mammalian cells can be stored frozen. I have kept cells from all the lines I have established either in a refrigerator at $-85\ °C$ or in liquid nitrogen ($-180\ °C$). *Antheraea eucalypti* cells are living after being frozen for 5 years and the *Aëdes aegypti* and *Bombyx mori* cell lines for 3 years. The percentage regeneration is usually quite high – between 70–90%.

One of the main differences between vertebrate and invertebrate cells is the temperature range at which the cells grow. Vertebrate cells will grow at temperatures between about 32–40 °C whereas invertebrate cells will grow only between about 16 and 32 °C.

17.3 Types of cell

Before a piece of tissue or a whole organism, such as an embryo, is set up in culture, either it is chopped into small pieces or the cells are dissociated, usually with trypsin. Although every type of cell within the tissue or organism is present at the time the culture is started, within a few days most types of cell disappear. One of the remarkable features of tissue culture is that, irrespective of whether the original tissues or cells were of vertebrate or invertebrate origin, the types of cells which eventually remain and grow in culture are morphologically similar. Insect cell cultures usually contain three main morphological types of cell, although I have shown (Grace 1968a) by cloning experiments (see later this section) that several distinct types of cell can be obtained from one cell line.

The most common type of cell in cultures is the fibroblast-like cell or spindle-shaped cell (fig. 17.1a). The term spindle-shaped is to be preferred as true fibroblasts are found in the vertebrates. These cells move actively but do not produce cytoplasmic processes. If they contact other cells they will move away or sometimes form a loose network of cells.

The second type of cell is the epithelial cell, (fig. 17.1b) which is usually rounder than the spindle-shaped cell. It adheres firmly to the surface of the culture vessel and may aggregrate into flat compact sheets. These cells are found in most insect tissue cultures and vary somewhat in shape and size. Mitsuhashi (1965) described several types of epithelial cells from tissues of a number of species of leafhopper.

The third main type of cell has been called various names – amoebocyte, phagocyte, amoeboid cell, wandering cell and plasmatocyte (Marks and Reinecke 1964; Hirumi and Maramorosch 1964a, b, c; Mitsuhashi 1965) (fig. 17.1C). As the names imply, these cells move actively and are phagocytic.

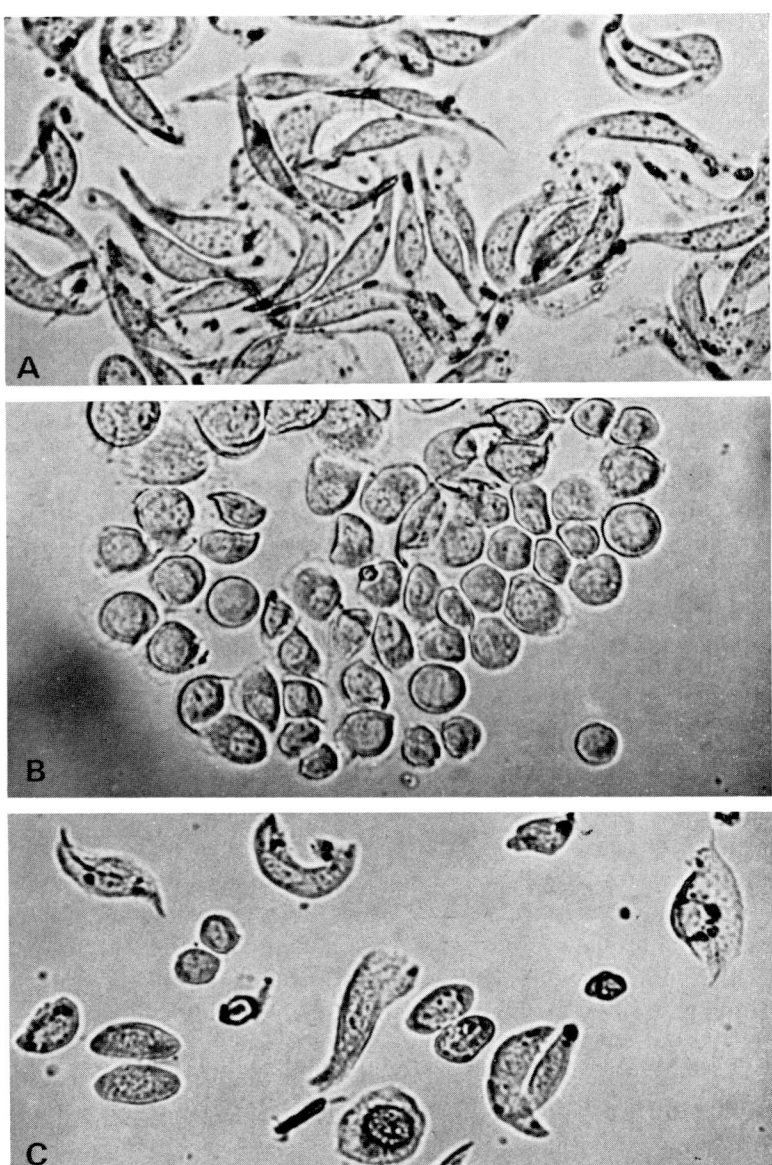

Fig. 17.1.   Light microscope photographs of the three main types of cell found in insect and tick cell cultures. A) spindle-shaped cells; B) epithelial-like cells; C) amoeboid (phagocytic) cells.

They usually spread out flat on the surface of the culture vessel and the cytoplasm is irregular at the edges. The cells do not aggregate and have no definite shape.

In cultures of vertebrate cells, 5 or 6 types of cell can often be distinguished. Three of the types – the epitheliocytes, mechanocytes (fibroblasts) and amoebocytes are very similar to the types found in invertebrate cultures. The other types are the nerve cells, lymphocytes and neuroglia cells. In invertebrate cultures cells corresponding to the last 3 types do not survive.

In most cultures, cells of all 3 types will be found but often one will predominate. In the *Antheraea eucalypti* cell line, for example, the spindle-shaped cell is the most common whereas in the *Aëdes albopictus* line (Singh 1967) most of the cells are round and epithelial. What determines the abundance or otherwise of each cell type in a culture is not clear. Neither is it clear from which type of cell in the insect the cultured cells originally arose. It is possible that the amoebocytes, for instance, may be blood cells.

Most of the tissues found in insects have been cultured but in only a few has it been possible to establish cell lines. Muscle cells, fat body, hypodermis, gut, and nerve tissue, have survived for short periods but with little growth (Schmidtmann 1925; Grace 1954; Judy 1969; Mitsuhashi 1965; Collier 1920; Gavrilov and Cowez 1941; Pfeiffer 1943; Seecof and Teplitz 1971).

17.3.1 Ultrastructural morphology

Very few studies have been made of the ultrastructure of the cells from any insect cell line. Grace and Mercer (1956) and White (1971) examined the *Antheraea eucalypti* cells in an electron microscope and Vago and Croissant (1963) studied the spindle-shaped cells from primary cultures of the silkworm *Bombyx mori*. The cells from the mosquito *Aëdes albopictus* have also been studied (R. S. Raghow and T. D. C. Grace, unpublished observations). The most interesting result of all these studies was that the cells, irrespective of their origin were very like each other and resembled undifferentiated cultured vertebrate cells. (Paul 1961; Willmer 1965, Kouri et al. 1971). The main characteristics of the cells are shown in fig. 17.2a–c. The nuclei in all the cells were large and usually contained more than one nucleolus (especially in the *Antheraea eucalypti* cells). The nuclei of the *A. eucalypti* and *Bombyx mori* cells were round but in the *Aëdes albopictus* cells were often lobed and on one side of the cell. All the cells have a large nuclear/cytoplasm ratio which is consistent with their being undifferentiated. The mitochondria may vary considerably in shape and size and may be very long, and may contain few

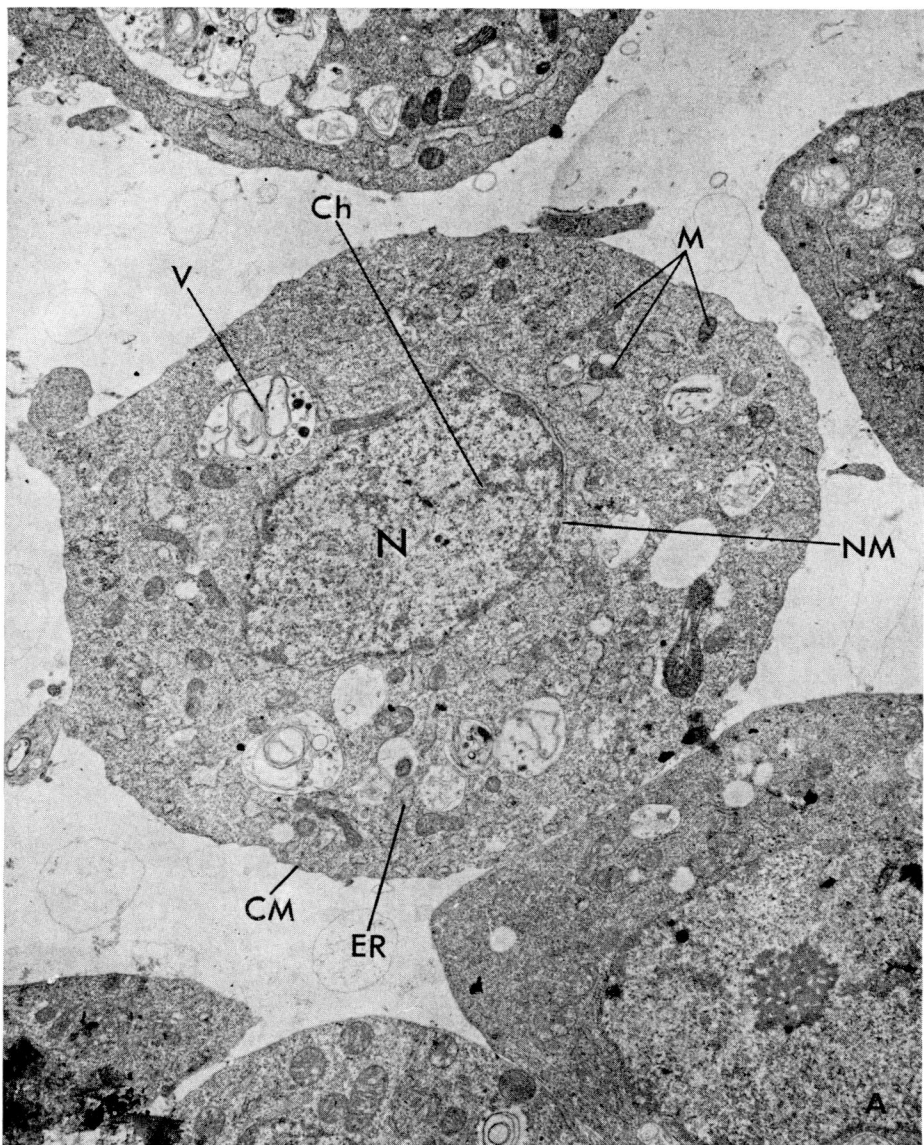

Fig. 17.2. Electron micrographs of sections through mosquito (*Aëdes albopictus*) cells (A and B); *Antheraea eucalypti* (C) and baby hamster kidney (BHK) cells (D). Ch = chromatin, CM = cell membrane, ER = endoplasmic reticulum; M = mitochondria; MT = microtubules; N = nucleus; NM = nuclear membrane; NP = nuclear pores; R = ribosomes; RER = rough endoplasmic reticulum; V = vacuole (lysozome).

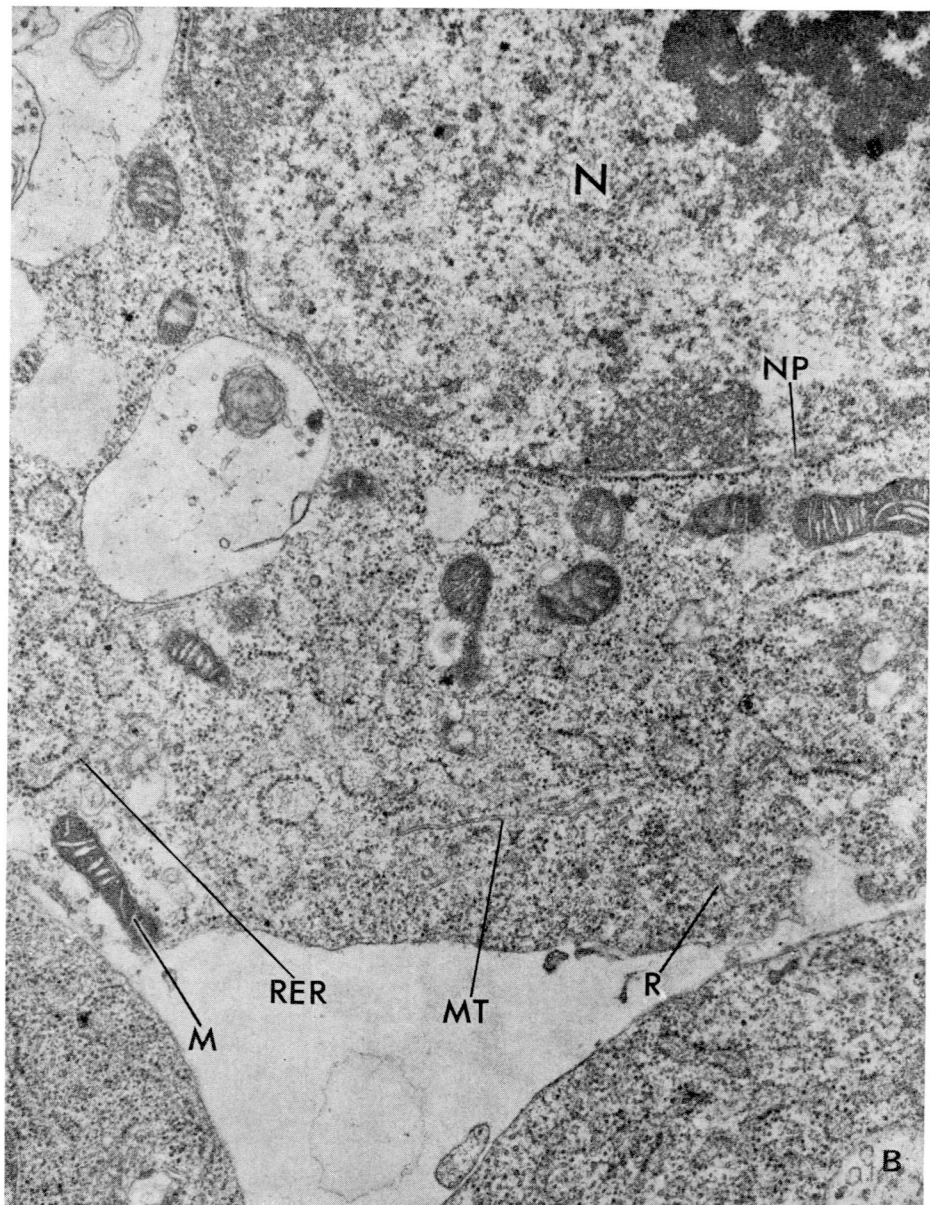

Fig. 17.2B.

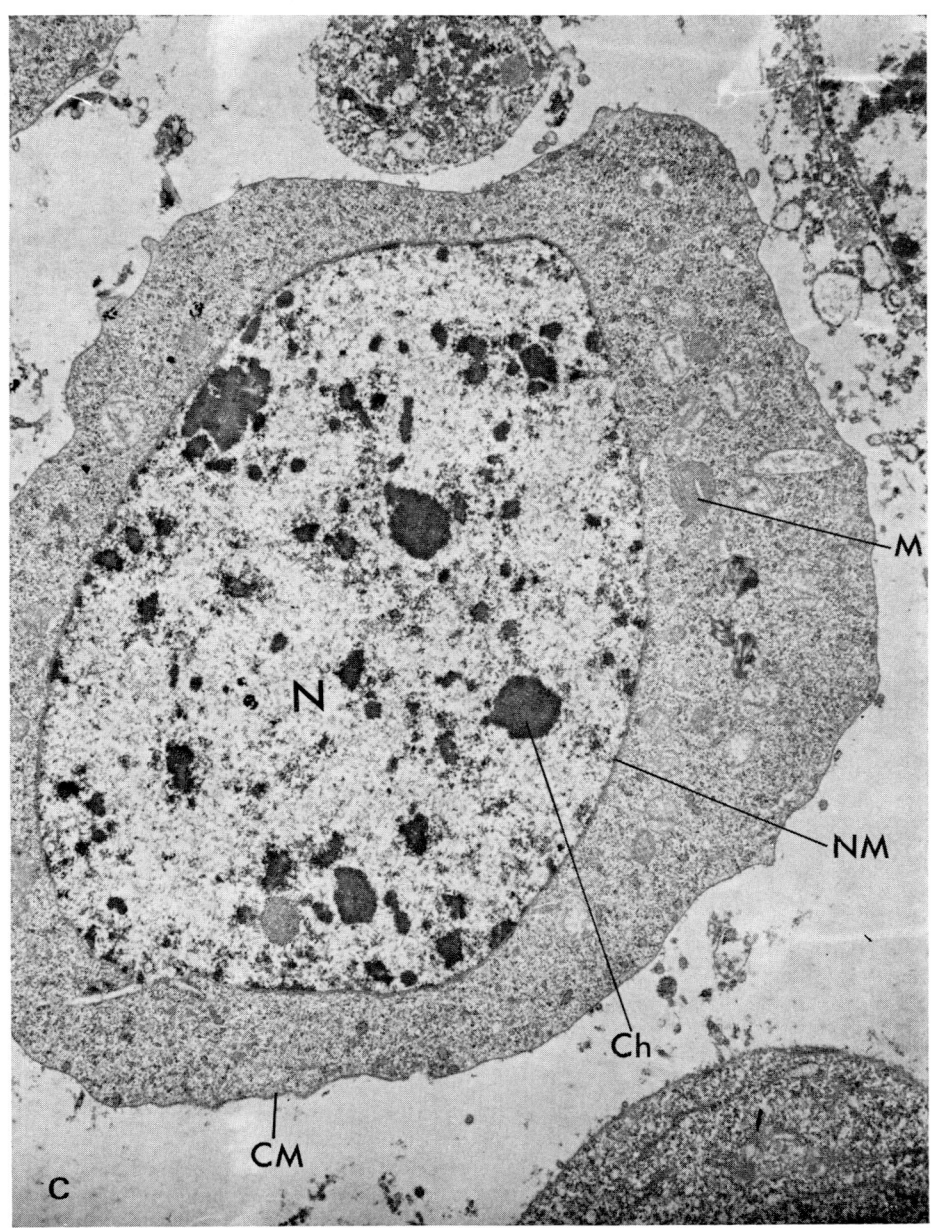

Fig. 17.2C.

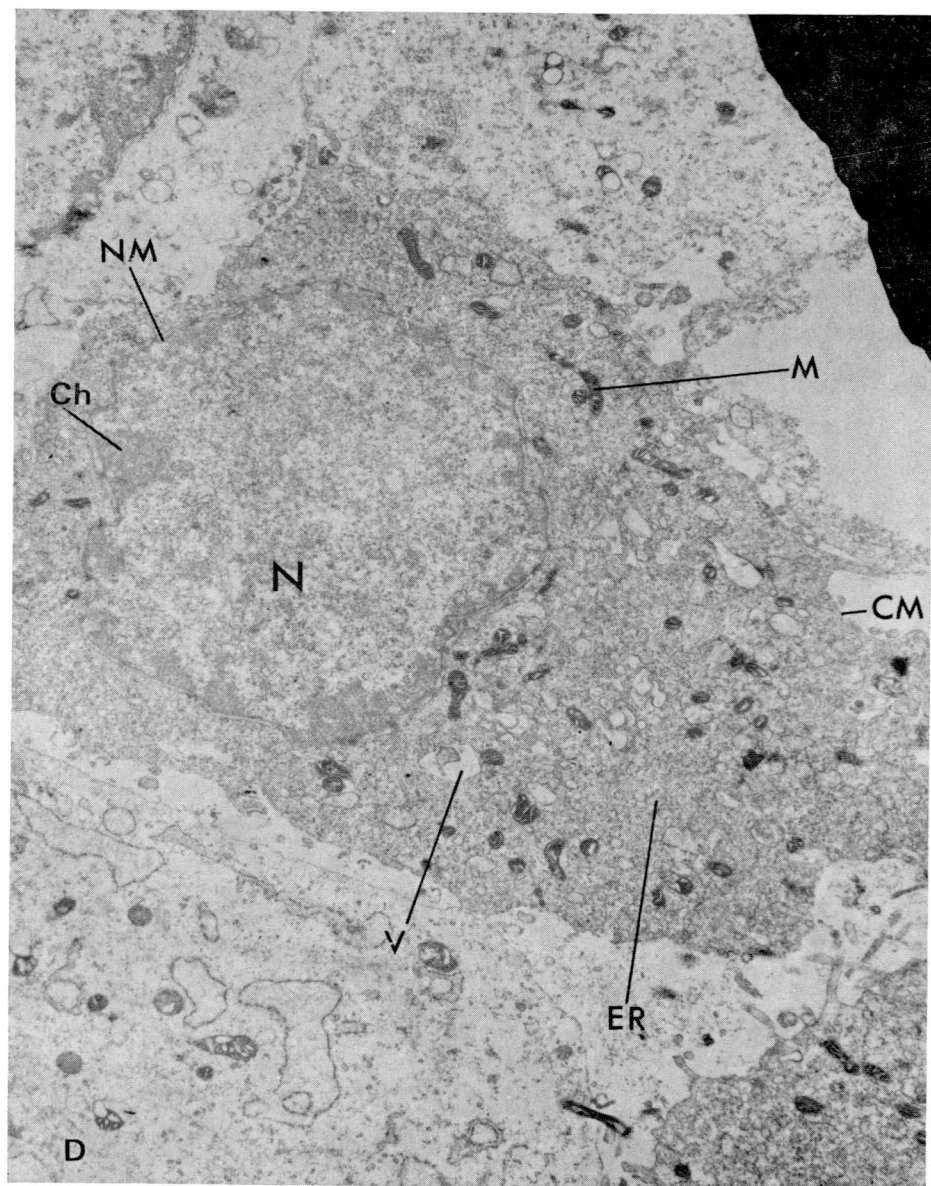

Fig. 17.2D.

cristae. In the cytoplasm of all the cells there were vesicles which appeared either as dark, dense bodies or contained cell debris. They were covered by a unit membrane and gave positive reactions to periodic acid-schiff reagent (PAS) and paraldehyde fuchsin (PAF); there is good evidence that they may be lysosomes (White 1971).

In many of the cells examined, from all species, pinocytotic vacuoles could be seen, especially around the nucleus. The endoplasmic reticulum, both smooth and rough (ribosome-covered) was prominent in all cells but there appeared to be less of it in *Antheraea eucalypti* cells. In the spindle-shaped cells, microtubules were plentiful and were generally oriented in the direction of the processes. The Golgi apparatus appeared normal in all cells and the cytoplasm was generally finely granular.

17.3.2 Lepidoptera cell cultures

Eleven lines of moth cells have been reported (see table 17.1). The three types of cell are present in all the lines but there are some interesting differences. In the line CP-1268, which is one of the two codling moth *(Carpocapsa pomonella)* lines obtained from macerated embryos (Hink and Ellis 1971) 61% of the cells are spindle-shaped, and 25–47 μ long and 9–16 μ wide. The next most common cell (23%) is round (13–19 μ in diameter). The rest of the cells (15%) have a single extension and are 9–10 μ wide by 24–35 μ long. In the other line, CP-169, the majority (70%) of the cells are round (11–20 μ diameter), 22% have a single extension (11–16 μ × 19–40 μ) and 7% are spindle-shaped, (10–13 μ × 27–31 μ). Not only are there differences in the proportion of each type, there are also differences in behaviour. The CP-1268 line has a lag phase of 24 hr, whereas it is 96 hr in line CP-169. The two lines also differ quite significantly in their rate of growth. Whereas CP-1268 reaches a maximum population of 10^7 cells/ml (a 40-fold increase) in 7–8 days CP-169 has a maximum population of 5×10^6 cells/ml (14-fold increase) which it reaches in 8–9 days. The chromosome contents of the cells also differ. About 5% of CP-1268 cells are diploid (2n = 54) and about 34% are tetraploid and some contain more than 250 chromosomes. In line CP-169, approximately 9% of the cells are diploid and about 72% contain more than 100 chromosomes.

In the cell line established from minced adult ovaries of the cabbage looper *(Trichoplusia ni)* (Hink 1970) the cells vary greatly in shape and size, some are round, others oval and some have one, two or three protoplasmic extensions. The cells are from 9 μ–59 μ in diameter. This line, like the

Antheraea eucalypti line (Grace 1962) and several others contain some binucleate cells. The cells have a population doubling time during their logarithmic phase of growth of about 16 hr which is comparable to the growth rate of vertebrate cell lines. One important characteristic of this line is that it is grown in a medium containing 8% foetal bovine serum, 8% whole chicken egg ultrafiltrate and 0.5% w/v crystallized bovine albumin, and never in a medium containing insect haemolymph, so that it is unlikely to have been contaminated with insect cells or viruses. Many attempts have been made over many years to culture haemocytes, however the results generally have been very poor. Only two lines of cells have been established from haemocytes and only one of these is still extant. In 1967, Mitsuhashi reported the establishment of a line of cells from the haemocytes of the rice stem borer *Chilo suppressalis*. Although there are seven types of haemocytes in *C. suppressalis* only the prohaemocytes and plasmatocytes continued to grow in culture. This cell line was extremely interesting as it was persistently infected with *Chilo* iridescent virus (CIV). Apparently the cultures had been accidentally infected with CIV about 15 months after they were started. All the cultures which were infected died except one, and although most cells in the surviving culture died some cells continued to multiply. The predominant cells were round (18 μ in diameter), and grew loosely attached to the culture vessel in an open network. When the cells were actively multiplying they did not show any iridescence (caused by masses of virus) but in dead cells and in masses of cells a bluish-green iridescence could often be seen when viewed by reflected light. Unfortunately, this line of cells died about 1968.

Recently, another line of haemocytes has been developed by Chao and Ball (1971), from the pupae of the cynthia moth *(Samia cynthia)*. Chao and Ball found that haemocytes from pupae which were 1–4 months old at first grew well in culture but died when subcultured. If, however, the haemocytes were removed from pupae which had been in diapause for 10 months or more they grew differently. They attached readily to the culture flask and formed a sheet of cells as they multiplied. Concurrently, small round cells (13–24 μ in diameter) appeared. These cells did not attach but floated freely in the medium, and were subcultured; it was not possible to isolate or subculture the attached cells. In old cultures of this line, spindle-shaped cells have appeared. Although it is not certain, it is believed that this cell line came, as did Mitsuhashi's, from prohaemocytes.

At the beginning of this section it was pointed out that there are three main morphological types of cell in all the cell lines. However, I (Grace 1968a) have shown by cloning experiments that there are in fact, many morphological types of cell in the *Antheraea eucalypti* cell line. In this line the most common cell is spindle-shaped (15) the next most frequent is of the epithelial type (20–40 μ in diameter) and there are also round cells (10–15 μ in diameter). By using the dilution method of cloning, based on the technique described by Paul (1961), I have isolated ten clones of cells. The clones, although morphologically distinct, can still be grouped into the three types originally described (Grace 1962). The cells in six of the clones were spindle-shaped. In two, the cells were round to spindle-shaped and grew in masses adhering to the glass. In one clone the cells were round and showed little variation in either shape or size. The cells of the tenth clone were very variable in shape and size. This variability was not eradicated by recloning.

The criterion used in cloning the cells was to separate all the cells of distinctly different shapes. Although this is the easiest characteristic to choose it may be the least useful as it may not be correlated with other cell characters (e.g. susceptibility to viruses). However, it did show that different shapes of cell are distinct types and not morphological variants, of the 3 basic forms.

There is little point in describing the characteristics of each of the other lines of moth cells, as in general they are similar to those that have been discussed.

17.3.3 Tick tissues

Many attempts have been made to culture tick cells (Rehacek 1958, 1962, 1965; Martin and Vidler 1962; Rehacek and Pesek 1960; Varma and Pudney 1967; Yunker and Cory 1967; Rehacek and Brzostowski 1969a, b) because of the importance of many species of ticks as vectors of viruses which cause serious diseases in man and animals. As stated previously, all the tissues and organs, have been used in culture. Developing adults or nymphs undergoing metamorphosis provide the best tissues. No cell lines have yet been established but primary cultures have survived for up to nine months (Rehacek and Brzostowski 1969b). The general method of preparing cultures is to remove the gut and Malpighian tubules and then to mince or cut the body into small fragments. The fragments are then set up in culture or dissociated into single cells with trypsin. Haemocytes can be separated if they are dissected very carefully. After the body has been fragmented the pieces are suspended in medium. After a short while, when the larger fragments have settled, the supernatant, containing the haemocytes is recovered and cultured separately.

The types of cell which are found most frequently in tick cell cultures are spindle-shaped or epithelial. Many of the spindle-shaped cells have long cytoplasmic processes at either end. In cultures of ovarian tissues from *Dermacentor andersoni* the cells were mainly of the epithelial type. The cytoplasm of many of the tick cells grown in vitro contain fat droplets. These are seen when the cells are first cultured and persist in most cells.

Primary cultures of tick and dipteran tissues, but not moth tissues often contain hollow vesicles. These are composed of a single layer of epithelial cells which is covered by a membrane on the outer surface, and can grow to quite a large size (2 mm in diameter). It is not understood why the vesicles form but sometimes they appear about one day after the cultures are started. Usually they are round, or oval and may either be attached to the original explant or to the culture vessel or float freely in the medium. Inside the vesicles haemocytes and other cells can be found as well as medium. The cells may float in the medium or form loose networks. The cells composing the vesicle multiply and the vesicles increase in size and gradually become darker, after which they degenerate. Usually the first few subcultures result in very good growth but in subsequent passages the vesicles become smaller but more numerous until eventually the cultures consist of many small vesicles with no cells.

17.3.4 Leafhopper and planthopper cells

Mitsuhashi (1969) is the only person so far who has attempted to culture planthopper cells. He cultured embryonic tissues from the planthopper *Laodelphax striatellus* which had not developed beyond blastokinesis. The embryos were dissociated with 0.1 % trypsin and cultured in Grace's medium which contained 20 % foetal bovine serum and 1 % egg extract from *L. striatellus*.

About a day after making the cultures spindle-shaped and epithelial cells migrated from the explants; also hollow vesicles, similar to those seen in the leafhopper and mosquito cells appeared. Mitsuhashi observed mitoses in the cells of the vesicles but not in the other cells. It was not possible to maintain the cultures beyond 40 days.

Although only four lines of leafhopper cells have been reported, this is not a true indication of the importance of these insects in the field of plant virology. Leafhoppers are vectors of many viruses and possibly mycoplasma-like organisms which cause serious damage to many plants of economic importance. Leafhoppers are also of interest to virologists because many of the disease agents multiply in their tissues as well as those of the plant host, although they do not have any obvious pathological effect on the insect.

The four lines were all established by Chiu and Black (1967) from embryos of 4 species of leafhopper – *Aceratagallia sanguinolenta, Agallia constricta, A. quadripunctata* and *Agalliopsis novella*. In addition to these lines, primary cultures have been made from ten other species of leafhopper (Hirumi 1971).

The media used for the leafhopper cultures are rather interesting. In their early experiments Chiu and Black used medium developed by Mitsuhashi and Maramorosch (1964) for the culture of leafhopper tissues but found it was not adequate. In the cultures they used Schneider's medium (1964) which was designed for growing *Drosophila* tissues. Chiu and Black included in their first media some *Antheraea pernyi* haemolymph but later found that it could be replaced with foetal bovine serum. More recently Chiu and Black (1969) have halved the concentrations of inorganic salts in their culture medium.

Tissues were cultured from leafhoppers at all stages of the life cycle but the most successful source, undoubtedly has been 7–8 day old embryos. Chiu and Black (1967) simply macerated the embryos before putting them in culture medium and did not use any trypsin. However, trypsin has been widely used to dissociate the cells and sometimes seems to have increased growth in the first few days of culturing.

The types of cell which grow in culture are mainly spindle-shaped or epithelial although there are some other forms which will be described. First to appear are spindle-shaped cells (40 × 8 μ) which usually grow very actively and in a matter of a few days form a network in which many cells may be seen dividing. Next epithelial cells (75–120 μ) migrate from the explants. These cells adhere to the surface of the culture vessel and form compact sheets. Wandering cells have been observed in cultures of *Macrosteles fascifrons* (Mitsuhashi and Maramorosch 1964) and in *Nephotettix cincticeps* (Mitsuhashi 1965).

In one line of *Agallia constricta* the cells are quite slender while in a subline from the same line there are two forms of epithelial cell, one being much larger than the other. The only other types of cell observed are 'phagocyte-like' cells which appeared about 24 hr after tissues of *Macrosteles fascifrons* were cultured. None of this type of cell was seen to divide.

Some of the fragments, especially muscles from various organs such as the gut, ovaries and legs may continue to contract in culture for many weeks.

17.3.5 Dipteran cells

As many species of mosquitoes are vectors of viruses, bacteria or protozoa which cause some of the most serious diseases in man and animals, it is not surprising that a great effort has been made to establish lines of cells from these important insects. The first line of dipteran cells was established in 1966 (Grace 1966) but there are now 27 lines (see table 17.1) and of these 21 are from mosquitoes and 6 from *Drosophila*. Primary cultures of tissues from the tsetse flies *Glossina morsitans* (Schneider 1971) and *Glossina palpalis* (Trager 1959), vectors of the trypanosome which causes the sleeping sickness, the flesh fly *Sarcophaga bullata* (Shinedling and Greenberg 1971) and two species of *Musca* – *M. domestica* and *M. sorbitans* – have been made but have not yet yielded cell lines.

17.3.5.1 Mosquito cells

Several media have been used for culturing the mosquito tissues, but only one was specifically designed for mosquito cells (Kitamura 1965). The others were designed for moth cell cultures (Grace 1962) or leafhopper cell cultures (Mitsuhashi and Maramorosch 1964). Scheinder (1969) used modified Grace's medium to culture tissues from *Anopheles stephensi*.

All the mosquito cell lines, except the one established by me (Grace 1966) have been initiated and maintained in media which contain no haemo-lymph, but, usually, foetal bovine serum.

The three types of cells already described are found in all the cell lines as well as the hollow vesicles already described above. In my original *Aëdes aegypti* line the majority of the cells were spindle-shaped and very large (40–50 μ long × 8–10 μ wide). There were also round cells 20–60 μ in diameter.

The way in which the mosquito and moth primary cultures developed into established lines seems to be different. When first cultured the moth tissues grew rapidly for several weeks, then the growth rate decreased and most cells died. The growth rate did not increase again for 3–12 months and was then maintained indefinitely. By contrast, in most of the mosquito cell lines there was little or no period in which the growth rate decreased, and the cells continued to grow rapidly practically from the time they were first put in culture medium.

Most lines of dipteran cells form monolayers and adhere quite firmly to the surface of the culture vessel. In some of the lines, such as the *A. aegypti* (Grace 1966) *A. vexans* and *Culiseta inornata* (Sweet and Dupree 1968) monolayers are not formed, the cells either float in the medium or are attached loosely to the substrate.

The *Aëdes aegypti* lines (Mos 29, Mos 20, Mos 20(A)) established by Varma and Pudney (1969) differ in their growth patterns from other mosquito cells. All were grown from minced first instar larvae in bottles or Leighton tubes with modified Kitamura's medium. In the Mos 29 no cells migrated and attached to the substrate until 25 days after the cultures were set up. About 7 days later hollow vesicles composed of epithelial cells appeared. The cells were first subcultured 67 days after they were started. Although the vesicles were transferred at each subculture they gradually decreased in number until by the 10th subculture none was found. The cells formed monolayers of spindle-shaped and epithelial cells. In the line Mos 20 the cells, soon after isolation were clearly epithelial and vesicles were never observed. In the third line Mos 20(A) vesicles appeared in the cultures 4 days after they were started and it was not until about 3–4 weeks later that epithelial cells appeared. At this stage the vesicles were discarded at each passage, and consequently the line eventually consisted of only epithelial cells. Occasionally vesicles appeared in the line but they were discarded.

The *Aëdes aegypti* line of Singh (1967) and the lines of *A. vittatus* (Bhat and Singh 1969, 1970), *Culex tritaeniorhynchus* and *C. salinarius* (Schneider 1971) are all similar in that they all contain some vesicles. The pattern of development was very similar. In each instance newly hatched larvae were macerated and the fragments put in culture medium. About 3 days later small vesicles appeared at the cut ends of the fragments. The vesicles, which were hollow and floated in the medium appeared to consist of a monolayer of epithelial cells. At this stage very few isolated cells were found. About two weeks after starting the cultures the larval fragments and vesicles were removed, cut into pieces and returned to the culture vessel. Two days later masses of epithelial cells attached themselves to the culture bottle and hollow vesicles continued also to form. Later the hollow vesicles were removed when, at 2 week intervals, the cells were subcultured. The cultures consisted mainly of epithelial cells but even after many subcultures hollow vesicles and tube-like structures still appear. In the two *Culex* lines (Schneider 1971) the predominant cells are epithelial but round and spindle-shaped cells, as well as the vesicles are also present.

Most of the mosquito lines, with the exception of my *Aëdes aegypti* line (Grace 1966) and the *A. vexans* and *Culiseta inornata* lines of Sweet and Dupree (1968) contain diploid ($2n=6$) cells. In the line of *Aëdes w-albus* (Singh and Bhat 1970) most of the cells are diploid but some are polyploid.

17.3.5.2 Drosophila cells

The establishment of lines of cells from *Drosophila* tissues has fulfilled the wishes of many geneticists. The first *Drosophila* lines were established by Echalier and Ohanessian (1970), from embryos 6–12 hr old. Further cell lines have been established by Kakpakov et al. (1969); Goosdev, Platova and Polukarova (1969), Schneider (unpublished) and Richard-Molard (unpublished). Most is known about the cells of Echalier and Ohanessian. They established eight cell lines. Most of them are diploid (2n = 8) although about 5–7% of the cells in some cultures have up to 32 chromosomes (tetraploid). There are some interesting genetic features of the cells which could be very useful in providing chromosomal markers. Besides there being normal female cells with two X chromosomes there were also cells containing only one X chromosomes (XO) and cells with one short heterochromatic body (possibly a Y chromosome) and an extra piece attached to one of the X chromosomes.

The majority of the cells appear to be epithelial although spindle-shaped and round cells are quite common. The shapes of the cells in some lines are quite distinct from the shapes found in other cell lines. For instance in diploid line 'K' the cells adhere loosely to the culture, are round or spindle-shaped and tend to remain separate. The nuclei are large and spherical and there is a distinct nucleolus. On the other hand in the diploid line 'C' the cells are spindle-shaped, they adhere firmly to the surface of the culture vessel and tend to form open networks of cells. The nuclei appear to be larger than in line K.

Recently, Richard-Molard (see Ohanessian 1971) has succeeded in establishing several lines of *Drosophila* cells homozygous for the 'refractory' gene 'ref'. The P^- variant of sigma virus (a rhabdovirus) cannot multiply in cells homozygous for this gene, however, the variant P^+ can multiply in the cells. Cells from refractory and non-refractory lines were inoculated with sigma virus strains P^+ and P^-. Both the virus variants multiplied in the non-refractory cell line but only the P^+ variant multiplied in the refractory line.

17.3.6 Aphid cells

Although there are no lines of cells from aphids their importance as vectors of viruses which cause diseases in plants warrants a short discussion of the efforts to culture their tissues and cells. Tokumitsu and Maramorosch (1966) cultured the cells of the pea aphid *Acyrthosiphum pisum*. The cells and tissues were obtained by rearing the aphids aseptically on pea plants *(Pisum sativum*

L) grown in tubes of agar medium. Parthenogenetic, apterous females were surface sterilized using 70% ethanol and placed on the pea plants. After a few days they produced nymphs which were used for tissue cultures. Besides the aseptic nymphs, ovaries, embryos and digestive tracts from sterilized females were also used. The insects or organs were cut into small pieces and placed in a modified Grace's medium.

The first cells to appear were spindle-shaped. These cells migrated over the surface of the vessel but none divided. In more recent experiments (Hirumi 1971) more active growth and mitotic figures were seen in cultured cells of the pea aphid in a medium consisting of 10 parts Schneider's medium, 10 parts Vago and Chastang's *Bombyx mori* medium 22, and 3 parts foetal bovine serum. Four different types of spindle-shaped cells, epithelial cells and large multinucleate cells grew for short periods.

Peters and Black (1970) cultured the ovarian tissues and embryos from the sowthistle aphid *Hyperomyzus lactucae L.* The tissues were obtained from apterous viviparous females, and were dissociated with pronase. Spindle-shaped cells with long processes were maintained for up to 10 days.

17.4 Identification of insect cells

One of the main dangers maintaining several lines of cells in one laboratory is that through carelessness or accident the lines will become mixed. This would be bad and not only if cells from one insect cell line became contaminated with those from another insect line but especially if an insect cell line and a vertebrate cell line became mixed. In recent years there have been numerous reports from several countries of vertebrate cell lines becoming contaminated with cells from another vertebrate cell line but so far there have been no reports of an insect cell line being contaminated with vertebrate cells or vice versa.

When there were only a few insect cell lines and they were grown in media more or less specifically designed for insect tissues, in a small number of laboratories, the dangers of contamination were quite small. Nowadays, however, when there are many lines that are grown in many laboratories, often alongside vertebrate cells, and in media which are, in some instances, modified vertebrate media, the danger of contamination is very great indeed. For this reason no vertebrate cells have ever been grown in my laboratory. If it is necessary to use vertebrate cells, e.g. to plaque assay for arboviruses on vertebrate cells, it is essential to arrange with another laboratory to use their cultures and facilities.

There are two factors which reduce the risk of insect cells and vertebrate cells contaminating each other. Firstly, insect cells grow best between 16–32 °C while vertebrate cells will only grow between about 32–40 °C. Secondly, although many of the media used, especially for the culture of mosquito and leafhopper cells, are based on vertebrate media (e.g. T.C. 199), various modifications are made to make them suitable for growing insect cells. Most important is the addition of protein hydrolysates to increase the free amino acid concentration which can be up to 20 g/l in some insects and is often between 8–15 g/l (Wyatt, Lougheed and Wyatt 1956). Often the salts in the media are adapted to reflect the concentrations found in insect haemolymph. In the haemolymph the content of magnesium is relatively high and chloride has largely been replaced by organic constituents both osmotically and in regard to cation balance. The mole ratio of sodium to potassium ranges from 0.1–2.5 in haemolymph; in human blood it is 29. Because of these changes in the media, vertebrate cells usually do not grow in the media used for growing insect cells. I (Grace 1963) have tried unsuccessfully to grow HeLa cells, in the medium used for *Antheraea eucalypti* cells; within a few hr after the medium was placed on the cells they rounded up, and came free from the glass, and after 48 hr every cell was dead. Similar experiments using *A. eucalypti* cells and Eagle's medium and T.C. 199 gave the same results. However, a recent report by Conover et al. (1971) shows that care must be taken when growing vertebrate and insect cells in the same laboratory. In experiments designed to produce human-mosquito somatic cell hybrids the authors had to grow mosquito cells (Grace 1966) and HeLa cells in the same media. They found that the HeLa cells grew slowly in Grace's medium at 37 °C but little or no growth occurred at 26–28 °C. By contrast the mosquito cells would not grow at 26 °C, 28 °C or 37 °C in Eagle's Minimum Essential Medium (MEM) (Lockhart and Eagle 1959) containing 10 % foetal bovine serum. When MEM and Grace's medium were mixed in equal parts the mosquito cells grew albeit slowly at 26–28 °C and the HeLa cells grew at 37 °C. It is thus possible that contamination could occur especially when cells were carried over in medium.

In 1963 I made extensive tests to determine whether the cells of *A. eucalypti* and *Aëdes aegypti* were different from each other and from vertebrate cells. I tested whether the cells would grow in vertebrate media (mentioned above) and also used gel diffusion serological tests. Five preparations of antigens were tested. 1) HeLa cells; 2) a strain of human epithelioma (Hep cells); 3) an acetone – dried preparation of normal mouse liver cells; 4) cells from the *Antheraea eucalypti* and *Aëdes aegypti* lines; and 5) fresh *Antheraea eucalypti* pupal haemolymph containing blood cells. An antiserum was prepared using *A. eucalypti* cells as immunogen. In the tests the only precipitin bands which formed were between the antiserum and the *A. eucalypti* cells and haemolymph. Further confirmatory evidence was provided from the study of the chromosome complements of the cells from the lines (Thomson and Grace 1963).

Greene and Charney (1971) have also tested the relationship of some of the insect cell lines to each other and to vertebrate cells. The insect cells were from Grace's *A. eucalypti* cell line, obtained from three different laboratories, two samples of Grace's *Aëdes aegypti* cell line obtained from two laboratories; and one sample each of Singh's *A. aegypti* and *A. albopictus* cells. The vertebrate antigens used were cells from rat, man, mouse, bovine kidney, fathead minnow and gekko lung. The antisera were prepared using Grace's *A. aegypti* and *Antheraea eucalypti* cells as immunogens. Greene and Charney made gel diffusion serological tests and also analysed the isoenzymes electrophoretically. The results of the agar gel immunodiffusion tests showed that Singh's *Aëdes aegypti* and *A. albopictus* cells are distinct from each other, from *Antheraea eucalypti* cells, from Grace's *Aëdes aegypti* cells and all the vertebrate cells tested. Also that Grace's *Antheraea eucalypti* cells are distinct from Singh's cell lines and from the vertebrate cells but indistinguishable from Grace's *Aëdes aegypti* cells.

The isoenzyme analysis was found to be an excellent method for identifying the different cell lines and distinguishing them from each other and vertebrate cells. It was possible to distinguish between human, mouse, moth and mosquito cells by identifying the patterns of glucose 6-phosphate dehydrogenase (G6PD) on polyacrylamide gel. The cells from the human, mouse and moth lines each formed a band for G6PD which were of different mobilities. No bands were formed by Singh's *A. aegypti* and *A. albopictus* cells but a band identical with the *Antheraea eucalypti* band was present in a clone of Grace's *Aëdes aegypti* cells developed by Suitor, Chang and Liu (1966) which once more confirmed that the *A. aegypti* cell line of Grace tested was a moth cell. The patterns of other isoenzymes-lactate dehydrogenase, malate dehydrogenase, 6-phosphogluconate dehydrogenase and acid phosphatase all distinguished between the moth cells, Singh's 2 lines of mosquito cells and the mammalian cells.

The results of these experiments indicate the need for testing the other insect cell lines now established. There is no doubt that rigorous precautions should be taken at all stages of cell culture to ensure that contamination does not occur and that cells should be tested at regular intervals. If contamination is found every effort should be made to notify all people who have the cell lines.

17.5 Conclusions

There is no doubt that the culture of invertebrate, especially insect and tick, tissues in vitro will play an ever increasing part in the study of viruses pathogenic to animals and plants. The rapid increase in the number of lines of mosquito cells and the awakening interests of geneticists to the potentialities of the *Drosophila* cell lines should lead, in the next few years to a great amount of research using invertebrate cell cultures.

With the ever increasing concern over the dangers and abuses of chemical insecticides, attention over the past decade has turned to the possibility of using insect viruses as specific insecticides. Consequently, techniques are being sought for growing large amounts of insect viruses. Although this can be done in living insects, there are a number of advantages in producing viruses in insect cells grown in vitro. 1) There is much greater control over the growth conditions, both physical and chemical; 2) Tissue cultures are generally free of other organisms; 3) The growth of large numbers of cells requires less space than growing a comparable mass of whole insects; 4) The cells can be stored very easily in a frozen state, when not in use; 5) There is much less trouble in purifying virus grown in cells than in whole insects; and 6) Only susceptible cells need be grown in culture.

It has been the purpose of this chapter to show that the culture of insect and tick tissues and cells in culture has progressed in the past 20 years from a very difficult and unrewarding technique to the point where it can be used quite effectively to solve a number of important problems in the relationship between insects, viruses, animals and plants.

Chiaroscura; the ecology of viruses and particular invertebrates

CHAPTER 18

Viruses and ticks

HARRY HOOGSTRAAL*

Contents

* This chapter is dedicated to my esteemed mentor Richard Moreland Taylor, M.D.

18.1 Introduction

Ten years ago, Smith (1962) listed 13 different viruses from 11 tick species. Five years ago, Burgdorfer and Varma (1967) reported 25 viruses transmitted by ticks. Now 68 different viruses (table 18.1) are recorded from more than 80 tick species (table 18.2) and certain other viruses are casually associated with ticks (sect. 18.4.2.11). Tickborne viruses known or believed to cause disease in man or domestic animals numbered 13 in 1962 and 20 in 1972 (marked by a plus (+) in table 18.1).

The 68 viruses (table 18.1) infecting about 20 argasid and 60 ixodid species (table 18.2) involve every major vertebrate animal group except marine ones. They occur on every continent and numerous islands. Ecologically, viral habitats range from arctic seabird breeding sites, through temperate conifer and mixed forests, plains, steppes, and mountain peaks, to equatorial savannas and primary and secondary forests. However, much of the world remains poorly explored for these viruses. No more than three viruses have been reported from South American ticks and one of these is probably biologically unimportant. Russian spring-summer encephalitis (RSSE) virus, but no other, has been recorded from the vast area of China. In the USSR, work with 'miscellaneous' viruses from ticks developed only within the past five years. As a result, several new viruses have been discovered in the USSR, knowledge of the geographical and host range of others has greatly increased, and numerous strains are under study. Madagascar, Australia, Japan, Philippines, China, and Central America are among the many important areas of the world still unexplored for tick viruses.

The universally accepted procedures for identifying and arranging these viruses into systematic physiochemical groups were developed during the last two decades. The present classification results chiefly from the investigation of physiological properties and the development of a serological classification of arboviruses by Jordi Casals and his colleagues at the Yale Arbovirus Research Unit (YARU) (Casals 1957, 1962, 1963, 1967, 1970, 1971; Clarke 1960, 1962, 1964a, b). Owing to the diversity of viruses from ticks, their characterization has been an especially important activity of YARU, which has elucidated the relationships and natural position of these and other arboviruses.

Most information concerning tick-virus interrelationships is derived from several post-World War II field and laboratory programmes devoted to unravelling epidemiological questions of the major group B tickborne infections, RSSE, European tickborne encephalitis (TBE), and Indian Kyasanur Forest

TABLE 18.1
Viruses isolated from ticks in nature*

Group B

(Russian spring-summer encephalitis (RSSE) complex)

+1. Russian spring–summer encephalitis (RSSE) (152)
Ixodes persulcatus, I. lividus, Haemaphysalis concinna, H. longicornis, H. japonica douglasi, Dermacentor silvarum, D. reticulatus (all from Siberia). *D. marginatus* (Kazakh SSR).

+2. Tickborne encephalitis (TBE) (154, also 153, 155)
Ixodes ricinus, Haemaphysalis inermis, H. punctata, H. concinna, Dermacentor marginatus, D. reticulatus (all from Europe).

+3. Louping ill (LI) (80)
Ixodes ricinus (Britain and Ireland; ?Bulgaria).

+4. Negishi encephalitis (NEGE) (101)
(Human corpse; tick association assumed, Japan).

+5. Omsk hemorrhagic fever (OHF) (156)
Dermacentor pictus, ?I. persulcatus, ?I. apronophorus (Siberia).

+6. Kyasanur Forest disease (KFD) (45)
Ornithodoros chiropterphila, Ixodes ceylonensis, I. petauristae, Haemaphysalis turturis, H. spinigera, H. kinneari, H. kyasanurensis, H. wellingtoni, H. bispinosa, H. minuta, H. cuspidata, Dermacentor auratus, Rhipicephalus turanicus (all Mysore, India).

+7. Langat encephalitis (LANE) (79)
Ixodes granulatus (Malaya).

+8. Royal farm (RF) (273)
Argas hermanni (Afghanistan).

+9. Powassan encephalitis (POWE) (94)
Ixodes spinipalpis, I. cookei, I. marxi, Dermacentor andersoni (N and W USA and Canada).

---

* Names of unregistered viruses in the text and tables are tentative and do not constitute original publication. Information on unpublished virus names derives from NAMRU-3 data, J. Casals (YARU), H. N. Johnson (Rockefeller Foundation), and C. E. Yunker and C. M. Clifford (Rocky Mountain Laboratory), to each of whom I am much indebted. Details regarding certain named viruses have been published only in the Catalogue of Arthropod-borne Viruses of the World. The registry number in the Catalogue is entered in parentheses following the name of the virus, or 'unreg.' if unregistered. (The practice of numbering newly registered viruses has recently been discontinued.) A plus sign (+) preceding the list number of a virus indicates that this virus has been recorded as infecting humans. Virus relationships are those demonstrated by complement fiation (CF) test results.

Group B (continued)

   (Other than RSSE complex)
   10. Kadam (KAD)
       Rhipicephalus pravus (Uganda).
   11. Tyuleniy (TYU) (275)
       Ixodes uriae (N USSR; USA, Oregon).
   12. West Nile (WN) (1) (generally mosquitoborne)
       Argas hermanni (Egypt). *Hyalomma m. marginatum* (SW USSR).

Hughes group

   13. Hughes (HUG) (161)
       Ornithodoros denmarki (USA, Florida. NW Mexico. Trinidad).
   14. Farallon (FAR) (unreg.)
       Ornithodoros sp. near *denmarki* (USA, California).
   15. Soldado (SOL) (180)
       Ornithodoros denmarki (Trinidad). *O. capensis* (Ethiopia).
(?+)16. Punta Salinas (PUNS)
       Ornithodoros amblus (Peru).
(?+)17. Zirqa (ZIR)
       Ornithodoros muesebecki (Arabian Gulf).
   18. Sapphire II (SAP II)
       Argas cooleyi (USA, Montana, Texas).

Quaranfil group

 +19. Quaranfil (QUA) (22)
       Argas arboreus (Egypt. Nigeria. South Africa). *A. hermanni* (Afghanistan. Nepal).
   20. Johnston Atoll (JA) (195)
       Ornithodoros capensis (mid-Pacific atoll and (as C5502) Australia).
   21. Abal (ABAL)
       Ornithodoros capensis (Australia).

Qalyub group

   22. Qalyub (QYB) (222)
       Ornithodoros erraticus (Egypt).
   23. Bandia (BDA) (205)
       Ornithodoros sonrai (Senegal).

Kemerovo group

   Chenuda subgroup
   24. Chenuda (CHE) (23)
       Argas hermanni (Egypt). *Ornithodoros peringueyi* (South Africa).

25. Baku (BAKU) (276)*
    Ornithodoros capensis (Azerbaijan SSR).
26. Mono Lake (ML)
    Argas cooleyi (USA, California).
27. Sixgun City (SC) (284)
    Argas cooleyi (USA, Texas, Colorado).
28. Huacho (HUA)
    Ornithodoros amblus (Peru).

Kemerovo subgroup

+29. Kemerovo (KEM) (157)
    Ixodes persulcatus (USSR, Siberia).
+30. Tribeč (TRB) (175)
    Ixodes ricinus (Czechoslovakia). *Haemaphysalis puntata* (Hungary).
31. Lipovnik (LIP) (186)
    Ixodes ricinus (Czechoslovakia).
32. Yaquina Head (YH) (281)
    Ixodes uriae (USA, Oregon, Alaska).
33. Great Island (GI)
    Ixodes uriae (Canada, Newfoundland).
34. Bauline (BAU)
    Ixodes uriae (Canada, Newfoundland).

Wad Medani subgroup

35. Wad Medani (WM) (35)
    Hyalomma marginatum isaaci (India). *H. a. anatolicum* (Pakistan). *Amblyomma cajennense* (Jamaica). *Rhipicephalus sanguineus* (Sudan). *Boophilus microplus* (Singapore and Malaya) (as SELETAR virus).

Uukuniemi group

36. Ponteves (PTV) (217)
    Argas reflexus (France).
37. Grand Arbaud (GA) (218)
    Argas reflexus (France). *A. hermanni* (Afghanistan).
38. Sumakh (unreg.)
    Ixodes ricinus (Azerbaijan and Ukrainian SSR).
39. Manawa (MWA) (206)
    Argas abdussalami, Rhipicephalus turanicus, R. ramachandrai (Pakistan).
40. Uukuniemi (UUK) (189)
    Ixodes ricinus (Finland. Czechoslovakia).

---

* The orthographic similarity between the Soviet 'Baku' virus (cat. no. 276) and the earlier 'Bakau' virus (cat. no. 114) will cause problems.

Dera Ghazi Khan group

    41. Kao Shuan (KAS)
       Argas robertsi (Taiwan. Java).
    42. Pathum Thani (PTH)
       Argas robertsi (Thailand. Ceylon).
    43. Abu Mina (ABUM)
       Argas streptopelia (Egypt).
    44. Abu Hammad (ABUH)
       Argas hermanni (Egypt).
    45. Dera Ghazi Khan (DGK) (212)
       Hyalomma dromedarii (Pakistan).

Crimean–Congo hemorrhagic fever group

    46. Crimean–Congo hemorrhagic fever (CCHF) (235)
       Hyalomma marginatum marginatum (SW USSR. SE Europe). *Dermacentor marginatus, Rhipicephalus rossicus, Boophilus annulatus* (RFSFR). *H. marginatum rufipes, H. truncatum, H. a. anatolicum, H. impeltatum, Amblyomma variegatum, Boophilus decoloratus.* [All from Africa (Nigeria, Senegal, East Africa). *H. a. anatolicum* from Nigeria. Central Asia (USSR), and Pakistan (Lahore)].
    47. Hazara (HAZ) (211)
       Ixodes redikorzevi (Pakistan, Himalayas).

Kaisodi group

    48. Lanjan (LJN) (194)
       Ixodes granulatus, Haemaphysalis nadchatrami, H. semermis, Dermacentor sp. (*D. compactus* or *D. atrosignatus*) (Malaya).
    49. Kaisodi (KSO) (164)
       Haemaphysalis spinigera (India, Mysore).
    50. Silverwater (SLW) (120)
       Haemaphysalis leporispalustris (Canada, Alberta, Ontario. USA, Wisconsin, Utah).

Ganjam group

   +51. Ganjam (GAN) (196) (also from mosquitoes)
       Haemaphysalis intermedia, H. wellingtoni (India, Orissa, Mysore, Gujerat, Kashmir).
   +52. Dugbe (DUG) (226)
       Hyalomma marginatum rufipes, H. truncatum, Amblyomma variegatum, Boophilus decoloratus (Nigeria). *A. lepidum* (Uganda).

Ungrouped

53. Nyamanini (NYM) (15)
    Argas arboreus (Egypt. Nigeria. South Africa). *A. robertsi* (Thailand. Ceylon).
54. Midway (MD) (unreg.)
    Ornithodoros capensis (Midway and Green Kure atolls).
55. Upolu (UPO) (229)
    Ornithodoros capensis (Australia).
56. Sakhalin (SAK) (277)
    Ixodes uriae (NE USSR. USA, Oregon).
57. Keterah (KET)
    Argas pusillus (Malaya).
58. Nepal (NEP) (unreg.)
    Ornithodoros piriformis (Nepal).
59. Matucare (MAT) (230)
    Ornithodoros boliviensis (Bolivia).
60. Sapphire I (SAP I)
    Ixodes howelli (USA, Montana).
61. African swine fever (ASF) (283)
    Ornithodoros moubata porcinus (Kenya. Tanzania).
+62. Colorado tick fever (CTF) (74)
    Otobius lagophilus, Haemaphysalis leporispalustris, Dermacentor andersoni, D. occidentalis, D. parumapterus (W USA).
+63. Bhanja (BHA) (197)
    Haemaphysalis intermedia (India, Orissa). *H. punctata* (Italy). *Hyalomma truncatum, Amblyomma variegatum, Boophilus decoloratus* (Nigeria).
64. Sawgrass (SAW) (191)
    Haemaphysalis leporispalustris, Dermacentor variabilis (USA, Florida).
65. Dhori (DHO) (208)
    Hyalomma dromedarii (India, Gujarat. Egypt). *H. m. marginatum* (as Astra virus: USSR, Astrakhan).
66. Wanowrie (WAN) (198)
    Hyalomma marginatum isaaci (India, Maharashtra). *H. impeltatum* (Egypt).
+67. Thogoto (THO) (138)
    Hyalomma a. anatolicum (Egypt). *Rhipicephalus* spp., *Boophilus decoloratus* (Kenya). *H. truncatum, Amblyomma variegatum, B. decoloratus* (Nigeria). *Rhipicephalus bursa* (Italy, Sicily).
68. Lone Star (LS) (239)
    Amblyomma americanum (USA, Kentucky).
+69. Nairobi sheep disease (NSD) (10)
    Rhipicephalus appendiculatus (Kenya. Uganda).
70. Jos (JOS) (unreg.)
    Amblyomma variegatum (Senegal. Nigeria).
71. (Rodent) encephalomyocarditis (EMC) (unreg.)
    Ixodes petauristae, Haemaphysalis spinigera (India, Mysore).

Miscellaneous

(Viruses surviving after ticks feed on a viraemic host but apparently not multiplying in or generally transmitted by ticks.)

Bakau group (mosquitoborne)
  Bakau (BAK) (114)
    Argas abdussalami (Pakistan, Lahore).
California encephalitis group (mosquitoborne)
  California encephalitis (CE) (99)
    Haemaphysalis leporispalustris, Dermacentor andersoni (USA, Montana, Colorado).
Tacaribe group (Arenovirus)
  Pichinde
    Ixodes tropicalis (Colombia).
Coxsackie virus
  'Coxsackie a-like virus'
    Haemaphysalis longicornis (Fiji).
Lymphocytic choriomeningitis
    Dermacentor andersoni (Canada). *Amblyomma variegatum, Rhipicephalus sanguineus* (Ethiopia).
Suckling mouse cataract agent
    Haemaphysalis leporispalustris (USA, Georgia).

TABLE 18.2

Tick families and species harboring viruses in nature

Family Argasidae

Virus	Virus group	Area	Remarks
Argas (Argas) reflexus (Fabricius)			
Ponteves	Uukuniemi	France (Rhone Delta)	pigeon house, rural
Grand Arbaud	Uukuniemi	France (Rhone Delta)	pigeon house, rural
Argas (Argas) hermanni Audouin			
Royal Farm	B(RSSE)	Afghanistan	pigeon house, rural and urban
West Nile	B	Egypt (Nile Delta)	pigeon house, rural
Quaranfil	Quaranfil	Afghanistan, Nepal	pigeon house, rural and urban
Abu Hammad	Dera Ghazi Khan	Egypt (Nile Delta)	pigeon house, rural
Chenuda	Kemerovo (Chen)	Egypt (Nile Delta)	pigeon house, rural
Grand Arbaud	Uukuniemi	Afghanistan	pigeon house, rural and urban
Argas (Argas) cooleyi Kohls and Hoogstraal			
Sapphire II	Hughes	USA (Texas, Montana)	swallow nests
Sixgun City	Kemerovo (Chen)	USA (Texas, California)	swallow nests
Mono Lake	Kemerovo (Chen)	USA (California)	swallow nests
Argas (Persicargas) arboreus Kaiser, Hoogstraat, Kaiser and Kohls			
Quaranfil	Quaranfil	Egypt. Nigeria. S. Africa	heron rookery, rural
Nyamanini	Ungrouped	Egypt. South Africa	heron rookery, rural
Argas (Persicargas) roberts: Hoogstraat, Kaiser and Kohls			
Kao Shuan	Dera Ghazi Khan	Taiwan. Java	heron and stork rookeries, rural and islands
Pathum Thani	Dera Ghazi Khan	Thailand. Ceylon	
Nyamanini	Ungrouped	Thailand. Ceylon	
Argas (Persicargas) streptopelia Kaiser, Hoogstraal and Horner			
Abu Mina	Dera Ghazi Kahn	Egypt (Western Desert)	palmtrees, oasis

Table 18.2 *(continued)*

Virus	Virus group	Area	Remarks
Argas (Persicargas) abdussalami Hoogstraal and McCarthy			
Manawa	Uukuniemi	Pakistan	vulture nest, rural
Bakau	Bakau	Pakistan	vulture nest, rural
Argas (Carios) pusillus Kohls			
Keterah	Ungrouped	Malaya	bat roost
Ornithodoros (Alveonasus) peringueyi Bedford and Hewitt			
Chenuda	Kemerovo (Chen)	South Africa	cliff swallow nests
Ornithodoros (Alectorobius) capensis Neumann			
Soldado	Hughes	Ethiopia	ibis and cormorant nests, lake
Johnston Atoll	Quaranfil	Mid-Pacific atoll	marine bird colony
Abal	Quaranfil	Australia	marine bird colony
Baku	Kemerovo (Chen)	Azerbaijan SSR	marine bird colony
Upolu	Ungrouped	Australia	marine bird colony
Midway	Ungrouped	Mid-Pacific atolls	marine bird colony
Ornithodoros (Alectorobius) denmarki Kohls, Sonnenshine and Clifford			
Hughes	Hughes	USA (Florida), Mexico, Trinidad	marine bird colony
Soldado	Hughes	Trinidad	marine bird colony
Ornithodoros (Alectorobius) species near denmarki			
Farallon	Hughes	USA (California)	marine bird colony
Ornithodoros (Alectorobius) amblus Chamberlain			
Punta Salinas	Hughes	Peru	marine bird colony
Huacho	Kemerovo	Peru	marine bird colony
Ornithodoros (Alectorobius) muesebecki Hoogstraal			
Zirqa	Hughes	Arabian Gulf	marine bird colony
Ornithodoros (Alectorobius) boliviensis Kohls and Clifford			
Matucare	Ungrouped	Bolivia	bat, house, lowland

Table 18.2 (continued)

Virus	Virus group	Area	Remarks
Ornithodoros (Reticulinasus) chiropterphila Dhanda and Rajagopalan			
Kyasanur Forest disease	B(RSSE)	India (Mysore)	bat cave, lowland
Ornithodoros (Reticulinasus) piriformis Warburton			
Nepal	Ungrouped	Nepal	bat cave, highland
Ornithodoros (Pavlovskyella) erraticus (Lucas)			
Qalyub	Qalyub	Egypt (Nile Delta and Valley)	rodent burrow, sandy cultivation
Ornithodoros (Pavlovskyella) somrai Sautet and Witkowski			
Bandia	Qalyub	Senegal	rodent burrow, savannah
Ornithodoros (Pavlovskyella) marocanus Velu			
African swine fever	Ungrouped	Spain	pigsties, rural
Ornithodoros (Ornithodoros) moubata porcinus Walton			
African swine fever	Ungrouped	Tanzania, Kenya	warthog burrow, savannah
Otobius lagophilus Cooley and Kohls			
Colorado tick fever	Ungrouped	USA (Nevada)	hares, ranch areas

Family Ixodidae

Ixodes (Ceratixodes) uriae White			
Tyuleniy	B	USSR (N), USA (Oregon)	
Sakhalin	Ungrouped	USSR (N), USA (Oregon)	marine bird colonies, temperate and arctic
Yaquina Head	Kemerovo (KEM)	USA (Oregon, Alaska)	
Great Island	Kemerovo (KEM)	Canada (Newfoundland)	
Bauline	Kemerovo (KEM)	Canada (Newfoundland)	

Table 18.2 (*continued*)

Virus	Virus group	Area	Remarks
Ixodes (Ixodes) ricinus Linnaeus			
Tickborne encephalitis	B(RSSE)	Europe (incl. W USSR)	wide variety of small and large vertebrate animals; temperate; chiefly mixed forests, taiga, and clearings
Louping ill	B(RSSE)	Britain and Ireland; ?Bulgaria	
Triheč	Kemerovo (KEM)	Czechoslovakia	
Lipovnik	Kemerovo (KEM)	Czechoslovakia	
Uukuniemi	Uukuniemi	Czechoslovakia, Finland	
Sumakh	Uukuniemi	Azerbaijan and Ukranian SSR	
Ixodes (Ixodes) persulcatus Schulze			
RSSE	B(RSSE)	USSR (Siberia)	
?Omsk hemorrhagic fever	B(RSSE)	USSR (Siberia)	
Kemerovo	Kemerovo (KEM)	USSR (Siberia)	
?*Ixodes (Ixodes) apronophorus* Schulze			
Omsk hemorrhagic fever	B(RSSE)	USSR (Siberia)	small mammal burrows, swampy steppe, temperate
Ixodes (Ixodes) granulatus Supino			
Langat	B(RSSE)	Malaya	small mammal burrows, tropical forest
Ixodes (Ixodes) redikorzevi Olenev			
Hazara	Crimean–Congo HF	Pakistan (Himalaya)	rodent, alpine meadow
Ixodes (Ixodes) petauristae Warburton			
Kyasanur Forest disease	B(RSSE)	India (Mysore)	rodents, tropical forest
Encephalomyocarditis	Ungrouped	India (Mysore)	rodents, tropical forest
Ixodes (Ixodes) spinipalpis Hadwen and Nuttall			
Powassan	B(RSSE)	USA (South Dakota)	rodents and other mammals, temperate forest and rural
Ixodes (Pholeoixodes) cookei Packard			
Powassan	B(RSSE)	Canada (Ontario). USA (New York)	
Ixodes (Pholeoixodes) marxi Banks			
Powassan	B(RSSE)	Canada (Ontario)	

Table 18.2 (continued)

Virus	Virus group	Area	Remarks
Ixodes (Pholeoixodes) lividus Koch			
RSSE	B(RSSE)	USSR (Siberia)	sand martin nests, temperate
Ixodes (Scaphixodes) howelli Kohls			
Sapphire I	Ungrouped	USA (Montana)	cliffsides (?host), temperate
Ixodes (Afrixodes) ceylonensis Kohls			
Kyasanur Forest disease	B(RSSE)	India (Mysore)	small mammals, tropical forests
Haemaphysalis (Gonixodes) leporispalustris (Packard)			
Silverwater	Kaisodi	Canada and USA (Wisconsin, Utah)	snowshoe hares, jackrabbits, other wild mammals
Colorado tick fever	Ungrouped	USA (Wyoming, Montana)	rabbits, hares, ranch areas
Sawgrass	Ungrouped	USA (Florida)	racoon, subtropical maple swamp
Haemaphysalis (Allophysalis) inermis Birula			
Tickborne encephalitis	B(RSSE)	E Europe	small and large mammals, temperate forests
Haemaphysalis (Aboimisalis) punctata Canestrini and Fanzago			
Tickborne encephalitis	B(RSSE)	Czechoslovakia, Bulgaria	domestic animals, rodents, birds; temperate rural and forest areas
Tribeč	Kemerovo (KEM)	Hungary	
Bhanja	Ungrouped	Italy (widely distributed)	
Haemaphysalis (Haemaphysalis) concinna Koch			
RSSE	B(RSSE)	USSR (Primorye)	various small and large vertebrates, temperate forests
Tickborne encephalitis	B(RSSE)	Austria	
Haemaphysalis (Haemaphysalis) japonica douglasi Nuttall and Warburton			
Haemaphysalis (Kaiseriana) longicornis Neumann			
RSSE	B(RSSE)	USSR (Primorye)	

Table 18.2 (continued)

Virus	Virus group	Area	Remarks
Haemaphysalis (Kaiseriana) semermis Neumann			
Haemaphysalis (Kaiseriana) nadchatrami Hoogstraal, Trapido and Kohls			
Lanjan	Kaisodi	Malaya	
Haemaphysalis (Kaiseriana) spinigera Neumann			
Kyasanur Forest disease	B(RSSE)	India (Mysore)	
Kaisodi	Kaisodi	India (Mysore)	
Encephalomyocarditis	Ungrouped	India (Mysore)	
Haemaphysalis (Kaiseriana) cuspidata Warburton			
Haemaphysalis (Kaiseriana) kinneari Warburton			
Haemaphysalis (Kaiseriana) bispinosa Neumann			various small and large vertebrates, tropical forest and rural areas
Haemaphysalis (Arbophysalis) kyasanurensis Trapido, Hoogstraal and Rajagopalan			
Haemaphysalis (–) minuta Kohls			
Haemaphysalis (–) turturis Nuttall and Warburton			
Kyasanur Forest disease	B(RSSE)	India (Mysore)	
Haemaphysalis (–) wellingtoni Nuttall and Warburton			
Kyasanur Forest disease	B(RSSE)	India (Mysore)	
Ganjam	Ganjam	India (Mysore)	
Haemaphysalis (–) intermedia Warburton and Nuttall			
Ganjam	Ganjam	India (widely distributed)	
Bhanja	Ungrouped	India (widely distributed)	

Table 18.2 (continued)

Virus	Virus group	Area	Remarks
Dermacentor (D.) reticulatus (Fabricius) (= *D. pictus* Hermann)			
RSSE	B(RSSE)	USSR (Siberia)	
Omsk hemorrhagic fever	B(RSSE)	USSR (Siberia)	various small and large mammals, temperate forest, steppe, and rural areas
Tickborne encephalitis	B(RSSE)	Czechoslovakia	
Dermacentor (D.) marginatus (Sulzer)			
RSSE	B(RSSE)	Kazakh SSR	
Tickborne encephalitis	B(RSSE)	Czechoslovakia, Bulgaria	
Crimean–Congo HF	Crimean–Congo HF	RSFSR	
Dermacentor (D.) silvarum Olenev			
RSSE	B(RSSE)	USSR (Primorye)	
Dermacentor (D.) auratus Supino			
Kyasanur Forest disease	B(RSSE)	India (Mysore)	tropical forest
Dermacentor (D.) sp. (compactus Neumann or *atrosignatus* Neumann)			
Lanjan	Kaisodi	Malaya	tropical forest
Dermacentor (D.) andersoni Stiles			
Powassan encephalitis	B(RSSE)	USA (Colorado, South Dakota)	rural, temperate
Colorado tick fever	Ungrouped	USA (Montana)	ground squirrels, other rodents, brush forests, temperate
Dermacentor (D.) variabilis (Say)			
Sawgrass	Ungrouped	USA (Florida)	racoon, subtropical maple swamp
Dermacentor (D.) occidentalis Marx			
Dermacentor (D.) parumapterus Neumann			
Colorado tick fever	Ungrouped	USA (Nevada)	hares, raneh areas
Hyalomma (H.) dromedarii Koch			
Dera Ghazi Khan	Dera Ghazi Khan	Pakistan	camels, semidesert
Dhori	Ungrouped	India (Kutch), Egypt	camels, semidesert

Table 18.2 *(continued)*

Virus	Virus group	Area	Remarks
Hyalomma (H.) impeltatum Schulze and Schlottke			
Wanowrie	Ungrouped	Egypt	camels, semidesert
Crimean–Congo HF	Crimean–Congo HF	Nigeria	cattle, dry savannah
Hyalomma (H.) anatolicum anatolicum Koch			
Wad Medani	Kemerovo (WM)	Pakistan (Lahore)	cattle, irrigated plain
Crimean–Congo HF	Crimean–Congo HF	USSR (Central Asia), Nigeria (as 'H. excavatum')	cattle, steppe and savannah
Thogoto	Ungrouped	Egypt	camels, market
Hyalomma (H.) a. anatolicum and Boophilus microplus (pooled ticks)			
Wad Medani	Kemerovo (WM)	West Pakistan (Lahore)	cattle, irrigated plain
Hyalomma (H.) marginatum marginatum Koch			
West Nile	B	USSR (SW)	marshy bird refuge
Dhori	Ungrouped	USSR (SW)	marshy bird refuge
Crimean–Congo HF	Crimean–Congo HF	USSR (SW), Bulgaria, Yugoslavia	birds and domestic animals, rural
Hyalomma (H.) marginatum isaaci Sharif			
Wad Medani	Kemerovo (WM)	India (Maharashtra)	} domestic animals, rural, tropical
Wanowrie	Ungrouped	India (Maharashtra)	
Hyalomma (H.) marginatum rufipes Koch			
Crimean–Congo HF	Crimean–Congo HF	Nigeria	
Dugbe	Ganjam	Nigeria	
Hyalomma (H.) truncatum Koch			
Crimean–Congo HF	Crimean–Congo HF	Nigeria	} domestic animals, tropical savannah
Dugbe	Ganjam	Nigeria	
Bhanja	Ungrouped	Nigeria	
Thogoto	Ungrouped	Nigeria	

Table 18.2 (*continued*)

Virus	Virus group	Area	Remarks
Rhipicephalus (R.) pravus Dönitz			
Kadam	B	Uganda	cattle, tropical savannah
Rhipicephalus (R.) sanguineus (Latreille)			
Wad Medani	Kemerovo (WM)	Sudan	goat, irrigated semidesert
Rhipicephalus (R.) appendiculatus Neumann			
Nairobi sheep disease	Ungrouped	Kenya, Uganda	sheep, tropical savannah
Rhipicephalus (R.) turanicus Pomerantzev			
Kyasanur Forest disease	B(RSSE)	India (Mysore)	tropical forest
Manawa	Uukuniemi	Pakistan (Hunza)	goat, mountain desert
(Under study, Mirzoeva et al.)		Azerbaijan SSR	hare, steppe
Rhipicephalus (R.) bursa Canestrini and Fanzago			
Thogoto	Ungrouped	Italy (Sicily)	domestic animals
(Under study, Mirzoeva et al.)		Azerbaijan SSR	cattle, steppe
Rhipicephalus (R.) rossicus Yakimov and Kohl-Yakimova			
Crimean–Congo HF	Crimean–Congo HF	USSR (SW)	domestic animals, temperate
Rhipicephalus (R.) ramachandrai Dhanda			
Manawa	Uukuniemi	Pakistan (Lahore)	grass gerbil nest, irrigated plains, tropical
Rhipicephalus spp. and *Boophilus decoloratus* (pooled ticks)			
Thogoto	Ungrouped	Kenya	domestic animals, tropical savannah

Table 18.2 *(continued)*

Virus	Virus group	Area	Remarks
Amblyomma (A.) variegatum (Fabricius)			
Crimean–Congo HF	Crimean–Congo HF	Nigeria, Uganda	
Dugbe	Ganjam	Nigeria, Uganda	
Bhanja	Ungrouped	Nigeria, Senegal	domestic mammals, tropical savannah
Thogoto	Ungrouped	Nigeria	
Jos	Ungrouped	Nigeria, Senegal	
Amblyomma (A.) lepidum Dönitz			
Dugbe	Ganjam	Uganda	
Amblyomma (A.) americanum (Linnaeus)			
Lone Star	Ungrouped	USA (Kentucky)	woodchuck burrow, temperate forest
Amblyomma (A.) cajennense (Fabricius)			
Wad Medani	Kemerovo (WM)	Jamaica	cattle, rural, tropical
Boophilus decoloratus (Koch)			
Dugbe	Ganjam	Nigeria	
Crimean–Congo HF	Crimean–Congo HF	Nigeria	domestic animals, tropical savannah
Thogoto	Ungrouped	Kenya, Nigeria	
Bhanja	Ungrouped	Nigeria	
Boophilus microplus (Canestrini)			
(see pooled *Hyalomma a. anatolicum* and *B. microplus*)			
Wad Medani (Seletar)	Kemerovo (WM)	Malaya, Singapore	cattle, tropical, rural
Boophilus annulatus (Say)			
Crimean–Congo HF	Crimean–Congo HF	USSR (SW)	domestic animals, temperate

disease (KFD). These investigations had an important byproduct, the discovery in ticks of numerous other viruses.

More recently, a few enthusiastic peripatetic researchers have travelled farther afield to collect an even greater variety of viruses from ticks. The new data, obtained either as epidemiological byproducts or from 'one-time only' field trips, have added numerous scattered details to knowledge of virus groups, and their distribution, ecology, and host relationships. Worthwhile as it is, this information has also raised perplexing scientific questions that may long remain unanswered.

Any attempt to generalize or to define operational systems among associations between ticks and their heterogeneous virus groups is marred by the uneven amount and variety of research within this field. Few biological characteristics of the well known tickborne group B infections of humans and lower animals are universally applicable, and biological information is lacking for the majority of viruses isolated from ticks. Therefore, I can offer the reader no neat package of positive data tied with integrating scientific principles. I present instead parts of a large jigsaw-puzzle scattered in an unfinished frame. Many of these pieces lie loose for lack of clues to where they belong within the frame. Some pieces fit to form intricate local scenes.

New laboratory and field investigations continue to modify long-cherished concepts. For instance, the vector capability of closely related tick species has generally been considered to be about equal. However, *Argas persicus* and *A. arboreus* diametrically differ in their ability to support and transmit Quaranfil virus (Kaiser 1966a, b). Nevertheless these ticks are so closely related and look so much alike that all African specimens had always been identified as *A. persicus*. *A. arboreus* was recognized as a structurally and biologically unique species, with virus vector capabilities distinctly different from those of *A. persicus*, only when inquiries were made into the unexplained absence of Quaranfil virus in some Cairo-area tick population samples and its presence in other samples (Kaiser, Hoogstraal and Kohls 1964). These species occur in nearby sites but they do not interbreed, even when placed together under optimum experimental conditions (Hafez, Abdel-Malek and Guirgis 1972).

Another concept, that tick-virus foci characteristically are stable in nature, probably applies chiefly to certain ecological situations described later as 'restricted' (18.3). Thus, active Omsk hemorrhagic fever virus has been difficult to rediscover in steppe ticks where this virus was first studied (Casals et al. 1970). Recent intensive efforts to reisolate Kemerovo virus from humans, ticks, and wild vertebrates in the Kemerovo area have also been unsuccessful (M. P. Chumakov and A. A. Maksimov, personal communications). The sudden outbreak of Kyasanur Forest disease in India (Work 1958; Indian Council for Medical Research 1964) is attributed to the recent human population explosion in Shimoga District and the altered conditions which favour virus spread (Boshell 1969). Powassan virus presents a similar threat in North America (Hoogstraal 1966). Crimean-Congo hemorrhagic fever virus has been spreading in the Astrakhan and Rostov areas despite diligent control efforts, but has disappeared from the Crimea, where it fulminated due to wartime upset of normal ecological balances (Casals et al. 1966, 1970; Hoogstraal 1967a).

Some important tick-virus associations were long overlooked owing to lack of inquiry and to ready acceptance of negative results from limited experimental investigations. The most recent and notable example is African swine fever virus, which is intimately associated with the *Ornithodoros* tick found in burrows of warthogs and certain other large African mammals (sect. 18.3.3.3).

18.2 Relationships in virus survival

18.2.1 Intrinsic factors

A perfectly adapted virus-tick relationship might be expected to include a high rate of transovarial transmission of the virus from one tick generation to the next. In rickettsia-tick associations, this phenomenon is the rule (Steinhaus 1947; Burgdorfer 1963; Hoogstraal 1967b; Balashov 1968). Certain group B and other viruses are known to pass from an infected tick through the eggs to the next generation (Hannoun, Corniou and Rageau 1970; Il'Enko, Gorozhankina and Smorodintsev 1970; Konrashova and Filippovets 1970; Řeháček 1965; Hoogstraal 1966; Burgdorfer and Varma 1967; Singh, Goverdhan and Bhat 1971) but most evidence is based on limited investigation. Transovarial transmission of Quaranfil virus apparently never occurs, even in its chief vector species, *Argas arboreus* (Kaiser 1966b; Taylor et al. 1966). However, transmission of African swine fever virus through the eggs of *Ornithodoros moubata porcinus* appears to be a natural maintenance mechanism (sect. 18.3.3.3). Extensive field and experimental research data are needed to evaluate the role of transovarial transmission in maintaining numerous tickborne viruses. Transstadial passage of virus from the larval to nymphal to adult stage is more frequent among ticks than among insects and, in the absence of transovarial passage, is essential for virus survival. Hematophagous insects generally feed on several hosts but ticks only on a single host in each developmental instar or stage. The lack of histolysis in most tick organs during molting may account for the transstadial phenomenon (Balashov 1968). The poorly understood mechanisms of stage-to-stage passage of viruses also offer a rich field for further investigation (Burgdorfer and Varma 1967).

The long tick life, commonly a year and often two or more years, involves tick-virus relationships in situations unique among hematophagous arthropods and arboviruses. The tick lifespan exceeds that of some of its hosts or at least involves successive breeding cycles of the host. Viruses surviving for months or years in ticks have a much enhanced potential in time and space for transmission to uninfected, susceptible vertebrate hosts. Many ixodid species have greatly dissimilar hosts for their immature and adult stages, hence the potential vertebrate reservoir range of the virus is greatly amplified when it multiplies being transmissible in each developmental stage of the tick.

The biology of tick parasitism also provides a special characteristic to circulation patterns of associated viruses. The milieu in which viruses survive and multiply within the tick is more distinct from that of insects than is generally appreciated. Interplay among various intrinsic and extrinsic factors is either ecologically equivalent to transovarial transmission or functions in nature together with this process to assure continuous virus circulation between ticks and hosts.

Homeostatic or intrinsic features that influence virus survival include numerous structural, physiological, and biochemical adjustments to the tick's enormous meals. Pools of various host fluids are imbibed in tremendous quantities and excess fluids are eliminated through argasid coxal organs or inoculated into the host from ixodid salivary glands. Long feeding periods are followed by slow digestive states and exceptionally slow physiological action and often by hibernation or diapause.

Tick feces (Hamdy 1972a, b, c) and argasid coxal fluid (Boné 1943; Lees 1946; Araman and Said 1972) are not known to be important in virus dissemination.

Injection of virus-infected tick saliva is the only known route of virus introduction from the tick into the vertebrate host. The nature of tick saliva and salivary gland function have been described by Gregson (1960, 1967, 1969) and Tatchell (1969a, b; Tatchell et al. 1972; Tatchell and Binnington 1972).*

---

* Ixodid saliva has a triple role; a cement is secreted to provide firm attachment for the feeding tick on the host, pharmacologically active substances aid in feeding, and excess

The mechanics of the way in which ticks transmit viruses via saliva to vertebrate animals is not included in this ecological review. However, the role of salivary transmission should be considered in arbovirus ecology, as it has epidemiological significance in certain tickborne infections. Lord (1970) summarized some vertebrate host aspects of arbovirus ecology as follows: 'Arboviruses seem to have evolved a dual system of propagation: the first, an enzootic or consuetudinal system defined as a gradually evolved, well-balanced habitual relationship of the parasite with its regular hosts, possibly but not necessarily including a vector; the second, an epizootic or amplifier host system favouring spectacular vectorborne spread, either in a long-lived alternate host population which has become non-immune or in a short-lived alternate host population during its annual peak'.

In this respect, I continue to be fascinated by the suggestion that arthropod transmission (the epizootic system) is a biological development of zoonotic infections (Casals et al. 1970), where virus is spread in infected vertebrate urine. In the zoonotic group, Machupo virus is shed in rodent urine for more than a year following inoculation and infant rodents acquire infection by contact with infected mothers (Johnson, Halstead and Cohen 1967). Lassa virus remained in the urine of a human patient for 23 days after treatment (Buckley and Casals 1970) and for 32 days after onset of illness in another patient (Leifer, Cocke and Bourne 1970).

In the epizootic system, scientific literature contains numerous more or less casual reports of the role of the vertebrate kidney in concentrating arboviruses and their appearance in urine; these reports pertain to practically every common arthropod transmitted virus. An example is the possibility of chronic infection by Powassan virus (group B) in the grey squirrel kidney, which might provide a means for virus survival in the absence of arthropod transmission (Timoney 1971). Omsk hemorrhagic fever (OHF) virus is isolated from shrew and vole urine (and brain and spleen tissue) (Kharitonova and Khadzhieva, 1966) and muskrat urine (and blood and brain tissue) (Fedorova 1966a, b). Infected urine may be responsible for spreading OHF virus to human handlers of muskrats (sect. 18.4.2.1).

Concentration and proliferation of virus in goat and bovine mammary glands, and human consumption of infected milk and cheese, is a significant factor in the epidemiology of tickborne encephalitis (sect. 18.4.2.1).

---

dietary water and electrolytes are eliminated by secretion into the host. Argasid nymphs and adults do not secrete cement but their watery saliva contains powerful cytolytic, anti-coagulant, and toxic components; argasids eliminate excess water and salts through coxal organs.

18.2.2 Extrinsic factors

Rodents and other common small vertebrate hosts of ticks often have a short lifespan and rapid physiological activity in distinct contrast with the generally long life and slow developmental cycle of ticks. Many wild rodents live for only six to twelve months and have two or three litters annually. Birds often nest twice a year. In the case of the cattle heron, the second summer brood allows Quaranfil virus to survive in the *Argas arboreus* population (sect. 18.3.2.4).

Within any tick-inhabited biotope, two distinct, ecologically determined patterns of habitat – tick – vertebrate host interrelationship influence virus circulation. The first pattern involves restricted vertebrate habitats and shelter-seeking ticks adapted to feed throughout their life only on sedentary animals nesting in burrows, trees, or ground-level colonies. The second pattern involves more generalized habitats where adult ticks feed on wandering hosts that usually are larger and ecologically distinct from the small or medium-sized hosts parasitized by the immature stages.

18.3 Restricted habitats, hosts, and ticks

18.3.1 Habitats

Caves and burrows, rookeries, bird colonies and bat roosts assure the annual return of hosts to a protected situation. Such sites harbor particular tick species (numerous *Argas* and *Ornithodoros* in the family Argasidae; some *Ixodes*, *Haemaphysalis*, and *Rhipicephalus*, but very few others in the family Ixodidae) which live their entire life cycle there. Their food is generally restricted to the host typical of the habitat. In many caves and burrows only a single adapted tick species is found, in certain others two or more tick species characteristically occur. Nesting bird colonies seldom, if ever, harbour more than one tick species.

Viruses from specialized ticks in these habitats are all those in the Hughes, Quaranfil, and Qalyub groups; the entire Chenuda subgroup of the Kemerovo group; Yaquina Head, Great Island, and Bauline of the Kemerovo subgroup; Ponteves and Grand Arbaud of the Uukuniemi group; Royal Farm and Tyuleniy of group B; others under study in the Dera Ghazi Khan group; and the ungrouped Nyamanini, Midway, Upolu, Sakhalin, Keterah, Nepal, Matucare, and African swine fever viruses.

18.3.2 Bird-associated ticks and viruses

18.3.2.1 Ecology

Bird hosts supporting dense populations of specially adapted ticks nest in holes in riversides (sand martins), or in nests constructed of mud (cliff swallows) or of large sticks in trees or on protected rock ledges (vultures). Others form large nesting colonies in trees (heron and open-bill stork rookeries) or on the ground (gulls and terns). Some are most numerous in man-made structures (domestic pigeons). (Notably, no viruses have yet been isolated from *Argas persicus*, a widely distributed parasite of domestic chickens).

Almost all argasid species and some ixodids parasitize only birds that are shelter- or colony-inhabiting. The seven *Argas*, seven *Ornithodoros*, and two *Ixodes* species which have been found to be virus-infected in these ecological situations are listed in the following sections and table 18.2. *Haemaphysalis* species that inhabit only bird (or mammal) nests have not been investigated. (No ticks of other genera are known to be confined to bird nests). At least two virus-infected tick species are known from each continent. The climatic range is from the subarctic *(Ixodes uriae)* to the tropics *(A. robertsi)*.

The ecology of viruses from ticks infesting bird nests appears to be relatively stable and easily investigated. Wild birds that nest in shelters or in colonies are available as tick hosts for only a few weeks during the annual breeding season. At other times the ticks retreat under nearby stones or in crevices (seldom in the nest) in relatively dry, often extremely warm or cold environments.

18.3.2.2 Domestic pigeon flocks

Domestic pigeons harbour *Argas reflexus* in southern Europe and *A. hermanni* from northwestern Africa to Nepal (Hoogstraal and Kohls 1960a, b). In domestic bird colonies, unlike those in nature, numerous hosts are available throughout the year. Viruses already isolated from these Palearctic *Argas* are West Nile (also mosquito borne) (Hoogstraal 1966) and Royal Farm (group B) (Williams et al. 1972), Quaranfil and Chenuda (Quaranfil and Kemerovo groups) (Taylor et al. 1966), Ponteves and Grand Arbaud (Uukuniemi group) (Hannoun, Corniou and Rageau 1970), and Abu Hammad (DGK group). This number and variety perhaps reflect the ease with which viruses may spread in and be isolated from large flocks of domestic birds and their ticks.

Some of the viruses from argasids parasitizing domestic pigeons also infect argasids associated with wild birds in protected nests or colonies and in different zoogeographical regions. Examples are Quaranfil virus from *A. hermanni* (domestic pigeons, Afghanistan) and from *A. arboreus* (wild herons, North to South Africa) and Chenuda virus from *A. hermanni* (domestic pigeons, Egypt) and from *Ornithodoros peringueyi* (cliff swallows, South Africa). The close association between domestic and wild pigeons and other wild birds that migrate great distances probably accounts for the extensive distribution of these viruses. Ticks and viruses from wild pigeon nests have not yet been investigated as they are difficult to obtain from deep rock wells or steep hill-side ledges.

18.3.2.3 Aquatic bird colonies

Marine ground-nesting birds in tropical and southern temperate zones fre-quently support virus-infected ticks. *Ornithodoros coniceps, O. capensis, O. denmarki, O.* sp. near *denmarki, O. amblus,* and *O. muesebecki* have yielded 10 viruses in the Hughes (5), Quaranfil (1), and Kemerovo (2) groups, and ungrouped (2). These viruses have been mostly recorded from only a single bird colony or from several colonies within a large zoogeographic region. However, Soldado virus (Hughes group) has been isolated in the Neotropical faunal region from *O. denmarki* inhabiting Trinidad tern nests and in the Ethiopian Region from *O. capensis* inhabiting ibis and cormorant nests in a mountain lake in Ethiopia.

Ixodes uriae replaces *Ornithodoros* in arctic and northern temperate zone marine bird nests. Current American and Soviet studies of *Ixodes uriae* reveal a variety of viruses, some of which are widely distributed. Tyuleniy virus (group B) infects *I. uriae* across northern USSR (L'Vov et al. 1971; Gaida-movich 1971), and on the Oregon coast of USA (Clifford et al. 1971). Sakhalin virus (ungrouped) also occurs in these ticks in northern USSR and Oregon. Three members of the Kemerovo group that infect *I. uriae* are Yaquina Head virus in Oregon and Alaska (C. M. Clifford and C. E. Yunker, personal com-munication) Great Island and Bauline in Newfoundland.

The biology of viruses of ticks from marine bird habitats has not been studied. Dense populations of birds and ticks, intense tick feeding activity, and rapid tick development during the host breeding season characterize many of these habitats.

The barren grounds where marine birds nest are usually inhabited by few if any other vertebrate animals and tick activity is limited to the nesting season. For example, where Zirqa virus infects *Ornithodoros muesebecki* on a small, rocky island in the hot Arabian Gulf, numerous cormorants nest close to one another in shallow depressions or slight gravel beds (Hoogstraal, Oliver and Guirgis 1970). The birds arrive in late August (late summer) and remain until early February (winter). Chicks hatch in early December and fly by the end of January. Ticks remain concealed beneath rocks while birds are absent but become increasingly active between late August and December–January, when nymphs run in search of a host. Humans venturing into the nesting grounds at this time can scarcely avoid being bitten. The fever, headache, itch, and erythema suffered following these bites is likely to be caused by Zirqa virus. The *Ornithodoros muesebecki* life cycle requires 43–85 days in the laboratory. Some ticks in nature might complete development within a single bird breeding season, others in two seasons.

In the temperate zone, Baku virus (L'Vov et al. 1971b) was isolated from samples of 1740 *O. coniceps* (17 strains from about 45 lots) and 72 herring gulls (1 strain from a fledgling) from a small island in the Caspian Sea (Gromashevsky, Sidorova and Tsirkin 1971). On this island with a warm semidesert climate and dry summer, several thousand herring gulls nest among sparse wormwood-saltwort vegetation (Aristova and Gostinshchikova 1971). Few other animals inhabit the island. *O. coniceps* ticks were recovered from 63 nests, especially those with thick litter, during the summer when gull eggs were incubating and fledglings were feeding. Fewer ticks, mostly nymphs, were found in the fall.

In subarctic USSR, *Ixodes uriae* was abundant in murre nests and absent where kittiwakes nested alone (Bekleshova, Terskikh and Smirnov 1971). Antibodies to Tyuleniy virus were equally frequent in murres from isolated and mixed colonies. Kittiwakes from mixed colonies, but not from unmixed colonies, showed antibody evidence of virus infection.

The environment in which *I. uriae* is infected by Tyuleniy and Sakhalin viruses on islands near the Oregon coast is described by Clifford et al. (1970). The murre is also the most commonly infested bird here. These islands are unique in harbouring two tick species parasitizing nesting birds: *I. uriae* and an argasid *(Ornithodoros* sp. (near *denmarki))*. No virus was isolated from the argasids.

18.3.2.4 Heron and stork rookeries

Argas ticks feeding on herons and storks in Africa and Asia usually find shelter under bark or in cracks of trees supporting the rookeries. However, in a heron rookery in a mangrove swamp on a small island in the Java Sea, I found *Argas* sp. (near *arboreus*) chiefly in the nests. The six or more viruses isolated from *A. arboreus* and *A.* sp. (near *arboreus*) that we collected from heron and stork rookeries are listed in table 18.2. Quaranfil and Nyamanini viruses are quite widely distributed (table 18.1).

Of viruses associated with ticks and birds having restricted ecological inter-relationships, only the ecology of Quaranfil virus has been studied in detail.

Forty-two (8.4%) of 500 pools (each five specimens) of *A. arboreus* collected at monthly intervals from an Egyptian rookery of the cattle heron, *Bubulcus i. ibis*, yielded Quaranfil virus throughout the year (Kaiser, 1966a).* In the spring, when the birds returned to nest after a winter absence, the incidence of the virus rose in the tick population. The incidence dropped soon afterward, remained low during the bird breeding season (summer), and rose again in the fall, but to a lower level than in the spring. The low summer incidence was attributed to a buildup of an immune bird population following fatalities of some young birds. Nesting birds are parasitized by hundreds of larval, nymphal, and adult *Argas arboreus*, but adult birds that have survived heavy tick infestations resist tick attachment (Kaiser 1966a; Hafez, Abdel-Malek and Guirgis 1972). Consequently, there was no viraemia in adult herons. Continued virus transmission depends on a continual supply of susceptible nestlings, especially the second summer brood, and on over-wintering survival of infected late-instar nymphs and adults of *A. arboreus*. Larvae and early-instar nymphs do not survive the winter (Guirgis 1971) and, as stated below, there is no transovarial transmission of the virus.

Quaranfil virus produced the highest viraemia levels in chicks four or five days following experimental inoculation (Kaiser 1966b). When *A. arboreus* nymphs fed on an infected chick, the virus persisted for up to 88 days, through several nymphal molts, to the adult stage. Infections in nymphs and adults generally required a 43–day extrinsic incubation period after the ticks fed on chicks. No evidence was obtained for transovarial transmission and attempts to transmit the virus orally to baby chicks were unsuccessful.

From these studies, Quaranfil virus appears to be a bird-adapted virus transmitted in nature by *A. arboreus*. The virus incubation period in the tick (about 43 days) appears to be an adaptation to the *arboreus* feeding pattern, in which most bloodmeals are more than 20 days apart. The absence of transovarial transmission suggests that Quaranfil virus is not as perfectly arthropod-adapted as are certain others (African swine fever, sect. 18.3.3.3). Quaranfil virus was first isolated from the blood of a febrile child living in a

---

* The same pools also yielded 76 strains of Nyamanini virus. No viruses were isolated from 2400 specimens of the closely related *Argas persicus* from nearby domestic chickens, and experimental studies (Kaiser 1966b) showed *A. persicus* to be a poor reservoir and vector of Quaranfil virus.

village near a heron rookery (Taylor et al. 1966), but the source of infection in humans is unknown.

18.3.2.5 Swallow and martin colonies

Several *Argas* and *Ornithodoros* species of the Nearctic, Palearctic, Ethiopian, and Australian faunal regions have a specialized association with cliff swallows and martins breeding in mud nest colonies on cliffs or buildings.

Although dense populations of *Argas lagenoplastis* infest Australian fairy martin nests (Hoogstraal and Kohls 1963), attempts to recover viruses from these ticks were unsuccessful (J. Řeháček, personal communication).

In western USA, *A. cooleyi* (Kohls and Hoogstraal 1960) is infected by one virus in the Hughes group and two in the Chenuda subgroup (Kemerovo group) (C. E. Yunker and C. M. Clifford, personal communication). The South African cliff swallow parasite *Ornithodoros peringueyi* is infected by Chenuda virus (Taylor et al. 1966) which, as already stated, is also common in *Argas hermanni* inhabiting Egyptian pigeon houses.

Little is known about ticks and viruses from sand martins but in Siberia, where *Ixodes lividus* parasitizes *Riperia riperia* nesting in holes in sandy bluffs (Glashinskaya-Babenko 1956), RSSE virus (group B) has been isolated from these ticks; this is unexpected as the virus was previously known from *I. persulcatus* and ground-feeding birds and mammals (Hoogstraal 1966).

18.3.2.6 Vulture nests

Vultures return annually to large permanent, *Argas*-infested nests on rock ledges (Hoogstraal, Kaiser and Kohls 1968) or in trees (Hoogstraal and McCarthy 1965; Hoogstraal and Kaiser 1970). In the only virological investigation of these *Argas*, Manawa virus (Uukuniemi group) was isolated from *A. abdussalami* infesting nests of the white-backed vulture near Lahore, Pakistan. Contrary to the characteristic pattern of viruses from ticks parasitizing birds in restricted situations, Manawa virus was also isolated from *Rhipicephalus turanicus* and *R. ramachandrai* near Lahore. *R. turanicus* is common from India to northern Africa; adults parasitize large domestic and wild mammals and immature stages feed on small mammals. All stages of *R. ramachandrai* are found only in burrows of grass gerbils or other rodents in the Indian subregion (Dhanda 1966). Thus the ecology of Manawa virus appears to differ from that of most others and deserves investigation.

18.3.3 Mammal-associated ticks and viruses

18.3.3.1 Bat caves

Caves, buildings, and trees in which bats rest are frequently infested by specialized bat-infesting *Argas*, *Ornithodoros*, or *Ixodes* ticks (Hoogstraal 1956). The ungrouped viruses recovered from these environments are Keterah from *Argas pusillus* taken on Malayan trees (A. Rudnick personal communication), Matucare from *Ornithodoros boliviensis* in thatch roofs of Bolivian houses (Kohls and Clifford 1964), and Nepal from *O. piriformis* in a cave near Katmandu. KFD virus (group B) infects bats and *O. chiropterphila* in Mysore, India (Dhanda and Rajagopalan 1971). KFD virus also circulates in numerous less specialized tick species (table 18.1) and, like RSSE virus in *Ixodes lividus* of swallow nests (sect. 18.3.4), may either have 'escaped' from its normal cycle or may originally have evolved in a protected habitat. The ecology of the ungrouped viruses from ticks inhabiting bat roosts has not been studied and many other ticks specific to bats remain to be investigated for virus infection.

18.3.3.2 Rodent burrows

Numerous argasid and some ixodid species spend their entire life cycle in burrows of rodents and shrews, but few have been investigated for viruses. In sandy cultivated areas of Upper and Lower Egypt, *Ornithodoros erraticus* (Hoogstraal, Salah and Kaiser 1954) inhabiting burrows of the Nile grass rat, *Arvicanthis n. niloticus*, is infected by Qalyub virus (Qalyub group) (Abdel Wahab, Williams and Kaiser 1970). In the dry savanna of Senegal, the closely related *O. sonrai* inhabiting burrows of the Nile grass rat and of the multimammate rat, *Mastomys* sp., is infected by Bandia virus, which is closely related to Qalyub virus (Bres, Cornet and Robin 1967; Robin 1971). These close relationships suggest a fascinating problem for more investigation.

Rhipicephalus ramachandrai, practically the only species of this large genus confined to rodent burrows, has already been mentioned in relation to Manawa virus (sect. 18.3.2.6).

Soviet workers postulate that *Ixodes apronophorus*, a parasite of rodent burrows in the marshy steppes of Omsk and Novosibirsk Oblasts, may be a vector of Omsk hemorrhagic fever virus (group B) (Casals et al. 1970).

Ixodes granulatus of tropical Asian forest rodent and shrew burrows (Hoogstraal et al. 1972) is infected by Langat virus (group B) (Smith 1956), which causes encephalitis in humans (Hoogstraal 1966). Notably, the closely related Langat and Royal Farm viruses are the only two of the nine in the group B RSSE complex (sect. 18.4.2.1) that are characteristic of specialized ticks from restricted habitats. *I. ceylonensis* and *I. petauristae*, infected by KFD virus (group B), may be restricted to rodent burrows and tree hole nests of squirrels (Rajagopalan 1969) but the biology of these species is poorly known.

18.3.3.3 Warthog burrows

Ornithodoros moubata porcinus (Hoogstraal 1956; Walton 1962; Van der Merwe 1968) and African swine fever (ASF) virus have an exceptionally intimate relationship. ASF is unrelated to any other known virus and is the only one mentioned in this chapter in which DNA has been demonstrated. Long associated with native African warthogs, ASF infection is asymptomatic in these animals but may be catastrophic among domestic swine in Africa or when it invades European piggeries (De Tray 1963). Recently, *O. moubata porcinus* was found to be abundant in warthog burrows in East African savannas and often naturally infected with ASF virus (Plowright, Parker and Peirce 1969a, b). Though the level of viraemia in infected warthogs may be inadequate to produce persistent infection in the tick, high rates of transovarial passage from naturally infected female ticks to their progeny and the ability of this generation to transmit the virus when biting susceptible pigs, assures a maintenance mechanism for ASF virus in nature (Plowright, Perry and Peirce 1970; Plowright et al. 1970). Thus we have, functionally, an arbovirus par excellence, and one satisfying every criterion of 'tickborne'. The ASF story is a cautionary tale, as there was earlier considerable inconclusive research on dissemination mechanisms, a few ticks were studied and no transmission obtained. Only with a fresh approach was it discovered that ticks are, in fact, efficient reservoirs and vectors of ASF virus.

18.4 Generalized habitats, hosts, and ticks

18.4.1 Ecological and epidemiological considerations

In generalized habitats, the adult tick parasitizes wandering hosts that have no fixed or specialized resting or nesting place. The immature stages of some species also feed on roaming hosts, but those of other species infest only hosts resting in burrows in the ground (ch. 5).

The immature and adult stages of all *Dermacentor* and *Amblyomma* species listed in table 18.2 infest hosts of greatly dissimilar sizes and families, as do all the *Hyalomma* species except *H. a. anatolicum* (each stage parasitizes large mammals), all the *Haemaphysalis* species except *H. wellingtoni* (each stage parasitizes ground-feeding birds), and all the *Rhipicephalus* species except *R. ramachandrai* (sect. 18.3.2.6). The *Ixodes* species with dissimilar hosts of immature and adult stages are those not mentioned in the discussion of restricted habitats (sect. 18.3). Each *Boophilus* species feeds on a single large, roaming, herbivorous mammal from the larval stage to the time the fed, fertilized female drops from the host to the ground to oviposit. *Otobius lagophilus*, the only argasid of generalized habitats now known to be virus-infected, usually parasitizes rabbits and hares. In summary, most adult and immature argasids occupy restricted habitats; the adult and often the immature stages of the more numerous ixodids occupy generalized habitats.

Habitats lacking special protection from the environment often have fewer but more varied hosts than restricted habitats. Two or more tick species frequently occur in such an area, thus enhancing the potential for virus dissemination and the complexity of virus circulation.

Various ticks may ingest virus from a viraemic animal. Whether a tick infests one or more hosts its feeding period usually overlaps that of other ticks. During the few weeks an immature rodent is in the nest, it may provide a viraemic bloodmeal to larvae, nymphs, or adults of one or more tick species. Some infected nymphs may feed the following year on medium size animals (such as carnivores) and as infected adults the third year on large animals (such as deer). Thus a complex variety of tick–host feeding interrelationships provides opportunities for virus circulation within a single season, from one season to the next, and from one year to the following year. The mobility of birds and of larger mammals adds space and distance to the time element of feeding ticks. In this connexion, it is notable that some arboviruses may circulate or recirculate in birds up to ten months after inoculation (Reeves 1961).

In tropical areas lacking a pronounced dry season, virus circulation may continue throughout the year. Elsewhere the virus lies dormant but viable during one or several dry seasons or winters till the tick takes a springtime bloodmeal. Even when the level of viraemia in the vertebrate host is low, the relatively long feeding period of most ticks and their enormous meals may allow sufficient virus intake to infect the parasite.

The population densities of birds and small mammals, and of immature ticks, are frequently high in semideserts, steppes, plains, savannahs, and forests, and even in mountainous alpine zones. Large mammals in these biotopes are usually fewer and more scattered. Thus relatively few adult ticks find hosts and survive. Immature ticks and smaller hosts then assume special importance in the virus cycle. Where there are domestic animals or herds of deer, antelopes, or other large animals as in nature preserves, the ratio between adult and immature ticks differs from that elsewhere and larger vertebrates may be more important in virus circulation.

The common fluctuation in rodent numbers over a period of years causes concurrent changes in incidence of RSSE and TBE viruses (literature reviewed in Hoogstraal 1966) and probably also in that of other viruses.

18.4.2 Ecology of virus groups

The viruses from ticks of generalized habitats represent every group except the Hughes, Quaranfil, and Qalyub groups and also include 11 of the 19 ungrouped viruses (table 18.1).

The literature on RSSE complex and other tickborne viruses causing diseases in humans was reviewed to early 1965 (Hoogstraal 1966) and contains numerous references to ecological factors, which are not repeated here.

18.4.2.1 Group B (RSSE complex)

Nine viruses comprise the RSSE complex. Negishi has been isolated only from human corpses and patients in Japan but no field investigations have been reported. Royal Farm (sect. 18.3.2.2) and Langat (sect. 18.3.3.2) are known only from ticks of restricted habitats. The other viruses infect several ixodid species (less often argasid species) of generalized habitats.

KFD and Langat occur in tropical forests of southern India and Malaya. Powassan is widely distributed in forests and rural areas of temperate North America. The remaining members of this complex are characteristic of certain areas of temperate Eurasia.

Omsk hemorrhagic fever (OHF) virus of the marshy steppes of Omsk and Novosibirsk Oblasts was originally reported from *Dermacentor reticulatus* (= *D. pictus*). It causes serious disease or death among muskrat handlers. The ecology and epidemiology of this virus, which is most closely related to KFD virus of tropical India, is unclear (Casals et al. 1970).

RSSE virus is found in the vast taiga forest in which *Ixodes persulcatus* is abundant (Korenberg et al. 1969). Tickborne encephalitis (TBE) is found in European forests and in rural areas occupied by *I. ricinus* (Korenberg, Dzyuba and Zhukov 1971). Louping ill (LI) virus is common in pastures and hedges where *I. ricinus* infests sheep in Great Britain and Ireland. The clinical features of these closely related viruses differ distinctly in larger vertebrate animals, including man.

I. persulcatus and *I. ricinus* live for two or more years and feed indiscriminately on any vertebrate animals they encounter, thus differing from most other tick species. Larvae and nymphs parasitize a wide variety of lizards, birds, and small mammals that crawl through or walk on ground litter. Adults, which quest for hosts from vegetation well above the ground, usually feed on medium and large size mammals. Dozens of resident and migrating bird species have been recorded as hosts of immature stages of both species.

Recent human exploitation of the taiga forest has exposed susceptible people to attack by infected *I. persulcatus* and has produced environmental changes that have altered the local incidence of RSSE virus in ticks, animals and man. Where the taiga forest is more humid and dense in the Far East (Primorye), tick species are more numerous and three additional *Haemaphysalis* and *Dermacentor* species are infected. Earlier suggestions that birds are hosts of *I. persulcatus* have recently been questioned or revised by Korenberg (1966, 1967, 1969), Korenberg and Ivanova (1967), and Korenberg et al. (1967).

Domestic animals and semi-wild deer in some parts of Europe maintain *I. ricinus* populations in surroundings much frequented by human beings and where numerous birds and small wild animals support larvae and nymphs. TBE was once considered to be caught by country people from infected goat milk and cheese. However this disease is now more frequently reported from towns people visiting the country.* The recovery of six different viruses from *I. ricinus* attests to the catholicity of its host associations.

The ecology of KFD virus (Hoogstraal 1966), which has recently caused an epidemic among the greatly increased human population of Shimoga District in tropical India (sect. 18.1), has continued to be carefully investigated (Boshell 1969, and papers by several authors in supplement, issue 4, volume 56, Indian Journal of Medical Research, 1968). There is a great number of tropical ticks, particularly in India, and many are involved in KFD ecology and epidemiology.

The langur monkey *Presbytis entellus* is highly susceptible to KFD virus which infects a high proportion of its tick parasites, chiefly *Haemaphysalis spinigera* and *H. turturis*, in which no transovarial passage has been demon-

---

* Some recent reports and reviews on TBE are those of Hoogstraal (1966); Blaškovič (1967, 1970), Blaškovič et al. (1967), and Blaškovič and Nosek (1972) on ecology and public health importance; Korenberg, Dzyuba and Zhukov (1971) on *Ixodes ricinus* distribution in western USSR; Riedl et al. (1971) on infection of *Haemaphysalis conncina* in Austrian marshes; Hannoun (1971) summarizing recent investigations throughout Europe; Hannoun et al. (1971) on the virus in France; Rosicky (1965, 1967) on birds and ecology.

strated in field and laboratory studies (Rajagopalan and Anderson 1970, 1971)*. However, the virus persists in nymphs of both species through the year. *H. spinigera* larvae are numerous on monkeys about a month earlier than *H. turturis* larvae. Nymphs are most important in transmitting virus to monkeys and human beings. The different stages of both species feed on various birds and mammals, and the host range of *H. spinigera* is especially wide. Although cattle resist KFD virus (Anderson and Singh 1971), they are hosts of adult *H. spinigera* and make tick control difficult (Drummond et al. 1969). The introduced American lantana bush has replaced hundreds of square miles of native forest and provides cover and food for tick hosts. More ticks are found on squirrels, monkeys, and birds than on shrews and smaller rodents in Shimoga District, but this does not seem to affect the incidence of KFD virus (Boshell 1969).

Langat encephalitis (LANE) virus has been recovered in nature only from *Ixodes granulatus*, a tick of restricted habitats (sect. 18.3.3.1), in a Malayan forest biotope in which several tick species of generalized habitats are common and Lanjan virus infects *Haemaphysalis semermis* and *H. nadchatrami* (Hoogstraal et al. 1972). Experimentally *Ixodes*, *Haemaphysalis*, and *Dermacentor* ticks can be infected by and transmit LANE virus (Varma and Smith 1962; Smith 1964). LANE and Royal Farm virus of *Argas hermanni* from pigeon houses (sect. 18.3.2.2) are closely related serologically and the restricted habitat of their vectors is unusual among RSSE viruses.

Powassan encephalitis (POWE) virus circulates among at least four tick species and numerous wild vertebrates in Canada and northern and western USA; the ecology and epidemiology of POWE virus have not been seriously investigated.

18.4.2.2 Group B (other than RSSE complex)
Three viruses infecting ticks and ecologically and biologically dissimilar are in this category.

Kadam virus is known in nature only by two strains from cattle-infesting *Rhipicephalus pravus* in Uganda (Henderson et al. 1970). *R. pravus* is a common parasite of various small and large mammals in certain drier savannas of

---

* Experimentally, KFD virus passes transstadially and transovarially in *Argas persicus* that have fed on viraemic chicks, and is transmitted when infected ticks bite virus-free chicks (Singh, Goverdhan and Bhat 1971). *A. persicus* infests domestic chickens in much of India but is not recorded from the KFD focus in Mysore.

East Africa (Hoogstraal 1956; Yeoman and Walker 1967). Experimentally infected *Dermacentor variabilis* transmitted the virus to baby mice three weeks afterward, but attempts to infect *Aëdes aegypti* mosquitoes and mosquito cell lines were unsuccessful (Mugo and Shope 1972).

Tyuleniy virus is associated with *Ixodes uriae* inhabiting restricted marine bird nests in northern temperate and subarctic zones of USA and USSR (sect. 18.3.2.3).

The widely distributed West Nile (WN) virus of Africa, southern Europe, and parts of Asia (including India), is generally mosquitoborne but may be tick borne. Unfortunately, an adequate review of WN ecology is impossible in this limited space. WN virus can be recovered throughout the year from *Argas hermanni* in Egyptian pigeon houses. It also infects immature *Hyalomma m. marginatum* from birds and hares in the Volga delta (Berezin et al. 1971), as well as numerous resident and migrating birds and some land mammals, bats, reptiles, and amphibia.

18.4.2.3 Kemerovo group (Kemerovo subgroup)

Three recently isolated members of the Kemerovo subgroup, Yaquina Head Great Island and Bauline, have been isolated from *Ixodes uriae* in marine bird nests of northwestern USA and eastern Canada (sect. 18.3.2.3).

Kemerovo (KEM) from *Ixodes persulcatus* in Siberia (and also from human beings, sect. 18.1), Tribeč (TRB) from *I. ricinus* and *Haemaphysalis punctata* in eastern Europe, and Lipovnik (LIP) from *Ixodes ricinis* in Czechoslovakia form a closely knit ecological group and are considered by some as variants of one virus (Libíková and Casals 1971). These authors identified the KEM strain isolated in September 1961, in Egypt (Schmidt and Shope 1971) from a southward migrating redstart, *Phoenicurus phoenicurus*, as the eastern (Siberian) type of the virus, and postulated that viraemia in this bird was induced by the stress of migration. In the Siberian taiga, 37% of the small birds tested (especially of the family Turdidae, which includes *P. phoenicurus*) had antibodies against KEM virus (Libíková et al. 1965). For other details of KEM and TRB viruses, see Hoogstraal (1966). In the Tribeč mountains, TRB circulates among rodents, goats, and different stages of *Ixodes ricinus* (Grešiková et al. 1970). Female *Haemaphysalis punctata* from cattle and sheep in Romania also yielded this virus (Topciu et al. 1968).

18.4.2.4 Kemerovo group (Wad Medani subgroup)

Wad Medani (WM) virus is a serotype of the Kemerovo group (Casals 1971) and its ecological characteristics remain an enigma. Since its isolation from *Rhipicephalus sanguineus* parasitizing goats in an irrigated semidesert of northeastern Sudan (Taylor, Hoogstraal and Hurlbut 1966), WM has been recovered from four other tick species characteristic of domestic animals in tropical regions; *Amblyomma cajennense* from Jamaica (J. Casals, personal communication), *Hyalomma marginatum isaaci* from India (V. Dhanda, personal communication), *H. a. anatolicum* from Pakistan (Begum, Wisseman and Casals 1970c), and *Boophilus microplus* from Singapore and Malaya (Rudnick, Marchette and Garcia 1967). I postulate that WM virus may have been introduced from Africa to Jamaica with *Amblyomma variegatum*, which is now widespread in the West Indies (Hoogstraal 1956). WM has not yet been discovered in Egyptian ticks.

18.4.2.5 Uukuniemi group

These five viruses are found in birds and their ticks. Ponteves and Grand Arbaud are known only from Palearctic *Argas* infesting pigeons (sect. 18.3.2.2). Manawa virus, from an *Argas* of vultures in Lahore, has spread into mammal-parasitizing *Rhipicephalus* ticks in the same area (sect. 18.3.2.5).

Uukuniemi (UUK) virus of *Ixodes ricinus* in southeastern Finland (Oker-Blom et al. 1965) has now been isolated from five species of passerine birds in Kumlinge and Uukuniemi parishes, but not from numerous mammals tested at the same time (M. Brummer, personal communication). This virus has also been recovered in Egypt from a southward migrating willow warbler, *Phylloscopus trochilus*. Since its discovery in Slovakia and the Tribeč Mountains (Kožuch, Grešíkova and Nosek 1968; Kožuch et al. 1970a, b), UUK has been recovered twice as often as TBE virus from unfed *Ixodes ricinus* from an oak and hornbeam forest area (Kolman and Husova 1971), and antibodies to UUK were found in 11 bird species (Ernek et al. 1971). In western Ukraine, UUK was isolated from *I. ricinus* infesting cattle grazing in woodlands (Gaidamovich, Vinograd and Obukhova 1971).

Although UUK and TBE viruses are both transmitted by *I. ricinus* their incidence in particular localities is not the same (Sekeyová, Grešíková and Stupalova 1970). The percentage of HI antibodies to UUK and TBE in human and bovine sera from localities in the Tribeč mountains were:

Sera	Locality	Antibodies (%)	
		UUK	TBE
Human	1	0.0	21.5
Human	2	3.9	15.2
Human	3	9.3	5.1

Human	4	5.8	6.2
Bovine	5	0.0	5.0
Bovine	6	2.0	25.0

Sumakh virus was recovered in Azerbaijan SSR from tissues of *Turdus merula*, a blackbird (Gaidamovich et al. 1971), which is frequently infested by *Ixodes ricinus*, the postulated vector of this virus.

18.4.2.6 Dera Ghazi Khan group
This group is also an enigma. Dera Ghazi Khan virus was isolated from a *Hyalomma dromedarii* infesting a camel in an irrigated desert area of Pakistan (Begum, Wisseman and Casals 1970b) but has not again been reported though several related viruses are under study from *Argas robertsi* collected in heron and stork rookeries in Java, Thailand, and Taiwan.

18.4.2.7 Crimean–Congo hemorrhagic fever group
The history, remarkably wide distribution, and severe clinical symptoms caused by Crimean–Congo hemorrhagic fever (CCHF) virus were reviewed by Casals et al. (1970) and Hoogstraal (1966, 1967a)*. In Bulgaria, the disease has been surprisingly common, with morbidity rates reaching 56% of the population in certain rural localities and mortality rates reaching 17% of the cases (Donchev et al. 1967). In Bulgaria, as in south-western USSR, CCHF is chiefly associated with agricultural workers and visitors who are bitten by *Hyalomma m. marginatum*. When ticks are dislodged partly fed from the host, they attach to any vertebrate animal, including man (see also Balashov 1968). Immature *H. m. marginatum* infest chiefly ground-feeding birds, but are especially numerous in the large rookeries of the rook, *Corvus frugilegus*, common in this area. Adults feed on domestic animals grazing near rookeries; thus the environment is ideal for developing large tick populations and important epizootic foci of the disease.

In Egypt, large numbers of *H. m. marginatum* larvae and nymphs are found on various birds migrating from Eurasia into Africa. *H. marginatum rufipes*, which is frequently infected in tropical Africa, is also commonly

---

* After this manuscript was completed, I received a book on Crimean hemorrhagic fever, Material of 3rd Regional Scientific-Practical Conference, Rostov on Don, May, 1970 (M. P. Chumakov, editor). This book contains many important papers, English translations of which can be obtained from my office.

found in Egypt on birds migrating from tropical Africa into Eurasia. Migration of tick-infested birds may be important in the extraordinarily wide distribution of CCHF virus.

The earlier reviews included reports of CCHF isolates from several different tick species parasitizing domestic animals in southeastern USSR and Central Asia. More recently, infected *Hyalomma a. anatolicum* have been recorded from Lahore (Begum, Wisseman and Casals 1970c). Many strains have been isolated from six species of cattle parasites in Nigeria (Causey et al. 1969, 1970, and personal communications), and four strains from *H. truncatum* and *H. marginatum rufipes* in Senegal (Y. Robin, personal communication). Thus, CCHF is now known in Eurasia from southeast Europe to northeast Pakistan and in tropical Africa from Kenya to Senegal. Domestic animals figure prominently in CCHF epidemiology everywhere and ground-feeding birds are important tick hosts in epizootic regions. As native fauna is replaced by species commensal with man, such as sparrows, crows, and weaver birds, epizootic foci may arise in pastoral areas where CCHF virus now circulates unnoticed.

Hazara virus, closely related to CCHF, infects *Ixodes redikorzevi* parasitizing rodents, *Alticola roylei,* in alpine meadows of Hazara District, northern Pakistan (Begum, Wisseman and Casals 1970a). *I. redikorzevi,* a parasite of birds and small and medium-size mammals, is distributed in a belt from Egypt through southern USSR and southwestern Asia to the lowlands of Nepal.

18.4.2.8 Kaisodi group

Lanjan virus was isolated from immature *Dermacentor* ticks (table 18.2) infesting rodents in a Malayan forest reserve (Smith et al. 1967) and later by N. Marchette in the same biotope from *Haemaphysalis semermis, H. nadchatrami,* and *Ixodes granulatus* (Hoogstraal et al. 1972). Kaisodi (KSO) virus is known by numerous strains, but only from *Haemaphysalis spinigera,* in the KFD focus of Mysore (Bhat et al. 1966; Pavri and Casals 1966). No evidence was obtained for KSO infection of the numerous vertebrate animals tested in this area. Silverwater (SLW) virus infects *Haemaphysalis leporispalustris* and snowshoe hares, *Lepus americanus,* in Ontario (McLean and Larke 1963), and Alberta and Wisconsin (Hoff et al. 1971a). Seropositive black-tailed jackrabbits, *Lepus californicus,* occur in Utah and red squirrels, chipmunks, coyotes, feral domestic rabbits, and cattle in Alberta (Hoff et al. 1971b). During a nine year study of SLW in Alberta, the last-named authors found neutralizing antibody in 31% and 17%, respectively, of adult and immature snowshoe hares. An increase in incidence in adults from 3% in 1961 to 47% in 1963 was associated with the hare's ten-year cycle of abundance.

18.4.2.9 Ganjam group

Ganjam (GAN) virus was first isolated from *Haemaphysalis intermedia* from goats in Orissa and antibodies were found in sheep and goats in different parts of India (Dandawate and Shah 1969). GAN also infects *Culex vishnui* complex mosquitoes and causes fevers in human beings (Dandawate et al. 1969a, b). Rajagopalan et al. (1970) collected infected nymphal *Haemaphysalis wellingtoni* from vegetation and red spurfowl, *Galloperdix spadicea*, in Mysore forests. Both *Haemaphysalis* species are common in Ceylon and India and the latter ranges through tropical Asia to Indonesia and Borneo. *H. intermedia* infests various birds and wild and domestic mammals. All stages of *H. wellingtoni* parasitize chiefly ground-feeding forest-dwelling birds and also small mammals.

Dugbe (DUG) virus was isolated from 4–17% of the cattle sampled at Ibadan, where the continuous introduction of susceptible cattle from the north allows the infection to persist in epizootic proportions (Kemp, Causey and Causey 1971). Incidence was greatest in the driest months of the year, when larval and nymphal *Amblyomma variegatum* are active. Causey and his colleagues (personal communication) obtained 654 DUG isolates between September 1964 and December 1968. Of these, two were from human patients with moderately severe fever, 185 from cattle, 314 from *Amblyomma variegatum*, 83 from *Boophilus decoloratus*, 65 from *Hyalomma marginatum rufipes*, and three from midges (*Culicoides* sp.). DUG has also been recovered from *Amblyomma variegatum* and *A. lepidum* in Uganda (Tukei et al. 1970).

18.4.2.10 Ungrouped

Ten of the 19 ungrouped viruses are exclusively from ticks of generalized habitats.

Wanowrie, Thogoto, and Dhori viruses have been recovered from ticks infesting camels in Egypt and from the same and other tick species characteristic of domestic animals in tropical Africa, Italy, southwestern USSR, and India (table 18.1) (Williams et al. 1973). These viruses are perhaps widely distributed with domestic animals over the caravan and marketing routes in steppe and semidesert regions of Asia and Africa; these animals are frequently infested by several tick species, each with individual biological and ecological properties. Dhori virus also appears to be associated with aquatic and shore birds and crows and rooks infested by immature *Hyalomma m. margi-*

natum, and with mosquitoes, in Astrakhan Oblast (Bashkirtsev et al. 1971).

Bhanja (BHA) virus, first isolated from *Haemaphysalis intermedia* from a paralysed goat in Orissa, infects goats and sheep in various parts of India (Shah and Work 1969). In Nigeria, Causey (personal communication) isolated 96 BHA strains, 12 from cattle, two from sheep, and 82 from *Hyalomma truncatum*, *Amblyomma variegatum*, and *Boophilus decoloratus* infesting domestic animals. In Italy, where *Haemaphysalis punctata* yielded 9 BHA strains (Sacca et al. 1969), antibodies to BHA are found chiefly in goats and sheep and also in cattle, humans, rodents, and birds (Verani, Balducci and Lopes 1970). Experimentally infected monkeys suffered from a central nervous system disease which killed some (Balducci et al. 1970).

Encephalomyocarditis (ECM) virus was isolated from Shimoga forest rodents and from unfed nymphs of *Ixodes petauristae* and *Haemaphysalis spinigera* (Paul et al. 1968). This finding suggests transstadial passage of ECM virus, but has not been further investigated.

Jos virus was isolated in Nigeria and Senegal from cattle-infesting *Amblyomma variegatum*. In the vicinity of Casamance, Senegal, where cattle are quite scattered, 38% of 144 bovine sera had antibodies to Jos (Robin et al. 1970).

Nairobi sheep disease virus, which causes considerable economic loss to sheep breeders in East Africa, is transmitted by *Rhipicephalus appendiculatus* and passes transovarially from the infected female to its progeny, which can transmit the virus for more than 100 days after hatching. Certain other tick species may be secondary vectors. Antibody surveys show evidence of up to 20% infection among humans near Entebbe. Other details are reviewed by Tukei et al. (1970) and Terpstra (1969).

Lone Star virus was isolated from a nymphal *Amblyomma americanum* parasitizing a marmot, *Marmota monax*, in Kentucky, USA (Kokernot et al. 1969).

Sawgrass virus is frequently isolated from *Haemaphysalis leporispalustris* and *Dermacentor variabilis* from wild animals in Florida, USA (Sather et al. 1970; Wellings, Lewis and Pierce 1972).

Colorado tick fever virus causes a denguelike, diphasic fever among people bitten by infected ticks in western USA, and circulates in nature among several tick species (table 18.1) and several mammals. The virus incidence in an important vector, *Dermacentor andersoni*, is high in humid mountain areas, where rodents and ticks are numerous, and low in arid sagebush areas where animals are more scattered. Transovarial passage of the virus apparently does not occur, but virus circulation persists in both ticks and mammals (Hoogstraal 1966).

18.4.2.11 Miscellaneous

At least five different viruses isolated from naturally infected ticks (table 18.1) apparently do not multiply in ticks, and there is no evidence that they are transmitted by ticks.

CHAPTER 19

Viruses and mites

J. T. SLYKHUIS

Contents

19.1 Introduction

Members of two families of plant feeding mites, the Tetranychidae or spider
mites, and the Eriophyidae including gall, bud and leaf curl mites have been
investigated as vectors of plant viruses. Reported evidence that any of the
spider mites is a vector lacks confirmation, but there is substantial evidence
that several of the eriophyid mites transmit plant pathogens, at least some
of which are viruses (Slykhuis 1965).

19.2 Spider mites as vector suspects

The Tetranychidae are 8 legged, medium-sized mites (up to 0.88 mm long),
oval or pear-shaped, varying in color through green, yellow or orange to
red. Most species spin a fine web over leaf surfaces, hence the name 'spider
mite' (Metcalf et al. 1962). Their piercing, sucking mode of feeding, free
mobility and wide host range appear to suit them well to be vectors of plant
viruses. Pale yellow to reddish-brown mottling caused by the feeding of
spider mites on the leaves of some plants has sometimes been mistaken for
symptoms of virus diseases.

The two-spotted spider mite *Tetranychus telarius* (L.) *urticae* (Koch)
(fig. 19.1) has been investigated as a vector of plant viruses. Schulz (1963)
reported that this mite transmitted potato virus Y from potato and *Nicotiana
glutinosa* to potato. Transmission appeared to be optimal when the mites

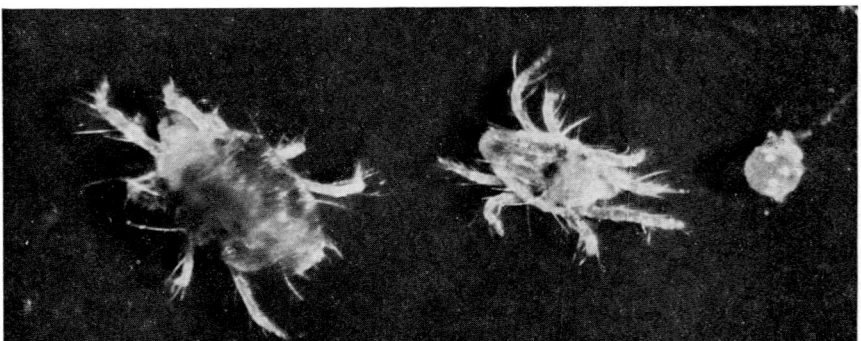

Fig. 19.1.   Adult, nymph and egg of the two-spotted spider mite (*Tetranychus* (*Telarius*)
urticae). Adults are up to 0.8 mm long. This mite has been a suspect as a vector of several
plant viruses, but none of the reports have been confirmed.

were fed for 5 min on the virus source plants, then for 5 min on the test plants. In such tests, 18 of 37 potato plants were infected with the virus. These results appear convincing but they have not been confirmed. Orlob (1968) tested collections of the same species of mite for transmission of certain strains of potato virus Y and other common and highly infectious viruses without evidence of transmission, even though some of the viruses could be recovered still infectious from the mite faeces. Also, infection resulted when mites were placed on leaves of plants sprayed with tobacco mosaic virus or potato virus X. In these special circumstances the mites assisted in inoculation, but to date there has been no confirmation that the two-spotted spider mite can transmit potato virus Y or any other virus under natural circumstances.

19.3 *Eriophyid mites and their effects on plants*

The Eriophyidae is a unique group of tiny (about 0.2 mm in length), four-legged, worm-like mites (Keifer 1952). These mites have specific and highly intimate host relations. Some are known only on closely related plant species. Others will infest plants in several genera. A few like *Aceria tulipae* (K.) (figs. 19.2, 19.3) can colonize plants in more than one family, but are limited to certain species within those families.

The development of some eriophyids from egg to adult may be completed in 6–14 days during warm weather. There are two nymphal instars, the second terminating in a resting period or pseudopupa during which the genitalia form and protrude through the body wall. Males are usually smaller than females and in some species are rarely found. Some species have two types of females, one specialized for hibernation. The eriophyids may be transported from plant to plant while clinging to the legs of bees, coccinellid beetles, aphids, thrips and other insects. Wind is of major importance in dispersing *A. tulipae*, the vector of wheat streak mosaic virus (Slykhuis 1955; Staples and Allington 1956). Under certain conditions the mites will swarm on the leaf surface, stand erect, then leap to be carried away by the wind. Some species have a waxy overcoat which may protect them from desiccation while wind-borne.

The eriophyids have piercing, sucking mouthparts, and although some are observed predominantly on leaves, most of them feed on the young tissues within buds, in leaf whorls or between flower parts. Some species have little apparent effect on the plants they colonize, but many have toxicogenic

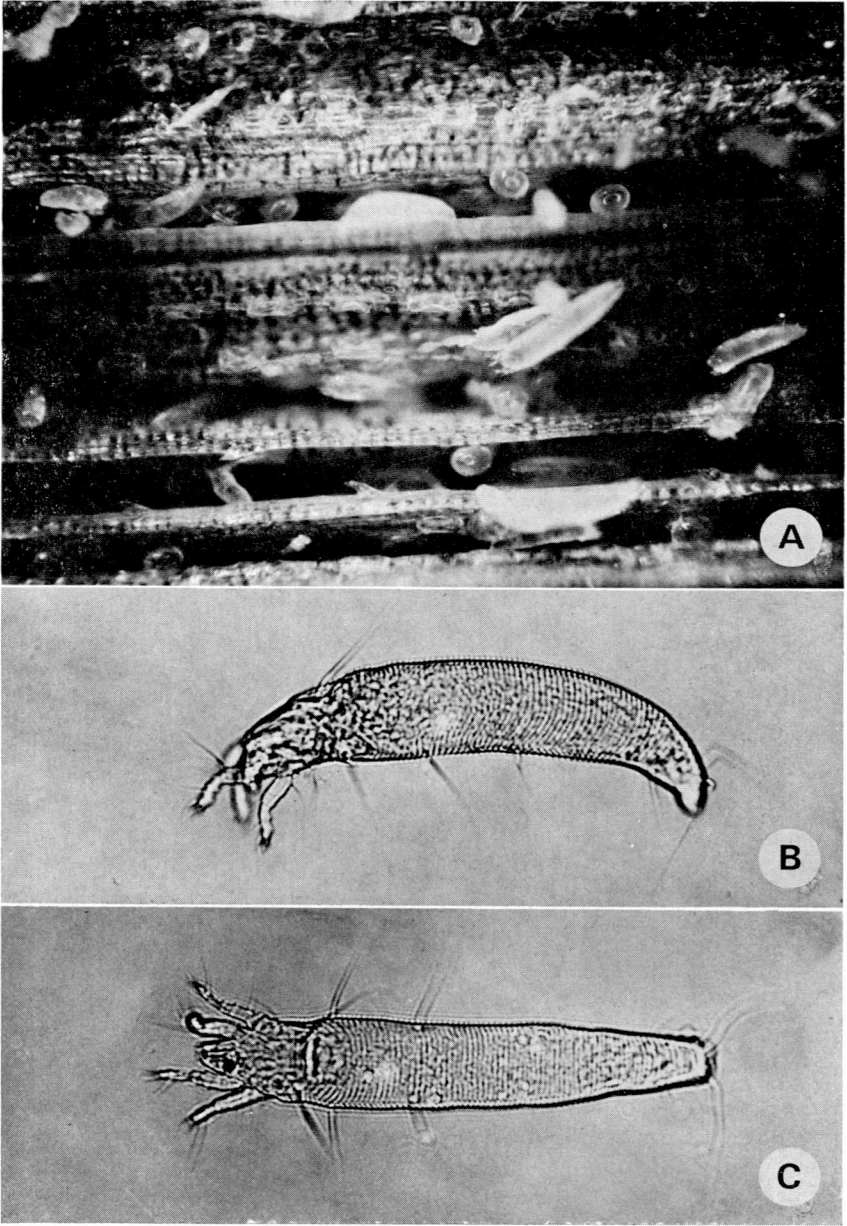

Fig. 19.2. *Aceria tulipae*, vector of wheat streak mosaic virus and wheat spot mosaic pathogen. A) eggs, nymphs, moulting mites and adults on the surface of a wheat leaf; B) lateral view; and C) ventral view of adults. All adults observed on wheat are females, varying from 175–285 μ long.

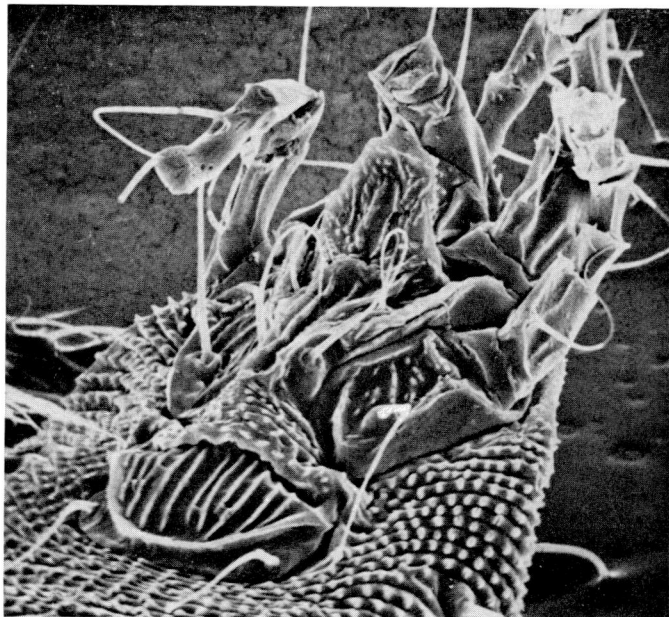

Fig. 19.3.   'Stereoscan' view of the front ventral area of *Aceria tulipae* showing the genital
flap and the position of the legs and the rostrum.

effects which result in leaf discoloration, bud deformation and blasting, gall
production, leaf puckering or rolling, and other abnormalities that may be
confused with symptoms that could be caused by viruses or other pathogens.

19.4 Infectious agents transmitted by eriophyid mites

When it is known that specific mites from diseased plants induce a disease
in test plants, other tests are required to determine whether the disease is
caused by an infectious agent carried by the mites, or if some or all of the
symptoms are caused by toxic feeding effects of the mites.

The continued development of symptoms after diseased plants have been
freed from the mites has generally been used to indicate that mite-induced
diseases were caused by infectious agents. However, this test cannot be
considered generally reliable because not all toxins have short-lived effects,
and some viruses produce transitory symptoms on some hosts.

Transmission by alternative means is one of the best indications that an infectious agent is present. Some viruses can be transmitted by rubbing juice from diseased plants on the leaves of test plants. Others can be transmitted by grafting scions from diseased onto healthy plants. For either test there should be precautions to enssure that live mites are not transferred during the transmission procedure.

Since the capacity to induce toxic symptoms is inherent, the ability to isolate non-symptom-inducing mites, by hatching eggs on healthy plants, can be used as evidence that symptoms induced by mites reared on diseased plants are caused by an infectious agent. This test would not be reliable for

TABLE 19.1

Evidence that disease symptoms associated with eriophyid mites are caused by infectious agents

Disease *Mite*	Symptoms continue after mites killed	Alternative transmission Juice Graft		Non-infective mites from eggs	Particle form and size
Currant reversion *Phytoptus ribis*	+	−	+	0	
Wheat streak mosaic *Aceria tulipae*	+	+	−	+	filamentous 19 × 700 nm
Wheat spot mosaic *Aceria tulipae*	+	−	−	+	
Fig mosaic *Averia ficus*	+	−	+	+	
Peach mosaic *Eriophyes insidiosus*	+	−	+	+	
Ryegrass mosaic *Abacarus hystrix*	+	+	−	+	filamentous 19 × 700 nm
Agropyron mosaic *Abacarus hystrix*	+	+	−	+	filamentous 19 × 700 nm
Pigeon pea sterility *Aceria cajani*	+	−	+	0	
Prunus latent virus *Vasates fockeui*	0	+	−	0	filamentous 750 nm
Rose rosette *Phyllocoptes fructiphilus*	+	−	+	0	

+ = tests with positive results; − = tests with negative results; 0 = no satisfactory tests a virus that is transmitted transovarially, but if non-infective mites are obtained, evidence for an infectious agent can be strengthened if the mites become able to induce the symptoms only after feeding on diseased plants.

Plant pathologists have been accustomed to designating as a virus any infectious agent not seen with a light microscope. Some leafhopper-transmitted plant pathogens that were assumed to be viruses now appear to be mycoplasma-like microorganisms. This discovery has introduced an unprecedented caution about designating any infectious agent as a virus unless it has been shown to have specific spherical, rod or thread-like particles. Of 10 plant diseases reported to be caused by mite-transmitted viruses, four have been shown to be caused by viruses; the other six are caused by infectious agents assumed to be viruses (table 19.1).

19.4.1 *Manually transmissible viruses with filamentous particles*

Three diseases of Gramineae, i.e., wheat streak mosaic, ryegrass mosaic and *Agropyron* mosaic, were known to be caused by manually transmissible viruses before mites were found to be the natural vectors. The particles of each of these viruses are filamentous (about 19×700 nm). Another virus with filamentous particles and that is manually transmissible has been isolated from apparently normal plum plants, but causes local necrotic lesions in some plants including *Chenopodium* spp. to which it has been transmitted by an eriophyid mite.

Wheat streak mosaic virus – There is more information on the association between this virus (WSMV) and its vector, *Aceria tulipae* (K.) than on any other virus–mite association. This is partly because of the considerable economic importance of the disease, and partly because of the relative facility with which many aspects of the association can be investigated. Recently Tosic (1971) reported that a new species of mite, *A. tosichella* Keifer, was a vector of WSMV in Yugoslavia.

Wheat streak mosaic was probably noted first in Nebraska in 1922, and the virus was transmitted manually in 1932 (McKinney 1937). Wheat is the principal economic host, but WSMV can infect most cereals and a number of annual grass weeds. In warm temperatures chlorotic dashes and streaks begin to show on wheat leaves 6–10 days after inoculation. The plants become stunted and often die prematurely. The disease has caused devastating losses in large acreages of wheat in the central great plains and western United States, and in Alberta, Canada. Since 1964 it has been found scattered through southwestern Ontario, and is widespread in maize and wheat in Ohio and Michigan. In the 1960's it was recognized in Rumania, different areas in the USSR, Yugoslavia and Jordan (Slykhuis 1967).

The discovery in 1952 that an eriophyid mite *(A. tulipae)*, previously overlooked on wheat, was an efficient vector of WSMV quickly lead to an increased understanding of the epidemiology and principles for control of the disease (Slykhuis 1955). The role of the mites was quickly confirmed by many investigators because both the virus and mites develop quickly, and the mites are efficient vectors (Staples and Allington 1956).

When reared on diseased wheat, 30–60% of the mites in all stages of development, except the eggs, were infectious. Non-viruliferous mites were obtained by transferring eggs to healthy wheat to hatch. The mites multiplied parthenogenetically, completing the life cycle in 7–10 days at 24–27 °C. Although the non-viruliferous mites caused an inrolling of the leaves of wheat and other Gramineae, and sometimes caused a chlorotic mottling on barley, only mites from diseased plants caused the chlorotic symptoms and stunting of plants, characteristic of manually transmitted wheat streak mosaic virus. More recently *A. tulipae* has been recognized as the cause of a reddish streaking on the kernels of maize (Nault et al. 1967), but none of the effects of the mites resemble the effects of WSMV.

The relationships of WSMV to *A. tulipae* include some characteristics of both non-persistent and persistent aphid-transmitted viruses. Like many non-persistent viruses, WSMV has long, flexuous particles and is readily transmitted manually. But, like persistent viruses it is acquired by the vector only during long feeding periods (15 min or longer) and is retained for long periods (ordinarily up to 9 days), including passage through molts. Although adults as well as nymphs can transmit the virus, infectivity is acquired only by nymphs.

Paliwal and Slykhuis (1967) found densely packed accumulations of WSMV particles in the midgut and hindgut of *A. tulipae* reared on diseased wheat. It was not interpreted that the particles multiplied in the mites, but that they accumulated during prolonged feeding. Since the gut contents could be regurgitated, ingested virus particles might get to the mouthparts and be introduced into plant cells while feeding. Alternatively, infection might result from particles in infective faeces. Takahashi and Orlob (1969) sometimes found a few virus-like particles in the parenchymatous cells around the mite's intestine. They speculated that transmission might occur through the intestinal wall into parenchyma tissue, then to the salivary glands, and into plant cells through salivary secretions while feeding.

Both WSMV and the mites require living hosts to survive. In some areas they are probably harbored together throughout the year on certain perennials, but they can be perpetuated on an overlapping succession of susceptible annuals grown in close proximity (Slykhuis et al. 1957). Where the disease is important its prevalence and severity are predominantly dependent on the presence and pattern of wheat culture. Wheat infected before winter carries the virus and mites over winter. During the warm weather the next spring and summer, the mites multiply rapidly. In a few weeks there may be hundreds on each diseased plant. They are dispersed by wind, carrying the virus to wheat and other susceptible plants nearby. The mites are such efficient vectors that in districts where WSMV occurs each virus-susceptible plant infested with mites is infected with the virus and is a source of infection for other plants. This includes any immature wheat, whether it be spring-sown crops, volunteer plants in fallow or at the edges of fields, or regrowth from hail-damaged crops. If such plants are still present when new crops of winter wheat are sown nearby, the annual cycle of infection is complete. In many areas infection of the autumn-sown crop can be avoided by eliminating all immature wheat, barley and other susceptible annuals in the vicinity before the new crops are sown. In some years this may necessitate a delay in seeding to permit maturation of a spring-sown crop. The growing of maize near wheat, as is done in parts of Ontario and Ohio, creates some problems. Maize may harbor infectious mites in immature ears until after the time winter wheat is normally sown (Gates 1970).

Ryegrass mosaic and Agropyron mosaic viruses – These viruses, RMV and AMV, like WSMV, have filamentous particles and are readily transmitted by the leaf rub method. Both have been reported to be transmitted by *Abacarus hystrix* (Nal.).

Mulligan (1960) found three species of Eriophydae on diseased ryegrass but only *A. hystrix* transmitted RMV. All instars of mites from diseased plants transmitted the virus. Virus-free mites acquired the virus in 2 hr on diseased ryegrass and the proportion that became infective increased with increasing feeding time up to 12 hr. Vectors lost infectivity within 24 hr on wheat which is immune to the virus.

There have been no further reports on the transmission of RMV by *A. hystrix*. Attempts to transmit RMV with *A. hystrix* in Canada have failed.

When *Agropyron* mosaic is found on *Agropyron repens* L. or wheat in Ontario, two eriophyid mites, *Abacarus hystrix* and *Aculus mckenziei* (K.), are usually also present. In vector tests, *Abacarus hystrix* but not *Aculus mckenziei* or *Aceria tulipae* transmitted the virus. Transmission of AMV by *Abacarus hystrix* is so inefficient that few test plants became infected even when large numbers of mites were transferred to them from diseased plants either manually or by blowing with a fan (Slykhuis 1969).

Although AMV and *A. hystrix* can be perpetuated on an overlapping succession of wheat in Ontario, much as WSMV and *Aceria tulipae* are perpetuated under similar circumstances in Alberta, the rate of transmission is so much slower that few plants are likely to be infected in commercial wheat fields.

Prunus latent virus – When the eriophyid mite *Vasates fockeui* (Nal.) was transferred from some diseased and symptomless plums in Germany to *Chenopodium foetidum* test plants, some of the latter developed ochre-colored local lesions surrounded by necrotic rings. The presence of a virus in the test plants was detected by mechanical transmission to other *C. foetidum* and *C. quinoa* test plants which also developed necrotic lesions. Extracts of the lesioned tissue contained flexuous particles 750 nm long (Proeseler and Kegler 1966).

Further tests showed that symptoms on *Chenopodium* sp. could be caused by single mites from plum, but more mites increased the infection rate and numbers of local lesions (Proeseler 1968). Infection occurred during feeding periods of 5 min, but more local lesions developed after longer feeding periods up to 24 hr. Larval and adult mites transmitted equally, and the virus persisted in them for at least four days. There was no mention of tests with non-viruliferous mites, or tests to show that the virus was not already present but latent in the test plants until incited by the mites.

19.4.2 Graft-transmissible pathogens assumed to be viruses

Three mite-associated diseases of woody perennials (currant reversion, pigeon pea sterility and rose rosette) are attributed to viruses as they are graft-transmissible and diseased plants continue to develop the symptoms after treatments to eliminate the mites. In addition to these criteria, evidence that the mites induce disease symptoms only after feeding on diseased trees shows that peach mosaic and fig mosaic were caused by infectious agents carried by mites. However, there have been no reports of virus-like particles being associated with these diseases.

Currant reversion was the first plant disease attributed to a mite-transmitted virus, but it has been one of the most difficult to investigate.

The black currant gall mite, *Phytoptus (Eriophyes) ribis* (Westw.) Nalepa, was recognized for many years as a serious pest of currants in England, causing swollen buds called 'big bud' (Massee 1928). Most of the affected buds dried up in early summer, or produced small, crumpled leaves. There were additional symptoms designated 'reversion' because the character of the leaves of affected black currants appeared to revert to the wild type. These occurred in association with big bud, but were differentiated from the direct effects of the mites. The leaves of bushes with reversion were narrower and smaller than normal, with the bases flattened. The leaf margins had fewer serrations and sub-main veins giving the leaves a coarser, rugose appearance. These symptoms were attributed to a virus when it was shown that the disease developed on normal plants after they were grafted with scions from reverted bushes. To avoid the possibility of transferring the gall mite with the scion material, some tests were done by cutting a length of stem from a diseased plant, stripping it of buds, then grafting it between a healthy rootstock and a healthy scion (Amos et al. 1927).

Earlier experiments on transmission of the pathogen by the mites were hampered by difficulties in establishing mites on new bushes, and slow development of symptoms (1–3 years) (Massee 1952). Smith (1962) found that the mites could be successfully transferred singly to the axils of new shoots of black currant seedlings during the normal period for mite migration in early summer. A single mite could start an infestation leading to bud galling and reversion, thus confirming an association between the mites and the disease.

Thresh (1964b) confirmed the presence of an infectious agent transmitted by the mites. He transferred mites from diseased bushes to small currant seedlings. After 4 days he dipped the seedlings in 0.05% Endrin solution to destroy the mites. There was good evidence that this treatment eliminated the mites. The plants did not develop big bud or other signs of mite injury, but some of them developed reversion symptoms. The presence of the reversion disease in these was confirmed by grafting patches of bark onto healthy bushes.

Although there have been no reports of successful experiments with non-infective mites, it appears well established that reversion is caused by an infectious agent carried by the mite *P. ribis*. However, the nature of the pathogen is still to be described.

The relationship between currant reversion and *P. ribis* is complex and mutually beneficial. The mites transmit the agent causing reversion and the disease increases the susceptibility of black currant bushes to the mites (Thresh 1964a). Hairs which impede the establishment of mites on healthy bushes are suppressed on diseased bushes, whereas vegetative growth is unaffected or even increased.

402   *J. T. Slykhuis*

Pigeon pea sterility, which causes partial sterility, reduction of flowering and sometimes mosaic on leaves of the pigeon pea *(Cajanus cajan* (Linn.) Millsp.) in India, was attributed to a graft-transmissible virus for many years before the association with mites was suspected (Seth 1962, 1965). Of three species of mites found on leaves of diseased plants, only the eriophyid mite *Aceria cajani* Channa Bosovanna induced the symptoms; 5–20 mites from diseased plants were placed on test plants, then after 4 days the plants were sprayed with 0.1 % Ekatox to kill the mites. In 3–5 weeks 31 of 136 of the test plants developed disease symptoms. The symptoms were not induced by mites from healthy plants. All pigeon pea varieties inoculated in the glasshouse, either by wedge grafting or by the mite vector, were found to be susceptible, but no hosts are known other than pigeon pea. There was no report of tests for acquisition of the virus by mites from healthy plants, or of rearing non-infective mites from eggs hatched on healthy plants.

Rose rosette is characterized by rosetting, mottling, and severe reduction in flowering of many species and varieties of roses. It has been serious in a breeding nursery near the South Platte River, Nebraska, and has been reported in Wyoming, California and Manitoba, usually in rural areas where roses are cultivated near wild roses which carry the disease. The cause was not transmitted by rubbing juice from diseased onto healthy rooted cuttings of *Rosa multiflora*, or seedlings of cucumber, squash or cowpea, but it was transmitted by grafting. Furthermore, it was transmitted by an eriophyid mite, *Phyllocoptes fructiphilus* Koch, usually found on diseased bushes (Allington et al. 1968). In a series of tests, 10–150 mites from diseased cuttings were transferred to the axils of leaves of small rose test plants. These plants and non-infested controls were kept in the same greenhouse, and after 35 days were sprayed with chlorobenzilate to control infestations of the two-spotted spider mite. In these trials, 11 of 38 test plants infested with *P. fructiphilus* developed the disease symptoms in 30–146 days. Some plants on which the mites survived did not develop symptoms. Rose rosette symptoms did not develop on any of the controls or on 59 other plants infested with the two-spotted spider mite.

The above results show that rose rosette is induced by *P. fructiphilus* from diseased roses. Attempts to culture the mites on healthy plants in the green-house failed, hence there has been no demonstration of the direct effects of the mites or of the acquisition of an infectious agent from diseased plants. The evidence for an infectious agent rests on graft transmission, but there was no assurance that mites were not present.

Fig mosaic and *peach mosaic*, though associated with different hosts and different mite species, have many similarities in history and characteristics. They were recognized and reported to be caused by graft transmissible viruses in the early 1930's (Condit and Horne 1933, Hutchins 1932). Both diseases caused a range of mottling, chlorotic blotching and deformation of leaves, and reduced grade and quality of fruit. Both diseases are widely distributed in many of the areas where their respective hosts are grown in the United States and in Europe.

Symptoms of both diseases were often associated with eriophyid mites which also caused toxicogenic leaf discolorations and deformations. In 1955 the pathogenic agent of each disease was reported in California to be transmitted by a specific eriophyid mite, *Aceria ficus* (Cotte), for fig mosaic (Flock and Wallace 1955), and a hitherto undescribed mite, *Eriophyes insidiosus*, for peach mosaic (Wilson et al. 1955; Keifer and Wilson 1956). When transferred singly from diseased leaves to susceptible test seedlings, 70% of the *Aceria ficus* mites transmitted fig mosaic, and 2.5% of *Eriophyes insidiosus* transmitted peach mosaic, some of them up to 2 days after removal from the diseased plants. The incubation period for symptoms was 10–90 days for fig mosaic and 14–100 days for peach mosaic. Plants infected with either disease continued to develop symptoms after treatment with acaricides that appeared to eliminate the mites. Graft transmission trials done with care to avoid transferring mites with the diseased scions confirmed the presence of infectious agents. Furthermore, mites in colonies originating from eggs hatched on healthy plants did not cause the mosaic symptoms.

Later experiments on the transmission of the fig mosaic pathogen showed that both larval stages and adults of *Aceria ficus* could acquire and transmit the pathogen (Proeseler 1969). The minimal acquisition and infection feeding periods were less than 15 min. Mites retained infectivity during a molt.

Recently Bradfute et al. (1970) reported double membrane bodies in the leaf tissue of mosaic-diseased figs. These bodies were accompanied by an open, tubular mass of electron dense, granular material which appeared to be structurally related.

19.4.3 A pathogen not artificially transmitted

Wheat spot mosaic was first detected when *A. tulipae* was recognized as a vector of WSMV in Alberta (Slykhuis 1956). When mites from naturally diseased wheat were tested singly on wheat for transmission of WSMV, some of the test plants that did not develop wheat streak mosaic developed chlorotic spots, severe chlorosis, stunting and necrosis, even though no manually transmissible virus was present. Mites transferred from diseased plants induced similar symptoms. Non-infective colonies, originating from eggs hatched on healthy plants could induce the symptoms only after they fed on diseased plants. When mites reared on diseased wheat were transferred to an immune host, some of them remained infective up to the 13th day.

Different isolates of the infectious agent originating from single mites caused symptoms that differed greatly in pattern and severity. The initial symptoms of most isolates included a chlorotic and necrotic spotting followed by a general yellowing and extensive chlorosis. Other isolates caused a mild mottling followed by extensive chlorosis. Field plants from which the pathogen was isolated were usually severely chlorotic and seldom had spotted leaves. Usually such plants were also infected with WSMV. In one test in which mites from a plant infected with both pathogens were tested singly, 65% transmitted the spot mosaic pathogen, while 34% transmitted WSMV, including some that transmitted both pathogens.

The only pathogens transmitted by *A. tulipae* but not by sap inoculation, and reported outside Alberta, are one causing a mild chlorosis of wheat in Ontario (Slykhuis 1961) and one in Ohio that appears identical to the wheat spot mosaic pathogen in all essential characteristics (Nault and Styer 1970). The host range and relations of the latter to *A. tulipae* were similar to those of the wheat spot mosaic pathogen. All stages of the mite except the egg could carry the pathogen, which persisted at least 8 days in mites kept on an artificial medium. This pathogen, like WSMV, was acquired by the nymphs but not by adult mites. It is not known if the wheat spot mosaic pathogen is acquired by adult mites.

Since there was no assurance that the Ohio pathogen was a virus it was referred to as 'wheat spot chlorosis pathogen'. The reluctance to call it a virus appears to be well justified. Electron microscopic studies of the ultrastructure of leaves from infected wheat, maize and barley revealed the presence of ovoid, double membrane bodies (0.1–0.2 μ) in the cytoplasm of parenchyma phloem and epidermis cells (Bradfute et al. 1970). The internal components of these bodies consisted of dispersed fibrils suggesting the presence of nucleic acid. Neither ribosomes characteristic of bacteria, mycoplasma and the psittacosis group of organisms, nor an electron dense nucleoid characteristic of many viruses was commonly evident. Similar bodies were found in fig leaves with fig mosaic. The double membrane bodies were not found in healthy leaf tissue, but they have not been positively identified as the infectious agent.

19.5 Conclusions

Eriophyid mites transmit 10 plant pathogens, four of which are viruses with filiform particles. Although the others were assumed to be viruses, their characteristics are undetermined.

CHAPTER 20

Viruses and Diptera

IAN D. MARSHALL

Contents

20.1 Introduction

The obligate association of haematophagous species of Diptera with verte-brates has incidentally allowed an association between these species and animal viruses. Many such viruses have evolved to an almost complete reliance upon the insect for transmission between vertebrate hosts, either by simple mechanical transfer on contaminated mouthparts or, as with the arboviruses, by a biological integration so complete that we do not know whether the agent is basically an insect virus which has adapted to parasitism in vertebrates, or an animal virus with an alternate cycle in invertebrates. The terminology used by animal and plant virologists differ for these types of transmission; stylet-borne (non-persistent) plant viruses are ecologically essentially similar to mechanically transmitted animal viruses, propagative (persistent) plant viruses to biologically transmitted animal viruses, whilst the equivalent of circulative (persistent) plant viruses, which enter the insect and are transmitted in the saliva without prior replication, has yet to be described for animal viruses. There are arguments in favour of animal virologists adopting the rather more precise terminology of the plant virologists and standardising the jargon of this common ground of the two fields.

It has long been recognized that Diptera are responsible for the perpetu-ation and dissemination of a large number of viral diseases of animals and man, but, until recently, seemed themselves plagued by few insect viruses. During the last few years the list of virus pathogens of Diptera larvae has been increasing steadily, but, with a few notable exceptions, there is little yet known of their natural history.

20.2 Viruses of Diptera

20.2.1 Nuclear and cytoplasmic polyhedrosis viruses

The virus pathogens of Diptera are principally parasites of larval forms. A nuclear polyhedrosis virus of *Culex tarsalis Coq.* studied under laboratory conditions has been shown to infect larvae by the oral route, and invariably kills them in the fourth instar (Kellen et al. 1963). A possible virus disease of larvae of *Anopheles subpictus* Grassi was reported by Dasgupta and Ray (1957) but the true nature of the pathogen was not resolved. A nuclear polyhedrosis virus producing symptoms superficially similar to the *A. sub-*

pictus agent was subsequently described in larvae of *Aëdes sollicitans* collected in Louisiana. Transmission rates per os to healthy larvae were low but all of those infected died before pupation (Clark et al. 1969).

A nuclear polyhedrosis virus of the crane fly *Tipula paludosa* Meig. differs from those of lepidopterous and hymenopterous larvae in pathogenesis and symptoms in the host, and in properties and shape of the polyhedra. The infection appears to be confined to the nuclei of blood cells and larvae live for a month or more, the larval period sometimes being extended beyond the normal duration (Smith and Xeros 1954; Smith 1967). Little is known of the natural history of the disease but presumably the polyhedral bodies from dead larvae remain infectious for long periods in the soil until ingested by Tipula larvae.

A cytoplasmic polyhedrosis virus has been recovered from *Culex salinarius* larvae collected in Louisiana and transmitted to colony reared larvae, but even heavily infected specimens are able to pupate and emerge as apparently healthy adult mosquitoes. Other species of mosquitoes have been infected in the laboratory (Clark et al. 1969).

20.2.2 Iridoviruses

Larvae of Diptera naturally infected with iridoviruses, or probable iridoviruses, have been recorded for the crane fly, *Tipula paludosa* (Xeros 1954), mosquitoes *Aëdes taeniorhynchus*, *A. vexans*, *A. fulvus pallens*, *Psorophora ferox* (Clark et al. 1965; Chapman et al. 1966) *Aëdes cantans* and *A. annulipes* (Weiser 1965), the blackfly *Simulium ornatum* (Weiser 1968) and *Chironomus plumosum* (Stoltz et al. 1968). Tipula iridescent virus (TIV) will infect a broad range of insect hosts (Smith et al. 1961) and might well be identical with iridoviruses recovered from larvae of other insects, but is not identical to mosquito iridescent virus (MIV). The fascination of the iridescence of the viral crystals in vivo and in vitro has led to extensive investigations of the properties of the virions (ch. 4), but attempts to unravel the natural history of any members of the group have been limited to MIV.

Despite the range of mosquito species yielding MIV in the field, the virus from *Aëdes taeniorhynchus* has been shown to be relatively species specific in laboratory colonies (Woodard and Chapman 1968), indicating that there might be a range of viruses with varying host specificities. The virus is ingested per os by larvae from contaminated breeding water or from infected larval cadavers. Time of onset of manifestations of the disease varies with the time of infection, but is usually within the span of the fourth instar, and death only occurs during this stage no matter when the virus is acquired. If larvae are infected later in life, particularly during the fourth instar, pupation proceeds and apparently healthy adults emerge. These transmit the virus transovarially, and, although only a relatively small proportion of normal larvae can be infected per os, it is probable that all the offspring of an infected adult will be carrying the virus and are fated to die in their fourth instar. Thus, although transmitted transovarially, the disease is not self perpetuating (Woodard and Chapman 1968; Linley and Nielsen 1968a).

Experiments and observations were extended from the laboratory to the field (Linley and Nielsen 1968b), and a tentative natural history of the disease developed. It is proposed that healthy larvae ingest virus directly from diseased cadavers during normal feeding activity. These larvae, already in their fourth instar when infected, will pupate and develop into infected adults, and these will transmit the virus transovarially to provide the next crop of infected eggs and doomed larvae. The implication of the cadaver as the key source of infection is required by the theory because of the failure to obtain significant experimental survival of virus for more than 24 hr when suspended in saline breeding waters or on grassy sods. However it is apparent that the postulated mechanism requires continuous breeding cycles throughout the year, as might well apply in Florida where the observations were made, or overwintering of infected adults or eggs in situations such as those reported in Czechoslovakia by Weiser (1968) where diseased larvae are first found in spring in 'marsh pools and pits in old stone quarries that are irregularly supplied with water from melting snows'. The overwintering mosquito-egg would be a far more hospitable environment for virus survival than saline waters and sods, and investigations of this aspect would be of interest.

The natural history of other iridoviruses might well be approached with these facts and postulations on MIV as a starting point.

20.2.3 Rhabdoviruses

Carbon dioxide is often used as a harmless anaesthetic for insects. In 1937 L'Heritier and Teissier observed that one of their laboratory strains of *Drosophila melanogaster* was sensitive to the gas; although the flies would recover from the induced anoxia, even the briefest exposure to high concentrations would produce subsequent paralysis and death within a few hours (L'Heritier and Teissier 1937). Characteristically with *Drosophila* this aberration was regarded as a genetic trait, and a large volume of information about the nature of its inheritance had accumulated before formal demonstration was forthcoming of a long suspected infectious agent (L'Heritier and Hugon de Scoeux 1947). The agent was named in a 'non-committal way', virus σ, and subsequently described defective variants, ρ, and ultra ρ. It is perhaps fortunate that investigations evolved in this way, for many of the subtle complexities of the interaction of host and parasite might well have eluded a purely virological approach.

Surveys in France indicate that about one third of *Drosophila* in the field are CO_2-sensitive and the condition has been found in wild populations in many countries. Indeed it appears to be rarer in laboratory than in wild colonies. As the only known consistent pathological effect is sensitivity to concentrations of CO_2 rarely found in nature, there is no apparent selection pressure against the defect, although Seecof (1964) has reported a slightly lowered survival of virus-infected progeny in a non-stabilized strain (q.v.) of *D. melanogaster*.

The virus is bullet-shaped (Berkaloff et al. 1965) and is therefore provisionally grouped with the rhabdoviruses (ch. 4), although no antigenic relationship with other members of the group has yet been observed. Two other rhabdoviruses, vesicular stomatitis and rabies viruses have been adapted to grow in *Drosophila* (Périès et al. 1966; Printz 1968; Plus and Atanasiu 1966), and the former produces delayed sensitivity to CO_2. Crude suspensions of σ virus inoculated into the haemocoele of *Drosophila* spp. undergo a characteristic growth cycle in which an initial eclipse or partial eclipse phase is followed by an exponential rise over the ensuing 14 days, and a plateau which remains remarkably constant for the life of the fly. The titre of virus attained at the plateau level correlates directly with the amount of inoculated virus. The development of CO_2 sensitivity in the inoculated fly is apparently a threshold effect, and its time of appearance varies with the virus strain and inversely with the dose of virus (L'Hèritier 1958). In apparent reflection of the neuronal site of action, the incubation period is shorter if the virus is inoculated at the front of the thorax than when inoculated intra-abdominally (Bussereau 1970a, b). In investigations directed at closer biochemical definition of the effect, acetylcholinesterase levels appear to be unrelated to sensitivity but other enzymes are being examined (Bussereau 1969).

Experimental reproducibility is most readily achieved at the relatively low temperature of 20 °C, a fact which is not entirely explained by the occurrence of temperature-sensitive virus mutants. Such mutants are incapable of replication if the flies are held from the outset at 30 °C, but if held for a short time at 20 °C after inoculation the virus enters an occult phase at 30 °C from which it can be rescued by return to 20 °C.

There is no evidence that σ virus will replicate in laboratory animals or in vertebrate cell cultures, but artificial inoculation has produced infection in four out of five dipteran species tested; *Musca domestica*, *Culex pipiens*, *Anopheles stephensi* and *Aëdes aegypti*. *Phormia terra novae* was refractory, as were all 8 species tested from 6 other orders (Jousset 1969). The virus is not transmitted during contact between infected and uninfected flies, although it appears that no deliberate attempts have been made to infect flies by feeding them virus-contaminated food, nor to assay the excretions and secretions of infected flies.

To this point σ virus could be regarded as behaving essentially like any other virus with restricted host range, albeit producing somewhat unexpected pathological effects in its natural host. The particular interest in the virus–host relationship resides in the mode of transmission, or more specifically, in the stabilized condition, which to date appears to be unique amongst viruses of organisms higher than bacteria.

When inoculated flies are mated, a proportion of the offspring will be found to be carrying the virus, the inheritance occurring only through the infected female after the virus has reached the plateau phase of replication. The proportion of infected eggs rapidly reaches a peak and then declines with resistant females, but a proportion of the sensitive females will produce only sensitive offspring no matter what the sensitivity status of the male partner. From the 'neostabilized' females in such broods are produced successive generations of stabilized CO_2-sensitive flies in which resistant individuals of either sex rarely occur. Stabilized females produce stabilized offspring no matter whether the male partner is sensitive or resistant. The offspring of stabilized males crossed with resistant females will include sensitives and resistants in proportions characteristic of the valency of the stabilized male; the valency is reproducible and, although not completely understood, appears to be dependent on virus strain and *Drosophila* genotype. As with the progeny of inoculated females, genetic experiments with such issue reveal potential non-stabilized sensitives, neostabilized sensitives, and resistants (L'Héritier 1958).

Infectious virus has not been detected in eggs freshly laid by resistant females mated to stabilized males, but it develops rapidly during embryonic and larval development to reach a plateau in the pupal stage which remains constant through metamorphosis and throughout the life of the imago. In contrast, about 50 infectious virus units are found in the eggs of stabilized females, but replication is slower and reaches a somewhat lower final titre than in non-stabilized flies (L'Héritier 1958).

Two types of defective σ viruses have been described, ρ and ultra ρ (L'Héritier 1958; Bernard 1970). Both conform to the same patterns of genetics inheritance as σ, but flies carrying these viruses are CO_2-resistant, although with ρ some sensitivity appears with ageing of the flies and small quantities of infectious virus are produced. This infectious virus is indistinguishable from σ virus and does not produce flies with ρ characteristics. The existence of these forms can be detected by the resistance of infected flies to superinfection by σ virus. The presence of ρ does not exclude infection by adapted strains of vesicular stomatitis virus (P. Printz, personal communication in Bernard 1970). The variant ultra ρ produces no infectious virus.

Trans-ovarian transmission is a recognized phenomenon for many insect, plant, and animal viruses, but it is usually accepted that this is a process involving infective virions. Although the evidence invoking some form of genetic inheritance of σ in stabilized flies was very strong, the presence of infective virus at all stages of the insect's development, including the egg, allowed of the possibility of an infectious process, the parameters of which eluded the experimental approach of the investigators. The demonstration of the ρ forms has proved that infective virions are not required for the perpetuation of the stabilized condition. Ultra ρ has been shown to be present in the imaginal wing discs of *Drosophila* larvae and, although mosaic forms are known, it seems reasonable to accept that all cells of stabilized flies carry the virus genome. That this is not incorporated into the *Drosophila* genome is indicated by the lack of inheritance through the stabilized male, so that inheritance is either cytoplasmic or extrachromosomal nucleic. The existence of similar mechanisms in other transovarially transmitted viruses must be suspected and can only be revealed by applying the methods of formal genetics to studies of the hosts concerned.

20.2.4 *Other viruses of Diptera*

Two other viruses have been recovered recently from *Drosophila*. P virus has been tentatively characterized as a parvovirus, and appears to be benignly present in low titres in many *Drosophila*. By artificial passage the titre is increased dramatically and lethal doses can be administered. Transmission, at least under experimental conditions, can be achieved by the association of susceptible with infected flies (Plus and Duthoit 1969).

The tradition of non-committal naming of *Drosophila* viruses has been astigmatically extended to ɩ virus. This virus was recovered from a suspension of triturated wild caught *D. immigrans*, and transmission to date has been only by serial artificial inoculation of *D. melanogaster*. Pathogenic expression is peculiarly restricted to male flies. These are rendered sensitive to CO_2, but their normal life span is also significantly reduced even in the absence of CO_2 challenge. The virus has yet to be visualized or serologically characterized (Jousset 1970).

20.3 *Viruses mechanically transmitted by Diptera*

20.3.1 *Plant viruses*

A leaf-miner fly, *Liriomyza langei* Frick, appears to be the only diptera species incriminated as a vector of plant viruses. Experimental transmission of tobacco mosaic virus has been achieved, most successfully between petunia plants (*Petunia hybrida* Vilm), and sowbane mosaic virus between sowbane plants (*Chenopodium murale* L.) where 60 successful transfers were made in 166 attempts (Costa et al. 1958).

Some of the leaf-miner flies, including *Liriomyza langei*, have an unusual manner of feeding, the ovipositor being used to lacerate the epidermis of a leaf prior to sucking out the cell contents. During egg laying the eggs are deposited beneath the epidermis through the laceration made by the ovipositor, and as virus has been shown to be transmitted during both ovipositing and feeding operations the mouthparts are not the only nor necessarily the most important vehicle of transmission. It is not suggested that this is the only mode of transmission of any of the plant viruses tested, and sowbane mosaic virus, which seems to be rare in nature, has been transmitted mechanically in the laboratory by other insects (Bennett and Costa 1961).

20.3.2 *Animal viruses*

Theoretically any insect which feeds on or associates with vertebrates or their excretions and secretions is capable of mechanically transmitting the viruses of the vertebrate under favourable conditions. In practice, effective mechanical transmission as a major mode of virus transmission is restricted to the skin piercers and scarifiers, occupations which are usually associated with blood feeding. Members of the superfamily *Muscoidea* are undoubted vectors of bacterial diseases and it is reasonable to assume that enteroviruses will at times be recovered from faeces and deposited on food destined for human or animal consumption, but the more usual modes of infection by these viruses are via the faecal and mucoid traces with which we smear our bodies and spice our food. Similarly, flies grazing on the exudates of poxes and rashes could subsequently introduce the causative viruses through cuts and abrasions to a new susceptible host, but again transfer directly or by fomites is more important.

For effective mechanical transmission the virus must be accessible to the probing mouthparts of the feeding insect; thus only those viruses which reach high titres in the skin or the blood are candidates. The virus must be resistant to inactivation, for it will be normally somewhat exposed to the elements during the period between blood meals taken by the vector. The only positive laboratory criteria of mechanical transmission are the capacity of the vector to transmit on interrupted feeding, the ability to sustain transmission rates during the early post-infection days when biologically transmitted viruses are undergoing their extrinsic incubation period (q.v.) and a decrease in transmission rates with time and with the number of post infection probes. As some biologically transmitted viruses pass through a relatively brief phase of mechanical transmission, the last two criteria are probably the most reliable indicators. The crucial demonstration of the absence of viral replication in the vector is fraught with the usual difficulties inherent in the proof of a negative proposition.

Relatively few animal viruses are primarily dependent upon mechanical transmission by insect vectors for survival, and most of these are viruses which produce viruliferous skin tumours or nodules. An exception appears to be equine infectious anaemia virus where the source of virus is believed to be the persistent high titred viraemia. Because of the difficulty of working with a virus for which there is no satisfactory host, the relative importance of mechanical transmission is largely inferential; incidence of the disease coincides with areas of mosquito abundance, epizootics occur during the mosquito season, the virus is readily transferred on contaminated hypodermic needles and mosquito transmission has been demonstrated experimentally. Other biting flies such as *Tabanus* spp. and *Stomoxys* spp. have also been incriminated (Hyslop 1966). The disease occurs in most countries and progress in control has been hampered by the high degree of specificity of the virus for the horse, the existence of carrier states, and the lack until recently of reliable and economic diagnostic methods (Kono et al. 1970; Hyslop 1966).

Coxsackie viruses have been recovered from pools of wild-caught mosquitoes (Taylor 1955; Maguire and Macnamara 1966). Attempts to demonstrate replication of the virus in mosquitoes have failed but Maguire obtained mechanical transfers from mouse to mouse by *Culiseta tonnoiri* up to 13 days after the infective feed and when virus could no longer be detected in triturated mosquitoes. There were no successful transfers in 46 attempts with *Aëdes australis* (Maguire 1970).

Mechanical transmission of reovirus type 3 has also been achieved in the laboratory. This virus is very resistant to inactivation and has been recovered from pools of wild caught mosquitoes on a number of occasions. Transfers of virus to susceptible mice were achieved up to 4 days after *Culex quinquefasciatus* had been fed on viraemic mice, and virus could be recovered from triturated mosquitoes of this and two other species up to 9 days after the acquisition feed. The importance of mechanical transmission in the widespread dissemination of this virus is difficult to assess (Miles and Stenhouse 1969).

Reoviruses are cosmopolitan in their range of vertebrate hosts but polio and many coxsackie viruses appear to be hosted almost exclusively by man. All are voided in human faeces, and, if the opportunity exists, coprophagous flies will inevitably ingest and also become superficially contaminated with these viruses. Coxsackie and polio viruses have been recovered from the faeces of flies, and there have been many recoveries from pools of flies trapped during but not before or after epidemics. However, virus has also been recovered from swabs of hands and soles of children in stricken households (Melnick 1949; Gelfand 1961). The faeces–fly–food route must produce some human infections with these viruses, but those communities which suffer the type of sanitation which would promote this route of infection usually live under conditions which would also maximise the opportunities for the direct bodily contact and faecal-oral routes currently favoured by epidemiologists.

With the exception, then, of equine infectious anaemia virus, mechanical transmission by *Diptera* of viruses of the vertebrate blood stream or faeces is, at best, incidental to some other more important means of spread. It is with those viruses which reach high concentrations in the skin of the vertebrate host that mechanical transmission attains primary importance. Of these, species of Diptera have been implicated in the transmission of the myxoma and avian pox subgroups of the poxvirus genus, and are potentially important in Yaba monkey tumor and Tana viruses, and in swinepox, where louse transmission has been demonstrated (ch. 4, table 4.1). Shope's rabbit papilloma is spread mechanically by mosquitoes and bugs and it is possible that some other members of the papillomavirus genus are also transmitted by this means (ch. 4).

The most devastating slaughter of an animal species by a viral disease, and probably by any other cause, was that of the European rabbit *(Oryctolagus cuniculus)* by myxomatosis, a disease transmitted mechanically by biting insects. As the vector–virus–vertebrate relationships have been studied in greatest depth myxomatosis will serve as a model for this type of transmission.

Myxoma virus is indigenous to the Americas and apparently to only two species of Sylvilagus; *Sylvilagus braziliensis* in central and South America and *S. bachmani* in California. Virus persists for long periods in these hosts in small tumours or nodules with no apparent detrimental general effect on the rabbit. If virus from these tumors is introduced into *Oryctolagus cuniculus*, whether wild or domesticated, a rapidly lethal disease is produced characterized by extensive skin tumors, inflammation of the muco-cutaneous junctions, swollen head and eye closure, and opalescent discharge from the eyes and nostrils, although the Californian virus from *Sylvilagus bachmani* tends to kill *Oryctolagus cuniculus* before the expression of these symptoms is fully developed. This dichotomic response of two host species to a given virus is a phenomenon familiar to plant virologists and is by no means unique in animal virology; myxomatosis is perhaps a peculiarly dramatic example because in its wild form *O. cuniculus* is superficially indistinguishable from many species of *Sylvilagus*.

The Moses strain of South American myxoma virus escaped from an experimental area on the Murray river in South-eastern Australia during the early summer of 1950 and a plot of the occurrence of the disease by the end of that summer would coincide fairly neatly with major watercourses and flood plains of eastern Australia. Transmission was achieved mainly by *Culex annulirostris*, a river haunting mosquito. The virus overwintered and its distribution in the summer of 1951–52 was augmented by large scale inoculation campaigns with laboratory produced Moses virus. Rather unexpectedly the virus spread through the thousands of square miles of semi arid country between the watercourses where the principal vector was found to be *Anopheles annulipes*, a species which shelters in the humid coolness of rabbit burrows throughout the dry country. Inestimable millions of rabbits died of the disease.

Two factors rapidly evolved to modify the virus–vertebrate–vector relationship during the early years of enzootic establishment in Australia; the dominant strains of myxoma virus circulating naturally in the field became less virulent than the introduced Moses strain of South American myxoma, and the intense selection pressure imposed on the rabbit population effectively increased the genetically controlled resistance of the species to this highly lethal disease (Fenner and Marshall 1957; Marshall and Fenner 1958, 1960; Marshall and Douglas 1961).

Transmission is mechanical and is achieved with varying efficiency by many species of mosquito, by fleas, Culicoides, mites and even the spines of thistles (Dyce 1961). A reproduction cycle has not been demonstrated in any vector species and the only possible selection of virus during its vector phase is by resistance to inactivation whilst on the proboscis of the insect, and no evidence of this has been forthcoming. The variants of the virus must arise during replication in the rabbit, and the variant which will survive and become dominant in the field is that best adapted for transmission, regardless of its effect on the vertebrate host. In laboratory investigations reliable transmission is achieved by mosquitoes only if the source of virus exceeds 10^7 LD_{20}/g. This titre is never reached in the blood during the relatively brief viraemia, or in areas of skin not actually over a tumour. The threshold titre is reached in the tumours produced by most naturally occurring strains of virus 6–7 days after initial infection of the rabbits and remains higher than the threshold until death or until late in the recovery phase of the rabbit. As rabbits infected with highly virulent strains such as the Moses virus die at 11 days, opportunity for the transmission of this type of virus only occurs for 4–5 days. The somewhat less virulent field virus strains such as prototype KM13, take about 23 days to kill 90% of genetically unselected rabbits, allowing about 17 days for transmission, and longer if the rabbit recovers. This fourfold increase in the length of time available for mosquitoes to acquire the virus is probably sufficient to explain the displacement of highly lethal strains in the field. Strains more attenuated than the KM13 type produce sufficient virus in the tumours for transmission, but the lesions break down and form scabs during the second or third week of infection. Although virus is present in the underlying tissue this is inaccessible to probing insects, so that the transmission period is longer than for highly virulent virus but significantly shorter than for the KM13 type, again giving a selective advantage to the latter strains. Viruses which infect the rabbit but do not produce sufficiently high titres to reliably contaminate the mouthpart of vectors probably occur naturally but have not so far been detected. Neuromyxoma, a laboratory-derived variant of the Moses virus, produces a benign infection of *Oryctolagus cuniculus* and transmission by mosquitos has rarely been achieved. No more than 10^5 ID_{50} of virus per gramme has ever been detected in the small tumours (Fenner and Ratcliffe 1965).

The overall effects of these considerations was assessed in the field by introducing a readily recognised South American strain (Lausanne virus) into a study area which experiences regular summer epizootics of myxomatosis. This strain produces large florid tumours in *O. cuniculus* in contrast to the smaller pale tumours of the Moses strain and the flat soft tumours of most attenuated enzootic strains. Virus samples were recovered from diseased rabbits throughout the summer and assessment of the type of virus carried out in the laboratory. A chance finding of a rabbit sick with the local endemic KM13 type virus during the inoculation period indicated that the introduced Lausanne virus would be competing, as expected, with a field selected variant. In the early stages of the epizootic Lausanne virus predominated, but weekly ratios gradually altered in favour of the local virus during the final 4 weeks of the 11 week epizootic, until only the local virus could be detected in the population. During the next epizootic the following summer no trace of Lausanne virus could be found, which also emphasizes the selection advantages of the more slowly evolving disease during the hazardous overwintering period (Fenner et al. 1957).

The field situation is certainly more complex than the reductionist approach would suggest; the evolution of the disease and the longevity of viruliferous tumours vary with the degree of innate resistance of the individual rabbit, the ambient temperature experienced by the population during an epizootic, the class of vector, and probably the nutritional and hormonal status of the stricken population, but in any circumstance mechanical transmission will favour the type of virus which produces the highest skin titres for the longest time.

The emergence of Australian races of *O. cuniculus* more resistant to the effects of the disease occurred rapidly, but it is still not known whether evolution will continue to the stage where myxomatosis is a trivial benign disease, as it is in *Sylvilagus braziliensis* and *S. bachmani*. Some of the parameters of the interactions of virus and vertebrate populations can be observed more clearly in the indigenous American situation where death of the host plays no part in the processes of selection.

Myxoma virus endemic in the South American tapeti *(S. braziliensis)* can be readily distinguished from myxoma virus endemic in the Californian brush rabbit *(S. bachmani)* by the symptoms produced in the European rabbit *(Oryctolagus cuniculus)*, although both are equally highly lethal in this host. This is not a casual divergence of characteristics due to evolution in isolation. Laboratory experiments with Californian virus in captured brush rabbits indicate that infection and mosquito transmission can be achieved as readily as in the European rabbit. Virus in skin slices from the tumours in brush rabbits reach a titre of 10^8 LD_{50}/g. However the Brazilian virus (Lausanne strain) produces a relatively transient tumour in the brush rabbit, mosquito transfer has not been achieved, and skin slice virus titres do not exceed $10^{4 \cdot 6}$

LD_{50}. KM13 virus, representing the dominant field type in wild European rabbits in Australia, gives minimal or negative responses in the brush rabbit (table 20.1) (Marshall and Regnery 1963).

Myxomatosis in North America appears to be confined to the area coincident with the natural range of the brush rabbit. The antigenically related Shope's fibroma virus occurs in *Sylvilagus* spp. in some eastern states, and this could presumably exclude the incursion of myxoma virus to these areas. However the range of the desert cottontail *(S. audubonii)* incorporates that of the brush rabbit and extends over a large area of western USA where myxomatosis has never been observed. Ranges closely neighbouring that of the brush rabbit include those of Nuttall's cottontail *(S. nuttallii)* and the pygmy rabbit *(S. idahoensis)*. The ranges of the brush rabbit and the tapeti are well separated, but are probably linked by the range of the desert cottontail through a tenuous overlap with the tapeti near the Gulf of Mexico. The eastern cottontail *(S. floridanus)*, one of the principal hosts of Shope's fibroma virus, ranges from Canada to central America and its distribution also overlaps that of the tapeti (Hall and Kelson 1959).

These six species of *Sylvilagus* have been tested for susceptibility to infection with Californian and Brazilian myxoma viruses and for their ability to produce tumours infective for mosquitoes. The results (table 20.1) suggest that adaptation of the Californian virus and the brush rabbit is so complete that the virus could not survive in nature in any other *Sylvilagus* spp. tested. Surprisingly infection was not achieved in the tapeti, the natural host of the Brazilian virus. The number so far tested is small but these were subsequently readily infected with Brazilian virus so their refractoriness was not due to immunity. The Californian virus infected and produced tumours in all other Sylvilagus species tested but in no case was transmission achieved, and where tested, virus in skin slices never reached the transmission threshold of 10^7 LD_{50}/g.

The Brazilian virus on the other hand, could presumably be sustained in nature by all *Sylvilagus* spp. tested except, significantly, by the brush rabbit. These experiments also indicated that resistance to the invasiveness of the virus is not a genetic character peculiar to Sylvilagus; the two specimens of Nuttall's cottontail infected with the Brazilian virus suffered severe generalized disease and death in reactions rather similar to KM13 type virus in *Oryctolagus cuniculus*. There is no apparent reason why the Brazilian or even KM13 types of myxoma could not become established in the *Sylvilagus* spp. of western USA with the exception of *S. bachmani* (Regnery and Marshall 1971).

I. D. Marshall

TABLE 20.1

Reactions of 5 species of *Sylvilagus* and *Oryctolagus cuniculus* to infection with 3 strains of myxoma virus

	S. bachmani		*S. brasiliensis*		*S. nuttallii*		*S. audubonii*		*S. floridanus*		*S. idahoensis*		*O. cuniculus*	
	Inf.	Trans.	Inf.	Trans.	Inf.	Trans.	Inf.	Trans.	Inf.	Trans.	Inf.	Trans.	Inf.	Trans.
California virus	+	T	−	−	+	−	+	−	+	−	+	−	+*	T
Brazilian virus (Lausanne)	+	−	+	T	+*	T	+	T	+	T	ND	ND	+*	T
Australian virus (KM13)	±	−	ND	ND	ND	ND	+	T	ND	ND	ND	ND	+*	T

+ Tumours produced
T Infection transmissible by mosquitoes
* Lethal disease
− No tumours produced or no transmission achieved
ND Not tested

Although mechanical transmission implies a passive role in the emergence of genetic variants of viruses, it in fact imposes a rigorous selection of available variants based upon titres of virus accessible to the mouthparts of the insect and the longevity of the source of virus in the vertebrate host. Myxoma in *Oryctolagus cuniculus* is still an unstable and evolving association which might eventually reach the rigorously selective association apparent between the Brazilian virus and *Sylvilagus brasiliensis* or the exquisitely balanced climax association between host and parasite as typified by the Californian virus and *S. bachmani*.

It seems likely that similar host-parasite interactions are occurring in the other virus diseases that are primarily maintained in nature by mechanical transmission.

20.4 *Viruses biologically transmitted by Diptera*

Diptera have not been incriminated as vectors in the biological transmission of plant viruses. Animal viruses maintained in this way not only have the facility to replicate in alternate cycles in the cells of vertebrates and invertebrates but, a priori, at relatively widely different temperatures. During the vertebrate cycle there must be a viraemia of a titre sufficiently high to ensure that an adequate amount of virus is imbibed in the small volume of the blood-meal. During the invertebrate cycle virus must be secreted in the saliva to ensure its transmission to a new vertebrate host. Viruses with these properties have been biologically grouped as 'arboviruses', a convenient epidemiological classification which ignores physico-chemical designations. Arboviruses include most currently accepted togaviruses, all the bluetongue-like viruses, some of the rhabdoviruses, and probably at least one picornavirus (ch. 4). Some are transmitted exclusively by ticks (chs. 5, 18), but most are transmitted by biting flies (tables 4.3A, 4.3B). Of these, most is known of toga-viruses transmitted by mosquitoes, and it is assumed that basic vertebrate–virus–vector relationships apply also to Culicoides-transmitted viruses such as bluetongue of sheep (Howell and Verwoerd 1971), and the insect-borne rhabdoviruses (Howatson 1970). Although arboviruses are common throughout the tropical and temperate regions, zoogeographical zones tend to harbour characteristic suites of viruses, and particular viruses are relatively infrequently found in more than one zone; this might be a reflection of the intimate association of the viruses with the distinctive zonal vertebrate and invertebrate fauna.

Arboviruses are probably exposed to fewer vicissitudes than most viruses
during the hazardous extra-cellular phase of their cycle, the only potentially
hostile environments being the blood of the vertebrate during viraemia, and
the gut of the insect. There is a higher degree of specificity in the virus–vector
relationship of biologically transmitted viruses than in mechanical transmission, a specificity which resides principally but not completely in the 'gut
barrier'. The blood imbibed by the fly is rapidly enshrouded in the semi-
permeable peritrophic membrane, and enzymic digestion begins at the
peripheral surface of the globule. Although there have been reports of
the almost immediate appearance of virus in the haemolymph (Boorman
1960; Aragao and Costa Lima 1929), it is more likely that the virus
normally attaches to receptors and replicates its way through the gut cells,
is released into the haemolymph and infects other organs and tissues of
the insect. During this phase there is a fairly high degree of virus–vector
selectivity. As far as is known all insect-borne viruses will replicate in all
species of mosquitoes if inoculated into the haemocoele. However if the
virus is imbibed with a blood meal each virus has a characteristic threshold of
infectivity for a given species of mosquito; the species which is a natural
vector of a particular virus usually has a low threshold for that virus. The
mechanism underlying this threshold effect is not known, but, largely by a
process of elimination, it seems to be related to the nature and number of
virus receptors on the gut cells (Chamberlain et al. 1954). Whatever the mechanism, this 'gut barrier' has a profound effect on the ecology of arboviruses; a given species of mosquito, for instance, might be completely refractory to natural infection by one arbovirus, have a high infection threshold
for another and yet be readily susceptible to infection by others. Particular
arboviruses vary in the range of vector species and genera that they can
readily infect, so that if the virus receptor theory is correct it seems that some
receptors occur more commonly than others over a range of insects. The time
that elapses between the infective blood meal and first transmission via the
salivary glands, the 'extrinsic incubation period', varies with the virus, the
vector, the infecting dose and the ambient temperature. Initially there is an
eclipse phase during which the titre of virus detectable in the fly falls below
that of the infecting dose, and, if the latter is small, virus will for a time be
below the level of detectability. The virus then replicates to a plateau level
which typically slowly declines over the life of the insect. Titration and fluorescent antibody staining of dissected mosquitoes indicate that initial replication is in the cells of the gut, followed by a viraemia in the haemolymph,
replication in most parts of the insect and finally virtually static titres in the

neurones and salivary gland and declining titres elsewhere (La Motte 1960; Thomas 1964; Doi et al. 1967). Although ovaries are infected, transovarian transmission has not been convincingly demonstrated in mosquitoes, remains to be finally confirmed in *Phlebotomus* spp. (Schmidt et al. 1971) and has not been reported for other dipteran vectors. A further barrier is sometimes detectable at the salivary gland; obviously this cannot occur in a vector species in nature and is mainly encountered during experimental infection by intra-haemocoele inoculation.

Apart from the gut and salivary gland barriers haematophagous Diptera appear to have no defence mechanisms against invading arboviruses, and can in fact be concurrently infected with at least two viruses. Sabin (1952) using carefully controlled doses of yellow fever virus detected a slightly higher threshold of infection in *Aëdes aegypti* already infected with dengue virus than in normal mosquitoes. Chamberlain and Sudia (1957) found that eastern and western equine encephalitis virus replicated as effectively in dually infected as in singly infected *Culex tarsalis* and were usually transmitted together. Lam and Marshall (1968a) used a range of related and unrelated alphaviruses and flaviviruses in various combinations and time intervals in *Aëdes aegypti* in an effort to determine whether mosquitoes secrete interferon-like substances in response to virus infection. In no instance did they detect any impairment to the characteristic growth cycles and transmission rates of the viruses used in dual compared to single infections, provided that the salivary glands remained undamaged (q.v.). On the other hand Altman (1963) found that the transmission rate of Murray Valley encephalitis by *Culex tritaeniorhynchus* was reduced if these mosquitoes were already infected with Japanese encephalitis virus, and Rozeboom and Kassira (1969) reported a similar phenomenon with two strains of West Nile virus in *Culex pipiens molestus*.

Adaptive defence mechanisms in insects are rudimentary at best (ch. 16), and contemporary arbovirus–vector relationship is such that there is now no apparent selection pressure favouring the development of such a mechanism, no matter what the evolutionary sequences were that led to the present commensal state. In the face of many negative reports it is generally accepted that the mosquito vector suffers no inconvenience in carrying a life-long infection by an arbovirus. There is only one recorded instance of the development of histopathologic lesions; with Semliki forest virus in *Aëdes aegypti* (Mims et al. 1966). This combination produces an abnormal evolution of transmission efficiency. There is a very short extrinsic incubation period and maximum transmission rates are reached 9 or 10 days after the acquisition blood meal.

Transmission rates then rapidly decline and in these experiments transmission was never achieved three weeks or longer after infection. This decline was associated with a general shrinkage of the salivary glands, altered consistency of the cytoplasm and disorientation of the cell array in the organ. No other abnormality was detected in the infected mosquitoes, and the virus content of whole mosquitoes remained at reasonably high levels. By subjective observation such mosquitoes appeared to experience difficulty in obtaining a blood meal, indeed it was this observation that led to more detailed investigation, but no positive objective measure of a deleterious effect could be obtained. In subsequent investigations Lam and Marshall (1968b) found that salivary glands already damaged by Semliki Forest virus were partially refractory to infection by a second arbovirus. If, however, Semliki Forest virus was introduced after another virus was already established in the gland, the gland was damaged in the usual way, but the transmission rates of the first virus continued at a characteristic level for the life of the mosquito. It was suggested that although an organ such as a salivary gland can be dually infected, the individual cells of that organ are not; at the cellular level there might be a mutual exclusion process operating.

Although arboviruses are the causative agents of many diseases of man and his domestic animals, characteristically the maintenance or primary cycle involves wildlife as vertebrate hosts, and although there must of necessity be a significant viraemia in this phase, disease is commonly inapparent. In this type of cycle man or domestic animals are infected tangentially; the vector acquires the virus during a blood meal on a viraemic wildlife host, and, after the extrinsic incubation period, usually transmits the virus during a subsequent feeding episode to another member of the wildlife community, but, according to the degree of host preference and availability, perhaps to man or domestic animal. If infection is successfully established in these latter hosts viraemia is usually of insufficient titre and duration to infect further vectors and thus no contribution is made to the perpetuation of the virus life-cycle. Manifestation of disease in these circumstances is also relatively infrequent; antibody surveys in man after an isolated epidemic of Murray Valley encephalitis in south eastern Australia indicated that only one clinically severe infection ensued from each 700 antibody inciting infections (Anderson et al. 1952), and with Japanese encephalitis the ratio of inapparent to clinical infections in children seems to be several thousands to one (Scherer et al. 1959). Such ratios will vary considerably with age-groups and probably with varying virulence of epidemic virus strains, and despite the limitations, arboviruses of this type are responsible for a number of diseases of significance in public and veterinary health such as western and eastern equine encephalitis and St. Louis, Japanese and Murray Valley encephalitis.

Yellow fever is the best documented example of an arbovirus maintained in nature in a primary sylvan cycle but which becomes established in persistent urban cycles involving man. In this case the primary cycle is perpetuated mainly but probably not exclusively by forest canopy mosquitoes with monkeys as the vertebrate hosts, but when introduced into an urban situation the highly domesticated mosquito, *Aëdes aegypti* is the intermediate host between man and man. It is likely that Chikungunya virus has similar primary and secondary cycles (McIntosh et al. 1964; DeMoor and Steffens 1970) and epidemic strains of Venezuelan equine encephalitis virus have equivalent cycles in rodents and horses.

The viruses of Phlebotomus fever and the suite of closely related arboviruses causing dengue and dengue haemorrhagic fever in man remain the only closely studied viruses of this group for which no primary wildlife cycle has been positively demonstrated. Antibodies to dengue viruses have been detected in monkeys, but usually in circumstances where it is equally probable that they have become involved in a secondary cycle initiated from the primary cycle in man, the reverse of the yellow fever sequence. *A. aegypti* is again the principal urban vector, although several other semi domesticated Aedine species have been implicated; *A. albopictus*, *A. scapularis*, *A. polynesiensis* and *A. scutellaris*. It is true that dengue viruses are only endemic in areas such as India and south-east Asia where there is a large sub-human primate population in addition to a high density human population, and the occurrence of dengue elsewhere tends to be of epidemic nature with no evidence of inter-epidemic persistence of the virus in the area. In these instances, as in the North American yellow fever epidemics of the past, it appears that the causative viruses are introduced into *A. aegypti* areas by viraemic humans or by transport of infected mosquitoes and the virus dies out with the waning of the epidemic as the density of susceptible human hosts declines, or the local vectors are controlled. Yellow fever has never been detected in Asia and this paradox is explained by some as due to the occupation of the appropriate niche by dengue viruses. It is more difficult to invoke man as the sustaining host of Phlebotomus fever, and there is some evidence that transovarial transmission of these viruses in the sandfly might be the perpetuating mechanism in endemic areas (Barnett and Suyemoto 1961; Schmidt et al. 1971).

The vast majority of arboviruses, then, are perpetuated in cycles involving insect vectors and wildlife and man is primarily involved when he associates with or alters the biotope supporting the nidus of the particular virus. In some instances human epidemics in temperate regions are regularly generated by transmission of virus from a wildlife source to an amplifying host which not only increases the accessibility of virus to insect vectors but also carries the reservoir of virus closer to human habitation. Thus the annual epidemics of Japanese encephalitis in the southern islands of Japan are at least partly due to initial virus activity in nesting herons followed by amplification in pigs and then widespread infection of humans (Buescher and Scherer 1959). The termination of epidemics in temperate regions results from the reduced availability of susceptible humans, or, more usually, by the reduced activity of vectors with the onset of lower ambient temperatures. Temperature and humidity may also have a direct effect on the development of the virus in the vector and its transmission (Derrick and Bicks 1958). No single mechanism can be invoked to explain the overwintering of arboviruses in temperate regions; continued transmission at a greatly reduced level, survival in infected hibernating mosquitoes, re-circulation of virus in birds or rodents induced by stress, trans-ovarian transmission in birds and persistence in ectoparasites have all been investigated with varying success, but none satisfactorily explain the observed tenacity of survival in the inter-epidemic period (Reeves 1961). The demonstration of re-circulation of western equine encephalitis virus in snakes after emergence from hibernation appears to provide a reliable mechanism for this virus, at least in certain areas (Thomas and Eklund 1962; Gebhardt et al. 1964), and other hibernating species such as bats might play a role in the overwintering of other arboviruses (Sulkin et al. 1965).

The apparent life-long commensal relationship between vector and virus and the capacity to support dual infections offers the ideal situation in nature for recombination between related viruses ingested at different times or between mutants arising during replication of one virus in the vector. Limited experiments with two temperature-sensitive mutants of Semliki Forest virus in *A. aegypti* failed to yield any detectable recombinants (Marshall, unpublished), in fact recombination of arboviruses as opposed to complementation has not been satisfactorily demonstrated in any system to date. Recombination requires the presence of the two components in one cell, and if the surmise is correct that the exclusion of a second virus at cellular rather than at tissue or organ level is absolute, recombination in the vector could not occur.

The vector does, however, play a role in the selection and survival of variants however these might arise. The almost universal practice of adapting field isolates to infant mice by serial intra-cerebral passage involves a selection of variants and the end product usually differs from the original in biological and occasionally antigenic characters. The properties of the field isolate are conserved more closely by mosquito to mosquito passage (Schaffer and Scherer 1971) or by alternate mosquito and mouse passages (Taylor and Marshall, in preparation). Some of the properties which best fit a virus to survive in the circumstances can be related directly to those that are important to mechanically transmitted viruses. Thus high titre and long duration of infectivity in the vertebrate host, in this case as a viraemia, will not only allow infection of a larger number of vectors but could also involve more species. However the fact of replication in the vector introduces restrictive subtleties of selection such as the ability to replicate efficiently at the widely differing temperatures and physiological conditions of vertebrate and insect, the ability to parasitise the intimate intracellular functions of cells from widely divergent phylogenetic orders, the ability to enter, replicate and be secreted by the insect's salivary glands without disturbing the physiological performance of the gland, and so on. Viraemia in the vertebrate is transient and even if death ensues the survival of an arbovirus is only indirectly affected by reducing the breeding potential of the vertebrate species, but death of the insect vector or even impairment of its normal functions before re-feeding will directly affect the survival of the virus. It is not surprising that the commensal relationship has become so highly developed, and, although some arboviruses have alternate modes of transmission, very few would survive without further adaptation if the ability of utilising an insect vector was lost. Any arbovirus variants which kill or seriously inconvenience their insect vector can only be rapidly eliminated.

CHAPTER 21

Viruses and Lepidoptera

J. F. LONGWORTH

Contents

21.1 Introduction

The devastating effects of viruses on laboratory and field populations of the Lepidoptera are well known. Detailed epidemiological studies have only been made with the occluded insect viruses, most commonly where the insect host is of some economic importance, this obviously imposes a limit on the content of this chapter.

There are two major baculovirus subgroups, in one, the nuclear polyhedrosis viruses (NPV), those members affecting the Hymenoptera develop only in the gut whereas those isolated from the Lepidoptera develop in the fat body, skin and blood cells. The granulosis viruses (GV) in the second subgroup are so far only known from the Lepidoptera. Some reference will be made in this chapter to the NPV of the Hymenoptera in view of their similarity to those of the Lepidoptera. All cytoplasmic polyhedrosis viruses (CPV), apart from one record from a Hymenopteran species and one from mosquitoes, are found in the Lepidoptera. Increasing numbers of entomo pox viruses are being recorded in the Lepidoptera and in other insect orders. Representatives of the iridovirus group have been isolated from the Lepidoptera, Diptera and Coleoptera and many are cross-transmissible to these and other insect orders. In recent years, several non-occluded viruses have been isolated from the Lepidoptera, but far too few to make comparisons with other animal viruses.

The larval stages are most susceptible to infection and there are numerous reports that with the occluded viruses, resistance to per os infection increases as larvae age. With the NPV of the diamond-backed moth, *Plutella maculipennis*, for example, the LD_{50} for first instar larvae was 46 polyhedra per larva and 211,300 polyhedra for third instar larvae (Zeya 1967). This type of response has generally been ascribed to a real increase in larval resistance, but could partly be explained by the normal increase in body weight which may serve to 'dilute' a constant virus dose (Ignoffo 1966).

Normally, virus infection of the adults arises from initial infection in the larval stage, either from a sub-lethal dose of virus or infection late in the larval stage, or some measure of resistance on the part of the host. Where infections in adults have been studied, their effects include reduced egg laying, reduced vigor and longevity and often deformation. Frequently however, the incidence of disease in the progeny of infected and uninfected adults is similar. Symptoms of infection and features of the replication of these viruses in larval and adult Lepidoptera are described in ch. 15.

The infection process may be affected by environmental conditions, particularly temperature. Larvae of the alfalfa caterpillar, *Colias eurytheme*, did not become infected when fed CPV polyhedra and reared at 35 °C whereas initial infection at 23 °C and subsequent transfer to 35 °C gave infected larvae (Tanada and Chang 1968). Similarly with the NPV of the bollworm, *Heliothis zea* and the NPV of the cabbage looper, *Trichoplusia ni*; when larvae were infected at low temperatures and then transferred to high temperatures, virus development was not inhibited, development was inhibited however if larvae were both infected and reared at high temperatures (Thompson 1959). Thus the mechanism of infection, but not the subsequent development of virus is temperature dependent.

There are often considerable differences in the susceptibility of different insect populations or individuals to infection, and in quantitative experiments in the laboratory some larvae may survive very large doses of virus. The information on resistance to virus infection in Lepidoptera and on the nature of resistance however is scanty. Certain stocks of larvae of the cabbage white butterfly, *Pieris brassicae* are resistant to infection with granulosis virus (David 1957; Rivers 1959), this effect was independent of the source of virus and persisted through many generations. The resistance was by no means absolute, and when given large doses of virus the resistant stocks merely had a slightly longer incubation period than normal stocks. Marked differences in mortality between resistant and susceptible stocks only occurred at low infecting doses. No studies on the nature of this resistance or its inheritance have been reported. Striking differences in the susceptibility of larvae of strains of the silkworm, *Bombyx mori* to infection with a CPV have been reported (Watanabe 1965). Fourth instar larvae of the Daizo strain had an ED_{50} of 1×10^9 polyhedra while for the Okusa strain the ED_{50} was $1 \times 10^{6.5}$. Watanabe suggested that the resistance in the Daizo strain was inherited as a complete dominance. Further, two susceptible silkworm strains were bred through several generations, the larvae were exposed to CPV in each generation and by breeding from the survivors a more resistant population was eventually selected (Watanabe 1967). Ten times more virus was needed to infect the selected strains than the two unselected strains in the seventh generation. Again, although these data indicate a measure of resistance to infection, probably genetically controlled, the mechanism of resistance remains unknown. The nature of resistance to virus infection in insects would appear to be a neglected problem, and one worthy of attention.

21.2 Transovarial transmission and latency

The large incidence of disease in early instars in laboratory and field populations of Lepidoptera has frequently been claimed to be a result of transmission of virus within the egg. Unequivocal demonstration of this phenomenon has generated much controversy, however, usually because external
contamination of the egg could rarely be ruled out. This subject has been
reviewed by Bergold (1958), Martignoni (1962), Martignoni and Milstead
(1962), and more recently by Smith (1967).

Martignoni and Milstead (1962) distinguished between transovarian transmission, that is from adult female to egg within the ovary, and transovum
transmission when eggs are contaminated outside the ovary, generally during
oviposition. These authors state that very few instances of transovarian
transmission have been verified experimentally; this is still the case.

I have found only two instances where unequivocal transovarian transmission seems to have been demonstrated. By far the most convincing example was described by Rivers (in Smith 1967) involving larvae of the atlas
moth, *Attacus atlas*, which failed to emerge from the egg and contained large
numbers of nuclear polyhedra. Further, surface sterilisation of eggs of the
cotton leafworm, *Prodenia litura* from two lines, one heavily infected with
NPV, did not reduce the incidence of disease in the developing larvae,
whether these were reared aseptically or not (Harpaz and Ben Shaked 1964).
Thus the larvae must have acquired the virus transovarially and they concluded therefore that transmission from generation to generation was not
merely a contamination of the egg surface or contents.

Transovum transmission has been demonstrated convincingly by many
workers, and indeed has been achieved experimentally by treating the genital
armatures of females of *Colias eurytheme* with a virus paste (Martignoni and
Milstead 1962). Furthermore, Bucher and Harris (1968) who studied a CPV
in *Calophasia lunula* showed that larvae infected late in life pupated successfully; the infected larval midgut lysed and was enclosed in the new adult
midgut where it contributed to the meconium. The meconium of infected
survivors brings a large amount of virus into contact with the eggs by direct
contamination either of the genital armature or of the eggs themselves. Part
of the meconium can be retained for several days so that infected adults can
contain polyhedra for most of their life. The CPV of the army worm, *Pseudaletia unipuncta* was described as a 'chronic disease of low morbidity' (Tanada
and Chang 1960). Even with large virus doses many larvae survived to produce adults in which polyhedra could be demonstrated in the midgut and

when these were voided they contaminated the eggs. Surface sterilisation of eggs from infected adults drastically reduced the incidence of disease in the progeny of the pink bollworm, *Pectinophora gossypiella*, infected with a CPV (Bullock et al. 1969) and in the progeny of the gypsy moth, *Porthetria dispar*, infected with an NPV (Doane 1969).

Despite many claims of transovarial transmission of NPVs and particularly CPVs in Lepidoptera one must conclude that the phenomenon has not been adequately demonstrated, because of the influence of contamination of the egg surface either by the meconium or the faeces. In most reports, no attempts were made to surface sterilise the eggs. An essential prerequisite for transovarian transmission is penetration of the oocytes by virus, resulting from an active infection in the ovarioles and this has not been demonstrated, though the observations of Rivers (loc. cit.) suggest that it occurs. Viruses that replicate in the tissues within the body cavity are most likely to be transovarially transmitted and it is difficult to see how this can arise with the CPVs, and with the NPVs of the Hymenoptera whose replication is confined to the gut. Although the gut is not the principal site of replication of the GV and NPV in the Lepidoptera this does not preclude transovum transmission. Indeed in the initial stages of infection, enveloped virus particles but not polyhedra are formed in the columnar cells of the gut before the fat body and other tissues are infected (Harrap and Robertson 1968; Summers 1969), and it is quite possible that virus may persist in the gut throughout metamorphosis, eventually contaminating the meconium and allowing transovum transmission.

It has been claimed that the NPVs of the Hymenoptera are transmitted transovarially but Neilson and Elgee (1968) working with the European spruce sawfly, *Diprion hercyniae*, state that transovum or transovarian transmission occur rarely if ever. They consider that there is little chance of the egg surface being contaminated during oviposition as the ovipositor is enclosed in a sheath when not in use, and also the eggs are laid singly in a slit in the spruce leaf. They conclude that virus is transmitted from one generation to the next on foliage contaminated either by diseased, or externally contaminated adults, or both.

There is no coherent explanation of latent virus infections in Lepidoptera. The confusion arises largely because of the considerable number of stress factors which, when applied to test larvae, may increase the incidence of overt virus infections. These factors include overcrowding, unsuitable diet, extremes of temperature and humidity, various chemicals, various heterologous non-infecting viruses, mechanical agitation and UV light. An infection may be latent when the initial virus dose is small and is received at a late stage of larval development (Smith 1967). For example, small doses of NPV were not fatal for later instars of Swaine's jack-pine sawfly, *Neodiprion swainei* (Smirnoff 1962), up to 60% survived and developed into normal adults which transmitted virus to their progeny. The interaction between host and virus which leads to latent or inapparent infections needs further investigation. Indeed in the majority of cases the effects of insect viruses are judged solely on the basis of larval mortality and it is probable that latent infections are merely inapparent infections which perhaps reflect a limited virus infection contained in some way by the host. That this has not been demonstrated may reflect the inadequacy of the techniques presently available. Perhaps the most pressing need is for a suitable assay system, completely independent of larval mortality. There is no evidence at present that Lepidoptera exhibit humoral response (ch. 16). Most insects however possess extensive defences against bacterial invasion and replication in the gut (Heimpel and Harshbarger 1965). In this context it would be useful to examine the response to virus particles injected into the hemocoele, for those populations in which resistance to per os infection has been shown.

21.3   Transmission of virus

21.3.1 Stability of inclusion bodies and virus particles

The principal means of transmission of virus to other individuals is contamination of food plants and soil from cadavers of infected larvae. Occluded viruses persist in the field despite bacterial putrefaction of the cadavers (Bergold 1963) and their infectivity is not markedly affected by temperatures in the range 40 to -20 °C. By contrast the iridescent viruses are thermolabile (Smith 1967), the infectivity of a dilute SIV solution was destroyed completely after 30 min at 50 °C.

Perhaps UV light is the most significant factor in reducing infectivity of virus in the field. Purified polyhedra of the NPV of *Trichoplusia ni* and inclusion bodies of the GV of the salt-marsh caterpillar *Estigmene acrea* were not infective after exposure to direct sunlight (Cantwell 1967). The NPV was gradually inactivated and was not infective after 3 hr, whereas the GV suddenly decreased in infectivity after 3 hr. It was concluded that the GV, with a single virus particle per inclusion, may have an 'all or none' response to exposure to UV whereas the NPV polyhedra, containing many virus particles

per inclusion may protect the more centrally located nucleocapsids. Purified preparations of the GV of *Pieris brassicae* are inactivated after 12–19 hr
exposure to direct sunlight on the upper surface of cabbage leaves (David et
al. 1968a). Crude preparations of GV inclusion bodies lost infectivity less
quickly however (David and Gardiner 1966), some being detected
after 16 weeks. This was attributed to the screening action of pigmented
materials and solids in the crude preparations. Thus inclusion bodies in cadavers in the field may remain a potent source of inoculum for long periods.
Certainly, remnants of larvae of the Great Basin tent caterpillar, *Malacosoma
fragile*, adhered to the host plant throughout the winter and they contained
infective virus (Clark 1956).

Polyhedra artificially applied to foliage as suspensions normally adhere
well and are not easily removed by rainfall. The CPV of the armyworm,
Thaumetyopiea pityocampa, persisted on pine needles when exposed to simu-
lated rainfall (Burgerjon and Grison 1965). This confirmed the report by Bird
(1964) that when spraying of virus was interrupted by heavy rains there was
no difference in mortality in areas sprayed before or after rain.

Virus reaching the soil, either in larval cadavers or in the faeces of parasites
and predators can provide a long-lasting source of inoculum within and between generations. When polyhedra of the NPV of *Trichoplusia ni* were applied to the soil, infectivity of the top inch of soil did not decline appreciably in
the first 90 weeks (Jaques 1964). Later tests however showed that after an
initial sharp decline infectivity remained constant for up to 231 weeks (Jaques
1969). Samples of dry soil or of soil watered weekly were equally infective
indicating that leaching of the polyhedra from the soil is not an important
factor. After heavy rain, leaves from cabbage plants grown on treated soils
were far more infective for test larvae than were leaves from plants grown on
non-treated soil. Thus virus in soil can be splashed onto foliage. Similarly,
the GV of *Pieris brassicae* showed little decline in infectivity after 2 years in
soil (David and Gardiner 1967). Furthermore, the incidence of GV in larval
populations of the small white butterfly, *Pieris rapae*, was studied in two plots
of crucifers, one with a long history of crucifer cropping, one with none
(Harcourt and Cass 1968). In the former site, incidence of virus was moderate
in the first generation each year and reached epizootic levels in the second
and third generations. At the second site, similar *P. rapae* population levels
were unaffected by virus in the first generation, but mortality increased subsequently although it never reached severe epizootic levels. These authors suggested therefore that soil-borne virus played an important part in the initi-
ation of the epizootics.

21.3.2 Transmission from larva to larva

Larvae undoubtedly become infected in the field by eating food contaminated with virus. The incubation period is determined by the age of the larva, the infecting dose, environmental conditions, particularly temperature, and other factors such as population density, the availability of food and possibly the incidence of other pathogens. Healthy larvae can be attracted to virus infected cadavers and become infected by feeding upon them. More usually however, the food plant may become contaminated either by exudates from dying infected larvae, or by regurgitated fluids, or the faeces. The migration of infected larvae is important for virus spread. Infected adults may contaminate the food plant when the gut and its contents are voided with the meconium. The normal behaviour of certain insects may change suddenly when they become infected with a virus disease (Smirnoff 1965). A well known example is that of larvae of the nun moth, *Lymantria monacha*, which gather at the tops of trees when infected with NPV. Similarly, virus-infected larvae of the poplar sawfly, *Trichiocampus viminalis*, and the willow sawfly, *Trichiocampus irregularis*, lose their gregarious habit and disperse, thus aiding dispersal of virus.

In some instances larval parasites have been shown to transmit virus from larva to larva, either during oviposition, or surface contamination of the foliage after feeding on freshly killed larvae. Under laboratory conditions the GV of *Pieris brassicae* can be transmitted from infected to healthy larvae by the hymenopterous parasite *Apanteles glomeratus* (David 1965). Predators and scavengers have been implicated in the spread of virus and dissemination through the faeces of these agents has been observed (Bergoin 1966; Vago and Bergoin 1968). NPV inclusion bodies and the non-occluded DNA virus of the greater wax moth, *Galleria mellonella*, retained a large part of their infectivity after passing through the alimentary canal of *Acheta domesticus*. The NPV, CPV and the non-occluded flacherie virus of *Bombyx mori* pass through the alimentary canal of the thrush, *Turdus cardis* without loss of virulence. Polyhedra of certain NPVs pass unchanged through the gut of a carnivorous bug, *Rhinoceris annulatus*, and the bird, *Erithacus rubella* (Franz et al. 1955). Larvae of the cabbage looper *Trichoplusia ni* move to the outer and more exposed surfaces of the plants at an advanced stage of infection with NPV, and thus become easy prey for the English sparrow, *Passer domesticus*. Aqueous suspensions of bird faeces collected from a cabbage field contained large numbers of viable inclusion bodies (Hostetter and Biever 1970).

Dispersal of the pathogen is also assisted by factors such as wind and rain

(Vago and Bergoin 1968). Bird (1961) found that viruses of insects in trees may be distributed by rain.

21.3.3 Transmission from generation to generation

Perhaps the most effective method of vertical transmission is the flight or migration of virus-carrying adults (Smith 1967). In species with more than one generation per year however, contamination of the environment may be sufficient to ensure transmission of virus between overlapping generations, despite UV light inactivating exposed inclusion bodies, for active residues of virus may persist on less exposed parts of the host plant and in cadavers (Jaques 1970). In univoltine species however, particularly on annual or deciduous hosts, transovum transmission may be more important, although in horticultural crops such as brassicas there must be significant carry-over of virus in the soil. Clark (1956) reported that virus smeared on twigs of *Cenothus cordulatus* and in remnants of *Malacosoma fragile* larvae killed by NPV was still viable after winter. The importance of this method of transmission in comparison with egg contamination is probably affected by host population density and may well differ from species to species. Indeed, Wellington (1962) found no evidence that the NPV of *Malacosoma pluviale* persisted between generations on contaminated hostplants. He considered that at low population density surviving adults from infected colonies were the only source of the small amount of disease appearing each year. As the population increased widespread disease appeared especially after a period of cool wet weather during the first instar. This stress was common to all localities and Wellington considered that the epizootic was caused by stress operating on latent virus infections throughout the population.

21.3.4 Transmission of virus between species

The specificity of insect viruses has been comprehensively reviewed by Ignoffo (1968). He emphasised the difficulty of reviewing this topic because most workers had not adequately considered the major variables arising in cross-transmission of virus, and in particular specific identification of the virus before and after the suspected transmission to another host. No useful purpose would be served here by enumerating the multitude of published records of cross-transmissions, but Ignoffo's conclusions are of interest. He stated that insect viruses are predominantly specific to the class Insecta, and that granulosis viruses were the most specific. Generic specificity does exist and species specificity may exist within each of the major types of insect viruses. Within the Nymphalidae for example, NPVs from four species are serologically and morphologically indistinguishable (Cunningham 1968). Moreover, three of these species feed on nettle and were observed together on the same host plant in one locality near Oxford, England. Furthermore, *Porthetria dispar*, of European origin and *P. obfuscata* from Kashmir shared a serologically and morphologically identical NPV, and three pierid species from Great Britain and the USA had identical GVs.

Cunningham and Longworth (1968) reported that CPV isolates from seven species representing four families of Lepidoptera were serologically indistinguishable. This is of particular interest in view of the wide cross-transmissibility of the CPVs within the Lepidoptera and suggests that many of the CPV isolates at present known by the specific name of the host from which they were isolated may be closely related or identical.

The densonucleosis virus of *Galleria mellonella* is extremely species specific, whereas a closely related virus from *Junonia coenia* (Spilling 1970) is transmissible per os and by injection to Lepidoptera from several families, but not *Galleria*.

There are few records of transmission of viruses of vertebrates to Lepidoptera, though Hurlbutt and Thomas (1960) report that 13 selected arboviruses were experimentally transmitted to 8 insect species representing 6 orders which included larvae of two Lepidoptera. The viruses reached high concentration without apparent damage to the host and persisted for long periods in a viable condition. Insect pathologists generally do not appear to have appreciated that viruses may replicate in unusual hosts without actually killing them. Most attempted cross-transmissions have been judged solely on whether the test insects have died.

The wide host range of the *Tipula* iridescent virus is well known and the iridescent viruses which have been recorded from the lepidopterous species are equally widely transmissible. The iridescent viruses of mosquitoes by contrast are apparently species specific, though mosquito larvae are susceptible to other iridescent viruses. The ecology of the iridescent viruses is of particular interest and has received surprisingly little attention. In New Zealand, iridescent viruses have been recorded from *Wiseana cervinata* (Kalmakoff and Robertson 1970) from *Witlesia sabulosella* and *Opogona osmocopa* (Robertson and Fowler unpublished) and from a Coleopteran, *Odontria* sp. (Pottinger unpublished). The first three isolates collected at the same time came from a pasture, and Robertson (unpublished) has shown that the *Wiseana* and *Witlesia* isolates are serologically and morphologically identical.

It is quite possible therefore that here and in other situations the iridescent viruses may affect a wide range of species within the same habitat. Any study of the ecology of viruses of this group therefore would be incomplete if considered only in the context of the species from which they were isolated. It is equally possible that the species from which an iridescent virus was isolated might not be the principal host. Because of this Tinsley and Kelly (1970) have suggested that the iridoviruses should not be given names based on hosts, but should be numbered like the adenoviruses, until more detailed comparative data are available. For example, it is likely that the *Wiseana* and *Witlesia* isolates are the same and this may well be the case with other iridescent viruses if these are isolated from sympatric species.

21.4 The influence of virus in changing natural population levels

So far, I have examined separately the factors operating in the dynamic interaction between lepidopterous host, virus, and environment. The periodic occurrence of disease among changing insect populations suggests that insect viruses are density dependent mortality factors (Steinhaus 1954). When large numbers of a susceptible host and adequately disseminated virus are present there is likely to be an epidemic, but even with a smaller host population the disease may help control host numbers. Epizootics are affected by environmental factors indirectly through their effects on the host, in that conditions favouring development of the host often also favour development of the disease. Canerday and Arant (1968) for example, reported that the incubation period expressed as a percentage of the duration of the larval period was relatively constant over a range of temperatures; thus the host–pathogen interaction was directly related to the host–environment response. Normally, epizootic disease greatly reduces the numbers of insects in a population without disturbing the overall effectiveness of parasites and predators.

Martignoni and Schmid (1961) state that virus epizootics occur at the edges of the range of distribution of the host, and decrease the density but increase resistance to virus infection. After an epidemic the remnant population contains more resistant individuals, so that further epidemics are less likely. Thus there is less selection for virus resistance and usually a return to a susceptibile heterogeneous population. These authors suggest that the three factors, tolerance, population density and the presence of virus and its inductors, explain recurring virus epizootics in insect populations.

David et al. (1968b) describe a *Pieris brassicae cheiranthi* stock which continued to 'carry' GV infection for 30 generations under laboratory conditions. Stringent conditions of hygiene were maintained and the incidence of overt infection in four of the generations was attributed to transmission of virus via the egg. Low population densities were maintained and a favourable environment, without stress was provided. Under these conditions overt infection usually did not occur, but in later generations, the percentage of larvae with inapparent infections increased, due possibly to the presence of more susceptible individuals in the population. They concluded that a virus epidemic could be started simply by an imbalance between the 'load' of latent virus and the host defense mechanisms without any help from adverse environmental factors. In a natural epidemic large amounts of virus are released into the environment and most individuals will die of the disease leaving only the most resistant individuals or those which escape contamination. In this

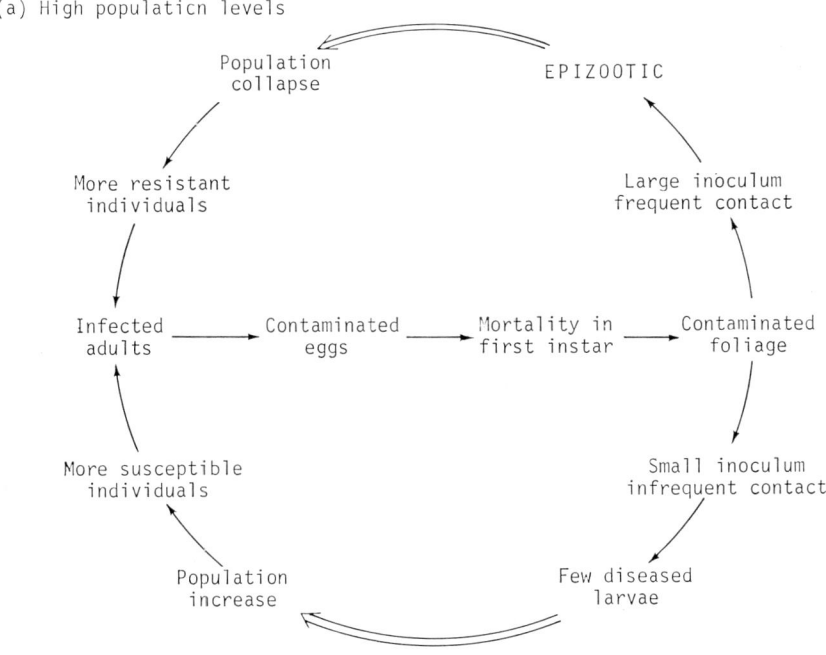

Fig. 21.1.   For explanation see text.

way the insect population returns to equilibrium with its virus and the progress towards imbalance begins again.

Finally, a virus–host system which has been examined in some detail is *Porthetria dispar* and its NPV in oakwoods and was described by Doane (1969, 1970). The virus is transmitted from generation to generation on the surface of the egg. The first instar larvae migrated to, and concentrated in the tops of the oak trees and any dying accumulated there. This small source of inoculum from dead first instar larvae was sufficient to cause an increasing rate of infection ending in a wave of mortality in the last instar, typical of the pattern of density dependent epizootics described by Steinhaus (1954). Doane's results, adapted to fit the hypothesis of Martignoni and Schmid (1961) are summarised in fig. 21.1. Doane points out that the final high mortality in the last instar was so impressive that it was easy to see why it might be described as a 'sudden' outbreak of disease. Similar situations in the literature which have been attributed to activation of occult virus may follow a similar pattern.

21.5 Conclusions

Viruses of Lepidoptera, like other insect viruses may be considered as density dependent mortality factors. Virus is transmitted from larva to larva within a generation mainly by contamination of foliage. Infected adults can transmit virus to their progeny by contamination of the surface of the egg or by contamination of the food material. Virus may persist in the environment, in the soil, on foliage and in the remains of cadavers, and be disseminated by wind and rain, and parasites and predators. Only two reports of transovarial transmission of virus have been noted in the literature; this must arise by active infection of the ovary, but this has not been demonstrated. Environmental factors may affect the host–virus interaction either by promoting the rapid development of high population densities or by modifying the response of infected larvae. Extreme environmental conditions can make the host susceptible.

There is little information on the nature and the incidence of resistance to infection with virus in the Lepidoptera, and work is urgently required on the establishment and survival of virus in sub-lethally infected larvae and on the incidence of infection in the progeny of such larvae. Without this information, there is little point in discussing latency. Indeed, previous work on resistance has centred on the levels of mortality in resistant and susceptible larvae after selected dosages of virus, perhaps more emphasis should have been placed on those larvae which survived infection!

Viruses affecting Lepidoptera are usually specific for that order, and whilst species specificity does exist, genus specificity may be more common than has been appreciated, and this will greatly affect the ecology of the viruses. Furthermore, the seemingly vast array of isolates at present described, need to be much more carefully compared, and descriptions of supposedly new isolates should be treated with scepticism.

The iridescent viruses, some other non-occluded viruses and perhaps the cytoplasmic polyhedroses seem to have wider host ranges, and detailed comparative studies on these viruses together with investigations into their effects upon the whole insect complex in particular ecosystems are warranted.

Finally, insect pathologists should be aware of the possibility that animal viruses, such as the arboviruses, may be found infecting non-vector species in the field.

CHAPTER 22

Viruses and Hymenoptera

L. BAILEY

Contents

The known viruses of Hymenoptera are few, but well illustrate the two main divisions of insect viruses. Those known in the sawflies (family Tenthredinidae) resemble in many ways the majority of known insect viruses – the polyhedrosis viruses (ch. 4) – and they are among the best known of these. By contrast those known in bees (family Apidae) have non-occluded particles; they resemble viruses of vertebrates and plants, and they, too, are the best known examples of their kind in insects.

22.1 Viruses of sawflies

22.1.1 Nuclear polyhedrosis viruses

These viruses attack many sawfly species, and kill young larvae. Bird (1961) found the infective dose for *Diprion hercyniae*, the European spruce sawfly, and *Neodiprion sertifer*, the European pine sawfly, was from 50–500 polyhedra, although susceptibility of larvae varies considerably according to age, with the youngest, at least of *N. swainei*, a jack pine sawfly (Smirnoff 1961), being the most susceptible. Bird (1961) found about 2×10^8 polyhedra developed per larva of *Diprion hercyniae* and *Neodiprion sertifer*, but Smirnoff (1961) reported that only about 6×10^6 developed per larva of *N. swainei*. Sick larvae change from their usual green to a yellow–green, their gut turns from a translucent green to an opaque milky–white and the yellow–green vomit produced by larvae when disturbed, becomes white. They exude a brown fluid from the anus that sticks larvae to the foliage, where they turn brown and eventually black. The sick larvae are flaccid and rupture easily. The incubation period ranges between 4 to about 16 days according to species and temperature (see Balch and Bird 1944; Bird and Whalen 1953). Infection changes the behaviour of larvae before they become noticeably sick. For example, larvae of *Trichiocampus viminalis*, the poplar sawfly, and *T. irregularis*, the willow sawfly, became solitary on both leaf surfaces instead of feeding as usual in single rows on the upper or lower edge respectively of leaves (Smirnoff 1965). Larvae of *Neodiprion swainei* also wander randomly instead of migrating in their usually orderly fashion in colonies of 40–70 individuals when their food is depleted: infection also seems to induce premature migration (Smirnoff 1960; 1961).

Sawfly nuclear polyhedrosis viruses multiply only in midgut tissue, in contrast to the many very similar viruses of Lepidoptera (ch. 21) which multiply in the epidermis, tracheal matrix, blood and fat-body cells. The sawfly viruses, therefore, seem potentially less destructive for their host than the Lepidoptera viruses, especially as they sometimes infect only a few gut cells (Smirnoff 1960). In fact, larvae that become infected during their late instars pupate successfully (Bird 1953; Smirnoff 1962), and the temporary pupal gut epithelium, of *Diprion hercyniae* at least, is immune. The new digestive cells that appear later in the pupal and adult stages are susceptible, however, and rapidly become infected in adults derived from infected larvae. Bird (1949) frequently found tumours, composed of cells from regenerative nidi of the midgut, protruding into the body cavity of such adults. However, Neilson and Elgee (1968a) found similar tumours also in about 10% of uninfected adults. The incidence of tumours clearly increased with virus infection but Neilson and Elgee found that infected adults with tumours transmitted infection to fewer progeny (see below) than did those without tumours. The tumours are either attached to the midgut or float free in the body cavity or in the gut lumen. They have a central core of partly broken down midgut cells, that are filled with polyhedra in infected adults, and are enclosed in a dense heavily pigmented layer, probably melanin, that is covered with many layers of flat elongated spindle-shaped cells. Polyhedra do not form in these cells. Frass remaining in some sawflies after their final moult also becomes surrounded by these tumours, which resemble those formed by gut cells against large parasites.

Larvae become infected by eating foliage contaminated with polyhedra. Rain disperses polyhedra in trees, according to Bird (1961), because disease spreads quicker down than up, although polyhedra seem impossible to remove with water once deposits from diseased larvae have dried on foliage (Bird and Burk 1961). Dispersal between trees is assumed to be by the wind, parasitic insects and birds, and to a lesser extent, by migratory larvae. These kinds of dispersal, which are probably more important between larvae within a generation ('horizontal' transmission) than between generations ('vertical' transmission), occur during summer when larvae grow and feed. Epizootic peaks, among *Diprion hercyniae* at least, occur towards the end of the season, suggesting a cumulative process. However, polyhedra deposited on foliage are not infective the next year, and few if any whole dead larvae remain on the foliage after the winter (Bird 1961). Therefore, transmission from year to year is probably by infected adults that overwinter as pupae in diapause, e.g. *D. hercyniae* and *Neodiprion lecontei*, the red-headed pine sawfly, or by eggs laid in autumn that hatch the next spring e.g. *N. sertifer* and *N. pratti bank-sianae*, a jack pine sawfly.

Bird (1949) found that infected adults of *Diprion hercyniae* transmitted polyhedrosis to more than 80% of their progeny. He considered (Bird 1961) that the virus was transmitted within eggs, because larvae do not chew their way out but split the shell by pressure and push their way out; hence he postulated that the embryos must be immune, as are prepupae. Larvae infected this way would soon die, and would easily be overlooked in the field, but viruses could spread from them to other larvae and so lead to obvious disease outbreaks later. Bird found no evidence to support the popular idea, which still lacks any acceptable evidence, that latent infections suddenly become activated in individual insects by unknown mechanisms, usually called 'stress', to cause widespread outbreaks of disease.

Neilson and Elgee (1968b) doubted that the virus of *Diprion hercyniae* was transmitted in eggs. They showed that infected adults crawling and ovipositing on foliage contaminated this with enough polyhedra to infect most larvae, including those from healthy sawflies, that ate it; they suggested that infection is usually transmitted this way. When populations were sparse, contamination by adults would be spread thinly, and many larvae would not encounter it till late in their development, so they would pupate to become infected adults. Neilson and Elgee considered that infection could persist in this way, inapparently and indefinitely. They failed to infect healthy larvae by feeding them with eggs from infected adults, hence their doubt that infection was transmitted in eggs. However, they fed third or fourth instar larvae an average of only one egg to every two larvae, so that each egg would have had to contain, on average, up to 10^3 polyhedra, or the equivalent amount of virus, to be infective. Newly hatched larvae may well be susceptible to much smaller doses than this, so Neilson and Elgee have not necessarily disproved Bird's hypothesis. The only obvious test for this seems to be to remove eggs laid by infected adults from the foliage in which they were laid, surface-sterilize them and allow them to hatch on uncontaminated foliage. However, Neilson and Elgee showed that contamination of foliage by adults, which become infected as late instar larvae, is probably a very important natural source of infection. In this respect, it is interesting that nuclear polyhedrosis spreads well the year after it becomes established among populations of *D. hercyniae* and *Neodiprion lecontei*, both of which overwinter as pupae in cocoons, whereas perennial transmission among populations of *N. sertifer* and *N. pratti banksianae*, which overwinter as eggs, is far less (Bird 1961). This could be explained by efficient transmission in pupae of virus, which multiplies in the adults that emerge during spring, and by loss of infectivity during winter of most virus deposited near and on eggs.

22.1.2 Cytoplasmic polyhedrosis viruses

The only reported cytoplasmic polyhedrosis virus (ch. 4) of sawflies attacks *Anoplonyx destructor*, the larch sawfly, in Britain (Longworth and Spilling 1970). Infected larvae cease feeding, become inactive after several days, turn from their healthy dark green to yellow, and die. The cadavers frequently stick to foliage and eventually become shrunken and brittle. Nothing is known of the ecology of the virus.

22.2 Viruses of honey bees

There are three known viruses of honey bees *(Apis mellifera)*; two attack the adults, one causing chronic bee-paralysis and the other acute bee-paralysis, and the third causes sacbrood of the larvae.

22.2.1 Chronic bee-paralysis virus

Chronic paralysis virus (ch. 4), causes the disease commonly known as 'paralysis', which has long been recognized. The median lethal dose of paralysis virus by injection is about 10^2 particles, but when sprayed on to the bees it is between 10^9 and 10^{11} particles, and when fed it is 10^{10} or more particles; the sprayed virus probably infects via the tracheae. The incubation period is about 5–7 days at 30–35 °C; cold prolongs the incubation period, although the virus multiplies less at 35 °C than at 30 °C. No other host has been found. Affected colonies of bees produce many flightless, moribund adult individuals, usually termed 'crawlers', many of which have bloated abdomens because their honey-sacs are distended with liquid (fig. 22.1). Paralytic bees occasionally appear dark and greasy. Severe attacks are usually noticed during summer when thousands of bees may crawl and die from a colony that seemed large and prosperous only a week or two previously, leaving the queen with a handful of attendant bees on neglected honey combs. The disease has been found in bees throughout the world; it is sometimes called 'Mal Noir' or 'Waldtrachtkrankheit' in Europe, but chronic bee-paralysis virus from all regions is serologically indistinguishable from local strains in Britain. All three castes of the adult honey bee are susceptible, seemingly at any age: young unmated queens, mated laying queens, workers and drones probably less than 24 hr old and even immature worker and drone pupae found mori-

Fig. 22.1. (A) Honey bee larvae: healthy individuals (A1); larvae recently killed by sacbrood (A2); and sacbrood larvae desiccated in natural conditions (A3), at intervals of about 1 week (left to right). (B) Honey-bee adults: healthy individual (left), and a chronically paralysed bee (right).

bund in or beneath paralytic colonies have contained as much chronic paralysis virus as paralysed mature adults (Bailey 1965a, 1971).

Between 10^{10} and 10^{11} particles can be extracted from one paralysed bee, and about half the number of particles are in the head (Bailey et al. 1968). Many are in the brain, and many are in the honey-sac in the abdomen. They probably are secreted into the honey-sac from the salivary and, possibly, the hypopharyngeal glands. However, although at least 70% of samples of crawling bees in Britain have paralysis, the proportion of colonies severely attacked is small, not more than one or two percent. Nevertheless, most colonies are infected, and 10% or more of the many tens of thousands of bees that die annually, from each apparently normal colony, die prematurely of paralysis (Bailey 1967). Moreover, many live, apparently healthy individuals are infected with the virus, which is localised in the salivary glands. Pollen collected in the field by normal foraging bees frequently contains infective virus that is probably secreted into the liquid added by bees to the pollen as they collect it (Bailey 1971). Therefore, most bees, although susceptible to infection, resist the multiplication of chronic paralysis virus enough to remain apparently healthy.

Susceptibility to paralysis seems to depend largely on genetic factors, but colonies may recover, and later, perhaps after a year, even with the same queen, they may suffer a severe relapse. This suggests that extraneous factors are involved, and Drescher (1964) found that strains of bees susceptible to paralysis show more disease when kept in uncultivated 'forest' regions in Germany than elsewhere. However, observations at Rothamsted for several years on colonies headed by queens bred from colonies with severe paralysis show that, although significantly more of these colonies have been attacked by paralysis than have colonies headed by normal queens, most remained apparently healthy even when kept in the same small area with the few that were severely diseased. A probable explanation of this is that queens occasionally produce flushes of susceptible workers, because they usually mate with several drones, the sperm of which do not become entirely mixed within the spermatheca (Taber 1955). Drones are produced from unfertilized eggs, so they should express recessive genes, and where only one or two genes are involved, the queen of a paralytic colony should produce many susceptible drones. However, observing paralytic colonies at Rothamsted for many years produced no evidence that proportionately very many more drones than workers die of paralysis. Moreover, many laboratory tests with apparently healthy larvae and newly emerged adult drones taken from colonies with many workers dying of paralysis have not shown many of the drones to be susceptible

(Bailey 1971). Susceptibility to paralysis may, therefore, be multifactorial with few chances of the many genes required coming together. These factors alone, or in some combinations, may be beneficial, for many apparently healthy colonies headed by queens reared from paralytic colonies are very productive.

22.2.2 Acute bee-paralysis virus

Acute paralysis virus (ch. 4) was discovered as a laboratory phenomenon during work on chronic paralysis (Bailey, Gibbs and Woods 1963). About 10^2 of the virus particles cause acute paralysis when injected into the haemolymph but about 10^{11} are needed to cause paralysis when ingested by a bee. The only alternative hosts known are bumble-bee *(Bombus)* species (Bailey and Gibbs 1964) (sect. 22.3.1). Bees suffering from acute paralysis have not been diagnosed in nature. In the laboratory acutely paralysed bees die within a day of becoming flightless, after an incubation period of five or six days at 30 °C.

Most apparently healthy bees during summer are infected with acute paralysis virus (Bailey 1971). When 1 μl of bacteria-free, concentrated extracts of whole apparently healthy bees, or of pollen collected by bees, e.g. 10 bees or 1 gram of pollen extracted in 1 ml of water, are injected into healthy bees most of these become flightless after about 5 or 6 days at 30 °C, and then die within about a day. Acute bee-paralysis virus has been detected in apparently healthy bees sent to Rothamsted from many parts of Britain, N. America and Italy (Bailey 1965b). Giauffret et al. (1969) isolated it from bees in France.

Acutely paralysed bees have much virus in several different tissues, such as the cytoplasm of the fat-body cells, the brain and especially the hypopharyngeal glands (Bailey and Milne 1969). About 10^{12} particles can be extracted from one acutely paralysed bee. Virus accumulates in various tissues of bees, shortly after they are fed sub-lethal doses, but then slowly decreases without causing apparent harm. It becomes systemic and lethal probably only when some particles enter the haemolymph, because antiserum prepared in rabbits against the virus and injected into the haemolymph protects bees equally against infection by mouth or by injection (Bailey and Gibbs 1964). Bees probably secrete acute paralysis virus from their glands, as they do chronic paralysis virus, adding it to the pollen they collect, and also passing it directly to other bees with food. Winter bees contain much less virus than summer bees (Bailey 1971), and there may be a relationship between the summer multiplication of paralysis and temperature because more than three times as much virus accumulates in living bees at 35 °C, which is the usual temperature within the summer cluster, than in bees killed by acute paralysis at 30 °C. These effects of temperature are the converse of those on the multiplication of chronic paralysis virus (sect. 22.2.1) (Bailey and Milne 1969).

22.2.3 Sacbrood virus

Larvae attacked by sacbrood virus (ch. 4) die when they are about seven days old, two days after they become sealed in their honeycomb cells, immediately before they would otherwise pupate. Two-day old larvae are most susceptible to infection, and the median lethal dose for them by mouth is between 10^7 and 10^8 particles. No other host has been found. Infected larvae continue to develop normally until their final moult, when they fail to shed their last skin. This becomes a transparent sac, around the pupal integument, and contains much ecdysial fluid, rich in infective sacbrood virus (fig. 22.1). The larva then soon dies and desiccates in a week or two to form a flat brown scale, now containing only inactivated virus, although adult bees frequently detect infected larvae within a day or so after they have been sealed in their cells, before they die, and sometimes eject many infected larvae before their cells are sealed (Bailey et al. 1964).

About 10^{13} virus particles can be extracted from one larva with sacbrood (Bailey 1969), but little is known of where the virus multiplies within the body of infected larvae. Lee and Furgala (1967) saw particles, sometimes in crystalline array, in the cytoplasm of fat, muscle and tracheal-end cells, 48 and 72 hr after 12–36 hr old larvae had been kept in the laboratory on food rich in virus. However, most of these larvae were classified as 'sick' or dead after 72 hr, at an age when they would still have been unsealed in their colony, so they were atypical of sacbrood larvae in nature, perhaps because they had ingested overwhelming amounts of virus. They may represent the usually few larvae that bees detect and eject before they are sealed in their cells. Fernando (1971) found the tissues of infected larvae appear normal almost until the time of pupation. The larval cuticle is then shed as usual and a normal cuticle forms beneath. However, the old larval endocuticle does not then dissolve in the ecdysial fluid, but remains to form the sac. Sacbrood larvae fail to secrete the enzymes needed to dissolve endocuticle, probably because of changes in the epidermis which is the probable source of virus in the ecdysial fluid.

In natural circumstances sacbrood usually disappears spontaneously during summer, and does not spread much even when combs containing many diseased larvae are placed in colonies (Hitchcock 1966). Despite this slight ability to spread during summer, the rapid inactivation of sacbrood virus in dead larvae and the lack of larvae during winter, sacbrood persists from year to year. It probably does so by multiplying in adult bees. Bees of any age become infected when injected with about 10^3 particles, and individuals younger than about 8 days become infected when they ingest 10^8 particles of sacbrood virus. They show no symptoms, but virus accumulates in their hypopharyngeal glands and, as they feed larvae with secretion – 'royal jelly' – from these glands, these seem a probable source of infection. The youngest bees are most susceptible to infection by ingestion of sacbrood virus and they are the ones most likely to ingest it in nature, because they clean honeycomb cells mostly during their first days of life. They detect and remove most sacbrood larvae within a day or two after the larvae die, when virus in the larvae is still infective, and they probably ingest ecdysial fluid when this is released from sacs ruptured during their removal. Less than 0.1 mg (i.e. 0.1 %) of a larva freshly killed by sacbrood contains enough virus to infect one of these bees, and much sacbrood virus collects in their hypopharyngeal glands within two days after they ingest an infective dose. However, although infected adults show no symptoms, they eat little or no pollen after infection, so bees that are infected when they are the youngest and most susceptible are the least likely to eat pollen (Bailey 1969). They will, therefore, secrete the least royal jelly and so presumably will not feed many larvae. Moreover, this lack of protein shortens their life and makes them least likely to survive the winter. Further, adults seem unable to infect each other with sacbrood virus when they exchange food.

Small colonies composed mostly of very young bees infected with sacbrood virus behave abnormally. They fail to rear many larvae, and the infected bees behave as though chilled, huddling together in the centre of the colony and not tending the queen. Some infected bees eventually forage normally, at least for pollen, and they perform apparently normal communication dances, indicating the source of their supply, to their colleagues. But recent tests show that they secrete sacbrood virus into pollen they collect, and this might be a source of reinfection for young bees. Although old bees are much more resistant to infection than young ones, it is comparatively easy to infect old bees from colonies that have been queenless for a time, probably because they have physiological characteristics of young bees. Such bees have a plentiful reserve of protein in their bodies, and are, therefore, well able to rear larvae, even after they are infected with sacbrood virus. Small colonies composed entirely of this kind of bee have in recent experiments lost from sacbrood almost 20% of the many larvae they reared. In usual circumstances, only very few adults that have already eaten pollen will later ingest sacbrood virus before they become immune, and so be able to infect larve. Many factors, therefore, seem to prevent the rapid spread of sacbrood.

Except for a possible alternative host species, which seems improbable, the only other obvious way that sacbrood virus might be transmitted is through the queen, but many attempts to show this at Rothamsted have failed (Bailey 1971). The virus was injected into laying queens or fed to young individuals, which successfully mated and produced larvae. None of the queens transmitted sacbrood, although infectivity and serological tests with extracts of their heads showed that sacbrood virus had multiplied in them.

Sacbrood virus multiplies readily in drones, especially in their brains, when it is injected into their haemolymph or when it is fed to them. Drones, like workers, are most susceptible to infection by mouth when young, and become immune to infection this way when they are about seven days old. However, the fact that sacbrood virus can multiply in drones is probably of no epidemiological significance because they are unlikely to secrete virus from their vestigial salivary glands, or to ingest an infective dose in nature because they do not clean combs. The particles of sacbrood and acute paralysis viruses, are physically similar and both multiply in adult bees, but their effects, especially on drones, are different. Acute paralysis kills drones at 30 °C within a week of injecting the virus, of which 10^{11} particles accumulates in the brain, whereas sacbrood virus, which multiplies in the brain and other tissues at least as much as does acute paralysis virus, causes no symptoms at any temperature and shortens the life of a drone by only a few days (Bailey 1971). A further difference is that whereas acute paralysis virus multiplies when injected into bumble-bees, sacbrood virus does not (Bailey unpublished).

None of this present information about the ecology of sacbrood virus helps to explain why severe outbreaks usually occur only during spring and early summer, but the proportion of susceptible young adults and larvae are then greatest because colonies are growing fastest.

22.3 *Viruses of other Hymenoptera*

22.3.1 *Bumble-bees*

Several species of bumble-bee – *Bombus agrorum*, *B. hortorum*, *B. lucorum*, *B. ruderarius*, and *B. terrestris* – are susceptible to acute paralysis (Bailey and Gibbs 1964). Concentrated extracts of apparently healthy bumble-bees, injected into further healthy individuals cause the disease, and the virus in the diseased bees is serologically closely related to that in honey-bees.

Pollen loads from bumble-bees foraging on red clover *(Trifolium pratense)* contained acute paralysis virus, so, as the pollen and anthers of the plant contained no acute paralysis virus, and as honey-bees, as usual, were not visiting the clover (Bailey 1971), it seems that bumble-bees, as honey-bees, are inapparently infected and secrete the virus from their salivary glands into the fluid they add to pollen as they collect it.

22.3.2 Ants

Steiger et al. (1968) described isometric virus-like particles of two sizes in the cytoplasm of nerve cells, glia and fat-body of *Formica lugubris*. Only one of several ants examined contained the small particles, 60–70 nm in diameter, but all contained the large particles, which resemble those of the iridescent viruses (ch. 4) and were often within the same cells as the small ones.

Nothing is known of the nature of these infections except that the individual worker ants, in which they were found, appeared healthy.

22.4 Conclusions

The polyhedra of sawflies resemble those found otherwise almost entirely among the Lepidoptera. This may be more than a coincidence, as the larvae of both groups of insects, which are not related, are phytophagous, often occupying similar habitats, so their viruses may have had common origins. However, the physical similarities of the polyhedra may no more indicate a relationship between the viruses than the similar, probably evolutionary convergent, appearances of the caterpillars do between the insect groups. There is, in fact, little evidence that the sawfly viruses themselves are related to one another. Smirnoff (1963) infected *Trichocampus irregularis*, the willow sawfly, by feeding them nuclear polyhedra of *Trichocampus viminalis*, the poplar sawfly, but only with difficulty; and the same virus did not infect *Pristiphora geniculata*, the mountain ash sawfly, or *Arge pectoralis*, a large sawfly of the family Cimbicidae. Smirnoff (1968) infected *Pristiphora erichsonii*, the larch sawfly, with a nuclear polyhedrosis of *Pristiphora geniculata*, but only a very few minute polyhedra formed in the foreign host. It seems improbable, therefore, that sawfly viruses will infect Lepidoptera or vice–versa. Sawflies are more closely related to bees than to Lepidoptera but sawfly viruses and bee viruses are obviously dissimilar.

All of this illustrates the difficulty of making useful, fundamental generalisations in insect pathology. Viruses of Hymenoptera presumably have some intrinsic uniformity at the biochemical level, but above this, and especially at the ecological level, each seems to be a special case. However, the host specificity of the viruses of Hymenoptera leads to at least one interesting similarity, which may be superficial but is, nevertheless, of practical importance. Honey-bee viruses, and probably other similar viruses of Hymenoptera, are perpetuated as inapparent infections, as most virus infections of vertebrates are 'subclinical or inapparent' (Fenner 1972); although most work on viruses of pest insects has been dominated, sometimes more from wishful thinking than evidence, by the concept of disease being acute and devastating, even the polyhedroses of sawflies are transmitted from year to year, at least, as a result of the inapparent infection of late instar larvae and the adults from these. This transmission of viruses from inapparently infected individuals to those that are susceptible to disease should be clearly distinguished from activation, or 'induction' (Smith 1967), of latent infections within the same inapparently infected individuals, an event that sometimes seems to occur under various unrelated or ill-defined circumstances, especially in the laboratory, and that may be of no significance in nature. Spectacular, natural outbreaks of virus diseases among populations of Hymenoptera, may be regarded as irregularities, depending, for example, on unusual or transient seasonal events leading to unusually large densities of susceptible individuals, or colonies, of young sawfly larvae, or to unusually large proportions of susceptible young bees, or on the rare associations of many genetic factors that together produce susceptible adult bees.

Viruses that were host specific, lethal and quick to spread would not survive in nature. Further, unless viruses can remain infective elsewhere than in their living host long enough and suitably placed to reinfect the next generation, as, for example, polyhedroses and granuloses of Lepidoptera persist in soil (ch. 29 sects. 2.1, 3.3), inapparent infection of individuals that produce the next generations, is their only means of survival. This is a relatively little studied aspect of the epizootiology of insect viruses, but it is important from the practical point of view of using them to control pests or of protecting useful insects from them.

CHAPTER 23

Viruses and Coccoidea

O. ROIVAINEN*

Contents

---

* This review is recommended to be read in conjunction with that of Leston, D., Entomology of the cocoa farm. Annual Review of Entomology 15, 273–294 (1970).

23.1 Introduction

Our knowledge of cocoa viruses transmitted by coccids is almost entirely based on the extensive research that has been done on the cocoa swollen shoot disease, one of the most economically damaging of all plant virus diseases. The disease was first discovered in Ghana, and Posnette (1940) showed that it was caused by a virus. Subsequently, cocoa viruses have been found in Nigeria, Ivory Coast, Togo, Sierra Leone, and also Trinidad and Ceylon. There are also reports from other countries but these have not been confirmed. Some aspects of cocoa viruses were reviewed by Thresh (1958a, b) and Thresh and Tinsley (1959), and a general account was prepared by Dale (1962). In addition to cocoa viruses, Heinze (1959) gives a comprehensive list of reports of coccid transmission of viruses from suspected virus infected plants but convincing evidence is lacking except for mealybug wilt of pineapple which appears to be a disease where virus is probably involved.

This disease was originally found in Hawaii and is linked with feeding of two mealybug species, *Dysmicoccus brevipes* (Cockerell) and *D. neobrevipes* Beardsley. The disease is now of less economical importance and is not discussed in the present paper because less progress has been made in the past decade, and adequate reviews of earlier work have been published elsewhere by Carter (1962) and Maramorosch (1963).

23.2 Cocoa viruses

Cocoa virus isolates obtained from different localities usually differ in symptoms and may also differ in host range, virulence, and mealybug transmissibility. There is little information on the relationships between West African isolates and those from other countries, and even the relationships between the various West African isolates themselves are not entirely clear. Thresh and Tinsley (1959) classified the West African mealybug transmitted isolates into the two groups, cocoa swollen shoot virus (CSSV) and cocoa mottle leaf virus (CMLV), on the basis of symptoms, host range and vector specificity. However, Kenten and Legg (1967, 1971) have shown that these two groups of viruses have many similarities and are serologically related and they have suggested that both groups should be referred to as CSSV. The status of cocoa (Trinidad) virus (CTV) is not clear because serological tests have not been done, however, some dissimilarities between CTV and the

West African viruses suggest that these are different viruses (Posnette 1944; Baker and Dale 1947; Kirkpatrick 1950). For sake of convenience here these three virus groups are treated as distinct viruses. Two other cocoa viruses are known but neither are mealybug-borne (Posnette 1950; Attafuah and Brunt 1960; Blencowe et al. 1963; Owusu 1971).

23.3 Transmission

At least fourteen distinct species of mealybugs (Pseudococcidae) are known to transmit one or more of the three cocoa viruses (Dale 1962). Only two species have been shown to be virus specific. Thus, *Ferrisia virgata* (Cockerell) is a vector of most of the virus isolates tested, but has consistently failed to transmit CMLV and Mampong isolate of CSSV. *Pseudococcus longispinus* (Targioni Tozzetti) has transmitted only Kpeve isolate of CMLV and Mampong isolate of CSSV (Posnette 1950; Attafuah and Brunt 1960).

Cocoa viruses can be transmitted mechanically using partially purified virus after chemical processing of infected leaf preparations, but crude sap is non-infective (Brunt and Kenten 1962). Transmission through seed or by dodder seems not to occur (Posnette 1947; Posnette, Robertson and Todd 1950).

23.4 Characteristic of transmission by vectors

The bulk of information on characteristics of transmission comes from the numerous contributions dealing with *Planococcoides njalensis* (Laing). Some other species, e.g. *Ferrisia virgata*, *Planococcus citri* (Risso), and *Dysmicoccus brevipes* (Cockerell), have also received attention, but to a lesser extent, and little is known about the transmission characteristics of the other vector species.

The process of virus transmission by mealybugs can be divided conveniently into different sequential sections or periods: a) period before virus acquisition feeding; b) virus acquisition feeding period on the infected source plant; c) period between cessation of acquisition feed and the next feed on a healthy plant and; d) period of feeding on a healthy plant. The length of each period and conditions during each period will determine whether the infection establishes in the healthy plant.

When searching for a new feeding site prior to acquisition feed, mealybugs may have to starve for different periods. After prolonged starvation the need to settle down to feed probably increases or perhaps the mealybugs are less fastidious about the new feeding site (Posnette and Robertson 1950; Roivainen, unpublished). Prolonged starvation before feeding also increases the rate of food uptake of mealybugs during the first two or three days of feeding (Roivainen, unpublished).

The chance of a mealybug becoming infective depends on the availability of virus in the tissues in which it is feeding. Generally, mealybugs can acquire virus from any part of a plant (Posnette and Strickland 1948), except the seed (Posnette 1947). However, there is evidence that virus is not always fully systemic in the plant. (Posnette and Robertson 1950). Young recently infected plant tissues with severe symptoms are good sources of virus and usually availability decreases with age of infected tissue (Posnette and Robertson 1950; Longworth 1964), and also with mildness of symptoms, but to some extent, virus can be acquired even from latently infected symptomless plants (Dale 1958b; Thresh 1958c).

It is known that *Ferrisia virgata* is more efficient as a vector than *Plano-coccoides njalensis* if it acquires virus from leaves. However, if the two species are fed on stems, *P. njalensis* is more efficient (Longworth 1964; Roivainen, unpublished). A very plausible explanation for this difference was given by Entwistle and Longworth (1963) who investigated the feeding behaviour of three mealybug species. In transverse sections of cocoa stems, stylets of *P. njalensis* were seen to end in the phloem more often than stylets of *Ferrisia virgata*. The third species, *Phenacoccus madeiriensis* Green, which is not a vector of cocoa viruses, fed differently; its stylets did not enter the phloem. Thus the inefficiency of virus transmission by *Ferrisia virgata* from stems can be explained by the infrequency its stylets reach the phloem. The low infection rate with *Planococcoides njalensis* from leaves may reflect the settling and feeding behaviour (Longworth 1964; Roivainen, unpublished). An obvious difference is that *Ferrisia virgata* prefers the leaf as a feeding site, whereas *Planococcoides njalensis*, although feeding to some extent on leaves, prefers leaf axils, crevices between veins and pulvinus, and scars on the stem.

It has been clearly established that virus is more readily available in mature sensitive than mature tolerant cocoa trees (Blencowe 1962; Owusu 1970). However, the availability of virus is much more variable in young cocoa (Dale 1958b; Longworth 1964; Igwegbe 1966; Owusu 1969). According to Posnette and Todd (1955), mealybugs can transmit a virulent virus strain and a mild strain equally from newly infected seedlings, but the mild strain becomes much less readily transmitted as the symptoms become less conspicuous with age.

The length of acquisition feeds on virus source plants affects greatly the infectivity of mealybugs; the infectivity increases with increasing length of acquisition feed (Posnette and Robertson 1950; Dale 1955; Roivainen, unpublished). Occasionally mealybugs become infective after an acquisition feed of one hr (Posnette 1951), but to obtain the largest proportion of infective mealybugs, the acquisition feed must continue up to about three days after which no further increase in infectivity can be observed (Roivainen, unpublished). If mealybugs have starved before the acquisition feed, acquisition feeds considerably shorter or longer than about three days result in a drop of infectivity. Roivainen (unpublished) suggests that this peak in the ability to transmit the virus results from increased amount of food and virus uptake per unit time by starved mealybugs during the first few days of acquisition feed. With longer feeds the infectivity does not increase so much because the amount of food and virus uptake decreases as the effect of starvation lessens, while the virus, originally acquired during the first few days, gradually loses its infectivity.

Because the virus does not persist for long, mealybugs leaving the source plant after acquisition feed may spread the virus only if they settle down to feed on a healthy plant in time. The loss of infectivity is approximately exponential with time; after every 8–10 hr half of the mealybugs have lost their capability to transmit the virus. Thus, after 3–4 days the infectivity is lost almost entirely (Roivainen unpublished).

Posnette and Strickland (1948) and Posnette and Robertson (1950) reported that infectivity was lost more rapidly after an acquisition feed if mealybugs were feeding rather than if they starved, and that a single mealybug seldom infected two or more healthy plants in a series. However, Roivainen (unpublished) could not demonstrate decreased loss of infectivity due to feeding on a healthy plant. Furthermore, young nymphs of *Planococcoides njalensis* which moulted after an acquisition feed still transmitted virus when transferred to cocoa bean test plants, thus indicating that the virus is transmitted in a circulative manner (Roivainen 1971). Thus it is rather difficult to explain how, after the acquisition feed, starving or feeding on a healthy plant would affect differently the subsequent mealybug infectivity as such, unless one could show that the salivary glands are less replete with virus after feeding than starving. The lower infectivity of satiated mealybugs could be because they delay settling and lose some of their infectivity.

The final stage, the transmission of virus by mealybugs into a healthy plant, takes 15 min to few hr, depending on how soon mealybugs settle down to feed (Posnette and Robertson 1950; Dale 1955). The actual penetration time for stylets to reach the phloem is about 10–30 min (Posnette and Robertson 1950; Entwistle and Longworth 1963), and it is probable that the virus must be deposited into phloem tissue in order to infect a plant.

The probability of transmission with single mealybugs can be easily maintained at about 0.3 and with more elaboration a probability of 0.5 or more is possible (Dale 1958a; Roivainen, unpublished). If the number of mealybugs is increased, the probability increases in accordance with mathematical expectation (Posnette and Robertson 1950; Roivainen, unpublished).

Posnette and Robertson (1950) suggest that adults of *Planococcus citri* are more efficient vectors of CSSV than nymphs. Kirkpatrick (1950), however, was unable to demonstrate any such difference with CTV, and it appears that there are no consistent differences in the transmission efficiency of vectors of varying age (Dale 1958a). Old adult females and males that may not feed at later developmental stages cannot therefore be vectors (Kirkpatrick 1950). There is no transovarial transmission of infectivity (Dale 1958a).

Posnette and Robertson (1950) report that one strain of *P. citri*, which was cultured from a single female, consistently failed to transmit CMLV under conditions which allowed transmission by a mixed progeny of five females. However, non-transmitting strains of *Planococcoides njalensis* were not found in tests with colonies reared from single individuals (Posnette and Robertson 1950; Lister 1953).

23.5 Ecology of cocoa viruses

There is no direct evidence about the origin of cocoa viruses. The experimental host range of these viruses is limited to some thirty species in the Bombacaceae, Malvaceae, Sterculiaceae and Tiliaceae (Posnette et al. 1950; Tinsley and Wharton 1958; Bald and Tinsley 1970). However, naturally infected indigenous plants are known from Ghana and Nigeria (Posnette et al. 1950; Attafuah and Tinsley 1958; Legg and Agbodjan 1969), and probably virus spread to cocoa in West Africa from these plants soon after introduction of the crop.

Whatever the importance of indigenous hosts in the past, the main reservoir of virus today is cocoa itself. Although mealybugs are always the sole agents of dissemination of virus at the final stage, some spread of virus may occur through infected pods which are moved at harvesting or through rooted cuttings if taken from infected trees (Thresh 1958b). In the pattern of natural spread from infected cocoa trees two distinct processes can be observed. A new outbreak of virus may start by discontinuous or jump spread from the existing primary outbreak. The subsequent enlargement of the new outbreak is by radial spread between the trees in contact (Thresh 1958b; Thresh and Lister 1960). Radial spread is slow by comparison with most other insect transmitted viruses but relentless, and inside the affected area trees seldom escape infection.

The pattern of spread coincides with the movements of mealybugs in cocoa. Thus, movement of mealybugs by walking from tree to tree through the canopy may spread the virus (radial spread) (Cornwell 1958), or passive dispersal by air currents over longer or shorter distances may occur (jump spread) (Strickland 1950; Cornwell 1960). Passage from tree to tree along the ground appears to be unimportant (Strickland 1951a; Cornwell 1956). Occasionally, ants carry mealybugs but this seems to happen only when an ant attended colony is disturbed (Cornwell 1955; Roivainen, unpublished).

The rate of virus spread in young cocoa plantations is very slow, obviously because the seedlings are not in contact which excludes radial spread and the 'target' for air-borne mealybugs migrating from elsewhere is smaller (Benstead 1951). Similarly, wide tree spacing may reduce spread of virus (Thresh 1958b).

Mealybugs are found in cocoa everywhere. They usually occur in small numbers of less than hundred per tree on an average, however, numbers in different trees can vary from zero to thousands (Strickland 1951b; Donald 1955). Because of their small numbers these insects can hardly be considered pests in their own right.

Some mealybugs such as *Ferrisia virgata* and *Planococcus citri* are only occasionally attended by ants. But in Ghana, where cocoa viruses are far more important than elsewhere, the most abundant and widespread vector, *Planococcoides njalensis*, is almost invariably attended by crematogasterine and pheidoline ants (Strickland 1951a, b). So close is this association that removal of ants eliminates *P. njalensis* almost completely (Taylor 1958; Entwistle 1959, 1960). There is no doubt that *P. njalensis* is protected against adverse environmental factors by the attending ants, and thus, this mealybug can occur at considerably higher population densities than some other mealybug species that are not associated with ants to the same extent (Strickland 1951b).

Usually less than 5% of *P. njalensis* or other species is parasitized, and this small incidence seems to not be influenced by attending ants (Strickland 1951b; Donald 1956). Predators may, therefore, be more important in controlling mealybugs (Donald 1956). Pathogenic fungi occur but their relative importance has not been studied in detail (Rojter, Bonney and Legg, 1966). Seasonal changes in density of *P. njalensis* were reported by Cornwell (1957) and he suggests that these trends may be brought about largely by changes in the abundance of predators and parasites.

23.6 *Some general considerations*

The ecology of cocoa viruses is such that, at least in Ghana and Nigeria, they could not be eradicated. Various attempts to control the mealybugs with insecticides have shown that repeated treatments are needed to suppress the mealybug populations to such a level that spread of virus is retarded, and this is not economical (Hanna, Judenko and Heatherington 1955; Hanna et al. 1959; Roivainen 1968, 1969). Indirectly, the mealybug populations can be decreased by controlling the mealybug attending ants, but again, repeated treatments are necessary and mealybug species not attended by ants would be unaffected (Entwistle 1959, 1960). A different approach altogether would be to protect cocoa against severe virus strains by deliberately infecting them with mild ones (Posnette and Todd 1955). Resistance and tolerance to virus infection was first shown by Posnette and Todd (1951), and it now appears that the use of resistant and/or tolerant cocoa varieties may provide the most practical method of combating the disease in future (Legg and Kenten 1970). But at present, the only general method of control is to remove all trees known to be infected and in this way to attempt to arrest spread by eradicating the obvious sources of infection.

CHAPTER 24

Persistent aphid-borne viruses

D. PETERS

Contents

24.1 *Introduction*

Plant viruses are transmitted by aphids in two quite different ways. Some viruses are carried by aphids for only a limited period, these will be discussed in ch. 25.

Other viruses are translocated after they have been ingested, passed through the vector's body and added to saliva. The aphid remains infective for a long period of its life, and does not lose its infectivity upon moulting. A small number of the plant viruses are transmitted in this way, and these are the subject of this chapter.

Some of these viruses seem only to circulate through the body of the vector apparently without multiplication, e.g. barley yellow dwarf virus and pea enation mosaic virus. Other viruses, like lettuce necrotic yellows virus and sowthistle yellow vein virus, replicate in the vector during their translocation. To distinguish between these, the latter are called propagative viruses, the former are referred to as circulative viruses.

Forty or more plant viruses are circulative or propagative in aphids (Gibbs 1969). These viruses mostly cause severe yellowing and leaf rolling symptoms and many, such as barley yellow dwarf virus and potato leafroll virus, are responsible for economically important crop diseases throughout the world. These viruses are acquired and transmitted most efficiently by aphids during long feeds. As a consequence their principal plant host is often also the principal host of their aphid vector. The spread of many of these viruses in crops can be controlled by treating the crops with insecticide (ch. 31).

This chapter will describe the route and the fate of these viruses during their transmission. Before doing this some of the characteristics of the viruses involved, will be briefly mentioned.

24.2 *The viruses*

The properties of most of these viruses are unknown. This is mainly so for two reasons. Firstly, they are difficult to study since they mostly occur in small quantities in infected plants and they seem often to be restricted to the phloem. Secondly, to demonstrate the infectivity of extracts, samples have to be fed to, or injected into virus-free vectors (fig. 24.2). These inoculated vectors are tested in turn for their infectivity on test plants. One circulative virus – pea enation mosaic virus – and one propagative virus – lettuce necrotic yellows virus – are readily transmitted mechanically from suspensions to plants, apparently because the epidermis and underlying tissues of their hosts are susceptible to these two viruses. Although sowthistle yellow vein virus can infect tissues other than the phloem of *Sonchus oleraceus* there is yet no evidence that this virus can be transmitted mechanically. A number of physical properties of the best studied viruses are summarized in table 24.1. Thus far, there is little information on the physical properties of the other circulative or propagative aphid-borne viruses, e.g. bean leafroll virus, beet mild yellows virus, carrot motley dwarf virus, filaree red leaf virus, and others.

TABLE 24.1
Physical properties of some circulative and propagative viruses

Virus	Morphology	Size (nm)	S20, w	Reference
Lettuce necrotic yellows	Bacilliform	227 × 66	940	a
Sowthistle yellow vein	Bacilliform	230 × 103		b
Potato leafroll	Isometric	24		c
Pea enation mosaic	Isometric	27	120, 100	d
Barley yellow dwarf	Isometric	30	115–118	e
Viruslike particle (24 nm)	Isometric	24	130, 56	f
Viruslike particle (29 nm)	Isometric	29		f

a) Harrison and Crowley 1965; b) Peters and Kitajima 1970; c) Peters 1967 a; d) Izadpanah and Shepherd 1966; e) Rochow and Brakke 1964; f) Peters 1965, 1967b, and unpublished results.

24.3 Acquisition

Viruses, circulative and propagative in aphids, are acquired from sap ingested from plants. As aphids feed mainly on the phloem their mouthparts have to pierce the plant tissue. The mouthparts consist of two pairs of needlelike stylets which are admirably adapted for penetrating to the vascular bundles (Forbes and MacCarthy 1969). A proteinaceous tube of saliva is produced around the stylets as they pierce through the plant tissues. The function of the sheath is uncertain. It has been proposed that it may act as a filter, preventing ingestion of bacteria and possibly also viruses. But a closed stylet sheath is difficult to reconcile with the rapid uptake of plant sap by some aphid species. Observations by Miles et al. (1964), and Kinsey and

466   *D. Peters*

McLean (1967) suggest that the sheath is open at its distal end. Carbon particles (from Chinese stick ink) in a sucrose solution were taken into the stomach; and stain did not diffuse through the wall of the sheath. In these experiments stylets were seen for several tens of microns beyond the distal end of the sheath, showing that the sheath is open at that end. Although all these experiments were done in artificial conditions, it is likely that food, and concurrently viruses, will be ingested through the open end of the sheath inside plants as well.

While feeding or an infected plant, aphids may acquire such a charge of virus that they become infective. Since aphids may interrupt feeding and need some time to penetrate to the vascular tissues, the period required by aphids to acquire virus, is called the acquisition access period. The minimum acquisition access period is approximately 5–15 min. In this period a small number of aphids are able to penetrate to a sieve tube and to ingest enough virus to make them infective. The efficiency of transmission is, however, definitely improved by increasing the acquisition access period to hours or even days as can be seen from fig. 24.1 for potato leafroll virus.

It seems that only part of the ingested virus is retained by the vector. Richardson and Sylvester (1965) showed that a part of the ingested virus was excreted by aphids feeding on plants infected with pea enation mosaic virus; the lumen of the alimentary tract was cleared of virus by feeding on healthy plants for a day or more. Identical results were obtained by Paliwal and

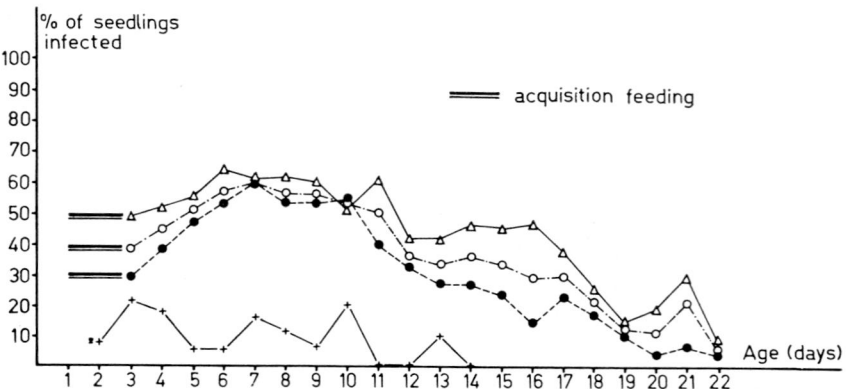

Fig. 24.1.   The daily transmission rate of potato leafroll virus by *Myzus persicae* after an acquisition access period of 4 hr (× —— ×) and of 2 days (△——△ ; ○——○ ; ●——●).
The aphids were tested daily until their death.

Sinha (1970) with barley yellow dwarf virus. These workers placed aphids for 6 hr on infected plants and maintained them then for 96 hr on healthy plants. The virus concentration in the gut following the acquisition access period remained about constant for the first 24 hr and then decreased to a low concentration by 96 hr. Thus the total virus acquisition time, sensu stricto, will exceed the acquisition access period, and after the acquisition access period virus will be absorbed from the lumen until it has been cleared of virus. Ponsen (1970) demonstrated this phenomenon in the transmission of potato leafroll virus; aphids *(Myzus persicae)* which had acquired the virus in a feeding period of 1 hr infected about 10% of *Physalis floridana* test seedlings on which they were fed for 7 days, whereas identical groups of aphids starved for 7 days at 4 °C and then tested for their infectivity, infected 70%. Ponsen concluded that during starvation virus, which would have been otherwise excreted, was absorbed from the lumen into the haemolymph, thus rendering the vector more infective. Thus the acquisition time is perhaps better defined not just from the time when virus is taken into the gut, but when it starts circulating through the body from the lumen.

Fig. 24.2.   Injection of aphids (Peters 1967b).

Viruses that are not transmitted by aphids maybe ingested but not transmitted by aphids. For example Hutchinson and Matthews (1963) demonstrated that substantial quantities of turnip yellow mosaic virus were taken up by the aphid *Hyadaphis brassicae*. Much of the ingested virus persists in the gut for days, and there is similar evidence of the occurrence of tobacco mosaic virus in the alimentary tract of aphids (Kikumoto and Matsui 1962). Thus the ingestion of viruses is non-specific and presumably the gut wall is the first barrier for the absorption of viruses into the vector and is the first site determining specificity.

Aphids not only acquire virus from plants but can also be made infectious by injecting virus samples into their haemocoel (fig. 24.2), and by feeding virus to them through a membrane. Rochow (1960) was the first who succesfully used the latter technique, and it was improved with the use of Parafilm membranes to study artificial diets for aphids (Mittler and Dadd 1962). Membrane feeding has been used in studies on beet western yellows virus (Duffus and Gold 1965), barley yellow dwarf virus (Rochow and Brakke 1964) and potato leafroll virus (Peters and Van Loon 1968). Duffus and Gold (1969), and Rochow (1970) used this technique in serum neutralisation tests to study the relationship between isolates of beet western yellows virus and of barley yellow dwarf virus, respectively.

24.4 The latent period

One of the characteristics of circulative and propagative viruses is the time taken for vectors to become infective after first acquiring the virus. This period is called the latent or incubation period. Its definition and especially its meaning has been the subject of many discussions. Severin (1931) defined it as 'the time for the infective principle to pass into mouthparts, alimentary canal, blood, salivary glands and out of mouthparts, in sufficient quantity to produce infection'. This definition emphasizes the concept of a minimum latent period. However, in certain experimental conditions no minimum incubation period can be observed, as for example with viruses with a latent period shorter than the combined periods of acquisition and inoculation. Such conditions are often met with the viruses of barley yellow dwarf, pea enation mosaic, and potato leafroll. To acquire these viruses aphids are mostly fed for approx 24 hr on the diseased plant and then tested for infectivity for the same period. Some aphids are able to infect plants in the first inoculation access period, though in subsequent inoculation feeding periods a larger proportion of the test plants becomes infected. This observation led Sylvester (1962) to define the latent period as a delay in the development of the maximum probability of transmission. However, this definition is not so useful since the exact length of such a latent period is hard to determine, and often some aphids in a group to be tested, do not become infective. Therefore, a new measure, the average latent period (LP 50) was suggested (Sylvester 1965.) The LP 50 being the time at which 50% of the vectors, which become eventually infective, can cause a first infection after the acquisition access. He used a cumulative measure based on the time of first transmissions to estimate this latent period.

The latent period was defined by Sylvester as 'the lapse between virus acquisition and the development of inoculativity', but was measured as 'the time occurring between the end of the acquisition access period and the end of the first positive inoculation access period'. The values of the LP 50s so measured are too small because the process to make the vector infective starts when the first virus is ingested. This may occur rather soon after the beginning of the acquisition access period since aphids reach the phloem in rather short periods (10–30 min) compared to the normal length of acquisition access periods. Thus a large part of the acquisition access period, whose length is arbitrarily chosen, is a part of the latent period.

The LP 50 of pea enation mosaic virus depends upon the temperature (Sylvester 1965). At 10, 20 and 30 °C the LP 50s were 70, 25 and 14 hr respectively. Estimates for the LP 50 have not yet been made for the other persistent viruses. But it is likely that the LP 50 for potato leafroll and barley yellow dwarf viruses will be longer than for pea enation mosaic virus; perhaps 2–4 days (Rochow 1963; Peters and Asjes, unpublished results).

Even longer latent periods have been observed in the transmission of sowthistle yellow vein virus and strawberry virus 3. In the former case this varied from 8–46 days according to the ambient temperature (Duffus 1963), and in the latter values ranged from 10–19 days.

24.5 *The fate of the virus in the vector*

No studies have been made on the sequence of events in the development of infection of viruses propagating in aphids, but there is no reason to assume that their infection cycle deviates much from that found for wound tumor virus in its leafhopper vector *Agallia constricta* (Sinha 1965).

Electron microscopy has revealed the presence of lettuce necrotic yellows virus particles, with and without their envelope in the vector *Hyperomyzus lactucae*. The virus was found in the cytoplasm of the cells, of the muscles, brain, fat body, mycetome, tracheae, epidermis, salivary glands and the alimentary canal (O'Loughlin and Chambers 1967). No particles were seen in the nucleus. As particles with and without membranes, which are considered to be different developmental stages of the virus, were found, it is likely that the virus multiplies in the vector. Moreover, virus was seen in some cells in bundles up to about 300 particles. Such cells were often surrounded by others in which no virus could be detected. O'Loughlin and Chambers believe that such a distribution is also indicative of virus multiplication in the vector.

A similar invasion has been shown for sowthistle yellow vein virus in the aphid *H. lactucae* (Sylvester and Richardson 1970). Virus was found in cells of the brain, suboesophageal ganglion, salivary glands, oesophagus, ventriculus, ovaries, fat body, mycetome and muscle. Uncoated virus particles were found in the nuclei of the infected tissues, whereas the particles with envelopes were found in the cytoplasm and perinuclear spaces. Particles have not been found in cells of the intestine or the hindgut. The results of this study suggest, therefore, that the virus tends to be concentrated in the cephalic and thoracic areas of the aphid.

Even less is known about the fate of circulative viruses. Bioassays have shown that these viruses occur in the haemolymph. Particles of pea enation mosaic virus have been seen in the cytoplasm of fat body cells (Shikata et al. 1966). Particles of this virus and also those of barley yellow dwarf virus (Paliwal and Sinha 1971), were also found in the lumen of the alimentary tracts of their vectors; these particles are those ingested by the vector.

An increase of the number of infectious particles after the acquisition is evidence for virus multiplication in the vector. Such evidence can be obtained by serially transferring virus-containing haemolymph to non-viruliferous aphids by injection, so that a dilution is attained in the transfers which exceeds the end-point dilution of the starting material. This method has been used for sowthistle yellow vein virus and potato leafroll virus. Sylvester and Richardson (1969) injected an estimated 10^5 particles of sowthistle yellow vein virus in the first recipient aphid, and in subsequent transfers made a 70-fold dilution, so that unless the virus multiplied, less than one virus particle was inoculated in the fourth passage. However, the virus transmitted equally efficiently at the first and at the sixth passage, when the experiment was stopped. Multiplication of this virus was also demonstrated by Peters and Black (1970) who infected primary cultures of cells of *H. lactucae*.

Infected cells could be detected by a fluorescent antibody technique 36 hr after inoculation. In experiments with potato leafroll virus Stegwee and Ponsen (1958) obtained evidence that the virus multiplied in the aphid *Myzus persicae*. They transferred haemolymph in 15 serial passages to obtain a total dilution of 10^{-21}, whereas the dilution end-point of haemolymph from the original donor aphids was 10^{-4}. However, in the same year but using different methods and a different virus isolate Harrison (1958) obtained no evidence of multiplication of potato leafroll virus in *M. persicae*. He found that extracts made from aphids immediately after they had fed for 24 or 48 hr on infected potato plants were infectious, whereas extracts made 1, 2 or 4 days later after the aphids had fed on plants immune for potato leafroll virus, were not infectious, even though aphids which have ingested potato leafroll virus for 24 or 48 hr, usually remain infective for their whole life. Presumably the amount of virus held in an infective aphid after its gut was cleared of virus by the passage of food from a virus-immune host, was too small to be detected by the assay technique used by Harrison. This suggests that a small part of the ingested virus retains the vector's infectivity.

Pea enation mosaic virus (Sylvester 1969) and barley yellow dwarf virus (Paliwal and Sinha 1970) could not be maintained in sequences of serial passages of haemolymph, suggesting that it is improbable that they multiply in the aphid.

Transovarial passage of virus from one generation to another has been well documented for a number of viruses transmitted by leafhoppers (Black 1959), but it seems to be rare among aphids. Miyamoto and Miyamoto (1971) reported the transmission of potato leafroll virus to parthenogenetic offspring of *M. persicae*. Similarly sowthistle yellow vein virus is transmitted to about 1% of the progeny of *Hyperomyzus lactucae* (Sylvester 1969). Transovarial passage does not prove multiplication in the vector (Black 1959), as a virus which passes through the vector tissues might also pass into the ovarial tissues.

24.6 Transmission

Virus is transmitted into the plant in saliva and regurgitated sap (ch. 25). In recent years there has been a considerable increase of knowledge of the transmission of viruses circulating in the vector, particularly the transmission of pea enation mosaic virus (Nault et al. 1964; Ehrhardt and Schmutterer 1965; Sylvester 1969); sowthistle yellow vein virus (Duffus 1963); and potato leafroll virus (Ponsen 1970). There are a number of transmission features common to or contrasting for all these viruses.

First, just as the probability of transmission is positively correlated with the duration of the acquisition access periods (fig. 24.1) so too the chance that an infective dose will be introduced is greater the longer the inoculation feeding period.

Sowthistle yellow vein virus is somewhat differently transmitted. The number of aphids becoming infectious is positively correlated with the acquisition access period, but if they become infective they are able to infect any *Sonchus oleraceus* test plant on which they feed for 24 hr. Thus the transmission efficiency for a single viruliferous vector is not dependent on the dose acquired. Duffus (1963) interpreted this as an evidence for multiplication. It may be questioned whether this is a general phenomenon for propagative viruses.

The transmission of potato leafroll virus is anomalous in that though multiplication in its vector has been demonstrated (Stegwee and Ponsen 1958), the vector's capacity to transmit depends on the dose acquired (fig. 24.1), as with the viruses which seem only to circulate in the aphid.

A second phenomenon which may distinguish circulative and propagative viruses can be observed in serial transmission experiments e.g. experiments in which viruliferous aphids are transferred at constant intervals to fresh test plants during their whole life. After acquisition aphids usually remain infective for their whole life. When nymphs acquire potato leafroll virus (MacCarthy 1954; Ponsen 1970; Peters and Asjes, unpublished results) or pea enation mosaic virus (Sylvester 1967), or other aphid circulative viruses, the transmission rate mostly shows an initial increase, then remains nearly constant for the larval stages and the first days of their imaginal stage, and finally shows a sharp decline (fig. 24.1). By contrast there is no final decline in the transmission of the sowthistle yellow vein virus, which is propagative.

The decline in efficiency could be caused by a loss of virus in the vector. Therefore, Sylvester and Richardson (1966) tried to re-establish the declining inoculative potential of aphids carrying pea enation mosaic virus by a second acquisition feed. However, the potential was only slightly restored, in that the decline was somewhat delayed. As the transmission efficiency of the recharged aphid decreased in the same way as that of the non-recharged aphids (fig. 24.3), it is unlikely that the decline was caused by a great loss of virus. Thus, other factors may be responsible too. Sylvester (1967) was able to correlate the decline of transmission with the rate of honey dew excretion. Working with *Aphis fabae* Banks and Macaulay (1964) observed that the

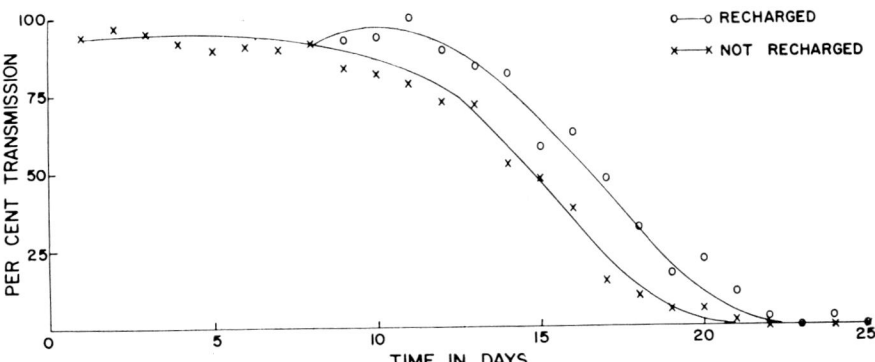

Fig. 24.3.   The daily transmission rate of pea enation mosaic virus by pea aphids at 25 °C. Aphids ($\times$———$\times$) given an acquisition access period of either 3 or 6 hr, and aphids ($\circ$———$\circ$) recharged for 24 hr in 8 days of transfer. All aphids were tested until their death. (Sylvester and Richardson 1966).

rate of feeding, which is about equal to the rate of excretion, decreased as the insect aged, particularly at the end of its reproductive life. If the production of saliva is likewise related to the intake of food it might explain the decline of infectivity with age.

   To infect plants with most of the viruses discussed the vector needs rather long inoculation feeding periods, and no transmission occurs after mere probing. This has been taken as evidence that only the phloem can be infected by these viruses. Pea enation mosaic virus, however, differs in that the virus can be transmitted by probing aphids and that tissues other than phloem can be infected (Nault and Gyrisco 1966). In veinal areas where the cells are tightly packed the rate of transmission is at least two times higher than in interveinal areas.

24.7   Specificity

The specificity of transmission may be obtained in different ways. As already discussed the wall of the intestinal tract is a site where specificity operates. The salivary glands are a second barrier for viruses as demonstrated by Rochow and Pang (1961). A strain of barley yellow dwarf virus, called MAV, is efficiently transmitted by the aphid *Macrosiphum avenae* and very inefficiently by the aphid *Rhopalosiphum padi* (Rochow 1969). However, the

virus can be detected in the haemolymph of *R. padi* feeding on plants infected with this isolate, by injecting samples of haemolymph from *R. padi*
into *Macrosiphum avenae*. These observations suggest that although the
virus reaches the haemolymph of *Rhopalosiphum padi*, it is not transmitted.

Other specificities of transmission are also known. For example not all
strains of a virus are equally efficiently transmitted by the same vector clone,
and different clones of one vector species can also differentially transmit a
single virus (Bath and Chapman 1967; Rochow 1969). The former authors
showed that a strain of *Acyrthosiphum pisum* from Arlington, Wisconsin,
did not transmit one pea enation mosaic virus isolate, but did transmit
another isolate, whereas a Californian strain of the aphid, isolated in El
Centro, California transmitted both virus isolates with a high frequency.

Thus far we have no understanding of the mechanisms controlling or
regulating specificity. Forbes (1964) suggested that virus particles pass
through cells by pinocytosis. This might seem a rather unspecific mechanism,
though as physical contact will be needed and when transporting the virus
there will be opportunities for affinities between the surface of the virus and
the cell to provide the specificity. The protein coat of the virus is likely to
have a function in this mechanism. Thus it is particularly interesting that
Rochow (1970) has recently demonstrated that the antigenic specificity and
vector specificity of barley yellow dwarf virus are correlated. As said above
R. padi rarely transmits the MAV isolate of barley yellow dwarf virus.
Though another isolate (RPV), which is serologically unrelated to MAV, is
transmitted efficiently by *R. padi*. However, this aphid transmits both strains
equally well from doubly infected plants, or from virus preparations purified
from these plants and fed to the aphids. The addition of MAV-antiserum to
the purified preparations did not diminish the transmission of MAV, though
when RPV antiserum was added *R. padi* failed to transmit either virus,
these results suggest that the infectious component of MAV has been
incorporated in the particles of RPV during mixed infections, and that the
protein coat of the particles controls specificity. This incorporation phenomenon is known as phenotypic mixing.

As early as 1946 Smith reported that there are two components in the
virus complex causing rosette disease in tobacco, one component (the mottle
virus) is dependent upon the presence in the plant of the other to be circu-
latively transmitted by aphids, yet is not by itself transmitted by aphids.
Several other complexes of 'helper' and 'helped' viruses have been found
since (e.g. carrot motley dwarf and groundnut rosette). But so far, this

phenomenon has not been explained, and phenotypic mixing is one possible explanation.

24.8 Conclusions

Circulative and propagative viruses will be ingested while the vector is feeding on an infected plant. Much, perhaps most, of the virus will be excreted with honey dew. The rest will be absorbed and initiate a process at the end of which the vector is infective. This takes place in the so-called latent period. In this period virus circulates and some viruses may propagate. As circulation is characteristic of both the circulative and propagative viruses, this period might be best called the circulation period.

There is some confusion as to which transmission features can be used to differentiate between circulative and propagative viruses. Several features described in this chapter, are often claimed to be evidence of multiplication of circulative virus in the vector. Sylvester (1969) considers the persistence of infectivity after a moult, the existence of a circulation period, the presence of virus in the haemolymph, the length of the circulation period being independent of the dose of virus acquired and the retention of the infectivity for a long period or for the vector's further life as evidence for such a hypothesis. However, both circulative and the propagative viruses share these features, and these features do not prove that the virus has multiplied in the vector. They merely result from the translocation of the virus through the vector.

An increase in virus concentration after acquisition is the only valid evidence of multiplication. Circumstantial evidence of multiplication is the occurrence of forms of the virus particles claimed to be developmental stages of the virus. At present it is accepted that there is convincing evidence that sowthistle yellow vein virus and lettuce necrotic yellows virus are propagative, however evidence of the multiplication of potato leafroll virus needs to be confirmed, particularly as the transmission kinetics of this virus in serial transmission experiments are similar to those of the circulative viruses, such as pea enation mosaic virus.

CHAPTER 25

Non-persistent aphid-borne viruses

R. G. GARRETT

Contents

25.1 Introduction

Non-persistent plant viruses are rapidly transmitted by aphids, and are rapidly lost by them. Indeed, the association of virus and vector is so tenuous that at first aphids were thought to merely contaminate their mouthparts with virus during probes on diseased plants, and to decontaminate them on healthy plants. Because recent reviewers have accepted that these viruses are transmitted by stylet-tip contamination of some kind (Smith 1965; Pirone 1969), the retention of the term 'non-persistent' (Watson and Roberts 1940) as opposed to 'stylet-borne' (Kennedy, Day and Eastop 1962) may seem a heresy. But the evidence for virus transmission by stylet contamination is not conclusive. Here, stylet contamination is discussed and an alternative transmission mechanism is proposed: that transmissible non-persistent viruses are carried in the foregut of their aphid vectors, rather than at the stylet-tips.

25.2 Transmission by stylet contamination

The idea that non-persistent viruses are transmitted on contaminated mouthparts makes sense for several reasons. The viruses are transmitted by aphids during very brief probes; they are lost when aphids shed exoskeleton, stylets, fore and hind-gut during ecdysis; they persist with aphids for only a few hours; and they persist less while aphids feed than while they do not feed.

Two properties in particular were inadequately explained with this transmission mechanism. Firstly, aphids transmitted virus less often when they had fed just *before* making access-probes on diseased plants than when they had been stored without food (Watson 1938; Watson and Roberts 1939). Virus transmission was thus affected by some change in the aphid or its behaviour. Secondly, an aphid species could be a better vector than another for one virus, and yet not for a second virus (Watson and Roberts 1939). To explain these effects, and the poor persistence of virus with aphids that fed *after* access-probes, Watson and Roberts (1939, 1940) suggested that aphids produced inhibitors while feeding and that these affected different viruses to different extents. However, such inhibitors have not been demonstrated and other explanations have been sought.

25.2.1 Secretion of saliva and stylet insertion

Saliva is secreted in two forms, one of which gels to form sheath material while the other remains fluid (Miles 1959). Before stylets are inserted, saliva is secreted onto the plant surface and the stylets are inserted through this into the plant; the salivary sheath is formed by the secretion of saliva in front of the advancing stylets, which penetrate it (Nault and Gyrisco 1966).

Stylet insertion as a mechanical process has been well studied, and details reported vary greatly. Some species secrete pectinase with their saliva and generally insert their stylets between cells, whereas others do not secrete pectinase and insert them through cells (Adams and McAllan 1956, 1958; McAllan and Adams 1961). However, the distinction between the two groups is not clear, for *Myzus persicae* (Sulz.) has been variously reported to insert its stylets mostly between cells (Smith 1926; Adams and McAllan 1958), or between and through them (Roberts 1940; Esau, Namba and Rasa 1961). Cell wall penetration is possible for all aphid species, as shown by the ultimate tapping of a cell for feeding. Also, Bradley and Cousin (1969) have recently shown that stylets are often inserted directly into cells when aphids probe briefly, and that virus may perhaps be acquired from the protoplast.

25.2.2 Virus acquisition

The ideas of how aphids acquire and carry transmissible virus must obviously be compatible with the nature of the sites in plants from which virus is acquired. But few attempts have been made to locate and characterise the sites without assuming that non-persistent virus is carried at the stylet-tips. During normal probes on intact plants, virus is usually acquired from the epidermis and inoculated into it. Recorded minimum access- or inoculation-probes can be as brief as 5 sec (see Sylvester 1954) and during this time aphids certainly make only shallow probes for they require about 1 min to insert their stylets beyond the epidermis (Roberts 1940; Bradley 1952). When access-probes last more than a few min aphids transmit virus less often, but this does not imply that less virus is available for acquisition from deeper tissues. For example, cucumber mosaic virus (CMV) can be acquired efficiently from the mesophyll after the epidermis has been removed (Namba 1962). It therefore seems that inefficient transmission after prolonged access-probes depends on the duration or the depth of these probes, or perhaps on some changes in aphid behaviour dependent on these.

Bradley (1952, 1961) and Sylvester (1954, 1962) proposed that more virus was removed from the outsides of the stylets after prolonged probes than after brief probes because the salivary sheath was hardened and longer. This was a particularly neat explanation of non-persistence, the inefficiency of prolonged access-probes and the poor persistence of the viruses while aphids fed. Sylvester (1962) further suggested that virus might be transmitted with a small salivary plug that sometimes adheres to aphid stylets after brief access-probes.

It was similarly thought that when aphids probed through 'Parafilm' membranes into plants, they transmitted less often because more virus was removed from the stylets (Bradley 1956). However, the wax component of the 'Parafilm', or an oil in it, may have affected transmission in this experiment. Aphids that probed wax after access- and before inoculation-probes transmitted virus less often than aphids that made intermediate probes on plants (Bradley, Wade and Wood 1962). Probing on plants brushed with mineral oil had effects similar to probes on wax. Because the oils were acquired by aphids when the rostrum touched the sprayed surface for only one sec, and could persist with aphids for several probes (Bradley 1963), it seems likely that the oils acted in low concentration. Mineral oils inhibit infection or affect plant susceptibility rather than act on virus carried at the stylet-tips (Loebenstein, Alper and Deutsch 1964). Support for this interpretation is provided by Külps (1968): aphids transmitted more often when they probed untreated leaves than when they probed leaves with the opposite surface treated with oil. Because of these effects on the plants it is impossible to demonstrate that 'Parafilm' or oils affect virus located at the stylet-tips.

Although some aspects of non-persistent virus transmission could be explained in terms of the development of the salivary sheath, the infrequent transmission by aphids that had fed just before making access-probes could not be explained in this way. Bradley (1952) showed that aphids which had not fed for about 15 min usually made only brief probes, of the sort normally associated with successful virus transmission. By contrast, aphids that had recently fed probed reluctantly, or tended to make prolonged probes of the kind usually ineffective for virus transmission. Also, when aphids withdrew their stylets after prolonged probes, they could not immediately re-ensheath them in the labial groove (Bradley 1961). The exposed stylets lack rigidity, and so a period without food is necessary for stylet re-ensheathment, and therefore for satisfactory probing and optimum virus transmission. Sylvester (1962) did not fully endorse this explanation, for when in his experiments aphids with exposed stylets were discarded and the remainder allowed only brief probes, an effect of pre-access feeding was still detected.

25.2.3 Attempts to locate virus on aphid stylets

The hypothesis that transmissible non-persistent virus is carried on the out-side of the stylet-tips is broadly compatible with the observations on sali-vation and probing. But these observations do not actually demonstrate that transmissible virus is carried there, and none actually denies the possibility that transmissible virus is carried in the aphid fore-gut. They merely provide explanations of particular features of transmission *if* non-persistent virus is carried at the stylet-tips. Attempts have been made to show that virus is carried at the stylet-tips, indirectly by applying treatments to the stylets, or by direct observation of the stylets or of virus released from them.

Firstly, most *M. persicae* failed to transmit virus if only their stylets were treated with 0.05 % formaldehyde after they had made access-probes on dis-eased plants. Also, treatment of the stylet-tips with water slightly reduced transmission frequencies (Bradley and Ganong 1955b). This was interpreted as evidence for virus inactivation on the stylet-tips, or the washing of virus from them. It was realised that the treatments could have affected virus carried within the stylets, perhaps at some distance from the tips. In a second method, ultra-violet irradiation was used as the stylet treatment (Bradley and Ganong 1955a). When only the terminal 15 µm of the stylets were irradiated after access-probes, most aphids did not transmit virus. This would be the most convincing evidence that virus was carried at the stylet-tips, were it not that irradiation of stylet-tips *before* access-probes prevented the aphids from ac-quiring and transmitting virus for at least 15 min after treatment (Bradley 1964 p. 169). In the presence of such effects, which must have been on the aphid, direct effects on transmissible virus cannot be demonstrated.

Nor has it been possible to detect virus on the stylet-tips, although it has been shown that the mandibular stylets are deeply ridged at the tips (Van der Want 1954) and could conceivably carry virus there. Although it has been shown that after probing infected plants, aphids can release virus into so-lutions they probe, or onto surfaces with secreted saliva (Hashiba and Misa-wa 1969), it was only assumed that virus was released from the stylet surfaces. It was in fact not known whether such virus had been carried on the stylet surfaces, in the food canal, or even in the fore-gut of the aphids.

25.3 Virus transmission by sap imbibition and ejection

Assuming only that non-persistent virus transmission depends on probing, aphids could carry transmissible virus on the outsides of the stylets or in any part of the fore-gut. Transmission from the stomach is less likely for two reasons: firstly, regurgitation has not been demonstrated and the oesophageal valve would seem to prevent it; and secondly, the stomach is retained during moulting whereas virus is lost.

The experimental evidence for stylet-contamination as the transmission mechanism does not in fact eliminate any of these ways for aphids to carry transmissible virus. In all instances, experimental methods have not allowed effects on transmissible virus to be separated from effects on either aphid behaviour or plant susceptibility, or they have not differentiated between virus carried outside the stylets and virus carried inside them, perhaps at some distance from the tips.

However, there are three important conclusions. Firstly, aphid stylets are extremely sensitive, and treatments applied to them can profoundly affect aphid behaviour (Bradley 1962; Mittler and Dadd 1965). Secondly, feeding or probing changes aphid behaviour. Aphids rarely insert their stylets to the phloem at the first attempt which often lasts less than a minute (Sylvester 1954) and it is these probes, of an apparently exploratory nature, that are effective for non-persistent virus transmission. Thirdly, the many experiments with stylet-tips have shown that they are the important route by which transmission takes place. But this could be because virus is carried there, or because receptors affecting aphid behaviour are there, or because there the alimentary canal starts.

25.3.1 The biological significance of probing

Although aphids transmit non-persistent viruses while probing rather than while feeding, there is scant information on the nature of probing, especially in relation to virus transmission. The explanation of transmission by stylet contamination has relied largely on the physical aspects of probing, and possible biological features have not been well considered.

Probing certainly plays a part in host recognition by aphids. When *Brevicoryne brassicae* probed cabbage (a host plant) or bean (not a host plant) it behaved differently (Wensler 1962) but on bean plants sprayed with a mustard-oil the aphids behaved as on cabbage. The reactions presumably resulted from sap-uptake and tasting, but it is likely that the four nerve dendrites of the two mandibular stylets would be able to detect only grossly unfavourable conditions at the stylet-tips. It is more probable that sap imbibed into the fore-gut would allow a reaction by the innervated hypodermal cells above the pharyngeal pump. Such uptake could play a significant part in host recognition by aphids.

Aphids have often been reported to imbibe sap while probing or feeding on solutions. Miles (1959) observed liquids secreted from the salivary canals move into and up the food canals of aphids as they recovered from anaesthesia, and aphids can be reared on artificial diets without the need for pressure to aid liquid uptake (Mittler and Dadd 1962; Auclair and Cartier 1963). Therefore, aphids imbibe liquids, presumably by the action of the pharyngeal pump. This does not mean that normal phloem feeding requires the action of the pharyngeal pump although it may serve to regulate uptake of food (Kennedy and Mittler 1953). But attempts to measure sap-uptake when aphids probed plants were at first unsuccessful. Ehrhardt (1961) and Day and Irzykiewicz (1953) could not detect sap-uptake by aphids that probed for short periods and it seems certain that sap-uptake by aphids when probing plays no part in nutrition (Hennig 1963). Watson and Nixon (1953) detected radioactive tracer in *Myzus persicae* that had probed for about 5 min but the aphids were thought to be superficially contaminated.

Using quite different techniques, McLean and Kinsey (1964, 1967, 1968) detected sap-uptake when aphids fed. Electrical conductivity between aphid and leaf was monitored during stylet insertion, and patterns of change in conductivity were correlated with salivation and feeding. A particular pattern which was thought to represent extension of the stylet-tips beyond the salivary sheath, correlated with virus transmission (McLean and Kinsey 1968). The possibility that sap-uptake occurred while the stylet-tips were exposed was not discussed although in earlier work (McLean and Kinsey 1967) they suggested that a particular pattern represented sap-imbibition and stylet-blockage followed by sap-ejection.

Recently, it was shown that when aphids probed for less than 10 min on plants heavily labelled with P^{32} they imbibed tracer (Garrett 1971). But they carried considerably less P^{32} after 6–8-min probes than after probes that lasted 3–5 min (fig. 25.1). Thus large proportions of the tracer must have been returned to the labelled plants. Furthermore, in 3-min inoculation-probes aphids could transfer to seedlings large proportions of the tracer they had acquired in earlier 5-min access-probes (fig. 25.2). The aphids were at first considered able to transmit tracer in any of several ways: by stylet contamination; by ingestion followed by perhaps regurgitation, excretion or assimilation and then secretion with saliva; or by sap-imbibition into the stylets or fore-gut and later sap-ejection.

However, the volumes of sap represented by the tracer were too large to be accommodated in the stylets, and in dismembered aphids only a little tracer was associated with the mouthparts, the large majority being carried with the

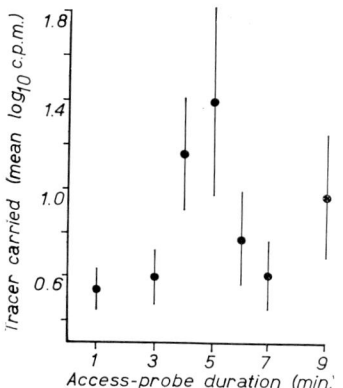
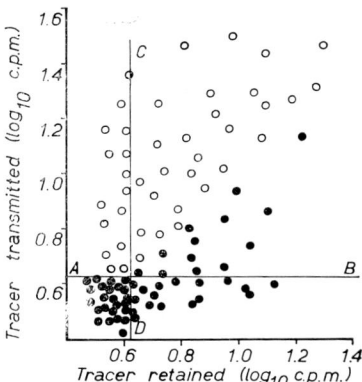

Fig. 25.1.   Amounts of ^{32}P carried by *Myzus persicae* that had probed labelled cucumber (mean and 95 % confidence limits for 5 aphids).

Fig. 25.2.   Comparison of ^{32}P transmitted and retained when 96 *Myzus persicae* probed labelled then unlabelled plants. Lines A–B and C–D represent 95 % confidence limits for background count-rate. Aphids that transmitted more than 50 % are marked thus (○).

bodies. Furthermore, because aphids could transfer large proportions of the tracer they carried (fig. 25.2) and sometimes all of it, most of the transferred tracer had certainly not been assimilated and secreted with saliva. In these experiments, the aphids did not excrete and regurgitation seems unlikely for reasons already mentioned. Therefore it was concluded that transferable tracer had been imbibed and carried in the fore-gut. Presumably, the brief probes that precede feeding involve imbibition which allows gustatory discrimination between plants (Wensler 1962) or between solutions (Mittler and Dadd 1962).

25.3.2 Transmission of tracer and virus compared

Using autoradiography for tracer assay, tracer and a non-persistent virus from radish were transmitted from separate plants (Nishizawa et al. 1959). Tracer was found to persist for longer than the virus in aphids as they probed. Although it was suggested that the virus was inactivated and so was less persistent in the aphids than tracer, an alternative explanation is that tracer assay was more sensitive than virus assay. Thus in successive probes, the end-point for virus detection may have been reached before that for tracer detection.

In 4 experiments Garrett (1971) compared the transmission of P^{32} and cucumber mosaic virus (CMV) by 27 groups of *M. persicae* and 27 groups of *Hyperomyzus lactucae* (L.). The same plant could not be used as a source of tracer and virus because the tracer inactivated the virus in vivo. Instead, comparisons were made between tracer and virus transfer from separate labelled and diseased plants to separate test-seedlings. Aphid behaviour was changed by storing aphids without food for up to 3 hr, and by allowing them access-probes lasting up to 9 min on labelled or virus infected plants. The access-probes were followed by single 3-min inoculation-probes on cucumber seedlings (for aphids from diseased plants) or on discs cut from cucumber cotyledons (for aphids from labelled plants). Estimates of the smallest volumes of sap transmitted by aphids when they transmitted virus, were obtained. In each experiment, this was done by choosing a sap-volume so that the total number of aphids that transmitted at least the selected volume from the labelled plants equalled the total number that transmitted virus from the diseased plants. This sap-volume ranged from 480 μm^3 to 841 μm^3 per aphid in the 4 experiments. The aphids that transmitted these volumes or more were distributed between the treatments in a similar way to those that transmitted virus (fig. 25.3 and 25.4). It was therefore suggested that sap and virus transmission were affected in the same way. Also, because the selected sap-volume was always at least 10 times the volume of the food-canal of the stylets, it was considered that the aphids transmitted virus by imbibing sap containing virus into the fore-gut during the access-probes and ejecting both during inoculation-probes.

25.3.3 *Interdependence between successive probes*

Several examples of serial transmission experiments have been described in which single aphids made one access-probe, then inoculation-probes on several successive plants. They are generally used to demonstrate non-persistence (Watson and Roberts 1939, 1940). However, they have also been used to test for interdependence between successive probes. Bradley (1952) used the over-all frequency of transmission to derive the numbers of aphids expected to transmit virus to different numbers of plants. He suggested that comparison of these with the observed numbers of aphids provides a test of interdependence between probes, and of the uniformity of virus charge carried by the aphids.

Different conclusions can be drawn from these experiments according to the kind of test used, and the attitude taken towards aphids that did not

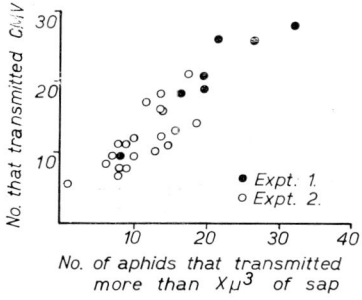 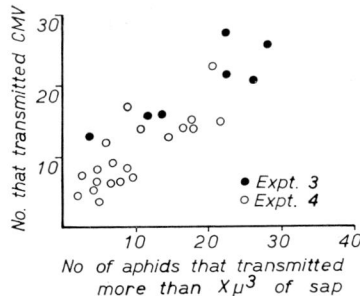

Fig. 25.3. Comparison of numbers of *Myzus persicae* that transmitted CMV with numbers of similarly treated aphids that transmitted certain volumes of sap. X is the estimated minimum sap-volume transmitted when aphids transmitted virus. X was 570 and 640 μ^3, and groups contained 40 and 30 aphids, in experiments 1 and 2 respectively. Fig. 25.4. Experiments 3 and 4 were similar to experiments 1 and 2, but *Hyperomyzus lactucae* was used instead of *Myzus persicae*. X was 480 and 841 μ^3 in experiments 3 and 4 respectively.

transmit to any plant. For example, McLean (1959) found that 12 of 50 aphids used did not transmit to any plant. When Bradley's (1952) test was applied, transmissions were shown to be independent events only if the 12 aphids were *excluded* on the argument that they had not acquired virus. However, Bradley's (1952) test was based on the distribution of aphids in only a few classes, and used only part of the information available in the experiment. An alternative test, based on the distribution of the infected seedlings, gives a different result. When each seedling is classified according to whether it became infected, and whether the preceeding plant became infected, then transmissions are shown to be independent events only if the 12 aphids that did not transmit virus are *retained*. The discrepancy probably arises because in this kind of experiment it is not known why some aphids failed to transmit virus, and so whether they should be included in the tests as aphids similar to those that transmitted virus. It could be that they had not acquired virus, that its location changed in aphids that failed to transmit, or that aphid behaviour changed. The various analyses do not show which interpretation is correct.

In another example, Sylvester (1955) found that although *M. persicae* transmitted lettuce mosaic virus to only a few plants, transmissions were apparently independent. In spite of the low frequency of transmissions, some aphids transmitted late in the series provided they had not transmitted earlier. This could be because they carried only small charges of virus. However, the hypothesis of virus transmission by stylet-tip contamination is difficult to reconcile with the ability of aphids to conserve small charges of virus during several successive inoculation-probes.

The questions raised by these experiments are answerable, especially if non-persistent virus is carried in the fore-gut of aphids. During inoculation-probes, sap-uptake would be expected as during access-probes. In such instances, the sap imbibed from healthy plants could impede the transmission of virus acquired during earlier probes and yet help to conserve it within the fore-gut for eventual transmission by the few aphids that later ejected large amounts of sap. This idea was tested by labelling with different viruses, the sap imbibed during successive probes (Garrett 1971).

Single *M. persicae* were successively allowed access-probes on plants infected with potato virus Y (PVY) then plants infected with CMV, followed by inoculation probes on *Physalis floridana* Rydb. seedlings (to detect PVY), then *Nicotiana tabacum* L. seedlings (to detect CMV and PVY). The tobacco seedlings were tested for the infecting virus by sap-inoculation to differential hosts, only if they developed symptoms. Aphids known to carry PVY (because they transmitted it to at least one plant) transmitted it to the *Physalis floridana* seedlings less often when they also later transmitted CMV to tobacco (4/27) than when they did not also transmit CMV (18/37). CMV did not inhibit infection by PVY, for aphids could transmit PVY as often after making second probes on CMV infected plants, as after second probes on healthy plants. It must therefore have been sap from the second access-probe that interfered with the transmission of PVY, presumably by displacing PVY within the aphid so that transmission was often delayed. Because such aphids eventually transmitted PVY the displacement was reversible. This seems more likely to happen within the stylets and fore-gut than at only the stylet-tips.

25.4 Some anomalies

Whether non-persistent viruses are considered to be transmitted by stylet contamination, or by sap-imbibition into and ejection from the fore-gut, there remain anomalies which have been unexplained for many years. They include semi-persistence, variations in efficiency of transmissions, specific requirements of some viruses for particular vector species or strains, and the inability of aphids to transmit some viruses which are easily transmitted by sap-inoculation.

25.4.1 Semi-persistent viruses

Semi-persistent viruses (Sylvester 1956) resemble non-persistent viruses in transmission properties because they do not have a latent period in the vector, and do not survive in aphids when they moult. On the other hand, they are transmitted more often when aphids feed on plants than when they probe them, and the viruses may persist in aphids for more than a day. Beet yellows and *Citrus* tristeza viruses are examples. These viruses are readily detected in phloem tissues (Price 1966; Esau 1967; Esau, Cronshaw and Hoefert 1967) and perhaps infect and multiply in them more readily than in other tissues. This is probably why aphids transmit them more often after long feeds than after brief probes. Similarly, the ability of *Brevicoryne brassicae* (L.) and *Myzus persicae* to transmit cauliflower mosaic virus after either brief or prolonged probes was explained in terms of a dual distribution of virus in the epidermis and in vascular tissue (Day and Venables 1961).

Stylet contamination is unlikely to be important unless these viruses survive longer on the stylets of feeding aphids than do non-persistent viruses. This is improbable as non-persistent viruses are usually more stable in plant extracts and are generally present in higher concentrations than semi-persistent viruses. However, if aphids transmit virus when imbibed sap is ejected, then persistence during feeding is the main anomaly. Although persistence for some hours in a few aphids could be because they did not feed, persistence for more than a day (Day and Venables 1961) is not likely to be explained in this way. Regurgitation is unlikely to be involved, especially after aphids had started to feed. Perhaps these viruses can survive in the fore-gut, possibly in the ducts to the hypodermal cells or perhaps absorbed to the fore-gut in the fashion that nematodes absorb viruses and retain them for long periods (Taylor and Robertson 1971). Alternatively, perhaps the cells of the epithelium to the fore-gut actually become infected (Watson and Plumb, 1972). However, as yet such explanations are at best speculative.

25.4.2 Efficiency and specificity

Some aphid species are better vectors than others, and some viruses are more often transmitted than others by aphids. Furthermore, different aphid species may transmit different viruses with different relative efficiencies. For these kinds of specificity to be attributed to an interaction between aphid and virus the various efficiencies must be measured at the same time, and ideally the viruses should be transmitted from the same plant. In this way it is possible to eliminate from the experiments many factors that are known to affect transmission frequencies to various extents, such as differences between aphid isolates (Stubbs 1955), virus isolates (Watson 1938), aphid growth forms (Hamlyn 1953), cultural conditions (Swenson 1962) and temperatures (Sylvester 1964). For example, Doncaster and Kassanis (1946) showed that *Myzus ascalonicus* Doncaster could transmit only cucumber virus 1 (CMV) from plants infected also with PVY, and only *Hyoscyamus* virus 3 from plants also infected with tobacco severe etch virus (TSEV). In tests with other aphid species, only *Myzus ascalonicus* consistently failed to transmit PVY and TSEV.

There is another form of specificity. Many viruses with very different properties attain high concentrations in sap, retain infectivity in extracts for a long time, but are not transmitted by aphids. Amongst these are tobacco mosaic virus (TMV), PVX, carnation mottle virus, and turnip yellow mosaic virus. Although Allard (1917) and Hoggan (1934) obtained transmission by aphids of viruses believed to be TMV, attempts to confirm these results have generally failed (Orlob 1963; Smith 1965; Pirone 1969). Although aphids may transmit TMV when they claw plants (Bradley and Harris, 1972) this does not explain their failure to transmit TMV when they probe plants.

The failure of aphids to transmit TMV has been attributed to virus inactivation or inhibition by saliva (Nishi 1958, 1969). Pirone (1970) was unable to demonstrate TMV inhibition, or inactivation, using saliva collected in solutions probed by aphids. Furthermore, aphids can cause infection when they insert their stylets into leaves sprayed with purified TMV (Teakle and Sylvester 1962), and can transfer virus from plants to solutions (Matsui et al. 1963) and infective virus from one solution to another (Pirone 1967). And yet the similar transmission of infective TMV from plant to plant has not been unequivocally shown.

An unusual form of specificity is shown by some viruses of the PVY group. Aided-viruses, such as potato virus C or potato aucuba mosaic virus, are transmitted by aphids only in the presence of another virus, a helper-virus, such as potato virus Y. The two viruses need not be acquired from the same plant, but the helper-virus must be acquired first (Kassanis and Govier 1971). Aphids could not transmit aided- or helper-viruses from extracts unless they had recently probed *intact* plants infected by the helper-virus, either untreated or which had been irradiated with UV light and so were a poor source of infective helper virus (Kassanis and Govier 1971). The role of the helper-virus, therefore, does not seem to involve either infective or non-infective virus particles. Presumably infection by a helper-virus changes the *intact* plant in some way so that aphids behave differently after probing them, or conditions within the aphid are changed so that the aided-virus can survive and be transmitted. These results show that interactions between virus and plant may prove as important in determining transmission as interactions between aphid secretions and viruses.

There is no adequately tested explanation of specificity. Of possible interactions between non-persistent virus and vectors, the action of saliva on virus is the most debated. Saliva for direct tests with virus is collected by allowing aphids to probe solutions or other substrates. But saliva contains a reacting mixture of phenolic compounds and polyphenoloxidase (Miles 1964) and reactions would be completed before the saliva is tested. Virus could be affected by being involved in the reaction, just as CMV is inactivated by the polyphenol-oxidase activity of tissue extracts (Pierpoint and Harrison 1963). In this case, only direct secretion of saliva into virus preparations maintaining high saliva to virus ratios, would show the effects. Alternatively, if virus is transmitted with imbibed sap the problem is less clearly defined. Conditions in the foregut are not known; secretions may be associated with functions of the hypodermal cells and may affect virus survival.

25.5 *Ecology of non-persistent and semi-persistent viruses*

There is no doubt that non-persistent and semi-persistent viruses can be brought into crops by alate aphids. The spread of these viruses within the crops can then lead to high incidences of infection. The most important aspects of the ecology of these viruses are therefore the behaviour of aphids during migration and the ways that the viruses can survive in the field from one season to the next.

25.5.1 *Aphid migration*

Mass migration of aphids may occur when most of the aphids produce winged offspring under the influence of meteorological conditions (Hille Ris Lambers 1966; Lees 1966) or changes in the physiology of the host (Rivnay 1937; Mittler 1958), or when large aphid populations build up in a crop (Dickson and Laird 1962; Lees 1966). But some winged aphids are continually produced and they may fly on any suitable day (Taylor 1965). When leaving their host plants, most aphid species are attracted to light of short wavelengths and are repelled by light of wavelengths longer than about 5000

Å (Taylor 1958). During migration their behaviour is reversed and aphids become attracted to the longer wavelengths (Moericke 1955) although the quality of light for maximum attraction may not be the same for all aphid species.

Aphids perhaps visually select young green leaves while flying over plants (Kennedy, Booth and Kershaw 1961). But they land as frequently on hosts as on non-host plants (Muller 1958a; Kennedy, Booth and Kershaw 1959) and host selection follows the initial probes they make. After probing briefly, most aphids move only a short distance (Muller 1958; Berry and Simpson 1967) where probing is repeated. This pattern of behaviour is the essence of the efficient transmission of non-persistent viruses by aphids. Because they probe rather than feed on the first plants they visit, migrating aphids can retain non-persistent viruses during several probes.

25.5.2 Virus spread by migrating aphids

Because non-persistent viruses can survive for some hr in aphids that are not feeding, long distance spread can occur. In this way, foci of infection are established for later spread within crops some kilometers from the original source of the virus. This kind of spread is especially important when virus is introduced into perennial crops, in which spread may occur in subsequent years. But in annuals, it is the movement of aphids within the crop and from nearby crops that is important for the spread of non-persistent viruses. In such crops virus is mostly spread over distances of a few metres because it is spread optimally by aphids making short flights (Cockbain et al. 1963).

Host selection occurs only after probing, and aphids only slightly favour their host plants (Muller 1958). Because of this behaviour, aphids can, while migrating, be vectors of viruses in crops they rarely colonise. Cucumber mosaic virus in gladiolus (Swenson and Nielson 1959), onion yellow dwarf virus in onion (Drake, Tate and Harris 1933) and tulip breaking virus in tulip are examples of viruses spread in this way. Spread is rapid in these crops and this is testimony of the importance of migrating alate aphids as vectors. In fact, non-persistent virus spread is generally attributable to the migration of alates, even in crops the aphids colonise. Although some spread presumably occurs when aphids walk from plant to plant, especially when the leaves of adjacent plants touch, this is generally late in the history of the crop and is not of great importance. Furthermore, in crops frequently sprayed with insecticides, in which apterous aphids are particularly well controlled, the spread of viruses may even be enhanced (Munster and Murbach 1952).

25.5.3 Perennation

Non-persistent viruses, because they do not survive long with their vectors, survive between seasons mainly with local infected plants. The source of infection for an annual crop can be either plants remaining after a previous season (seed crops, nearby crops of similar kinds, root storage clamps, volunteer plants or weed hosts) or diseased materials introduced at planting (seed, budwood, tubers, etc.). This leads directly to the classical method of disease control: to interrupt the natural cycle of the disease in the field.

Some dramatic control measures have been developed using planting materials certified as free of disease. Among examples are the use of lettuce seed with less than 0.1% infection to control lettuce mosaic virus (Tomlinson 1962), the seed potato scheme in the U.K. and the control of potato virus Y (Todd 1961) and the isolation of beet seed-crops from the main cropping district (Hull 1965).

However, disease control in this way is difficult to apply in some instances, especially when perennial crops are involved. *Citrus* tristeza is semi-persistent in its vector and may spread over large distances. Because these crops cannot be replaced frequently, small rates of spread of the virus into the crops and subsequent spread within them, lead to high disease incidences. A similar problem arises when the multiplication rate of a crop is slow, as with tulips and narcissi. Stocks of bulbs must be quite free of disease if they are to remain relatively disease-free for the many years they are retained.

Non-persistent viruses have been controlled by severe roguing combined with insecticidal sprays (Bagnall 1953; Adams 1962). But studies on the ecology of non-persistent and semi-persistent viruses, and especially studies on aphid migration, have led to some interesting approaches to virus control.

Firstly, by using mulches that reflect daylight, migrating aphids have been repelled from crops and the incidence of infection with non-persistent viruses greatly reduced (Smith and Webb 1969). Secondly, the growing of carnations under glass has allowed the replacement of diseased plants by healthy ones, under conditions that protect the carnations from infection by migrating aphids. This approach will prove useful for maintaining nuclear stocks of field crops when geographical isolation from diseased crops is not possible. Thirdly, some non-persistent viruses have been controlled in crops sprayed with mineral oil suspensions (Bradley, Moore and Pond 1966). This is of particular interest. The oils apparently affect the susceptibility of the plants to infection at the site the aphid probes. It is the only type of field spray that can be used to reduce the rate of re-infection of perennial crops. Variations on this theme would seem worth investigating, in particular because the oils have been found to impede the transmission of at least one semi-persistent virus (Dutrecq and Vanderveken 1969).

25.6 *Conclusions*

The case for stylet-contamination has been presented by Bradley (1961, 1964), Sylvester (1954, 1962) and Pirone (1969) whereas Bawden (1964) discussed the question of the transmission mechanism but did not draw conclusions. Watson and Plumb (1972) favour an internal site for transmissible non-persistent virus in aphids. Their arguments are based largely on consideration of specificity. Certainly there are too many ill-explained features remaining for the mechanism to be considered understood. But specificity, and the non-transmission of TMV and other viruses, are not likely to be explained only in physical terms. Aphids presumably transmit viruses by transmitting sap containing them. It may therefore be possible to explain many aspects of virus transmission in terms of sap transmission, without explaining how some viruses fail to survive or be transmitted with it.

If such a mechanism is to be considered, I think that one based on sap inbibition into and ejection from the fore-gut is more useful than one based on stylet-tip contamination. In particular, the evidence that uptake of sap during inoculation-probes affects virus transmission is an important development. This allows explanation of several features ill-explained in terms of only stylet-tip contamination. These include the conservation of small virus charges and their subsequent transmission, interdependence of transmissions, faster loss of virus by feeding than by non-feeding aphids, non-persistence (by displacement or dilution as well as by loss during transmission), semi-persistent virus transmission, and the generally low frequencies of transmission by single aphids. Sap-sampling in this way is regarded here as a facet of the exploratory nature of probing and so is affected by environment, plant or aphid species used, previous aphid behaviour, and treatments applied to sensitive parts. Also, sap-sampling is a behavioural pattern which is often inhibited by feeding and by prolonged probes. So virus transmission is affected by these factors.

Nevertheless, whatever transmission mechanism is proposed, it will be considered inadequate until we understand better such problems as specificity and semi-persistence.

CHAPTER 26

Viruses and leafhoppers*

R. C. SINHA

Contents

---

* Contribution No. 712 from C.B.R.I., Canada Agriculture, Ottawa.

26.1 Introduction

The relationships of 'plant viruses' to their leafhopper (Cicadallidae) or planthopper (Delphacidae) vectors, in general, are more intimate and complex than to other vectors, such as aphids, thrips, mealybugs, beetles, mites, nematodes or fungi. These viruses can be readily transmitted to plants only by means of their vectors, with the exception of potato yellow dwarf virus which can also be transmitted by sap inoculation. The hopper-transmitted viruses, after being sucked from the infected plants into the vector's gut, are released into the hemolymph, carried to the salivary glands and then are injected back into the healthy plants while the insect feeds. Indications for such 'circulative' movement of viruses in their hopper vectors came from the early work on maize streak and sugar beet curly top viruses. Storey (1933) was the first to demonstrate that maize streak virus could be transmitted to its leafhopper vector (*Cicadulina mbila* Naude) by introducing the inoculum into a hole made in the abdomen by a finely pointed needle. Using this method, and preparing inocula from viruliferous hoppers, he showed that the virus was present in the thorax, the abdomen and in the hemolymph. Bennett and Wallace (1938) showed that sugar beet curly top virus occurred in the hemo-lymph, gut and salivary glands of viruliferous leafhoppers, *Circulifer tenellus* (Baker). This was demonstrated by allowing the virus-free leafhoppers to feed through a membrane, on various inocula prepared from different tissues of hoppers carrying the virus, and then testing their ability to transmit the virus to healthy plants. Certain features of transmission of these two viruses suggest that they may not multiply in their hopper vectors.

Rice tungro is the only leafhopper borne virus, so far discovered, that does not appear to circulate in its leafhopper vector, *Nephotettix impicticeps* (Ishihara). Ling (1966) reported that the infective nymphs failed to transmit the virus after moulting into adults, and suggested that tungro virus may be carried only on the stylets of its vector. On the other hand several other viruses have been shown not only to circulate but also to multiply in their hopper vectors and to invade them systemically, and thus they could be considered both plant and insect viruses. Many of these 'propagative' viruses show a close resemblance in structure with certain viruses of mammals, and they bridge the gap between plant and animal viruses. In this review I shall discuss the behaviour of such viruses in their leaf- or planthopper vectors, their similarities with some viruses infecting mammals, and virus growth in vectors as well as in cultured cells.

For several years the causal agents of 'yellows'-type diseases of plants,

TABLE 26.1
Morphology of some viruses transmitted by hoppers

Shape	Common name	Size (nm)	Virus particles observed in	References
Large polyhedral	wound tumor	60	purified preparations, sections of plant and vector tissues	Bils and Hall 1962; Shikata and Maramorosch 1965
	rice dwarf	70	purified preparations, sections of plant and vector tissues	Fukushi et al. 1962; Shikata 1966
	maize rough dwarf	60–70	purified preparations, sections of plant and vector tissues	Gerola et al. 1966; Lovisolo et al. 1967
Small polyhedral	oat blue dwarf	28–30	purified preparations and plant sections	Banttari and Zeyen 1969; Zeyen and Banttari 1969;
	rice tungro	30–33	purified preparations	Galvez 1967
Bacilliform	potato yellow dwarf	380 × 75	sections of plants and virus pellets	MacLeod et al. 1966
	maize mosaic	242 ± 40 × 48 ± 10	purified preparations, sections of plant and vector tissues	Herold et al. 1960; Herold and Munz 1965
	wheat striate mosaic	250–260 × 75–80	purified preparations and plant sections	Lee 1964; Ahmed et al. 1970; Sinha 1971
	rice transitory yellowing	325 × 93	plant sections	Shikata and Chen 1969
	northern cereal mosaic	500–600 × 40 300–350 × 60	partially purified preparations, sections of plant and vector tissues	Shikata and Lu 1967
Long flexuous threads	hoja blanca	variable in length × 10	sections of plant and vector tissues	Shikata and Galvez 1969

transmitted by leafhoppers, were believed to be viruses, although they were never morphologically identified. Doi et al. (1967) were the first to provide electron micrographs of infected plant tissues which suggested that such diseases are caused by microorganisms resembling *Mycoplasma*, and not by viruses. This was the first time that mycoplasmas were clearly implicated as disease causing agents in plants. Since then, the association of such organisms with several other plant diseases has been demonstrated. The behaviour of mycoplasmas in hopper vectors resembles in many ways that of viruses and will also be discussed briefly.

26.2 Morphology of viruses

Until 1962, only two hopper-transmitted viruses had been photographed in the electron miscroscope. The particles of wound tumor virus were found to be polyhedral in shape and about 60 nm in diameter (Brakke et al. 1954; Bils and Hall 1962), whereas those of potato yellow dwarf virus in partially purified preparations were of variable shape and size (Brakke et al. 1951), though later examination of sections of the virus infected leaves revealed the presence of bacilliform particles with average dimensions of 380×70 nm (MacLeod et al. 1966). The particles of bacilliform viruses are labile and disrupt when extracted from infected tissues but if they are fixed in situ or after extraction from infected tissues with glutaraldehyde, the particles retain their true bacilliform morphology (MacLeod 1968; Ahmed et al. 1970).

In the last ten years, the morphology of several other hopper-transmitted viruses have been studied in purified preparations and/or in sections of infected plants or vector tissues. The particles of all such viruses, so far, have been found to be either polyhedral (ranging from 28–70 nm) or bacilliform (ranging from 250–380 nm in length) with the one exception of hoja blanca virus, which apparently has long flexuous threadlike particles (table 26.1, figs. 26.1–4). Recently, Black (1969) made an interesting observation that the

Fig. 26.1.   Wound tumor virus particles in a purified preparation from infected plants. (Courtesy L. M. Black, G. J. Hills and R. Markham.)

Fig. 26.2.   Wheat striate mosaic virus particles in a partially purified preparation from infected plants. (Sinha 1971.)

Fig. 26.3.   Oat blue dwarf virus particles in a purified preparation from infected plants. (Courtesy E. E. Banttari.)

Fig. 26.4.   Flexuous threadlike particles of hoja blanca virus found in lumen of the gut of an infective hopper. (Courtesy E. Shikata.)

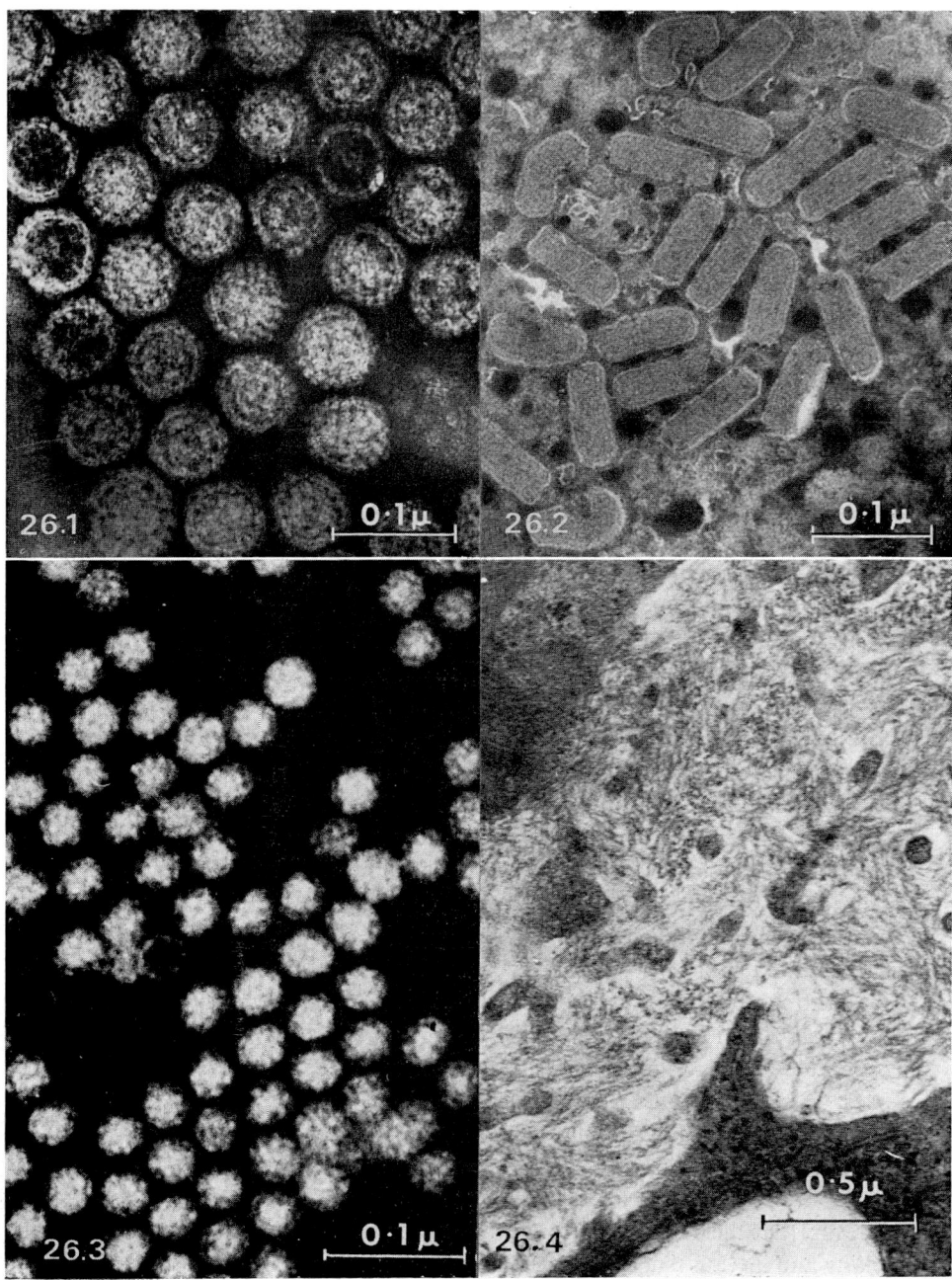

viruses which have been definitely shown to multiply in both plants and hopper vectors are either as large as wound tumor (60 nm) or larger, and suggested that it is possible that this unusual ability cannot be incorporated into the small viruses.

26.3 Virus–vector relationships

Whether or not certain viruses are capable of multiplying in their leafhopper vectors was discussed and argued in several papers published between 1928 and 1950. The history of research on this problem and the evaluation of kinds of evidence for and against virus multiplication in insect vectors is well documented (Black 1959). Several tests can provide evidence indicating virus multiplication, such as persistence of virus in the vector for a long period, frequency of virus transmission by single insects to plants being independent of the period they had spent on infected plants, a cytopathogenic effect in the vector, and cross protection of strains of virus in the insects. There are two methods, however, which can provide unequivocal evidence of virus multiplication in a vector. The first is demonstration of transovarial passage of the virus from generation to generation until such time that the progeny insects contain in total much more virus than could possibly have been present in the original female from which the line was initiated. Of course, in this kind of experiment it must be ensured that the progeny insects do not increase their virus concentration due to fresh acquisition of virus through the plants on which they are maintained. This is done either by using plants that are immune to the virus or, if susceptible plants have to be used, by transferring the insects to fresh plants before they can acquire more virus. Fukushi (1935) was the first to demonstrate the inheritance of rice dwarf virus to the sixth generation of a single *Nephotettix cincticeps* (Uhler), and concluded that this virus multiplies in its leafhopper vector. Bawden (1950), however, argued that although Fukushi's experiments provide strong presumptive evidence of virus multiplication, they did not rule out the possibility that the virus initially contained in the original female leafhopper was diluted among the progeny. Later, tests similar to that of Fukushi were done by Black (1950) with clover club leaf virus in *Agalliopsis novella* (Say), but he extended his experiments to 21 generations over 5 years and established beyond doubt that this virus indeed multiplies in its leafhopper vector. It should be pointed out here, however, that the evidence of viral etiology for clover club leaf disease is still lacking and its causal agent has never been purified or characterized morphologically. Maramorosch et al. (1970) in a recent review speculate that this disease may be caused by mycoplasma, and quote that Black (as personal communication) also shares their suspicion. The second method that can provide unequivocal evidence of virus multiplication in a vector is serial passage of the virus from insect to insect, using the injection technique, until the dilution attained exceeds the dilution end point of the starting inoculum (Maramorosch 1952). This method, which provides direct evidence for virus multiplication in insects has been used to demonstrate the multiplication of at least 5 viruses in their hopper vectors, namely, wound tumor (Black and Brakke 1952), rice dwarf (Kimura 1962), wheat striate mosaic (Sinha and Chiykowski 1967b), rice stripe (Okuyama et al. 1968) and northern cereal mosaic (Yamada and Shikata 1969).

26.3.1 *Some factors affecting virus transmission*

The age at which hoppers are allowed to acquire virus from infected plants can sometimes affect their subsequent ability to transmit viruses. For example, more planthoppers (*Delphacodes pellucida* Fabricius) were able to transmit European wheat striate mosaic virus if they acquired it as nymphs rather than as adults (Sinha 1960). Similar results were obtained for wound tumor and potato yellow dwarf viruses with their common vector *Agallia constricta* Van Duzee (Sinha 1963). Individual hoppers, irrespective of their age, may also differ in their ability to transmit viruses. Storey (1933) isolated two races of *Cicadulina mbila* which, though indistinguishable morphologically, could be separated by the fact that hoppers of one race (active) were able to transmit maize streak virus and the other (inactive) were not. Such inactive races within a single species of leafhopper are rare but vectors of several other viruses, such as curly top, potato yellow dwarf, rice stripe, European wheat striate mosaic, and wound-tumour have been shown to occur in strains that vary significantly in their ability to transmit the virus to plants (see Sinha 1968).

There is some evidence to suggest that the differences in ability of hoppers to transmit viruses could be due to the difficulty a virus encounters in passing through the gut wall of the insects into the hemolymph. This hypothesis was first suggested by Storey (1933) who found that when abdomens of hoppers from the inactive race of *C. mbila* were punctured with a fine needle, either just before or soon after acquisition of maize streak virus from infected plants, they became able to transmit the virus to plants. Similarly, when the abdomens of adult vectors of European wheat striate mosaic, wound tumor and potato yellow dwarf viruses were punctured, their ability to transmit increased almost to the degree expected of nymphs, suggesting that the permeability of the gut wall to these viruses decreases with increasing age of vectors (Sinha 1960, 1963). Susceptibility of gut cells to virus infection and the extent to which the virus can spread in this tissue may also be affected by the age of hoppers at the time of acquisition access period. For example, wound tumor virus infection, as determined by immunofluorescent technique, spread gradually throughout the gut when the vector insects acquired the virus as nymphs, but if they acquired the virus as adults the infection remained restricted to the filter chamber of the gut, suggesting that cells of other parts of the gut become resistant to virus infection as the insect ages (Sinha 1967). Also, in most leafhoppers that acquired the virus as adults the virus antigens were not detected in the hemolymph, indicating that the virus was not able to pass through the gut wall into the body cavity.

Movement of viruses inside their hopper vectors may also be prevented by exposing insects to high temperatures, and thereby making them non-infective. Translocation of wound tumor virus from the vector's filter chamber to other parts of the gut, and thence to the hemolymph and salivary glands was prevented by exposing the hoppers to 36 °C (Sinha 1967). The multiplication of maize rough dwarf virus in its planthopper vector, *Laodelphax striatellus* (Fallén), was also affected by high temperatures (36 °C), and under such heat conditions the hoppers were unable to transmit the virus to plants (Klein and Harpaz 1970).

Transmission of viruses by hoppers to their progeny may also depend on the age of the mother at the time of acquisition. If female *Delphacodes pellucida* acquired European wheat striate mosaic virus when adult, there was little or no transovarial transmission of the virus. However, if the mother acquired the virus as a nymph and the incubation period of the virus was completed in the mother before egg laying, not only did most progeny insects inherit the virus but they also started transmitting it to plants soon after hatching (Watson and Sinha 1959; Sinha 1960). There seems to be a critical time at which the virus infection must be established in the ovaries for successful transovarial transmission. For successful transmission of wound tumor virus to progeny insects, the minimum period was 14–21 days before eggs were laid. The progeny insects that inherited the virus also needed a minimum incubation period of 6–9 days before they were able to transmit the virus to plants (Sinha and Shelley 1965).

How viruses penetrate into the eggs of hoppers is not known, but work reported by Nasu (1965) on rice dwarf virus in *Nephotettix cincticeps* suggests that mycetocytes may play an important role. Examination of ultrathin sections of ovaries of viruliferous hoppers showed the presence of virus particles at the surface of certain symbiotes present in the cytoplasm of mycetocytes. As each ovariole contained a mycetocyte, which migrates to an oocyte, the author suggested that the virus is transported by the symbiotes of mycetocytes into the oocytes, resulting in transovarial transmission of the virus.

26.3.2 Vector specificity

Vector specificity of hopper-transmitted viruses cannot be explained solely on the basis of the ability of a virus to pass through the gut wall into the hemolymph of an insect, because Storey (1933) failed to make non-vector leafhoppers transmit maize streak virus either by puncturing their abdomens or by introducing the virus directly into their body cavity. Since the multiplication of several viruses in their hopper vectors has been demonstrated, vector specificity might be correlated with the ability of viruses to multiply only in certain species of hoppers. However this would not explain the vector specificity of viruses, such as curly top, that circulate but evidently do not multiply in their vectors and occur in several strains which can be transmitted only by a specific species of hoppers (Costa 1952). Irrespective of whether or not a circulative virus multiplies in its hopper vector, it has to pass through the gut wall into the hemolymph and enter the salivary glands in order to be injected back into the plants. Failure of any one of these steps could prevent a hopper from acting as a vector. Of course, the first step involving passage of the virus from the gut into the hemolymph can be bypassed experimentally by introducing the virus mechanically into the body cavity of hoppers by injection, but the virus must also survive in the hemolymph so that it can be transported to the salivary glands which in turn must be permeable to the virus. How viruses penetrate different tissues of hoppers is not known. These viruses are too large to pass directly across cell membranes, but may cross by a pinocytosis-type mechanism.

There is no information at all as to how viruses actually replicate in hopper cells. The separation of nucleic acid from its protein is probably the first step of virus infection in hoppers since this is an essential step in all known viral replication processes but experimental evidence for such a happening in hoppers is lacking. It is true that the infective RNA of one hopper-transmitted virus, namely rice dwarf, has been isolated (Yoshii and Kiso 1959), but as yet no work has been reported that would provide any clues on the process of virus multiplication in hopper vectors. Future work on these lines may provide a better understanding of vector specificity, at least for propagative viruses.

26.3.3 Growth of viruses in vectors

There are only a few reports on the quantitative measurements of virus increase in insect vectors. The methods available for virus assay in hoppers, which weigh only about 1–2 mg, are not only laborious but also inaccurate.

Serological techniques have been employed to study the synthesis of wound tumor virus and its soluble antigen in the leafhopper vector, *Agallia constricta*. The non-infective soluble antigen is a low molecular weight protein with a sedimentation coefficient of about 5S (Black 1959). It remains in the supernatant after the virus is pelleted from extracts of viruliferous hoppers by high speed centrifugation, and reacts specifically with the antiserum prepared against the virus. When virus-free hoppers were injected with a massive dose of virus, soluble antigen was first detected in the insects on the 4th day and thereafter its titre rose sharply reaching a maximum between days 8 and 10 (Whitcomb and Black 1961). Later, it was demonstrated that the amount of soluble antigen synthesized in hoppers was directly proportional to the amount of virus injected, and the production of virus in insects followed the same pattern as that of soluble antigen (Reddy and Black 1966). The relative virus concentration in hoppers, after the insects acquired the initial virus by feeding on infected plants for 1 day, rose steeply between days 7 and 30 and increased about 1000-fold during this period, but thereafter declined to about one-tenth of the peak concentration by day 45. Gamez and Black (1968) used a particle-counting technique to estimate the absolute concentration of the virus in extracts of viruliferous hoppers. They also found that the virus concentration in insects reached a peak on day 30 but then declined rather sharply. The average number of virus particles per infected hopper were calculated to be about $10^{6.6}$, $10^{9.26}$, and $10^{8.7}$ at days 6, 30 and 40, respectively. The reason or reasons for this decline in virus concentration is not known.

The relative concentration of wheat striate mosaic virus in its leafhopper vector, *Endria inimica* (Say), at various intervals after the insects had acquired the initial virus by feeding on infected plants for 1 day, was studied using infectivity bioassays (Sinha and Chiykowski 1969). The method involved injection of virus-free hoppers with various dilutions of extracts of viruliferous insects and then testing the injected hoppers singly for their ability to transmit the virus to plants. The number of insects that became infective was found to be roughly proportional to the relative virus concentration in various inocula. The virus concentration in hoppers increased by about 5000-fold between days 1 and 7, and thereafter remained at approximately the same level up to 42 days.

26.3.4 Distribution and some sites of virus multiplication

Until 1962 very little, if anything, was known about the multiplication sites of viruses in their hopper vectors. Fukushi et al. (1962), working with rice dwarf virus in its vector *Nephotettix cincticeps*, were first to visualize virus particles in various tissues of the hoppers. Since then, distribution of several propagative viruses in their vectors have been studied, by means of infectivity tests, serology or electron microscopy, in order to determine possible sites of virus multiplication or accumulation.

Clusters of rice dwarf virus particles, ranging from small groups to large masses, were observed in cells of the gut, Malpighian tubules, hemolymph, ovaries, salivary glands, mycetomes and fat body tissues of viruliferous hoppers (Fukushi et al. 1962; Nasu 1965). The particles were found in the cytoplasm but not in the nuclei of infected cells. It was suggested that the virus probably multiplies in several tissues of the vector.

Distribution of wound tumor virus in its vector *Agallia constricta* has been studied by means of both serology and electron microscopy. Fluorescent antibody technique was used to study the localization of virus antigens in various tissues of the insect vectors (Sinha 1965a). The technique involves staining virus antigens with specific antibodies that have been labelled with fluorescein isothiocyanate. The reaction produces a microprecipitate which gives a yellow-green fluorescence when observed in a microscope fitted with an ultraviolet light source. Wound tumor virus antigens were detected in hemocytes, gut, Malpighian tubules, female reproductive organs, fat body tissues, mycetome, brain and salivary glands of viruliferous hoppers, by using fluorescent antibody technique. This technique also made it possible to study the distribution of virus antigen within a particular organ (Sinha 1965a). Electron microscopic studies revealed the presence of virus particles in all the tissues mentioned above, and also in muscle, trachea and epidermal cells, thus providing evidence for the systemic invasion of the hopper vectors by the virus (Shikata and Maramorosch 1965). These authors also suggested that the virus was assembled in electron dense areas designated as 'viroplasms', that develop in the cytoplasm of infected cells.

Sequence of infection in hoppers by wound tumor virus was studied by using fluorescent antibody technique (Sinha 1965a). The virus antigens were first detected on day 4 in a specific corner of the filter chamber of the gut of insects that had acquired the virus by feeding on infected plants for 1 day. As the incubation period of the virus in hoppers progressed, the virus spread to other parts of the gut and by day 12 the virus antigens could also be detected in the hemolymph. On day 14, the virus antigens were found in the fat body tissues, brain, Malpighian tubules, and on the 17th day in anterior lobes of the salivary glands. These results suggested that the gut of the hoppers was the initial site of virus multiplication and its filter chamber the primary focus of infection. The virus presumably also multiplies in other susceptible sites, such as hemocytes, fat body cells, mycetome and salivary glands. The virus probably does not multiply equally well in all the lobes of the salivary glands because virus antigens were found restricted mostly to the anterior lobes.

That maize rough dwarf virus is also capable of systemically invading its planthopper vector, *Laodelphax striatellus*, was recently shown by the presence of virus particles in several tissues of viruliferous insects (Vidano 1970). Although the virus particles were observed in various tissues of the vector, the organs most frequently found infected were gut and salivary glands.

Particles of other viruses, such as maize mosaic (Herold and Munz 1965), northern cereal mosaic (Shikata and Lu 1967) and rice hoja blanca (Shikata and Galvez 1969), have also been observed in certain organs of their hopper vectors. It is not yet known if these viruses invade their vectors systemically.

Distribution of potato yellow dwarf virus in its vector *Agallia constricta* was studied by using infectivity bioassays (Sinha 1965b). The extracts, prepared from various internal organs of viruliferous hoppers, were injected into virus-free insects and the latter were then tested for their infectivity on plants susceptible to the virus. The virus was recovered from hemolymph and all the internal organs, showing that the virus invades the vector systemically. Northern cereal mosaic virus has been shown to occur in the gut and salivary glands of viruliferous hoppers but not in the fat body tissues, indicating that this virus perhaps multiplies only in certain tissues (Yamada and Shikata 1969).

Wheat striate mosaic virus was recovered, using infectivity bioassays, from hemolymph, hemocytes, gut, salivary glands, fat body tissues, brain and mycetomes, but not from ovaries, testes or Malpighian tubules, of viruliferous hoppers *Endria inimica* (Sinha and Chiykowski 1969). The approximate concentration of the virus in certain tissues of hoppers, at various times after the insects had an acquisition access period of 1 day, was also determined. The virus was first recovered from the gut on day 2, from hemolymph,

hemocytes, and the salivary glands by day 4. The virus concentration increased rapidly in the gut, reached a peak by day 8, and then declined slightly
by day 32. In hemolymph, hemocytes and salivary glands, the virus attained
a plateau level between 6 and 8 days. These results suggested that the gut of
the vector is the primary site of virus multiplication and that the virus probably is synthesized in hemocytes and salivary glands also, but that it may not
multiply but only accumulate in other tissues of the vector. However, when
the same experiments were done with injected insects virus was recovered
from all the tissues except the gut (Sinha, unpublished results). It seems that
the virus cannot infect the gut from the hemolymph. Since the virus multiplies in hoppers irrespectively whether they receive it by injection or by
feeding on infected plants, tissues other than the gut must be capable of
supporting virus synthesis.

26.3.5 *Effect of viruses on vectors*

Several viruses have been reported to produce deleterious effects on their
hopper vectors, and the subject has been reviewed critically (see Jensen 1963;
Sinha 1968). However, several of these reports dealt with diseases which are
now suspected to be caused by mycoplasmas and not by viruses as had been
assumed earlier.

It has been suggested that European wheat striate mosaic virus may be
injurious to its vector, *Delphacodes pellucida* (Watson and Sinha 1959). Viruliferous females were found to produce 40% fewer progeny insects than
non-viruliferous ones. Examination of eggs laid by viruliferous females
showed that some embryos died in the egg at a late stage of their development,
indicating that the virus may be pathogenic to them. Similar results were later
reported with rice stripe virus in its vector *Delphacodes striatella* Fall (Nasu
1963). Kisimoto and Watson (1965), however, found that some of the eggs of
D. pellucida from the first inbred generation were abnormal and failed to
hatch, irrespective of whether the eggs were laid by viruliferous or non-viruliferous females.

Yoshii (1959) reported that the oxygen consumption, respiratory quotient
and oxidative phosphorylation rate were higher in hoppers carrying rice
dwarf virus than in those that were virus-free. Viruliferous hoppers in Yoshii's
experiments were obtained by caging insects on infected plants for a long
period of time, and therefore, it is quite probable that the metabolic abnormalities in infective hoppers could have been due to malnutrition. Nasu
(1963) reported that this virus adversely affects the fecundity of hopper vec-

tors. Viruliferous females laid 28–68 % fewer eggs as compared to non-viruliferous ones. Also, the survival rate of nymphs that received the virus through inheritance was found to be 15 % less than that of progeny insects from virus-free females. These differences in longevity and mean fecundity were, however, not statistically significant (Nakasuji and Kiritani 1970). Reduction in fertility and longevity of hoppers *Sogatodes oryzicola* (Muir) carrying hoja blanca virus has also been reported (Jennings and Peneda, 1970). It could be that such deleterious effects in hoppers are due to virus infection, but the possibility that they are the result of malnutrition cannot be ruled out.

The presence of wound tumor virus in the nervous system of *Agallia constricta* was found to be associated with degenerating ganglion cells (Hirumi et al. 1967). The membrane structures of the cytoplasmic organelles such as endoplasmic reticulum, the nuclear membrane and the Golgi apparatus were indistinct in the degenerating cells. Also, no normal mitochondria were observed in these cells. However, no macroscopic or behavioural disorder was noted in viruliferous hoppers. At present, there seems to be no critical evidence to show that these cytological changes are pathological effects of the virus.

So far, unequivocal evidence showing that any virus is actually harmful to its hopper vector is lacking.

26.4 *Tissue culture*

Since a separate chapter on insect tissue culture appears elsewhere in this book (ch. 17), this section will only deal briefly with the multiplication of viruses in their hopper vector cells grown in vitro.

Embryonic tissues of hoppers, *Nephotettix cincticeps*, carrying rice dwarf virus were cultivated in vitro (Mitsuhashi 1965). Electron microscopic examination of epithelial cells that grew from the explants showed, by the 27th day of cultivation, typical virus particles in almost all the cells. However, when healthy embryonic tissue cultures were inoculated with the virus, only a few normal virus particles (70 nm) were observed in some cells, but small virus-like particles (30 nm) were found in almost all the cells. The author suggested that the small particles might be one of the forms of virus because they have also been seen in the cells of intact viruliferous hoppers (Nasu 1965). Later, the multiplication of the normal (70 nm) virus in primary cultures was demonstrated after they were inoculated with the inoculum prepared from other primary cultures established from viruliferous hoppers (Mitsuhashi and Nasu 1967).

Chiu et al. (1966) were able to infect healthy embryonic tissue cultures of *Agallia constricta* with wound tumor virus, and found virus antigens to be localized only in the cytoplasm of inoculated cells, as determined by immuno-fluorescent technique. It was also demonstrated, by infectivity bioassays and serial passage, that the virus multiplied in these cells. Minimum concentration of the virus that produced infection in monolayers of cultured cells was estimated to be about 10^6 particles per ml of inoculum (Gamez and Chiu 1968). Later, it was found that the number of cells that showed specific fluorescence had a linear relationship with the relative virus concentration in the inoculum, thereby making it possible to use a fluorescent cell counting method for assaying the infectivity of the virus in cell cultures (Chiu and Black 1969).

Cell cultures of non-vector hoppers have also been shown to be susceptible to wound tumor virus, although the virus multiplied only to a limited extent in such cells (Chiu and Black 1969).

Potato yellow dwarf virus has been shown to multiply in monolayers of vector cells after they were inoculated with the virus obtained from infected plants (Chiu et al. 1970). Fluorescent antibody and electron microscopic techniques were employed to demonstrate the presence of virus antigens or particles in the inoculated cells. In early stages of infection, specific fluorescence was observed only in the nuclei of infected cells, but later some virus antigens were detected in the cytoplasm also.

Successful development of monolayer cultures of hopper cell lines and their inoculation with viruses provides an important method for studying the molecular biology of hopper-transmitted viruses.

26.5 *Similarities with viruses of other animals*

Several physical and chemical characteristics of wound tumor virus resemble those of reoviruses occurring in mammals. The particles of both viruses are polyhedral, about 60 nm in diameter, and they consist of 92 subunits (Bils and Hall 1962; Vasquez and Tournier 1962). It has also been demonstrated that RNA of wound tumor virus (Black and Markham 1963; Tomita and Rich 1964) as well as that of reovirus (Gomatos and Tamm 1963) is double stranded.

Streissle and Maramorosch (1963) reported that wound tumor virus and reovirus are serologically related, as determined by the complement fixation test. However, Gomatos and Tamm (1963) did not find any common antigens between wound tumor virus and three prototypes of reoviruses. Later, Gamez et al. (1967), by using the passive hemagglutination inhibition test, which is one of the most sensitive methods of detecting antigens that can be bound to erythrocytes, also failed to find any evidence of serological relationship between the two viruses. As yet, therefore, there is no evidence showing that these viruses have any antigens in common.

Maize mosaic was the first hopper-transmitted virus that was shown to have bacilliform particles (Herold et al. 1960). Since then, several other viruses with similar particles have been described (see table 26.1). These viruses resemble in structure several vertebrate and insect viruses, such as, vesicular stomatitis, rabies, and sigma (see Howatson 1970). The term 'rhabdovirus' has been suggested as a generic name to denote this group of viruses. The distinguishing feature of the group is the bullet-shaped or bacilliform particles with a cross striated viral nucleocapsid, which has a pitch of 4–5 nm. In this group, all vertebrate viruses have been found to be bullet-shaped, whereas particles of hopper-transmitted viruses are reported to be bacilliform and where bullet-shaped particles are found in vitro preparations, they are believed to be the result of damage of bacilliform particles. Recently, however, both bullet-shaped and bacilliform particles have been observed in sections of plants infected with wheat striate mosaic virus (Sinha 1971).

Not much is known about the properties of these hopper-transmitted viruses. Until their biochemical and biophysical properties are studied in detail, it is still a matter of discussion whether these viruses should be included under the same genus as vertebrate viruses.

26.6 Mycoplasma–vector relationships

Organisms resembling mycoplasmas have been found associated with at least 40 plant diseases (see Maramorosch et al. 1970). It should be realized, however, that proof that mycoplasma cause these diseases can only come by satisfying Koch's postulates. Although evidence of multiplication of the agents, so far, has been reported for only 3 diseases, namely, white leaf of sugar cane (Lin and Lee 1969), corn stunt (Chen and Granados 1970), and European clover phyllody (Giannotti et al. 1971), it has been assumed that the other diseases are caused by mycoplasmas also. No specific names have yet been assigned to these 'plant-pathogenic mycoplasmas' and, therefore, for convenience, the various diseases they cause will be referred to by their common names followed by the word 'mycoplasma'.

Most of the diseases caused by mycoplasmas are spread from plant to plant by means of hoppers but other vectors have been reported also (Hibino et al. 1971). Behaviour of these organisms in their hopper vectors resembles in many ways that of viruses, and this was one of the reasons that several diseases, now suspected to be caused by mycoplasmas, were erroneously thought to be viral in nature, in spite of the fact that their causal agents were never morphologically identified. Like viruses, mycoplasmas have been shown to circulate and multiply in their hopper vectors, to undergo an incubation period before an insect can transmit them to plants, to be inactivated by high temperatures, and to occur in strains which exhibit the interference phenomenon. As a matter of fact, some of the viruses often quoted as classic examples of 'virus multiplication in leafhoppers' have now been placed in the mycoplasma category, for example aster yellows.

Mycoplasmas are pleomorphic organisms, have no cell wall, are bound by a membrane and contain typical ribosomes in their cytoplasm (fig. 26.5).

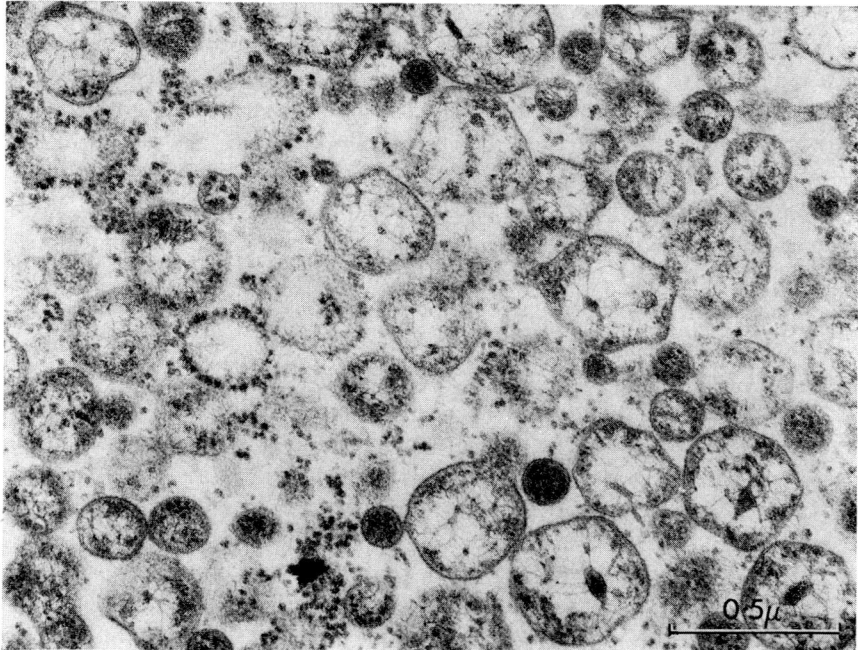

Fig. 26.5.   Clover phyllody mycoplasma in an acinus of salivary glands of an infective hopper.

Small forms of mycoplasma, commonly referred to as 'elementary bodies', ranging from 60–150 nm in size, are spherical in shape and are filled with ribosomes in a dense matrix. Forms larger than elementary bodies, ranging from 150–900 nm in diameter, have also been observed in tissues of infective hoppers. Some of the larger forms of mycoplasma cells show a nuclear area with DNA-like fibrils in the center and ribosomes at the periphery. Occasionally, the large forms of mycoplasma give rise to elementary bodies by a budding process, which presumably is one of the methods of multiplication of mycoplasmas in vector insects (Sinha and Paliwal 1970).

Corn stunt mycoplasma has been observed in the salivary glands, Malpighian tubules, gut, brain, fat body cells of infective *Dalbulus elimatus* (Ball) (Granados et al. 1968). The occurrence of clover phyllody mycoplasma has been reported in the gut and salivary glands of the vector *Euscelis plebejus* Fall. from Europe (Giannotti et al. 1968) and of *Macrosteles fascifrons* (Stål) from Canada (Sinha and Paliwal 1970). Hirumi and Maramorosch (1969) reported the presence of aster yellows mycoplasma in the salivary glands of *M. fascifrons*. Western X mycoplasma has been observed in the gut and brain of infective leafhoppers, *Colladonus montanus* (Van Duzee).

Distribution of aster yellows mycoplasma has been studied in the vector *Macrosteles fascifrons*, using infectivity bioassays (Sinha and Chiykowski 1967c). The extracts prepared from hemolymph, gut, ovaries and salivary glands of hoppers were infective, but those from mycetomes, Malpighian tubules, fat body tissue, testes or brain were not. Studies on sequential infection of tissues of the vector suggested that the mycoplasma multiplies in the gut and hemolymph of hoppers, after the insects acquire the organism by feeding on infected plants. Similar results were reported for clover phyllody mycoplasma in its vector *M. fascifrons* (Sinha and Chiykowski 1968).

It has been demonstrated that aster yellows mycoplasma not only can be acquired by non-vector hoppers but is also retained by them for several days, and to multiply to a limited extent in the gut of non-vector hoppers (Sinha and Chiykowski 1967a). Vector specificity of aster yellows mycoplasma, therefore, cannot be explained solely on the basis of its ability to multiply only in a vector. It seems that several factors discussed earlier in relation to vector specificity of viruses may be valid for mycoplasmas also.

The work on Western X mycoplasma has shown that this organism is injurious to its hopper vector, *Colladonus montanus* (Jensen 1959). The average length of life of infective hoppers was found to be reduced from about 50–20 days. Moreover, histopathological effects have been observed in several tissues of hoppers carrying the mycoplasma (Whitcomb et al. 1967, 1968).

Certain cytological abnormalities have also been observed in fat body cells of hoppers carrying aster yellows mycoplasma (Littau and Maramorosch 1960). At present, there is no critical evidence to suggest that these cytological changes are pathological and it is possible that they may be artifacts caused by staining and embedding procedures used.

CHAPTER 27

Viruses and nematodes

B. D. HARRISON

Contents

27.1 Introduction

Although most of the recent interest in nematodes as virus vectors has come from work with plant viruses, the earliest evidence to implicate them in the spread of virus disease was obtained in research with mammals. Thus Shope (1941, 1943) detected swine influenza virus not only in pig lungworms *(Metastrongylus elongatus)* obtained from virus-infected pigs, but also in lungworm ova from pig faeces, suggesting that the lungworms might retain the virus during the phase of their life cycle in which they parasitize earth-worms (ch. 12), and then transmit the virus back to pigs. Whether the virus multiplied in *M. elongatus* was not established. In 1947, Syverton et al. showed that *Trichinella spiralis* larvae acquired lymphocytic choriomeningitis virus in virus-infected guinea pigs, and could transmit it to healthy animals. Also, several instances are known in which nematodes can act synergistically with mammalian viruses (Woodruff 1968). But despite these observations, it seems unlikely that the spread of these mammalian viruses usually involves nematode vectors. By contrast, there are several plant viruses that are rarely spread in any other way, and these are the subject of this chapter. I shall, however, not attempt to provide an exhaustive account of this rapidly growing subject, but simply to describe the main types of phenomena found. Other examples and details are given in reviews by Taylor and Cadman (1969) and Taylor (1971).

Grapevine fanleaf was the first plant virus shown to have a nematode vector (Hewitt et al. 1958). At the time this work was done, the virus could otherwise be transmitted only by grafting, but soon afterwards it was shown to be transmissible by inoculation of sap and to be related to arabis mosaic virus (Cadman et al. 1960), which in turn was already considered a member of a group of sap-transmissible viruses that have isometric particles, and many properties in common. All these viruses have now been found to have nematode vectors.

Various other plant viruses have long been known to spread via the soil (Harrison 1964a). Some of these are now known to have fungus vectors, but at least one further group, which have straight tubular particles, are dependent on nematodes for their spread. Indeed the list of plant viruses with nematode vectors continues to grow steadily and already includes several viruses of major economic importance. Grapevine fanleaf virus occurs in most grapevine-growing areas of the world, especially where rootstocks are used and, in Europe, some cultivars are totally infected; with one cultivar grown on different rootstocks the loss of yield was estimated to be 18–43 %

(Gallay et al. 1955). Tobacco rattle virus can infect more than 400 species. It is prevalent on sandy soils in Western Europe, including Britain, occurs in North and South America, and causes economically important diseases in potato and various bulbous ornamentals. Other nematode-borne viruses cause locally important diseases in crops as diverse as cherry, cucumber, leek, lettuce, mulberry, pea, raspberry, strawberry and sugar beet, to name a few examples. Nematode-borne viruses are reported from most well-developed countries of the world and will very probably also be found in less-developed regions in due course.

27.2 *Nematode vectors of plant viruses*

Nearly all the plant viruses reported to have nematode vectors fall into two groups, called nepoviruses and tobraviruses (previously netuviruses) (Harrison et al. 1971). Similarly, the vector species belong to two corresponding groups, the Dorylaimina and Diphtherophorina respectively (ch. 12). Within the Dorylaimina only species of *Xiphinema* and *Longidorus* are known to act as vectors, and within the Diphtherophorina only *Trichodorus* species (table 27.1). These nematodes not only transmit the viruses in laboratory conditions but also are associated with disease outbreaks in the field. Also listed in table 27.1 are a few 'other viruses' that have isometric particles and are reported to be transmitted by nematodes in laboratory experiments,

TABLE 27.1
Nematode-transmitted plant viruses and their vectors

Virus	Vector	Reference
Nepoviruses		
Arabis mosaic	*Xiphinema diversicaudatum*	Harrison and Cadman 1959; Jha and Posnette 1959
	X. coxi	Fritzsche 1964
Cherry leaf roll	*X. diversicaudatum*	Fritzsche and Kegler 1964
	X. coxi	Fritzsche 1964
Grapevine fanleaf	*X. index*	Hewitt et al. 1958
	X. italiae	Cohn et al. 1970
Mulberry ringspot	*Longidorus martini*	Yagita and Komuro 1970
Raspberry ringspot type strain	*L. elongatus*	Taylor 1962

Table 27.1 *(continued)*

Virus	Vector	Reference
English strain	*L. macrosoma*	Harrison 1964b
	L. elongatus	Taylor and Murant 1969
cherry strain	*L. macrosoma*	Fritzsche and Kegler 1968
	Xiphinema diversicaudatum	Fritzsche and Kegler 1968
Strawberry latent		
ringspot	*X. diversicaudatum*	Lister 1964
	X. coxi	Putz and Stocky 1970
Tobacco ringspot	*X. americanum*	Fulton 1962
	X. coxi	van Hoof 1971
Tomato black ring		
English strain	*Longidorus attenuatus*	Harrison et al. 1961
Scottish strain	*L. elongatus*	Harrison et al. 1961
Tomato ringspot	*Xiphinema americanum*	Breece and Hart 1959

Tobraviruses

Pea early browning		
Dutch isolates	*Trichodorus pachydermus*	van Hoof 1962
	T. teres	van Hoof 1962
English isolates	*T. anemones*	Harrison 1967a
	T. primitivus	Harrison 1966
	T. viruliferus	Gibbs and Harrison 1964
Tobacco rattle		
European isolates	*T. anemones*	van Hoof 1968
	T. cylindricus	van Hoof 1968
	T. nanus	van Hoof 1968
	T. pachydermus	Sol and Seinhorst 1961
	T. primitivus	Harrison 1961; Sänger 1961
	T. similis	Cremer and Kooistra 1964
	T. teres	van Hoof 1968
	T. viruliferus	van Hoof 1968
N. American isolates	*T. allius*	Jensen and Allen 1964
	T. christiei	Walkinshaw et al. 1961
	T. porosus	Ayala and Allen 1966

Other viruses

Brome mosaic	*Xiphinema diversicaudatum*	Schmidt, Fritzsche and Lehmann 1963
	Longidorus macrosoma	Fritzsche 1970
Carnation ringspot	*Xiphinema diversicaudatum*	Fritzsche and Schmelzer 1967
Prunus necrotic		
ringspot	*Longidorus macrosoma*	Fritzsche and Kegler 1968

although it is not known whether these species have any importance as vectors in nature. In general, viruses that are transmitted by nematodes seem not to have alternative vectors among other groups of organisms. When infective soil is fractionated and the vector species picked out, these nematodes transmit virus to healthy bait plants, whereas no infection occurs among bait plants grown in the residue (mainly soil particles and root fragments) which contains many other organisms. Also, when naturally infested soil is dried at about 20 °C for two weeks, the vector nematodes are killed and the infectivity of the soil abolished. These two kinds of test are among the most useful for implicating nematodes as vectors. However, despite these generalizations, tobacco ringspot virus is reported to be transmitted in laboratory tests not only by *Xiphinema americanum*, but also by spider mites *(Tetranychus* sp.) (Thomas 1969) and by *Thrips tabaci* (Messieha 1969).

The nepoviruses are a group of viruses with isometric particles about 30 nm in diameter. Typically the particles are of three types, all with the same protein shell; in one type the shell is empty and in the other two it contains different amounts of RNA. Many of the nepoviruses have wide host ranges and cause ringspot diseases and, unlike most plant viruses, all are transmitted through seed. Among the nepoviruses there is considerable specificity between virus and vector. For example, *Xiphinema diversicaudatum* readily transmits arabis mosaic virus but does not transmit tomato black ring virus, or even grapevine fanleaf virus, which is serologically distantly related to arabis mosaic virus. Sometimes the specificity is a matter of degree only. Thus *Longidorus elongatus* transmits the English strain of tomato black ring virus on rare occasions but the Scottish strain readily, and it transmits the English strain of raspberry ringspot virus only slightly less efficiently than the type strain (Taylor and Murant 1969).

The tobraviruses have straight tubular particles and an RNA genome in two parts, which are found in particles of two corresponding lengths. The RNA in the longer particles is infective on its own, replicates in plants and induces symptoms, but without producing the usual tubular particles, because the RNA in the shorter particles includes the coat protein cistron. This smaller RNA does not replicate in plants or cause symptoms on its own but cooperates with the larger RNA to produce new RNA-containing tubular particles of both lengths. The tobraviruses have wide host ranges, tobacco rattle virus having more known hosts than any other virus. In general, the specificity between virus and vector seems somewhat less well developed in the tobraviruses than in the nepoviruses. For instance, several species of

Trichodorus transmit tobacco rattle virus in the Netherlands (van Hoof 1968). Some specificity exists nevertheless, because *T. anemones* transmitted an English but not a Dutch strain of pea early-browning virus (Harrison 1967a), and *T. pachydermus* transmitted only one out of five Dutch isolates of tobacco rattle virus (van Hoof 1968).

27.3 Transmission processes

27.3.1 Acquisition and inoculation of virus

The efficiency of nematode transmission of nepoviruses and tobraviruses increases with increasing times of access to both infected source and healthy bait plants, though in most tests fewer than half the individuals transmit. The nematodes feed intermittently and many take time to find suitable feeding points on the roots, so that the increase in transmission efficiency with increasing access times undoubtedly reflects the increasing proportion of nematodes that have fed. Whether individual nematodes become more efficient vectors with increasing times of feeding has not been ascertained.

Among the nepoviruses, tomato ringspot virus can be acquired by *Xiphinema americanum* in 1 hr and inoculated in 1 hr, but increasing the periods of access to a day or more greatly increases the number of nematodes that transmit (Teliz et al. 1966). Grapevine fanleaf virus can occasionally be acquired and inoculated in as little as 15 min (Das and Raski 1968). Similar trends are found with tobraviruses. Tobacco rattle virus can be acquired in 1 hr and inoculated in 1 hr, but again efficiency increases with the duration of the access periods (Ayala and Allen 1968).

The vectors seem usually to acquire and inoculate the viruses near the root tips, because they feed mainly just behind the root cap, in the zone of cell elongation, and usually not as far from the root tip as the root hair zone. The feeding itself can induce the production of galls, or kill cells and stop the roots growing, quite independently of virus infection. The nematodes may not feed equally on, or be equally pathogenic to, different types of root. For instance on apple, *Trichodorus viruliferus* feeds mainly on the white extending roots and little on the finer feeder roots (Pitcher and Flegg 1965).

Nepoviruses and tobraviruses can be transmitted by both adult and juvenile forms of their vector species. All juvenile stages (except the first which was not tested), and adults, of *Xiphinema index* transmitted grapevine fanleaf virus (Raski and Hewitt 1960) and of *X. americanum* transmitted tomato ringspot virus (Teliz et al. 1966), and all stages except the first juvenile one of *Trichodorus allius* transmitted tobacco rattle virus (Ayala and Allen 1968). If, as seems very possible, the viruses are not retained by nematodes through the moult (sect. 3.2), these results would suggest that each of the stages tested can both acquire and inoculate virus. There is evidence that, in some conditions, different stages differ in efficiency as vectors. For instance, adult female *Xiphinema diversicaudatum* transmitted arabis mosaic virus more often than adult males (Harrison 1967b), and in one series of tests tomato black ring virus was transmitted by juvenile but not by adult *Longidorus elongatus* (Harrison et al. 1961); later work however showed that adults could also transmit (Yassin 1968). By contrast adult *Trichodorus allius* transmitted tobacco rattle virus more efficiently than juveniles (Ayala and Allen 1968).

27.3.2 Behaviour of virus in the vector, and mechanism of transmission

The period for which nematodes can retain virus has been tested using handpicked nematodes kept in previously sterilized medium without plants (the method least subject to error), nematodes kept in naturally infested soil without plants or nematodes kept on plants apparently immune to the virus. All these kinds of test show that at about 20 °C the viruses can persist for several weeks or more in a transmissible form. Some of the tests also suggest that the viruses are retained for shorter periods in *Longidorus* than in *Xiphinema* or *Trichodorus* vectors. Thus raspberry ringspot and tomato black ring viruses occasionally persisted up to 12 weeks in *Longidorus elongatus* (Murant and Lister 1967; Taylor 1970), whereas *Xiphinema americanum* transmitted tobacco ringspot virus after 49 weeks (Bergeson and Athow 1963), and *Trichodorus pachydermus* transmitted tobacco rattle virus after about 2 years (van Hoof 1970) without plants.

What little evidence there is, suggests that nepoviruses and, possibly, tobraviruses are not retained by nematodes through a moult, and do not pass through nematode eggs. The tests were however on too small a scale to eliminate the possibility of occasional transmission (Harrison and Winslow 1961; Taylor and Raski 1964; Ayala and Allen 1968).

Excellent progress has recently been made in locating virus particles within nematodes by means of electron microscopy of ultrathin sections. Virus-like particles were found in *Longidorus elongatus* carrying raspberry ringspot virus and carrying tomato black ring virus (Taylor and Robertson 1969), *Xiphinema diversicaudatum* and *X. index* carrying arabis mosaic and grapevine fanleaf viruses respectively (Taylor and Robertson 1970a), *X. americanum* carrying tobacco ringspot virus (McGuire et al. 1970) and *Trichodorus pachydermus* carrying tobacco rattle virus (Taylor and Robertson 1970b). These particles were not found in virus-free nematodes, and there can be little doubt that they are those of the viruses, especially with tobacco rattle virus, which have easily recognizable particles; and in each instance the particles were found only in a specific part of the nematode's body. The viruses transmitted by *Longidorus elongatus* are associated with the inner surface of the stylet guiding sheath, where they are found in a layer one particle thick, together with mucous-like material. The viruses transmitted by *Xiphinema* spp. are not associated with the guiding sheath, but are found on the inner surface of the cuticle lining the oesophagus, from the anterior end of the odontophore to the posterior end of the oesophageal bulb (fig. 12.2). The particles are in a single layer, sometimes in a hexagonally close-packed array (fig. 27.1), and interspersed with mucous-like material. Tobacco rattle virus is found in a similar position to that of the viruses in *Xiphinema* spp., namely in association with the cuticle lining the pharynx and oesophagus of *Trichodorus pachydermus* (figs. 12.2; 27.1). Thus the apparent difference in persistence in their vectors between *Longidorus*-transmitted viruses on the one hand, and *Xiphinema*- and *Trichodorus*-transmitted viruses on the other, is paralleled by a difference in the site of virus retention in the nematode. Virus-like particles were not found in other parts of these nematodes using thin-section methods, and it seems reasonable to conclude that the main sites of virus retention and accumulation were located.

It is thought that as material passes from the plant to the nematode's intestine virus particles are selectively adsorbed to the guiding sheath or oesophageal cuticle (Taylor and Robertson 1969, 1970a, b). Viruses that are not transmitted seem not to be retained at these sites. For example, arabis mosaic virus particles were not found on the guiding sheath of *Longidorus elongatus* after feeding on infected plants (Taylor and Robertson 1969). Vector specificity therefore seems to reflect specificity of adsorption, which in turn presumably depends on the detailed structure of the surfaces of the nematode and the virus particle. Indeed the possible importance of the virus particle surface in virus transmission was suggested by Harrison (1964b) to explain the fact that with tomato black ring and raspberry ringspot viruses vector specificity is correlated with serological relatedness. In this instance therefore, the difference between strains in particle surface causes both a difference in serological behaviour and, presumably, a difference in ability to adsorb to the guiding sheath of the nematodes. The precise features of

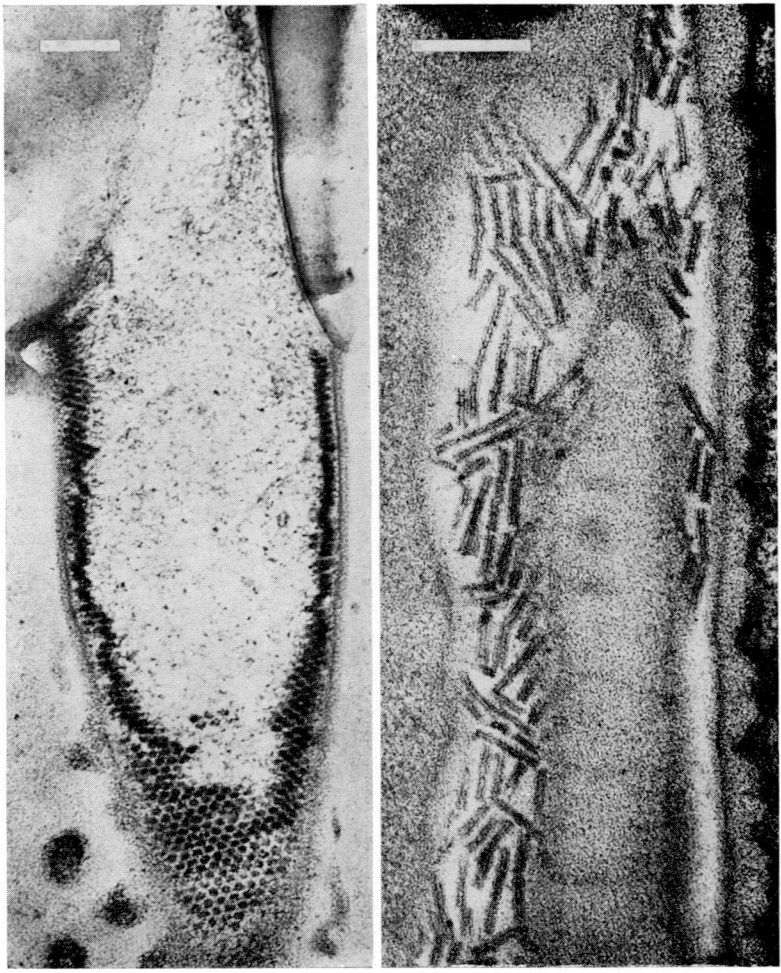

Fig. 27.1. Left, oblique longitudinal section of the junction between the stylet (upper part) and odontophore (lower part) in *Xiphinema diversicaudatum*; particles of arabis mosaic virus line the lumen of the odontophore but not that of the stylet (Taylor and Robertson 1970a). Right, longitudinal section of the lumen of the oesophageal bulb of *Trichodorus pachydermus*, showing particles of tobacco rattle virus; the anterior part is uppermost (Taylor and Robertson 1970b). The bars represent 200 nm.

the virus surface and nematode surface that are important for transmission are not known.

The suggested mechanism of virus transmission fits well with other observations. Thus injecting grapevine fanleaf virus into *Xiphinema index* did not make the nematodes infective (Betto and Raski 1966), and although infective virus can sometimes be extracted from cut nematodes, probably from their intestine (Sänger et al. 1962; Raski and Hewitt 1963; Taylor and Murant 1969), such virus is unlikely to be transmitted because the oesophago-intestinal valve seems designed to prevent regurgitation from the intestine. Indeed there is no direct evidence that any of the viruses multiply in nematodes, and although Roggen (1966) reported some morphological and physiological differences between *X. index* raised on virus-infected and virus-free plants, these may be dietary effects.

The lengthy persistence of the viruses in nematodes that are allowed to feed on plants would suggest that the virus particles are eluted from the nematode surfaces less readily than they are adsorbed. Thus when nematodes were moved to fresh healthy bait plants every day or two, some individuals of *Trichodorus pachydermus* (van Hoof 1964), *Xiphinema diversicaudatum* (Harrison 1967b) and *Longidorus attenuatus* (Harrison 1969) transmitted viruses to several plants in the series; these plants were either consecutive or not consecutive. The apparent failure of the viruses to be retained through the moult is easily explained. During moulting, the lining of the stylet guiding sheath is cast with the nematode's outer cuticle, and with it the viruses carried by *L. elongatus*. Not only is the lining of this anterior part of the food canal shed, but also the cuticular lining of the oesophagus, which seems, from electron microscopy of sections of *Xiphinema*, to pass back into the intestine (Taylor and Robertson 1970a). Thus viruses carried by *Longidorus* and *Xiphinema* are removed from the surfaces from which they can be transmitted. Moulting in *Trichodorus* has still to be studied in similar detail.

The way in which virus transmission by a process of adsorption and elution might occur during the normal feeding of the nematodes could until recently only be surmised. However, Wyss (1971) has made a detailed cinematographic record of feeding by *Trichodorus*, and his analysis, which extends that of Chen and Mai (1965) provides answers to many of the questions that arise. Wyss separates the feeding process in *Trichodorus* into five phases: probing, penetration, salivation, ingestion and retraction. First the stylet is thrust repeatedly into the wall of a root cell, and soon penetrates it, often in less than a minute. Even after penetration the stylet continues to make short movements, and saliva from the oesophageal bulb (where the salivary ducts join the oesophagus) is injected into the cell by pulsations of the bulb, perhaps aided by the stylet movements. During this salivation phase, which may last for several (perhaps 10) min the oesophago-intestinal valve is closed; and inside the root cell, cytoplasm accumulates around the puncture hole. Finally the stylet makes deep thrusting movements through the cytoplasmic mass, the oesophago-intestinal valve opens, and the cell contents are ingested, a process that takes perhaps 20 sec, with the stylet still pulsating and probably now pumping in the reverse direction, helped by contractions of the oesophagus. The stylet is then retracted and the nematode moves to another cell, often leaving attached to the cell wall a salivary tube, which was possibly formed within the nematode's stoma.

Virus adsorption presumably occurs when sap is being ingested from an infected plant, and the mouthparts and oesophageal lining are bathed in virus-containing plant extract. Detachment of virus from its site of retention would then occur during the salivation phase of subsequent feeds, possibly because the saliva provides conditions in which adsorption is less strong, or because it acts on some material that binds the virus to the appropriate nematode surface. Inoculation would occur when the virus is carried into the plant cell with the saliva although, for the plant to become infected, the inoculated cell must not be damaged too seriously. Wyss observed, however, that ingestion of the cell contents was often very incomplete and that cytoplasmic streaming resumed in some cells after they were fed on, so this requirement seems satisfied, for *Trichodorus* at least.

27.4 Nematode behaviour and virus ecology

27.4.1 Nematode occurrence and soil type

Many of the vector species occur predominantly on particular types of soil. Thus, in Britain, *Longidorus attenuatus* occurs on sandy soils, *L. elongatus* mainly on light loams and *L. macrosoma* on heavy soils or in river valleys where the water table is high. In Scotland, Cooper (1971) found that *Trichodorus pachydermus* was the commonest vector species of tobacco rattle virus occurring on sandy soils, whereas *T. primitivus* was the most prevalent on light loams. Moreover the species were not evenly distributed among different soil series of a soil type, some series being almost invariably infested and a few not at all. Neither *T. pachydermus* nor *T. primitivus* occurred on heavy soils.

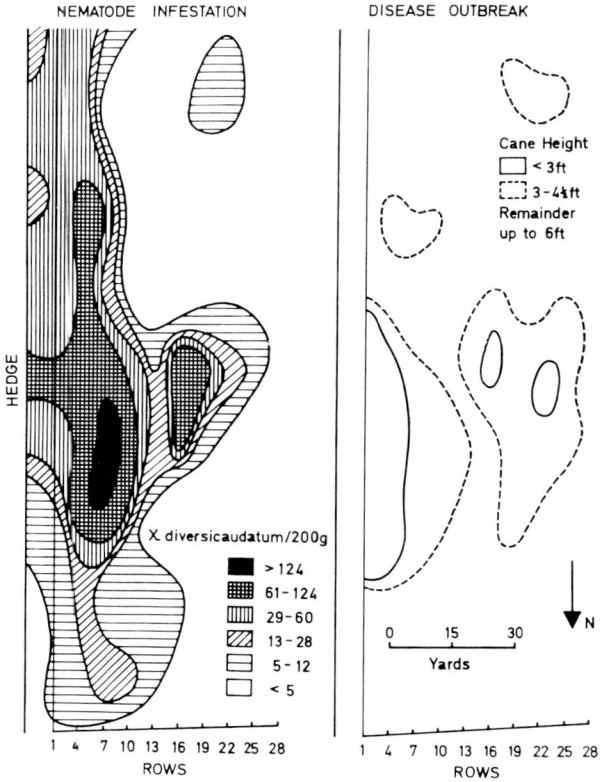

NEMATODE INFESTATION   DISEASE OUTBREAK

HEDGE

X. diversicaudatum/200g

>124
61–124
29–60
13–28
5–12
< 5

Cane Height
< 3ft
3–4½ft
Remainder
up to 6ft

0   15   30
Yards

N

1 4 7 10 13 16 19 22 25 28   1 4 7 10 13 16 19 22 25 28
ROWS   ROWS

Fig. 27.2.   Diagram showing the distribution of *Xiphinema diversicaudatum* at an outbreak
of strawberry latent ringspot virus in a ten-year-old plantation of Malling Jewel raspberry.
(Taylor and Thomas 1968.)

In many of the fields where they occur, the vector species are patchily
distributed and the patches often correspond to patches of virus-infected
plants in a crop (Harrison and Cadman 1959; Harrison et al. 1961; Taylor
and Thomas 1968; fig. 27.2). These patches recur in successive crops, and
usually expand only slowly, presumably because the vectors do not move
far. For example, in uncultivated land Harrison and Winslow (1961) estimated that the edge of an infestation of *Xiphinema diversicaudatum* advanced
only about 30 cm per year. Indeed transport of nematodes with soil during
cultivation may be as important in enlarging infestations in cultivated land
as movement of the nematodes themselves.

27.4.2 Frequency of transmission

Many factors affect the speed with which a crop becomes infected with virus when it is planted on land infested with virus-carrying nematodes. These include the availability of virus sources, the susceptibility of the crop plants to virus infection, their attractiveness to the nematodes and the density of the vector populations. However, it is not necessary that a nematode should be able to reproduce on a plant for that plant to become infected by vector-borne virus. For example raspberry will not maintain *Longidorus elongatus* populations (Taylor 1967) but nevertheless frequently becomes infected with raspberry ringspot virus. In assessing the status of a plant as a host for a nematode species it is therefore important to test whether the nematode can reproduce, can grow and moult, or can merely probe or feed sporadically on it.

Various soil factors affect nematode activity, especially soil moisture and temperature. Temperature has different effects on the transmission of different viruses by different vectors (Fulton 1962; Harrison 1966, 1969; Das and Raski 1968) though these effects are difficult to analyse because so many interacting elements are involved. Soil moisture is important because for optimum nematode mobility the soil particles should be covered by a film of water of a thickness that depends on the diameter of the nematode (Wallace 1963). Thus in a sandy soil infested with *Trichodorus* spp. carrying tobacco rattle virus, transmission ceased abruptly when the moisture content decreased below 10%. Also, the incidence of tobacco rattle virus infection of potato tubers in Scottish crops seems related to the amount of rainfall during the period when the tubers are small, presumably because virus transmission is favoured in wet soil (Cooper and Harrison 1971).

27.4.3 Persistence of virus at sites

Many nematode-borne viruses have a wide host range including numerous weed species, and thus can survive when a succession of crops immune to the virus are grown. These viruses are perhaps best considered viruses of wild plants, that infect crop plants incidentally when they are planted on infested land. Grapevine fanleaf virus is probably an exception to this rule, because it seems rarely to infect weeds in the field, vineyard plantings are long-lived, and after a planting is grubbed living roots can remain in the soil for several years (Taylor and Raski 1964).

The viruses transmitted by *Xiphinema* and *Trichodorus* are retained long enough to pass the winter, or longer, in their vectors. By contrast, tomato black ring and raspberry ringspot viruses are usually not retained through the winter by *Longidorus elongatus*. Instead these viruses survive in perennating plants and in dormant weed seeds. Both viruses are frequently seed-borne in the field, and are reacquired by *L. elongatus* in the spring from young infected weed seedlings (Murant and Taylor 1965; Lister and Murant 1967; Murant and Lister 1967). Seed transmission also occurs with the viruses transmitted by *Xiphinema* and *Trichodorus* spp., and may play a role in their survival, but it seems less essential and occurs less frequently in the field than with the viruses transmitted by *Longidorus* spp. (Murant and Lister 1967).

27.4.4 Spread of virus into new areas

Spread of virus can occur either by the spread of virus-carrying nematodes into new areas or by the dissemination of infected plant material to sites that may or may not already be infested with vector species. As already mentioned, the vectors can travel only short distances on their own but they may also be moved over either short or long distances in soil carried on farm machinery, wheels, the feet of birds and other animals, the roots of transplanted ornamental or crop plants, etc. Most of these methods are inefficient, but the last is probably mainly responsible for the worldwide distribution of *Xiphinema index* carrying grapevine fanleaf virus. Some ecological niches serve as reservoirs of virus-carrying vectors, which spread from them into adjacent areas. For example, in parts of Britain, old woody hedgerows harbour arabis mosaic virus and *X. diversicaudatum*, which spreads from these into adjoining fields (Harrison and Winslow 1961; Pitcher and Jha 1961).

Another way in which viruses can become established at new sites involves the dissemination of virus-containing seed, particularly weed seed, to areas already colonized by virus-free vectors. This method is thought to be important for raspberry ringspot, tomato black ring and probably other viruses (Murant and Lister 1967).

This brief outline of nematode behaviour and the ecology of nematode-borne viruses indicates the main factors that must be considered in formulating control measures, of which the most successful have been the growing of disease-escaping crop cultivars, and the treatment of soil with nematicides (ch. 31).

CHAPTER 28

Other invertebrates

A. J. GIBBS

Contents

28.1 Introduction

I mentioned in the preface that most known viruses have been isolated from
a small proportion of the different phyla of organisms. The ecology of viruses
in each of the principal invertebrate phyla that are associated with viruses
has been reviewed in the preceding chapters of this section. However, viruses
and virus-like particles have been found in other phyla of invertebrates,
ranging from 'phycomycetaceous protozoa' to tunicates, and these will be
reviewed in this chapter.

28.2 Protozoa

There are several reports of viruses in protozoa. For example Diamond et al.
(1972) and Mattern et al. (1972) found transmissible virus-like agents in four
out of six cultures of *Entamoeba histolytica* isolated from human patients
with amoebic dysentry. The virus-like agents were obtained from *Entamoeba*
cultures that lysed spontaneously, they were disrupted by freezing and
thawing, added to 'healthy' *Entamoeba* cultures and caused lysis. Entamoebae
infected with the commonest agent contained isometric particles about
70–80 nm in diameter. These were found in the cytoplasm, but never in
intact nuclei, whereas entamoebae infected with another agent contained
clusters of fine filaments in their nuclei.

   Another fascinating example of a virus–invertebrate association is that of
the kappa killer factor of the ciliate *Paramecium aurelia*, and it is a story
that is still incomplete. Certain strains of *P. aurelia* contain endosymbiotic
Gram negative bacteria, which are directly or indirectly responsible for
killing strains of *P. aurelia* that do not contain the same endosymbiont.
Several of these endosymbionts are known (Beale et al. 1969), the best
known is Kappa. About a quarter of the individuals in each population of
Kappa contain one or more refractile (R) bodies. Those that contain R
bodies do not reproduce, and also contain a number of bacteriophage-like
particles, whose properties have been studied by Preer et al. (1971). It seems
that Kappa are released by paramecia and ingested by others, and those
Kappa that contain R bodies kill Kappa-free sensitive paramecia. The R
body is a rolled tape, which when unwound is 13 nm thick, 200–500 nm wide
and up to 20 µm long. The phage-like particles are isometric, tailless and
mostly 50–60 nm in diameter, and they have a molecular weight of about
250×10^6 dalton of which about 45% is DNA with a G+C content of 45%.

There has been much speculation on how Kappa kill sensitive paramecia, and the role, if any, of the R bodies and phage-like particles in the killing. Much interesting work has been done on the genetic interactions of the paramecia and their endosymbionts but little is known of the ecology of the endosymbionts and their effects on their hosts.

Virus-like particles have been found in the nuclei of an oyster pathogen *Labyrinthomyxa marina* (Perkins 1969), in *Naegleria gruberi* a free living amoeboflagellate occasionally isolated from human beings with encephalitis (Schuster and Dunnebacke 1971), and also in species of *Plasmodium* the malarial parasite (Terzakis 1969; Davies and Howells 1971). Dunnebacke and Schuster (1971) also found that extracts of *Naegleria* contain a transmissible agent that causes cytopathic effects in chick embryo fibroblasts, but did not show that the cytopathic effects were caused by the virus-like particles.

Recently virus-like particles have been found in *Aphelidium* (Schnepf et al. 1970), which is considered by some to be a phycomycete and by others a protozoon. *Aphelidium* was found parasitizing cultures of *Scenedesmus armatus*, a green alga. Some of the *Aphelidium* protoplasts inside *Scenedesmus* cells were disorganized and contained large numbers of angular polyhedral particles which were about 200 nm in diameter and resembled those of some of the mosquito iridoviruses. These particles have not been transmitted experimentally, and although it is probable that they are virus particles it is not known whether the virus infected the *Aphelidium* when it was in the zoospore stage before the *Aphelidium* attacked the *Scenedesmus*, or whether it was latent in either the *Aphelidium* or the *Scenedesmus*.

28.3 Annelids

Dougherty et al. (1963) have found abnormal individuals in axenic cultures of the microannelid *Enchytraeus fragmentosus*. In some cultures individuals developed tumour-like swellings, and in others most individuals lysed. Extracts of lysed colonies contained numerous irregular rod-shaped particles which appeared to be jointed, with one end swollen and the other narrow.

28.4 Arthropods

Virus-like pathogens have been found in crustaceans. Vago found virus-like particles in paralysed crabs *(Macropippus depurator)* from the French Mediterranean coast. The particles were either isometric and 50–60 nm in diameter (Vago 1966) or ovoid and 150–300 nm in size with an amorphous core and surface subunits packed hexagonally (Bonami and Vago 1971). The prevalence of individuals containing these particles was not reported.

Bang (1971) has also found a transmissible disease of crabs, but in *Carcinus maenas* from the Brittany coast of France. The disease was obtained from one of 679 crabs. It stopped blood clotting, and affected the number, behaviour and distribution of blood amoebocytes in the animal. Diseased amoebocytes contained isometric virus-like particles 95–125 nm in diameter.

There are other groups of insects not reviewed in previous chapters of this section, and with which viruses are associated. For example viruses closely resembling the enteroviruses of mammals and bees (chs. 4 and 22) have been isolated from crickets (Meynadier 1966; Reinganum et al. 1970) and termites (Gibbs et al. 1970). These viruses are widespread and common in both types of host in Australia, but seem not to be associated with any obvious disease in nature, though the cricket virus causes paralysis and death of young crickets in experimental colonies. The only other virus so far reported from orthopterans is a nuclear polyhedrosis of North American grasshoppers (Henry and Jutila 1966).

One of the most widespread plant viruses is tomato spotted wilt virus (ch. 4), it is common in temperate and subtropical regions thoughout the world (Best 1968; Ei 1970), and it causes mottling and mosaic symptoms in a wide range of plants. Its vectors are various species of thrips (ch. 13), and, although there is no direct evidence, it is likely that the virus multiplies in the thrips, as thrips that have fed on a virus-infected plant are not immediately able to infect healthy plants. This so-called latent period is long compared with the life span of thrips, so that only thrips that fed on infected plants when they were larvae are able to infect healthy plants, but usually not before they become adults (Sakimura 1963).

28.5 Molluscs

Rungger et al. (1971) have recently described a lethal and common virus disease of octopus in the Bay of Naples. Infected individuals develop edematous tumours in the tentacle muscles. Cells in the tumours contain groups of particles about 100–130 nm in diameter and of the iridovirus type (ch. 4). Although nothing is yet known of the epidemiology of this virus and others resembling iridoviruses, it is notable how prevalent they are in hosts with moist environments.

28.6 Tunicates

Virus-like particles of the enterovirus type (ch. 4) are widespread but not plentiful in extracts of the cunjevoi, an Australian littoral tunicate *(Pyurus stoloniferum)*. Attempts to transmit these particles by injection into the haemocoel and increase their numbers, as can be done with bee enteroviruses (ch. 22) have been unsuccessful (Gibbs and Macdonald, unpublished results).

Triptych; the control of viruses and invertebrates

CHAPTER 29

Control of invertebrates by viruses

L. BAILEY

Contents

29.1 Introduction

A newcomer to entomology might consider that the use of diseases to control insects and mites has become possible only recently, because he will find many review articles and research papers published in the past few years that might well impress him with their optimism, especially about novel virus diseases. However, the announcements of 'new viruses', often reported 'for the first time', reflect increasing research, not a new natural phenomenon. They are reminiscent of similar excitements half a century or more ago about novel bacterial pathogens. The idea of biological control of pest insects with diseases came even earlier than this, after the work of Pasteur on serious diseases of the domesticated silkworm; it was also much stimulated by the disasters caused by disease to laboratory cultures of field insects, and by the discovery of diseases in the honey bee. These facts were indisputable evidence that insects suffer diseases, and they continue to be given as examples by some who advocate the use of pathogens to control insect pests. However, they are three of the worst illustrations that can be selected to suggest that diseases can adequately control populations of insects in the field. The silkworm, *Bombyx mori*, is a domesticated insect – the only domesticated insect species – and may well be more susceptible to diseases than wild types, possibly because it lacks the genetic diversity of natural populations that allows resistant strains to appear quickly, but especially because of the way it is kept confined by man. Similarly, laboratory populations of field insects are more susceptible to their enzootic diseases than natural populations because their unusually crowded conditions much increase chances of contagion. Honeybees, admittedly, are field insects, no less than insect pests of crops, but, contrary to popular belief, they are usually strikingly resistant to the worst effects of their many known pathogens, especially when they are in environments where they can prosper (Bailey 1968). They have been scrutinized, as have silkworms, for disease far more than other insects, and the many parasites that have consequently been found are considered seriously because the insects are of direct benefit to man. Accordingly, any losses they suffer are important; but they are usually slight when compared with the devastating effects required to control insect pests successfully.

   None of this is to deny that diseases take a substantial toll of pest insects, or that pathogens may be applied in artificial ways to good effect, but, in case hopes for pest control with viruses and other pathogens go beyond the bounds of possibility, an assessment of the achievements and the difficulties experienced so far might be useful.

Early enthusiasm for control of insects by bacterial pathogens, which eventually had some success in the field, was soon subdued by the striking achievements with synthetic insecticides from about 1940. But interest in diseases of invertebrates continued, owing much to the enthusiasm of the late Edward A. Steinhaus (1947, 1949, 1963) and, so far as viruses are concerned, to the demonstration with the electron microscope by Bergold (1947) of virus particles within the long recognized polyhedral bodies in insects killed by the 'polyhedroses'. More recently, the association of viruses with molecular biology has given invertebrate virology a fashionable interest, but the revival about biological control owes much to concern about the rapid development of strains of some pests resistant to insecticides and, perhaps most, to the fear of the harmful effects of these materials on the environment, e.g. 'The most important factor in favour (of insect viruses for biological control) is the necessity to reduce environmental pollution' (Cameron 1968).

There have been outstanding and permanent successes from time to time during the last seventy years or so with the use of some parasitic and predatory insects against insect pests and weeds, under certain conditions (see DeBach 1964). These successes in biological control have supported hopes for similar results with viruses, although, more recently, much of the effort has been devoted to attempts to use viruses more in the manner of insecticides than as agents of biological control in the usually accepted sense. The use of viruses to control insect pests may conveniently be considered from these two points of view as follows: 1) their use in the manner of biological control either a) by reintroducing them to exotic pests that have escaped virus-free from their native environment to regions where they have become pests; or b) by introducing exotic viruses into native pests; 2) their application as insecticides. Before discussing either of these applications it may be instructive to consider first the effects of spontaneous enzootic infections.

29.2 *Enzootic virus diseases*

Pest insects are sometimes killed spontaneously by enzootic virus infections in the field. Table 29.1 lists some of these: there are many other reports of enzootic infections in insect populations but they contain no quantitative information. The examples are too few for comparisons to be made safely between different kinds of viruses, but it seems that the severest infections are by nuclear polyhedrosis viruses, which sometimes destroy well over half the populations. However, this is not enough to control them. When, for ex-

TABLE 29.1
Examples of the incidence of enzootic virus diseases

Host	Locality	Type of virus	infection (%)	Reference
Panonychus citri (Citrus red mite)	California	non-occluded	25	Shaw and Beavers 1970
Chironomus plumous (midge)	Wisconsin	iridescent	40	Stolz et al. 1968
Chironomus luridus (midge)	Germany	insect-pox	20–40	Huger et al. 1970
Trichoplusia ni (cabbage looper)	California	cytoplasmic polyhedrosis	40	Vail and Gough 1970
Malacosoma alpicola (alpine tent caterpillar)	Switzerland	nuclear polyhedrosis	53	Benz 1962
Hemerocampa pseudotsugata (Douglas fir tussock moth)	California	nuclear polyhedrosis	61	Dahlsten and Thomas 1969
Porthetria dispar (gypsy moth)	Connecticut	nuclear polyhedrosis	80	Doane 1969

ample, a pair of insects has twenty offspring, which is very few for most species, then 90% of the offspring must die, before they can reproduce, if the population is to remain the same size. The death of only 80% would allow the population to double at each generation.

29.2.1 *Effect of host population density*

Some of the observations recorded in table 29.1 were on insect populations that were unusually dense, and there have been several other reports suggesting that the percentage of infected individuals is positively correlated with the population density. For example, an epizootic of a granulosis in *Eucosoma griseana*, the larch bud moth, coincided with the peak of the population density of the insect, which had been multiplying rapidly and consistently for five years in Switzerland (Martignoni 1957). Other examples include nuclear polyhedroses of *Colias eurytheme*, the alfalfa caterpillar (Thompson and Steinhaus 1950), and of sawflies (Bird 1955; Neilson and Elgee 1968), non-occluded virus diseases of *Panonychus ulmi*, the European red mite, in Canada (Putman 1970) and of *Panonychus citri*, the citrus red mite in California

(Shaw et al. 1968b; Shaw and Beavers 1970) and three kinds of occluded virus disease in *Pseudaletia unipuncta* in Hawaii. The incidence of nuclear polyhedrosis of *Neodiprion sertifer*, the European pine sawfly, is proportional to the number of colonies, which are the infectible units, of larvae per tree (fig. 29.1).

Sometimes the effect of crowding is slow to appear; for example, the nuclear polyhedrosis of *Malacosoma disstria*, the forest tent caterpillar, does not become epizootic in Canada till 4–6 years after the first year of dense populations (Stairs 1964a). Sometimes, the effect is strikingly localized; for example Benz (1962) observed that the infected individuals in a single population of *M. alpicola*, the alpine tent caterpillar, were found on mountain slopes only above 1850 m where the population had become most dense (fig. 29.2): the density of the residual population of uninfected individuals was similar to that of the virus-free part of the population at lower altitudes.

By contrast with these epizootics in dense populations a nuclear polyhedrosis of *Spilonota ocellana*, the eye-spotted bud moth, is rare in apple orchards of Nova Scotia because the insects are sparse and solitary (Jaques and Stultz

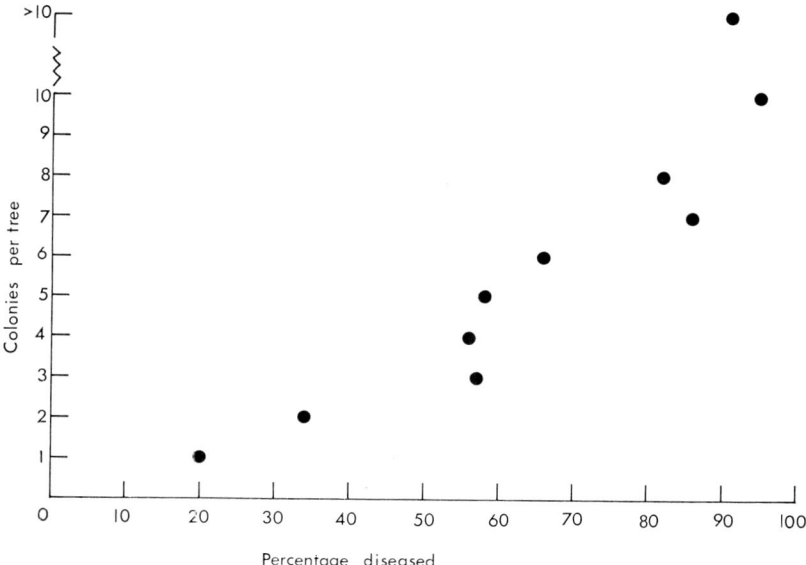

Fig. 29.1. The percentage of colonies of *Neodiprion sertifer* infected with nuclear polyhedrosis virus. (Each value was derived from the number of colonies on 22–76 trees.) (From Bird 1961.)

1966). Probably there are many similar examples that, because of their nature, go unnoticed.

There has been much controversy about the responses of populations of parasitic insects and predators that attack insect pests to the densities of their host populations (see DeBach 1964). This is no reason for supposing that insect-virus relationships are similarly difficult to analyse, but many workers seem to wish to make a mystery by classifying crowding as 'stress', a term often favoured as an explanation of disease outbreaks when no obvious cause can be found. It seems clear, however, that when viruses are transmitted by eating food contaminated with virus shed by infected individuals, as viruses pathogenic for insects often are, then crowded populations that are enzootically infected will be more likely to suffer epizootics than those that are uncrowded: the insects multiply till a previously unnoticed infection can spread quickly by contact, as for example the nuclear polyhedrosis of *Diprion*

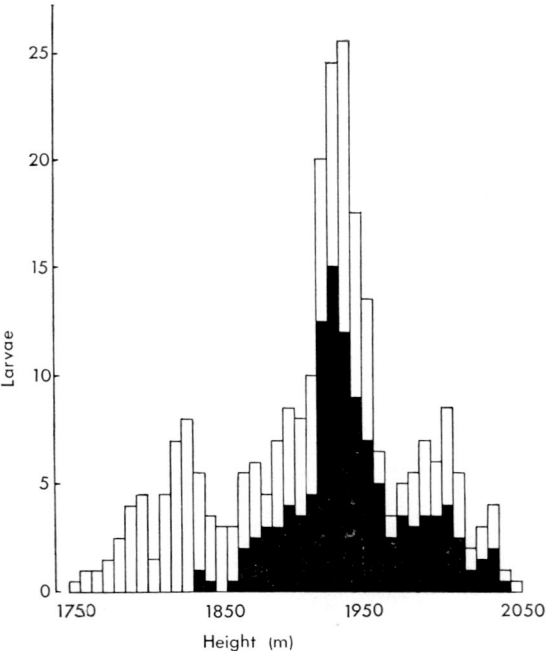

Fig. 29.2 The number of larvae of *Malacosoma alpicola* per 10 m² on alpine slopes at different altitudes. Solid columns indicate the number of larvae with nuclear polyhedrosis. (From Benz 1962.)

hercyniae, the European spruce sawfly (Bird 1961; Neilson and Elgee 1968). As a result of crowding, insects in a group may simultaneously run short of food; for example, epizootics of nuclear polyhedrosis occurred among populations of *Trichoplusia ni*, the cabbage looper, when they had destroyed most of their host plants (Wolfenbarger 1965). Such epizootics are not necessarily caused by food shortage, as is often supposed.

A simple, but interesting, exception to the association of disease outbreaks with dense populations is when polyhedrosis viruses persist for years in soil and accumulate there, contaminating low growing plants successfully enough to infect a new generation of pests each year before their populations become very dense, such as occurs with *T. ni* on cabbage (Jaques 1970a). Crop rotation will diminish the effect of this cumulative kind of enzootic infection.

Epizootics that depend on crowding and the ecology of which may not be simple include one of *Panonychus citri*, the citrus red mite, which seems to be more susceptible to a particular non-inclusion virus when the virus is fed to it, and it is placed in colonies of healthy mites rather than when it is kept alone or with only one other mite (Gilmore and Munger 1965). Similarly, larvae of *Trichoplusia ni* were more susceptible to a nuclear polyhedrosis virus when fed in groups of five than when kept alone (Ignoffo and Garcia 1969). The reasons for these effects are unknown, but to call them stress does not advance our understanding of them.

29.2.2 *Other factors*

Clearly, factors additional to virus infections are required to keep insect populations small. Benz (1962) considered that a nuclear polyhedrosis of *Malacosoma alpicola* was important but pointed out that Hymenopterous parasites and rodents are also needed effectively to control the insects. Holling (1959) concluded that in Canada populations of *Neodiprion sertifer*, the European pine sawfly, increased in spite of a nuclear polyhedrosis virus, when not attacked by small mammal predators, which eat the pupae on the forest floor. The mammals themselves, however, could not cope with the very dense populations of the sawfly that existed before the nuclear polyhedrosis was introduced in 1952 (see below). In other words, several kinds of enemy are necessary in combination to control insect pests. The multiplication of parasitic insects, many of which have only one generation per year, lags behind that of the host, but the parasites can very efficiently seek out their prey when these are few. Small mammals can reproduce quickly, increasing rapidly with prey densities, and also have a so-called 'functional response' when they eat more than usual of a particular prey when its population density increases (Holling 1959). Pathogens such as viruses have no searching abilities or responses, so their effectiveness is most simply related to host densities, and as they multiply quickest of all the enemies, they are potentially able to spread fastest between crowded individuals or crowded colonies of individuals. Each link in the chain of parasites, predators and pathogens may well be necessary to effectively minimize the population density of an insect pest. When the principal limiting factors are few and are only, for example, food and pathogens, the population may fluctuate wildly and population crashes can then be followed by even more serious outbreaks of the pest because parasitic insects and predators are greatly depleted (e.g. Ullyett and Schonken 1940), in the same way that insecticides can cause 'pest upsets' (see DeBach 1964).

29.3 The application of viruses

29.3.1 Exotic pests

The successful control of the European spruce sawfly, *Diprion hercyniae*, in Canada marked the beginning of intense interest in the use of viruses for the biological control of insect pests. The sawfly became well established in Canada about 1930 and its nuclear polyhedrosis was discovered there about 1936, first in laboratory cultures and then in the field. The virus was probably imported from Europe accidentally with various parasitic insects between 1934 and 1939 in attempts to control the sawfly. The incidence of this spontaneous infection was then estimated by Balch and Bird (1944). Almost all the larvae sampled in 1939, 1940 and 1941 were diseased and, simultaneously, the density of the insect population fell dramatically (fig. 29.3). Balch and Bird (1944) very reasonably concluded that their findings illustrated 'fairly clearly the importance of the disease in bringing about the end of the outbreak' of the sawfly. The disease must have had great effect, but the importance of other factors has often been overlooked.

According to Prebble (1941), populations of *D. hercyniae*, which has two generations each year in New Brunswick, decreased there in 1939 and 1940 from 'a combination of circumstances, all unfavourable to the insect'. For obscure reasons in these years, unusually few pupae of the first generations, which are less severely attacked by polyhedrosis than the second (Bird and Burk 1961), went into diapause, which can last up to seven years or more. The increased second generations were severely attacked by polyhedrosis and, as the virus is transmitted through the winter within the cocoons of infected diapausing pupae (ch. 22 sect. 1.1), unusually many of these may well have provided foci of infection early the next year.

Balch and Bird (1944) made it very clear that 'by 1938, several species of parasitic insects had become well established'; and Bird and Elgee (1957) stated that of many parasitic species brought from Europe during 1934 to 1939, two were multiplying rapidly and exerting considerable control of sawfly populations by 1938; and they 'might have raised the average generation mortality from the already estimated 85–95% to the 97.5% necessary to give a stable population'. Seemingly the virus did this instead, but the two established parasitic insect species were 'apparently eliminated as control factors during the severe virus epizootics' (Bird and Elgee 1957) and have not been seen since. Fortunately, two other species of parasitic insects became established and since 1944 'appear to have been chiefly responsible for maintaining sawfly populations at an extremely low level until 1951' (Bird and

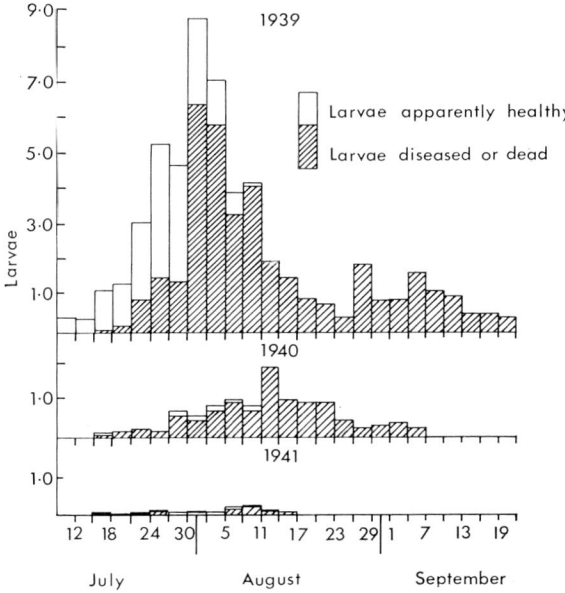

Fig. 29.3. The number of 5th instar larvae, with and without nuclear polyhedrosis, of *Diprion hercyniae*, collected per unit area beneath trees in New Brunswick. (From Balch and Bird 1944.)

Elgee 1957), and they also seemed to be mostly responsible for controlling the comparatively small outbreak of sawflies that occurred then, because the polyhedrosis did not spread. Bird and Elgee thought that this failure of virus to spread in 1951 may have reflected an increased resistance of sawflies to infection, but later evidence does not support this view (Neilson and Elgee 1960).

Later Bird and Burk (1961) found an area of 4 square miles of spruce near Sault Ste. Marie, about 100 miles west of the known distribution of *D. hercyniae*, that was infested with the sawfly. The insects seemed free from virus and 'practically free from introduced parasites'. About 10^9 polyhedra were sprayed on each of seven trees in 1950 and surveys were then made till 1959. The virus became established and spread to other trees, but the sizes of the samples of sawflies, which reflected the population density, at the end of the period were as great as at the beginning (fig. 29.4). Moreover, these samples suggested that the population density of sawflies was similar to that in New Brunswick in 1939, when polyhedrosis apparently began to decrease the population to the low levels reached by 1941 (fig. 29.3). Bird and Burk could not explain why the sawfly populations at Sault Ste. Marie were not diminished. They believed that the virus decreased the population to levels that did not cause 'serious defoliation' although the percentage of sawfly larvae

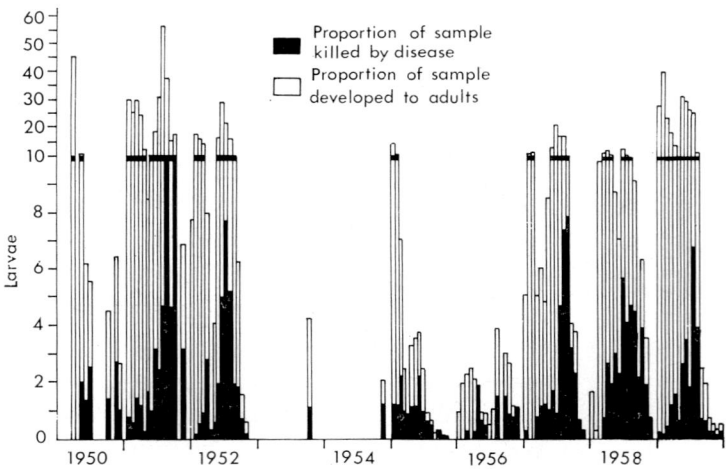

Fig. 29.4. The number of larvae, with and without nuclear polyhedrosis, of *Diprion hercyniae*, collected per unit area beneath trees at Sault Ste. Marie. (From Bird and Burk 1961.)

killed by the virus was usually very much smaller than the 97.5% estimated to give a stable population (fig. 29.4). This may have depended on predation of cocoons by small mammals, which Holling (1959) and others have shown to be important and about which Bird and Burk had no information. The absence of insect parasites would seem to have been a serious deficiency in the ecosystem in view of their great effect in areas such as New Brunswick where the sawfly had become well established.

Another species of sawfly, *Neodiprion sertifer*, the European pine sawfly, was accidentally taken to N. America, seemingly without its nuclear poly-hedrosis virus, and was discovered first in New Jersey in 1925. First reported in Canada in 1939, the sawfly seemed to be free from disease, so the nuclear polyhedrosis virus of *N. sertifer* in Sweden was imported to Canada in 1949 and sprayed from aircraft or smeared on tree trunks, from where it spread. However, the virus does not spread without help readily enough to control the insect, which continues to cause severe damage unless the virus is disseminated artificially (Bird 1953, 1955).

The only other documented example of biological control of exotic pests with viruses is the work of Marschall (1970) who introduced the virus discovered by Huger (1966) in Malaya, of *Oryctes rhinoceros*, the rhinoceros beetle, to Western Samoa. Here the beetle has become a serious pest of coconut palms since its fairly recent introduction to the islands of the South Pacific. Marschall, who had found no virus 'during 5 years of investigation of the beetle population', macerated 1500 grubs killed by the virus in decomposed sawdust and then distributed this over rotting coconut logs, where the beetles multiply, in two islands. He collected the first diseased larvae 18 months later when, at the same time, he observed a 'conspicuous decrease', presumably compared with other islands, of damage to palms in some areas. After two years, between 57.5% and 73% of grubs collected in the field died of virus. This is a substantial proportion of the 90% mortality required to maintain a stable population, if the female beetles lay about 20 viable eggs, as they do in other parts of S.E. Asia (Goonewardena 1958).

29.3.2 Exotic viruses

Nothing equivalent to the spread of myxomatosis among populations of the European rabbit has been known to occur among insects, and little interest has been shown in the possibility. Perhaps most workers are discouraged by the apparent host specificity of virulent viruses or the comparative lack of virulence of the less specific ones, such as the cytoplasmic polyhedroses. More important perhaps from the aspect of spontaneous biological control, is that as most insect pathogens are not vector borne, very virulent insect viruses would be self-eliminating as they would kill their hosts before these had time to wander, or become adult and transmit virus to their offspring. The results of the few investigations that have been made with exotic viruses have not been encouraging. Chautani et al. (1968) tested a nuclear polyhedrosis virus of *Heliothis armigera*, the cotton bollworm, from the Sudan and the Ivory Coast against *H. zea* in Texas but the native virus was significantly more virulent than the African strains. David and Gardiner (1965) found strains of granulosis virus from the Canary Islands and Britain were equally virulent for *Pieris brassicae*, although *P. brassicae* from the Canaries were more suscepti- ble than British strains of the insect to virus from either source. Ossowski (1960) found the nuclear polyhedrosis virus of *Heliothis armigera* less effec- tive when fed to *Kotochalia junodi*, the wattle bagworm, in Africa than a similar local virus of the bagworm, although virus of *K. junodi* from distant plantations killed significantly more bagworms than did local viruses. He may, however, have been dealing with different strains of bagworms (Came- ron 1967). Stairs (1964b) found that nuclear polyhedrosis viruses from *Mala- cosoma* spp. in N. America were all about equally virulent for *Malacosoma disstria* in Ontario, but a similar virus from *M. alpicola* in Europe killed *M. disstria* more slowly than the American strains. Tanada and Hukuhara (1968) found an Oregon strain of granulosis virus more infective for *Pseudaletia uni- punctata*, the armyworm, also from Oregon, than a similar virus from the same insect species in Hawaii. All the observations suggest that exotic strains of virus are usually less infective than native strains and that much effort might be needed to find unusually virulent types. Unsuitable though these might be for the spontaneous spread without vectors among insect popu- lations, they would be a great attraction for those who wish to use viruses in the style of insecticides.

29.3.3 The artificial dissemination of enzootic viruses

When classical biological control methods are used some pest damage must be expected. When this is unacceptable, the only solution is to destroy the pest, and many attempts have been made to do this using enzootic viruses as insecticides. The nuclear and cytoplasmic polyhedroses have been tried most because, apart from being the best known, they seem to stay infective for long periods within their crystalline protein inclusion body or capsule. For example, the nuclear polyhedra of *Diprion hercyniae* stay infective for many years in the air-dried cadavers of their host (Bird 1955; Neilson and Elgee 1960); nuclear polyhedra of *Neodiprion swainei* resist oil and benzene, which are used in the formulation of fluid suspensions suitable for spraying (Smirnoff 1961); polyhedra of *N. sertifer* stay infective in stomach contents of birds (Bird 1955); and polyhedra of *Trichoplusia ni* kept in wet soil out of doors maintain all their infectivity for two years and retained 25% after five years (Jaques 1964, 1967a, b), and they were unaffected by proteolytic bacteria and fungi (Jaques and Huston 1969). Granulosis particles are similarly resistant: those of *T. ni* lost no infectivity after two years in water at 4 °C (Paschke et al. 1968) and those of *Pieris brassicae* lost very little infectivity after two years in soil or sand in an unheated shaded greenhouse (David and Gardiner 1967).

Spray techniques, with sticking, 'surfactant' and spreading agents, similar to or the same as those suitable for insecticides, have been used to good effect with virus preparations (Akutsu 1967; Bird 1953; Wolfenbarger 1965). Crude extracts of diseased insects seem more efficient than purified virus preparations (Ignoffo 1964; Magnoler 1968), perhaps because they are protected from damaging light by coloured and solid impurities (see sect. 4.1) and possibly because they attract insects as do plant extracts. For example, Montoya et al. (1966) found in laboratory and field tests that extracts of parts of plants such as green beans, tomatoes and corn attracted *Heliothis zea* and *H. virescens* when they were mixed and sprayed with viruses, which then killed more caterpillars than when sprayed alone. Allen and Pate (1966) obtained similar results, increasing the kill of *H. zea* from 20–48% on cotton leaves by mixing an extract of corn silks and kernels with the virus suspensions.

The entire crop must be sprayed with viruses to get the quickest results, but attempts have been made to find cheaper methods of distribution. Smirnoff (1961) for example, found it cheapest to spread a nuclear polyhedrosis of *Neodiprion swainei* by spraying small areas of forest with dilute virus; in this way many infected adults were produced, which transmitted disease to between 33% and 95% of their offspring. Martignoni and Milstead (1962) applied nuclear polyhedra in an ophthalmic ointment to the genital armature of female adult moths of *Colias eurytheme* which successfully transmitted disease to many of their offspring. Elmore and Howland (1964), however, who mechanically contaminated adults of *Trichoplusia ni* in a similar way so that they infected up to 51% of their progeny, considered that spraying virus directly on foliage was a more effective way of disseminating viruses.

Some tests have shown that results approaching or similar to those with insecticides can be achieved by spraying crops with polyhedra, especially nuclear polyhedra, and with granulosis capsules. Results of different workers cannot be compared with confidence because their preparations were not standardised, and their reports are often incomplete, especially about virus concentrations in sprays, and the volumes of spray applied per crop unit.

Table 29.2 shows some results with nuclear and cytoplasmic polyhedrosis viruses and with granulosis viruses. Cameron (1967) said that virus from about 12–25 insects was usually enough to treat one hectare, but this seems minimal, and often very inadequate according to the results of many workers.

TABLE 29.2

Results of applying viruses on insect pests in the field

Pest	Dose*	Result †	Reference
(a) Nuclear polyhedrosis viruses			
Heliothis zea (cotton bollworm; corn earworm)	$1.5–2.5 \times 10^{12}$ (250 l.e.)/ha	'compared favourably with insecticides in current use'	Ignoffo 1965
Heliothis virescens (tobacco budworm)		98% kill when sprayed on eggs and young larvae	Ignoffo 1966a
H. zea	3×10^{11}/ha	Almost as good as 'toxaphene' + DDT	Allen et al. 1967
H. virescens			
H. zea	2×10^8/g of dust brushed on corn silk		Tanada and Reiner 1962
Trichoplusia ni (cabbage looper)	1.5×10^{12} (125 l.e.)/ha	As good as 'mevin-phos' at 560g/ha	Jaques 1970b
T. ni	1.25×10^{11} + 'surfactants'/ha	Better than 'endrin'	Wolfenbarger 1965
T. ni	10^{11}/ha	No control	McEwen and Hervey 1959
Colias eurytheme (alfalfa caterpillar)	5×10^{10}/ha		Thompson and Steinhaus 1950
Mamestra brassicae (cabbage armyworm)	2.5×10^9– 25×10^{10}/ha	70–100% kill	Akutsu 1967
Orgyia pseudotsugata (Douglas fir tussock moth)	250 l.e./2 m tree	90% kill	Morris 1963
Porthetria dispar (gypsy moth)	1.25×10^{13}/ha	Significant decrease in defoliation of oak	Rollinson et al. 1965
Choristoneura fumiferana (spruce budworm)	0.6 g/3 m tree	26% kill	Stairs and Bird 1962

Table 29.2 *(continued)*

Pest	Dose*	Result†	Reference
Malacosoma disstria (forest tent caterpillar)	1.5×10^9/tree	92% kill	Stairs 1964a
Malacosoma fragile (Great Basin tent caterpillar)	10^8–10^{10}/1 m of branch	84–98% kill	Stelzer 1965
Neodiprion sertifer (European pine sawfly)	10^8 (1 l.e.)/tree 2.5×10^{10}/ha	90% kill	Bird 1953
Neodiprion swainei (jack pine sawfly)	5×10^{10} (5×10^3 l.e.)/ha	100% kill	Smirnoff 1961
Galleria mellonella (greater wax-moth)	10^8 (1 l.e.)/0.13 m² of comb	Prevented new infestations	Bailey 1964
Achroia grisella (lesser wax-moth)	10^8/0.13 m² of comb	Did not prevent serious damage	Bailey 1964

(b) Cytoplasmic polyhedrosis viruses

Pectinophorae gossypiella (pink bollworm)	4.25×10^{14}/ha		Ignoffo and Adams 1966
Dendrolimus spectabilis (pine caterpillar)	10^{11} (200 l.e.)/ha		Koyama and Kzumasa 1967
Thaumetopoea pityocampa (pine processionary caterpillar)	7.5×10^{11}/ha (larvae from 1300 nests for 320 ha)		Franz 1961

(c) Granulosis viruses

Pieris rapae (cabbage white butterfly; common cabbage worm)	300 g of diseased larvae/ha	As good as 'endrin'	Akutsu 1967
P. rapae	5×10^{13}/ha	As good as DDT + 'mevinphos'	Jaques 1970b; Fox and Jaques 1966
P. rapae	12.5 l.e./ha	No control	McEwen and Hervey 1959
Choristoneura murinana (fir-shoot roller)	2.5×10^{12}/ha	100% kill	Schonherr 1969

* Number of polyhedra or granulosis capsules/crop unit (l.e. = larval equivalent: the amount of virus from 1 diseased larva).

† Satisfactory control was claimed when no information to the contrary is given.

Most effort with a non-occluded virus has been made using a virus with isometric particles against *Panonychus citri*, the citrus red mite, in California. The most effective results were when live infected mites, either 10,000 laboratory reared individuals, or, more effectively, up to 5000 infected mites from the field were distributed on each tree. This increased enzootic infection from about 1 % to about 9 % for several months (Shaw et al. 1968a) and increased the proportion of infected females from about 25 % to about 50 % in dense populations (Shaw and Beavers 1970). Spraying the virus was less effective, probably because of its instability (see sect. 4.1). Suspensions of triturated diseased mites, at the rate of 5 µg of mite in 2 ml on each lemon or of 100 mg of mites in 1 litre on each tree were respectively, only 'marginally effective' (Gilmore and Munger 1963) or increased the proportion of infected female mites from the existing 0–13 % by only 2 % or 3 % (Shaw et al. 1968a). Shaw et al. (1969) suppressed mites by using sprays containing 50–1000 mg of diseased individuals/100 ml adjusted to pH 6.0, in greenhouses in coastal regions, but not inland, even by repeated spraying, when temperatures were higher. A virus with rod-shaped particles, of *P. ulmi*, the European red mite, in Ontario, also was ineffective when sprayed, whereas live infected mites, placed in peach trees that had been artificially infested, caused infection in 15–27 % of the established population after one month. After two months nearly all the mites of the infected populations had gone whereas plenty remained on control trees that had not been infected (Putman 1970).

Some tests have been made with a non-occluded virus, 'dense-nucleus disease' virus, of *Galleria mellonella*. Lavie et al. (1965) stated they used it successfully, although they gave no details; Giauffret (1966) said it was less effective against *G. mellonella* in bee-hives than in laboratory tests. I found that the virus from about 35 desiccated larvae dusted over 0.5 m^2 of honeycomb, which is about $^1/_{10}$ to $^1/_{20}$ of the comb occupied by a bee colony, was necessary to prevent serious damage by new infestations (Bailey 1970).

Little use has been made of the iridescent viruses despite their wide host range and the academic interest they have attracted, perhaps because none seems especially virulent. For example, Woodward and Chapman (1968) infected only 16–20 % of *Aëdes taeniorhynchus* larvae with an iridescent virus when they exposed early instars to 'substantial amounts' of virus for 48 hr, and when the insects were exposed similarly to infection in their 4th instar, only about 20 % of their progeny became diseased.

29.4 Difficulties with viruses

29.4.1 Instability

Although the infectivity of polyhedra persists for many years in some conditions (see sect. 3.3), it is severely decreased by light. For example, nuclear polyhedra of *Trichoplusia ni* lost infectivity after one month on foliage out of doors, ultraviolet light soon inactivating the virus, especially when wet (Jaques 1967a). Polyhedra of *Heliothis zea* lost infectivity on leaves in sunlight within a day (Allen 1967; Bullock 1967); those of *Trichoplusia ni* lost infectivity after 3 hr in direct sunlight and capsules of the granulosis of *Estigmene acrea*, the salt marsh caterpillar, lost infectivity within 3 hr in the same conditions (Cantwell 1967) as did capsules of the granulosis of *Pieris brassicae* within 12 hr (David et al. 1968). When granulosis capsules are exposed to sunlight in films of water they lose infectivity more slowly when the water contains coloured and solid impurities, or when the film is allowed to dry before exposure to sunlight (David 1969).

Non-occluded viruses are usually less stable than occluded ones. One with rod-shaped particles from *Panonychus ulmi* loses infectivity almost at once in water and retains some infectivity for only about a week in natural deposits from infected mites (Putman 1970). A virus with isometric particles from *Panonychus citri* stays infective for only a few hours in dried preparations; and although natural deposits from infected mites stay infective for several days, they are inactivated by freezing and thawing or by 30 min of sunlight (Gilmore and Munger 1963). Some infectivity remains in intact dead mites kept at 21–27 °C for about 28 days, but aqueous suspensions of triturated mites lost infectivity in about two days. Giauffret (1966) found unpurified preparations of 'dense-nucleus' disease virus of *Galleria mellonella* infective after at least 3 months in dilute honey or in pollen or on honeycomb. However, I found this virus soon lost most infectivity in similar preparations at about 18 °C. After 3 months the equivalent of one larva contained about 10^5 median lethal doses – 0.1 % of the original infectivity – by injection (Bailey 1970), and enough to kill only about 100 newly hatched larvae by mouth sufficiently quickly to prevent them causing serious damage.

29.4.2 Timing the application of viruses

The instability of viruses, especially in sunlight, makes careful timing of application essential to obtain the best possible results. Also, as insects are

very susceptible to virus infection only when young (e.g. Smirnoff 1961;
Morris 1962; Stairs 1964a; Ignoffo 1966b; Doane 1967; Bailey 1968), the
most effective time to spray viruses is when the insect eggs are hatching. The
increased resistance with age of insects to virus infections is probably largely
responsible for the shallowness of the slopes of the dose/mortality curves,
observed by Bucher (1958), for virus and other infections of several insect
species. Comparing them with the steepness of slopes for insecticides, Bucher
concluded that it is 'frequently impractical to apply doses of pathogens to kill
95–99 % of the population'.

Thompson and Steinhaus (1950) found that once damage to alfalfa by
Colias eurytheme became apparent it was usually too late to control the insect
with nuclear polyhedrosis virus, and that well supervised spraying programmes were needed to ensure virus was applied at the right time because 'one
day may mean the difference between economic loss and satisfactory control'.
Similar conclusions have been reached about other viruses, e.g. that *Neodiprion sertifer* should be sprayed at egg-hatching time, otherwise growers follow up with chemicals (Cameron 1967). Growers, moreover, 'demand 100 %
control in a short time' and the 'delayed action of *Heliothis* nuclear polyhedrosis is a limiting factor when it is used on certain plants, e.g. corn, tomatoes and tobacco' (Allen 1967). *Heliothis virescens* for example, caused
severe damage to tobacco even though all the insects were killed with nuclear
polyhedrosis virus in 13 days (Chamberlin and Dutky 1958). Damage to some
crops can be tolerated, e.g. on cotton (Allen 1967) or cattle food, but larvae
killed by viruses often stick to crops and even cattle will not eat alfalfa when
it is contaminated with dead caterpillars (Cameron 1967).

Another kind of timing problem is posed when introducing live virus-
infected individuals into populations of *Panonychus citri* (see sect. 3.3) because infected mites cease to transmit virus within two days after a five day
incubation period (Gilmore and Tashiro 1966).

29.4.3 *Effects on parasites*

The elimination of some parasitic insects from populations of *Diprion hercyniae* by an epizootic of nuclear polyhedrosis has already been mentioned
(sect. 3.1). Thompson and Steinhaus (1950) found that *Apanteles* spp. of
parasitic insects killed many young larvae of *Colias eurytheme*, so, as nuclear
polyhedrosis virus had to be applied to such larvae to be effective, 'the possibility of control by *Apanteles* is forfeited'. Harcourt (1966) observed that a
granulosis of *Pieris rapae* in Canada killed the caterpillars before Tachinid

parasites in them could complete their development, and Steinhaus (1954) refers to similar parasites that do not attack *Lymantria monacha*, the nun-moth, when this is already infected with polyhedrosis virus. Kaya (1970) found that a granulosis virus from Hawaii caused proteinaceous material to form in *Pseudaletia unipunctata* that was harmless for the caterpillar but that killed the larvae of insect parasites within its body.

29.4.4 *Resistance of insects to virus infections*

Advocates of the use of viruses as insecticides emphasize the lack of evidence that resistant strains of insects have been selected (Cameron 1967). Allen (1967) found strains of *Heliothis* showed no increase in resistance to nuclear polyhedrosis after they were exposed to infection for 18 generations; and Ignoffo (1971), who said that resistance to insecticides is induced by exposing 15–20 generations of insects to the chemicals, considered there was only frag-mentary evidence showing that equivalent resistance to 'microbial insecti-cides', including viruses, could occur, and that it took 30–40 generations to develop. The infectivity of the nuclear polyhedrosis virus of *Diprion hercyniae* remained constant for at least 10 years after it became widespread in Canada (Neilson and Elgee 1960), although the continued sensitivity of this sawfly is not surprising because it reproduces parthenogenetically: such insects do not show increasing resistance towards insecticides (Helle 1968). However, the nuclear polyhedrosis of *Neodiprion sertifer* in Canada was as infective in 1955 as it was in 1949 when it was introduced (Bird 1955), and Franz (1961) found the infective dose of the virus the same for *N. sertifer* in Germany, where the sawfly and virus are native, as in Canada (Bird 1953).

Nevertheless, there is plenty of evidence that individuals of field popu-lations of some insect species differ in their resistance to virus infections. Mar-tignoni (1957) found the LD_{50} for larvae of *Eucosoma griseana*, the larch bud moth, in Switzerland 1 year after an outbreak of granulosis was about 40 times the LD_{50} at the time of the outbreak; and Martignoni and Schmid (1961) observed a similar change during the decline of populations of *Phryga-nidia californica*, the Californian oakworm, that were enzootically infected with a nuclear polyhedrosis virus. Resistant strains of insects have also been selected in the laboratory (David and Gardiner 1960, 1965; Watanabe 1967).

In view of the common occurrence of virus infections among field insects (sect. 2) their populations probably always have many resistant individuals and further striking selection for resistance will not happen readily. For example, most individuals in honey bee colonies are resistant to enzootic viruses when these are applied artificially (Bailey 1971).

29.4.5 Culture of viruses

The production of much of a particular insect virus still needs very many insect hosts. Although simple in principle, especially now that semi-synthetic foods for several species can be used, the rearing of insects on an industrial basis is not as easy to manage as the manufacture of insecticides. Moreover, the specificity of viruses means that different hosts and, therefore, different methods are needed for different viruses. The yield of virus in some instances would clearly be unprofitable, e.g. Matta and Lowe (1970) could multiply their inoculum of an iridescent virus of an *Aëdes* sp. by a factor of only 3 at the most.

Insect tissue can be cultured in vitro but, again, the yield of virus from cultures is small. In fact, the mass production of viruses has 'yet to be commercially demonstrated' (Ignoffo 1971). Perhaps novel hosts, easily and cheaply grown, such as fungal cells, may be induced to support the multiplication of insect viruses. It has been claimed, for example, that the nucleic acid of silkworm polyhedrosis virus will multiply in mammalian cells and produce polyhedra and virus particles that are indistinguishable from those in the silkworm (Himeno et al. 1967).

29.5 Conclusions

Strong beliefs prevail about infectious diseases, because we fear their effects on ourselves, our crops and our animals. However, the magnitude of these effects rarely approaches the destruction that must be done to populations of insect pests so that they do not seriously damage crops. It may well be unreasonable, therefore, to expect to control insect populations easily with viruses. Should it become possible to produce viruses very cheaply, then enough might be spread to control some pests; but the effects, even with 'biological' agents, can then have undesirable consequences, no less than with insecticides, of depriving parasitic insects of their hosts, and allowing other pests, immune to the viruses, to multiply in place of the ones destroyed.

Viruses may perhaps be used with advantage on crops where some damage is tolerable and where insecticides are best avoided. However, enzootic infections by viruses are important in helping to suppress unusually dense populations of insects, and may thereby enable other biotic factors, that work well on less dense populations, to become useful. The application of insect viruses may, therefore, continue to be most rewarding in combination with other agents in classical biological control programmes, when they help to suppress pests that have broken free from their native diseases in a new environment.

CHAPTER 30

Control of viruses spread by invertebrates to animals

C. E. GORDON SMITH and G. SURTEES

Contents

30.1 Introduction

The general principles governing the transmission of arthropod-borne virus diseases are similar to those which apply to other infections but the interpolation of an arthropod host between successive infections in vertebrate hosts adds complexity. Susceptibility to infection with a particular virus varies with species, both arthropod and vertebrate, and often only a few species of each are susceptible. Whether or not one of a range of susceptible species actually plays an important role in the transmission of a particular infection in a specific area depends critically on its overall ecology and in particular on the interacting patterns of behaviour and population dynamics of all the species involved. The hosts, vertebrate and arthropod, can be classified conveniently into 'maintenance hosts' on which the infection depends for its continued existence in the area, and 'incidental hosts' which become infected but play no important role in maintaining the infection (Smith 1964). Rates of transmission depend on the frequency of 'effective contact' between infected and susceptible hosts, contact with a refractory or immune host is ineffective. Effective contact between an infected vertebrate and a susceptible arthropod host depends on viraemia of a sufficient intensity in the former during the taking of a blood meal. The probability of such contact is dependent on the duration of effective viraemia and on the number of arthropods biting. The role of the arthropod may be passive (mechanical) in which it acts merely as a 'flying pin' carrying the virus on its contaminated mouthparts (as with myxomatosis in rabbits), or active (biological) when the virus is ingested and invades the arthropod's tissues including the salivary gland so that virus is secreted when a further blood meal is taken. The time taken for this process from infection to infectivity in the arthropod is known as the extrinsic incubation period and its duration is dependent on the environmental temperature because virus multiplication rates are temperature dependent. This period also differs with species (Chamberlain et al. 1954): a 7-day difference was found between individual strains of *Culex pipiens fatigans* and *C. p. pipiens* infected with St. Louis encephalitis virus.

The arboviruses are a heterogeneous collection of some 320 viruses, not all of them normally or perhaps even occasionally transmitted by arthropods (Yaru*, 1969). With a few possible exceptions (e.g. dengue, urban yellow fever) they are zoonoses, infections maintained in a vertebrate other than

---

* The information, obtained from the Yale Arbovirus Research Unit, 1969 Annual Report, is quoted with the kind permission of the Director.

man which when transmitted to him, cause a disease. Very many of them are, however, 'viruses in search of a disease' and the pathogenicity of many of them for man is unknown. The arboviruses comprise 41 groups based on antigenic relationships (Casals 1957) together with some 60 viruses not yet grouped with others. 28 groups contain viruses known or believed to be transmitted in nature by mosquitoes, 9 groups contain known or probable tick-borne viruses, 3 groups Culicoides-borne viruses, 2 groups Phlebotomus-borne viruses and 1 group may be mite-borne. From the point of view of human disease the more important groups are* A (20 viruses all mosquito-borne), B (26 mosquito-borne, 8 tick-borne, 8 unknown), C (11 mosquito-borne), Bunyamwera (BUN: 14 mosquito-borne), California (CAL: 10 mosquito-borne), arena-viruses (AR: 9 possibly mite-borne), Phlebotomus (PHL: 15 Phlebotomus-borne), and Congo (CO: 2 tick-borne); the most important groups causing veterinary diseases are A, B, vesicular stomatitis, (VSV: 6 mosquito-borne), African Horse Sickness (AFS: 3 *Culicoides*-borne) and the ungrouped Nairobi sheep disease, African swine fever and Rift Valley fever viruses. Arbovirus infections are worldwide stretching from Finland in the north to the Murray Valley of Australia in the south. Of zoogeographical regions, 45 viruses have been found in the Nearctic, 114 in the Neotropical, 88 in the Ethiopian, 45 in the Palaearctic, 46 in the Oriental, and 29 in the Australasian. Viruses in groups A and B are known in all regions: Bunyamwera group viruses in all except the Australasian: California group viruses in all except the Australasian and Oriental Regions. Group C is so far confined to the New World, so far as virus isolations are concerned. Disease is advantageous to a parasite only when it increases the probability of effective contact (e.g. cough in a respiratory infection). In their vertebrate hosts, arthropod-borne viruses rely only on viraemia to ensure that subsequent transmission is successful. There is no selection in favour of disease-producing host-parasite relationships. Thus, arboviruses often cause no recognizable disease in their maintenance hosts; disease is usually confined to incidental vertebrate hosts which (except in a few infections) include man. With the exception of one laboratory relationship (Mims et al. 1966), there is no evidence that arboviruses cause disease in their arthropod hosts. Arbovirus diseases in man range from a mild febrile illness, with or without rash, to severe and commonly fatal encephalitis or haemorrhagic fever. Many of these diseases are biphasic with a mild febrile illness (often unrecognised)

---

* Throughout the text the group of a virus will be indicated in parenthesis after its name.

as the first (viraemic) phase; this may or may not be followed by one of the more serious types of illness by which time viraemia may have ceased and immunological responses including antibody production have occurred. In any large arbovirus epidemic predominantly of a febrile illness (e.g. Venezuelan encephalitis (A) or West Nile Fever (B)) there is usually a small proportion of patients who develop encephalitis. A high proportion of infections with many arboviruses are however inapparent: for example in Korea, between 600 and 1000 people are infected with Japanese encephalitis virus (B) with little or no illness for each one which develops encephalitis (Chang et al. 1965).

The main veterinary arbovirus diseases are encephalitis affecting horses (A, B), sheep (B), pheasants (A), turkeys (B), vesicular stomatitis, African horse sickness, African swine fever, Nairobi sheep disease and Rift Valley fever.

30.2 *Arbovirus diseases*

30.2.1 *Mosquito-borne human diseases*

Chikungunya (A) and O'nyong-nyong (A) are very closely related viruses which have caused large epidemics of similar, generally non-fatal, febrile illnesses with sudden onset of crippling joint pains, particularly in the back, which are described by their dialect names. Chikungunya virus has caused several epidemics in Africa (Lumsden 1955; McIntosh et al. 1963) but more recently there has been a series of large epidemics of high prevalence in the major cities of South India and Ceylon: in Madras with a population of 1.8 million, there were an estimated 380,000 cases (Sharma et al. 1965). In urban situations the main mosquito responsible has been *Aëdes aegypti* (Reuben 1967) but the virus has also been repeatedly isolated from *Aë. africanus* in Africa. O'nyong-nyong virus caused an epidemic involving some 2 million people in East Africa between 1959 and 1962 (Shore 1961). It started in north west Uganda and spread across the country to Kenya, Tanzania and Malawi. In the worst areas as much as 70% of the population was affected within a short period. It was transmitted to man by the malaria mosquitoes *Anopheles gambiae* and *funestus* (Williams et al. 1965). The vertebrate maintenance hosts of these two viruses (other than man in epidemics) are unknown. Venezuelan encephalitis virus (A) has caused very large epidemics of 'influenza-like' illness and a small proportion of encephalitis cases in tropical America (Sellers et al. 1965, Sanmartin and Arbelez 1963), where evidence of infection is widespread. It has frequently been isolated from *Aëdes serratus* and *Aë. taeniorhynchus* and from both rodents and birds. In a recent epidemic in Venezuela and Colombia there were over 30,000 cases, with 200 deaths in man, and many deaths in horses and mules. This epidemic was associated with flooding and hence both increased mosquito breeding and the crowding of animals onto islands. At least in Central America the virus appears to be maintained by rodents (Srihongse et al. 1967). Eastern encephalitis virus (A) occurs over most of the east of the Americas from Argentina to the United States: outbreaks of encephalitis in man have been small and sporadic (Beadle 1966). *Culiseta melanura* appears to be the main mosquito involved in infection in birds (Wallis and Whitman 1967). Western encephalitis virus (A) is much more widespread in the Americas and has caused larger epidemics of encephalitis in the United States and Canada, mainly in the west (Beadle 1966). It has been isolated from many bird species, from *Culex tarsalis*, its main mosquito host in the western United States, and from *Culiseta melanura* in the east (Hess and Hayes 1967).

At least 4 closely related dengue (B) viruses are endemic throughout much of the tropics, particularly in Asia, the Pacific and Caribbean, and periodically cause large epidemics of the classical type of disease in subtropical or even temperate areas where populations of *Aëdes aegypti* have become sufficiently dense. Dengue viruses have recently been demonstrated in West Africa (Carey et al. 1971). An epidemic of some million cases occurred in Greece in 1927–28 and in an epidemic in Brisbane in 1905 nearly 90% of tramway workers, of hospital workers and of the staff of a large food retail concern were affected in a very short period. In 1963–4 a large epidemic swept the Caribbean islands and caused more than 10,000 cases in Venezuela (Briceno-Rossi 1964). Clinically the disease is not unlike those caused by Chikungunya and O'nyong-nyong viruses. Urban dengue is transmitted by *Aë. aegypti* but other *Stegomyia* are also involved: for example *Aëdes albopictus* in rural south east Asia, *Ae. polynesiensis* in many Pacific Islands. Recently, however, a much more serious disease caused by dengue viruses has appeared in epidemics in cities of south Asia and the western Pacific again with *Aë. aegypti* breeding in close association with man. This is haemorrhagic dengue fever which has caused large epidemics and high mortality among children under 14, first in Manila in 1953, then in Bangkok (1958 to date) and later in Calcutta (1963–4). This severe disease which seldom occurs over 15 years of age, is characterized by haemorrhages and shock (Halstead 1965, 1966) and appears to be due to multiple infection with dengue viruses with a critical interval (perhaps of the order of 6 months) between a dose of virus which sensitizes the patient and a later one which causes an immunological disaster (Halstead 1970). Dengue haemorrhagic fever has always been associated with a much larger number of cases of the classical type of illness – for example in Bangkok in 1952 there were between 150,000 and 200,000 cases of mild disease and over 8000 of severe disease (Halstead 1966). The reader's attention is drawn to the extensive discussion of this disease in the Yale Journal of Medicine Science and Biology, volume 42 part 5. In general West Nile virus (B) causes a disease similar to classical dengue. The virus is widely distributed but epidemics have been recorded only in Israel. Sporadic cases have occurred in Egypt, elsewhere in Africa and in Southern France where the disease situation has been created by a vast increase in the population of *Culex modestus* because of desalinization of surface waters by irrigated rice culture (Hannoun et al. 1964, 1969). The virus is probably maintained by *Culex* mosquitoes and birds and in Egypt it has been isolated from *Argas* ticks associated with bird colonies (Taylor et al. 1966).

Yellow fever (B) is the only arbovirus infection subject to International Health Regulations. It is enzootic in the large tropical rain forests of Africa and America maintained by monkeys in both continents, by *Aëdes africanus* in Africa and by *Haemagogus* and *Sabethes* species in America. All aspects of this disease were reviewed by Strode (1951). In urban yellow fever epidemics in West Africa and America transmission from man to man has been by *Aëdes aegypti*. *Aëdes aegypti* has also been involved in epidemics outside the tropics, in southern Europe and in the United States as far north as Philadelphia. However the situation has been more variable in Eastern Africa: in the Nuba Mountains epidemic of 1940 *Aëdes vittatus*, *Aë. metallicus* and *Aë. taylori* were thought to be responsible; in the very extensive 1962–3 epidemic in Ethiopia, *Aë. simpsoni* was mainly responsible (Serie et al. 1964) and in the west of Uganda *Aë. simpsoni* acts as a link host in banana plantations biting foraging monkeys from the *Aë. africanus* forest maintenance cycle and man from villages and towns where the virus could be transmitted by *Aë. aegypti*. In recent years there have been outbreaks in Nigeria (1951–4), Congo (1958) and Central America (1958). In Ethiopia in 1962–63 there were more than 30,000 cases and 2000 deaths, and in Senegal (1965) there may have been as many as 20,000 cases and at least 216 deaths (Chambon et al. 1967). In 1969, there were outbreaks in Ghana, Mali, Nigeria, Togo and Upper Volta, the latter probably involving thousands of cases. The disease is a severe febrile illness with serious liver damage and haemorrhages. African monkeys suffer no detectable disease but it is fatal in some South American species.

Several group B viruses epidemic in temperate areas and endemic in the tropics cause encephalitis. Most important are Japanese and St. Louis encephalitis viruses. Japanese encephalitis occurs from the far eastern USSR south to Borneo and the Indian subcontinent. Epidemics in Korea, China and Japan and other offshore islands are often annual and sometimes involve thousands of cases mainly children (Fan et al. 1962). In Japan the seasonal cycle starts in spring with intense transmission amongst nestling herons by *Culex tritaeniorhynchus* which also bites pigs; the latter act as amplifier hosts infecting a population of *C. tritaeniorhynchus* close to man (Buescher and Scherer 1959). Recent work suggests that bats may be very important hosts of Japanese encephalitis virus in Japan and they may play a role in the over-wintering of the virus (Sulkin et al. 1970). Epidemics can be predicted by about 2 weeks by monitoring the infection rate in pigs (Konno et al. 1967). In tropical areas the disease is sporadic but infections very common: in Sarawak infection rates of between 6 and 10 % per annum appear to occur in man. It is associated mainly with ricefields and *C. tritaeniorhynchus* but other *Culex* species are involved notably the predominantly pig and cattle biting species, *C. gelidus* (Hill et al. 1969; Simpson et al. 1970). St. Louis encephalitis occurs from Brazil to the United States in which it is the most important mosquito-borne disease. It is sporadic in South and Central America but in the United States an urban epidemic occurs somewhere in most years and the mortality is particularly high in the aged. In a Florida outbreak in 1962 largely affecting older people there were 200 cases and 43 deaths (Dow et al. 1964). In Houston, Texas in 1962 there were 300 cases (Beadle 1966). The disease is most prevalent in poorer areas where poor sanitation provides higher populations of the main mosquito involved, *C. pipiens fatigans (quinquefasciatus)* (Philips and Melnick 1965; Bond et al. 1966). *C. nigripalpus* which has been incriminated in Jamaica and *C. tarsalis*, the most important mosquito host in the western United States, are both bird biting species. Birds are the principal maintenance hosts: sparrows (in which infection rates closely parallel those in man) probably act as amplifier hosts (Hayes et al. 1967). In Trinidad and Panama, the virus has been isolated from *Culex, Psorophora* and *Sabethes* species.

Of the Bunyamwera viruses (Casals and Whitman 1960) three (Bunyam-wera, Ilesha, Germiston) caused febrile illnesses in man in Africa, one (Guaroa) in America, one (Batai) in Malaysia and one (Calovo) in Czecho-slovakia. Bunyamwera virus has been isolated from *Aëdes circumluteolus* and *Aë. pembaensis.* The American Guaroa virus has been repeatedly isolated from man in Colombia and Brazil (Causey et al. 1962).

Group C viruses (Casals and Whitman 1961) are widely distributed in South and Central America and 9 have been incriminated in mild febrile disease mainly in forest workers. They have been variously isolated from rodents, marsupials and a sloth and from a range of mosquito species including *Culex* and *Sabethes* (see Smith 1968).

The California group of viruses (Hammon and Sather 1966) occurs in both North America and Europe and has caused sporadic cases of human encephalitis throughout North America. They appear to have lagomorphs as their vertebrate maintenance hosts.

30.2.2 Tick-borne human diseases

In terms of human disease the most important tick-borne viruses are those in group B (Smith 1962; Clarke 1964) which in Eurasia cause diseases ranging from the Siberian form with severe paralytic encephalitis transmitted mainly by *Ixodes persulcatus*, through the biphasic and less paralytic encephalitis of Central Europe transmitted by *I. ricinus*, to louping ill transmitted in the British Isles by *I. ricinus* which for ecological reasons rarely infects man but is mainly a disease of sheep. The Central European virus has also caused outbreaks in man through the infected milk of goats (Popov 1967) infected by ticks. In central Asia another member of the group causes Omsk haemorrhagic fever (Clarke 1964; Casals et al. 1966). These viruses have a wide range of wild vertebrate hosts but ground-living small mammals are probably their main maintenance hosts (see Smith 1968). All except louping ill are associated with forests. Powassan is the comparable and closely related virus in North America (McLean and Donohue 1959; Thomas et al. 1960) and has caused sporadic encephalitis in man. 2 of this group cause disease in south Asia: Kyasanur Forest disease first occurred in Mysore in 1956 as an epidemic of about 500 cases of a disease mainly resembling typhoid fever with some 50 deaths; it was associated with a fatal epizootic in monkeys (Work 1958). The disease has since remained seasonably endemic in this area. *Haemaphysalis spinigera* is the main tick responsible for human disease (Work 1958). Langat virus is maintained by *Ixodes granulatus* and forest rodents but man is rarely infected for ecological reasons (Smith 1962). The remaining important human infections are Colorado tick fever and diseases due to Congo viruses. The former is a non-fatal but distressing disease mainly occurring in the Rocky Mountains area of North

America and transmitted by *Dermacentor* species (Florio et al. 1950).
Crimean haemorrhagic fever (Co) occurs in Bulgaria, Crimea and central
USSR and is transmitted by *Hyalomma* ticks. Congo viruses (Simpson et al.
1967) are widespread throughout tropical Africa and can cause a fatal illness
in man. They also occur in Pakistan, probably transmitted by *Hyalomma*
ticks, and Nigeria where they have been isolated from cattle, a goat, a
hedgehog and from *Hyalomma*, *Amblyomma* and *Boophilus* ticks (Causey
et al. 1970).

30.2.3 Other human diseases

Phlebotomus-borne sandfly-fever viruses have caused large epidemics of
dengue-like illness in the Mediterranean area transmitted by *Phlebotomus
papataci*. In Serbia in 1948 an epidemic affected 75% of a population of
1.2 million (Guelmino and Jevtic 1955). The disease has also been reported
from the Sudan, the Middle East, Pakistan, India and the southern USSR.
Sergentomyia species may be important in some areas.

The arenaviruses (see Smith 1968) contains the viruses of Argentinian
and Bolivian haemorrhagic fevers. Although they may be mite-borne, transmission to man, at least of Bolivian haemorrhagic fever virus, is probably
from the urine of rodents in which it is excreted for long periods. *Calomys
callosus* can be chronically infected and act as a maintenance host (Johnson
1965; Kunz 1965). Since 1958, 300 to 1000 cases of Argentinian haemorrhagic
fever (Ruggiero et al. 1964) have been reported annually, mainly from the
north east of Buenos Aires Province. The disease, previously mainly
among agricultural workers has shown signs of spread to urban areas.
Bolivian haemorrhagic fever has caused a single severe epidemic affecting a
small town (Mackenzie et al. 1964): in a population of 2500 there were 650
cases and 115 deaths.

30.2.4 Mosquito-borne veterinary diseases

Venezuelan encephalitis virus (A) causes large epizootics (with associated
human epidemics) of encephalitis among Equidae in Central and South
America. Eastern encephalitis virus (A) causes horse epizootics along the
eastern margin of America; further west, encephalitis in horses is caused by
western encephalitis virus (A). Rift Valley fever is an economically important
disease in sheep, goats and cattle in Africa (Daubney et al. 1931). An epizootic in 1950–51 in South Africa killed some 100,000 sheep and cattle and
caused 20,000 human cases.

30.2.5 Other veterinary diseases

The Culicoides-borne viruses are important and include African horse sickness and blue-tongue of sheep. African horse sickness was confined to Southern Africa until 1959 when it broke out in epizootic form in Iran spreading to West Pakistan and Afghanistan, and in 1960 to India, Iraq, Syria and Cyprus causing thousands of deaths in Equidae (Maurer 1961). In 1965–6 the same virus strain caused an epizootic in Algeria (Pilo-Moron et al. 1966). In India a mortality rate just under 50% was recorded in cavalry horses (Shah 1964). Blue tongue occurs mainly in Africa, especially in the south-east. It has spread in recent years to Cyprus, Palestine, Turkey, Spain, Portugal, Pakistan, Japan and the United States. Virus has been isolated from *Culicoides* species, *C. variipennis* in the United States. Wild animals may act as reservoirs in Africa (Andrewes 1964). Nairobi sheep disease virus causes acute gastroenteritis in sheep in East Africa, Mozambique and Botswana. It is transmitted by the tick *Rhipicephalus appendiculatus* (Daubney and Hudson 1931).

30.3 Ecology

The majority of arboviruses have been isolated from arthropods in situations where there was no evidence that they were causing any disease in any species. As disease confers no advantage on the virus, such a symbiotic relationship is probably the normal state of maintenance of these agents. Thus it is likely that many potentially pathogenic viruses, as yet unknown or not yet the cause of naturally-occurring disease, exist ecologically isolated from man. But as he ventures into their habitats, and particularly if he interferes with them, new viruses and new diseases become apparent.

Stable maintenance of an arthropod-borne virus requires a dynamic equilibrium in the whole ecosystem including climate, botanical and zoological species composition, population turnovers and behavioural patterns. Disturbance will lead to a new equilibrium, to an epizootic or to the disappearance of the infection from the area. For the infection to persist, the chain of transmission must never be broken but continue very close to an average rate such that each viraemic maintenance vertebrate infects a sufficient number of arthropod maintenance hosts that enough of them will survive long enough to bite and infect one susceptible vertebrate maintenance host. Clearly the viraemia has to persist at or above an infective level sufficiently

long for there to be a high probability of the animal being bitten by arthropods of the relevant species (Smith 1964). The number of arthropods infected
must be such that a sufficient proportion will survive the extrinsic incubation
period and subsequently bite. With Diptera which bite frequently, even if
90% survive each 24 hr, only 1 in 3 will survive 10 days. With ticks the
extrinsic incubation period is often short compared with the frequency of
biting and the next blood meal may be months or a year away. In a stable
tick population only perhaps 1 in 1000 larvae survive to adult life. The
infecting bite must be on a maintenance host so that if the 'host preferences'
are such that only (say) 50% of a species bites the maintenance host the
probability is that only 1 in 4 (0.5^2) will also bite a maintenance host at the
infective feed. The longevity of arthropods is markedly affected by ambient
temperature (as is inversely the length of the extrinsic incubation period)
(Bates and Garcia 1946; Chamberlain et al. 1954) and by relative humidity
(Lewis 1933; Ingram 1954). The latter is particularly important in the resting
places, hence the great importance of vegetation in survival. Because infection
leads to immunity or death, the infection rate in the vertebrate maintenance
hosts must be such that it does not outstrip their replacement (i.e. their
reproduction rate), so that the probability of an infected arthropod biting a
susceptible host falls below the critical level for maintenance. For instance
it has been found that urban yellow fever epidemics (where man is a temporary maintenance host) die out when between a half and two thirds of the
population has become immune. This indicates that, in this situation, from
each infectious man the probable number of potential transmissions to other
men is between two and three (Smith 1964). This rate will differ with circumstances but it is likely that the proportion of vertebrates becoming immune
and rates of reproduction are the main controlling factors in maintaining
the overall equilibrium in infections borne by Diptera. Thus in a study of
Japanese encephalitis (B) infections in pigs in Sarawak, the infection rate was
17% per month from the age of 4 months upwards but, because of the high
reproduction rate of pigs, a relatively constant level of susceptibility was
maintained among pigs (Simpson et al. 1970). The factors controlling the
equilibrium of tick-borne infections may however be different. To summarise
for Diptera-transmitted viruses, the rate of transmission of an arbovirus is
directly proportional to the number of arthropod hosts of the virus biting
each vertebrate host of the virus; to the effective duration of viraemia; to
the proportion of vertebrates susceptible and, to the proportion of arthropod
hosts biting vertebrate hosts. The rate of transmission varies inversely with
the length of the interval between arthropod blood meals. The relationships

to the extrinsic incubation period and to the longevity of the arthropod are more complex – the rate varies as $p^n / -\log_e p$, where p is the probability of the arthropod surviving 24 hr and n is the duration of the extrinsic incubation period (Smith 1964). The exponential relationships indicate the exceptional importance of these factors and that, in particular, environmental and control factors influencing arthropod longevity may have profound effects.

Because of the short life-span of Diptera, the average intervals between vertebrate infections cannot be longer than the difference between the average longevity of the arthropod and the extrinsic incubation period – this difference is seldom longer than a few days. With such a high frequency of transmission, diptera-borne infections can remain static only if the vertebrate maintenance host has a reproduction rate high enough to provide a continuous supply of new susceptible animals – most likely if the maintenance hosts are smaller species of birds or rodents. In other circumstances the infection must move continuously as a wave through the population and not return to an immunized population until it has regenerated. This was very graphically illustrated by the monkey epizootic in Central America between 1949 and 1956 in which the infection spread unidirectionally and finally burnt out at about the limit of suitable ecology because of the linear geography of the area. It may be that only the great forests of the Congo and Amazon basins are large enough to allow a primate-maintained virus infection like yellow fever (B) to wander indefinitely never reaching a dead end in terms of susceptible vertebrate hosts.

Tick-borne infections characteristically behave quite differently: ticks usually have a well defined botanical habitat which limits the distribution of the infection (Smith 1962). For example *Ixodes persulcatus*, vector of tick-borne encephalitis (B) predominates only in marshy taiga (Siberian forest) while *Dermacentor pictus*, vector of Omsk haemorrhagic fever (B) predominates in secondary forest and in the northern forest steppe. *Ixodes persulcatus* can occur in different types of vegetation if there is deep shade and constantly wet soil providing a suitable microclimate. Tick-borne infections persist for very long periods in fixed foci and the problem of availability of susceptible vertebrate hosts shrinks into insignificance because the intervals between tick blood-feeds are very much longer than with Diptera. The life-span of an infected tick may be a year or longer and infection is transmitted transstadially. Some arbovirus infections are also transmitted transovarially in ticks but not in Diptera with the possible exception of *Phlebotomus*. Tick-borne infections are probably spread from an infected to a new focus by longer-ranging vertebrates carrying either the virus or infected ticks.

The seasonal arbovirus infections of the temperate zones present the special problem of how the virus persists through the winter. Tick-borne infections probably overwinter mainly in the tick (Burgdorfer and Varma 1967) but the problem is much more controversial with mosquitoes. A number of possible mechanisms have been demonstrated.

A) Over-wintering of infected adult mosquitoes. Shestakov and Polenova (1965) have found large numbers of *Culex tritaeniorhynchus* hibernating in places with high relative humidity (storehouses, wells, basements) in the Maritime Territory of the USSR. Bellamy et al. (1967) found that under outdoor conditions *C. tarsalis* remained infective by bite for up to 8 months after infection with western encephalitis virus by feeding on chicks. Mosquitoes infected in October transmitted the infection in the succeeding January, February, April and May. It was estimated that 3–4 mosquito–chick cycles would be required to maintain the virus round the year;

B) Hibernating vertebrates – if a vertebrate goes into hibernation immediately after infection, virus multiplication and immunological responses to it are suspended but are switched on again when the animal awakes; thus viraemia may occur in the spring in animals infected in the autumn – hedgehogs, garter snakes and bats have been shown to behave in this way (Burton et al. 1966; Gebhardt et al. 1966). Recent work suggests that bats may be hosts of arboviruses, such as Japanese encephalitis (B), Rio Bravo (B), Dakar bat virus and Entebbe bat virus (Sulkin et al. 1970, and see Smith 1968);

C) Annual re-introduction in spring by viraemic migrant birds – for example Western encephalitis (A) in California or Japanese encephalitis (B) in Japan. It may be that these and other mechanisms combine but it is at present difficult to account entirely for the very rapid build up of infection in the spring.

Undoubtedly the most potent influence on the epidemiology of arbovirus infection is man's behaviour which impinges in many different ways.

30.3.1 Intrusion

Whenever man ventures into the biocenosis of an arbovirus, he risks infection. Thus there were several outbreaks of Russian spring–summer encephalitis (B) during the development and settlement of the Siberian forest (Zakorkina 1958; Groshkova et al. 1959) and today the victims of tick-borne encephalitis (B) in the mixed forests of Central Europe are foresters, 'mushroom'-pickers, and campers (Blascovic et al. 1967). Similarly it was

when the villagers gathered firewood in the forest that they were infected
with Kyasanur Forest disease (B) (Work 1958). Thus occupation, leisure
activities and the agricultural or economic development of new areas all may
carry risks of arbovirus disease in various parts of the world.

30.3.2 Agricultural activities

Most important is irrigation especially that associated with rice culture
(Surtees 1970a). Notable examples are St. Louis (B) and western encephalitis
(A) transmitted by *C. tarsalis* in California in areas otherwise arid and where
80% of the breeding places are man-made pools (Reeves and Hammon 1962).
Japanese encephalitis (B) is associated throughout its distribution with the
rice-field-breeding *C. tritaeniorhynchus* (Buescher and Scherer 1959, Macdonald et al. 1967). West Nile fever recently emerged in the Camargue where
a huge increase in its mosquito host, *C. modestus*, resulted from the desalinization of surface waters for rice-growing (Hannoun et al. 1964). In east
African rice-fields *Anopheles gambiae* and *An. funestus* breed prolifically and
are hosts of arboviruses (notably onyongnyong (A)) as well as malaria and
filariasis (Grainger 1947; Webbe 1961; Surtees 1970b). Channels in irrigated
cotton plantations in Egypt breed *An. gambiae* while *An. coustani* occupies
the same niche in west African banana plantations (Toumanoff and Simond
1956; Lewis 1966). *Psorophora confinnis* breeds in the irrigation channels of
date ranches in California (Al-Azawi and Chew 1959). Excess water from the
irrigation pastures in Nebraska breeds extensive populations of *Culex tarsalis*
and *Culiseta inornata* (Edmunds 1958) and similar conditions in Wyoming
yield large seasonal populations of *Aëdes dorsalis* (Owen and Gebhardt 1957)
and *Ae. vexans* which similarly occur in the same habitat in Czechoslovakia
(Novak 1965). Domestic animal populations may be important e.g. pigs as
amplifier hosts of Japanese encephalitis (B) or cattle and goats transmitting
tick-borne encephalitis (B) to man in their milk. Domestic animals which
are incidental hosts may also be important because of their influence on
arthropod populations. For example the outbreak of Kyasanur Forest
disease (B) from a latent forest maintenance cycle was probably mediated
by the introduction of grazing cattle (incidental hosts) to the forest and the
consequent vast increase in the population of the tick *Haemaphysalis
spinigera*.

30.3.3 Urbanization

The major mosquito-borne disease problems in urban environments arise from inadequate sewage disposal or from water storage in domestic containers in the tropics, and elsewhere from sewage systems and accumulations of water-retaining domestic rubbish (Surtees 1971). Although many new towns in the tropics are being built with piped water supplies, at least in their central areas, the consistent pattern is still that water for domestic use is collected daily from a supply point, hydrant or stream, and stored in or around individual houses (Rao 1967; Surtees 1970c). Disease problems due to *Aëdes aegypti*, host of urban yellow fever and dengue (Strode 1951), are essentially urban in that the mosquito breeds in domestic containers in close association with man (Christophers 1960, Sheppard et al. 1969). In tropical Asia and on the east African coast, rapid urban growth associated with inadequate sewage disposal has been followed by a dramatic increase in *C. fatigans* and in Bancroftian filariasis in man (see Surtees 1971). Although this species is not known to transmit a virus in these regions, very closely related species are important virus hosts in America. Mosquito-borne virus diseases are also associated with sewage disposal in more temperate areas: in the United States *Culex tarsalis*, *Aëdes vexans*, *Anopheles quadrimaculatus*, *Culiseta inornata* and *Psorophora confinnis* all breed in sewage stabilization ponds. One or more of them transmit St. Louis encephalitis (B), Western encephalitis (A), California encephalitis (CAL), Tensaw (BUN), Cache Valley (BUN) or Ilheus (BUN) viruses to man. Up-to-date sewage systems maintain breeding populations of *Culex pipiens fatigans (quinquefasciatus)* host of St. Louis encephalitis virus (Curtis 1963; Smith 1969b; Sudia et al. 1967).

Water-retaining debris is a problem the world over. With the advent of piped water supplies the opportunity arose to overcome the domestic container-breeding mosquito problem, but with rising standards of living, tinned foods became a feature of life and discarded tins now provide prolific breeding sites for mosquito hosts of yellow fever, dengue and the New-World encephalitides. In the United States, such water-retaining rubbish is an acute problem in areas of low socio-economic status. In some areas the size of *Aëdes aegypti* populations is a direct function of the number of potential breeding sites provided by man, and infection with St. Louis encephalitis virus has been closely correlated with debris-breeding *Culex fatigans quinquefasciatus* (Fletcher and Flanigan 1957; Surtees 1969, 1970c; Henderson et al. 1970).

30.3.4 Engineering

Reservoirs and hydroelectric dams may be prolific sources of mosquitoes (Carreker 1965; Van Tongeren 1965; Lewis 1966) and, like irrigation, can have an additional important effect by increasing relative humidity and hence arthropod longevity. Improper disposal of industrial waste waters may have similar effects (Horsfall 1956). An outbreak of St. Louis encephalitis in Kentucky was directly attributed to *C. fatigans (quinquefasciatus)* breeding in industrial effluent (Ranzenhoffer et al. 1957).

30.3.5 Travel

Travel has long been important in the movement of disease and its arthropod hosts. *Aëdes aegypti* is most notable among arthropod travellers, having spread east from Africa to most of south and south east Asia, as far as Hawaii and the Pacific followed by a train of dengue epidemics; and west to the Caribbean, tropical and sub-tropical America causing urban yellow fever and dengue epidemics. *Culex fatigans* was introduced into Australia in the late 19th century by sailing ship and, by the same means, *Aëdes aegypti* and *Aë. albopictus* were taken to Hawaii and *Anopheles gambiae* to Brazil. Freight trains have carried 6 different genera of mosquitoes from Mexico to North America. During a recent survey at Nairobi airport, 13 species of mosquito were taken from aircraft and one, an American species, *Aëdes taeniorhynchus*, was found on an aircraft which had started from Rome. It has also been established that the yellow fever mosquito could survive, bite and breed even in southern England during the summer (Surtees et al. 1971). Recently two notable new diseases appeared, each of which might have become more widespread than it did. The first of these was due to Marburg virus and caused 31 cases of a serious illness with 7 deaths among laboratory workers and the hospital staffs who cared for them (Smith 1971). The mode of transmission was contact with blood or tissues although one case may have been infected venereally. The origin of this disease was vervet monkeys wild-caught in northern Uganda and imported for laboratory use. As the disease is uniformly fatal in monkeys it is unlikely that they are the maintenance hosts which remain unknown. The second, Lassa (AR) disease, is likewise of unknown origin in nature (Frame et al. 1970). It caused serious illness and death in at least 12 patients in Nigeria. One of them, a nurse, was flown on a normal commercial flight to New York where one virologist died of the disease and another narrowly escaped death after working on specimens from the patient. In both these situations the means of transmission and the prevailing circumstances were such that general spread did not occur. We may be less fortunate next time.

30.4 Control

Effective control of the spread of diseases by arthropods depends on an understanding of the ecological relationships involved and on their subsequent exploitation. This is not to say that chemical insecticides have not in the past contributed greatly to the control of mosquito-borne disease. For instance, in 1947 about 300 million people suffered from malaria, some 3 million dying annually. By 1964 it was claimed that insecticides, mainly DDT, had eradicated malaria from areas with a population of over 400 million people. However, the haphazard dissemination of a toxic chemical which often results before long in a resistant species (Busvine 1967), and a polluted environment is now no longer permissible. Insecticides should now be used only in carefully planned control programmes. In 1969–70 the annual cost of mosquito control in the State of California rose to $ 10 million and *Culex tarsalis* resistant to each of all the licensed insecticides could be found. The vital ecological relationships may be summarised as follows: the arthropod host requires suitable microhabitats for breeding and resting, it must reproduce in sufficient numbers to maintain biological transmission, it must be able to find and feed on a viraemic vertebrate host in numbers sufficient that enough will survive long enough to become infective and then bite a susceptible host. The basis for successful control is breaking this sequence of ecological events by a sufficient reduction in effective contact or in numbers or in both.

30.4.1 Reducing effective contact

Probably the simplest and most widely used method of reducing contact is mosquito screening. Secondly it is important in any urban development within the tropics and subtropics, that periurban scrub and forest should be cleared so as to reduce opportunities for arthropod hosts of viruses to feed on both wild vertebrate hosts and man. This may incidentally increase the dependence of mosquitoes on non-human hosts thus ensuring a degree of zooprophylaxis (Hess and Hayes 1970). In Siberia tick-borne encephalitis has been kept out of suburban areas by maintaining a cleared zone, inhospi-

table to *Ixodes persulcatus* (a forest species) around them. Such measures also
tend to prevent ticks being carried into dwellings by domestic animals such
as cats and dogs. In the Crimea, tick hosts of viruses have been found to be
introduced into villages by birds nesting in trees and by small mammals
taking winter shelter in barns. In irrigation schemes such as ricefields, sugar
and cotton plantations, peripheral vegetation provides fringe habitats for
contact between man, arthropods and wild animals, especially birds attracted
to an irrigated area. This will tend to widen the spectrum of disease agents
to which man is exposed. Clearing of such vegetation also removes resting
sites favouring mosquito survival and so reduces the probability of contact
between man and infective mosquitoes. In irrigation or agricultural extension
schemes, associated towns should be sited beyond the flight-range of medi-
cally important mosquitoes from the irrigation breeding sites. For instance,
the flight range of *Anopheles gambiae* (which transmits filariasis, malaria and
arboviruses) is about 1 km (Gillies and Demeillon 1968) and villages should
be sited at least 1 km from well defined major breeding places such as ricefields. House structure is also important: in Sarawak, *Culex tritaeniorhynchus*
tends to enter better constructed village houses less than more open ones
(Bendell 1970); and in the United States, infection rates with St. Louis
encephalitis virus tend to be greater in poor, less well protected dwellings
(Henderson et al. 1970). Thus in an area with irrigated agriculture, well-to-do
families in houses with tight walls, floors and roof and adequate screening
were less likely to become infected with encephalitis virus than migratory
agricultural workers living in tents or crude cabins (Reeves and Hammon
1962). Certain insecticides such as pyrethrum have an irritant effect on
mosquitoes and when used to spray houses may reduce entry and biting,
hence reducing contact.

30.4.2 Reduction in populations

There are a number of ways of reducing populations sufficiently in degree
and duration to diminish the frequency of effective contact and thus break
transmission. Efforts can be directed at either the arthropod or vertebrate
hosts of the virus. Arthropod populations may be reduced by ecological
methods, such as suitably controlled irrigation in the case of mosquitoes,
by the clearance of vegetation and by drainage in the case of ticks, by
chemical methods including insecticides, oils or inert compounds applied as
monomolecular layers or by the use of predators, insect diseases or population manipulation (e.g. sterile males). Control of vertebrate hosts may be

directed either to reduction of animal populations or to their effective
reduction by immunization in the case of man and his domestic animals.

30.4.3 Habitat control

Virtually every crop requiring irrigation supports mosquito populations,
many of them species of medical importance (Surtees 1970a). The resultant
mosquito-borne diseases are not confined to the tropics but occur seasonally
for instance in North America, France, Czechoslovakia and Japan.

Reduction of numbers in ricefields has been achieved by environmental
manipulation in a number of ways, including weed-free irrigation channels to
allow free flow of water, proper banking to prevent overflow and the formation of marshy areas, grading of fields to eliminate depressions favouring
weed growth, a shortened period of irrigation, intermittent irrigation designed
to strand larvae, and careful preparation of land so as not to encourage
breeding in isolated pools (see Surtees 1971a). Growing rice in trenches with
a continual flow of water to wash away mosquito larvae (Peredo Reyes 1945),
and recently the use of water flow through nursery plots to reduce the larval
population, have been proposed (Surtees 1970d). Natural fertilisers and
animal manure have been used to discourage mosquito breeding while the
burning of stubble as opposed to cutting it down has been suggested so as
to avoid producing pools favouring massive egglaying.

The long term study by Reeves and Hammon (1962) in California indicates how environmental manipulation can reduce the intensity of breeding:
Culex tarsalis breeding was largely the result of excess water and could have
been reduced by better irrigation practice while, if areas above the valley
bottom were used, rapid run-off of water would prevent intense breeding.
Transforming an arid area into a rich agricultural one, created not only ideal
conditions for mosquitoes but also attracted large populations of birds which
helped to maintain western (A) and St. Louis (B) encephalitis viruses.

Environmental control of virus diseases transmitted by container-breeding
mosquitoes is largely a matter of 'good housekeeping', and of improvement
of water supplies and sewage disposal. These mosquitoes will be effectively
reduced only by the provision of piped water supplies together with legislation enpowering health officials to prevent container storage of water.
Mosquito breeding in domestic rubbish can be diminished by efficient refuse
disposal supported by suitable legislation (Surtees 1971). Mosquito populations in sewage lagoons can be reduced by good shore line maintenance,
prevention of algal growth and, above all by control over marginal vegetation

which may provide an ideal milieu for breeding. Although sewage lagoons have relatively low capital cost, subsequent maintenance must be of a high order if they are not to create a public health hazard.

Reservoirs and other large water impoundments also endanger health by supporting mosquito breeding, but the hazard can be reduced by good design and management. The shore line should be vertical as shallow edges support marginal vegetation favourable to mosquito breeding. If a constant level is maintained, wave action in shallow parts may produce sand bars enclosing small pools which breed mosquitoes. Fluctuation of the water level strands larvae and helps to reduce the population. Plants with floating leaves or freely floating vegetation extend favourable margin conditions into deeper water thus enlarging the breeding area, particularly of the *Mansonia* species which attach, in their larval stages, to the roots of floating plants (see Surtees 1971a).

30.4.4 *Control by population manipulation*

Genetic control is the use of any condition or treatment that reduces the reproductive potential of populations by altering or replacing hereditary material. Genetically defined breeding methods can produce adult mosquitoes that are reproductively competitive but sexually sterile. Several genes conferring sex-linked sterility have been isolated in *Aëdes aegypti*. In one strain, the males are completely fertile and the females viable and fecund but their eggs die 12 hr after fertilization due to a defective shell (Bhalla and Craig 1964). Another mutation confers sterility on the males when reared at 30 °C but not when reared at 27 °C. In *Aë. albopictus* another host of dengue virus in tropical Asia, a gene exists which converts the labella of the female into tarsi so that she cannot take a blood meal (Bat-Miriam and Craig 1966). A further genetic factor in *Aë. aegypti* distorts the sex ratio; males carrying this factor produce only male offspring (Hickey 1965). Experimental populations were established to study this effect over a number of generations. Females were exposed to various proportions of male-producing and normal males and populations were allowed to breed continuously. Combinations of 5:1 and 10:1 of mutant to normal males were equally effective in maintaining distorted sex ratios containing only 10% females. Crossing numbers of allopatric populations of the *Culex fatigans* complex sometimes results in infertile off-spring due to cytoplasmic incompatibility. In an incompatible cross, the sperm are blocked before they can fuse with the haploid egg nucleus. Twenty different crossing types are known in this complex, some of which are incompatible in this sense. Female mosquitoes do not discriminate between normal and potentially incompatible males. An excellent example of the application of this principle was reported by Laven (1967). During 1967, males of a strain of *C. fatigans* with cytoplasm from a Paris strain and genome from a Californian strain, were released in a village in Burma. The natural population was estimated at between 4000 and 20,000. 5000 incompatible males were released daily from March 16 to May 6. 5 weeks later, nearly 20% of the egg rafts were non-viable. This level rose to 40% at the 8th week, to 70% in the 10th week, and by the 12th week to 100%. Genetic control of *Anopheles gambiae* was attempted by Davidson, (Odetoyinbo, Colussa and Coz 1970) in a pilot scheme in Haute Volta using sterile males produced by crossing two sibling species, male *An. gambiae* B and female *An. melas*. This experiment gave disappointing results but has led to further work which suggests that the method has promise.

Dominant lethal mutations can be induced in insects by electromagnetic radiations through the whole range of gamma- and X-rays as well as by particular radiations such as neutrons, alpha and beta particles from ingested isotopes. The mutagenic capability of some chemicals (La Chance 1967) has been used to induce sterility in insects by producing dominant lethal mutations in reproductive cells. Most of those investigated have been alkylating agents (Ross 1962) for example, ethyleneimine analogues possessing from 1–6 aziridinyl groups including tepa, thiotepa and apholate, chloromethyl ethers, epoxides and diazo alkanes.

The use of insects to control populations of their own kind through the transfer within natural populations of selected or deliberately damaged genetic material, has not yet been fully exploited in mosquito control. The prospects for control of *Aëdes aegypti* are discussed by Smith (1967). These autocidal methods all make use of dominant lethal effects in sperm or ova, the mutant insects then being used to transmit lethal effects to normal individuals in a natural population with the aim of systematically reducing its intrinsic rate of increase. Two main methods may be used: firstly mosquitoes can be reared, sterilized (or selected) and released in large numbers, or secondly a natural population can be exposed to a chemical sterilant, sterilized adults then competing for mates within the population. It has been estimated that a natural population, subject to control by insects treated so that 90% of each generation is sterilized, will tend to extinction by the 5th generation (Knipling 1967). The ease with which *Aë. aegypti*, *Culex fatigans*, *C. tarsalis* and *C. tritaeniorhynchus* can be mass reared (Smith 1967) suggests that control by autocidal methods is feasible and should be given greater research priority. *Aëdes aegypti* can be produced at a rate of half a million per week by a simple pilot plant operated by two men and the technique is amenable to factory-scale expansion (Morlan et al. 1963). Attention should also be paid to the use of alkylating agents against natural populations of larvae, in conjunction with a food bait and possibly a light attractant such as a beta-light (Bertram et al. 1970), a long-term method which may prove suitable for large scale irrigation schemes.

30.4.5 *Control by other organisms*

Fish have been successfully used to control mosquito breeding in ricefields in Czechoslovakia, central Asia, United States, Russia and parts of south east Asia. The species most commonly used are *Gambusia affinis*, *Lebistes reticulatus* and *Cyprinus carpio* (see Surtees 1971a). Analysis of their stomach contents has revealed that as much as 20–30% of their food may consist of mosquito larvae. Although even high densities of fish will not eliminate some larval populations, detailed studies of the control of *Culex tarsalis* in the United States have shown that reductions of 95 and 99% could be achieved with 80 and 400 fish per hectare respectively. Feeding may be selective: both *Gambusia affinis* and *Culex carpio* fed on *C. tarsalis* but only *Gambusia affinis* fed on *Anopheles freeborni*.

The main groups of microorganisms that have been considered for the control of arthropods are insect viruses, sporeforming bacteria, protozoa and fungi (for references in this section, see Surtees 1971c). Their possible advantages over chemicals are their very low mammalian toxicity, no residues, perhaps less likelihood of the evolution of resistance, and in some cases, the ability to establish a persistent infection in field populations. They are also highly species specific in their mode of action.

Mosquito iridescent virus (MIV) is highly specific for mosquito larvae and has been reported only from this group of insects. MIV has been isolated from *Aëdes taeniorhynchus*, *Aë. fulvus pallens*, *Aë. vexans* and *Psorophora ferox*, all species of medical importance. Larvae usually become infected by feeding on infected cadavers and those infected in the early larval instars usually do not pupate. Death generally does not occur before the final larval instar regardless of the age of infection. The virus does not retain infectivity in soil but can be maintained in a population by transovarial transmission. The rate of cross infection between four known hosts of MIV was always less than 4%.

The most important group of protozoa which act as insect pathogens are the sporozoa, most of which are obligatory species-specific parasites. The most promising microbial agent for the control of mosquito larvae is the fungus *Coelomomyces*. 22 species are found on mosquito larvae and over 45 species of mosquito have been recorded as being parasitized. The fungus has an almost world wide distribution: east Africa, India, Malaysia, Philippine Islands, southern United States, Pacific Islands and Tasmania. Although the fungus mainly attacks the larval stages, adults of *Anopheles gambiae*, *An. funestus*, *An. hyrcanus*, *Culex tritaeniorhynchus* and *Aëdes taeniorhynchus* have been found infected. A study by Laird (1960) demonstrated the potential of this agent for mosquito control. *Aë. polynesiensis* transmits the filarial worm *Wuchereria bancrofti* and probably dengue fever to man on the Tokelau Islands; it breeds in tree holes and similar small water-retaining containers. Up to 1956 *Coelomomyces* fungi were absent from the islands and in 1958, the three islands in the group were used for a unique study in biological control. *Coelomomyces stegomyia* from Singapore was introduced to 761 larval habitats on one island, dieldrin-cement briquettes were used on a second, while the third was left untreated. Examination two years later showed that the fungus was well established in the *Aëdes* population with an overall incidence five to seven times greater than in Singapore and it had also appeared in old coconut shells used as breeding sites near to experimentally infected tree holes.

Invertebrate predators of mosquitoes include beetles, dragonfly larvae, water bugs and carnivorous mosquito larvae of the genus *Toxorhynchites* and *Culex tigripes*. Larvae of *Toxorhynchites splendens* have been recorded as destroying 14 fully grown larvae of *Aëdes aegypti* per day and members of the genus have from time to time been used in attempts to control other mosquitoes.

30.4.6 *Control by chemical insecticides*

Chemical insecticides fall into the broad categories of stomach poisons and contact insecticides. Stomach poisons such as Paris green (calcium aceto-arsenite) are mainly used to kill arthropods with chewing mouthparts and are dusted directly on to their food. Contact insecticides, while they may be ingested, typically act on the central nervous system. They are divided into four main types: inorganic, natural organic, organochlorine and organo-phosphorous. Inorganic substances such as sulphur dust are used for killing mites and scale insects, while natural organic compounds, such as nicotine sulphate, pyrethrum and rotenone (derris) are widely used in arthropod control. The remaining groups have been most used against disease-carrying arthropods, particularly mosquitoes: organochlorine compounds include DDT, benzene hexachloride and the dieldrin, aldrin, endrin group. Examples of organophosphates are parathion, malathion and diazinon.

Resistance to insecticides arises by a number of mechanisms (Brown 1967). In *Aë. aegypti* for example, DDT-resistance is due to detoxification which does not occur in susceptible strains, probably by the enzyme DDT-dehydro-chloriase (Kimura and Brown 1964). Other species respond to the presence of DDT by excreting large quantities of peritrophic membrane (Abedi and Brown 1961). Larvae of dieldrin-resistant mosquitoes contain 25% more phospholipid than susceptible strains (Khan 1964). Strains resistant to organophosphorous compounds usually show lower levels of absorption rather than the acquisition of degrading enzymes (Matsumura and Brown 1963). In the Trinidad strain of *Aë. aegypti* inheritance is due to a single gene allele (Coker 1958; Qutubuddin 1958), hybrids normally being exactly intermediate (Brown and Abedi 1962).

DDT is dissolved in kerosene for spraying, and concentrations of 3–5% are usually effective; it is also used in paints, when its effectiveness depends on residual activity; wettable or water-dispersable powders are also sprayed on to walls as adulticides. Methoxychlor is a water-insoluble analogue of DDT which is used as an oil-based insecticide. The gamma-isomer of benzene hexachloride is commonly used to spray walls and has a vapour effect in closed interiors. Dieldrin has a prolonged residual effect, is water-insoluble and is used against both larvae and adults; it is more toxic to man than DDT and its analogues. Heptachlor and aldrin can be used against DDT-resistant species. Of the organophosphorous insecticides, parathion can be used as a wettable powder but is highly toxic to man and dichlorous is a volatile insecticide especially useful in enclosed spaces.

Insecticides (for references see Surtees 1971a) can be used against either the larval stages of mosquitoes in water or the adults. Applications against larvae are used particularly in domestic water containers or in irrigated areas, while adults are more commonly attacked in buildings. Insecticides for use in domestic water containers must have very low mammalian toxicity. Methods for larval control in irrigated areas include introducing chemicals into the water, dispersal of fog or granulated compounds from ground level or from the air, or treatment of the soil before flooding. Control of *Psorophora confinnis* was achieved in ricefields with abate and fenthion used at 0.5 k/ha in conjunction with a herbicide; no residues of either compound were found in the plant tissues. Very low concentrations have been used in ricefields: 90% control of *Culex tritaeniorhynchus* was attained with only 0.006 k/ha fenthion applied to the water as an emulsified concentrate. In the control of *Aëdes melanion* and *Ae. nigromaculis* in irrigated pastures emulsified concentrates of abate, parathion, dursban and fenthion have been used by drip feed near the main pump outlet. Complete control was obtained in plots of approximately 5–12 ha with 0.17 ppm. abate, 0.14 ppm. parathion, 0.07 ppm. dursban and 0.2 fenthion. Following an application rate of 0.1 ppm., dursban residues in the water were 0.04 ppm. at 29–60 m from the outlet after 10 days, but had fallen below 0.01 ppm. after 20 days. Parathion has also been used at a pump outlet and was effective against *Psorophora confinnis* at 0.8 km distance down an irrigation canal and 119 m across a planted field.

Chlorothion and malathion failed in ricefields because of percolation into the soil followed by degradation. Effective control of *Aëdes dorsalis* was achieved in ricefields by spraying them before flooding with emulsified DDT at 0.18–0.72 k/ha. The fields were flooded 9 days after treatment and 100% kills were obtained for up to 2 weeks. Similarly 0.54 k/ha DDT applied as a 3% emulsion before flooding gave control of *Psorophora confinnis* for only 4 weeks but dieldrin at 0.18 k/ha as a 1% emulsion gave control for up to 10 weeks. Aircraft application of an emulsion of DDT at 0.06 k/ha controlled *Aëdes dorsalis*, *Aë. nigromaculis* and *Aë. vexans* for 2 weeks. Relatively small amounts of insecticide may be effective in aerial spraying: using a Husman–Longcoy spray unit on a Piper Cub aircraft, an aqueous emulsion of DDT at 5–10% killed *Anopheles* mosquitoes and *Psorophora confinnis* at dosages of about 0.02–0.04 k/ha when the aircraft was laying down a 15 m swathe in a cross wind. However, poor control was obtained if the fields were treated before flooding.

Thermal fogs applied from a truck-mounted generator using Baygon, raled and dursban were effective up to 90 m against caged adults of *Anopheles albimanus*, *Culex quinquefasciatus* and *Aëdes aegypti* when the compounds were used at 1.3–4 g/l and at a discharge rate of 180 l/h with the truck moving at 8 k/h. In irrigated areas such methods for control of mosquitoes fail if the breeding sites are protected by marginal broad-leaved vegetation. However *Culex tarsalis* was controlled in such a habitat with a 44 micron particle parathion dust, gravity released from an aircraft distributing 0.4–1.0 k/ha from 15 m at 144 k/h in 50–60 m swathes. The resultant dosage was 0.01 k/ha.

Ultra low volume (ULV) techniques are being used increasingly both in irrigated agricultural areas and against epidemics in urban situations. These techniques use undiluted insecticides or high concentration formulations delivered by compressed air or electric pumps (Glancy et al. 1970). Using ground level ULV application of technical malathion the best control of *Aëdes taeniorhynchus* was obtained with 6.4–10.8 μm droplets at a dosage of 0.009 k/ha (Mount et al. 1970). ULV techniques have also been used against *Aë. simpsoni* breeding in banana axils *(Musa ensetta)* using malathion at a rate of 0.36 l/ha: populations were reduced by 93–100% but with 0.13 l/ha 76–86% reduction was achieved. Thus the canopy formed by *M. ensetta* was not an effective barrier to the penetration of the spray droplets (Brooks et al. 1970). Trials against *Aëdes aegypti* in three villages near Bangkok indicated that at least 0.13 l/ha was required to control adult mosquitoes. The aircraft flew at 40 m at 161 k/h using 95% technical grade malathion and swathes 23 m wide. Tests at a river port near Bangkok using two treatments of 0.13 k/ha applied 4 days apart over an urban area reduced the number of adult mosquitoes coming to human bait by 88–99% during the following 10 days. However, in these or similar tests only fair control of *Culex fatigans* was obtained and populations of all species monitored quickly built up again (Lofgren et al. 1970, Kilpatrick et al. 1970).

Monomolecular films offer the possibility of larval control without toxic hazard to the environment. *Anopheles* larvae cannot cling to the water surface and respire when the surface tension falls below 27 dynes per cm. Using aliphatic amines, mostly cationic surfactants, *Culex fatigans* shows a high mortality: 95% of pupae were killed when surface tension was reduced using the non-ionic surfactant, nonyl-phenoxy polyethoxy ethenol. Amyl alcohol also kills both larvae and pupae while extracts of natural lipids kill pupae and prevent the emergence of adults (Mulla and Chaudhury 1968, McMullen and Hill 1971).

Control has been attempted of breeding mosquitoes in woodland pools using abate briquettes containing 0.6–0.8 g abate, 1 briquette per 10 ft^3 water: good control of *Aëdes canadensis* and *Culex restuens* was obtained (Barnes and Webb 1968).

Oils are still used extensively for distributing insecticides over the greatest possible surface area of water. In thin films, oil blocks the tracheae and kills mosquito larvae by suffocation. Oil may also act as a stomach poison. The most effective, such as gas oils and diesels, are those with a mid-range boiling point: sodium alkyl sulphate improves their spreading properties.

Many attempts at mosquito control by chemical insecticides result in an immediate high kill but the populations rapidly return to pre-treatment densities. Studies of an isolated village population of *Aëdes aegypti* in secondary forest in southern Nigeria (Surtees 1959) showed that the species was restricted to the village, that the majority of breeding was in domestic water pots inside houses and that the population was at a minimum (largely represented by drought resistant eggs) from February to March. Aerial spraying caused very little mortality but fogging of individual houses did. However, as the tests were carried out in November, pre-treatment population levels were quickly re-attained. An attack in February–March would probably have been more successful and more prolonged in effect. The only sound basis for control is a good background of ecological data which must include information on the whereabouts of the mosquitoes, on the best method by which they can be reached and on when the population is most vulnerable.

In terms of tick control, Russian workers (see Smith 1968) have described the use of aerially distributed granules of 10% DDT at a density of 20 kgm/ha in early spring on top of the snow. In a large focus of tick-borne encephalitis near Tomsk almost all adult ticks were killed and there was a significant effect on immature stages. This use of granules appears to have a considerable advantage in economy over the use of 10% DDT dust which was used in a similar manner at 50 kgm/ha: 73–100% of adult ticks were killed and these reduced levels persisted for 3 years. If limited areas are treated with an acaricide then the rate of re-introduction of ticks is a limiting factor.

30.4.7 *Control of vertebrate hosts*

As a possible method of control, reduction of populations of vertebrate hosts of arboviruses is mainly confined to rodents and perhaps birds, particularly in urban environments. The most spectacular effect of rodent control was the termination of the Bolivian haemorrhagic fever (AR) epidemic in San Joaquin by destruction of the rodent, *Calomys* (Mackenzie et al. 1964). In attempts to control foci of tick-borne encephalitis (B), Russian workers (see Smith 1968) have reported on the use of aerially distributed zinc phosphide in oats at a density of 2 kgm/ha, and on the use of similar measures in a focus of haemorrhagic fever with renal syndrome: there was a marked reduction in rodent populations (especially *Clethrionomys glareolus*) over 3 months and the case incidence fell to about 10% of that in untreated areas. The effect was not, however, long-lasting. They also describe the use of rodenticides aerially distributed in forest during the summer and the use of poisoned zones around habitations against the autumn migrations of small mammals to shelter. The possible risks of such blunderbuss treatments in our present state of knowledge were demonstrated by Russian workers who found that after acaricide treatment *Ixodes trianguliceps* appeared to replace *I. persul-*

catus and that there was an apparent increase in haemorrhagic fever with renal syndrome to offset the reduction in tick-borne encephalitis. As with the mosquitoes, studies are required to give us a much more profound understanding of the bionomics of tick and vertebrate maintenance hosts before and during control measures.

In Florida, St. Louis encephalitis (B) occurs in dense urban foci and the virus has been isolated repeatedly from mourning doves, house sparrows, jays, pigeons and swifts. Mosquitoes such as *Culex tarsalis*, *C. pipiens* and *C. fatigans* from which St. Louis virus has been isolated, bite birds as well as man (e.g. Kokernot et al. 1969). Nestling house sparrows may be particularly important in the epidemiology of Western (A) and St. Louis encephalitis. During one epidemic, the weekly infection rates in *C. tarsalis* and nestling house sparrows ran closely parallel. House sparrows appear to be important amplifiers of the risk of epidemics (Hayes et al. 1967). In such circumstances substantial reduction of the sparrow population might make a considerable contribution to control of these virus infections.

Domestic animals play an important role in some arbovirus diseases (e.g. pigs in Japanese encephalitis (Simpson et al. 1970), cattle and goats in tick-borne encephalitis) and certain measures may help, such as housing pigs away from close proximity to man or pasteurization of milk. These and other domestic animals (e.g. horses, dogs) are naturally infected by a number of the arboviruses but in most cases their epidemiological importance is unknown because, unless they have viraemia of a sufficient height and duration to infect the biting arthropod host, they are incidental hosts. For instance the horse has inadequate viraemia with eastern encephalitis virus to infect several potentially important mosquito species including the probable maintenance species *Culiseta melanura* (Schaeffer and Arnold 1954) – even though the virus causes fatal encephalitis in the horse. It has been suggested that domestic animals, such as cattle or dogs could be used for zooprophylaxis (Hess and Hayes 1970) but their use should be assessed with care as, for instance, dogs may be involved as hosts of Venezuelan encephalitis or Japanese encephalitis (Sidwell et al 1967, Simpson et al. 1970).

Vaccination may be of value when domestic animals are amplifier hosts. For example vaccination of a pig population in a Japanese village resulted in fewer isolations of the virus from mosquitoes (Takahashi 1968). However in domestic animals as in man, most vaccines are used to prevent disease; they can have no real controlling effect on the continued maintenance of the infection unless used in major maintenance hosts. Vaccines are in use in domestic animals against eastern, western and Venezuelan encephalitis (A), louping ill (B), African horse sickness, blue-tongue, and Rift Valley fever (Smith 1969a).

30.4.8 *Vaccination in man*

The vaccines available or under development for human use against arbovirus diseases were reviewed by Smith (1969). The only one which has been both widely and very successfully used is yellow fever vaccine which gives long lasting immunity and has prevented epidemic yellow fever wherever it has been given to a sufficiently high proportion of the population – at least 70%. Vaccination programmes, particularly against such diseases as yellow fever which can be transmitted from man to man, require maintenance by regular vaccination of young children because, with high birth rates and now decreasing neonatal and infantile death rates in many cities in developing countries, the proportion of susceptibles in their populations builds up quickly (Smith 1970). In some areas of west Africa where vaccination programmes were disrupted or allowed to lapse in the late 1950s or early 1960s epidemics have followed. Yellow fever vaccine is required for international travel to and from endemic areas and is particularly effective in urban yellow fever where man is the maintenance host. A vaccine against Venezuelan encephalitis virus (Alevizatos et al. 1967) was recently widely used in both man and Equidae in the large epidemic/epizootic which swept Central America during recent years (Escalona et al. 1969, Morilla-Gonzalez and Mucha Macias 1969). Few other arbovirus vaccines have had more than tentative trials mainly because arbovirus diseases occur either sporadically or in epidemics which are unpredictable both in time and space. Moreover most of them occur in economically poor areas of the world where there are many other important disease problems and where there is no economic incentive for commercial vaccine development and production. A number of tick-borne encephalitis vaccines have been widely used in the USSR with varying degrees of success (Smith 1969a).

While there must be no slackening of the efforts to make new and better vaccines, their deployment is likely to be slow mainly because of the unpredictable behaviour of arbovirus diseases. Tick-borne infections, with their relatively fixed distribution are more amenable to disease control by vaccination. Some vaccines would be useful in occupations at high risk such as foresters in areas with serious tick-borne infections. Some vaccines have been used experimentally with military personnel for similar reasons and much of the development of arbovirus vaccines has been for the possible protection of troops who might have to operate in areas with enzootic arbovirus infections. Further progress can be made on prototype arbovirus vaccines by developing veterinary vaccines against representative viruses in the several groups which cause veterinary diseases of economic importance.

30.5 Diseases warranting control

The costs of research and development and of the implementation of disease control methods are high and, particularly in developing countries, must compete with many other demands for often scarce human and financial resources. Priority may be determined in terms of mortality, high morbidity or the danger of epidemics. But, particularly in developing countries, the economic importance of disease is a significant factor, whether it affects a particular industry or impairs output of an important agricultural product. While diseases that kill tend to be given prominence in mass media, temporary incapacitating illnesses which affect a large proportion of a population are probably of greater economic significance. Classical dengue fever for example can cause very high morbidity rates: over one million persons were affected in the United States in 1922 and over half a million in Australia in 1925–26. During an epidemic in Brisbane in 1905, 90% of the city's transport workers and hospital staff were put out of action within a short space of time. The economic importance of veterinary arbovirus infections is much easier to demonstrate. As mentioned above, African horse sickness caused havoc among Equidae in India, Pakistan and the Middle East in 1959–60; bluetongue is an important disease in the Mediterranean area, the Middle East and parts of the United States; Rift Valley fever killed some 100,000 sheep and cattle in South Africa in 1950–51; Venezuelan encephalitis has caused very large epizootics of equine encephalitis in South and Central America; and in some years louping ill kills as much as 50% of the lambs on some Scottish hill farms (Smith et al. 1964).

30.6 International control problems

Probably the greatest achievement in the control of an arbovirus disease was the elimination of urban yellow fever from South America which illustrates some of the problems associated with disease control at an international level (Soper 1967). In 1923, the Rockefeller Foundation organized a yellow fever eradication programme in northern Brazil based on the elimination of *Aëdes aegypti* by systematic inspection and environmental hygiene. The object was not so much to eradicate the species but rather the disease, by breaking its man-*aegypti* maintenance cycle. There was however an outbreak of yellow fever in one locality after 6 years freedom from the disease, apparently due to the persistence of the insect at an undetectable level. In 1930, Dr. Fred Soper joined the yellow fever service in Brazil determined to eradicate the disease by developing an effective *aegypti* surveillance system: working areas were mapped out block by block and house to house itineraries prepared; all work was recorded and cross-checked by supervised supervisors. Larval indices were checked against independent adult catches and, to avoid breeding in the same loci from week to week, a mixture of fuel oil and kerosene was applied to containers found with larvae. Routes were laid down for each inspecting team and every block of buildings marked with a number on an outside wall: a set routine of inspection was adhered to for each block. As a result, *Aë. aegypti* was eradicated from 8 cities by 1933. At this point a budgetary problem arose because all workers were automatically receiving bonuses for perfect work as there were no mosquitoes to miss. The monitoring of clean areas was therefore reduced to monthly and three quarters of the staff released to move outwards into the suburbs and surrounding villages. Eventually, eradicated areas were put on to a 3-monthly, then a 6-monthly or annual inspection basis and the eradication of *Aë. aegypti* progressively expanded at the periphery of each area.

At an international level however there had been less progress. By 1946, *Aë. aegypti* control in Brazil had progressed so far that re-infestation from adjacent territories was the most serious problem. The Pan American Health Organization carried on with *Aë. aegypti* eradication throughout South and Central America. The United States, recognising the effectiveness of yellow fever vaccine and DDT could not be persuaded to undertake eradication of *Aë. aegypti* until 1964, after Mexico had claimed eradication of the species. However this programme made only limited progress and *Aë. aegypti* has adapted to tree-hole breeding in the southern U.S. and is probably now ineradicable for practical purposes.

30.7 Conclusions

There is no doubt about the urgency with which new methods for arthropod control must be sought – especially methods highly specific for dangerous

species, methods which do not cause persistent or cumulative toxic contamination of the environment, and perhaps most important, methods
against which resistance (biochemical or behavioural) is not readily developed. However there is equally no doubt about the enormous advantages
to health and agriculture which have resulted from the use of insecticides,
particularly DDT. In countries where arthropod-borne diseases are a prevalent cause of serious ill-health, we must not be panicked into abandoning
DDT before a satisfactory alternative is found because its accumulation in
the tissues of man and other animals may be seriously harmful. A careful
balance must be sought between the advantages and disadvantages of
abandoning such an effective insecticide. In many developing countries the
balance must at present come out in favour of its continued use at least
against malaria and certain agricultural pests. Any toxic chemical must
however be used by the most economical and effective methods consistent
with minimal contamination of the environment, and wherever possible only
with understanding of the behaviour of the species to be controlled. Where,
because of resistance, an insecticide has become largely or wholly ineffective
for control of disease it should of course be abandoned; in some situations,
such as the control of *Culex tarsalis* in California (Reeves 1970) there may
soon be no effective insecticides left available and completely new methods
may be required. This sort of outcome may be avoidable if arthropod
control measures are devised at the planning stage of agricultural and engineering schemes in areas at risk. Wherever possible, crops which pose the
least risk of disease problems should be chosen.

New control methods, whether chemical or biological and however applied, must derive from a fuller understanding of the behaviour, ecology,
genetics, physiology etc. of the species to be controlled. Moreover control
programmes must be very carefully designed and rigorous attention given
to detail in their execution – a point so well made by Soper's eradication
programme for *Aë. aegypti*. Most of the infections discussed in this chapter
are maintained in wild vertebrates thus, although disease in man and domestic animals could be controlled by vaccination (were the vaccines available), the infection in the maintenance hosts would remain unaffected. Such
infections are ineradicable and probably have to be lived with in many parts
of the world. However they normally remain a cause of only sporadic disease
unless their seriousness is aggravated by one or more of the many types of
human interference with the environment. The growth of towns and of the
risk of urban arbovirus epidemics continues to increase and, because of the
unpredictability of these events, there is a special need for the development

of very rapidly applicable methods for the reduction of mosquito populations over large urban areas, probably using ultra low volume aerial spray techniques. Large scale interference with the environment is essential if sufficient food is to be grown for the world's escalating population. Urban development and the need for more living space in towns increase the need for alternative and additional agricultural land. However, if serious consequences are to be avoided, urgent and careful attention must be given in all large scale development plans to studies of possible adverse consequences which may arise and to methods which minimize disease risks.

The advanced countries have made great strides in the control of lethal and important infectious diseases but find themselves recognizing new frontiers to conquer: infective hepatitis, dental caries, slow viruses, infections made important by the use of immunosuppressives and so on. The advances so far have been achieved over many decades with much trial and error. The developing countries start with a much more complex disease situation, with galloping population problems, widespread malnutrition and an industrial revolution. They are impatient for rapid progress and neither they nor their more prosperous neighbours can afford unnecessary and expensive errors. Indeed with the rapid increase in population movement of all types, our developing neighbours come closer and closer, not least in terms of their transmissible diseases (Dorolle 1968). There is thus the greatest urgency to help them solve their problems by well-designed aid both in research and development and in technical and expert assistance.

Efforts must be made to make existing knowledge of the control of infectious disease more available and practicable in the developing countries and also to aid and promote research, development and production of all the means of disease control specifically needed by the developing countries. Much of this will have to be furnished by the advanced nations but they should recognize the clear self-interest which they have in doing it. 'In this age of jet planes and soon of supersonic transport, the only way of preventing the old plagues, and some new ones, from spreading from continent to continent and from country to country is to help the poorest nations of the world to reach such a level of economic and technical development that it will be possible for them to combat the evil at its source' (Dorolle 1968).

CHAPTER 31

Control of viruses spread by invertebrates to plants

G. D. HEATHCOTE

Contents

31.1 Introduction

Few virus diseases kill plants but they can harm them in many ways and decrease the amount of food, fibres, flowers etc. they produce. Unless viruses are controlled it is impossible to grow many crop plants satisfactorily. The value of the crop determines how much can be spent on treatments to decrease virus spread. Using biocides to control the virus vectors is often the easiest and most effective way to protect crops, but it is often expensive and can only be used on high-value crops therefore. Also it is increasingly obvious that the residues of biocides left in the crop or environment may have long-lasting deleterious effects. Less harmful, and of more use for low-value crops are methods based on plant breeding and agronomy, such as choice of the most suitable time or density of sowing, and it is the aim, even with high-value crops, to integrate the control measures and combine the minimal use of chemicals with the maximum plant and crop hygiene, the best varieties and best agronomic techniques.

Many of these plant viruses have a complex disease cycle, involving at least the spread of the virus from an infected plant to a vector and back again to a susceptible host plant, and the aim of control measures is to attack the vulnerable place in such a disease cycle. Many methods have been used to break disease cycles and control virus spread, but not all are completely successful, and some have actually increased spread (Broadbent 1957a). Control measures are often aimed at the virus vectors, which, as plant parasites, should be controlled anyway as they cause loss when they feed irrespective of whether they are also transmitting viruses. The methods used to control invertebrates acting as vectors of plant viruses differ little from those used to control them as pests. However, their efficiency in these two roles will be different as, for example, pests only cause feeding damage when present in large numbers, whereas when acting as vectors they may be few yet cause much damage. Thus a particular method may kill enough of the population of a particular invertebrate to stop it damaging a crop directly, yet leave more than enough individuals to spread virus to all plants in a crop. Also, although most of the methods devised to control crop pests are also used against virus vectors, a few techniques are obviously not applicable. For example, the use of sex 'attractants' or the release of sterile males (techniques being developed against pink bollworm and other moths) would never succeed against aphids, which reproduce asexually throughout most of the year.

Obviously very different techniques are needed to control the spread of soil-borne viruses (i.e. those with vectors such as nematodes which occur in the soil and which infect plants through their roots) and air-borne viruses (i.e. those carried by vectors such as aphids, whitefly or thrips that fly or are blown by the wind and which infect plants through the leaves or stems). For example, nematode-borne viruses are usually spread only short distances and the aim of control measures is to kill all the viruliferous (virus-carrying) nematodes in soil before a crop is planted, but air-borne viruses often spread widely and control measures might attempt to control the viruses at source or might merely aim to prevent aphids or other winged vectors from spreading the disease within a crop.

A detailed knowledge of the ecology of the vectors is often needed before the most effective control measures can be devised but, armed with that knowledge, simple changes in farming practice or in the choice of plant variety may give economic control of a virus disease. Most control measures are not absolutely successful, but that does not make them useless. The epidemic of a disease, and the growth rate of a crop can be represented diagramatically by sigmoid curves, and the area under these curves is proportional to yield loss and gain respectively. Thus a great economic advantage can come merely from delaying the course of an epidemic (e.g. sugar-beet plants lose up to 5% of their potential yield for every week they are infected with virus yellows (Watson et al. 1946) and virus spread late in the season is much less important than early spread).

The control methods described here are mainly those used now in the more highly-developed countries, with a temperate climate, but the principles involved should be generally applicable. The methods described are grouped according to the point at which they attempt to break the disease cycle.

31.2 Preventing infection from sources within the crop

One way this can be done is by modifying the farming methods, and this tactic should be adopted whenever possible as undesirable 'side-effects' are likely to be small compared with those resulting from, for example, chemical methods of control (i.e. no pollution of the environment, no accumulation of toxic residues or taint of edible crops), and it is often inexpensive. Agronomic methods of control will therefore be considered first.

31.2.1 Crop rotation

Fields may contain virus-infected plants or viruliferous vectors and these must be destroyed before a crop is planted. Crop rotation affects soil-borne viruses less than might be hoped because most of the viruses and their vectors have wide host ranges. For example, tomato black ring virus and its vector, *Longidorus elongatus* (de Man), persisted through a 4-year ryegrass-white clover ley and affected leeks and onions in the following year (Calvert and Harrison 1963).

31.2.2 Groundkeepers and infected plant debris

Virus-infected plant material remaining in a field after harvest may carry infection to following crops, and infected plants may persist for several years. For example, a break of at least five years before replanting vines is needed to decrease the incidence of fanleaf virus as the vector, *Xiphinema index* (Thorne and Allen), can survive under a winter grain-summer fallow rotation for at least $4\frac{1}{2}$ years (Raski et al. 1965), and can acquire virus from pieces of grapevine root which may survive in the soil for several years before they decay. Similarly, small roots of sugar beet which escape the harvester or fall from the root-cleaning machinery may survive mild winters if partially buried in soil, even better than large ones (Wallis 1967), also virus-infected sugar-beet plants can sometimes be found in grass roadside verges in Britain where they they have fallen from lorries carrying beet to the processing factory, and all are sources of virus.

Doncaster and Gregory (1948) showed that large numbers of potato tubers were left in the soil after harvest and grew into plants in the following years. Such plants even persisted for five years. In Britain today all cereal crops (which usually follow potato crops) are treated with herbicides which destroy groundkeepers together with the weeds and there is no longer a danger of virus spread. Infected plant material of this type can also be destroyed by cultivation.

31.2.3 Infected weeds and weed seeds

Almost all viruses that infect crop plants also have weed hosts, which may show no symptoms, and some viruses are seed-borne in weeds. Infected weed seeds can introduce viruses into fields and help perpetuate them, especially if the vector also feeds on the weed host. Cadman (1963) suggested that efficient weed control may be as beneficial as the use of nematicides to control some soil-borne viruses, as most nepoviruses are transmitted through the seed of some of their weed hosts. Weed control may also be important to limit the spread of air-borne viruses. Tomlinson et al. (1970) have found that annual weeds and their seeds are a major overwintering source of cucumber mosaic virus infecting lettuce crops in Britain; the virus is transmitted to from 5–8% of the seeds of infected *Stellaria media* (L.) Vill. (chickweed), a common and widespread weed in temperate regions. The relationship between weeds, herbicides and plant virus diseases was reviewed by Heathcote (1971).

31.2.4 Special cultivations

Cultivation of the soil can affect survival of soil-borne viruses as it may affect the population of some nematodes, including virus-vector species. Oostenbrink (1964) showed that mixing soil with a rotary cultivator decreased a population of *Trichodorus flevensis* Kuiper and Loof from 1210 to 160 per l of soil, but some nematodes are too deep in the soil to be affected by this treatment.

A period of fallow will not necessarily decrease nematode populations as some can live for long periods without food. Harrison and Hooper (1963) extracted live *Longidorus elongatus* from samples of a light loam soil that appeared to contain no living roots, and which had been kept in polythene bags in darkness at room temp. for 29 months. However, if soil is allowed to dry for more than one week at 20 °C nematodes are killed (Cadman 1963).

Flooding by water is an age-old technique for weed control in paddy rice culture, and may also control nematodes and soil fungi, but I do not know of it being used to control a soil-borne virus disease.

31.2.5 Partial soil sterilisation

Partial soil sterilisation (fumigation) may greatly increase crop yield, but the biocides used are non-selective and kill friend (e.g. parasites and predators of virus vectors) as well as foe. Such treatments have diverse effects; weeds are killed, nitrification is retarded and mineralisation of soil nitrogen is increased (Gasser and Peachey 1964), damage to roots caused by ectoparasitic nematodes, fungi and other organisms decreased, and, in addition, spread of virus by nematodes is decreased.

Soil may be sterilised by putting volatile liquids such as 'D-D' (dichloro-propane-dichloropropene) or chloropicrin into the soil, or by pumping a gas such as methyl bromide under a temporary cover of rubber or plastic sheeting. These sterilants are lost from the soil by evaporation when the soil is cultivated, and by leaching, absorption and microbial activity, and rarely cause a residue problem. Machines are now available for injecting liquids or gas into soil at a required depth on a field scale (fig. 31.1).

Fig. 31.1.   A tractor-mounted machine for injecting chemicals such as D-D into the soil. (Photograph Broom's Barn Experimental Station.)

Fig. 31.2.   Control of the spread of arabis mosaic virus by *Xiphinema diversicaudatum*. Plot in lower left treated with D-D, plot in upper right not treated. (After Harrison, Peachy and Winslow 1963.)

If 'D-D' is applied at 820 l/ha (2 lb/100 sq. ft) it will kill over 90% of *Xiphinema diversicaudatum* (Micol.) to a depth of 0.7 m (28 in), and almost stop the spread of arabis mosaic virus in strawberry crops; methyl bromide at a similar rate is equally effective (Harrison et al. 1963) (fig. 31.2). Although this treatment is effective with shallow-rooting crops, it may be less so with deeper-rooting crops such as vine, and although Vuittenez (1960) achieved some control of grapevine fan leaf virus by soil fumigation, the nematode vector has been found in soil at a depth 2.1 m (7 ft) (Hewitt et al. 1962).

Only in certain conditions is it possible to fumigate soil efficiently, it must be neither too wet nor too dry, nor too cold, and 'capping' the soil by com-pacting or wetting the surface may improve efficiency of fumigation. The technique is expensive but the cost is justified for a high-value crop, or when reinfestation of the soil by nematodes takes several years. Nematodes usually spread very slowly. For example, at one site *Xiphinema diversicaudatum* in-vaded uncultivated soil at a rate of only 0.3 m (1 ft) per year (Harrison and Winslow 1961).

31.3 *Preventing the introduction of virus (primary spread)*

31.3.1 *Quarantine and certification*

It is often difficult to detect viruses in seed or planting material as they rarely show symptoms. Even when growing plants are imported from another dis-trict or country they often lack leaves (e.g. young fruit trees or sugar beet 'stecklings', which are seed plants in their first year ready for transplanting). Therefore the plants should be grown in quarantine in vector-free conditions and checked for virus infection. Plants may be held in quarantine in the im-porting country, or even in an intermediate country. For example, rooted cuttings of cacao from South or Central America are quarantined in Britain before they are imported into west Africa (Thresh 1960).

In Britain, 'seed' potato crops have long been grown in isolation to prevent virus infection, and 'seed' is then sold to growers of 'ware' potatoes (those for consumption) in one of three grades. The growing crop is examined by the certifying authority and a certificate is issued of purity of type and freedom from disease. The highest quality seed is from fields which contained not more than 4 plants per acre with potato leaf roll or potato virus Y and 0.25% of plants with potato virus X. The next grade is from fields with not more than 0.5% of plants with leaf roll or virus Y, and 2% with virus X, and the lowest grade of seed has not more than 2% with leaf roll or virus Y. Similarly, tristeza virus of citrus trees is controlled in Argentina by a certification scheme for imported budwood (Fernandez Valiela 1963), and a similar certification scheme provides virus-free top-fruit trees in Britain (Posnette 1962).

31.3.2 Restricted planting

Legislation may restrict the growing of crops that are likely to become infected with virus disease, or act as sources of disease for other crops, and crops may have to be registered. Such restrictions in Australia, the Philippines and elsewhere have been used to prevent the spread of bunchy top of banana and abaca, which are vegetatively propagated, and where latent infection is a particular problem (Magee 1961). In the main sugar-beet growing areas of Britain sugar-beet seed crops can only be grown near sugar-beet root crops if the stecklings are certified as containing few or no virus-infected plants.

31.3.3 Virus-free seed and planting material

Happily, few viruses are transmitted through the seed of infected plants, but up to 15% of the viable seeds of a lettuce plant infected with lettuce mosaic virus may carry infection (Kramer et al. 1945) and so lettuce seed must be collected from virus-free plants. Several virus diseases of legumes are seed-borne also (e.g. alfalfa mosaic, bean common mosaic and broad bean stain viruses) and stocks should be checked for virus.

To ensure that planting material is virus-free, sap from sample leaves should be inoculated to test plants that produce characteristic symptoms when infected with the viruses suspected to be present, or serological tests made. In the well-known slide agglutination tests for potato viruses, drops of sap from test leaves are mixed with a drop of suitably diluted antiserum and an agglutination of the chloroplasts in the sap indicates the presence of the virus. However, this test can only be used with some viruses, especially those that attain large concentrations in the sap and have anisometric particles. If the virus induces callose formation in the phloem or some other pathological change, a staining technique can sometimes be used to detect it (e.g. the Ingel-Lange/resorcin test for potato leaf roll (Keller and Bérces 1966) and the ninhydrin test for asparagine in maize plants infected with maize rough dwarf virus (Harpaz and Applebau 1961)).

31.3.4 Meristem culture and heat treatment

Plants that are vegetatively propagated, such as potato, sugar cane and many ornamental plants, can sometimes be freed from virus by heat treatment or by growing plants from their apical meristems or shoot tips (Kassanis 1950).

Clones from plants produced by meristem culture may yield more than virus-infected ones, but care must be taken to stop them being reinfected. Kassanis grew the excised apical meristems of sprouts from potato of the varieties 'King Edward' and 'Arran Victory', infected respectively with para-crinkle virus or potato virus S, and obtained virus-free plants. The virus-free stock of King Edward was grown in virus-free areas in Northern Ireland until there was sufficient to supply potato-seed growers. This clone outyields the virus-infected commercial stocks by an average of 10% and produces a larger proportion of marketable tubers (Kassanis 1957, 1961).

Virus-free plants are also obtained by heating infected plants, but only certain virus–plant combinations are suitable for this type of treatment. The plants are either heated for short periods to high temperatures, as for example when sugar cane planting material is heated for 8 hr at 54 °C to free it from ratoon stunt virus, or the plants are maintained for long periods at a lower temperature, e.g. 16 days at 38 °C. The subject has been reviewed by Hollings (1965). Such treatments are very successful; for example, heat-treated and virus-free 'Baccara' roses yielded 13.5% more blooms than untreated plants (Pool et al. 1970). Often heat treatment and meristem tip culture methods are combined, and tips are cultured from heat-treated plants. Obviously the successful application of these techniques to field problems depends on keeping the stocks virus-free. This can only be done by efficient 'nuclear stocks' or 'budwood' schemes.

31.4 *Control of incoming vectors*

31.4.1 Isolation from infected crops or weeds

It is never possible to specify the degree of isolation necessary to prevent the spread of an air-borne virus from one crop to another, as the disease gradient around a source is exponential. Thus the slope of the exponential determines the distance that will be needed to provide a given amount of isolation, the slope may vary from one year to the next, and will depend on other factors such as the type of the intervening crop, but obviously the greater the isolation the smaller the risk. Martin and Kantack (1960) found that 0.09 km (100 yds) isolation decreased the incidence of aphid-transmitted internal cork virus of sweet potato from 11–23 % to 1 %, and Heathcote and Cockbain (1966) found that most spread of virus yellows from mangold clamps was within 0.5 km (440 yds) of the clamp site, and so short distances may prevent spread sufficiently.

31.4.2 Cages, barriers and cover crops

Occasionally a valuable crop may be caged to keep out insect vectors of viruses or mycoplasmas (e.g. aster plants caged against aster yellows in the USA) but often a partial barrier gives sufficient protection. Physical barriers such as tarred paper or coarse string netting have been tried to protect brassica seedlings from aphid-borne non-persistent viruses, but a narrow barrier of barley is more effective and cheaper, perhaps because incoming viruliferous aphids feed on the barley barrier and lose the virus they are carrying before they move to the brassica seedlings (Broadbent 1957b). Similarly, a barrier of maize and cowpea has been used successfully to decrease spread of virus to *Capsicum* (Ramallo and Garciá 1968).

Another alternative, often used with biennial seed crops, is to grow the plants in the first year under a crop, often a cereal. Such 'cover crops' have been used in Britain for many years to protect sugar-beet seed plants in their first year of growth from aphid-borne virus yellows. Barley is generally used, and the barley is harvested, although the crop may be a poor one. Mustard is equally effective as a cover crop in preventing infection by virus yellows, but provides a microclimate within the sugar-beet crop which encourages the growth of sugar beet downy mildew, a fungus disease. Similarly, bananas are grown under coconuts as they are then less frequently attacked by the banana aphid, *Pentolonia nigronervosa* Coq., and infected with the bunchy-top virus it carries than when they are grown in recently cleared forest (Magee 1967).

31.4.3 Reflective surfaces and repellants

The use of light-reflecting surfaces under plants to repel aphids that, having flown, are negatively phototactic and plant-seeking, was first suggested by Kring (1964). Strips of aluminium foil laid between rows of plants have been shown to decrease the incidence of aphid-borne viruses in gladiolus, turnips, lettuce and sugar beet (Heinze 1967a, b; Heathcote 1968a; Johnson et al. 1967) but the method is expensive in material and labour. No chemicals have been found which can be sprayed on living plants and which will then prevent aphids or other air-borne vectors from landing.

31.4.4 *Virus inhibitors and inactivators*

Attempts have been made to inhibit transmission of viruses by spraying the leaves of crop plants with a wide range of substances. Vanderveken (1968) reported that spraying sugar-beet plants with mineral oils, corn oil or whole milk inhibited the transmission of beet mosaic virus, but only mineral oil affected the transmission of beet yellows virus (a semi-persistent virus) by aphids. Doubt has been cast on this technique by Crane and Calpouzos (1969) who have shown that plant resistance to beet yellows virus is localised and temporary, and that transmission and movement of virus through the plant is not prevented, although symptoms are suppressed for several weeks.

31.4.5 *Chemical control of vectors at source*

Although many immigrant vectors can be killed by treating the crop, most pesticides do not act sufficiently quickly to prevent infection of a plant by a vector already carrying virus. Better control may be achieved by treating the source of the disease, if it is known. For example, Simons et al. (1959) found that eliminating the weed hosts in an area before planting peppers was more effective in checking virus spread than using insecticides on the crop, provided all growers in the region co-operated in weed control. Similarly, desert areas in south west USA where leafhoppers (*Circulifer tenellus* (Baker)) congregate on Russian thistle and other weeds in winter may be sprayed to prevent curly-top being carried to fields of sugar beet (Cook 1967).

   Local control of incoming vectors may be important if they do not move far. In central Europe the beet leaf bug, *Piesma quadrata* Fieb., overwinters under trees and in hedges and moves into beet fields in spring, bringing with it beet Krauselkrankheit. It is controlled by treating only the borders of the fields with insecticidal dust, although the entire field may be treated once (Scheibe 1955). This technique has replaced 'trap cropping', in which control was achieved by sowing an early crop of sugar beet round the edges of fields and ploughing it in when the bugs had moved on to the beet plants, and before the main crop of sugar beet had germinated (Hull 1960). Trap cropping meant sowing the main crop undesirably late, and wasted land or resulted in a mixed crop in the field.

   The pesticides applied to crops to prevent virus spread will be discussed in more detail later (sect. 5.4).

31.4.6 Virus-resistant and tolerant varieties

The plant breeder has a duty to avoid releasing a new cultivar which is particularly susceptible to attack by any widespread virus disease, and it is obviously desirable that any new 'variety' should resist virus vectors and virus disease, but the degree of such resistance is seldom a major factor in the choice of a particular variety. If sources of virus disease and the vectors are numerous a resistant or tolerant variety may give a satisfactory crop which would otherwise be unobtainable, e.g. of sugar beet where curly-top is widespread.

The plant breeder may introduce material with greater resistance or tolerance from foreign or other localities, select strains with improved resistance or tolerance to disease but also with desirable agronomic qualities, hybridize between resistant species or varieties with those that are not resistant or tolerant but otherwise desirable, or graft susceptible but otherwise desirable types on resistant or tolerant strains (Snelling 1941).

Some varieties of crop plants may resist or repel virus vectors more than others, but resistance is never complete enough to prevent trial feeding and the transmission of aphid-borne non-persistent viruses. The transmission of aphid-borne persistent viruses takes longer and may sometimes be affected. Plants may resist or discourage vectors such as aphids in diverse ways, e.g. by the biochemical nature of their sap, the physical properties or the thickness of the waxy cuticle (as in some brassicas), or the density of glandular hairs on a leaf (as in tobacco). Even the length of hairs on a leaf may affect the resistance of plants to leaf-feeding vectors; a high density of long hairs on the cotton lamina decreases the attack by *Empoasca devastans* Dist., but the length of hairs on the midrib is less important (Sikka et al. 1966). These factors may vary with the age of the plant and the conditions under which it is grown. For example, the preference shown by *Therioaphis maculata* (Buckton) for 'aphid resistant' or 'aphid susceptible' alfalfa varieties varies with temperature (Schalk et al. 1969) and the toxicity of tobacco varieties to *Myzus persicae* (Sulz.) is related to leaf age, and parallels the amount of exudate from leaf hairs, which is toxic and sticky (Abernathy and Thurston 1969), or the alkaloid content of leaves (Thurston et al. 1966). Even the width of the space between the leaf veins of cereals seems to affect the transmission of cocksfoot mottle virus; the mouthparts of the beetle vector (*Lema melanopa* (L.)) are too large to be inserted for feeding between the veins of certain cultivars (Shade and Wilson 1967).

Some varieties of plants are extremely resistant to infection with virus diseases and in the field remain virus-free. Other varieties may be susceptible to infection, but tolerate the effects of infection so that they lose less of their potential yield when infected than more susceptible varieties. Hypersensitivity to infection can also prevent virus spread in crops, for plants that react severely may not become systemically infected and may eventually recover, or die before they can act as sources of virus.

Varieties of plants almost immune to some nematode-transmitted viruses have been found. Losses caused by arabis mosaic virus in raspberry crops can be avoided by growing the cultivar 'Malling Jewel', however, the variety is susceptible to strawberry latent ringspot virus, which is a very closely related virus with the same vector (Taylor and Thomas 1968). Similarly, some varieties of pea are highly tolerant or resistant to nematode-borne pea early browning virus (Hubbeling and Kooistra 1963).

Less success has been achieved with aphid-borne viruses, but Jones and Catherall (1970) found a barley variety tolerant to a number of isolates of barley yellow dwarf virus, and seed of the lettuce cultivar 'Gallega de Invierno' does not carry lettuce mosaic virus (Marrou 1969). The breeding of varieties of sugar beet tolerant to hopper-borne curly-top virus is a success story. It is well known that there is a great range of susceptibility to curly-top in varieties of sugar beet; some varieties are very susceptible and greatly injured and others show few or no symptoms when infected (Bennett 1957), and without these tolerant varieties it would be impossible to grow sugar beet in parts of North America. Encouraging though such examples may be, the use of cultivars resistant to infection with a particular strain of a virus may not be the answer to the whole problem because they may be susceptible to, and hence select out, novel, perhaps more virulent strains of the virus.

31.5 Preventing spread within the crop (secondary spread)

Virus spread within the field can be decreased by many agronomic, chemical or biological means, and where little virus is brought into the crop initially control of secondary spread may be vital.

31.5.1 Agronomic methods

Even the choice of the field in which it will be grown will affect the spread of virus within a crop. If a choice is possible large open fields should be selected. As Van der Plank (1948) pointed out the edge of a crop which is often most heavily infected by immigrant vector-borne viruses forms a greater proportion of a small field than of a larger one, and the infection entering from nearby sources per unit area is approximately inversely proportional to the square root of the field area. Damage by other pests and diseases that come into crops may be decreased by having fewer and larger fields. Most pests are commonest on the edges of fields and air-borne vectors are most likely to be deposited near windbreaks such as hedges, trees or tall buildings (Lewis 1967).

In groundnuts, sugar beet, field beans and other row crops, losses from air-borne viruses have been shown to be less if the plants are densely spaced than if widely spaced, as a smaller proportion of the plants become infected, and sometimes the number of plants infected per unit area is smaller also (A'Brook 1964; Way and Heathcote 1966; Heathcote 1970a). There should be as few gaps in a crop as possible, as plants on the edge of gaps, like plants on the edge of fields, are particularly liable to infection with air-borne viruses.

Time of sowing and of harvesting crops may affect virus incidence. In India early-sown crops of tobacco are less likely to be affected by whitefly-transmitted leaf curl than later-sown crops (Patel and Patel 1966) and in Britain late-sown crops of sugar beet are more likely to be affected with yellowing viruses than early-sown ones (Heathcote 1970a). Harpaz (1961) reported an interesting situation in Israel where the incidence of maize rough dwarf virus was decreased from 45–3% by delaying sowing until late May. The vector is a delphacid planthopper *Calligypona marginata* Fab. which overwinters on wild grasses and moves to irrigated maize fields when the grass withers at the end of the rainy season. The planthopper does not breed on the maize crop and is scarce during the hot weather from May to September. Similarly the spread of wheat streak mosaic virus in North America can be decreased by delaying sowing of winter wheat until early September when the previous wheat crop has been harvested and volunteer wheat plants carrying the vector, an eriophyid mite, *Aceria tulipae* Keifer, have been destroyed by cultivation (Slykhuis 1955).

Older plants are more resistant to virus infection than younger ones, and may be less attractive hosts to the vector, and so, where possible, crops should be sown so that they are well-grown before migrant vectors are likely to invade them. It may even be possible to harvest crops before there is much spread of virus. Potato crops in Scotland are harvested early, and the foliage is destroyed before any virus introduced late in the season can pass to the tubers, and in the Netherlands seed potatoes are not certified unless they have been lifted before a date fixed annually by the Nederlandse Algemene Keuringsdienst (N.A.K.) (Hille Ris Lambers 1955). It takes from 8–16 days for potato leaf roll virus to move from leaves to the tubers (Bradley and Ganong 1953), so the aerial parts are often destroyed by spraying with dilute sulphuric acid before harvest, though pulling and removing them is safer. It is important that the tubers are not allowed to sprout, as the sprouts are particularly attractive to aphids and susceptible to virus (Keller and Weiss 1956).

Spread of viruses cannot be prevented by changing fertilisers. The only nutrient that greatly affects virus vectors is nitrogen. Large dressings of nitrogen generally make a plant more attractive to aphids, and their fecundity may be increased (Van Emden 1966a). Conversely, salt can depress populations of *Aphis fabae* on sugarbeet plants by as much as 30%, but the effect is not economically significant (Watson et al. 1951).

If a relatively small proportion of plants is infected with a virus disease, 'roguing', removing infected plants, may be economically justified to control spread. When swollen shoot disease of cacao was first discovered in Nigeria the removal of infected trees was the only possible means of control, and between 1946 and 1950, one and a half million trees were destroyed, but this failed to eradicate the disease (Thresh 1959). Initially, only obviously infected trees were removed, but later all trees within 27.4 m (30 yds) of each outbreak, including apparently healthy ones, were removed. Roguing is regularly carried out in potato seed crops, removing genetically undesirable plants as well as those infected with virus diseases, and may be done with banana or abaca crops (Magee 1967). It is seldom completely successful as there is always a delay before virus-infected plants show symptoms, the disease may remain latent for a long period, and the method selects mild or symptomless strains of the virus. It is usually better to destroy infected plants with a herbicide (which may also kill the vectors) rather than to try and dig it up and risk disturbing active vectors.

Occasionally extra plants are grown to allow for the destruction of part of the crop to eradicate virus-diseased plants. Kromnek disease of Virginian tobacco (tomato spotted wilt virus) in South Africa is transmitted by thrips, *Thrips tabaci* Lindeman, which settle at random on the crop but do not move again from plant to plant. Van der Plank and Anderssen (1944) reported that where tobacco plants were planted in pairs, and later thinned (and at the same time diseased plants were removed), an epidemic which would have destroyed 40% of the crop set out in the usual way with one plant per hill was effectively controlled.

31.5.2 Biological control

This is usually taken to mean the release or encouragement of predators or parasites of the pest which have a host range limited to the pest species and its near relatives, so that they do not endanger beneficial species. This type of control has roused great public and scientific interest in recent years, but so far, seems to be of less value in controlling virus spread than in controlling plant pests. A small number of virus-vectors may cause much damage, and successful biological control techniques do not completely eradicate the pest species. This is because usually the populations of host and parasite vary cyclically and slightly out of phase; first the host pest multiplies rapidly and it is then attacked by predators and parasites, which in turn multiply rapidly and are present in large numbers after the pest population has declined. Some individuals of this generation of parasites or predators will survive until the pest numbers increase again, but the cycle depends on the survival of the pest. If all the pests are killed, the predators and parasites will starve and die and it will be necessary to release more in the future to control any pests that are reintroduced.

Orchards and glasshouses may provide a stable environment where parasites and predators of virus vectors can become established, but short-term field crops seldom provide the opportunity for them to do so.

Probably 40–60% of the aphids eaten by predators are consumed by coccinellid beetles (ladybirds), and as these beetles are relatively easy to introduce into crops artificially, their use to control aphid-borne viruses is often considered. Unfortunately, in the field, coccinellids only multiply once there is an established aphid population, and under favourable conditions aphids multiply faster than their predators. Late in the season the host plant may be damaged by aphids feeding, and it may become less suitable as food for aphids as it matures (Atwal and Sethi 1963), also intra-specific competition may slow aphid reproduction so that predators have a greater effect on their population. The effectiveness of aphidophagous insects in decreasing aphid populations has been reviewed by Van Emden (1966b).

About 30 species of Hymenoptera are known to attack *Myzus persicae*, the most important aphid vector species, although few are specific. The release of adult parasites is thought less likely to succeed than providing a contributing source of parasites by putting plants infested with parasitized aphids into the field (Starý, 1968). Similarly, Wyatt (1970) has suggested that *Myzus persicae* could be controlled in glasshouses by introducing stabilised populations of heavily parasitized aphids on chrysanthemum cuttings.

Fungi, including *Entomophthora* spp., can also cause heavy mortality of aphids under humid conditions (i.e. about r.h. 80%) but infection is generally negligible at r.h. 50% (Ramaseshiah 1967). *Entomophthora* spp. are difficult to culture, although they can be kept for one year in airtight tubes (Prasertphon 1967), but *Cephalosporium* spp., other parasitic fungi, can be cultured on sorghum grains and used to control aphids or mealybugs under humid conditions. Suspensions of this fungus sprayed on heavily-infested cacao plants considerably decreased the population of *Planococcoides njalensis* (Laing), the vector of swollen shoot (Rojter et al. 1966).

Predators of all groups of virus vectors are known, including nematodes, which are attacked by tardigrades, collembolids and by other nematodes, and even by some fungi which form a trap with an adhesive network of mycelial threads (Shepherd 1956; Southey 1965). However, predators and parasites of aphids have received particular attention.

31.5.3 *Indirect control of the vectors by chemicals*

Chemicals are not always used directly against air-borne virus vectors. As has alreay been pointed out (4.5), herbicides may be used to destroy the weed hosts of viruses, or defoliants sprayed on the winter hosts of the vectors to prevent egg laying. The latter method shows promise in an integrated control programme for *Myzus persicae* on peach trees described by Tamaki and Weeks (1968); the trees are defoliated in autumn before the sexual adults are able to lay overwintering eggs. There is also the possibility of destroying symbionts associated with some virus vectors, for example, *Crematogaster* ants construct protective 'tents' of vegetable material around *Planococcoides njalensis*, the mealybug vector of swollen shoot, and they do not thrive without the ants (Hanna et al. 1956).

Antibiotics can be used to slow virus multiplication or reproduction of the vector. Terramycin (0.2% concentration) produces complete sterility of *Aphis fabae* by the second generation when sprayed on infested *Vicia faba* plants (Jayaraj et al. 1967). A plant growth retardant, 'Cycocel', (2-chloroethyl) tremethylammonium chloride, also decreases the number of nymphs produced by *Brevicoryne brassicae* (L.) (Van Emden 1964), but, interesting though these findings are, there seems no advantage in using antibiotics or growth-regulating substances rather than pesticides intended to kill virus-vectors immediately.

31.5.4 Chemical control of the vectors

Considerable ingenuity has been shown by chemical manufacturers, agricultural engineers and plant pathologists in producing the various pesticides in different forms and devising different methods of applying them. Many can be applied without special equipment, as a seed dressing, or added to a drench of liquid fertiliser, as has been done to control *Myzus persicae* in glasshouse (Tonks 1968), but often the use of special equipment is economically justified.

The danger of undesirable side effects from using seed dressings is slight as only small quantities are used, and here the chemical can be used as an insurance against an attack by a pest that only occasionally causes crop loss. Large seed can be coated with a pesticidal dust, or systemic pesticides can be sprayed onto the seed where it is absorbed by the plant and translocated with the sap through the tissues after the seedlings have emerged through the soil. Sugar-beet 'seed' (or more correctly fruits) have been treated with disulfoton or phorate in the USA to control the leafhopper vector of curly top (Hills et al. 1960), and the technique has been tried in Britain to decrease the spread of yellowing viruses in sugar beet. If seed is pelleted in inert material to make it easier to drill, pesticides, such as menazon, can be incorporated in the pellet, but the germination of the seed may be decreased (Heathcote 1968b, 1970b). Potato 'seed' has also been experimentally treated with menazon, and the spread of potato leaf roll virus decreased by the treatment (Broadbent et al. 1964).

Systemic organophosphorous pesticides are sometimes used in the form of granules; adsorbed on particles of inert material such as pumice or charcoal. The granules may be used to treat soil, before or at sowing, or applied between the rows after the emergence of seedlings, or on the foliage of larger plants. Provided the granules are free from dust, they are easier and safer to handle than liquid formulations, the pesticide persists longer, and when applied to the soil is unlikely to harm beneficial insects or birds. Granules can protect emerging seedlings from pest attack when they are difficult to spray adequately, and if applied with the seed make it unnecessary for more machinery to be driven through the crop. Spray machinery generally causes some damage to crops, and may compact the soil. Pesticide granules are sometimes applied with fertiliser granules (Ridgeway et al. 1967) but the chemicals may not be compatible, and if they are of a different size or specific gravity will not be applied uniformly.

Pesticide granules decrease the damage done by curly top to sugar beet

(Malm and Finkner 1968), by potato leaf roll virus to potato crops (Burt et al. 1960) and by barley yellow dwarf virus to barley (Stern 1967).

The disadvantages of using granules are their cost (often equivalent to the cost of two sprays of a similar pesticide), their phytotoxicity (sugar-beet seedlings are very susceptible to injury for example) and that the decision to apply granules to the soil must be made early in the season when it is uncertain if pests will attack the crop, though late in the season granules can be applied from aircraft.

Pesticide dusts are seldom used today to control spread of viruses (their only advantage over sprays is that they are light to transport) and suspensions of contact insecticides (which require mixing in the field) are only a little more popular. Contact insecticides such as DDT may stimulate vectors to move from treated plants and increase virus spread, and, although organophosphorous systemic insecticides are less persistent than contact ones, the conservationist may consider it desirable to use an insecticide with a very limited period of activity. Most systemic insecticides are more dangerous to handle than contact ones, but they are degraded in plant tissues and do not leave toxic residues or taint edible crops, nor cause much harm to beneficial insects. Several European countries have imposed at least a partial ban on the use of persistent organochlorine insecticides such as dieldrin and DDT.

Pesticide sprays are generally diluted with water, but, because of the cost of moving water, 'low volume' (i.e. with little water) or even 'ultra low volume' (less than 5 l/ha) sprays rather than 'high volume' sprays are often used. By using a rotating disc device instead of more conventional spray nozzles (Burt et al. 1966) it is possible to spray undiluted insecticides. often in oil, which are as effective as standard emulsion-in-water sprays even when a much smaller quantity is applied per unit area (fig. 31.3).

Provided the area to be sprayed is large enough it may be cheaper to spray from the air using aircraft, helicopters or hovercraft than from the ground, as it can be done more quickly and without causing physical damage to crops by the passage of machinery, but most crops are sprayed from the ground, either using knapsack sprayers (1–3 gals capacity) or heavier, tractor-mounted sprayers (usually 40–100 gals capacity). All spraying requires skill and care, both to protect the operator from toxic chemicals and to apply droplets of suitable size onto the plants to ensure that the correct dose of pesticide is applied to each plant.

Crops may be sprayed at regular intervals to prevent virus spread but the number of sprays should be as small as possible, not only because of cost, but to prevent selection of strains of the vector which are resistant to the pesticide.

Fig. 31.3. A hand-held, rotary disc, ultra-low volume sprayer (Micron Sprayers Ltd.) operated by electric torch batteries. (Photograph Broom's Barn Experimental Station.)

Myzus persicae resistant to organophosphorous insecticides are known to occur in glasshouses treated regularly with these insecticides (Russell 1965). The timing of the application of a spray is most important, and with some high-value crops, the growing of which is organised centrally, the growers may be advised when crops should be sprayed to prevent virus spread. Thus the British Sugar Corporation field staff warns sugar-beet growers in Britain when aphids are likely to spread yellowing viruses in their areas, basing the warning on the number of aphids found by fieldmen of the Corporation on beet plants checked regularly when aphids are likely to colonize the plants (Hull 1968). A similar scheme exists in Denmark and other European countries (Stapel 1965). In the Netherlands, seed potato growers are warned by the N.A.K. when to spray their crops to prevent virus spread (Hille Ris Lambers 1955), and in New Zealand cereal growers are warned when to spray to decrease the spread of barley yellow dwarf virus (Lamb 1958).

31.6 Integrated control

No one method of control is likely to keep crops entirely free from virus infection and as many preventive measures should be taken as are economically justified. One of the main aims when integrating control measures is to avoid changing the 'balance of nature', so that, for example, there is the minimum possibility of the vectors becoming resistant to the control measures. New genes may arise giving insects resistance to what are normally very toxic chemicals. If using a pesticide is only one of several measures in a control programme, the pest needs to resist several adverse factors at the same time, and if less of the pesticide is used, the selection pressure exerted will be less also. For this reason the pesticide should also be changed from time to time, for example from organophosphorus to carbamate, to decrease the chance of resistant mutants being selected. After studies of an analogous system Jones et al. (1967) published mathematical models to simulate the population changes of the potato cyst-nematode infesting eelworm-resistant and susceptible potato varieties, and they showed that the useful life of the resistant variety was prolonged when the varieties were grown alternately.

31.7 Forecasting

It is always difficult to forecast the extent to which viruses will spread at any particular site in any one season, but recent work suggests that useful predictions can be made with some diseases; the use of weather information to predict plant disease epiphytotics has been reviewed by Austin Bourke (1970), and forecasting the occurrence of plant pests by Buhl and Schütte (1971).

Outbreaks of virus diseases carried by aphids can sometimes be related to weather data. In Britain, a weather factor computed from the number of days in January, February and March when temperatures are below freezing together with the mean weekly temperature in April is correlated closely with the percentage of sugar-beet plants that are infected with yellowing viruses in the following August, and thus can be used to predict yellows incidence (Watson 1966) (fig. 31.4). It seems likely that the same weather data can be used to predict severe outbreaks of carrot motley dwarf virus in carrot in Britain, although it has a different aphid vector. Freezing usually kills weeds and aphids living on them, and in mild winters these plants survive and are important as sources of virus as well as of aphids which move to crops in early spring. By contrast, aphid eggs laid on trees or shrubs are usually unaffected by cold winters (some contain glycerol, which accounts for their cold-hardiness (Sømme 1969)) but these alternative host plants do not carry the viruses that affect the crop plants. On the European mainland winter temperatures are lower than in Britain and aphids seldom overwinter on herbaceous hosts, but Gabriel (1965) found that spread of potato leaf roll and potato Y viruses depends upon late winter and early summer temperatures in Poland. In Denmark and in Britain the incidence of virus yellows of sugar beet is correlated with the number of aphid-infested mangold clamps unused during the winter (Stapel 1965) but this is also a reflection of the winter weather.

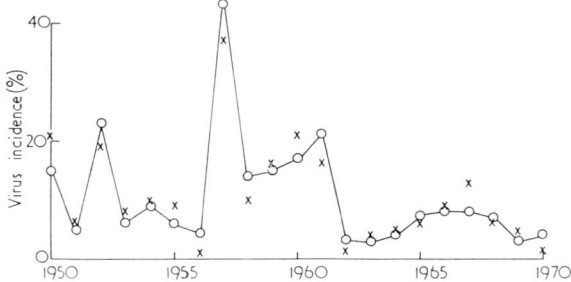

Fig. 31.4. Observed values for percentage of sugar-beet plants with yellowing viruses at the end of August in England for the years 1950–1970 (open circles) compared with the calculated values (crosses) from the multiple regression equation for dependence of yellows incidence on the number of freezing days in January to March and deviation from the mean temperature in April.

Unfortunately virus incidence cannot be predicted further ahead, for example, the incidence of virus disease in one year is usually independent of the incidence in the previous year (as shown by Gibbs (1966) with virus yellows of beet), and counts of eggs laid by aphids on their winter (primary) host have so far proved of little value in forecasting because, due to the attack of predators, there is no simple relation between their number and the number of individuals hatching from them and reaching maturity (Hille Ris Lambers 1955).

31.8 Conclusions

Control of spread of air- or soil-borne viruses is often difficult because spread can be initiated by small numbers of vectors, which may be too few to be considered as pests otherwise. Control of virus spread without the use of pesticides is desirable, as it avoids tainting food crops and pollution of the environment, but it is difficult to achieve and may not be as cheap as chemical control, and simple agronomic methods of control are seldom effective alone. Crops can be isolated geographically or in time from infected crops or weeds, and cover crops or barriers used to keep out the vectors once the area has been freed from sources of infection such as 'groundkeepers', stored bulbs or roots in clamps.

It may be possible to select cultivars that are resistant or tolerant to attack by virus diseases, or are resistant to the vectors, but plant breeding is a slow process. Control of virus vectors by releasing their parasites, predators or pathogens alone offers little promise, as the method depends upon the survival of a reservoir of the controlling species, and in turn this means that a small population of the vector must survive.

Little success has been achieved in finding chemicals to repel vectors or inhibit the transmission of viruses, but pesticides can be applied in many different forms and in many different ways to limit virus spread. Best control of viruses spread by invertebrates is obtained from well-designed 'integrated' programmes based on an understanding of the biology of the vector and of the virus, using cultural control measures and the minimum amount of pesticide applied at carefully chosen times. Considerable increases in yield can often be obtained for little cost when crops are kept free from virus diseases.

References

In parentheses after each reference is the number of the chapter(s) in which the reference(s) appear(s).

ABEDI, Z. H. and A. W. A. BROWN, 1961, Ann. Entomol. Soc. Am. *54*, 539–542. (30).

ABERNATHY, C. O. and R. THURSTON, 1969, J. Econ. Entomol. *62*, 1356–1359. (31).

ABDEL-WAHAB, K. S. E., R. E. WILLIAMS and M. N. KAISER, 1970, Folia Parasitol. *17*, 355–358. (18).

ABDEL-WAHAB, K. S. E., J. D. ALMEIDA, F. W. DOANE and D. M. MCLEAN, 1964, Can. Med. Assoc. J. *90*, 1068–1072. (14).

ABOUL-EID, H. Z., 1969, Nematologica *15*, 451–463. (12).

A'BROOK, J., 1964, Ann. Appl. Biol. *54*, 199–208. (31).

ACHESON, N H. and I. TAMM, 1967, Virology *32*, 128–143. (14).

ACHESON, N. H. and I. TAMM, 1970a, Virology *41*, 306–320. (4).

ACHESON, N. H. and I. TAMM, 1970b, Virology *41*, 321–329. (14).

ACTON, R. T., P. F. WEINHEIMER and E. E. EVANS, 1969, J. Invert. Pathol. *13*, 463. (16).

ADAM, A. V., 1962, Plant Disease Reptr. *40*, 366–370. (25).

ADAMS, J. B. and J. W. MCALLAN, 1956, Can. J. Zool. *34*, 541–543. (25).

ADAMS, J. B. and J. W. MCALLAN, 1958, Can. J. Zool. *36*, 305–308. (25).

ADLDINGER, H. K., S. S. STONE, W. R. HESS and H. L. BOEHRACH, 1966, Virology *30*, 750–752. (4).

ADLER, S. and O. THEODOR, 1957, Ann. Rev. Entomol. *2*, 203–226. (11).

AFIFI, S. A., 1968, Bull. Brit. Museum Nat. Hist. (Entomol. Suppl.) *7*, 1–210. (10).

AHMED, M. E., R. C. SINHA and R. M. HOCHSTER, 1970, Virology *41*, 768–771. (26).

AITKEN, T. H. G., C. B. WORTH, A. H. JONKERS, E. S. TIKASINGH and W. G. DOWNS, 1968, Am. J. Trop. Med. Hyg. *17*, 237–252. (11).

AIZAWA, K., 1959, J. Insect Pathol. *1*, 67–74. (14).

AIZAWA, K., 1962a, J. Insect Pathol. *4*, 72–76. (15).

AIZAWA, K., 1962b, J. Insect Pathol. *4*, 122–127. (15).

AIZAWA, K., 1963, *In:* Steinhaus, E. A., ed., Insect Pathology, Vol. 1 (Academic Press) pp. 381–412. (14, 15).

AIZAWA, K. and S. IIDA, 1963, J. Insect Pathol. *5*, 344–348. (4).

AIZAWA, K. and T. KAWARABATA, 1963, J. Invert. Pathol. *5*, 356–360. (14).

AKUTSU, K., 1967, *In:* The U.S.-Japan Committee on Scientific Cooperation, Proceedings of the Joint United States–Japan Seminar on Microbial Control of Insect Pests, Fukuoka, pp. 43–49. (29).

AL-AZAWI, A. and R. M. CHEW, 1959, Ann. Entomol. Soc. Am. *52*, 345–347. (30).

ALEVIZATOS, A. C., R. W. MCKINNEY and R. W. FEIZIN, 1967, Am. J. Trop. Med. Hyg. *16*, 762–768. (30).

ALLARD, H. A., 1917, J. Agric. Res. *10*, 615–632. (3, 25).

ALLEN, G. E., 1967, *In:* The U.S.–Japan Committee on Scientific Cooperation, Proceedings of the joint United States–Japan Seminar on Microbial Control of Insect Pests, Fukuoka, pp. 37–41. (29).

ALLEN, G. E. and T. L. PATE, 1966, J. Invert. Pathol. *8*, 129–131. (29).

ALLEN, G. E., B. G. GREGORY and T. L. PATE, 1967, J. Invert. Pathol. *9*, 40–43. (29).

ALLEN, G. E. and C. M. IGNOFFO, 1969, J. Invert. Pathol. *13*, 378–381. (14, 15).

ALLINGTON, W. B., R. STAPLES and G. VIEHMEYER, 1968, J. Econ. Entomol. *61*, 1137–1140. (19).

ALLISON, A. C. and D. C. BURKE, 1962, J. Gen. Microbiol. *27*, 181–194. (4).

ALTMAN, R. M., 1963, Am. J. Trop. Med. Hyg. *12*, 425–434. (20).

ALTMAN, R., M. GOLDFIELD and O. SUSSMAN, 1967, Med. Clinics of North America *51*, 661. (1).

AMOS, J., R. G. HATTON, R. C. KNIGHT and A. M. MASSEE, 1927, Ann. Rep. E. Malling Res. Sta., Kent, Suppl. *11*, 126–150. (19).

ANANTHAKRISHNAN, T. N., 1970, Indian Thysanoptera, C.S.I.R. Monograph No. 1. New Delhi. 171 pp. (13).

ANDERSON, C. R. and K. R. P. SINGH, 1971, Indian J. Med. Res. *59*, 195–198. (18).

ANDERSON, E. S., J. A. ARMSTRONG and J. S. F. NIVEN, 1959, *In:* Isaacs, A. and B. W. Lacey, eds., Symp. Soc. Gen. Microbiol. *9*, 224–255. (14).

ANDERSON, J. F., 1970, J. Invert. Pathol. *15*, 219–224. (4).

ANDERSON, J. R. and S. C. AYALA, 1968, Science *161*, 1023–1025. (11).

ANDERSON, R. C., 1959, Proc. 6th Internat. Congr. Trop. Med. Malaria *2*, 444–449. (11).

ANDERSON, S. G., M. DONNELLY, W. J. STEVENSON, N. J. CALDWELL and M. EAGLE, 1952, Med. J. Australia *1*, 110–114. (20).

ANDO, H., 1910, J. Japan Agr. Soc. *347*, 1–3. (3).

ANDREWES, C. H., 1957, *In:* The evolution of some animal tissues. Biological aspects of the Transmission of Disease, (Oliver and Boyd). (1).

ANDREWES, C. H., 1964, *Viruses of Vertebrates* 1st Ed. (Bailliere, Tindall & Cox. London) 401 pp. (30).

ANDREWES, C. H. and H. G. PEREIRA, 1967, *Viruses of Vertebrates* 2nd Ed. (Bailliere, Tindall & Cassell, London). (4).

APPEL, O., 1906, Jahresber. Angew. Botan. *3*, 122–136. (3).

ARAGAO, H. de B., and A. DA COSTA LIMA, 1929, Mem. Inst. Osw. Cruz. *10* (suppl.) 253–254. (20).

ARAI, K., Y. DOI, K. YORA, and H. ASUYAMA, 1969, Ann. Phytopathol. Soc. Japan *35*, 10–15. (4).

ARAMAN, S. F. and A. SAID, 1972, J. Parasitol. *58*, 348–353 (18).

ARISTOVA, V. A. and G. V. GOSTINSHCHIKOVA, 1971, Tezisy Dokl. Vop. Med. Virus., Inst. Virus. Imeni Ivanovsky, D.I., Akad. Med. Nauk SSSR (October 19–21), 123. (18).

ARNOTT, H. J. and K. M. SMITH, 1968a, J. Ultrastruct. Res. *21*, 251–268. (2, 14).

ARNOTT, H. J. and K. M. SMITH, 1968b, J. Ultrastruct. Res. *22*, 136–158 (4, 14).

ARNOTT, H. J., K. M. SMITH and S. L. FULLILOVE, 1968, J. Ultrastruct. Res. *24*, 479–507. (2, 4, 14).

ARTHUR, D. R., 1960, *Ticks, A Monograph of the Ixodoidea. Part V. On the genera Derma-*

centor, Anocentor, Cosmiomma, Boophilus and Margaropus, (Cambridge at the University Press) 251 pp. (5).

ARTHUR, D. R., 1963, *British Ticks*, (Butterworths, London) 213 pp. (5).

ARTHUR, D. R., 1965, *Ticks of the Genus Ixodes in Africa*, (The Athlone Press, University of London) 348 pp. (5).

ARTHUR, D. R. and R. P. CHAUDHURI, 1965, Parasitology 55, 391–400 (5).

ARUGA, H., N. YOSHITAKE and H. WATANABE, 1963a, J. Insect Pathol. 5, 72–77. (14).

ARUGA, H., T. HUKUHARA, S. FUKUDA and Y. HASHIMOTO, 1963b, J. Insect Pathol. 5, 415–421. (14).

ARUGA, H., T. HUKUHARA, N. YOSHITAKE and I. N. AYUDHYA, 1961, J. Insect Pathol. 3, 81–92. (14).

ASAHINA, S. and Y. TSURUOKA, 1968, Kontyu 36, 190–202. (8).

ASHBURN, P. M. and C. F. CRAIG, 1907, Philippine J. Sci. 2, 93. (1).

ATTAFUAH, A. and A. A. BRUNT, 1960, Ann. Rep. W. Afr. Cocoa Res. Inst., 1958–59, 16–17. (23).

ATTAFUAH, A. and T. W. TINSLEY, 1958, Ann. Appl. Biol. 46, 20–22. (23).

ATWAL, A. S. and S. L. SETHI, 1963, J. Animol. Ecol. 32, 481–488. (31).

AUCLAIR, J. L., 1963, Ann. Rev. Entomol. 8, 439–490. (7).

AUCLAIR, J. L. and J. J. CARTIER, 1963, Science 142, 1068–1069. (25).

AULMANN, G., 1913, *Psyllidarum Catalogus* (W. Junk, Berlin) 92 pp. (7).

AUSTIN BOURKE, P. M., 1970, Ann. Rev. Phytopathol. 8, 345–370. (31).

AUZINS, I. and D. ROWLEY, 1962, Australian J. Exptl. Biol. Med. Sci. 40, 283. (16).

AVSATTHI, B. L. and L. S. HIREGAUDAR, 1971, Gujvet 5, 43–46. (5).

AYALA, S. C., 1970a, J. Parasitol. 56, 387–388. (11).

AYALA, S. C., 1970b, J. Parasitol. 56, 417–442. (11).

AYALA, A. and M. W. ALLEN, 1966, Nematologica 12, 87. (27).

AYALA, A. and M. W. ALLEN, 1968, J. Agr. Univ. Puerto Rico 52, 101–125. (27).

AYALA, A., M. W. ALLEN and E. M. NOFFSINGER, 1970, J. Agr. Univ. Puerto Rico 54, 341–369. (12).

BABOS, S., 1964, *Die Zeckenfauna Mitteleuropas*, (Akad. Kiado, Budapest) 410 pp. (5).

BACON, J. S. D. and B. DICKINSON, 1957, Biochem. J. 66, 289–297, (7).

BAGNALL, R. H., 1953, Can. J. Agr. Sci. 33, 509–519. (25).

BAILEY, L., 1964, Wax moth. Rothamsted Exptl. Sta. Rept. for 1963, p. 167. (29).

BAILEY, L., 1965a, J. Invert. Pathol. 7, 132–140. (15, 22).

BAILEY, L., 1965b, J. Invert. Pathol. 7, 167–169. (22).

BAILEY, L., 1967, Ann. Appl. Biol. 60, 43–48. (22).

BAILEY, L., 1968, Ann. Rev. Entomol. 13, 191–212. (29).

BAILEY, L., 1969, Ann. Appl. Biol. 63, 483–491. (22).

BAILEY, L., 1970, Wax moth. Rothamsted Exptl. Sta. Rept. for 1969, p. 261. (29).

BAILEY, L., 1971, Virus diseases of the Honey Bee. Rothamsted Exptl. Sta. Rept. for 1970, pt. 2, pp. 171–183. (22, 29).

BAILEY, L. and A. J. GIBBS, 1964, J. Insect Pathol. 6, 395–407. (16, 22).

BAILEY, L., A. J. GIBBS and R. D. WOODS, 1963, Virology 21, 390–395. (15, 22).

BAILEY, L., A. J. GIBBS and R. D. WOODS, 1964, Virology 23, 425–429. (22).

BAILEY, L., A. J. GIBBS and R. D. WOODS, 1968, J. Gen. Virol. 2, 251–260. (4, 22).

BAILEY, L. and R. G. MILNE, 1969, J. Gen. Virol. 4, 9–14 (22).

BAILEY, S. F., 1957, Bull. Calif. Insect Surv. *4*, 141–220. (13).

BAILEY, S., B. J. MILLER and E. L. COOPER, 1971, Immunology *21*, 81. (16).

BAKER, R. E. D. and W. T. DALE, 1947, Ann. Appl. Biol. *34*, 60–65. (23).

BALACHOWSKY, A. S., 1963, Entomologie appliquée à l'agriculture I, Coléoptères Vol. 2. Paris. (9).

BALASHOV, YU. S., 1968, *Bloodsucking Ticks (Ixodoidea) – Vectors of Diseases of Man and Animals*, (Akademiya Nauk SSSR, Zoologicheske Institut, Leningrad) pp. 319 (1967). (5, 18).

BALCH, R. E. and F. T. BIRD, 1944, Sci. Agr. *25*, 65–80. (22, 29).

BALD, J. G. and T. W. TINSLEY, 1970, Virology 40, 369–378. (23).

BALDUCCI, M. P. VERANI, M. C. LOPES and F. NARDI, 1970, Acta Virol., Prague (English edition) *14*, 237–243. (18).

BALL, E. D., 1909, U.S.D.A. Entomol. Bull. *66*, 33–52. (3).

BANERJEE, K. and K. R. P. SINGH, 1968, Indian J. Med. Res, *56*, 812–814. (14).

BANG, F. B., 1961, Biol. Bull. *121*, 57. (16).

BANG, F. B., 1966, J. Immunol. *96*, 960. (16).

BANG, F. B., 1971, Infect. Immunol. *3*, 617–623. (28).

BANKS, J. C. and E. D. M. MACAULEY, 1964, Ann. Appl. Biol. *53*, 229–242. (24).

BANTARI, E. E. and R. J. ZEYEN, 1969, Phytopathology *59*, 183–186. (26).

BARNES, W. W. and A. B. WEBB, 1968, Mosquito News. *28*, 458–461. (30).

BARNETT, H. C., 1962, *In:* Maramarosch, K., ed., *Biological Transmission of Disease Agents*, (New York & London: Academic Press). (11).

BARNETT, H. C. and W. SUYEMOTO, 1961, Trans. N.Y. Acad. Sci. Ser. II *23*, 609–617. (20).

BARTLETT, B. R. and J. C. BALL, 1966, Ann. Entomol. Soc. Am. *59*, 42. (16).

BARWISE, A. H. and I. O. WALKER, 1970, FEBS Letters *6*, 13–16. (4).

BASHKIRTSEV, V. N., M. P. CHUMAKOV, V. V. BEREZIN and A. M. BUTENKO, 1971, Tezisy Dokl. Vop. Med. Virus., Inst. Virus. Imeni Ivanovsky, D. I., Akad. Med. Nauk SSSR (October 19–21), 139–140. (In Russian, Engl. trans.: NAMRU3-T511). (18).

BATES, M. and M. ROCA-GARCIA, 1946, Am. J. Trop. Med. *26*, 585–605. (30).

BATH, J. E. and R. K. CHAPMAN, 1967, Virology *33*, 503–506. (24).

BAT-MIRIAM, M. and G. B. CRAIG, 1966, Mosquito News. *26*, 13–22. (30).

BAWDEN, F. C., 1939, *Plant Viruses and Virus Diseases*, 1st Ed. (Chronica Botanica, Waltham, Mass.). (3).

BAWDEN, F. C., 1943, *Plant Viruses and Virus Diseases*, 2nd Ed. (Chronica Botanica, Waltham, Mass.). (3).

BAWDEN, F. C., 1950, *Plant Viruses and Virus Diseases*, 3rd Ed. (Chronica Botanica, Waltham, Mass.). (1, 10, 26).

BAWDEN, F. C., 1964, *Plant Viruses and Virus Diseases*, 4th Ed. (Ronald Press, New York). (25).

BAWDEN, F. C. and N. W. PIRIE, 1937, Proc. Roy. Soc. (London) *B123*, 274–320. (3).

BAWDEN, F. C. and N. W. PIRIE, 1938, Brit. J. Exptl. Pathol. *19*, 251–263. (3).

BAWDEN, F. C., N. W. PIRIE, J. D. BERNAL and I. FANKUCHEN, 1936, Nature *138*, 1051. (3).

BEADLE, J. D., 1966, Mosquito News. *26*, 483–486. (30).

BEALE, G. H., A. JURAND and J. R. PREER, 1969, J. Cell Sci. *5*, 65–91. (28).

BEARDSLEY, J. W., 1959, Proc. Hawaiian Entomol. Soc. *17*, 29–37. (10).

BEARDSLEY, J. W., 1960, Proc. Hawaiian Entomol. Soc. *17*, 199–243. (10).

BEASLEY, A. R., W. LICHTER and M. M. SIGEL, 1961, Arch. Ges. Virusforschung *10*, 672–683. (14).

BEAVERS, J. B. and R. B. HAMPTON, 1971, Ann. Entomol. Soc. Am. *64*, 804–806. (6).

BEGUM, F., C. L. WISSEMAN jr. and J. CASALS, 1970a, Am. J. Epidemiol. *92*, 192–194. (18).

BEGUM, F., C. L. WISSEMAN jr. and J. CASALS, 1970b, Am. J. Epidemiol. *92*, 195–196. (18).

BEGUM, F., C. L. WISSEMAN jr. and J. CASALS, 1970c, Am. J. Epidemiol. *92*, 197–202. (18).

BEIJERINCK, M. W., 1898, Centr. Bakteriol. Abt. II *5*, 27–33. (4).

BEIRNE, B. P., 1956, Canad. Entomol. *88*, Suppl. 2, 1–180. (8).

BEKLESHOVA, A. YU., I. I. TERSKIKH and V. A. SMIRNOV, 1971, Tezisy Dokl. Vop. Med. Virus., Inst. Virus, Imeni Ivanovsky, D. I., Akad. Nauk. SSSR (October 19–21), 117–118. (In Russian.). (18).

BELLAMY, R. E., W. C. REEVES and R. P. SCRIVANI, 1967, Am. J. Epidemiol. *85*, 282–296. (30).

BELLETT, A. J. D., 1965a, Virology *26*, 127–131. (2, 14).

BELLETT, A. J. D., 1965b, Virology *26*, 132–141. (14).

BELLETT, A. J. D., 1968, Advan. Virus Res. *13*, 225–246. (2, 4, 14, 15).

BELLETT, A. J. D., 1969, Virology *37*, 117–123. (4, 14).

BELLETT, A. J. D. and R. B. INMAN, 1967, J. Mol. Biol. *25*, 425–432. (4).

BELLETT, A. J. D. and E. H. MERCER, 1964, Virology *24*, 645–653. (14).

BELLETT, A. J. D. and F. FENNER, 1968, J. Virol. *2*, 1374–1380. (4).

BENDELL, R. J. E., 1972, Trans. R. Soc. Trop. Med. Hyg. *64*, 497–502. (30).

BEN-ISHAI, Z., N. GOLDBLUM and Y. BECKER, 1968, J. Gen. Virol. *2*, 365–375. (14).

BENNETT, C. W., 1934, J. Agr. Res. *48*, 665–701. (3).

BENNETT, C. W., 1953, *In:* Plant Diseases, the Year Book of Agriculture. U.S. Dept. Agr. (3).

BENNETT, C. W., 1957, J. Am. Soc. Sugar Beet Tech. *9*, 553–565. (31).

BENNETT, C. W. and H. E. WALLACE, 1938, J. Agr. Res. *56*, 31–50. (26).

BENNETT, C. W. and A. S. COSTA, 1961, Phytopathology *51*, 546–550. (3, 20).

BENNETT, G. F., 1961, Can. J. Zool. *39*, 17–33. (11).

BENNETT, G. F., P. C. C. GARNHAM, and A. M. FALLIS, 1965, Can. J. Zool. *43*, 927–932. (11).

BENNETT, H. S., 1966, J. Biophys. Biochem. Cytol. *2*, 4, Suppl., 99. (16).

BENSTEAD, R. J., 1951, In: Rep. Cocoa Conf. London 1951. The Cocoa, Chocolate and Confectionery Alliance Ltd. pp. 111–116. (23).

BENZ, G., 1962, Schweiz. Entomol. Gesch. *34*, 382–392. (29).

BENZ, G., 1963, J. Insect Pathol. *5*, 215–241. (15).

BEREZIN, V. V., M. P. CHUMAKOV, V. N. BASHKIRTSEV and B. F. SEMENOV, 1971, Tezisy Dokl. Vop. Med. Virus., Inst. Virus. Imeni Ivanovsky, D. I., Akad. Med. Nauk SSSR (October 19–21), 137–138. (In Russian, Engl. trans.: NAMRU3-T510). (18).

BERGE, T. O., R. E. SHOPE and T. H. WORK, 1971, Am. J. Trop. Med. Hyg. *19*, 1082–1160. (11).

BERGESON, G. B. and K. L. ATHOW, 1963, Phytopathology *53*, 871. (27).

BERGOIN, M., 1966, Entomophaga *11*, 253–259. (21).

BERGOIN, M., G. DEVAUCHELLE and C. VAGO, 1971, Virology *43*, 453–467. (4).

BERGOLD, G. H., 1947, Z. Naturforsch. *2b*, 122–143. (2, 29).

BERGOLD, G. H., 1948, Z. Naturforsch. *3b*, 338–342. (2).

BERGOLD, G. H., 1958, *In:* Hallauer, C. and K. F. Meyer, eds., *Handbuch der Virusforschung*, Vol. 4 pp. 60–142. (Springer, Vienna). (4, 21).

BERGOLD, G. H., 1963a, *In:* Steinhaus, E. A., ed., Insect Pathology. An Advanced Treatise, vol. *1* Academic Press, New York. pp. 413–456. (4, 21).

BERGOLD, G. H., 1963b, J. Ultrastruct. Res. *8*, 360–378. (4).

BERGOLD, G. H. and J. WEIBEL, 1962, Virology *17*, 554–562. (14, 15).

BERGOLD, G. H., O. M. SUAREZ and K. MUNZ, 1968, J. Insect Pathol. *11*, 406–428. (14).

BERKALOFF, A., J. C. BREGLIANO and A. OHANESSIAN, 1965, Compt. Rend. *260*, 5956–5959. (4, 20).

BERNARD, J., 1970, J. Gen. Virol. *8*, 209–218. (4, 20).

BERRY, R. E. and R. G. SIMPSON, 1967, Colo. Agr. Expt. Sta. Tech. Bull. *92*, 34 pp. (25).

BERTRAM, D. S., K. UNSWORTH and R. M. GORDON, 1964, Ann. Trop. Med. Parasitol. *40*, 223–254. (6).

BERTRAM, D. S., M. G. R. VARMA, R. C. PAGE and O. H. U. HEATHCOTE, 1970, J. Med. Entomol. *7*, 267–270. (30).

BEST, R. J., 1968, Adv. Virus. Res. *13*, 65–146. (13, 28).

BETTO, E. and D. J. RASKI, 1966, Nematologica *12*, 453–461. (27).

BHALLA, S. C. and G. B. CRAIG, 1964, Bull. Entomol. Soc. Am. *10*, 173. (30).

BHAT, U. K. M. and K. R. P. SINGH, 1969, J. Med. Entomol. *6*, 71–74. (17).

BHAT, U. K. M. and K. R. P. SINGH, 1970, Current Sci. *39*, 388–390. (17).

BHATT, P. N., K. G. KULKARNI, J. BOSHELL, P. K. RAJAGOPALAN, A. P. PATIL, M. K. GOVERDHAN and K. M. PAVRI, 1966, Am. J. Trop. Med. Hyg. *15*, 958–960. (18).

BILS, R. F. and C. E. HALL, 1962, Virology *17*, 123–130. (26).

BIRD, F. T., 1949, Nature *163*, 777. (22).

BIRD, F. T., 1953a, Can. J. Zool. *31*, 300–303. (22).

BIRD, F. T., 1953b, Can. Entomologist *85*, 437–446. (22, 29).

BIRD, F. T., 1955, Can. Entomologist *87*, 124–127. (29).

BIRD, F. T., 1959, J. Insect Pathol. *1*, 406–430. (15).

BIRD, F. T., 1961, J. Insect Pathol. *3*, 352–379. (21, 22, 29).

BIRD, F. T., 1962, Can. J. Microbiol. *8*, 533–534. (14).

BIRD, F. T., 1964, Can. J. Microbiol. *10*, 49–52. (14).

BIRD, F. T., 1964, Intern. Colloq. Insect Pathol. et Lutte Microbiol., 2e, Paris, 1962, pp. 465–473. (21).

BIRD, F. T., 1965, Can. J. Microbiol. *11*, 497–501. (14).

BIRD, F. T., 1966, Can. J. Microbiol. *12*, 337–339. (14).

BIRD, F. T. and J. M. BURK, 1961, Can. Entomologist *93*, 228–238. (22, 29).

BIRD, F. T. and D. E. ELGEE, 1957, Can. Entomologist *89*, 371–378. (29).

BIRD, F. T. and M. M. WHALEN, 1953, Can. Entomologist *85*, 433–437. (22).

BIRD, G. W., 1971, J. Nematol. *3*, 50–57. (12).

BISHOPP, F. C. and H. L. TREMBLEY, 1945, J. Parasit. *31*, 1–54. (5).

BLACK, L. M., 1943, Genetics *28*, 200–209. (3).

BLACK, L. M., 1950, Nature *166*, 852–853. (3, 26).

BLACK, L. M., 1953, Phytopathology *43*, 466. (15).

BLACK, L. M., 1959, *In:* Burnet, F. M. and W. M. Stanley, eds., *The Viruses*, Vol. II, (Academic Press, New York) pp. 157–185. (24, 26).

BLACK, L. M., 1969, Ann. Rev. Phytopathol. *7*, 73–100. (17, 26).

BLACK, L. M., 1970, C.M.I./A.A.B. Descrip. Plant Viruses No. *35*. (4).

BLACK, L. M. and BRAKKE, M. K., 1952, Phytopathology *42*, 269–273. (15, 26).

BLACK, L. M. and R. MARKHAM, 1963, Neth. J. Plant Pathol. *69*, 215. (26).

BLAIR, C. A. and J. R. GROVES, 1952, J. Hort. Sci. *27*, 14–43. (6).

BLASKOVIC, D., 1967, Bull. World Health Organ. *36*, 5–13. (18).

BLASKOVIC, D., 1970, Arch. Environ. Health *21*, 453–461. (18).

BLASKOVIC, D. and J. NOSEK, 1972, Progr. Med. Virol. *14*, 275–320. (18).

BLASKOVIC, D., G. PUCEKOVA, L. KUBINYI, S. STUPALOVA and V. ORAVCOVA, 1967, Bull. World Health Organ. *36*, 89–94. (18, 30).

BLENCOWE, J. W., 1962, In: Rep. Cocoa Conf. London 1961. The Cocoa, Chocolate and Confectionery Alliance Ltd. pp. 141–145. (23).

BLENCOWE, J. W., A. A. BRUNT, R. H. KENTEN and N. K. LOVI, 1963, Trop Agr. *40*, 233–236. (23).

BLOWERS, J. R. and V. C. MORAN, 1967, J. Entomol. Soc. S. Africa *30*, 75–81. (7).

BOL, J. F., L. VLOTEN-DOTING and E. M. J. JASPERS, 1971, Virology *46*, 73–85. (4).

BOLLE, J., 1894, Atti e Memoire dell'i.T. Societa agraria, Gorz. (2).

BONAMI, J. R. and C. VAGO, 1971, Experientia *27*, 1363–1364. (28).

BOND, J. O., D. T. QUICK, A. L. LEWIS, W. MCD. HAMMON and G. E. SATHER, 1966, Am. J. Epidemol. *83*, 564–570. (30).

BONDAR, G., 1923, *Aleyrodideos do Brasil* (Secretaria da Agricultura, Bahia) 182 pp. (13).

BONDAR, G., 1924, Chacaras e quintaes *30*, 216–218. (13).

BONE, G. J., 1943, Ann. Soc. R. Zool. Belg. *74*, 16–31. (18).

BOORMAN, J., 1960, Trans. Roy. Soc. Trop. Med. Hyg. *54*, 362–365. (20).

BORATYNSKI, K. and R. G. DAVIES, 1971, Biol. J. Linnean Soc. *3*, 57–102. (10).

BOS, L., 1970, C.M.I./A.A.B. Descrip. Plant Viruses No. *40*. (4).

BOS, L., and E. M. J. JASPERS, 1971, C.M.I./A.A.B. Descrip. Plant Viruses No. *46*. (4).

BOSHELL, J., 1969, Am. J. Trop. Med. Hyg. *18*, 67–70. (18).

BOWNE, J. G. and R. H. JONES, 1966, Virology *30*, 127–133. (4, 15).

BOX, H. E., 1945, Nature (London) *155*, 608. (3).

BOYCE, A. M., 1948, In: Batchelor, L. D. and H. J. Webber, eds., *The Citrus Industry*, Vol. 2, (Univ. California Press) pp. 665–812. (6).

BRADFUTE, O. E., R. E. WHITMOYER, and L. R. NAULT, 1970, 28th Ann. Proc. EMSA, 178–179. (19).

BRADLEY, R. H. E., 1952, Ann. Appl. Biol. *39*, 78–97. (25).

BRADLEY, R. H. E., 1956, Can. J. Microbiol. *2*, 539–547. (25).

BRADLEY, R. H. E., 1959, IXth International Botanical Congress, Montreal.

BRADLEY, R. H. E., 1961, In: *Recent Advances in Botany*, Vol. 1 (University of Toronto Press, Toronto) pp. 528–533. (3, 25).

BRADLEY, R. H. E., 1962, Can. Entomol. *94*, 707–722. (25).

BRADLEY, R. H. E., 1963, Can. J. Microbiol. *9*, 369–380. (25).

BRADLEY, R. H. E., 1964, In: Corbett, M. K. and H. D. Sisler, eds., *Plant Virology* (University of Florida Press, Gainesville) pp. 148–174. (25).

BRADLEY, R. H. E. and M. T. COUSIN, 1969, Virology *39*, 338–342. (25).

BRADLEY, R. H. E. and R. Y. GANONG, 1953, Can. J. Botany *31*, 143–144. (31).

BRADLEY, R. H. E. and R. Y. GANONG, 1955a, Can. J. Microbiol. *1*, 775–782. (3, 25).

BRADLEY, R. H. E. and R. Y. GANONG, 1955b, Can. J. Microbiol. *1*, 783–793. (25).

BRADLEY, R. H. E. and K. F. HARRIS, 1972, Virology *50*, 615–618. (25).

BRADLEY, R. H. E., C. A. MOORE and D. D. POND, 1966, Nature London, *209*, 1370–1371. (25).

BRADLEY, R. H. E., C. V. WADE and F. A. WOOD, 1962, Virology *18*, 327–328. (25).

BRAKKE, M. K., 1971, C.M.I./A.A.B. Descrip. Plant Viruses. No. *48*. (4).

BRAKKE, M. K., L. M. BLACK and R. W. G. WYCKOFF, 1951, Am. J. Botany *38*, 332–342. (26).

BRAKKE, M. K., A. E. VATTIER and L. M. BLACK, 1954, Brookhaven Symp. Biol. *6*, 137–156. (26).

BRČÁK, J., 1957, Phytopathol. Z. *30*, 414–428. (9).

BREECE, J. R. and W. H. HART, 1959, Plant Disease Reptr. *43*, 989–990. (27).

BREESE, S. S. and C. J. DE BOER, 1966, Virology *28*, 420–428. (4).

BRÈS, P., M. CORNET and Y. ROBIN, 1967, Ann. Inst. Pasteur Paris *113*, 739–747. (18).

BRICENO ROSSI, A. L., 1964, Rev. Venez. Sanid. Asist. Soc. *29*, 351–437. (30).

BRIGGS, J. D., 1958, J. Exptl. Zool. *138*, 155. (16).

BRIGGS, J. D., 1964, *In:* Rockstein, M., ed., *The physiology of insecta*, vol. 3 (Academic Press, N.Y.) p. 259. (16).

BRITTAIN, W. H., 1923, Proc. Arcad. Entomol. Soc. *8*, 23–42. (7).

BROADBENT, L., 1957a, Ann. Rev. Entomol. *2*, 339–354. (31).

BROADBENT, L., 1957b, *Agri. Res. Council Rep. Series No. 14*. Cambridge University Press. 31).

BROADBENT, L., P. E. BURT and G. D. HEATHCOTE, 1964, Exptl. Hort. *11*, 40–50. (31).

BROOKS, G. D., P. NERI, N. G. GRATZ and D. B. WEATHERS, 1970, Bull. World Health. Organ. *42*, 37–54. (30).

BROOKS, M. A. and T. J. KURTTI, 1971, Ann. Rev. Entomol. *16*, 27–52. (17).

BROOKS, W. M., 1969, *In:* Jackson, G. J., R. Herman and I. Singer, eds., *Immunity to Parasitic Animals*, vol. 1 (North-Holland, Amsterdam) pp. 149–171. (16).

BROWN, A. W. A., 1967, Bull. World Health Organ. *36*, 578–580. (30).

BROWN, A. W. A. and Z. H. ABEDI, 1962, Canad. J. Genet. Cytol. *4*, 319–332, (30).

BRUNT, A. A., 1970, C.M.I./A.A.B. Descrip. Plant Viruses No. *10*. (4).

BRUNT, A. A., 1971, C.M.I./A.A.B. Descrip. Plant Viruses, No. *51*. (4).

BRUNT, A. A. and R. H. KENTEN, Virology *16*, 199–201. (23).

BUCHER, G. E., 1958, *In:* Becker, E. C., ed., Proc. 10th Int. Cong. Ent. Montreal, 1956, Vol. 4. pp. 695–702. (29).

BUCHER, G. E. and P. HARRIS, 1968, J. Invert. Pathol. *10*, 235–244. (21).

BUCKLEY, S. M., 1969, Proc. Soc. Exptl. Biol. Med. *131*, 25–30. (14).

BUCKLEY, S. M., 1971, Trans. Roy. Soc. Trop. Med. Hyg. *65*, 535–536. (14).

BUCKLEY, S. M. and J. CASALS, 1970, Am. J. Trop. Med. Hyg. *19*, 680–691. (18).

BUHL, C. and F. SCHÜTTE, 1971, *Prognose wichtiger Pflanzenschädlinge in der Landwirtschaft*. (Parey, Berlin) 364 pp. (31).

BUESCHER, E. L. and W. F. SCHERER, 1959, Am. J. Trop. Med. Hyg. *8*, 719–722 (20, 30).

BURGDORFER, W., 1963, Exptl. Parasitol. *14*, 152–159. (18).

BURGDORFER, W. and M. G. R. VARMA, 1967, Ann. Rev. Entomol. *12*, 347–376. (18, 30).

BURGE, B. W. and J. H. STRAUSS, 1970, J. Mol. Biol. *47*, 449–466. (14).

BURGERJON, A. and P. GRISON, 1965, J. Invert. Pathol. *7*, 281–284. (21).

BURRELL, C. J., E. M. MARTIN and P. D. COOPER, 1970, J. Gen. Virol. *6*, 319–323. (14).

BURT, P. E., L. BROADBENT and G. D. HEATHCOTE. 1960, Ann. Appl. Biol. *48*, 580–590. (31).

BURT, E. C., D. B. SMITH and E. P. LLOYD, 1966, J. Econ. Entomol. *59*, 1487–1489. (31).

BURTON, A. N., J. R. MCLINTOCK and J. G. REMPEL, 1966, Science *154*, 1029–1031. (30).

BUSNEL, R. C. and M. CHEVALIER, 1938, Rev. Zool. Agric. Appl. *37*, 180–187. (9).

BUSSEREAU, F., 1969, Compt. Rend. *269*, 2620–2623. (20).

BUSSEREAU, F., 1970a, Ann. Inst. Pasteur. *118*, 367–385. (20).

BUSSEREAU, F., 1970b, Ann. Inst. Pasteur. *118*, 626–645. (20).

BUSVINE, J. R., 1967, Bull. World Health Organ. *37*, 287–292. (30).

BULLOCK, H. R., 1967, J. Invert. Pathol. *9*, 434–436. (29).

BULLOCK, H. R., C. L. MAGNUM and A. A. GUERRA, 1969, J. Invert. Pathol. *14*, 271–273. (21).

CADMAN, C. H., 1963, Ann. Rev. Phytopathol. *1*, 143–172. (31).

CADMAN, C. H., H. F. DIAS and B. D. HARRISON, 1960, Nature, Lond. *187*, 577–579. (27).

CAIRNS, J., 1960, Virology *11*, 603–623. (14).

CALDWELL, J. S., 1938, Bull. Ohio Biol. Surv. *34*, 228–281. (7).

CALDWELL, J. S. and L. F. MARTORELLA, 1951, Ann. Entomol. Soc. Am. *44*, 603–613. (7).

CALVERT, E. L. and B. D. HARRISON, 1963, Hort. Res. *2*, 115–120. (31).

CAMERON, G. R., 1932, J. Pathol. Bact. *35*, 933. (16).

CAMERON, G. R., 1934, J. Pathol. Bact. *38*, 441. (16).

CAMERON, J. W. M., 1967, *In:* Van der Laan, P. A., ed., Proceedings of the International Colloquium on Insect Pathology and Microbial Control, Wageningen, 1966, pp. 182–196, (29).

CAMERON, J. W. M., 1968, Proc. Entomol. Soc. Ontario *99*, 73–79. (29).

CAMIN, J. H., 1953, Chicago Acad. Sci., spec. Publs. No. *10*, 1–75. (6).

CAMPBELL, K. G., 1964, J. Entomol. Soc. Australia *1*, 3–4. (7).

CAMPBELL, R. N., 1971, C.M.I./A.A.B. Descrip. Plant Viruses. No. *43*. (4).

CAMPBELL, R. N. and W. M. COLT, 1967, Phytopathology *57*, 502–504. (9).

CANERDAY, T. D. and F. S. ARANT, 1968, J. Invert. Pathol. *12*, 344–348. (14, 21).

CANTWELL, G. E., 1967, J. Invert. Pathol. *9*, 138–140. (21, 29).

CANTWELL, G. E., R. M. FAUST and H. K. POOLE, 1968, J. Invert. Pathol. *10*, 161–162. (14).

CAPENER, A. L., 1970, J. Entomol. Soc. S. Afr. *33*, 195–200. (7).

CAREY, D. E., O. R. CAUSEY, S. REDDY and A. R. COOKE, 1971, Lancet (I), *7690*, 105–107. (30).

CARNEGIE, J. W. and G. S. BEAUDREAU, 1969, J. Virol. *4*, 311–312. (14).

CARREKER, J. R., 1965, Trans. Am. Soc. Agri. Eng. *8*, 161–166. (30).

CARSNER, E., 1919, Phytopathology *9*, 413–421. (3).

CARTER, W., 1927, J. Agric. Res. *34*, 449–453. (3).

CARTER, W., 1962, Insects in relation to plant disease, (Interscience Publishers, New York and London) pp. 238–265. (23).

CARTER, W., 1963, Ann. N.Y. Acad. Sci. *105*, 741–764. (3, 10).

CARTWRIGHT, K. L. and D. C. BURKE, 1970, J. Gen. Virol. *6*, 231–248. (14).

CASALS, J., 1957, Trans. N.Y. Acad. Sci. *19*, 219–235. (18, 30).

CASALS, J., 1962, Proc. Symp. Biol. Viruses Tick-borne Enceph. Complex (Smolenice, October 11–14, 1960) pp. 53–66. (18).

CASALS, J., 1963, Anals Microbiol. *11*, 13–34. (18).

CASALS, J., 1967, Japan. J. Med. Sci. Biol. *20*, 119–129. (18).

CASALS, J., 1970, Misc. Publ. Entomol. Soc. Am. *6*, 327–329. (18).

CASALS, J., 1971, (Intr. Lect. Proc. Symp.) Int. Symp. Tick-borne Arboviruses (Excluding Group B) (Smolenice, September 9–12, 1969) pp. 13–20. (18).

CASALS, J. and D. H. CLARKE, 1965, *In:* Horsfall, F. L. and I. Tamm, eds., *Viral and Rickettsial Infections of Man.*, (Lipincott Co, Philadelphia and Toronto) pp. 583 and 603. (14).

CASALS, J., B. E. HENDERSON, H. HOOGSTRAAL, K. M. JOHNSON and A. SHELOKOV, 1970, J. Infect. Diseases *122*, 437–453. (18).

CASALS, J., H. HOOGSTRAAL, K. M. JOHNSON, A. SHELOKOV, N. H. WIEBENGA and T. H. WORK, 1966, Am. J. Trop. Med. Hyg. *15*, 751–764. (18, 30).

CASALS, J. and L. WHITMAN, 1960, Am. J. Trop. Med. Hyg. *9*, 73–77. (30).

CASALS, J. and L. WHITMAN, 1961, Am. J. Trop. Med. Hyg. *10*, 250–258. (30).

CATHERALL, P. L., 1970, C.M.I./A.A.B. Descrip. Plant Viruses. No. *23*, (4).

CATHERALL, P. L., 1970, Plant Pathol. *19*, 101–103. (9).

CATLING, H. D., 1969, J. Entomol. Soc. S. Afr. *32*, 191–223. (7).

CAUSEY, O. R., G. E. KEMP, M. H. MADBOULY and T. S. DAVID-WEST, 1970, Am. J. Trop. Med. Hyg. *19*, 846–850. (18, 30).

CAUSEY, O. R., G. E. KEMP, M. H. MADBOULY and V. H. LEE, 1969, Bull. Soc. Pathol. Exotique *62*, 249–253. (18).

CAUSEY, O. R., R. E. SHOPE and A. RODRIGUES, 1962, Rev. Serv. Esp. Saude Publ. *12*, 55–59. (30).

CHADWICK, J. S., 1967, Federation Proc. *26*, 1675. (16).

CHADWICK, J. S., 1970, J. Invert. Pathol. *15*, 455. (16).

CHAIN, M. M. T., F. W. DOANE and D. M. MCLEAN, 1966, Can. J. Microbiol. *12*, 895–900. (14).

CHAMBERLAIN, R. W., E. C. CORRISTAN and R. K. SIKES, 1954, Am. J. Hyg. *60*, 269–277. (30).

CHAMBERLAIN, R. W., D. B. NELSON and W. D. SUDIA, 1954, Am. J. Hyg. *60*, 278–285. (20).

CHAMBERLAIN, R. W. and W. D. SUDIA, 1957, J. Infect. Diseases *101*, 233–236. (20).

CHAMBERLAIN, R. W. and W. D. SUDIA, 1961, Ann. Rev. Entomol. *6*, 371–390. (11, 14).

CHAMBERLIN, F. S. and S. R. DUTKY, 1958, J. Econ. Entomol. *51*, 560. (29).

CHAMBERS, V. C., 1957, Virology, *3*, 62–75. (14).

CHAMBON, L. and I. WONE, 1967, Bull. World Health Organ. *36*, 113–114. (30).

CHANG, I. C., K. W. HONG and K. S. WHANG, 1965, Korean Med. J. *10*, 77. (30).

CHANTANAO, A. and H. J. JENSEN, 1969, J. Nematol. *1*, 216–218. (12).

CHAO, J. and G. H. BALL, 1971, Current Top. Microbiol. Immunol. *55*, 28–31. (17).

CHAPMAN, H. C., T. B. CLARK, D. B. WOODARD and W. R. KELLEN, 1966, J. Invert. Pathol. *8*, 545–546. (20).

CHAUTHANI, A. R., D. CLAUSSEN and C. S. REHNBORG, 1968, J. Invert. Pathol. *12*, 335–339. (29).

CHEN, T. A. and R. R. GRANADOS, 1970, Science *167*, 1633–1636. (26).

CHEN, T. A. and W. F. MAI, 1965, Phytopathology *55*, 128. (27).

CHENG, T. C., E. RIFKIN and H. W. F. YEE, 1968, J. Invert. Pathol. *11*, 302. (16).

CHITWOOD, B. G. and M. B. CHITWOOD, 1950, *An Introduction to Nematology* (B. G. Chitwood, Baltimore) 213 pp. (12).

CHITWOOD, M. B., 1969, *In:* Florkin, M. and B. T. Scheer, eds., *Chemical Zoology*, Vol. 3 (Academic Press, New York and London) pp. 223–244. (12).

CHIU, R. J. and L. M. BLACK, 1967, Nature, London *215*, 1076–1078. (17).

CHIU, R. J. and L. M. BLACK, 1969, Virology *37*, 667–677. (17, 26).

CHIU, R. J., H. Y. LIU, R. MACLEOD and L. M. BLACK, 1970, Virology *40*, 387–396. (15, 26).

CHIU, R. J., D. V. R. REDDY and L. M. BLACK, 1966, Virology *30*, 562–566. (26).

CHRISTOPHERS, R., 1960, *Aedes aegypti* (L). *The Yellow Fever Mosquito, its Life History, Bionomics and Structure.* (Cambridge University Press, England) 739 pp. (30).

CHRISTYAKOVA, A. V., 1967. Dokl. Akad. Nauk. SSSR. Bot. Sci. Sect. *173*, 215–216. (4).

CLARK, E. C., 1956, Ecology *37*, 728–732. (21).

CLARK, L. R., P. W. GEIER, R. D. HUGHES and R. F. MORRIS, 1967, *The ecology of insect populations in theory and practice*, (Methuen, London) (4).

CLARK, T. B., H. C. CHAPMAN and T. FUKUDA, 1969, J. Invert. Pathol. *14*, 284–286. (20).

CLARK, T. B., W. R. KELLEN and P. T. M. LUM, 1965, J. Invert. Pathol. *7*, 519–521. (14, 20).

CLARKE, D. H., 1960, J. Exptl. Med. *111*, 21–32. (18).

CLARKE, D. H., 1962, Proc. Symp. Biol. Viruses Tick-borne Enceph. Complex (Smolenice, October 11–14, 1960), pp. 67–75. (18).

CLARKE, D. H., 1964a, Bull. World Health. Organ. *31*, 45–56. (18, 30).

CLARKE, D. H., 1964b, Anais Microbiol. *11*, 143–148. (18).

CLIFFORD, C. M., G. M. KOHLS and D. E. SONENSHINE, 1964, Ann. Entomol. Soc. Am. *57*, 429–437. (5).

CLIFFORD, C. M., C. E. YUNKER. E. R. EASTON and J. E. KEIRANS, 1970, J. Med. Entomol. *7*, 438–445. (18).

CLIFFORD, C. M., C. E. YUNKER, L. A. THOMAS, E. R. EASTON and D. CORWIN, 1971, Am. J. Trop. Med. Hyg. *20*, 461–468. (18).

COCKBAIN, A. J., 1971, *Rept. Rothamsted Exptl. Sta. for 1970*, pp. 184–185. (9).

COCKBAIN, A. J., A. J. GIBBS and G. D. HEATHCOTE, 1963, Ann. Appl. Biol. *52*, 133–142. (25).

COHIC, F., 1970, Ann. Univ. d'Abidjan E II (2) 1969, 1–156. (13).

COHN, E., 1969, Nematologica *15*, 179–192. (12).

COHN, Z. A. and M. E. FEDORKO, 1969, *In:* Dingle, J. T. and H. B. Fell, eds., *Lysosomes in Biology and Pathology.* (North-Holland, Amsterdam), pp. 43–63. (16).

COHN, E. and M. MORDECHAI, 1969, Nematologica *15*, 295–302. (12).

COHN, E., E. TANNE and F. E. NITZANY, 1970, Phytopathology *60*, 181–182. (27).

COKER, W. Z., 1958, Ann. Trop. Med. Parasitol. *52*, 443–455. (30).

COLLIER, W. A., 1920, Z. Wiss. Insektbiol. *16*, 1–5. (17).

CONDIT, I. J. and W. T. HORNE, 1933, Phytopathology *23*, 887–889. (19).

CONOVER, J. H., H. D. ZEPP, K. HIRSCHORN and H. L. HODES, 1971, Curr. Top. Microbiol. Immunol. *55*, 85–92. (17).

CONVERSE, J. L. and S. C. NAGLE, jr., 1967, J. Virol. *1*, 1096–1097. (14, 17).

COOK, W. C., 1967, U.S. Dept. Agri. Tech. Bull. *1365*, 1–122. (8, 31).

COOLEY, R. A. and G. M. KOHLS, 1944, Am. Midl. Nat. Monogr. No. 1, pp. 152. (5).

COOLEY, R. A. and G. M. KOHLS, 1945, Natl. Inst. Health Bull. *184*, 246. (5).

COOMANS, A. and L. DE CONINCK, 1963, Nematologica *9*, 85–96. (12).

COOMANS, A., and P. A. A. LOOF, 1970, Nematologica *16*, 180–196. (12).

COOPER, E. L., 1968, Transplantation *6*, 322. (16).

COOPER, E. L., 1971, Transplantation Proc. *3*, 214. (16).

COOPER, E. L., R. T. ACTON, P. F. WEINHEIMER and E. E. EVANS, 1969, J. Invert. Pathol. *14*, 402. (16).

COOPER, E. L. and L. M. RUBILOTTA, 1969, Transplantation *8*, 220. (16).

COOPER, J. I., 1971, Plant. Pathol. *20*, 51–58. (12, 27).

COOPER, J. I. and B. D. HARRISON, 1971, Rep. Scott. Hort. Res. Inst. for 1970, p. 52. (27).

COOPER, J. I. and P. R. THOMAS, 1970, Plant Pathol. *19*, 197. (12).

CORBETT, G. H., 1935, J. Fed. Malay St. Mus. *17*, 722–852. (13).

CORNALIA, E., 1856, Rend. 1st Lombardo Sci. Letters. Mem. I, pp. 348–351. (2).

CORNWELL, P. B. ,1955, *In:* Proc. W. Afr. Int. Cacao Res. Conf. 1953, (Crown Agents, London). pp. 8–11. (23).

CORNWELL, P. B., 1956, Bull. Ent. Res. *47*, 137–166. (23).

CORNWELL, P. B., 1957, Bull. Ent. Res. *48*, 375–396. (23).

CORNWELL, P. B., 1958, Bull. Ent. Res. *49*, 613–630. (23).

CORNWELL, P. B., 1960, Bull. Entomol. Res. *51*, 175–201. (23).

COSTA, A. S., 1952, Phytopathology *42*, 396–403. (26).

COSTA, A. S., 1969, *In:* Maramorosch, K., ed., *Viruses, Vectors and Vegetation* (Interscience Publishers, New York), pp. 95–119. (13).

COSTA, A. S., D. M. DE SILVA and J. E. DUFFUS, 1958, Virology *5*, 145–149 (20).

COTT, H. E., 1956, Univ. Calif. Publs. Entomol. *13*, 1–210. (13).

COTTEN, J., J. J. M. FLEGG and A. M. POPHAM, 1971, Nematologica *16*, 584–590. (12).

CRANE, G. L. and L. CALPOUZOS, 1969, Phytopathology *59*, 697–698. (31).

CRAWFORD, D. L., 1919, Phillippine J. Sci. *15*, 139–207. (7).

CRAWFORD, D. L., 1920, Philippine J. Sci. *17*, 353–359. (7).

CRAWFORD, D. L., 1925, Broteria, ser. Zool. *22*, 56–74. (7).

CRAWFORD, D. L., 1928, Entomol. Mitteil. *17*, 425–426. (7).

CRAWFORD, L. V., E. A. C. FOLLETT, M. G. BURDON and D. J. McGEOGH, 1969, J. Gen. Virol. *4*, 37–46. (4).

CREMER, M. C. and G. KOOISTRA, 1964, Nematologica *10*, 69–70. (27).

CROWSON, R. A., 1954, *The Natural Classification of the Families of Coleoptera* (Nathaniel Lloyd, London). (9).

CUENOT, L., 1914, Arch. Zool. Exptl. Gen. *54*, 267. (16).

CUNNINGHAM, J. C., 1968, J. Invert Pathol. *11*, 132–141. (4, 21).

CUNNINGHAM, J. C., 1971, Can. J. Microbiol. *17*, 69–72. (14).

CUNNINGHAM, J. C. and J. F. LONGWORTH, 1968, J. Invert. Pathol. *11*, 196–202. (4, 15, 21).

CUNNINGHAM, J. C., and T. W. TINSLEY, 1968, J. Gen. Virol. *3*, 1–8. (4).

CURTIS, L. C., 1963, Proc. Entomol. Soc. B.C. *60*, 22–23. (30).

DAHLSTEN, D. L. and G. M. THOMAS, 1969, J. Invert. Pathol. *13*, 264–272. (29).

DALE, W. T., 1953, Ann. Appl. Biol. *40*, 385–392. (9).

DALE, W. T., 1955, Ann. Rep. W. African Cocoa Res. Inst. 1954–55, 33–35. (23).

DALE, W. T., 1958a, Ann. Rep. W. African Cocoa Res. Inst., 1956–57, 22–24. (23).

DALE, W. T., 1958b, Ann. Rep. W. African Cocoa Res. Inst. 1956–57, 27–30. (23).

DALE, W. T., 1962, *In:* J. Brian Wills, ed., *Agriculture and Land Use in Ghana* (Oxford University Press, London), pp. 286–316. (23).

DALES, S., P. J. GOMATOS and K. C. HSU, 1965, Virology *25*, 193–211. (4).

DALES, S. and E. H. MOSBACH, 1968, Virology *35*, 564–583. (5).

DALMASSO, A., 1970, C.R. Acad. Sci. Paris *270*, 824–827. (12).

DALMAT, H. T., 1958, J. Exptl. Med. *108*, 9–20. (4).

DANDAWATE, C. N., P. K. RAJAGOPALAN, K. M. PAVRI and T. H. WORK, 1969a, Indian J. Med. Res. *57*, 1420–1426. (18).

DANDAWATE, C. N. and K. V. SHAH, 1969, Indian J. Med. Res. *57*, 799–804. (18).

DANDAWATE, C. N., T. H. WORK, J. K. G. WEBB and K. V. SHAH, 1969b, Indian J. Med. Res. *57*, 975–982. (18).

DAS, S. and D. J. RASKI, 1968, Nematologica *14*, 55–62. (27).

DASGUPTA, B. and H. N. RAY, 1957, Parasitology *47*, 194–195. (20).

DAUBNEY, R. and J. R. HUDSON, 1931, Parasitology *23*, 507. (30).

DAUBNEY, R., J. R. HUDSON and P. C. C. CARNHAM, 1931, J. Pathol. Bacteriol. *34*, 345. (30).

DAVID, W. A. L., 1957, Z. Pfl. Krankh. Pfl. Pathol. Pfl. Schutz. *64*, 572–577. (21).

DAVID, W. A. L. 1965, Ann. Appl. Biol. *56*, 331–334. (21).

DAVID, W. A. L., 1969, J. Invert. Pathol. *14*, 336–343. (29).

DAVID, W. A. L. and B. O. C. GARDINER, 1952, Proc. Roy. Entomol. Soc. London A *27*, 54. (2).

DAVID, W. A. L. and B. O. C. GARDINER, 1960, J. Insect Pathol. *2*, 106–115. (29).

DAVID, W. A. L. and B. O. C. GARDINER, 1965, J. Invert. Pathol. *7*, 285–291. (29).

DAVID, W. A. L. and B. O. C. GARDINER, 1966, J. Invert. Pathol. *8*, 180–183. (21).

DAVID, W. A. L. and B. O. C. GARDINER, 1967, J. Invert. Pathol. *9*, 342–348. (21, 29).

DAVID, W. A. L., B. O. C. GARDINER and S. E. CLOTHIER, 1968a, J. Invert. Pathol. *12*, 238–244. (21).

DAVID, W. A. L., B. O. C. GARDINER and M. WOOLNER, 1968b, J. Invert. Pathol. *11*, 496–501. (21, 29).

DAVIDSON, G., J. A. ODETOYINBO, B. COLUSSA and J. COZ, 1970, Bull. World Health Organ. *42*, 55–67. (30).

DAVIES, E. E. and R. E. HOWELLS, 1971, Trans. R. Soc. Trop. Med. Hyg. *65*, 13–14. (28).

DAVIS, N. C., 1932, Am. J. Hyg. *16*, 163–176. (14).

DAY, M. F., 1955, Exptl. Parasitol. *4*, 387–418. (11).

DAY, M. F. and M. L. DUDZINSKI, 1966, Australian J. Biol. Sci. *19*, 481–493. (14).

DAY, M. F., J. L. FARRANT and C. POTTER, 1958, J. Ultrastruct. Res. *2*, 227–238. (14).

DAY, M. F. and T. D. C. GRACE, 1959, Ann. Rev. Entomol. *4*, 17–38. (17).

DAY, M. F. and H. IRZYKIEWICZ, 1953, Australian J. Biol. Sci. *6*, 98–108. (25).

DAY, M. F., H. IRZYKIEWICZ and A. MCKINNON, 1952, Australian J. Sci. Res. (B) *5*, 128–142. (8).

DAY, M. F. and A. MCKINNON, 1951, Australian J. Sci. Res. (B), *2*, 125–135. (8).

DAY, M. F. and E. H. MERCER, 1964, Australian J. Biol. Sci. *17*, 892–902. (14).

DAY, M. F. and D. G. VENABLES, 1961, Australian J. Biol. Sci. *14*, 187–197. (25).

DEBACH, P., 1964, *Biological control of insect pests and weeds.* (Chapman and Hall, London) 844 pp. (29).

DE FOLIART, G. R., R. O. ANSLOW, R. P. HANSON, C. D. MORRIS, O. PAPADOPOULOS and G. E. SATHER, 1969, Am. J. Trop. Med. Hyg. *18*, 440–447. (11).

DELGADO-SANCHEZ, S. and R. G. GROGAN, 1970, C.M.I./A.A.B. Descrip. Plant Viruses. No. 37. (4).

DELONG, D. M., 1971, Ann. Rev. Entomol. *16*, 179–210. (8).

DE MOOR, P. P. and F. E. STEFFENS, 1970, Trans. Roy. Soc. Trop. Med. Hyg. *64*, 927–934. (20).

DENNEL, R., 1942, Phil. Trans. Roy. Soc. London, Ser. B. *231*, 247–259. (9).

DERRICK, E. H. and V. A. BICKS, 1958, Australasian Ann. Med. *7*, 102–107. (20).

DETINOVA, T. S., 1968, Ann. Rev. Entomol. *13*, 427–450. (11).

DETRAY, D. E., 1963, Advan. Vet. Sci. *8*, 299–333. (18).

DEVAUCHELLE, G., C. VAGO, J. GIANOTTI and J. M. QUIOT, 1969, Entomophaga *14*, 457. (2).

DHANDA, V., 1966, J. Parasitol. *52*, 1025–1031. (18).

DHANDA, V., S. M. KULKARNI and P. PRATT, 1972, J. Parasitol. *58*, in press. (5).

DHANDA, V. and P. K. RAJAGOPALAN, 1971, Oriental Insects *5*, 135–143. (18).

D'HERDE, J. and J. VAN DEN BRANDE, 1964, Nematologica *10*, 454–458. (12).

DIAMOND, L. S., C. F. T. MATTERN and I. L. BARTGIS, 1972, J. Virol. *9*, 326–341. (28).

DICKSON, R. C. and E. F., LAIRD jr., 1962, J. Econ. Entomol. *55*, 501–504. (25).

DOANE, C. C., 1967, J. Invert. Pathol. *9*, 376–387. (29).

DOANE, C. C., 1969, J. Invert. Pathol. *14*, 199–210. (21, 29).

DOANE, C. C., 1970, J. Invert. Pathol. *15*, 21–33. (21).

DOBOS, P. and P. FAULKNER, 1970, J. Virol. *6*, 145–147. (14).

DOBREANU, E. and C. MANOLACHE, 1961, Insecta *8*, 1–376. (7).

DOERR, R., 1908, Berlin Klin. Weschr. *45*, 1857. (1).

DOHERTY, R. L., 1958, Virology *6*, 575–583. (14).

DOHERTY, R. L., H. A. STANDFAST, R. DOMROW, E. J. WETTERS, R. H. WHITEHEAD and J. G. CARLEY, 1971, Trans. Royal Soc. Trop. Med. Hyg. *65*, 504–513. (14).

DOI, R., A. SHIRASAKA and M. SASA, 1967, Japan J. Exptl. Med. *37*, 227–238. (14, 20).

DOI, Y., M. TERANAKA, K. YORA and H. ASUYAMA, 1967, Ann. Phytopathol. Soc. Japan *33*, 259–266. (26).

DONALD, R. G., 1955, Ann. Rep. W. African Cocoa Res. Inst. 1954–55, 102–104. (23).

DONALD, R. G., 1956, J. W. African Sci. Assoc. *2*, 48–60. (23).

DONCASTER, J. P. and P. H. GREGORY, 1948, Agr. Res. Council Rep. *7*, London, H.M.S.O. (31).

DONCASTER, J. P., and B. KASSANIS, 1946, Ann. Appl. Biol. *33*, 66–68. (25).

DONCHEV, D., G. KEBEDZHIEV and M. RUSAKIEV, 1967, Bulgar. Akad. Nauk Mikrobiol. Inst., 1. Kongr. Mikrobiol. (1965), 777–784. (In Bulgarian, Engl. trans.: NAMRU3-T465). (18).

DOOLITTLE, S. P., 1916, Phytopathology *6*. 145–147. (3).

DOOLITTLE, S. P. and M. N. WALKER, 1928, Phytopathology *18*, 143 (abstr.). (3).

DOROLLE, P., 1968, Brit. Med. J. *4*, 789–792. (30).

DOUGHERTY, C., J. FERRAL, B. M. BRODY and M. L. GOTTHOLD, 1963, Nature, London *198*, 973–975. (28).

DOW, R. P., P. H. COLEMAN, K. E. MEADOWS and T. H. WORK, 1964, Am. J. Trop. Med. Hyg. *13*, 462–468. (30).

DOWNES, J. A., 1958, Ann. Rev. Entomol. *3*, 249–266. (11).

DOWNS, W. G., T. H. G. AITKEN, C. B. WORTH, L. SPENCE and J. H. JONKERS, 1968, Am. J. Trop. Med. Hyg. *17*, 224–236. (11).

DRAKE, C. J., H. D. TATE and H. M. HARRIS, 1933, J. Econ. Entomol. *26*, 841–846. (25).

DRESCHER, W., 1964, Z. Bienenforsch. *7*, 116–124. (22).

DRUMMOND, R. O., P. K. RAJAGOPALAN, M. A. SREENIVASAN and P. K. B. MENON, 1969, J. Med. Entomol. *6*, 245–251. (18).

DUFFUS, J. E., 1963, Virology *21*, 194–202. (15, 24).

DUFFUS, J. E. and A. H. GOLD, 1965, Virology *27*, 388–390. (24).

DUFFUS, J. E. and A. H. GOLD, 1969, Virology *37*, 150–153. (4, 24).

DULBECCO, R., 1952, Proc. Natl. Acad. Sci., U.S. *38*, 747. (17).

DUNNEBACKE, T. H. and F. L. SCHUSTER, 1971, Science *174*, 516–518. (28).

DUNSTAN, A. G., 1927, J. Econ. Entomol. *20*, 68–75. (7).

DUPRAT, P., 1967, Ann. Inst. Pasteur, Paris *113*, 867. (16).

DU TOIT, R. and G. THEILER, 1964, Sci. Bull. S. Afr. Dep. Agric. Tech. Serv. No. *364*, 28. (5).

DUTRECQ, A. and J. VANDERVEKEN, 1969, Bull. Res. Agron. Gembloux *4*, 66–75. (25).

DYCE, A. L., 1961, C.S.I.R.O. (Australia) Wildlife Res., *6*, 88–90. (20).

EAVES, G. and T. H. FLEWETT, 1955, J. Hyg. *53*, 102–105. (4).

EBLE, A. F., 1966, Am. Zool. *6*, 339. (16).

ECHALIER, G. and A. OHANESSIAN, 1968, Intern. Colloq. Tissue Culture Invert., 2nd, Tremezzo-Como, Italy, 1967. (17).

ECHALIER, G. and A. OHANESSIAN, 1970, In Vitro. *6*, 162–172. (17).

EDMUNDS, L. R., 1958, Mosquito News *18*, 23–26. (30).

EDWARDSON, J. R., 1966, Am. J. Botany *53*, 359–363. (4).

EHRHARDT, P., 1961, Experimentia *17*, 461–463. (25).

EHRHARDT, P. and H. SCHMUTTERER, 1965, Phytopathol. Z. *52*, 73–88. (24).

EIDE, P. E. and T. H. CHANG, 1969, Exptl. Cell Res. *54*, 302–308. (17).

ELBL, A. and G. ANASTOS, 1966, Ann. Mus. R. Afr. Cent., Ser. 8v0, s. Sci. Zool. Vols. 1–4. (5).

ELMER, C. H., 1922, Science *56*, 370. (10).

ELMER, C. H., 1925, Res. Bull. Iowa Agri. Exptl. Sta. *82*, 39 (10).

ELMORE, J. C. and A. F. HOWLAND, 1964, J. Insect Pathol. *6*, 430–439. (29).

ELS, H. L. and D. W. VERWOERD, 1969, Virology *38*, 213–219. (4).

ENDERS, J. F., T. H. WELLER and E. C. ROBBINS, 1949, Science *109*, 85. (17).

ENGSTROM, A. and R. KILKSON, 1968, Exptl. Cell Res. *53*, 305–310. (14).

ENTWISTLE, P. F., 1959, Rep. W. African Cocoa Res. Inst. 1957–58, 31–36. (23).

ENTWISTLE, P. F., 1960, Rep. W. African Cocoa Res. Inst. 1958–59, 30–32. (23).

ENTWISTLE, P. F., 1972, *Pests of cocoa*, (Longmans) 799 pp. (10).

ENTWISTLE, P. F. and J. F. LONGWORTH, 1963, Ann. Appl. Biol. *52*, 387–391. (10, 23).

ERLANDSON, R. A., V. I. BABCOCK, C. M. SOUTHAM, R. B. BAILEY and F. H. SHIPKEY, 1967, J. Virol. *1*, 996–1009. (14).

ERNEK, E., O. KOZUCH, M. SEKEYOVA, K. HUDEC and C. FOLK, 1971, Acta Virol., Prague (English edition) *15*, 335. (18).

ESAU, K., 1967, Phytopathology *5*, 45–76. (25).

ESAU, K., J. CRONSHAW and L. L. HOEFERT, 1967, J. Cell Biol. *32*, 71–87. (25).

ESAU, K., R. NAMBA and E. A. RASA, 1961, Hilgardia *30*, 517–529. (25).

ESCALONE, A. S., L. T. FINOL and S. RYDER, 1969, Invest. Clin. *22*, 45–57. (30).

ESTES, Z. E. and R. M. FAUST, 1965, J. Invert. Pathol. *7*, 259–261. (4).

ESTES, Z. E. and R. M. FAUST, 1966, J. Invert. Pathol. *8*, 145–149. (15).

EVANS, G. O., J. G. SHEALS and D. MACFARLANE, 1961, *The Terrestrial Acari of the British Isles, Vol. 1, Introduction and Biology*, (British Museum (Nat. Hist.), London.) 1–219. (6).

EVANS, G. O. and W. M. TILL, 1965. Bull. Brit. Museum (Nat. Hist.) Zool. *13*, 249–294. (6).

EZZAT, Y. M. and H. S. MCCONNEL, 1956, Bull. Med. Agr. Expt. Sta. *A 84*, 108 pp. (10).

FAIRCHILD, G. B., G. M. KOHLS and V. J. TIPTON, 1966, *In:* Wenzel, R. L. and V. J. Tipton, eds., *Ectoparasites of Panama*, (Field Museum of Natural History, Chicago, Illinois) pp. 167–219. (5).

FALCON, L. A., 1965, Bull. Entomol. Soc. Am. *11*, 84. (2).

FAN, K. Y., T. C. HSU and C. L. CHEN, 1962, J. Formosan Med. Assoc. *61*, 421–428. (30).

FAULKNER, P., 1962, Virology *16*, 479–484. (4).

FAUST, R. M. and J. R. ADAMS, 1966, J. Invert. Pathol. *8*, 526–530. (14, 15).

FEDOROVA, T. N., 1966a, *In:* Maksimov, A. A. and G. I. Netsky, eds., *The Muskrat of Western Siberia*. (Biol. Inst., Akad. Nauk SSSR, Sibirsk, Otd., Novosibirsk) pp. 131–135. (In Russian, Engl. trans.: NAMRU3-T555). (18).

FEDOROVA, T. N., 1966b, *In:* Maksimov, A. A. and G. I. Netsky, eds., *The Muskrat of Western Siberia*. (Biol. Inst. Akad. Nauk SSSR, Sibirsk. Otd., Novosibirsk) pp. 136–140. (In Russian, Engl. trans.: NAMRU3-T556). (18).

FEIDER, Z., 1965, *Fauna Republicii Populare Romane. Arachnida, Vol. 5, fasc. 2. Acaromorpha, Suprafamilia Ixodoidea (Capuse)*, (Editura Academiei Republicii Populare Romane, Bucuresti) 404 pp. (5).

FENG, S. Y., 1965, Biol. Bull. *129*, 95. (16).

FENG, S. Y., 1967, Federation Proc. *26*, 1685. (16).

FENNER, F., 1972, *The Biology of Animal Viruses*. 2nd Ed., (Academic Press, London and New York). (4, 22).

FENNER, F., M. F. DAY and G. M. WOODROOFE, 1953, Australian J. Exptl. Biol. Sci. *30*, 139. (1).

FENNER, F. and I. D. MARSHALL, 1957, J. Hyg. *55*, 149–191. (20).

FENNER, F., W. E. POOLE, I. D. MARSHALL and A. L. DYCE, 1957, J. Hyg. *55*, 192–206. (20).

FENNER, F. and F. N. RATCLIFFE, 1965, *Myxomatosis*, (Cambridge University Press, London and New York). (4, 20).

FERNANDEZ VALIELA, M. V., 1963, Bol. Divulg. Delta de. Paraná *3*, 3–37. (31).

FERNANDO, E. F. W., 1971, Sacbrood. Rothamsted Exptl. Sta. Rept. for 1970. (22).

FIFE, J. M. and V. L. FRAMPTON, 1936, J. Agr. Res. *53*, 557–580. (3).

FILIPPOVA, N. A., 1966, Argasid Ticks (Argasidae). Fauna SSSR, Paukoobraznye *4*, 255 pp. (In Russian, Engl. trans.: NAMRU3-T600). (5).

FILSHIE, B. K. and J. REHACEK, 1968, Virology *34*, 435–443. (14).

FINCH, J. T. and A. KLUG, 1965, J. Mol. Biol. *13*, 1–12. (4).

FISHER, R. C., 1961, J. Exptl. Biol. *38*, 605. (16).

FLEGG, J. J. M., 1968, Nematologica *14*, 197–210. (12).

FLETCHER, O. K. and L. FLANIGAN, 1957, Mosquito News *17*, 29–32. (30).

FLOCK, R. A. and J. M. WALLACE, 1955, Phytopathology *45*, 52–54. (19).

FLORIO, L., M. S. MILLER and E. R. MUGRAGE, 1950, J. Immunol. *64*, 257. (30).

FORBES, A. R., 1964, Mem. Entomol. Soc. Can. *36*, 1–74. (7, 24).

FORBES, A. R., 1969, Can. Entomol. *101*, 31–41. (7).

FORBES, A. R. and H. R. MACCARTHY, 1969, *In:* Maramorosch, K., ed., *Viruses, Vectors and Vegetation* (Interscience Pubs.) pp. 211–234. (7, 24).

FORBES, A. R. and D. B. MULLICK, 1970, Can. Entomol. *102*, 1074–1082. (7).

FOX, C. J. S. and R. P. JAQUES, 1966, Can. J. Plant Sci. *46*, 497–499. (29).

FRAME, J. D., J. M. BALDWIN, D. S. GOCKE and J. M. TROUP, 1970, Am. J. Trop. Med. Hyg. *19*, 670. (30).

FRANCKI, R. I. B. and J. W. RANDLES, 1970, C.M.I./A.A.B. Descrip. Plant Viruses. No. 26. (4).

FRANZ, J. M., 1961, Ann. Rev. Entomol. *6*, 183–200. (29).

FRANZ, J., A. KRIEG and R. LANGENBUCH, 1955, Z. Pfl. Krankh. Pfl. Schutz *62*, 721–726. (21).

FREDERICKSON, L. E. and L. THOMAS, 1965, Publ. Health. Rep. U.S. *80*, 495. (1).

FREITAG, J. H., 1936, Hilgardia *10*, 305–342. (3).

FREITAG, J. H., 1956, Phytopathology *46*, 73–81. (9).

FRIEDMAN, R. M., 1968a, J. Virol. *2*, 26–32. (14).

FRIEDMAN, R. M., 1968b, J. Virol. *2*, 547–552. (14).

FRIEDMAN, R. M., 1968c, J. Virol. *2*, 1076–1080. (14).

FRIEDMAN, R. M. and I. K. BEREZESKY, 1967, J. Virol. *1*, 374–383. (14).

FRIEDMAN, R. M., H. B. LEVY and W. B. CARTER, 1966, Proc. Natl. Acad. Sci. U.S. *56*, 440–446. (14).

FRIST, R. H., I. J. BENDET, K. M. SMITH and M. A. LAUFFER, 1965, Virology *26*, 558–566. (4).

FRITZSCHE, R., 1964, Wiss. Z. Univ. Rostock. Math.-Naturwiss. Reihe *13*, 343–347. (27).

FRITZSCHE, R., 1970, Zeszyty Probl. Postep. Nauk. Roln. *92*, 293–300. (27).

FRITZSCHE, R. and H. KEGLER, 1964, Naturwissenschaften *51*, 299. (27).

FRITZSCHE, R. and H. KEGLER, 1968, Deut. Landwirtsch. *97*, 289–295. (12, 27).

FRITZSCHE, R. and K. SCHMELZER, 1967, Naturwissenschaften *54*, 498–499. (27).

FUJII-KAWATA, I., K. I. MIURA and M. FUKE, 1970, J. Mol. Biol. *51*, 247–253. (4).

FUKAYA, M. and S. NASU, 1966, Appl. Entomol. Zool. *1*, 69–72. (4).

FUKUSHI, T., 1933, J. Fac. Agr. Hokkaido Imp. Univ. *37*, 41–64. (3).

FUKUSHI, T., 1935. Proc. Imp. Acad. (Tokyo) *11*, 301–303. (26).

FUKUSHI, T., 1940, J. Fac. Agr. Hokkaido Imp. Univ. *45*, 83–154. (3).

FUKUSHI, T., E. SHIKATA and I. KIMURA, 1962, Virology *18*, 192–205. (15, 26).

FUKUSHI, T., E. SHIKATA and I. KIMURA, 1963, Virology *21*, 503–505. (15).

FULTON, J. P., 1962, Phytopathology *52*, 375. (27).

FURMAN, D. P., 1959, Am. J. Trop. Med. Hyg. *8*, 5–12. (6).

GABRIEL, W., 1965, Ann. Appl. Biol. *56*, 461–475. (31).

GAIDAMOVICH, S. YA, E. E. MEL'NIKOVA, A. YU. BEKLESHOVA, I. I. TERSKIKH and V. R. OBUKHO-VA, 1971. Tezisy Dokl. Vop. Med. Virus., Inst. Virus. Imeni Ivanovsky, D. I., Akad. Med. Nauk SSSR (October 19–21), 68–69. (In Russian, Engl. trans.: NAMRU3-T492). (18).

GAIDAMOVICH, S. YA, L. P. NIKIFOROV, V. L. GROMASHEVSKY, V. R. OBUKHOVA, G. T. KLISENKO, V. I. CHERVONSKY and E. E. MELNIKOVA, 1971, Acta Virol., Prague (English edition) *15*, 155–160. (18).

GAIDAMOVICH, S. YA., I. A. VINOGRAD and V. R. OBUKHOVA, 1971, Acta Virol., Prague (English edition) *15*, 333. (18).

GALINDO, P. and H. TRAPIDO, 1955, Am. J. Trop. Med. Hyg. *4*, 543–546. (11).

GALLAY, R., W. WURGLER, R. BOVEY, M. STAEHELIN and H. LEYVRAZ, 1955, Revue Romande Agr. Vitic. Arboric. *11*, 17–24. (27).

GALVEZ, G. E., 1967, Virology *33*, 357–359. (26).

GAMEZ, R. and L. M. BLACK, 1968, Virology *34*, 444–451. (26).

GAMEZ, R., L. M. BLACK and R. MACLEOD, 1967, Virology *32*, 163–165. (4, 26).

GAMEZ, R. and R. J. CHIU, 1968, Virology *34*, 356–357. (26).

GARRETT, R. G., 1971, *The mechanism of transmission of non-persistent viruses.* Doctoral thesis, University of Adelaide, Australia. (25).

GASSER, J. K. R. and J. E. PEACHEY, 1964, J. Sci. Food. Agr. *15*, 142–146. (31).

GATES, L. F., 1970, Can. Plant Diseases Surv. *50*, 59–62. (19).

GAVRILOV, W. and S. COWEZ, 1941, Ann. Parasit. Hum. Comp. *18*, 180. (17).

GAW, Z-Y., N. T. LIU and T. V. ZIA, 1959, Acta Virol. *3*, (suppl.) 55–60. (14).

GEBHARDT, L. P., G. J. STANTON, D. W. HILL and G. C. COLLETT, 1964, New Engl. J. Med. *271*, 172–177. (20).

GEBHARDT, L. P., G. J. STANTON and L. de ST. JEON, 1966, Proc. Soc. Exptl. Biol. Med. *123*, 233–235. (30).

GELFAND, H. M., 1961, Progr. Med. Virol. *3*, 193–244. (4, 20).

GEORGE, J. E., 1971, J. Med. Entomol. *8*, 461–479. (6).

GEROLA, F. M., M. BASSI, O. LOVISOLO and C. VIDANO, 1966, Phytopathol. Z. *56*, 97–99. (26).

GERSHENSON, S. M., 1960, Probl. Virol. (USSR) (Engl. Transl.) *6*, 720–725. (2).

GERSHENSON, S. M., I. P. KOK, K. I. VITAS, G. N. DOBROVOL'SKAYA and I. N. SKURATOVSKAYA, 1963, Proc. 5th. Int. Cong. Biochem. Moscow 1961 Vol. 9 p. 150 (Pergamon Press, Oxford). (4).

GHIRADELLA, H. T., 1965, Biol. Bull. *128*, 77. (16).

GIANNOTTI, J., C. VAGO, G. DEVAUCHELLE and G. MARCHOUX, 1968, Entomol. Exptl. Appl. *11*, 470–474. (26).

GIANNOTTI, J., C. VAGO, J. SASSINE and D. CZARNECKY, 1971, C.R. Acad. Sci. Paris, Serie D. *272*, 1776–1778. (26).

GIAUFFRET, A., 1966, Bull. Apicole *9*, 35–42. (29).

GIAUFFRET, A., J. L. DUTHOIT, F. POUTIERS and M. J. TOSTAINCAUCAT, 1969, Bull. Apicole *12*, 13–22. (22).

GIBBS, A. J., 1966, Plant Pathol. *15*, 150–152. (31).

GIBBS, A. J., 1969, Adv. Virus Res. *14*, 263–328. (4).

GIBBS, A. J., F. J. GAY and A. H. WEATHERLEY, 1970, Virology *40*, 1063–1065. (28).

GIBBS, A. J. and B. D. HARRISON, 1964, Ann. Appl. Biol. *54*, 1–11. (27).

GIBBS, A. J. and B. D. HARRISON, 1970, C.M.I./A.A.B. Descrip. Plant Viruses. No. *1*. (4).

GIBBS, A. J., B. D. HARRISON, D. H. WATSON and P. WILDY, 1966, Nature, London *209*, 450–454. (4).

GILBERT, L. L., 1967, Adv. Insect Physiol. *4*, 69–211. (7).

GILLIES, M. T. and B. DE MEILLON, 1968, St. African Inst. Med. Res. Publ. No. *54*. (30).

GILMORE, J. G. and F. MUNGER, 1963, J. Insect Pathol. *5*, 141–152. (29).

GILMORE, J. G. and F. MUNGER, 1965, J. Invert. Pathol. *7*, 156–161. (29).

GILMORE, J. G. and H. TASHIRO, 1966, J. Invert. Pathol. *8*, 334–340. (29).

GINGRICH, R. E., 1964, J. Insect Physiol. *10*, 179. (16).

GLANCEY, B. M., H. R. FORD and C. S. LOFEREN, 1970, Mosquito News *30*, 174–180. (30).

GLASHINSKAYA-BABENKO, L. V., 1956, Mater. Pozn. Fauny Flory SSSR, n. s, Otd. Zool., vypusk 34 (XLIX); Ektoparazity, Fauna, Biol., Prakt. Znachenie, 21–105. (In Russian). (18).

GLASS, E. H., 1944, Tech. Bull. Va. Agri. Expt. Sta. *95*, 16. (10).

GLICK, P. A., 1960, U.S. Dept. Agri. Tech. Bull. *1222*, 1–16. (8).

GLITZ, D. G., G. J. HILLS and C. F. RIVERS, 1968, J. Gen. Virol. *3*, 209–220. (4, 14).

GOLDSCHMIDT, R., 1915, Proc. Natl. Acad. Sci., U.S. *1*, 220–222. (17).

GOMATOS, P. J. and I. TAMM, 1963, Proc. Natl. Acad. Sci. U.S. *50*, 878–885. (26).

GOODWIN, R. H. and B. K. FILSHIE, 1969, J. Invert. Pathol. *13*, 317–329. (4).

GOONEWARDENA, H. F., 1958, Trop. Agri. Ceylon *114*, 39–60. (29).

GOTHILF, S. and S. D. BECK, 1966, J. Econ. Entomol. *59*, 489–490. (10).

GOVIER, D. A. and R. D. WOODS, 1971, J. Gen. Virol. *13*, 127–132. (4).

GRACE, T. D. C., 1954, Nature, London *194*, 187. (17).

GRACE, T. D. C., 1958, Science *128*, 249–250. (14).

GRACE, T. D. C., 1962a, Nature, London *195*, 788–789. (14, 17).

GRACE, T. D. C., 1962b, Virology *18*, 33–42. (14).

GRACE, T. D. C., 1963, Doctoral thesis, Australian National University, Canberra. (17).

GRACE, T. D. C., 1966, Nature, London *211*, 366–367. (14, 17).

GRACE, T. D. C., 1967, Nature, London *216*, 613. (17).

GRACE, T. D. C., 1968a, Exptl. Cell Res. *52*, 451–458. (17).

GRACE, T. D. C. 1968b, In Vitro. *3*, 104–117. (17).

GRACE, T. D. C., 1969, Adv. Virus. Res. *14*, 201–220. (17).

GRACE, T. D. C. and E. M. MERCER, 1965, J. Invert. Pathol. *7*, 241–244. (4, 17).

GRAINGER, W. E., 1947, E. African Med. J. *24*, 16–22. (30).

GRANADOS, R. R., K. MARAMOROSCH and E. SHIKATA, 1968, Proc. Natl. Acad. Sci. U.S. *60*, 841–844. (26).

GRANADOS, R. R. and D. W. ROBERTS, 1970, Virology *40*, 230–243. (2, 4).

GRANADOS, R. R., L. S. WARD and K. MARAMOROSCH, 1968, Virology *34*, 790–796. (15).

GRANOFF, A., 1969, Current Top. Microbiol. Immunol. *50*, 107–137. (4).

GRASSE, P., 1951, *Traité de Zoologie*, X *(II)*, (Masson et Cie, Paris) pp. 1805–1869. (13).

GRASSE, P.,-P., 1965, *Traité de Zoologie. Anatomie, Systematique, Biologie. Vol. 4, parts 2 and 3*, (Masson et Cie, Paris) 1497 pp. (12).

GRAVITZ, N. and C. WILLSON, 1968, J. Econ. Entomol. *61*, 1458–1459. (10).

GREENBERG, B., 1971, *Flies and Disease. Vol. I. Ecology, Classification and Biotic Associations*, (Princeton. Univ. Press). (11).

GREENE, A. E. and J. CHARNEY, 1971, Currents Top. Microbiol. Immunol. *55*, 51–61. (17).

GREGSON, J. D., 1956, Publ. Dept. Agr. Canada, No. *930*, pp. 92. (5).

GREGSON, J. D., 1960, Acta Trop. *17*, 47–79. (18).

GREGSON, J. D., 1967, Parasitology *57*, 1–8. (18).

GREGSON, J. D., 1969, Proc. 2nd. Int. Congr. Acarol. (Sutton Bonington, July 19–25, 1967), pp. 329–339. (18).

GRESIKOVA, M., E. ERNEK. O, KOZUCH and J. NOSEK, 1970, Folia Parasitol. *17*, 379–382. (18).

GRESSIT, J. L. and S. KIMOTO, 1961, Pacific Insects Monograph 1A. The Chrysomelidae (Coleopt.) of China and Korea. Part 1. Honolulu. (9).

GRIFFIN, G. D. and H. M. DARLING, 1964, Nematologica *10*, 471–479. (12).

GRIMLEY, P. M., I. K. BEREZESKY and R. M. FRIEDMAN, 1968, J. Virol. *2*, 1326–1338. (14).

GRIMLEY, P. M. and R. M. FRIEDMAN, 1969, Bacteriol. Proc. 185. (14).

GRIMSTONE, A. V., S. ROTHERAM, and G. SALT, 1967, J. Cell Sci. *2*, 281. (16).

GROMASHEVSKY, V. L., G. A. SIDOROVA and YU. M. TSIRKIN, 1971, Tezisy Dokl. Vop. Med. Virus, Inst. Virus. Imeni Ivanovsky, D.I., Akad. Med. Nauk SSSR (October 19–21), pp. 122. (In Russian). (18).

GROSHKOVA. I. M., M. S. PAVLOVA, V. M. POPOV and M. K. TIUSHNAKOVA, 1959, Probl. Virol. *4*, 67. (30).

GUELMINO, D. J. and M. JEVTIC, 1955, Bibl. Hig. Inst. Srbje. *8*, 1–71. (30).

GUIRGIS, S. S., 1971, J. Med. Entomol. *8*, 407–414. (18).

HADDOW, A. J., 1948, Bull. Entomol. Res. *39*, 188–213. (11).

HADDOW, A. J., 1965, Trans. Roy. Soc. Trop. Med. Hyg. *59*, 436–440. (30).

HADDOW, A. J., P. S. CORBET, J. D. GILLETT, I. DIRMHIRN, T. H. E. JACKSON and K. W. BROWN, 1961, Trans. Roy. Entomol. Soc. London *113*, 249–368. (11).

HAFEZ, M., A. A. ABDEL-MALEK and S. S. GUIRGIS, 1972, J. Med. Entomol. *9*, 19–29. (18).

HALDANE, J. B. S., 1960, Nature *187*, 879. (14).

HALL, C. C., 1967a. Ann. Entomol. Soc. Am. *60*, 91–94. (6).

HALL, C. C., 1967b. Kans. Univ. Sci. Bull. *47*, 601–675. (6).

HALL, E. R. and K. R. KELSON, 1959, *The Mammals of North America*, Vol. 1, (The Ronald Press Co., New York). (20).

HALSTEAD, S. B., 1970, Yale J. Biol. Med. Sci. *42*, 350–362. (30).

HALSTEAD, S. B., 1966, Bull. World. Health Organ. *35*, 3–15. (30).

HAMDY, B. H., 1972, J. Med. Entomol. *9*, 346–350. (18).

HAMDY, B. H., 1973a, J. Med. Entomol. *10*, 53–57. (18).

HAMDY, B. H., 1973b, J. Med. Entomol. *10*. (18).

HAMLYN, B. M. G., 1953, Ann. Appl. Biol. *40*, 393–402. (25).

HAMMON W. MCD., 1968, J. Am. Med. Ass. *203*, 647. (1).

HAMMON, W. MCD and G. E. SATHER, 1966, Am. J. Trop. Med. Hyg. *15*, 199–204. (30).

HANNA, A. D., W. HEATHERINGTON, H. R. MAPOTHER and R. WICKENS, 1959, Bull. Entomol. Res. *50*, 209–225. (23).

HANNA, A. D., E. JUDENKO and W. HEATHERINGTON, 1955, Bull. Ent. Res. *46*, 669–710. (23).

HANNA, A. D., E. JUDENKO and W. HEATHERINGTON, 1956, Bull. Ent. Res. *47*, 219–226. (31).

HANNOUN, C. 1971, Bull. Inst. Pasteur, Paris *69*, 241–278. (18).

HANNOUN, C., J. CHATELAIN, S. KRAMS and J. M. GUILLON, 1971, C.R. Hebd. Seanc. Acad. Sci., s.D. *272*, 766–768. (18).

HANNOUN, C., B. CORNIOU and J. RAGEAU, 1970, Acta Virol., Prague (English edition) *14*, 167–170. (18).

HANNOUN, C., R. PANTHIER and R. CORNOU *1969*, *In:* Arboviruses of the California complex and Bunyamwera group. Slovak. Acad. Sci. (30).

HANNOUN, C., R. PANTHIER, J. MOUCHET and J. P. EOUZAN, 1964, C.R. Hebd. Seanc. Acad. Sci. Paris *259*, 4170–4172. (30).

HANSON, R. P., 1968, Am. J. Epidemiol. *87*, 264–266. (11).

HANSON, W. J., 1961, Ann. Entomol. Soc. Am. *54*, 317–322. (11).

HARCOURT, D. G., 1966, Can. Entomologist *98*, 653–662. (29).

HARCOURT, D. G. and L. M. CASS, 1968, J. Invert. Pathol. *11*, 142–143. (21).

HARDY, J. L., R. P. SCRIVANI, R. N. LYNESS, R. L. NELSON and D. R. ROBERTS, 1970, Am. J. Trop. Med. Hyg. *19*, 552–563. (11).

HARKER, J. E., 1958, J. Exptl. Biol. *35*, 251. (16).

HARPAZ, I., 1961, F.A.O. Plant. Prot. Bull. *9*, 144–147, (8, 31).

HARPAZ, I. and S. W. APPLEBAU, 1961, Nature London *192*, 780–781. (31).

HARPAZ, I. and Y. BEN SHAKED, 1964, J. Insect Pathol. *6*, 127–130. (21).

HARRAP, K. A., 1969, *In:* Melnick J. L., ed., Proceedings of the First International Congress of Virology, Helsinki 1968, (S. Karger, Basel). Intern. Virol. *1*, 281. (15).

HARRAP, K. A., 1970, Virology *42*, 311–318. (14, 15).

HARRAP, K. A. and B. E. JUNIPER, 1966, Virology *29*, 175–178. (4).

HARRAP, K. A. and J. S. ROBERTSON, 1968, J. Gen. Virol. *3*, 221–225. (2, 14, 15, 21).

HARRISON, B. D., 1958, Virology *6*, 265–277. (24).

HARRISON, B. D., 1961, Rept. Rothamsted Exptl. Sta. for 1960, p. 118. (27).

HARRISON, B. D., 1964a, *In:* Corbett, M. K. and H. D. Sisler, eds. *Plant Virology*, (Florida University Press, Gainesville) pp. 118–147. (27).

HARRISON, B. D., 1964b, Virology *22*, 544–550. (27).

HARRISON, B. D., 1966, Ann. Appl. Biol. *57*, 121–129. (27).

HARRISON, B. D., 1967a, Rept. Rothamsted Exptl. Sta. for 1966, p. 115. (27).

HARRISON, B. D., 1967b, Ann. Appl. Biol. *60*, 405–409. (27).

HARRISON, B. D., 1969, Zentbl. Bakt. ParasitKde *123*, 226–229. (27).

HARRISON, B. D., 1970, C.M.I./A.A.B. Descrip. Plant Viruses. No. *12*. (4).

HARRISON, B. D. and C. H. CADMAN, 1959, Nature, London *184*, 1624–1626. (27).

HARRISON, B. D. and N. C. CROWLEY, 1965, Virology *26*, 297–310. (24).

HARRISON, S. C., A. DAVID, J. JUMBLATT and J. E. DARNELL, 1971, J. Mol. Biol. *60*, 523–528. (4).

HARRISON, B. D., J. T. FINCH, A. J. GIBBS, M. HOLLINGS, R. J. SHEPHERD, V. VALENTA and C. WETTER, 1971, Virology *45*, 356–363. (4, 127).

HARRISON, B. D. and D. J. HOOPER, 1963, Nematologica *9*, 159–160. (31).

HARRISON, B. D. and R. D. WINSLOW, 1961, Ann. Appl. Biol. *49*, 621–633. (27, 31).

HARRISON, B. D., W. P. MOWAT and C. E. TAYLOR, 1961, Virology *14*, 480–485. (27).

HARRISON, B. D., J. E. PEACHEY and R. D. WINSLOW, 1963, Ann. Appl. Biol. *52*, 243–255. (31).

HASHIBA, T. and T. MISAWA, 1969, Tohoko J. Agric. Res. *20*, 159–171. (25).

HAUPT, H., 1935, Tierwelt Mitteleuropas *4*, 221–252. (7).

HAY, A. J., J. J. SKEHEL and D. C. BURKE, 1968, J. Gen. Virol. *3*, 175–184. (14).

HAYASHI, Y., 1970, J. Invert. Pathol. *16*, 442–450. (14).

HAYASHI, Y. and F. T. BIRD, 1968a, J. Invert. Pathol. *11*, 40–44. (4).

HAYASHI, Y. and F. T. BIRD, 1968b, J. Invert. Pathol. *12*, 140 (4).

HAYASHI, Y. and S. KAWASE, 1964, Virology *23*, 612–614. (4).

HAYASHI, Y. and A. RETNAKARAN, 1970, J. Invert. Pathol. *16*, 150–151. (14).

HAYES, R. O., L. C. LAMOTTE and P. HOLDEN, 1967, Am. J. Trop. Med. Hyg. *16*, 675–687. (30).

HEATHCOTE, G. D., 1968a, Plant Pathol. *17*, 158–161. (31).

HEATHCOTE, G. D., 1968b, Ann. Appl. Biol. *62*, 113–118. (31).

HEATHCOTE, G. D., 1970a, Plant Pathol. *19*, 32–39. (31).

HEATHCOTE, G. D., 1970b, J. Inst. Intern. Recherche Betterave *5*, 42–51 (31).

HEATHCOTE, G. D., 1971, *Proc. 10th Br. Weed Control Conf. 1970 3*, 934–941. (31).

HEATHCOTE, G. D. and A. J. COCKBAIN, 1966, Ann. Appl. Biol. *57*, 321–336. (31).

HEIE, O., 1967, Spolia Zool. Mus. Haun. *26*, 274. (7).

HEIE, O., 1969, Mitt. Geol. Palaeont. Inst. Univ. Hamburg *38*, 143–151. (7).

HEIMPEL, A. M. and J. R. ADAMS, 1966, J. Invert Pathol. *8*, 340–346. (14).

HEIMPEL, A. M. and J. C. HARSHBARGER, 1965, Bact. Rev. *29*, 397–405. (15, 21).

HEINZE, K., 1959, *Phytopathogene Viren und ihre Überträger*, (Duncker & Humblot, Berlin) pp. 254–258. (13, 23).

HEINZE, K., 1967a, Mitt. Biol. Bund Anst. Ld-u. Forstw. *121*, 132–139. (31).

HEINZE, K., 1967b, Nachbr. Bl. Deut. Pfl. Schutzdienst. *19*, 150–153. (31).

HELLE, W., 1968, Meded. Rijksfakt. Landbouwwetensch. Ghent. *33*, 621–628. (29).

HENDERSON, B. E., C. A. PIGFORD, T. WORK and R. D. WENDE, 1970, Am. J. Epidemiol. *91*, 87–98. (30).

HENDERSON, B. E., P. M. TUKEI, A. W. R. MCCRAE, Y. SSENKUBUGE and W. N. MUGO, 1970, E. African Med. J. *47*, 273–276. (18).

HENNIG, E., 1963, Entomol. Exptl. Appl. *6*, 326–336. (25).

HENRY, J. E. and J. W. JUTILA, 1961, J. Invert. Pathol. *8*, 417–418. (28).

HENRY, J. E., B. P. NELSON and J. W. JUTILA, 1969, J. Virol. *3*, 605–610. (15).

HEROLD, F., G. H. BERGOLD and J. WEIBEL, 1960, Virology *12*, 335–347. (26).

HEROLD, F. and K. MUNZ, 1965, Virology *25*, 412–417. (26).

HERRIN, C. S. and D. E. BECK, 1965, Brigham Young Univ. Sci. Bull., Biol. s. *6*, 1–19. (5).

HESS, A. D. and R. O. HAYES, 1967, Am. J. Med. Sci. *253*, 333–348. (30).

HESS, A. D. and R. O. HAYES, 1970, Am. J. Trop. Med. Hyg. *19*, 327–334. (30).

HEWITT, W. B., G. MARTELLI, H. F. DIAS and R. H. TAYLOR, 1970, C.M.I./A.A.B. Descrip. Plant Viruses. No. *28*. (4).

HEWITT, W. B., A. C. GOHEEN, D. J. RASKI and G. V. GOODING, 1962, Vitis *3*, 57–83. (31).

HEWITT, W. B., D. J. RASKI and A. C. GOHEEN, 1958, Phytopathology *48*, 586–595. (27).

HIBINO, H., G. H. KALOOSTIAN and H. SCHNEIDER, 1971, Virology *43*, 34–40. (26).

HICKEY, W. A., 1965, XII Int. Congress Entomol. (London) 1964. (30).

HIGASHI, N., A. MATSUMOTO, K. TABATA and Y. NAGATOMO, 1967, Virology *33*, 55–69. (14).

HILGARD, H. R. and J. H. PHILLIPS, 1968, Science *161*, 1243. (16).

HILL, M. N., M. G. R. VARMA, S. MAHADEVAN and P. O. MEERS, 1969, J. Med. Entomol. *6*, 398–406. (30).

HILLE RIS LAMBERS, D., 1955, Ann. Appl. Biol. *42*, 355–360. (31).

HILLE RIS LAMBERS, D., 1966, Ann. Rev. Entomol. *11*, 47–78. (7, 25).

HILLS, G. J. and R. N. CAMPBELL, 1968, J. Ultrastruct. Res. *24*, 134–144. (4).

HILLS, G. J. and K. M. SMITH, 1959, J. Insect Pathol. *1*, 121–128. (4).

HILLS, O. A., A. C. VALCARCE, H. K. JEWELL and D. C. COUDRIET, 1960, J. Am. Soc. Sugar Beet Tech. *11*, 15–24. (31).

HIMENO, M., F. SAKAI, K. ONODERA, H. NAKAI, T. FUKUDA and Y. KAWADE, 1967, Virology *33*, 507–512. (14, 29).

HINK, W. F., 1970, Nature, London *226*, 466–467. (17).

HINK, W. F. and J. D. BRIGGS, 1968, J. Insect Physiol. *14*, 1025. (16).

HINK, W. F. and B. J. ELLIS, 1971, Current Top. Microbiol. Immunol. *55*, 19–28. (17).

HINK, W. F. and C. M. IGNOFFO, 1970, Exptl. Cell Res. *60*, 307–309. (17).

HIROYUKI, H. and H. SCHNEIDER, 1970, Phytopathology *60*, 499. (3).

HIRUMI, H., 1971, Current Top. Microbiol. Immunol. *55*, 170–196. (17).

HIRUMI, H., T. A. CHEN, K. J. LEE and K. MARAMOROSCH, 1968, J. Ultrastruct. Res. *24*, 434–453. (12).

HIRUMI, H. and K. MARAMOROSCH, 1964a, Exptl. Cell Res. *36*, 625–631. (17).

HIRUMI, H. and K. MARAMOROSCH, 1964b, Contrib. Boyce. Thompson Inst. *22*, 343–352. (17).

HIRUMI, H. and K. MARAMOROSCH, 1964c, Science *144*, 1465–1467. (17).

HIRUMI, H. and K. MARAMOROSCH, 1964d, Contrib. Boyce Thompson Inst. 22, 259–268.(17).

HIRUMI, H. and K. MARAMOROSCH, 1969, J. Virol. *3*, 82–84. (26).

HIRUMI, H., R. R. GRANADOS, and K. MARAMOROSCH, 1967, J. Virol. *1*, 430–444. (15, 26).

HITCHCOCK, J. D., 1966, J. Econ. Entomol. *59*, 1154–1156. (22).

HODEK, I., 1967, Ann. Rev. Entomol. *12*, 79–104. (7).

HOFF, G. L., J. O. IVERSEN, T. M. YUILL, R. O. ANSLOW, J. O. JACKSON and R. P. HANSON, 1971a, Am. J. Trop. Med. Hyg. *20*, 320–325. (18).

HOFF, G. L., T. M. YUILL, J. O. IVERSEN and R. P. HANSON, 1971b, Am. J. Trop. Med. Hyg. *20*, 326–330. (18).

HOFFMAN, A., 1962, Rev. Soc. Mex. Hist. Natl. *23*, 191–307. (5).

HOFFMAN, J. A., A. PORTE and P. JOLY, 1968, C.R. Acad. Sci. Paris *267*, 776. (16).

HOGGAN, I. A., 1934, J. Agri. Res. *49*, 1135–1142. (3, 25).

HOLDEN, P., 1955, Proc. Soc. Exptl. Biol. *88*, 607. (1).

HOLLING, C. S., 1959, Can. Entomologist *91*, 293–320. (29).

HOLLINGS, M., 1965, Ann. Rev. Phytopathol. *3*, 367–396. (31).

HOLOWEZAK, J. A. and W. K. JOKLIK, 1967, Virology *33*, 717–725. (4).

HOOGSTRAAL, H., 1956, *African Ixodoidea. I. Ticks of the Sudan.* (Dept. Navy, Bur. Med. Surg., Washington, D.C.) 1101 pp. (5, 18).

HOOGSTRAAL, H., 1966, Ann. Rev. Entomol. *11*, 261–308. (18).

HOOGSTRAAL, H., 1967a, Exptl. Parasitol. *21*, 98–111. (18).

HOOGSTRAAL, H., 1967b, Ann. Rev. Entomol. *12*, 377–420. (5, 18).

HOOGSTRAAL, H., 1970, Misc. Publ. Entomol. Soc. Am. *6*, 359–363. (6).

HOOGSTRAAL, H., 1973, Proc. 2nd Symp. Med. Vet. Acaroentomology (Gdansk October 1971). Wiad. Parazyt. (in press). (5).

HOOGSTRAAL, H. and D. HEYNEMAN, 1969, Am. J. Trop. Med. Hyg. *18*, 1091–1210. (11).

HOOGSTRAAL, H. and M. N. KAISER, 1970, Ann. Entomol. Soc. Am. *63*, 205–210. (18).

HOOGSTRAAL, H. and M. N. KAISER, 1972, Ann. Entomol. Soc. Am. *66*, 1–3. (5).

HOOGSTRAAL, H., M. N. KAISER and G. M. KOHLS, 1968, Ann. Entomol. Soc. Am. *61*, 744–751. (18).

HOOGSTRAAL, H., M. N. KAISER and R. M. MITCHELL, 1970, Ann. Entomol. Soc. Am. *63*, 1576–1585. (5).

HOOGSTRAAL, H., M. N. KAISER, M. A. TRAYLOR, S. GABER and E. GUINDY, 1961, Bull. World Health Organ. *24*, 197–212. (5).

HOOGSTRAAL, H., M. N. KAISER, M. A. TRAYLOR, E. GUINDY and S. GABER, 1963, Bull. World Health Organ. *28*, 235–262. (5).

HOOGSTRAAL, H. and G. M. KOHLS, 1960a, Ann. Entomol. Soc. Am. *53*, 611–618. (18).

HOOGSTRAAL, H. and G. M. KOHLS, 1960b, Ann. Entomol. Soc. Am. *53*, 743–755. (18).

HOOGSTRAAL, H. and G. M. KOHLS, 1963, Ann. Entomol. Soc. Am. *56*, 577–582. (18).

HOOGSTRAAL, H., B. L. LIM and G. ANASTOS, 1969, J. Parasitol. *55*, 1075–1077. (5).

HOOGSTRAAL, H., B. L. LIM, M. NADCHATRAM and G. ANASTOS, 1972, Bull. Brit. Museum Zool. *22*, 153–172. (18).

HOOGSTRAAL, H. and V. C. MCCARTHY, 1965, Ann. Entomol. Soc. Am. *58*, 756–762. (18).

HOOGSTRAAL, H., R. M. OLIVER and S. S. GUIRGIS, 1970, Ann. Entomol. Soc. Am. *63*, 1762–1768. (18).

HOOGSTRAAL, H., A. A. SALAH and M. N. KAISER, 1954, J. Egypt. Publ. Health Assoc. *29*, 127–138. (18).

HOOGSTRAAL, H. and R. J. TATCHELL (eds.), 1972, Misc. Publs. Entomol. Soc. Amer. *8*, 161–376. (18).

HOOGSTRAAL, H., M. A. TRAYLOR, S. GABER, G. MALAKATIS, E. GUINDY and I. HELMY, 1964, Bull. World Health Organ. *30*, 355–367. (5).

HORIKAWA, M. and A. S. FOX, 1964, Science *145*, 1437–1439. (17).

HORSEFALL, W. R., 1956, J. Econ. Entomol. *49*, 416. (30).

HORZINEK, M. and M. MUSSGAY, 1969, J. Virol. *4*, 514–520. (4).

HOSTETTER, D. L. and K. D. BIEVER, 1970, J. Invert. Pathol. *15*, 173–176. (21).

HOWATSON, A. F., 1970, Adv. Virus Res. *16*, 195–256. (2, 20, 26).

HOWELL, P. G. and D. W. VERWOERD, 1971, Virol. Monogr. *9*, 35–74. (20).

HOWELL, C. J., 1970, Sci. Bull. Dept. Agr. Tech. Serv. No. *389*, pp. 16. (5).

HSU, S. H., 1971, Current Top. Microbiol. Immunol. *55*, 140–148. (14).

HSU, S. H., H. H. LIU and E. C. SUITOR, 1969, Mosquito News *29*, 439–446. (17).

HSU, S. H., W. H. MAO and J. H. CROSS, 1970, J. Med. Entomol. *7*, 703–707. (17).

HU, S. M. K., 1939, Am. J. Hyg. *29*, D67. (16).

HUBBELING, N. and E. KOOISTRA, 1963, Euphytica *12*, 258–260. (31).

HUFF, F. H., 1963, J. Appl. Meteorol. *2*, 39–43. (8).

HUGHES, K. M., 1952, J. Bacteriol. *64*, 375–380. (14).

HUGHES, R. D. and N. GILBERT, 1968, J. Anim. Ecol. *37*, 553–563. (7).

HUGHES-SCHRADER, S., 1948, Adv. Genet. *2*, 127–203. (10).

HUGER, A., 1963, *In:* Steinhaus, E. A., ed., *Insect Pathology*. Vol. 1 (Academic Press. New York and London). (14).

HUGER, A. M., 1966, J. Invert. Pathol. *8*, 38–51. (15, 29).

HUGER, A. M., A. KRIEG, P. ENSCHERMAN and P. GÖTZ, 1970, J. Invert. Pathol. *15*, 253–261. (29).

HUKUHARA, T., 1962, J. Insect Pathol. *4*, 132–135. (2).

HUKUHARA, T. and Y. HASHIMOTO, 1966, J. Invert Pathol. *8*, 184–192. (14).

HUKUHARA, T. and Y. HASHIMOTO, 1966, J. Invert Pathol. *8*, 234–239. (4).

HUKUHARA, T. and Y. HASHIMOTO, 1967, J. Invert Pathol. *9*, 278–281. (14).

HULL, R., 1960, Sugar beet diseases. Min. Agric. Fisheries & Food Bull. *142*, Her Majesty's Stationery Office, London. (31).

HULL, R., 1965, Ann. Appl. Biol. *56*, 345–347. (25).

HULL, R., 1965, J. Royal. Agri. Soc. *113*, 86–102. (25).

HULL, R., 1968, Plant Pathol. *17*, 1–10. (31).

HURLBUT, H. S., 1965, Adv. Virus Res. *11*, 277–292. (15).

HURLBUT, H. S. and J. I. THOMAS, 1960, Virology *12*, 391–407. (14, 21).

HUTCHINS, L. M., 1932, Science *76*, 123. (19).

HUTCHINSON, P. B. and R. E. F. MATTHEWS, 1963, Virology *20*, 169–175. (24).

HYMAN, L. H., 1951, *The Invertebrates: Acanthocephala, Aschelminthes and Entoprocta. The Pseudocoelomate Bilateria.* Vol. 3, (McGraw Hill Book Co., New York). (12).

HYSLOP, N. ST. G., 1966, Vet. Rec. *78*, 858–864. (20).

IGNOFFO, C. M., 1964, J. Insect Pathol. *6*, 318–326. (2, 29).

IGNOFFO, C. M., 1965, J. Invert. Pathol. *7*, 227–236. (29).

IGNOFFO, C. M., 1966a, J. Invert. Pathol. *8*, 531–537. (29).

IGNOFFO, C. M., 1966b, J. Invert. Pathol. *8*, 279–282. (21, 29).

IGNOFFO, C. M., 1968, Bull. Entomol. Soc. Am. *14*, 265–276. (15, 21).

IGNOFFO, C. M., 1971, *In:* Burges, H. D. and N. W. Hussey, eds., *Microbial Control of Insetcs and Mites* (Academic Press, London). (29).

IGNOFFO, C. M. and J. R. ADAMS, 1966, J. Invert. Pathol. *8*, 59–67. (29).

IGNOFFO, C. M. and C. GARCIA, 1969, J. Invert. Pathol. *14*, 282–284. (29).

IGNOFFO, C. M. and W. F. HINK, 1971, *In:* Burges, H. D. and N. W. Hussey, eds., *Microbial Control of Insects and Mites*, (Academic Press, London and New York). (14).

IGWEGBE, E. C. K., 1966, Ann. Rep. Cocoa Res. Inst. Nigeria 1964–65, 62–65. (23).

IMMS, A. D., 1957, *A General Textbook of Entomology.* 9th Ed., revised (Methuen, London). (10, 13).

INDIAN COUNCIL OF MEDICAL RESEARCH, 1964, Kyasanur Forest disease 1957–1964. Publ. Indian Couns. Med. Res. *30*. (18).

INGRAM, R. L., 1954, Am. J. Hyg. *60*, 169–185. (30).

IL'ENKO, V. I., T. S. GOROZHANKINA and A. A. SMORODINTSEV, 1970, Med. Parazit., Moskva *39*, 263–269. (18).

ISHAAYA, I., I. MOORE and D. JOSEPH, 1971, J. Insect Physiol. *17*, 945–953. (9).

ISHII, T., 1966, Res. Bull. Hokkaido Nat. Agri. Expt. Sta. *89*, 49–54. (8).

ISHII, T., 1967, Japan. J. Appl. Entomol. Zool. *11*, 191–192. (8).

ISHIMORI, J. 1934, Compt. Rend. Soc. Biol. *116*, 1169. (2).

IZADPANAH, K. and R. J. SHEPHERD, 1966, Virology *28*, 463–476. (24).

JACOBSON, M. F. and D. BALTIMORE, 1968, Proc. Natl. Acad. Sci. U.S. *61*, 77–84. (14).

JACOBY, M., 1908, *The Fauna of British India. Coleoptera (Chrysomelidae 1)*, (Taylor Francis, London). (9).

JAGGER, I. C., 1916, Phytopathology *6*, 148–151. (3).

JAMES, H. C., 1937, Bull. Entomol. Res. *28*, 429–461. (10).

JANOFF, A. and E. HAWRYLKO, 1964, J. Cell Comp. Physiol. *63*, 267. (16).

JANZEN, H. G., A. J. RHODES and F. W. DOANE, 1970, Can. J. Microbiol. *16*, 581–586. (14, 15).

JAQUES, R. P., 1962, J. Insect Pathol. *4*, 433–445. (15, 21).

JAQUES, R. P., 1964, J. Insect Pathol. *6*, 251–254. (21, 29).

JAQUES, R. P., 1967a, Can. Entomologist *99*, 785–794. (29).

JAQUES, R. P., 1967b, Can. Entomologist *99*, 820–829. (29).

JAQUES, R. P., 1969, J. Invert. Pathol. *13*, 256–263. (21).

JAQUES, R. P., 1970a, Can. Entomologist *102*, 36–41. (29).

JAQUES, R. P., 1970b, J. Invert. Pathol. *15*, 328–340. (21, 29).

JAQUES, R. P. and F. HUSTON, 1969, J. Invert. Pathol. *14*, 289–290. (29).

JAQUES, R. P. and H. T. STULTZ, 1966, Can. Entomol. *98*, 1036–1045. (29).

JAYARAJ, S., P. EHRHARDT and H. SCHMUTTERER, 1967, Ann. Appl. Biol. *59*, 13–21. (31).

JEANNEL, R., 1949, *In:* Grassé, P. P., ed., *Traité de Zoologie, Anatomie, Systematique. Biologie. Tome 9*, (Masson et Cie, Paris). (9).

JELLISON, W. L., E. J. BELL, R. J. HUEBNER, R. R. PARKER and H. H. WELSH, 1948, Publ. Health. Rep. U.S. Publ. Health. Serv. *63*, 1483–1489. (5).

JENKIN, C. R. and D. ROWLEY, 1970, Australian J. Exptl. Biol. Med. Sci. *48*, 129. (16).

JENNINGS, P. R. and A. PINEDA, 1970, Phytopathology *61*, 142–143. (26).

JENSEN, D. D., 1951, Hilgardia *20*, 299–324. (7).

JENSEN, D. D., 1959, Virology *8*, 164–175. (2, 26).

JENSEN, D. D., 1963, Ann. N.Y. Acad. Sci. *105*, 685–712. (26).

JENSEN, H. J., 1967, Plant Disease Reptr. *51*, 98–102. (12).

JENSEN, H. J. and T. C. ALLEN, 1964, Plant Disease Reptr. *48*, 333–334. (27).

JENSEN, H. J. and J. O. STEVENS, 1969, J. Nematol. *1*, 293. (12).

JHA, A. and A. F. POSNETTE, 1959, Nature, London *184*, 962–963. (27).

JOHNSON, B., 1958, Anim. Behav. *6*, 9–26. (8).

JOHNSON, C. G., 1969, *Migration and Dispersal of Insects by Flight*, (Methuen, London). (4, 8).

JOHNSON, G. V., A. BING and F. F. SMITH, 1967, J. Econ. Entomol. *60*, 16–19. (31).

JOHNSON, J., 1925, Wis. Agri. Expt. Sta. Res. Bull. *63*. (3).

JOHNSON, K. M., 1965, Am. J. Trop. Med. Hyg. *14*, 816–818. (30).

JOHNSON, K. M., S. B. HALSTEAD and S. N. COHEN, 1967, Progr. Med. Virol. *19*, 105–158. (18).

JOHNSON, P. T. and F. A. CHAPMAN, 1970, J. Invert. Pathol. *16*, 127. (16).

JOKLIK, W. K., 1968, Ann. Rev. Microbiol. *22*, 359–390. (4).

JONES, A. T. and P. L. CATHERALL, 1970, Ann. Appl. Biol. *65*, 147–152. (31).

JONES, B. M., 1966, Cells and Tissues in Culture. *3*, 397. (17).

JONES, E. K. and C. M. CLIFFORD, 1972, Ann. Entomol. Soc. Am. *65*, 730–740. (5).

JONES, F. G. W. and M. G. JONES, 1964, *Pests of Field Crops*, (Edward Arnold, London) 406 pp. (12).

JONES, F. G. W., D. M. PARROTT and G. J. S. ROSS, 1967, Ann. Appl. Biol. *60*, 151–171. (31).

JONES, J. C., 1962, Am. Zool. *2*, 209. (16).

JONKERS, A. H., 1967, Am. J. Epidemiol. *86*, 286–291. (11).

JONKERS, A. H., L. SPENCE, W. G. DOWNS, T. H. C. AITKEN and C. B. WORTH, 1968, Am. J. Trop. Med. Hyg. *17*, 285–297. (11).

JOUSSET, F. X., 1969, Compt. Rend. *269*, 1035–1038. (20).

JOUSSET, F. X., 1970, Compt. Rend. *271*, 1141–1144. (20).

JUDY, K. J., 1969, Science *165*, 1374–1375. (17).

KAISER, M. N., 1966a, Am. J. Trop. Med. Hyg. *15*, 964–975. (18).

KAISER, M. N., 1966b, Am. J. Trop. Med. Hyg. *15*, 976–985. (18).

KAISER, M. N. and H. HOOGSTRAAL, 1969, Ann. Entomol. Soc. Am. *62*, 885–890. (5).

KAISER, M. N., H. HOOGSTRAAL and G. M. KOHLS, 1964, Ann. Entomol. Soc. Am. *57*, 60–69. (18).

KAKPAKOV, V. T., V. A. GOOSDEV, T. P. PLATOVA and L. G. POLUKAROVA, 1969, Genetika 5, 67–75. (17).

KALMAKOFF, J., L. J. LEWANDOWSKI and D. R. BLACK, 1969, J. Virol. 4, 851–856. (4, 14).

KALMAKOFF, J. and J. S. ROBERTSON, 1970, Proc. Otago Med. School 48, 16–18. (4, 21).

KALMAKOFF, J. and J. H. TREMAINE, 1968, J. Virol. 2, 738–744. (4).

KASSANIS, B., 1950, Ann. Appl. Biol. 37, 339–341. (31).

KASSANIS, B., 1957, Ann. Appl. Biol. 45, 422–427. (31).

KASSANIS, B., 1961, Rept. Rothamsted Exp. Sta. for 1960, p. 117. (31).

KASSANIS, B. and D. A. GOVIER, 1971, J. Gen. Virol. 10, 99. (25).

KASSANIS, B. and D. A. GOVIER, 1971, J. Gen. Virol. 13, 221–228. (25).

KAWASE, S., 1964, J. Invert. Pathol. 6, 156–163. (4).

KAWASE, S. and T. HUKUHARA, 1967, J. Invert. Pathol. 9, 273–274. (5).

KAWASE, S. and I. KAWAMORI, 1968, J. Invert. Pathol. 12, 395–404. (14).

KAWASE, S. and S. MIYAJIMA, 1968, J. Invert. Pathol. 11, 63–69. (2, 4, 14).

KAWASE, S. and S. MIYAJIMA, 1969, J. Invert. Pathol. 13, 330–336. (14).

KAYA, H., 1970, Science 168, 251–253. (29).

KEIFER, H. H., 1952, Bull. Calif. Insect Surv. 2. (19).

KEIFER, H. H., 1964, Eriophyid studies B-11. Bureau of Entomology, California Dept. Agr., 1–20. (5).

KEIFER, H. H. and N. S. WILSON, 1956, Bull. Calif. Dept. Agr. 44, 145–146. (19).

KEIRANS, J. E., C. M. CLIFFORD and J. M. CAPRILES, 1971, Ann. Entomol. Soc. Am. 64, 1410–1413. (5).

KELLEN, W. R., T. B. CLARK and J. E. LINDEGREN, 1963, J. Insect Pathol. 5, 98–103. (20).

KELLEN, W. R., T. B. CLARK, J. E. LINDEGREN and J. D. SANDERS, 1966, J. Invert. Pathol. 8, 390–394. (14).

KELLER, E. R. and S. BÉRCES, 1966, European Potato J. 9, 1–12. (31).

KELLER, E. R. and R. WEISS, 1956, Mitt. Schweiz. Landwirtsch. 6, 97–104. (31).

KELLY, R. B., J. L. GOULD and R. L. SINSHEIMER, 1965, J. Mol. Biol. 11, 562–575. (14).

KEMP, G. E., O. R. CAUSEY and C. E. CAUSEY, 1971, Bull. Epiz. Dis. Afr. 19, 131–135. (18).

KENNEDY, J. S., C. O. BOOTH and W. J. S. KERSHAW, 1959, Ann. Appl. Biol. 47, 410–423. (25).

KENNEDY, J. S., C. O. BOOTH and W. J. S. KERSHAW, 1961, Ann. Appl. Biol. 49, 1–21. (25).

KENNEDY, J. S., M. F. DAY and V. F. EASTOP, 1962, A conspectus of aphids as vectors of plant viruses. Commonwealth Inst. of Entomol., London. (7, 8, 15, 25).

KENNEDY, J. S. and T. E. MITTLER, 1953, Nature, London 171, 528. (25).

KENTEN, R. H. and J. T. LEGG, 1967, J. Gen. Virol. 1, 465–470. (23).

KENTEN, R. H. and J. T. LEGG, 1971, Ann. Appl. Biol. 67, 195–200. (23).

KETTLE, D. S., 1969, Acta Trop. 26, 235–248. (11).

KHAN, M. A. Q., 1964, PhD Thesis, University of West Ontario, London, Ontario. (30).

KHARITONOVA, N. N. and T. M. KHADZHIEVA, 1966, In: Maksimov, A. A. and G. I. Netsky, eds., *The Muskrat of Western Siberia*. (Biol. Inst., Akad. Nauk SSSR, Sibirsk. Otd., Novosibirsk). pp. 136–140. (18).

KIKUMOTO, T. and C. MATSUI, 1962, Virology 16, 509–510. (24).

KILPATRICK, J. W., R. J. TONN and S. JATANASEN, 1970, Bull. World Health Organ. 42, 1–14. (30).

KIMURA, I., 1962, Ann. Phytopathol. Soc. Japan 27, 197–203. (26).

KIMURA, T. and A. W. A. BROWN, 1964, J. Econ. Entomol. 57, 710–716. (30).

KINSEY, M. G. and D. L. MCLEAN, 1967, Ann. Entomol. Soc. Am. 60, 1263–1265. (24).

KIRKPATRICK, T. W., 1927, Bull. Dept. Agri., Kenya. *18*. (10).

KIRKPATRICK, T. W., 1931, Bull. Entomol. Res. *22*, 323–363. (3).

KIRKPATRICK, T. W., 1950, Bull. Entomol. Res. *41*, 99–117. (10, 23).

KIRKPATRICK, T. W., 1953, Rept. Cacao Res. 1952, 62–71. (10).

KIRKWOOD, A. C., 1968, Entomol. Exptl. Appl. *11*, 315–320. (6).

KISLEV, N., I. HARPAZ and A. ZELCER, 1969, J. Invert. Pathol. *14*, 245–257. (14).

KISIMOTO, R., 1958, Japan. J. Appl. Entomol. Zool. *2*, 128–134. (8).

KISIMOTO, R., 1959a, Japan. J. Appl. Entomol. Zool. *3*, 49–55. (8).

KISIMOTO, R., 1959b, Japan. J. Appl. Entomol. Zool. *3*, 200–207. (8).

KISIMOTO, R., 1965, Bull. Shikou Agri. Expt. Sta. *13*, 1–106. (8).

KISIMOTO, R., 1971. *In:* Proc. Symp. Rice Insects. Trop. Agric. Res. Centre, MAF, Tokyo, (201–216). (8).

KISIMOTO, R. and M. A. WATSON, 1965, J. Invert. Pathol. *7*, 297–305. (26).

KITAMURA, S., 1965, Kobe J. Med. Sci. *11*, 23–30. (17).

KITAMURA, S., 1970, Kobe J. Med. Sci. *16*, 41–50. (17).

KLEIN, M. and I. HARPAZ, 1970, Virology *41*, 72–76. (26).

KLIGLER, I. J. and M. ASHNER, 1929, Brit. J. Exptl. Pathol. *10*, 347–352. (4).

KLIMASZEWSKI, S. M., 1964, Ann. Zool. Warsaw *22*, 139–156. (7).

KLIMASZEWSKI, S. M., 1969, Klucze Ozhacz. Kreg. Pol. *17*, 1–89. (7).

KLOET, S. K. and W. D. HINCKS, 1964, Handbooks Ident. Brit. Inst. *11*, 65–66. (7).

KLUG, A., R. E. FRANKLIN and S. P. F. HUMPHREYS-OWEN, 1959, Biochim. Biophys. Acta. *32*, 203–219. (4).

KNIPLING, E. F., 1967, *In:* Wright, J. W. and R. Pal, eds., *Genetics of Insect Vectors of Disease*, (Elsevier, Amsterdam). (30).

KNUDSON, D. and R. MACLEOD, 1972, Virology *47*, 285–295. (4).

KOHLS, G. M., 1960, Ann. Entomol. Soc. Am. *53*, 855–856. (5).

KOHLS, G. M. and C. M. CLIFFORD, 1964, J. Parasitol. *50*, 792–796. (18).

KOHLS, G. M., D. E. SONENSHINE and C. M. CLIFFORD, 1965, Ann. Entomol. Soc. Am. *58*, 331–364. (5).

KOKERNOT, R. H., C. H. CALISHER, L. J. STANNARD and J. HAYES, 1969, Am. J. Trop. Med. Hyg. *18*, 789–795. (18).

KOKERNOT, R. H., J. HAYES, R. L. WILL, C. H. TEMPLIS, D. H. M. CHAN and B. RADIVOJEVIC, 1969, Am. J. Trop. Med. Hyg. *18*, 750–761. (30).

KOLMAN, J. M. and M. HUSOVA, 1971, Folia Parasitol. *18*, 329–335. (18).

KOMAREK, J. and V. BREINDL, 1924, Z. Angew. Entomol. *10*, 99–162. (2).

KONDRASHOVA, Z. N. and R. V. FILIPPOVETS, 1970, Vop. Virus. *15*, 703–708. (18).

KONICEK, D. E. and H. J. JENSEN, 1961, Proc. Helminthol. Soc. Wash. *28*, 216–218. (12).

KONHO, J., K. ENDO and N. ISHIDA, 1967, Proc. Soc. Exptl. Biol. Med. *124*, 73–75. (30).

KONO, Y., T. YOSHINO and Y. FUKANAGA, 1970, Arch. Gesch. Virusforsch. *30*, 252–256. (4, 20).

KOPEC, S., 1911, Arch. Entwicklungsmech. Organ. *33*, 1. (16).

KORENBERG, E. I., 1966, Zool. Zh. *45*, 245–260. (In Russian, Engl. trans.: NAMRU3-T212). (18).

KORENBERG, E. I., 1967, Byull. Mosk. Obshch. Ispyt. Prirody, Otd. Biol. *72*, 32–36. (18).

KORENBERG, E. I., 1969, Zool. Zh. *48*, 356–363. (18).

KORENBERG, E. I., M. I. DZYUBA and V. I. ZHUKOV, 1971, Zool. Zh. *50*, 41–50. (18).

KORENBERG, E. I. and L. M. IVANOVA, 1967, Med. Parazit., Moskva *36*, 270–275. (18).

KORENBERG, E. I., V. V. KUCHERUK, L. I. POGORELENKO, L. G. SUVOROVA and YU. V. KOVA-LEVSKY, 1967, Mater 13. Sess. Inst. Aktual. Probl. Virus. Spetsifich. Profilakt. Virus. Zabolev... (June 27–30, 1967). 143–144. (18).

KORENBERG, E. I., V. I. ZHUKOV, A. V. SHATKAUSKAS and L. K. BUSHUEVA, 1969, Zool. Zh. *48*, 1003–1014. (18).

KOURI, J., G. KOURI, E. BRAVO, P. RODRIGUEZ and A. AGUILERA, 1971, J. Microscopie *11*, 331–338. (17).

KOVACS, E. and B. BUCS, 1967, Life Sciences *6*, 347–348. (2).

KOYAMA, R. and K. KAZUMASA, 1967, *In:* The U.S.–Japan Committee on Scientific Co-operation, Proceedings of the joint United States–Japan Seminar on Microbial Control of Insect Pests, Fukuoka. (29).

KOZLOV, E. A. and I. P. ALEXEENKO, 1967, J. Invert. Pathol. *9*, 413–419. (4).

KOZUCH, O., M. GRESIKOVA and J. NOSEK, 1968, Acta. Virol., Prague *12*, 475. (18).

KOZUCH, O., M. GRESIKOVA, J. NOSEK and J. CHMELA, 1970a, Folia Parasitol. *17*, 337–340. (18).

KOZUCH, O., J. RAJCANI, M. SEKEYOVA and J. NOSEK, 1970b, Acta Virol., Prague *14*, 163–166. (18).

KRAMER, M., A. ORLANDO and K. M. SILBERSCHMIDT, 1945, Biologico *11*, 121–134. (31).

KRANTZ, G. W., 1970, *A Manual of Acarology*, (O.S.U. Book Stores, Oregon), 1–335. (6).

KREUTZBERG, V. E., 1940, Dokl. Acad. Sci. U.S.S.R. *27*, 614–617. (13).

KREUTZBERG, V. E., 1955, Entomol. Obozr. *34*, 95–98. (13).

KRIEG, A. and A. HUGER, 1960, J. Insect Pathol. *2*, 274–288. (15).

KRIEG, A. and A. M. HUGER, 1969, J. Invert. Pathol. *13*, 272–279. (14).

KRING, J. B., 1964, Frontiers Plant Sci. *17*, 6–17. (31).

KRUG, R. M. and P. J. GOMATOS, 1969, J. Virol. *4*, 642–650. (4).

KRYWIENCZYK, J., 1963, J. Invert. Pathol. *5*, 309–317. (4, 14).

KRYWIENCZYK, J. and G. H. BERGOLD, 1960a, Virology *10*, 308–315. (4).

KRYWIENCZYK, J. and G. H. BERGOLD, 1960b, J. Immunol. *84*, 404–408. (4).

KRYWIENCZYK, J. and G. H. BERGOLD, 1961, J. Insect Pathol. *3*, 15–28. (4).

KRYWIENCZYK, J., Y. HAYASHI and F. T. BIRD, 1969, J. Invert. Pathol. *13*, 114–119. (4).

KRYWIENCZYK, J. and S. S. SOHI, 1967, J. Invert. Pathol. *9*, 568–570. (4, 14).

KUIPER, K. and P. A. A. LOOF, 1962, Verslag Plantenziektek. Dienst Wageningen *136*, 193–200. (12).

KULPS, G., 1968, Z. Pfl. Krankh. Pfl. Path. Pfl. Schutz *74*, 213–217. (25).

KUNKEL, L. O., 1926, Am. J. Botany *13*, 646–705. (3).

KUNKEL, L. O., 1937, Am. J. Botany *24*, 316. (3).

KUNZ, M. L., 1965, Am. J. Trop. Med. Hyg. *14*, 813–816. (30).

KURIBAYASHI, K., 1931, Nagano Agri. Expt. Sta. Bull. *2*, 45–69. (8).

KURSTAK, E., S. BELLONCIK and C. BRAILOVSKY, 1969, C.R. Acad. Sci., Paris *269D*, 1716–1719. (4).

KURSTAK, E. and J. R. CÔTÉ, 1969, C.R. Acad. Sci., Paris *268D*, 616–619. (4).

LA CHANCE, L. E., 1967, *In:* Wright J. W. and R. Pal., eds., *Genetics of Insect Vectors of Disease*, (Elsevier, Amsterdam). (30).

LAIRD, M., 1960, *In:* Conference on Biological Control of Insects of Medical Importance. Am. Inst. Biol. Sci. Tech. Rept. (30).

LAM, K. S. K. and I. D. MARSHALL, 1968a, Am. J. Trop. Med., Hyg. *17*, 625–636. (20).

LAM, K. S. K. and I. D. MARSHALL, 1968b, Am. J. Trop. Med. Hyg. *17*, 637–644. (14, 20).

LAMB, K. P., 1958, N.Z.J. Sci. *1*, 579–589. (31).

LA MOTTE, L. C. Jr., 1960, Am. J. Hyg. *72*, 73–87. (14, 15, 20).

LANGRIDGE, R. and P. J. GOMATOS, 1963, Science *141*, 694–698. (4).

LASCANO, E. F., M. I. BERRIA and J. G. BARRERA ORO, 1969, J. Virol. *4*, 271–282. (14).

LAUDEHO, Y. and A. AMARGIER, 1963, Reo. Pathol. Veg. Entomol. Agr. Fr. *42*, 207–210. (14).

LAUTERER, P., 1965, Casopia Moravskeho Musea *50*, 171–190. (7).

LAVEN, H., 1967, Nature, London *216*, 383–384. (30).

LAVIE, P., J. FRESNAYE and C. VAGO, 1965, Ann. Abeille *8*, 321–323. (29).

LEE, P. E., 1964, Virology *23*, 145–151. (26).

LEE, P. E. and B. FURGALA, 1967, J. Invert. Pathol. *9*, 178–187. (15, 22).

LEES, A. D., 1946, Parasitology *37*, 172–184. (18).

LEES, A. D., 1953, Ann. Appl. Biol. *40*, 449–486. (6).

LEES, A. D., 1966, Adv. Insect. Physiol. *3*, 207–277. (7, 25).

LEGG, J. T. and F. X. AGBODJAN, 1969, Ann. Rep. Cocoa Res. Inst. Ghana 1967–68, 23–25. (23).

LEGG, J. T. and R. H. KENTEN, 1970, Ann. Appl. Biol. *65*, 425–434. (23).

LEIFER, E., D. J. GOCKE and H. BOURNE, 1970, Am. J. Trop. Med. Hyg. *19*, 677–679. (18).

LESPERON, L., 1937, Arch. Zool. Exptl. Gen. *79*, 1. (16).

LESSEPS, R. J., 1965, Science *148*, 502–503. (17).

LEUTENEGGER, R., 1964, Virology *24*, 200–204. (14).

LEUTENEGGER, R., 1967, Virology *32*, 109–116. (14).

LEVIN, J. G. and R. M. FRIEDMAN, 1971, J. Virol. *7*, 504–514. (14).

LEWANDOWSKI, L. J., J. KALMAKOFF and Y. TANADA, 1969, J. Virol. *4*, 857–865. (4, 14).

LEWIS, D. J., 1933, Bull. Entomol. Res. *24*, 363–372. (30).

LEWIS, D. J., 1966, *In:* Lowe-Connell, X., ed., *Man-made Lakes*, (Academic Press, London and New York). (30).

LEWIS, T., 1964, Ann. Appl. Biol. *53*, 165–170. (13).

LEWIS, T., 1967, Ann. Appl. Biol. *60*, 23–31. (31).

LEWIS, T. and L. R. TAYLOR, 1967, *Introduction to experimental ecology*, (Academic Press, London and New York). (4).

L'HERITIER, PH., 1958, Adv. Virus. Res. *5*, 195–245. (20).

L'HERITIER, PH. and F. HUGON DE SCOEUX, 1947, Bull. Biol. France Belg. *81*, 70–91. (20).

L'HERITIER, PH. and G. TEISSIER, 1937, Compt. Rend. *205*, 1099–1101. (20).

LIBIKOVA, H. and J. CASALS, 1971, Acta Virol. *15*, 65–78. (18).

LIBÍKOVÁ, H., J. ŘEHÁČEK and V. MAYER, 1965, Proc. Symp. Theor. Quest. Nat. Foci Diseases, (Prague, November 26–29, 1963), 429–437. (18).

LIMA, M. B., 1968, C.r. 8e Symp. Int. Nematol. Antibes, Sept. 1965, pp. 30. (12).

LIN, S. C. and C. S. LEE, 1969, Sugarcane Pathol. Newsletter *3*, 2–3. (26).

LINDBERG, H. and F. OSSIANNILSSON, 1960, Fauna Fenn. *8*, 1–23. (7).

LINDSTEN, K., 1961, Kungl. Lantbrukshögsk. Ann. *27*, 199–271. (8).

LING, K. C., 1966. Phytopathology *56*, 1252–1256. (26).

LINLEY, J. R. and H. T. NIELSEN, 1968a, J. Invert. Pathol. *12*, 7–16. (20).

LINLEY, J. R. and H. T. NIELSEN, 1968b, J. Invert. Pathol. *12*, 17–24. (20).

LISON, L., 1942, Mem. Acad. Roy. Belg. Cl. Sci. (8°) *19*, 1. (16).

LISTER, R. M., 1953, Ann. Rep. W. African Cacao Res. Inst. 1952–53, *9*. (23).

LISTER, R. M., 1964, Ann. Appl. Biol. *54*, 167–176. (27).

LISTER, R. M. and C. E. BRACKER, 1969, Virology *37*, 262–275. (4).

LISTER, R. M. and A. F. MURANT, 1967, Ann. Appl. Biol. *59*, 49–62. (27).

LITTAU, V. C. and K. MARAMOROSCH, 1960, Virology *10*, 483–500. (26).

LIVINGSTON, L. G., 1935, Am. J. Botany *22*, 75–87. (25).

LLOYD, D. C., 1952, Can. Entomol. *84*, 308–310. (10).

LOCKART, R. Z. Jr., 1960, Virology *10*, 198–210. (14).

LOCKART, R. Z. and H. EAGLE, 1959, Science *129*, 252–254. (17).

LOEBENSTEIN, G., M. A. ALPER and M. DEUTSCH., 1964, Phytopathology *54*, 960–962. (25).

LÖFFLER, F. and P. FROSCH, 1898, Zent. Bakt. I. Orig. *23*, 371. (1, 4).

LOFGREN, C. S., H. R. FORD, R. J. TONN and S. JATANASEN, 1970, Bull. World Health Organ. *42*, 15–25. (30).

LOGINOVA, M. M., 1960, Tr. Vses. Entomol. Obshch. *47*, 53–93. (7).

LOGINOVA, M. N., 1964a, Tr. Zool. Inst. *34*, 52–112. (7).

LOGINOVA, M. N., 1964b, *In:* Bey-Bienko, G. Y., *Classification Keys to the insects of the European part of the U.S.S.R.* Vol. 1. 437–482. (Acad. Sci. Moscow and Leningrad). (7).

LONGWORTH, J. F., 1964, Ann. Rep. Nigerian Cocoa Res. Inst. 1962–63, 28–33. (10, 23).

LONGWORTH, J. F. and J. C. CUNNINGHAM, 1968, J. Invert. Pathol. *10*, 361–367. (14, 15).

LONGWORTH, J. F. and K. A. HARRAP, 1968, J. Invert. Pathol. *10*, 139–145. (15).

LONGWORTH, J. F. and C. R. SPILLING, 1970, J. Invert. Pathol. *15*, 276–280. (14, 22).

LONGWORTH, J. F., T. W. TINSLEY, A. H. BARWISE and I. O. WALKER, 1968, J. Gen. Virol. *3*, 167–174. (4).

LOOF, P. A. A., 1965, Verslag Meded. Plantenziektenk. Dienst Wageningen *142*, 132–136. (12).

LOPEZ-ABELLA, D., F. JIMENEZ-MILLAN and F. GARCIA-HIDALGO, 1967, Nematologica *13*, 283–286. (12).

LOTMAR, R., 1941, Mitt. Schweiz. Entomol. Ges. *18*, 372–373. (2).

LOVISOLO, O., E. LUISONI, M. CONTI and C. WETTER, 1967, Naturwiss. *54*, 73–74. (26).

LOVISOLO, O., 1971, C.M.I./A.A.B. Descrip. Plant Viruses. No. *72*. (4).

LUMSDEN, W. H. R., 1955, Trans. Roy. Soc. Trop. Med. Hyg. *49*, 33–57. (30).

L'VOV, D. K., V. L. GROMASHEVSKY, G. A. SIDOROVA, YU. M. TSIRKIN, V. I. CHERVONSKY, G. V. GOSTINSHCHIKOVA and V. A. ARISTOVA, 1971a, Vop. Virus. *16*, 434–438. (18).

L'VOV, D. K., A. A. TIMOFEEVA, V. I. CHERVONSKY, V. L. GROMASHEVSKY, G. A. KLISENKO, G. V. GOSTINSCHCHIKOVA and I. N. KOSTYRKO, 1971b, Am. J. Trop. Med. Hyg. *20*, 456–460. (18).

MACCARTHY, H. R., 1954, Phytopathology *53*, 1161–1163. (24).

MACDONALD, W. W., C. E. G. SMITH, P. S. DAWSON, A. CANAPATHIPILLAI and S. MAHADEVAN, 1967, J. Med. Entomol. *4*, 146–157. (30).

MACKENZIE, R. B., H. K. BEYE, C. L. VALVERDE and H. GARRON, 1964, Am. J. Trop. Med. Hyg. *13*, 620–625. (30).

MACKIN, J. G., 1951, Bull. Marine Sci. Gulf Carribbean *1*, 72. (16).

MACLEOD, J., M. H. COLBO and S. BEK-PEDERSEN, 1970, Bull. Epiz. Diseases Afr. *18*, 355–358. (5).

MACLEOD, R., 1968, Virology *34*, 771–777. (26).

MACLEOD, R., L. M. BLACK and F. H. MOYER, 1966, Virology *29*, 540–552. (26).

MAESTRI, A., 1856, *In:* Frammenti Anatomici Fisiologici e Patologici sul Baco da Seta; pp. 117–120, (Fusi, Pavia). (2).

MAGEE, C. J., 1961, Rep. 6th Commonwealth Mycol. Conf. 1960, pp. 117–119. (31).

MAGEE, C. J., 1967, Tech. Pap. S. Pacif. Commn. No. *150*, 1–13. (31).

MAGGENTI, A. R., 1971, *In:* Zuckerman, B. M., W. F. Mai and R. A. Rohde, eds., *Plant Parasitic Nematodes* Vol. I, (Academic Press, New York and London). (12).

MAGNIN, J., 1953, Agron. Trop. Nogent-Sur-Marne *8*, 292–299. (10).

MAGUIRE, T., 1970, J. Hyg. *68*, 625–630. (5, 20).

MAGNOLER, A., 1968, J. Invert. Pathol. *11*, 326–328. (29).

MAGUIRE, T. and F. N. MACNAMARA, 1966, J. Hyg. *64*, 451–456. (4, 20).

MAGUIRE, T. and J. A. R. MILES, 1965, Arch. Ges. Virusforsch. *15*, 457–474. (14).

MALM, N. R. and R. E. FINKNER, 1968, J. Am. Soc. Sugar Beet Tech. *15*, 246–254. (31).

MARAMOROSCH, K., 1952, Phytopathology *42*, 59–64. (3, 26).

MARAMOROSCH, K., 1963, Ann. Rev. Entomol. *8*, 400–401. (23).

MARAMOROSCH, K., R. R. GRANADOS and H. HIRUMI, 1970, Adv. Virus Res. *16*, 135–193. (2, 26).

MARCHALONIS, J. J. and G. M. EDELMAN, 1968, J. Mol. Biol. *32*, 453. (16).

MARKHAM, R. and K. M. SMITH, 1949, Parasitology *39*, 330–342. (4, 9).

MARKS, E. P. and J. P. REINECKE, 1964, Science *143*, 961. (17).

MARROU, J., 1969, Ann. Phytopathol. *1*, 213–218. (31).

MARSCHALL, K. J., 1970, Nature *225*, 288–289. (29).

MARSHALL, I. D. and G. W. DOUGLAS, 1961, J. Hyg. *59*, 117–122. (20).

MARSHALL, I. D. and F. FENNER, 1958, J. Hyg. *56*, 288–302. (20).

MARSHALL, I. D. and F. FENNER, 1960, J. Hyg. *58*, 485–488. (20).

MARSHALL, I. D. and D. C. REGNERY, 1963, Am. J. Hyg. *77*, 213–219. (20).

MARSHALL, J., 1959, Proc. Entomol. Soc. B.C. *56*, 69–71. (7).

MARTIGNONI, M. E., 1957, Mitt. Schweiz. Anst. Forst. Versuchwesen *32*, 371–418. (29).

MARTIGNONI, M. E., 1962, A list of papers concerning the transovum transmission of insect viruses. No. 5. Mimeographed series. University of California, Berkeley, U.S.A. 4 pp. (21).

MARTIGNONI, M. E. and J. E. MILSTEAD, 1962, J. Insect Pathol. *4*, 113–121. (21, 29).

MARTIGNONI, M. E. and J. E. MILSTEAD, 1964, J. Insect Pathol. *6*, 517–531. (15).

MARTIGNONI, M. E. and R. J. SCALLION, 1961, Nature, London *190*, 1133–1134. (14).

MARTIGNONI, M. E. and P. SCHMID, 1961, J. Insect Pathol. *3*, 62–74. (21, 29).

MARTIN, E. M. and J. A. SONNABEND, 1967, J. Virol. *1*, 97–109. (14).

MARTIN, H. M. and B. O. VIDLER, 1962, Exptl. Parasitol. *12*, 192–203. (17).

MARTIN, W. J. and E. J. KANTACK, 1960, Phytopathology *50*, 150–152. (31).

MASSEE, A. M., 1928, Bull. Entomol. Res. *18*, 297–309. (19).

MASSEE, A. M., 1952, Ann. Rep. E. Malling Res. Sta, Kent, 1951, 162–165. (19).

MATHAD, S. B., C. M. SPLITTSTOESSER and F. L. MCEWAN, 1968, J. Invert. Pathol. *11*, 456–464. (14).

MATTHEWS, R. E. F., 1970a, *Plant Virology*, (Academic Press, New York and London). (4).

MATTHEWS, R. E. F., 1970b, C.M.I./A.A.B. Descrip. Plant Viruses No. 2 (4).

MATHUR, R. N., 1935, Indian Forest Records *1*, 35–71. (7).

MATSUI, C., T. SASAKI and T. KIKUMOTO, 1963, Virology *19*, 411–412. (25).

MATSUMURA, F. and A. W. A. BROWN, 1963, Mosquito News *23*, 26–31. (30).

MATTA, J. F., 1970, J. Invert. Pathol. *16*, 157–164. (4).

MATTA, J. F. and R. E. LOWE, 1970, J. Invert. Pathol. *16*, 38–41. (15, 29).

MATTERN, C. F. T., L. S. DIAMOND and W. A. DANIEL, 1972, J. Virol. *9*, 342–358. (28).

MATTINGLY, P. F., 1965, Symp. Br. Soc. Parasitol. *3*, 29–45. (11).

MATTINGLY, P. F., 1969, *The Biology of Mosquito-borne Disease* (Allen and Unwin, London). (11).

MAULIK, S. 1919, *The Fauna of British India: Coleoptera, Chrysomelidae (Hispinae and Cassidinae)*, (Taylor Francis, London). (9).

MAULIK, S., 1936a, *The Fauna of British India: Coleoptera, Chrysomelidae (Chrysomelinae and Halticinae)*, (Taylor Francis, London). (9).

MAULIK, S., 1936b, *The Fauna of British India: Coleoptera, Chrysomelidae (Galerucidae)*, (Taylor Francis, London). (9).

MAURER, F. D., 1961, J. Am. Vet. Med. Ass. *138*, 15–16. (30).

MAYO, M. A., A. F. MURANT and B. D. HARRISON, 1971, J. Gen. Virol. *12*, 175–178. (4).

MCALLAN, J. W. and J. B. ADAMS, 1961, Can. J. Zool. *39*, 305–310. (25).

MCDADE, J. E. and M. R. TRIPP, 1967, J. Invert. Pathol. *9*, 531. (16).

MCEWEN, F. L. and G. E. R. HERVEY, 1959, J. Insect Pathol. *1*, 86–95. (29).

MCGEE-RUSSELL, S. M. and G. GOSZTONYI, 1967, Nature, London *214*, 1204–1206. (14).

MCGUIRE, J. M., K. S. KIM and L. B. DOUTHIT, 1970, Virology *42*, 212–216. (27).

MCINTOSH, B. M., R. H. HARWIN, H. E. PATERSON and M. L. WESTWATER, 1963, Central African J. Med. *9*, 351–359. (30).

MCINTOSH, B. M., H. E. PATTERSON, G. MCGILLIVRAY and J. DE SOUSA, 1964, Ann. Trop. Med. Parasitol. *58*, 45–51. (20).

MCKAY, D. and C. R. JENKIN, 1969, Immunology *17*, 127. (16).

MCKAY, D. and C. R. JENKIN, 1970a, Australian J. Exptl. Biol. Med. Sci. *48*, 139. (16).

MCKAY, D. and C. R. JENKIN, 1970b, Australian J. Exptl. Biol. Med. Sci. *48*, 599. (16).

MCKAY, D. and C. R. JENKIN, 1970c, Australian J. Exptl. Biol. Med. Sci. *48*, 609. (16).

MCKENZIE, H. L., 1967, *Mealybugs of California* (University of California Press). (10).

MCKINNEY, H. H., 1937, U.S. Dept. Agr. Circ. *442*, 23pp. (19).

MCLEAN, D. L., 1959, J. Econ. Entomol. *52*, 1057–1062. (25).

MCLEAN, D. L. and M. G. KINSEY, 1964, Nature, London *202*, 1358–1359. (25).

MCLEAN, D. L. and M. G. KINSEY, 1967, Ann. Entomol. Soc. Am. *60*, 400–406. (25).

MCLEAN, D. L. and M. G. KINSEY, 1968, Ann. Entomol. Soc. Am. *61*, 730–739. (25).

MCLEAN, D. M., 1953, Australian J. Exptl. Biol. Med. Sci. *31*, 481–490. (14).

MCLEAN, D. M., 1955, Australian J. Exptl. Biol. Med. Sci. *33*, 53–66. (14).

MCLEAN, D. M. and W. L. DONOHUE, 1959, Can. Med. Ass. J. *80*, 708. (30).

MCLEAN, D. M. and R. P. B. LARKE, 1963, Can. Med. Ass. J. *88*, 182–185. (18).

MCLINTOCK, J. A. and L. B. SMITH, 1918, J. Agr. Res. *14*, 1–60. (3).

MCMULLEN, A. I. and M. N. HILL, 1971, Nature, London *234*, 51–52. (30).

MECS, E., J. A. SONNABEND, E. M. MARTIN and K. H. FANTES, 1967, J. Gen. Virol. *1*, 25–40. (14).

MELNICK, J. L., 1949, Am. J. Hyg. *49*, 8–16. (4, 20).

MERCER, E. H. and M. F. DAY, 1965, Biochim. Biophys. Acta. *102*, 590–599. (4).

MESSIEHA, M., 1969, Phytopathol. *59*, 943–945. (27).

METCALF, C. L., W. P. FLINT and R. L. METCALF, 1962, *Destructive and Useful Insects*, 4th Ed., (McGraw-Hill, New York). (19).

MEYNADIER, G., 1966, C.R. Acad. Sci. Paris. Serie D *263*, 742–744. (28).

MEYNADIER, G., C. VAGO, G. PLANTEVIN and P. ATGER, 1964, Rev. Zool. Agri. Appl. *63*, 207. (4).

M'FADYEAN, J. A., 1900, J. Comp. Pathol. *13*, 1 (1).

MICHELSON, E. H., 1961, Am. J. Trop. Med. Hyg. *10*, 423. (16).

MICHELSON, E. H., 1963, Ann. N.Y. Acad. Sci. *113*, 486. (16).

MICHELSON, E. H., 1964, Am. J. Trop. Med. Hyg. *13*, 36. (16).

MILES, J. A. R. and A. C. STENHOUSE, 1969, Am. J. Trop. Med. Hyg. *18*, 427–432. (4, 20).

MILES, P. W., 1959, Nature, London *183*, 756. (25).

MILES, P. W., 1964, J. Insect Physiol. *10*, 121–129. (25).

MILES, P. W., D. L. MCLEAN and M. G. KINSEY, 1964, Experientia *20*, 582. (24).

MILNE, R. G., 1970, J. Gen. Virol. *6*, 267–276. (4).

MILZER, A., 1942, J. Infect. Diseases *20*, 152–172. (4).

MIMS, C. A., M. F. DAY and I. D. MARSHALL, 1966, Am. J. Trop. Med. Hyg. *15*, 775–784. (1, 14, 15, 20, 30).

MINKIEWICZ, S., 1927, Mem. Inst. Polon. Ec. Rur. Pulawy, A *8*, 457–528. (7).

MIRCHAMSY, H., A. HAZRATI, S. BAHRAMI and A. SHAFYI, 1970, Am. J. Vet. Res. *31*, 1755–1761. (14).

MIRZOEVA, N. M., I. G. KANBAY, Z. D. SULTANOVA, E. I. SOKOLOVA and N. M. KULIEVA, 1971, Tezisy Dokl. Vop. Med. Virus. Inst. Virus. imeni Ivanovsky, D.I., Akad. Med. Nauk SSSR (October 19–21), 65–66. (Engl. trans.: NAMRU3–T491). (18).

MISKO, I. S., 1968, B.Sc. Honours Thesis, Australian National University Canberra. (16).

MITSUHASHI, J., 1965, Japan. J. Appl. Entomol. Zool. *9*, 137–141. (26).

MITSUHASHI, J., 1965, Japan. J. Appl. Entomol. Zool. *9*, 217–224. (17).

MITSUHASHI, J., 1967, Nature, London *215*, 863–864. (14, 17).

MITSUHASHI, J., 1968, Appl. Entomol. Zool. *3*, 1–4. (1, 7).

MITSUHASHI, J., 1969, Appl. Entomol. Zool. *4*, 105–113. (17).

MITSUHASHI, J. and K. KOYAMA, 1969, Appl. Entomol. Zool. *4*, 185–193. (8).

MITSUHASHI, J. and K. MARAMOROSCH, 1964, Contrib. Boyce, Thompson Inst. *22*, 435–460. (17).

MITSUHASHI, J., and S. NASU, 1967, Japan. J. Appl. Entomol. Zool. *2*, 113–114. (26).

MITTLER, T. E., 1958, J. Exptl. Biol. *35*, 74–84. (25).

MITTLER, T. E. and R. H. DADD, 1962, Nature, London *195*, 404. (24, 25).

MITTLER, T. E., and R. H. DADD, 1965, Entomol. Exptl. Appl. *8*, 107–122. (25).

MIURA, K., I. FUJII, T. SAKAKI, M. FUKE and S. KAWASE, 1968, J. Virol. *2*, 1211–1222. (2, 4).

MIURA, K., I. FUJII-KAWATA, H. IWATA and S. KAWASE, 1969, J. Invert. Pathol. *14*, 262–265. (4).

MIYAJIMA, S. and S. KAWASE, 1968, J. Invert. Pathol. *12*, 329–334. (14).

MIYAMOTO, S. and Y. MIYAMOTO, 1971, Sci. Rep. Fac. Agri. Kobe Univ. *9*, 59–70. (24).

MIYATAKE, Y. 1963, J. Fac. Agri. Kyushu Univ. *12*, 323–357. (7).

MIYATAKE, Y., 1964, J. Fac. Agri. Kyushu Univ. *13*, 1–37. (7).

MIYATAKE, Y., 1965, Kontyu *33*, 171–189. (7).

MIYATAKE, Y., 1969, Bull. Osaka Museum Nat. Hist. *22*, 63–83. (7).

MOCHIDA, O., 1971, Trans. Royal Entomol. Soc. London (In press). (8).

MOERICKE, V., 1950, Z. Tierpsychol. *7*, 265–274. (25).

MOERICKE, V., 1955, Z. Pfl. Krank. Pfl. Path. Pfl. Schutz. *62*, 588–592. (25).

MONTOYA, E. L., C. M. IGNOFFO and R. L. MCGARR, 1966, J. Invert. Pathol. *8*, 320–325. (29).

MOORE, K. M., 1970, Austr. Zool. *15*, 248–376. (7).

MOORHOUSE, D. E., 1966, J. Med. Entomol. *3*, 168–171. (6).

MORAN, V. C. and J. R. BLOWERS, 1967, J. Entomol. Soc. S. Afr. *30*, 96–106. (7).

MORGAN, C., C. HOWE and H. M. ROSE, 1961, J. Exptl. Med. *113*, 219–234. (14).

MORILLA-GONZALEZ, A. and J. DE MUCHA-MACIAS, 1969, Rev. Invest. Salud. Publ. *29*, 3–20. (30).

MORLAN, H. B., R. O. HAYES and H. F. SCHOOF, 1963, Publ. Health. Rept. (Washington). *78*, 711–719. (30).

MORREN, E., 1869, Acad. Royal Belgique *28*, 434–442. (3).

MORRIS, O. N., 1962, J. Insect Pathol. *4*, 207–215. (15, 29).

MORRIS, O. N., 1963, J. Insect Pathol. *5*, 401–415. (29).

MORRIS, O. N., 1968a, J. Invert. Pathol. *10*, 28–38. (14, 15).

MORRIS, O. N., 1968b, J. Invert. Pathol. *11*, 476–486. (15).

MORTON, H. V. and V. G. PERRY, 1968, Nematologica *14*, 11. (12).

MOUND, L. A., 1962, Entomol. Exptl. Appl. *5*, 99–104. (13).

MOUND, L. A., 1963, Proc. Roy. Entomol. Soc. London (A). *38*, 171–180. (13).

MOUND, L. A., 1965a, Empire Cotton Growing Rev. *42*, 33–40. (13).

MOUND, L. A., 1965b, Bull. Brit. Museum Entomol. *17*, 115–160. (13).

MOUND, L. A., 1965c, Empire Cotton Growing Rev. *42*, 290–294. (13).

MOUND, L. A., 1966, Bull. Brit. Museum Entomol. *17*, 397–428. (13).

MOUND, L. A., 1971, Bull. Entomol. Res. *60*, 547–548. (13).

MOUNT, G. A., C. S. LOFGREN, N. W. PIERCE and K. F. BALDWIN, 1970, Mosquito News *30*, 48–51. (30).

MUGO, W. N. and R. E. SHOPE, 1972, Trans. Roy. Soc. Trop. Med. Hyg. *66*, 300–304. (18).

MULDREW, J. A., 1953, Can. J. Zool. *31*, 313. (16).

MULLA, M. S. and M. F. B. CHAUDHURY, 1968, Mosquito News *28*, 187–191. (30).

MÜLLER, H. J., 1956, Handb. Pflanzenkr. *5*, 306–331. (7).

MÜLLER, H. J., 1957, Zool. Jahrb. Syst. *85*, 317–430. (8).

MÜLLER, H. J., 1958a, Entomol. Exptl. Appl. *1*, 66–72. (25).

MÜLLER, H. J., 1958b, Zool. Anzeig. *160*, 294–312. (8).

MULLIGAN, T. E., 1960, Ann. Appl. Biol. *48*, 575–579. (19).

MUNSTER, J. and R. MURBACH, 1952, Rev. Romande Agr. Viticult. Aboricult. *8*, 41–43. (25).

MURANT, A. F., 1970, C.M.I./A.A.B. Descrip. Plant Viruses No. *16*, (4).

MURANT, A. F., R. A. GOOLD, I. M. ROBERTS and J. CATHRO, 1969, J. Gen. Virol. *4*, 329–341. (4).

MURANT, A. F. and R. M. LISTER, 1967, Ann. Appl. Biol. *59*, 63–76. (27).

MURANT, A. F. and C. E. TAYLOR, 1965, Ann. Appl. Biol. *55*, 227–237. (27).

MURPHY, F. A., E. C. BORDEN, R. E. SHOPE and A. HARRISON, 1971, J. Gen. Virol. *13*, 273–288. (4).

MURPHY, F. A., A. K. HARRISON and W. K. COLLIN, 1970, Lab. Invest. *22*, 318–328. (14).

MURPHY, F. A., A. K. HARRISON, G. W. GARY Jr., S. G. WHITFIELD and F. T. FORRESTER, 1968c, Lab. Invest. *19*, 652–662. (14).

MURPHY, F. A., A. K. HARRISON and T. TZIANABOS, 1968a, J. Virol. *2*, 1315–1325. (4, 14).

MURPHY, F. A., W. F. SCHERER, A. K. HARRISON, H. W. DUNNE and G. W. GARY Jr., 1970, Virology *40*, 1008–1021. (4, 11).

MURPHY, F. A. and S. G. WHITFIELD, 1970, Exptl. Mol. Pathol. *13*, 131–146. (14).

MURPHY, F. A., S. G. WHITFIELD, P. H. COLEMAN, C. H. CALISHER, E. R. RABIN, A. B. JENSON, J. L. MELNICK, M. E. EDWARDS and E. WHITNEY, 1968b, Exptl. Mol. Pathol. *9*, 44–56. (4, 14).

MUSSGAY, M., 1964, Prog. Med. Virol. *6*, 193–267. (14).

MUSSGAY, M. and J. WEIBEL, 1962, Virology *16*, 52–62. (14).

NAGAYAMA, A., B. G. T. POGO and S. DALES, 1970, Virology *40*, 1039–1051. (4).

NAGLE, S. C., W. C. CROTHERS and N. L. HALL, 1967, Appl. Microbiol. *15*, 1497–1498. (17).

NAITO, A. and J. MASAKI, 1967a, Japan. J. Appl. Entomol. Zool. *11*, 50–56. (8).

NAITO, A. and J. MASAKI, 1967b, Japan. J. Appl. Entomol. Zool. *11*, 150–156. (8).

NAKASUJI, F. and K. KIRITANI, 1970, Appl. Entomol. Zool. *5*, 1–12. (26).

NAMBA, R., 1962, Virology *16*, 267–270. (25).

NASU, S., 1963, Bull. Kyushu Agr. Sta. *8*, 153–349. (8, 26).

NASU, S., 1965, Japan. J. Appl. Entomol. Zool. *9*, 225–237. (8, 26).

NAULT, L. R., M. L. BRIONES, L. E. WILLIAMS and B. D. BARRY, 1967, Phytopathology *57*, 986–989. (19).

NAULT, L. R. and W. E. STYER, 1970, Phytopathology *60*, 1616–1618. (19).

NAULT, L. R., G. C. GYRISCO and W. F. ROCHOW, 1964, Phytopathology *54*, 1269–1272. (15, 24).

NAULT, L. R. and G. C. GYRISCO, 1966, Ann. Entomol Soc. Am. *59*, 1185–1197. (24, 25).

NEILSON, M. M. and D. E. ELGEE, 1960, J. Insect Pathol. *2*, 165–172. (29).

NEILSON, M. M. and D. E. ELGEE, 1968a, J. Invert. Pathol. *10*, 70–76. (22).

NEILSON, M. M. and D. E. ELGEE, 1968b, J. Invert. Pathol. *12*, 132–140. (21, 22, 29).

NEITZ, W. O., 1956, Ann. N.Y. Acad. Sci. *64*, 56–111. (5).

NELSON-REES, W. A., 1960, J. Exptl. Zool. *144*, 111–137. (10).

NEWTON, W., 1953, F.A.O. Plant Protect. Bull. *2*, 40 (10).

NEWTON, W. L., 1952, J. Parasitol. *38*, 362. (16).

NIELSON, M. W., 1968, U.S. Dept. Agr. Tech. Bull. *1382*, 1–386. (8).

NIJVELDT, W., 1969, *Gall Midges of Economic Importance.* Vol. *VIII: Gall Midges-Miscellaneous*, (Crosby Lockwood). (7).

NISHI, Y., 1958, Ann. Phytopathol. Soc. Japan *23*, 185–188. (25).

NISHI, Y., 1969, *In:* Maramorosch, K., ed., *Viruses, Vectors and Vegetation*, (Wiley-Interscience, New York). (25).

NISHIMURA, A., and Y. HOSAKA, 1969, Virology *38*, 550–557. (4).

NISHIZAWA, T., Y. NISHI and T. KIMURA, 1959, Virus *9*, 130–133. (25).

NIVEN, J. S. F., J. A. ARMSTRONG, C. H. ANDREWES, H. G. PEREIRA, and R. C. VALENTINE, 1961, J. Pathol. Bacteriol *81*, 1–14. (4).

NOVAK, D., 1965, Cah. ORSTOM. (Entomol. Med.) *3*, 35–38. (30).

NUTMAN, F. J., 1969, Ann. Appl. Biol. *64*, 188. (3).

NUTTALL, G. H. F., 1911, *In:* Nuttall, G. H. F. and C. Warburton, *Ticks. A Monograph of the Ixodoidea. Part II. Ixodidea*, (Cambridge University Press, London). (5).

NUTTING, W. L., 1951, J. Morphol. *89*, 501. (16).

OBERG, B. and L. PHILIPSON, 1971, J. Mol. Biol. *58*, 725–737. (14).

OHANESSIAN, A., 1971. Current Top. Microbiol. Immunol. *55*, 230–231. (17).

OHKUBO, N., 1968, Ann. Meet. Japan. Soc. Appl. Entomol. Zool. *6*, (8).

OHKUBO, N. and R. KISIMOTO, 1971, Japan. J. Appl. Entomol. Zool. *15*, 8–16. (8).

OKADA, T., 1969, Appl. Entomol. Zool. *4*, 147–148. (14).

OKER-BLOM, N., A. SALMINEN, M. BRUMMER-KORVENKONTIO, L. KÄÄRIÄINEN and P. WECK-STRÖM, 1965, Proc. Symp. Theor. Quest. Nat. Foci Dis. (Prague, November 26–29, 1963) 441–447. (18).

OKUYAMA, S., K. YORA and H. ASUYAMA, 1968, Ann. Phytopathol. Soc. Japan *34*, 255–264. (26).

OLITSKY, P. K., 1925, Science *62*, 442. (10).

OLIVEIRA, A. R. and M. B. PONSEN, 1966, Neth. J. Plant. Pathol. *72*, 259–264. (14).

OLIVER, J. H. JR., 1971, Am. Zool. *11*, 283–299. (5).

O'LOUGHLIN, G. T. and T. C. CHAMBERS, 1967, Virology *33*, 262–271. (15, 24).

OMAN, P., 1969, *In:* Maramorosch, K., ed., *Viruses, Vectors and Vegetation*, (Interscience, New York). (8,15).

ONODERA, K., M. HIMENO, F. SAKAI and I. MORISHIMA, 1968, J. Biochem. (Tokyo) *64*, 649–655. (14).

ONODERA, K., T. KOMANO, M. HIMENO and F. SAKAI, 1965, J. Mol. Biol. 13, 532–539. (4).

OOSTENBRINK, M., 1964, Nematologica *10*, 49–56. (31).

OORTWIJN BOTJES, J. G., 1920, Diss. Wageningen, The Netherlands. (3).

ORLANDO, A. and K. SILBERSCHMIDT, 1946, Arq. Inst. Biol. São Paulo *17*, 1–36. (3).

ORLOB, G. B., 1963, Phytopatholgy *53*, 822–830. (9, 25).

ORLOB, G. B., 1968, Virology *35*, 121–133. (19).

ORTON, W. A., 1914, U.S.Dept. Agr. Bull. *64*, 44–48. (3).

OSSIANNILSSON, F., 1949, Opuscula Entomol. Suppl. *10*, 1–142. (8).

OSSIANNILSSON, F., 1963, Entomologist, London. *96*, 249–257. (7).

OSSIANNILSSON, F., 1970, Entomologia Scand. *1*, 135–144. (7).

OSSOWSKI, L. L. J., 1960, J. Invert. Pathol. *2*, 35–44. (29).

OTA, Z., 1965, Virology *25*, 372–378. (14).

OWEN, W. B. and R. W. GERHARDT, 1957, Univ. Wyoming Publ. *21*, 71–141. (30).

OWUSU, G. K., 1969, Ann. Rept. Cocoa Res. Inst. Ghana. 1967–68. 32–34. (23).

OWUSU, G. K., 1970, Ann. Rept. Cocoa Res. Inst. Ghana 1968–69, 31–33. (23).

OWUSU, G. K., 1971, Trop. Agr. *48*, 133–139. (23).

PAILLOT, A., 1926, Compt. Rend. *182*, 180–182. (2).

PAILLOT, A., 1934, Compt. Rend. *198*, 204–205. (2).

PAILLOT, A., 1937, Compt. Rend. *205*, 1264–1266. (2).

PALIWAL, Y. C. and R. C. SINHA, 1970, Virology *42*, 668–680. (15, 24).

PALIWAL, Y. C. and J. T. SLYKHUIS, 1967, Virology *32*, 344–353. (19).

PARKER, L., E. BAKER and N. F. STANLEY, 1965, Australian J. Exptl. Biol. Med. Sci. *43*, 167–170. (4).

PASCHKE, J. D., R. E. LOWE and R. L. GIESE, 1968, J. Invert. Pathol. *10*, 327–334. (29).

PATEL, V. C. and H. K., PATEL, 1966, Indian J. Entomol. *28*, 339–344. (31).

PAUL, J., 1960, *Cells and Tissue Culture*, 2nd Ed., (The Williams and Wilkins Co. Baltimore). (17).

PAUL, S. D. and K. R. P. SINGH, 1969, Current Sci. *38*, 241–242. (14).

PAUL, S. D., K. M. PAVRI, L. V. D'LIMA, P. K. RAJAGOPALAN, M. GOVERDHAN and K. R. P. SINGH, 1968, Indian J. Med. Res. *56*, 264–274. (18).

PAVRI, K. M. and J. CASALS, 1966, Am. J. Trop. Med. Hyg. *15*, 961–963. (18).

PEACOCK, P. B., 1958, S. African Med. J. *32*, 201–202. (5).

PEDERSEN, I. R., 1971, Nature, London *234*, 112–114. (4).

PELEG, J., 1965, Am. J. Trop. Med. Hyg. *14*, 158–164. (11).

PELEG, J., 1968a, Am. J. Trop. Med. Hyg. *17*, 219–223. (14).

PELEG, J., 1968b, Virology *35*, 617–619. (14).

PELEG, J., 1969a, J. Gen. Virol. *5*, 463–471. (14).

PELEG, J., 1969b, Nature *221*, 193–194. (4, 14, 17).

PERIES, J., P. PRINTZ, M. CANIVET and J. C. CHUAT, 1966, Compt. Rend. *262*, 2106–2107. (20).

PERKINS, F. O., 1969, J. Invert. Pathol. *13*, 199–222. (28).

PETERS, D., 1965, Virology *26*, 159–161. (24).

PETERS, D., 1967a, Virology *31*, 46–54. (24).

PETERS, D., 1967b, Meded. Fonds Landbouw Export Bureau 1916/1918, Wageningen. No. *45*, 1–100. (24).

PETERS, D. and L. M. BLACK, 1970, Virology *40*, 847–853. (15, 17, 24).

PETERS, D., and E. W. KITAJIMA, 1970, Virology *41*, 135–150. (4, 24).

PETERS, D. and L. C. VAN LOON, 1968, Virology *35*, 597–600. (4, 24).

PETERS, D. H. A. and G. MÜLLER, 1963, Virology *21*, 266–269. (4).

PETRE, Z. and P. PLOAIRE, 1969, Experientia *25*, 842–844. (14).

PFEFFERKORN, E. R. and R. L. CLIFFORD, 1964, Virology *23*, 217–223. (14).

PFEFFERKORN, E. R. and H. S. HUNTER, 1963, Virology *20*, 446–456. (14).

PFEIFFER, H. H., 1943, Naturwissenschaften *31*, 47. (17).

PHILLIPS, J. H., 1963, *In:* Dougherty, E. C., Z. N. Brown, E. D. Hanson and W. D. Hartman eds., *The lower Metazoa. Comparative Biology and Phylogeny.* (University of California Press, Los Angeles). pp. 425–431 (16).

PHILLIPS, C. A. and J. L. MELNICK, 1965, J. Am. Med. Assoc. *193*, 207–211. (30).

PIENKOWSKI, P. L. and J. T. MEDLER, 1964, Ann. Entomol. Soc. Am. *57*, 588–591. (8).

PIERPOINT, W. S. and B. D. HARRISON, 1963, J. Gen. Microbiol. *32*, 429–440. (25).

PILO-MORON, E., J. VINCENT and P. SUREAU, 1966, Arch. Inst. Pasteur, Alg. *44*, 78–101. (30).

PIRONE, T. P., 1967, Virology *31*, 569–571. (15, 25).

PIRONE, T. P., 1969, *In:* Maramorosch, K., ed., *Viruses, Vectors and Vegetation*, (Wiley-Interscience, New York). (25).

PIRONE, T. P., 1970, Phytopathology *60*, 1657–1659. (25).

PITCHER, R. S., 1968, Nematologica *13*, 547–557. (12).

PITCHER, R. S. and J. J. M. FLEGG, 1965, Nature, London *207*, 317. (27).

PITCHER, R. S. and A. JHA, 1961, Plant Pathology *10*, 67–71. (27).

PITCHER, R. S. and D. G. MCNAMARA, 1970, Nematologica *16*, 99–106. (12).

PITTENDRIGH, C. S., 1950, Evolution *4*, 64–78. (11).

PLANK, J. E. VAN DER, 1948, Empire J. Exptl. Agr. *16*, 134–142. (31).

PLANK, J. E. VAN DER and E. E. ANDERSSEN, 1944, Bull. Dept. Agr. S. Africa *240*, 1–6. (31).

PLOWRIGHT, W., J. PARKER and M. A. PEIRCE, 1969a, Nature, London *221*, 1071–1073. (18).

PLOWRIGHT, W., J. PARKER and M. A. PEIRCE, 1969b, Vet. Rec. *85*, 668–674. (4, 18).

PLOWRIGHT, W., C. T. PERRY and M. A. PEIRCE, 1970, Res. Vet. Sci. *11*, 582–584. (18).

PLOWRIGHT, W., C. T. PERRY, M. A. PEIRCE and J. PARKER, 1970, Arch. Gesch. Virusforsch. *31*, 33–50. (4,18).

PLUS, N. and P. ATANASIU, 1966, Compt. Rend. *263*, 89–92. (20).

PLUS, N. and J. L. DUTHOIT, 1969, Compt. Rend. *268*, 2313–2315. (20).

POGO, B. G. T., S. DALES, M. BERGOIN and D. W. ROBERTS, 1971, Virology *43*, 306–309. (4).

POINAR, G. O., 1969, *In:* Jackson, G. J., R. Herman and I. Singer, eds., *Immunity to Parasitic Animals.* Vol. 1, (North-Holland, Amsterdam) pp. 173–210 (16).

POLLARD, D. G., 1969, Proc. Roy. Entomol. Soc. London (A). *44*, 173–185. (10).

PONSEN, M. B., 1965, Neth. J. Plant Pathol. *71*, 54–56. (4).

PONSEN, M. B., 1970, Neth. J. Plant Pathol. *76*, 234–239. (24).

PONSEN, M. B. and D. J. DE JONG, 1964, J. Insect Pathol. *6*, 373. (16).

POOL, R. A. F., H. K. WAGNON and H. E. WILLIAMS, 1970, Plant Disease Reptr. *54*, 825–826. (31).

POPOV, V. F., 1967, Medskaya Parazit. *36*, 288–289. (30).

PORTERFIELD, J. S., 1959, Nature, London *183*, 1069–1070. (14).

POSNETTE, A. F., 1940, Trop. Agr. *17*, 98. (23).

POSNETTE, A. F., 1944, Trop. Agr. *21*, 105–106. (23).

POSNETTE, A. F., 1947, Ann. Appl. Biol. *34*, 388–402. (23).

POSNETTE, A. F., 1950, Ann. Appl. Biol. *37*, 378–384. (23).

POSNETTE, A. F., 1951, Trop. Agr. *28*, 133–142. (23).

POSNETTE, A. F., 1962, Rept. E. Malling Res. Sta., 1961, 125–127. (31).

POSNETTE, A. F., and J. MCA. TODD, 1951, Ann. Appl. Biol. *38*, 785–800. (23).

POSNETTE, A. F. and J. MCA. TODD, 1955, Ann. Appl. Biol. *43*, 433–453. (23).

POSNETTE, A. F. and N. F. ROBERTSON, 1950, Ann. Appl. Biol. *37*, 363–377. (23).

POSNETTE, A. F. and A. H. STRICKLAND, 1948, Ann. Appl. Biol. *35*, 53–63. (3, 23).

POSNETTE, A. F., N. F. ROBERTSON and J. MCA. TODD, 1950, Ann. Appl. Biol. *37*, 229–240. (23).

POYARKOFF, E., 1910, Archs. Anat. Microsc. *12*, 333–474. (9).

PRASERTPHON, S., 1967, J. Invert. Pathol. *9*, 140–142. (31).

PREBBLE, M. L., 1941, Can. J. Res. D. *19*, 245–454. (29).

PREER, J. R., L. B. PREER, B. RUDMAN and A. JURAND, 1971, Mol. Gen. Genet. *111*, 202–208. (28).

PRICE, W. C., 1966, Virology *29*, 285–294. (25).

PRICE, W. C., 1970, C.M.I./A.A.B. Descrip. Plant Viruses. No. *33*, (4).

PRIESNER, H., 1964a, Ordnung Thysanoptera, Bestimmungsbucher zur Bodenfauna Europas, Lief 2 Akademie-Verlag, Berlin. (13).

PRIESNER, H., 1964b, Pubis. Inst. desert Egypte (1960) *13*, 1–549. (13).

PRINTZ, P., 1967, C.R. Acad. Sci. Paris. *264*, 169–172. (4).

PRINTZ, P., 1968, Compt. Rend. Soc. Biol. *162*, 372–373. (20).

PRITCHARD, A. E. and E. W. BAKER, 1955, Pacific Coast Entomol. Soc. Mem. Ser. *2*, San Francisco, 1–472. (6).

PROESELER, G., 1968, Phytopathol. Z. *63*, 1–9. (19).

PROESELER, G., 1969, Z. Zentralbl. Bakteriol. Parasitenk. Infekt. Hyg. *123*, 288–292. (19).

PROESELER, G. and H. KEGLER, 1966, Monatsberichte der Deut. Akad. der Wiss. zu Berlin. *8*, 472–476. (19).

PROWSE, R. H., and N. N. TAIT, 1969, Immunology *17*, 437. (16).

PU, M. H., 1963, Acta. Entomol. Sin. *12*, 117–136. (8).

PUDNEY, M. and M. G. R. VARMA, 1971, Exptl. Parasitol. *29*, 7–12. (17).

PUTMAN, W. L., 1970, Can. Entomologist *102*, 305–321. (29).

PUTZ, C. and G. STOCKY, 1970, Ann. Phytopathol. *2*, 329–347. (27).

QUAINTANCE, A. L. and A. C. BAKER, 1913, Bull. U.S. Bur. Entomol. *27*, 1–109. (13).

QUAINTANCE, A. L. and A. C. BAKER, 1917, Proc. U.S. Nat. Museum *51*, 335–445. (13).

QUANJER, H. M., 1913, Meded. Landb. Hogesch. Wageningen. *6*, 41–80. (3).

QUANJER, H. M., 1931, Phytopathology *21*, 577–613. (3).

QUANJER, H. M., 1936, Plziekte *42*, 45–54. (3).

QUTUBUDDIN, M., 1958, Bull. World Health. Organ. *19*, 1109–1111. (30).

RAATIKAINEN, M., 1967, Ann. Agr. Fenn. *6*, Suppl. *2*, 1–149. (8).

RADEWALD, J. D. and D. J. RASKI, 1962, Phytopathology *52*, 748. (12).

RAJAGOPALAN, P. K., 1969, Proc. Rodent. Symp. (Calcutta, December 8–11, 1966), 220–223. (18).

RAJAGOPALAN, P. K. and C. R. ANDERSON, 1970, Indian J. Med. Res. *58*, 1184–1187. (18).

RAJAGOPALAN, P. K. and C. R. ANDERSON, 1971, Indian J. Med. Res. *59*, 847–860. (18).

RAJAGOPALAN, P. K., M. A. SREENIVASAN and S. D. PAUL, 1970, Indian J. Med. Res. *58*, 1195–1196. (18).

RAMALLO, J. C. and A. E. GARCÍA, 1968, Misc. Univ. Nac. Tucuman *25*, 1–16. (31).

RAMASESHIAH, G., 1967, J. Invert. Pathol. *9*, 128–130. (31).

RAMIREZ GOMEZ, C., 1956a, Bol. Soc. Esp. Hist. Nat. (Biol.) *53*, 151–217. (7).

RAMIREZ GOMEZ, C., 1956b, Bol. Soc. Esp. Hist. Nat. (Biol.), *54*, 63–106. (7).

RAMIREZ GOMEZ, C., 1960, Bol. Soc. Esp. Hist. Nat. (Biol.), *57*, 5–87. (7).

RAND, F. V. and L. C. CASH, 1920, Phytopathology *10*, 133–140. (9).

RAND, F. V. and E. M. A. ENLOWS, 1916, J. Agr. Res. *6*, 417–434. (9).

RAND, F. V. and W. D. PIERCE, 1920, Phytopathology *10*, 189–231. (3).

RANZENHOFER, E. R., E. R. ALEXANDER, L. D. BEADLE, A. BERNSTEIN and D. C. PICKARD, 1957, Am. J. Hyg. *65*, 147–161. (30).

RAO, R. T., 1967, Bull. World Health. Organ. *36*, 547–551. (30).

RAO, S. R., 1943, Current Sci. (India) *12*, 208. (10).

RASKI, D. J. and W. B. HEWITT, 1960, Nematologica *5*, 166–170. (27).

RASKI, D. J. and W. B. HEWITT, 1963, Phytopathology *53*, 39–47. (27).

RASKI, D. J., W. B. HEWITT, A. C. GOHEEN, C. E. TAYLOR and R. H. TAYLOR, 1965, Nematologica *11*, 349. (31).

RASKI, D. J., N. O. JONES and D. R. ROGGEN, 1969, Proc. Helminthol. Soc. Washington *36*, 106–118. (12).

REDDY, D. V. R. and L. M. BLACK, 1966, Virology *30*, 551–561. (26).

REED, W., J. CARROLL, A. AGRAMONTE and J. W. LAZEAR, 1900, Phila. Med. *6*, 790. (1).

REEVES, W. C. and W. MCD. HAMMON, 1962, Univ. Calif. Publ. Health Vol. 4. (30).

REEVES, W. C., 1961, Progr. Med. Virol. *3*, 59–78. (18, 20).

REEVES, W. C., 1970, Proc. Calif. Mosq. Control Assoc. *38*, 6–9. (30).

REEVES, W. C., R. E. BELLAMY and R. P. SCRIVANI, 1958, Am. J. Hyg. *67*, 78. (1).

REGNERY, D. C. and I. D. MARSHALL, 1971, Am. J. Epidemiol. *94*, 508–513. (20).

ŘEHÁČEK, J., 1958, Acta Virol. *2*, 253–254. (17).

ŘEHÁČEK, J., 1962, Acta Virol. *6*, 188. (17).

ŘEHÁČEK, J., 1965, Acta Virol. *9*, 332–337. (14, 17).

ŘEHÁČEK, J., 1965, Ann. Rev. Entomol. *10*, 1–24. (18).

ŘEHÁČEK, J., 1968a, Acta Virol. *12*, 241–246. (14).

ŘEHÁČEK, J., 1968b, Acta Virol. *12*, 340–346. (14).

ŘEHÁČEK, J., 1971, Current Top. Microbiol. Immunol. *55*, 113–126. (17).

ŘEHÁČEK, J. and H. W. BRZOSTOWSKI, 1969a, J. Insect Physiol. *15*, 1431–1436. (17).

ŘEHÁČEK, J. and H. W. BRZOSTOWSKI, 1969b, J. Insect Physiol. *15*, 1683–1686. (17).

REHÁČEK, J. and J. PESEK, 1960, Acta Virol. *4*, 241–245. (17).

REINGANUM, C., G. T. O'LOUGHLIN and T. W. HOGAN, 1970, J. Invert. Pathol. *16*, 214–220. (28).

REISS-GUTFREUND, R. L., J. ANDRAL and C. SÉRIÉ, 1962, Ann. Inst. Pasteur *102*, 36–43. (4).

RENNIE, J., 1923, Proc. Royal. Phys. Soc. Edinburgh, *A20*, 265–267. (2).

REUBEN, R., 1967, Indian J. Med. Res. *55*, 1–12. (30).

RICHARDSON, J. and E. S. SYLVESTER, 1965, Virology *25*, 472–475. (24).

RICHARDSON, J. and E. S. SYLVESTER, 1968, Virology *35*, 347–355. (15).

RICHARDSON, J. and E. S. SYLVESTER, 1970, Virology *42*, 1023–1042. (15).

RIDGWAY, R. L., H. J. WALKER, R. L. HANNA and W. L. OWEN, 1967, J. Econ. Entomol. *60*, 592–594. (31).

RIEDL, H., O. KOZUCH, W. SIXL. E. SCHMELLER and J. NOSEK, 1971, Arch. Hyg. Bakt. *154*, 610–611. (18).

RIVERS, C. F., 1959, Trans. 1st. Int. Conf. Insect. Pathol. Biol. Control, Praha. 1958. (21).

RIVNAY, E., 1937, Bull. Entomol. Res. *28*, 173–179. (25).

ROBERTS, F. H. S., 1970, *Australian Ticks*, (Commonwealth Scientific and Industrial Research Organization, Australia, Melbourne). (5).

ROBERTS, F. M., 1940, Ann. Appl. Biol. *27*, 348–358. (25).

ROBIN, Y., 1971, Rapp. Fonct. Tech. Inst. Pasteur, Dakar. 69–80. (18).

ROBIN, Y., J. L. CAMICAS, P. BRÈS and G. HERY, 1970, Folia Parasitol. *17*, 345–348. (18).

ROBINSON, L. E., 1926, *Ticks, A Monograph of the Ixodoidea. Part IV. The genus Amblyomma*, (Cambridge University Press) pp. 302. (5).

ROCHOW, W. F., 1960, Virology *12*, 223–232. (24).

ROCHOW, W. F., 1963, Phytopathology *53*, 355–356. (24).

ROCHOW, W. F., 1969, *In:* Maramorosch, K., ed., *Viruses, Vectors and Vegetation*, (Interscience Publishers, New York). pp. 175–198. (24).

ROCHOW, W. F., 1970, Science *167*, 875–878. (24).

ROCHOW, W. F., 1970, C.M.I./A.A.B. Descrip. Plant Viruses. No. *32*. (4).

ROCHOW, W. F. and M. K. BRAKKE, 1964, Virology *24*, 310–332. (24).

ROCHOW, W. F. and E. PANG, 1961, Virology *15*, 382–384. (24).

ROGGEN, D. R., 1966, Nematologica *12*, 287–296. (27).

ROGGEN, D. R., D. J. RASKI and N. O. JONES, 1967, Nematologica *13*, 1–16. (12).

ROHDE, R. A. and W. R. JENKINS, 1957, Phytopathology *47*, 29. (12).

ROIVAINEN, O., 1968, Ann. Rept. Cocoa Res. Inst. Ghana 1966–67, 39. (23).

ROIVAINEN, O., 1969, Ann. Rept. Cocoa Res. Inst. Ghana 1967–68, 44. (23).

ROIVAINEN, O., 1971, Proc. Third Int. Cocoa Res. Conf. Ghana. 518–521. (23).

ROJTER, S., J. K. BONNEY and J. T. LEGG, 1966, Ghana J. Sci. *6*, 110–114. (23, 31).

ROLAND, G., 1936, Sucr. Belge. *55*, 213. (3).

ROLLINSON, W. D., F. B. LEWIS and W. E. WATERS, 1965, J. Invert. Pathol. *7*, 515–517. (29).

ROSICKÝ, B., 1965, Proc. Symp. Theor. Quest. Nat. Foci Dis. (Prague, November 26–29, 1963), 151–164. (18).

ROSICKÝ, B., 1967, *In:* Cockburn, A., ed. *Infectious Diseases*, (Charles C. Thomas, Springfield, Ill.). (18).

ROWE, W. P., F. A. MURPHY, G. H. BERGOLD, J. CASALS, J. HOTCHIN, K. M. JOHNSON, F. LEHMANN-GRUBE, C. A. MIMS, E. TRAUB and P. A. WEBB, 1970. J. Virol. *5*, 651–652. (4).

ROZEBOOM, L. E. and E. N. KASSIRA, 1969, J. Med. Entomol. *6*, 407–411. (20).

RUBIN, H., M. BALUDA and J. E. HOTCHIN, 1955, J. Exptl. Med. *101*, 205–212. (14).

RUDNICK, A., N. J. MARCHETTE and R. GARCIA, 1967, Abstr. Pap. I. SE. Asian Reg. Semin. Trop. Med., 3, Conf. Parasitic Dis., Semin. Malar. (Bangkok, August 7–11, 1967), 40–41. (18).

RUGGIERO, H. R., A. S. PARODI, H. G. RUGGERIO, N. METTLER, M. BOXACA, A. L. DEGUERRERO, A. CINTORA, C. MAGNONI, H. MILANI, F. MAGLIO, C. GONZALEZ CHAMBACERES, L. ASTARLOA, G. SQUASSI, D. FERNANDEZ and A. GIACOSA, 1964, Bull. Hyg. *39*, 1254. (30).

RUNGGER, D., M. RASTELLI, E. BRAENDLE, R. G. MALSBERGER, 1971, J. Invert. Pathol. *17*, 72–80. (28).

RUSSELL, G. E., 1965, Bull. Entomol. Res. *56*, 191–196. (31).

RUSSELL, G. E., 1970, C.M.I./A.A.B. Descrip. Plant Viruses No. *13*, (4).

RUSSELL, G. J., E. A. C. FOLLETT, J. H. SUBAK-SHARPE and B. D. HARRISON, 1971, J. Gen. Virol. *11*, 129–138. (4).

RUSSELL, L. M., 1948, Misc. Publs. U.S. Dept. Agr. 635. (13).

SABIN, A. B., 1952, Am. J. Trop. Med. Hyg. *1*, 30–50. (20).

SACCÀ, G., M. L. MASTRILLI, M. BALDUCCI, P. VERANI and M. C. LOPES, 1969, Ann. Ist. Sup. Sanità. *5*, 21–28. (18).

SAKIMURA, K., 1962, *In:* Maramorosch, K., ed., *Biological Transmission of Disease Agents*, (Academic Press, New York). (13).

SAKIMURA, K., 1963, Phytopathology *53*, 412–415. (28).

SAKIMURA, K., 1970, Pacif. Insects *11*, 761–762. (13).

SALAMA, H. S. and M. R. SALCH, 1971, Proc. 13th Int. Congr. Entomol. Moscow, 1968, 434–435. (10).

SALT, G., 1955, Proc. Roy. Soc. (London) B. *144*, 380. (16).

SALT, G., 1960, Proc. Roy. Soc. (London) B *151*, 446. (16).

SALT, G., 1961, *In:* Ramsay, J. A., and V. B. Wigglesworth, eds., *The cell and the organism*, (Cambridge University Press). pp. 175–192. (16).

SALT, G., 1963, Parasitology *53*, 527. (16).

SALT, G., 1966, Proc. Roy. Soc. (London) B *162*, 303. (16).

SALT, G., 1970, Proc. Roy. Soc. (London) B *176*, 105. (16).

SAMUEL, G. and BALD, J. G., 1931, Nature, London *128*, 494. (3).

SANGER, H. L., 1961, Proc. 4th Conf. Potato Virus Dis. Braunschweig, Germany 1960, 22–28. (27).

SANGER, H. L., M. W. ALLEN and A. H. GOLD, 1962, Phytopathology *52*, 750. (27).

SANMARTIN, C. and N. ARBELAEZ, 1963, Bol. of Santi. Pan-Am *59*, 516–525. (30).

SANTOS DIAS, J. A. T., 1963, Mem. Estud. Mus. Zool. Univ. Coimbra. No. *285*, 34. (5).

SATHER, G. E., A. L. LEWIS, W. JENNINGS, J. O. BOND and MCD. W. HAMMON, 1970, Amer. J. Trop. Med. Hyg. *19*, 319–326. (18).

SATURNO, A., 1963, Virology *21*, 131–133. (14).

SCHAEFER, H. A., 1949, Mitt. Schweis. Entomol. Gesch. *22*, 1–96. (7).

SCHAEFFER, M. and E. H. ARNOLD, 1954, Am. J. Hyg. *60*, 231–241. (30).

SCHAFFER, R. A. and W. F. SCHERER, 1971, Am. J. Epidemiol. *93*, 68–74. (20).

SCHALK, J. M., S. D. KINDLER and G. R. MANGLITZ, 1969, J. Econ. Entomol. *62*, 1000–1003. (31).

SCHÄLLER, G., 1961, Entomologia Exptl. Appl. *4*, 73–85. (7).

SCHEELE, C. M. and E. R. PFEFFERKORN, 1969a, J. Virol. *3*, 369–375. (14).

SCHEELE, C. M. and E. R. PFEFFERKORN, 1969b, J. Virol. *4*, 117–122. (14).

SCHEIBE, K., 1955, Zucker 8, 371–372. (31).

SCHELL, S. C., 1952, Trans. Am. Microsc. Soc. *71*, 293. (16).

SCHERER, W. F., E. L. BUESCHER and H. E. MCCLURE, 1959, Am. J. Trop. Med. Hyg. *8*, 689. (1).

SCHERER, W. F. and H. S. HURLBURT, 1967, Am. J. Epidemiol. *86*, 271–285. (4).

SCHERER, W. F., J. E. VERNA and G. W. RICHTER, 1968, Am. J. Trop. Med. Hyg. *17*, 120–128. (15).

SCHERER, W. F., M. KITAOKA, T. OKUNO and T. OGATA, 1959, Am. J. Trop. Med. Hyg. *8*, 707–715. (20).

SCHLESINGER, R. W., 1971, Current Top. Microbiol. Immunol. *55*, 241–245. (14).

SCHMIDT, H. B., R. FRITZSCHE and W. LEHMANN, 1963, Naturwissenschaften *50*, 386. (27).

SCHMIDT, J. R., M. L. SCHMIDT and M. I. SAID, 1971. Am. J. Trop. Med. Hyg. *20*, 483–490. (20).

SCHMIDT, J. R. and R. E. SHOPE, 1971, Acta Virol. Prague *15*, 112. (18).

SCHMIDTMANN, M., 1925, Z. Gesch. Exptl. Med. *45*, 714. (17).

SCHMUTTERER, H., 1961, Z. Angew. Entomol. *47*, 277–301; 416–439. (9).

SCHNEIDER, F., 1969, Ann. Rev. Entomol. *14*, 103–124. (7).

SCHNEIDER, I., 1964, J. Exptl. Zool. *156*, 91–104. (17).

SCHNEIDER, I., 1969, J. Cell Biol. *42*, 603–606. (14, 17).

SCHNEIDER, I., 1971, Current Top. Microbiol. Immunol. *55*, 1–12. (17).

SCHNEIDER, I., 1972, J. Embryol. Exptl. Morphol. *27*, 353–365. (17).

SCHNEPF, E., C. J. SOEDER and E. HEGEWALD, 1970, Virology *42*, 482–487. (28).

SCHÖNHERR, J., 1969, Entomophaga *14*, 251–260. (29).

SCHULZ, J. T., 1963, Plant Disease Reptr. *47*, 594–596. (19).

SCHUSTER, F. L. and T. H. DUNNEBACKE, 1971, J. Ultrastruct. Res. *36*, 659–668. (28).

SCHUSTER, M. F., 1963, Pl. Dis. Reptr. *47*, 510–511. (9).

SCHWARTZ, P. H. and B. G. TOWNSHEND, 1968, J. Invert. Pathol. *12*, 288. (16).

SCOTT, M. T., 1971, Transplantation *11*, 78. (16).

SEAMAN, G. R. and N. L. ROBERT, 1968, Science *161*, 1359. (16).

SEECOF, R. L., 1964, Virology *22*, 142–148. (20).

SEECOF, R. L. and R. L. TEPLITZ, 1971, Curr. Top. Microbiol. Immunol. *55*, 71–75. (17).

SEINHORST, J. W., 1970, Nematologica *16*, 330. (12).

SEKEYOVÁ, M., M. GREŠÍKOVÁ and S. STÚPALOVÁ, 1970, Folia Parasitol. *17*, 341–343. (18).

SELLERS, R. F., G. H. BERGOLD, O. M. SUAREZ and A. MORALES, 1965, Am. J. Trop. Med. Hyg. *14*, 460–469. (30).

SERIE, C., L. ANDRAL, A. LINDREC and P. NERT., 1964, Bull. World Health Organ. *30*, 299–319. (30).

SERJEANT, E. P., 1967, Ann. Appl. Biol. *59*, 31–38. (9).

SERRANO, F. M. H., 1964, Pecuaria, Loanda 37–42. (5).

SETH, M. L., 1962, Indian Phytopathol. *15*, 225–227. (19).

SETH, M. L., 1965, Indian Phytopathol. *18*, 317–319. (19).

SEVERIN, H. H. P., 1931, Hilgardia *6*, 253–276. (24).

SHADE, R. E. and M. C. WILSON, 1967, Ann. Ent. Soc. Am. *60*, 493–496. (31).

SHAH, K. V., 1964, Indian J. Vet. Sci. Anim. Husb. *34*, 1–14. (30).

SHAH, K. V. and T. H. WORK, 1969, Indian J. Med. Res. *57*, 793–798. (18).

SHAPIRO, D., W. E. BRANDT, R. D. CARDIFF and P. K. RUSSELL, 1971, Virology *44*, 108–124. (14).

SHAPIRO, M., 1967, J. Invert. Pathol. *9*, 19. (16).

SHARMA, H. H., C. A. K. SHANMUGHAM, S. P. IYER, A. RAMACHANDRA RAO and S. A. KUPPUS-WAMI, 1965, Indian J. Med. Res. *53*, 720–728. (30).

SHARMA, R. D., 1965, Meded. Landb. Hoogesch. Ghent *30*, 1437–1443. (12).

SHATKIN, A. J., 1969, Adv. Virus. Res. *14*, 63–87. (4).

SHATOURY, H. H. el, 1956, Arch. Entwicklungsmech. Organ. *148*, 391. (16).

SHATOURY, H. H. el. and C. H. WADDINGTON, 1957, J. Embryol. Exptl. Morphol. *5*, 143. (16).

SHAW, J. G. and J. B. BEAVERS, 1970, J. Econ. Entomol. *63*, 850–853. (29).

SHAW, J. G., D. L. CHAMBERS and H. TASHIRO, 1968a, J. Econ. Entomol. *61*, 1352–1355. (29).

SHAW, J. G., H. TASHIRO and E. J. DIETRICH, 1968b, J. Econ. Entomol. *61*, 1492–1495. (29).

SHAW, J. G., J. B. BEAVERS, J. L. PAPPAS and R. B. HAMPTON, 1969, J. Econ. Entomol. *62*, 1154–1156. (29).

SHAW, M. J. P., 1970, Ann. Appl. Biol. *65*, 191–212. (7).

SHEPHERD, A. M., 1956, Friesia, Copenhagen *5*, 396–408. (31).

SHEPHERD, R. J., 1970a, C.M.I./A.A.B. Descrip. Plant Viruses No. 24. (4).

SHEPHERD, R. J., 1970b, C.M.I./A.A.B. Descrip. Plant Viruses. No. 25. (4).

SHEPHERD, R. J. and D. E. PURCIFULL, 1970, C.M.I./A.A.B. Descrip. Plant Viruses. No. *55*, (4).

SHEPHERD, R. J., R. J. WAKEMAN and R. R. ROMANKO, 1968, Virology *36*, 150–152. (4).

SHEPPARD, P. M., W. W. MACDONALD, R. J. TONN and R. GRAB, 1969, J. Anim. Ecol. *38*, 661–702. (30).

SHESTAKOV, V. I. and I. N. POLENOVA, 1965, Zool. Zh. *44*, 1871. (30).

SHIGEMATSU, H. and A. NOGUCHI, 1969, J. Invert. Pathol. *14*, 301–307. (14).

SHIKATA, E., 1966, J. Fac. Agr. Hokkaido Univ. *55*, 1–110. (26).

SHIKATA, E. and M. J. CHEN, 1969, J. Virol. *3*, 261–264. (26).

SHIKATA, E. and G. E. GALVEZ, 1969, Virology *39*, 635–641. (26).

SHIKATA, E. and Y. T. LU, 1967, Proc. Japan. Acad. *43*, 918–923. (26).

SHIKATA, E. and K. MARAMOROSCH, 1965, Virology *27*, 461–475. (26).

SHIKATA, E. and K. MARAMOROSCH, 1967, Virology *32*, 363–377. (15).

SHIKATA, E., K. MARAMOROSCH and R. R. GRANADOS, 1966, Virology *29*, 426–436. (15, 24).

SHINEDLING, S. T. and B. GREENBERG, 1971, Current Top. Microbiol. Immunol. *55*, 12–19. (17).

SHINKAI, A., 1966, Ann. Phytopathol. Soc. Japan *32*, 317. (8).

SHINKAI, A., 1967, Ann. Phytopathol. Soc. Japan *33*, 318. (8).

SHOPE, R. E., 1940, Arch. Gesch. Virusforsch. *1*, 457–467. (4).

SHOPE, R. E., 1941, J. Exptl. Med. *74*, 41–68. (1, 4, 12, 27).

SHOPE, R. E., 1943, J. Exptl. Med. *77*, 111–126. (27).

SHOPE, R. E., 1958, J. Exp. Med. *107*, 609–622. (1, 12).

SHOPE, R. E., 1958, J. Exp. Med. *108*, 159. (1).

SHOPE, R. E., R. MANGOLD, L. G. MACNAMARA and K. R. DUMBELL, 1958, J. Exptl. Med. *108*, 797–802. (4).

SHORE, H., 1961, Trans. Roy. Soc. Trop. Med. Hyg. *55*, 361–373. (30).

SHOTTS, E. B., J. W. FOSTER, M. BRUGH, H. E. JORDAN and J. L. MCQUEEN, 1968, J. Exptl. Med. *127*, 359–369. (12).

SIDOR, C., 1960, Virology *10*, 551–552. (15).

SIDWELL, R. W., L. P. GEBHARDT and B. D. THORPE, 1967, Bacteriol. Rev. *31*, 65–81. (30).

SIKES, R. K. and R. W. CHAMBERLIN, 1954, J. Parasitol. *40*, 691–697. (6).

SIKKA, S. M., V. M. SAHNI and D. K. BUTANI, 1966, Euphytica *15*, 383–388. (31).

SIMONS, J. N., J. R. ORSENIGO, R. E. STALL and P. L. THAYER, 1959, Mimeogr. Rep. Everglades Exptl. Sta. Univ. Fla. (31).

SIMPSON, D. I. H., E. M. KNIGHT, G. COURTOIS, M. C. WILLIAMS, M. P. WEINBREN and J. W. KIBUKAMSOKE, 1967, E. African Med. J. *44*, 87–92. (30).

SIMPSON, D. I. H., E. T. W. BOWEN, G. S. PLATT, H. WAY, C. E. G. SMITH, S. PETO, S. KAMATH, B. L. LIM and T. W. LIM, 1970, Trans. Roy. Soc. Trop. Med. Hyg. *64*, 503–510. (30).

SIMPSON, R. W. and R. F. HAUSER, 1966, Virology *29*, 654–667. (4).

SINGH, K., 1931, Mem. Dept. Agr. India Entomol. *12*, 1–98. (13).

SINGH, K. R. P., 1967, Current Sci. *36*, 506–508. (14, 17).

SINGH, K. R. P., 1968, Parasitology *58*, 461–463. (5).

SINGH, K. R. P., M. K. GOVERDHAN and U. K. M. BHAT, 1971, Indian J. Med. Res. *59*, 213–218. (18).

SINGH, K. R. P. and S. D. PAUL, 1968, Current Sci. *37*, 65–67. (14).

SINHA, R. C., 1960, Virology *10*, 344–352. (26).

SINHA, R. C., 1963, Phytopathology *53*, 1170–1173. (3, 15, 26).

SINHA, R. C., 1965a, Virology *26*, 673–686. (8, 15, 24, 26).

SINHA, R. C., 1965b, Virology *27*, 118–119. (26).

SINHA, R. C., 1967, Virology *31*, 746–748. (26).

SINHA, R. C., 1968, Adv. Virus. Res. *13*, 181–223. (26).

SINHA, R. C., 1971, Virology *44*, 342–351. (26).

SINHA, R. C. and L. N. CHIYKOWSKI, 1967a, Virology *31*, 461–466. (26).

SINHA, R. C. and L. N. CHIYKOWSKI, 1967b, Virology *32*, 402–405. (26).

SINHA, R. C. and L. N. CHIYKOWSKI, 1967c, Virology *33*, 702–708. (26).

SINHA, R. C. and L. N. CHIYKOWSKI, 1968, Acta Virol. *12*, 546–550. (26).

SINHA, R. C. and L. N. CHIYKOWSKI, 1969, Virology *38*, 679–684. (26).

SINHA, R. C. and Y. C. PALIWAL, 1970, Virology *40*, 665–672. (26).

SINHA, R. C. and S. SHELLEY, 1965, Phytopathology *55*, 324–327. (26).

SKALIY, P. and W. J. HAYES, 1949, Am. J. Trop. Med. *29*, 759–772. (6).

SKEHEL, J. J. and W. K. JOKLIK, 1969, Virology *39*, 822–831. (4).

SLACK, S. A. and J. P. FULTON, 1971, Virology *43*, 728–729. (9).

SLACK, S. A. and H. A. SCOTT, 1971, Phytopathology *61*, 538–540. (9).

SLYKHUIS, J. T., 1955, Phytopathology *45*, 116–128. (3, 19, 31).

SLYKHUIS, J. T., 1956, Phytopathology *46*, 682–687. (19).

SLYKHUIS, J. T., 1961, Can. J. Plant Sci. *41*, 304–308. (19).

SLYKHUIS, J. T., 1962, *In:* Maramorosch, K., ed., *Biological Transmission of Disease Agents* (Academic Press, New York and London), pp. 41–61. (6).

SLYKHUIS, J. T., 1965, Adv. Virus. Res. *11*, 97–137. (3, 19).

SLYKHUIS, J. T., 1967, Rev. Appl. Mycol. *46*, 401–429. (19).

SLYKHUIS, J. T., 1969, Phytopathology *59*, 29–32. (19).

SLYKHUIS, J. T., J. E. ANDREWS and U. J. PITTMAN, 1957, Can. J. Plant Sci. *37*, 113–127. (19).

SMIRNOFF, W. A., 1960, Can. Entomol. *92*, 957–958. (22).

SMIRNOFF, W. A., 1961, J. Insect Pathol. *3*, 29–47. (22, 29).

SMIRNOFF, W. A., 1962, J. Insect Pathol. *4*, 192–200. (21, 22).

SMIRNOFF, W. A., 1963, J. Insect Pathol. *5*, 104–111. (22).

SMIRNOFF, W. A., 1965, J. Invert. Pathol. *7*, 387–388. (15, 21, 22).

SMIRNOFF, W. A., 1968, J. Invert. Pathol. *10*, 436–438. (22).

SMITH, B. D., 1962, Ann. Rep. Agr. Hort. Res. Sta., Long Ashton, Bristol, 1961, 170–172. (19).

SMITH, C. E. G., 1956, Nature, London *178*, 581–582. (18).

SMITH, C. E. G., 1962, Symp. Zool. Soc. London 199–221. (18, 30).

SMITH, C. E. G., 1964, Symp. Br. Soc. Parasitol. *2*, 1–31. (11).

SMITH, C. E. G., 1964, *In: Host-Parastic Relationships in Invertebrate Hosts.* (Blackwell Scientific Publications, Oxford). (18).

SMITH, C. E. G., 1964, Sci. Basis. Med. Ann. Rev. 1964. (30).

SMITH, C. E. G., 1968, Abstr. Hyg. *43*, 1398–1436. (30).

SMITH, C. E. G., 1969, Brit. Med. Bull. *25*, 142–147. (30).

SMITH, C. E. G., 1970, Proc. R. Soc. Med. *63*, 1181–1189. (30).

SMITH, C. E. G., 1971, Sci. Basis Med. Ann. Rev., 1971. (30).

SMITH, C. E. G., D. A. MCMAHON and E. T. W. BOWEN, 1967, Nature, London. *214*, 1154–1155. (18).

SMITH, C. N., 1967, Bull. World Health Organ. *36*, 633–635. (30).

SMITH, F. F. and R. E. WEBB, 1969, *In:* Maramorosch, K., ed., *Viruses Vectors and Vegetation*, (Wiley-Interscience, New York). (25).

SMITH, K. M., 1926, Ann. Appl. Biol. *13*, 109–139. (10, 25).

SMITH, K. M., 1931, Proc. Roy. Soc. (London) B *109*, 251–266. (3).

SMITH, K. M., 1937, *Textbook of Plant Virus Diseases*, 1st Ed., (J. & A. Churchill, London). (3).

SMITH, K. M., 1941, Parasitology *33*, 110–116. (9).

SMITH, K. M., 1946, Parasitology *37*, 131–134. (24).

SMITH, K. M., 1958, Parasitology *48*, 459–462. (2).

SMITH, K. M., 1963, *In:* Steinhaus, E. A., ed., *Insect Pathology* Vol. 1, (Academic Press, London and New York). (14).

SMITH, K. M., 1965, Adv. Virus. Res. *11*, 61–96. (9, 25).

SMITH, K. M., 1967, *Insect Virology*, (Academic Press, New York). (4, 9, 15, 20, 21, 22).

SMITH, K. M. and G. J. HILLS, 1959, J. Mol. Biol. *1*, 277–280. (14).

SMITH, K. M. and G. J. HILLS, 1962, Proc. 11th Entomol. Congr. Vienna, 1960, *2*, 823–827. (4).

SMITH, K. M., G. J. HILLS, F. MUNGER and J. GILMORE, 1959, Nature, London *184*, 70. (4).

SMITH, K. M., G. J. HILLS and C. F. RIVERS, 1961, Virology *13*, 233–241. (14, 20).

SMITH, K. M. and R. W. G. WYCKOFF, 1950, Nature, London *166*, 861. (2).

SMITH, K. M. and N. XEROS, 1953a, Nature, London *172*, 670–671. (14).

SMITH, K. M. and N. XEROS, 1953b, Parasitology *43*, 178–185. (2).

SMITH, K. M. and N. XEROS, 1954a, Nature. London *173*, 866–867. (20).

SMITH, K. M. and N. XEROS, 1954b, Parasitology *44*, 71–80. (14).

SMITH, L. W., 1969, Mosquito News *29*, 556–563. (30).

SMITH, R. E. and P. BONQUET, 1951a, Phytopathology *5*, 103–107. (3).

SMITH, R. E. and P. A. BONQUET, 1915b, Phytopathology *2*, 335–342. (3).

SMITH, R. E., H. J. ZWEERINK and W. K. JOKLIK, 1969, Virology *39*, 791–810. (4).

SNELLING, R. O., 1941, J. Econ. Entomol. *34*, 335–340. (31).

SOGAWA, K., 1965, Japan. J. Appl. Entomol. Zool. *9*, 275–290. (8).

SOGAWA, K., 1967a, Appl. Entomol. Zool. *2*, 13–21. (8).

SOGAWA, K., 1967b, Appl. Entomol. Zool. *2*, 195–202. (8).

SOGAWA, K., 1968a, Appl. Entomol. Zool. *3*, 13–25. (8).

SOGAWA, K., 1968b, Appl. Entomol. Zool. *3*, 67–73. (8).

SOGAWA, K., 1970, Jap. J. Appl. Entomol. Zool. *14*, 101–106. (8).

SOHI, S. S., 1968, Can. J. Zool. *46*, 11–13. (17).

SOHI, S. S., 1969, Can. J. Microbiol. *15*, 1197–1200. (17).

SOHI, S. S. and C. SMITH, 1970, Can. J. Zool. *45*, 427–432. (17).

SOHI, S. S., 1971, Can. J. Zool. *49*, 1355–1358. (17).

SOL, H. H. and J. W. SEINHORST, 1961, Tijdschr. Pl. Ziekten *67*, 307–309. (27).

SØMME, L., 1969, Norsk Entomol. Tidsskr. *16*, 107–111. (31).

SONENSHINE, D. E., C. M. CLIFFORD and G. M. KOHLS, 1966, Ann. Entomol. Soc. Am. *59*, 92–122. (5).

SONNABEND, J. A., E. M. MARTIN and E. MECS, 1967, Nature, London *213*, 365–367. (14).

SONNABEND, J. A., L. DALGARNO, R. M. FRIEDMAN and E. M. MARTIN, 1964, Biochem. Biophys. Res. Comm. *17*, 455–460. (14).

SOPER, F. L., 1967, Bull. World Health Organ. *36*, 521–527. (30).

SOUTHAM, C. M., F. H. SHIPKEY, V. I. BABCOCK, R. BAILEY and R. A. ERLANDSON, 1964, J. Bacteriol. *88*, 187–199. (14).

SOUTHEY, J. F., 1965, Tech. Bull. Min. Agr. Fish. Food. *7*, H.M.S.O. London. (12, 31).

SPILLING, C. R., 1970, D.Phil. Thesis, University of Oxford, England, 1970. (21).

SPRENT, J. F. A., 1969, *In:* Jackson, G. J., R. Herman, and I. Singer, eds., *Immunity to Parasitic Animals.* Vol. 1, (North-Holland, Amsterdam) pp. 3–26. (16).

SREEVALSAN, T. and R. Z. LOCKHART, 1966, Proc. Natl. Acad. Sci. U.S. *55*, 974–981. (14).

SREEVALSAN, T., R. Z. LOCKHART, M. L. DODSON and K. A. HARTMAN, 1968, J. Virol. *2*, 558–566. (14).

SRIHONGSE, S., W. F. SCHERER and P. GALINDO, 1967, Am. J. Trop. Med. Hyg. *16*, 519–524. (30).

SRIVASTAVA, V. S., 1959, Proc. Roy. Entomol. Soc. London *34*, 57–62. (9).

STACE-SMITH, R., 1970, C.M.I./A.A.B. Descrip. Plant Viruses. No. *18*. (4).

STAIRS, G. R., 1964a, Can. Entomol. *96*, 1017–1020. (29).

STAIRS, G. R., 1964b, J. Insect Pathol. *6*, 164–169. (29).

STAIRS, G. R., 1965a, J. Invert. Pathol. *7*, 5–9. (15).

STAIRS, G. R., 1965b, J. Invert. Pathol. *7*, 427–429. (15).

STAIRS, G. R., 1970, J. Invert. Pathol. *15*, 60–62. (14).

STAIRS, G. R. and F. T. BIRD, 1962, Can. Entomol. *94*, 966–969. (29).

STAIRS, G. R., W. B. PARRISH and M. ALLIETTA, 1968, J. Invert. Pathol. *12*, 359–365. (14).

STANLEY, W. M., 1935, Science *81*, 644–645. (3).

STANLEY, W. M., 1936, Phytopathology *26*, 305–320. (3).

STANNARD, L. J., 1968, Bull. Ill. Nat. Hist. Surv. *29*, 210–552. (13).

STAPEL, C., 1965, Tidsskr. Landkøn. *5*, 173–198. (31).

STAPLES, R. and W. B. ALLINGTON, 1956, Univ. Neb. Agr. Exptl. Sta. Res. Bull. *178*. (19).

STAPLES, R., 1968, J. Econ. Entomol. *61*, 1378–1380. (7).

STARKOFF, O., 1958, *Ixodoidea d'Italia. Studio Monografico, Il Pensiero,* (Scientifico Editore, Roma) pp. 385 (5).

STARÝ, P., 1966, *Aphid parasites of Czechoslovakia,* (Academia, Prague). (7).

STARÝ, P., 1968, Acta Entomol. Bohemoslov. *65*, 76–77. (31).

STEGWEE, D. and M. B. PONSEN, 1958, Entomol. Exptl. Appl. *1*, 291–300. (3, 15, 24).

STEPHENS, W. F. and M. R. L. JOHNSTON, 1966, J. Ultrastruct. Res. *15*, 543–554. (4).

STEIGER, U., H. E. CAMPORTER, C. SANDRI and K. AKERT, 1968, Arch. Gesch. Virusforsch. *26*, 271–282. (22).

STEINHAUS, E. A., 1947, Science *106*, 323. (2).

STEINHAUS, E. A., 1947, *Insect Microbiology, An Account of the Microbes Associated with Insects and Ticks with Special Reference to the Biologic Relationships Involved,* (Comstock Publishing Company, Inc. Ithaca, New York). (18, 29).

STEINHAUS, E. A., 1949, *Principles of Insect Pathology* (McGraw-Hill, New York). (29).

STEINHAUS, E. A., 1954, Hilgardia *23*, 197–261. (21, 29).

STEINHAUS, E. A., 1963, *Insect Pathology,* Vols. I and II, (Academic Press, New York). (29).

STELZER, M. J., 1965, J. Invert. Pathol. *7*, 122–126. (29).

STEPHENS, J. M., 1959, Can. J. Microbiol. *5*, 203. (16).

STEPHENS, J. M., 1962, Can. J. Microbiol. *8*, 491. (16).

STEPHENS, J. M., 1964, *In:* Steinhaus, E. A., ed., *Insect Pathology: An advanced Treatise* Vol. 1, (Academic Press New York and London). pp. 273–297. (16).

STEPHENS, J. M. and J. H. MARSHALL, 1962, Can. J. Microbiol. *8*, 719. (16).

STERN, V. M., 1967, J. Econ. Entomol. *60*, 485–490. (31).

STEVENS, T. M., 1970, Proc. Soc. Exptl. Biol. Med. *134*, 356–361. (14).

STILES, G. W., 1944, Am. J. Vet. Res. *5*, 318–319. (5).

STOLLAR, V., 1969, Virology *39*, 426–438. (14).

STOLLAR, V., R. W. SCHLESINGER and T. M. STEVENS, 1967, Virology *33*, 650–658. (14).

STOLLAR, V., T. M. STEVENS and R. W. SCHLESINGER, 1966, Virology *30*, 303–312. (14).

STOLTZ, D. B. and HILSENHOFF, W. L., 1969, J. Invert. Pathol. *14*, 39–48. (14, 15).

STOLTZ, D. B., W. L. HILSENHOFF and H. F. STICH, 1968, J. Invert. Pathol. *12*, 118–128. (4, 20, 29).

STONE, A., K. L. KNIGHT and H. STARCKE, 1959, A Synoptic Catalog of the Mosquitoes of the World. Thomas Say Foundation, No. 6. Washington: Entomological Society of America. (11).

STOREY, H. H., 1932, Proc. Roy. Soc. (London) B. *112*, 46–60. (3, 15).

STOREY, H. H., 1933, Proc. Roy. Soc. (London) B. *113*, 463–485. (3, 15, 26).

STOREY, H. H., 1938, Proc. Roy. Soc. (London) *125*, 455–477. (3).

STRAUSS, J. H., B. W. BURGE and J. E. DARNELL, 1969, Virology *37*, 367–376. (14).

STRAUSS, J. H., B. W. BURGE and J. E. DARNELL, 1970, J. Mol. Biol. *47*, 437–448. (14).

STRAUSS, J. H., B. W. BURGE, E. R. PFEFFERKORN and J. E. DARNELL, 1968, Proc. Natl. Acad. Sci. U.S., *59*, 533–537. (4, 14).

STREAMS, F. A. and L. GREENBERG, 1969, J. Invert. Pathol. *13*, 371. (16).

STREISSLE, G. and K. MARAMOROSCH, 1963, Science, N.Y. *140*, 996–997. (26).

STRICKLAND, E. H., 1938, Can. Entomol. *70*, 200–206. (7).

STRICKLAND, A. H., 1945, J. Animal Ecol. *14*, 1–11. (10).

STRICKLAND, A. H., 1950, Proc. Roy. Entomol. Soc. London (A). *25*, 1–9. (23).

STRICKLAND, A. H., 1951a, Bull. Entomol. Res. *41*, 725–748. (10, 23).

STRICKLAND, A. H., 1951b, Bull. Entomol. Res. *42*, 65–103. (10, 23).

STRODE, G. K., 1951, *Yellow Fever*, (McGraw-Hill Book Co., New York). (1, 30).

STRONG, F. E., 1963, Hilgardia *34*, 43–61. (7).

STRONG, F. E., 1967, Ann. Entomol. Soc. Am. *60*, 668–676. (7).

STUART, A. E., 1968, J. Pathol. Bacteriol. *96*, 401. (16).

STUBBS, L. L., 1955, Australian J. Biol. Sci. *8*, 68–74. (25).

STURHAN, D., 1967, Mitt. Biol. BundAnst. Ld-u. Forstw. *121*, 146–151. (12).

SUDIA, W. D., P. H. COLEMAN, R. W. CHAMBERLAIN, J. S. WISEMAN and T. H. WORK, 1967, J. Med. Entomol. *4*, 32–36. (30).

SUITOR, E. C., 1966, Virology *30*, 143–145. (14).

SUITOR, E. C., 1969, J. Gen. Virol. *5*, 545–546. (14).

SUITOR, E. C. and F. J. PAUL, 1969, Virology *38*, 482–485. (14).

SUITOR, E. C., L. L. CHANG and H. H. LIU, 1966, Exptl. Cell Res. *44*, 572–578. (17).

SULKIN, S. E., R. ALLEN, T. MIURA and E. TOYOKANA, 1970, Amer. J. Trop. Med. Hyg. *19*, 77–87. (30).

SULKIN, S. E., A. RAE, R. SIMS and S. K. TAYLOR, 1965, Am. J. Public Health *55*, 1376–1385. (20).

SUMMERS, D. F. and J. V. MAIZEL, 1968, Proc. Natl. Acad. Sci. U.S. *59*, 966–971. (14).

SUMMERS, M. D., 1969, J. Virol. *4*, 188–190. (14, 15, 21).

SUMMERS, M. D., 1971, J. Ultrastruct. Res. *35*, 606–625. (14, 15).

SUMMERS, M. D. and D. L. ANDERSON, 1972, J. Virol. *9*, 710–713. (4).

SUMMERS, M. D. and H. J. ARNOTT, 1969, J. Ultrastruct. Res. *28*, 462–480. (14).

SURTEES, G., 1959, Bull. Entomol. Res. *50*, 681–686. (30).

SURTEES, G., 1969, J. Med. Entomol. *6*, 317–320. (30).

SURTEES, G., 1970a, Int. J. Envir. Studies *1*, 35–42. (30).

SURTEES, G., 1970b, J. Med. Entomol. *7*, 509–517. (30).

SURTEES, G., 1970c, J. Med. Entomol. *7*, 273–276. (30).

SURTEES, G., 1970d, Bull. Entomol. Res. *60*, 275–283. (30).

SURTEES, G., 1971a, J. Trop. Med. Hyg. *74*, 255–259 (30).

SURTEES, G., 1971b, Abs. Hyg. *46*, 121–134. (30).

SURTEES, G., 1971c, Intern. J. Envir. Studies. *2*, 195–201. (30).

SURTEES, G., M. N. HILL and J. BROADFOOT, 1971, Survival of a tropical mosquito, *Aedes aegypti* in southern England. Bull. World Health. Organ. *44*, 707–709. (30).

SWEET, B. H. and L. T. DUPREE, 1968, Mosquito News *28*, 368–373. (17).

SWEET, B. H. and J. S. MCHALE, 1970, Exptl. Cell Res. *61*, 51–63. (17).

SWENSON, K. G., 1962, Australian J. Biol. Sci. *15*, 468–482. (25).

SWENSON, K. G. and R. L. NELSON, 1959, J. Econ. Entomol. *52*, 421–425. (25).

SWIRSKI, E., 1954, Ktavim *4*, 61–68. (7).

SYLVESTER, E. S., 1954, Hilgardia *23*, 53–98. (25).

SYLVESTER, E. S., 1955, Phytopathology *45*, 357–370. (25).

SYLVESTER, E. S., 1956, J. Econ. Entomol. *49*, 789–800. (25).

SYLVESTER, E. S., 1962, *In:* Maramorosch, K. ed., *Biological transmission of disease agents* (Academic Press, New York) pp. 11–31. (3, 24, 25).

SYLVESTER, E. S., 1962, J. Econ. Entomol *49*, 789–800. (3).

SYLVESTER, E. S., 1964, J. Econ. Entomol. *57*, 538–544. (25).

SYLVESTER, E. S., 1965, Virology *25*, 62–67. (24).

SYLVESTER, E. S., 1967, Virology *32*, 524–531. (24).

SYLVESTER, E. S., 1969, Virology *38*, 440–446. (24).

SYLVESTER, E. S., 1969, *In:* Maramorosch, K., ed., *Viruses, Vectors and Vegetation.* (Interscience Publ. New York) pp. 159–173. (3, 24).

SYLVESTER, E. S. and J. RICHARDSON, 1966, Virology *30*, 592–597. (24).

SYLVESTER, E. S. and J. RICHARDSON, 1969, Virology *37*, 26–31. (24).

SYLVESTER, E. S. and J. RICHARDSON, 1970, Virology *42*, 1023–1042. (24).

SYVERTON, J. T., O. R. MCCOY and J. J. KOOMEN, 1947, J. Exptl. Med. *85*, 759–769. (4, 12, 27).

TABER, S., 1955, J. Econ. Entomol. *48*, 522–525. (22).

TAJIMA, M., H. NAKAJIMA and Y. ITO, 1969, J. Virol. *4*, 521–527. (4).

TAKAHASHI, K., 1968, Trop. Med. *10*, 181–194. (30).

TAKAHASHI, R., 1963, Mushi *37*, 167–190. (13).

TAKAHASHI, Y. and G. B. ORLOB, 1969, Virology *38*, 230–240. (19).

TAKAMI, N., 1901, J. Japan. Agr. Soc. *241*, 22–30. (3).

TAKATA, K., 1895, J. Japan. Agr. Soc. *171*, 1–4. (3).

TAMAKI, G. and R. E. WEEKS, 1968, J. Econ. Entomol. *61*, 431–435. (31).

TANADA, Y., 1954, Ann. Entomol. Soc. Am. *47*, 553–574. (4).

TANADA, Y. and G. Y. CHANG, 1960, J. Insect Pathol. *2*, 201–208. (21).

TANADA, Y. and G. Y. CHANG, 1962, J. Insect Pathol. *4*, 361–370. (15).

TANADA, Y. and G. Y. CHANG, 1968, J. Invert. Pathol. *10*, 79–83, (21).

TANADA, Y. and T. HUKUHARA, 1968, J. Invert. Pathol. *12*, 263–268. (15, 29).

TANADA, Y. and R. LEUTENEGGER, 1968, J. Invert. Pathol. *10*, 39–47. (14).

TANADA, Y. and R. LEUTENEGGER, 1970, J. Ultrastruct. Res. *30*, 589–600. (14).

TANADA, Y. and C. REINER, 1962, J. Insect Pathol. *4*, 139–155. (29).

TANADA, Y. and A. M. TANABE, 1965, J. Invert. Pathol. *7*, 184–188. (14).

TANADA, Y., T. HUKUHARA and G. Y. CHANG, 1969, J. Invert. Pathol. *13*, 394–409. (14).

TARJAN, A. C., 1964, Proc. Helminthol. Soc. Wash. *31*, 65–76. (12).

TARR, S. A. J., 1951, *Leaf Curl Disease of Cotton*, (Commonw. Myc. Inst. London). pp. 13–32. (3).

TARSHIS, I. B. and W. D. OMMERT, 1961, J. Am. Vet. Med. Ass. *138*, 665–669. (5).

TATCHELL, R. J., 1969a, Parasitology *59*, 93–104. (18).

TATCHELL, R. J., 1969b, J. Insect Physiol. *15*, 1421–1430. (18).

TATCHELL, R. J. and K. C. BINNINGTON, 1973, Proc. 3rd. Intern. Congr. Acarol. Prague, 000–000. (18).

TATCHELL, R. J., R. CARNELL and D. H. KEMP, 1972, Z. Parasitenk. *38*, 32–44. (18).

TAYLOR, C. E., 1962, Virology *17*, 493–494. (27).

TAYLOR, C. E., 1967, Ann. Appl. Biol. *59*, 275–281. (27).

TAYLOR, C. E., 1970, Zesz. Probl. Postep. Nauk. Roln. *92*, 283–289. (27).

TAYLOR, C. E., 1971, *In:* Zuckerman, B. M., W. F. Mai and R. A. Rohde, eds., *Plant Parasitic Nematodes* (Academic Press, New York and London). (12, 27).

TAYLOR, C. E. and C. H. CADMAN, 1969, *In:* Maramorosch, K., ed., *Viruses, Vectors and Vegetation*, (Interscience Publishers, New York). (27).

TAYLOR, C. E. and A. F. MURANT, 1969, Ann. Appl. Biol. *64*, 43–48. (27).

TAYLOR, C. E. and D. J. RASKI, 1964, Nematologica *10*, 489–495. (27).

TAYLOR, C. E. and W. M. ROBERTSON, 1969, Ann. Appl. Biol. *64*, 233–237. (12, 27).

TAYLOR, C. E. and W. M. ROBERTSON, 1970a, Ann. Appl. Biol. *66*, 375–380. (12, 25, 27).

TAYLOR, C. E. and W. M. ROBERTSON, 1970b, J. Gen. Virol. *6*, 179–182. (27).

TAYLOR, C. E. and P. R. THOMAS, 1968, Ann. Appl. Biol. *62*, 147–157. (27, 31).

TAYLOR, C. E., P. R. THOMAS, W. M. ROBERTSON and I. M. ROBERTS, 1970, Nematologica *16*, 6–12. (12).

TAYLOR, D. J., 1958, *In: Rep. Cocoa Conf. London 1957.* (The Cocoa, Chocolate and Confectionery Alliance Ltd.) pp. 125–131. (23).

TAYLOR, J., 1965, Virology *25*, 340–349. (14).

TAYLOR, K. L., 1960, Australian J. Zool. *8*, 383–391. (7).

TAYLOR, L. R., 1958, Proc. Linnean Soc. London *169*, 67–73. (25).

TAYLOR, L. R., 1965, Proc. N. Central Branch Entomol. Soc. Am. *20*, 9–19. (25).

TAYLOR, R. M., 1955, Proc. 6th. Intern. Cong. Microbiol. Rome, *3*, 236–240. (20).

TAYLOR, R. M., 1967, Public Health Service Publication No. 1760, U.S. Govt. Printing Office, Washington, D.C. (4, 11).

TAYLOR, R. M., H. HOOGSTRAAL and H. S. HURLBUT, 1966, Am. J. Trop. Med. Hyg. *15*, 75. (18).

TAYLOR, R. M., H. S. HURLBUT, T. H. WORK, J. R. KINGSTON and H. HOOGSTRAAL, 1966, Am. J. Trop. Med. Hyg. *15*, 76–86. (18, 30).

TCHUBIAISHVILI, C. A., 1970, Acta Virol. *14*, 75–77. (2).

TEAKLE, D. S. and E. S. SYLVESTER, 1962, Virology *22*, 520–538. (3, 25).

TEAKLE, R. E., 1969, J. Invert. Pathol. *14*, 18–27. (14).

TELIZ, D., R. G. GROGAN and B. F. LOWNSBERY, 1966, Phytopathology *56*, 658–663. (27).

TERPSTRA, C., 1969, Doctoral Thesis. Diergeneeskunde Rijksuniversiteit, Elinkwijk, Utrecht. (18).

TERZAKIS, J. A., 1969, Mill. Med. *134*, 916–921. (28).

THATCHER, V. E., 1968, Ann. Entomol. Soc. Am. *61*, 1141–1143. (11).

THEILER, G., 1962, Rep. Dir. Vet. Serv. Onderstepoort, Proj. *S.9958*, 255. (5).

THEILER, M. and H. H. SMITH, 1937, J. Exptl. Med. *65*, 757–787. (1).

THOMAS, C. E., 1969, Phytopathology *59*, 633–636. (27).

THOMAS. L. A., 1963, Am. J. Hyg. *78*, 150–165. (14, 20).

THOMAS. L. A. and C. M. EKLUND, 1960, Proc. Soc. Exptl. Biol. (N.Y.). *105*, 52. (1).

THOMAS, L. A. and C. M. EKLUND, 1962, Proc. Soc. Exptl. Biol. Med., *109*, 421–424. (20).

THOMAS, L. A., R. C. KENNEDY and C. M. EKLUND, 1960, Proc. Soc. Exptl. Biol. N.Y. *104*, 355. (30).

THOMAS, P. R., 1969, Nematologica *15*, 582–590. (12).

THOMAS, R. S., 1961, Virology *14*, 240–252. (4).

THOMAS, R. S. and R. C. WILLIAMS, 1961, J. Biophys. Biochem. Cytol. *11*, 15–29. (4).

THOMPSON, C. G., 1959, J. Insect Pathol. *1*, 189–190. (21).

THOMPSON, C. G. and E. A. STEINHAUS, 1950, Hilgardia *19*, 411–445. (29).

THOMPSON, W. R. and F. J. SIMMONDS, 1964, A catalogue of the parasites and predators of insect pests. Section 3. Commonwealth Agricultural Bureaux, London (7).

THOMSON, J. A. and T. D. C. GRACE, 1963, Australian J. Biol. Sci. *16*, 869–876. (17).

THRESH, J. M., 1957, Rept. W. African Cocoa Res. Inst. 1955–56, 83. (10).

THRESH, J. M., 1958a, Techn. Bull. W. African Cocoa Res. Inst. *4*, 1–36. (23).

THRESH, J. M., 1958b, Techn. Bull. W. African Cocoa Res. Inst., *5*, 1–36. (23).

THRESH, J. M., 1958c, Ann. Rep. W. African Cocoa Res. Inst., *1956–57*, 78–81. (23).

THRESH, J. M., 1959, Trop. Agr. Trin. *36*, 35–44. (31).

THRESH, J. M., 1960, F.A.O. Plant Protection. Bull. *8*, 89–92. (31).

THRESH, J. M., 1964a, Nature, London, *202*, 1028. (19).

THRESH, J. M., 1964b, Nature, London, *202*, 1085–1087. (19).

THRESH, J. M. and R. M. LISTER, 1960, Ann. Appl. Biol. *48*, 65–74. (23).

THRESH, J. M. and T W. TINSLEY, 1959, Techn. Bull. W. African Cocoa Res. Inst. *7*, 1–32. (23).

THURSTON, R., W. T. SMITH and B. P. COOPER, 1966, Entomol. Exptl. Appl. *9*, 428–432. (31).

TIMONEY, P., 1971, Acta Virol., Prague *15*, 429. (18).

TINSLEY, T. W., 1971, J. Appl. Ecol. *8*, 491–495. (10).

TINSLEY, T. W. and D. C. KELLY, 1970, J. Invert. Pathol. *16*, 470–472. (15, 21).

TINSLEY, T. W. and A. L. WHARTON, 1958, Ann. Appl. Biol. *46*, 1–6. (23).

TODD, J. M., 1961, European Potato. J. *4*, 316–329. (25).

TOKUMITSU, T. and K. MARAMOROSCH, 1966, Exptl. Cell. Res. *44*, 652–665. (17).

TOMITA, K. and A. RICH, 1964, Nature, London *201*, 1160–1163. (26).

TOMLINSON, J. A., 1962, Plant Pathol. *11*, 61–66. (25).

TOMLINSON, J. A., 1970, C.M.I./A.A.B. Descrip. Plant Viruses. No. 8. (4).

TOMLINSON, J. A., A. L. CARTER, W. T. DALE and C. J. SIMPSON, 1970, Ann. Appl. Biol. *66*, 11–16. (31).

TOMPKINS, G. J., J. R. ADAMS and A. M. HEIMPEL, 1969, J. Invert. Pathol. *14*, 343–357. (14).

TONKS, N. V., 1968, Can. J. Plant Sci. *48*, 332–334. (31).

TOPCIU, V., N. ROȘIU, L. GEORGESCU, D. GHERMAN, P. ARCAN and N. CSÁKY, 1968, Acta Virol., Prague *12*, 287. (18).

TŐSIČ, M., 1971, Phytopathol. Z. *70*, 145–162. (19).

TOUMANOFF, C. and M. SIMOND, 1956, Bull. Soc. Pathol. Exot. *49*, 667–674. (30).

TOWNSEND, C. O., 1915, Phytopathology *5*, 282. (3).

TRAGER, W., 1935, J. Exptl. Med. *61*, 501–513. (14).

TRAGER, W., 1959, Ann. Trop. Med. Parasitol. *53*, 473–491. (17).

TRAPIDO, H. and P. GALINDO, 1957, Am. J. Trop. Med. Hyg. *6*, 114–152. (11).

TRAVERSI, B. A., 1949, Rev. Invest. Agr. Buenos Aires *3*, 345–351. (13).

TRENT, D. W. and A. A. QURESHI, 1971, J. Virol. *7*, 379–388. (14).

TRENT, D. W., C. C. SWENSEN and A. A. QURESHI, 1969, J. Virol. *3*, 385–394. (14).

TRIPCONEY, D., 1970, J. Invert. Pathol. *15*, 268–275. (4).

TRIPP, M. R., 1960, Biol. Bull. *119*, 273. (16).

TRIPP, M. R., 1961, J. Parasitol. *47*, 745. (16).

TRIPP, M. R., 1966, J. Invert. Pathol. *8*, 478. (16).

TRIPP, M. R., 1969, *In:* Jackson, G. J., R. Herman, and I. Singer, eds., *Immunity to parasitic animals*, vol. 1, (North-Holland, Amsterdam) pp. 111–128. (16).

TUKEI, P. M., M. C. WILLIAMS, L. G. MUKWAYA, B. E. HENDERSON, G. W. KAFUKO and A. W. R. MCCRAE, 1970, E. African Med. J. *47*, 265–272. (18).

TUTHILL, L. D., 1943a, Coll. J. Sci. *17*, 443–660. (7).

TUTHILL, L. D., 1943b, Occ. Pap. B.P. Bishop Mus. *17*, 221–228. (7).

TUTHILL, L. D., 1944a, Entomol. News *55*, 93–96. (7).

TUTHILL, L. D., 1944b, J. Kansas Entomol. Soc. *17*, 143–149. (7).

TUTHILL, L. D., 1945a, J. Kansas Entomol. Soc. *18*, 1–29. (7).

TUTHILL, L. D., 1945b, Entomol. News *56*, 235–238. (7).

TUTHILL, L. D., 1952, Pacific Sci. *6*, 83–125. (7).

TUTHILL, L. D., 1959, Rev. Peruana Entomol. Agric. *2*, 1–27. (7).

TUTHILL, L. D., 1964, Insects of Micronesia *6*, 353–376. (7).

TUTHILL, L. D. and K. L. TAYLOR, 1955, Australian J. Zool. *3*, 227–257. (7).

UICHANCO, L. B., 1921, Philippine J. Sci. *18*, 259–288. (7).

UILENBERG, G., 1965, Rev. Élev. Méd. Vét. Pays Trop. n.s. *18*, 89–94. (5).

ULLYETT, G. C. and D. B. SCHONKEN, 1940, Union S. Africa Dept. Agr. Forestry Sci. Bull No. *218*. (29).

VAGO, C., 1963, J. Insect Pathol. *5*, 275–276. (2, 4).

VAGO, C., 1966, Nature, London *209*, 1290. (28).

VAGO, C. and M. BERGOIN, 1963, Entomophaga *8*, 253–261. (14).

VAGO, C. and M. BERGOIN, 1968, Adv. Virus Res. *13*, 247–303. (9, 15, 21).

VAGO, C. and S. CHASTANG, 1962, Entomophaga *7*, 175–179. (17).

VAGO, C. and O. CROISSANT, 1963, Ann. Epiphyties *14*, 43. (17).

VAGO, C., G. MEYNADIER and J. L. DUTHOIT, 1964, Ann. Epiphysies *15*, 475–479. (2).

VAGO, C., P. ROBERT, A. AMARGIER and J. L. DUTHOIT, 1969, Zentr. Mikro. Forsch. Methodik *25*, 378–386. (2).

VAGO, C. and L. VASILJEVIC, 1956, Microskopie *11*, 136. (16).

VAIL, P. V. and D. GOUGH, 1970, J. Invert. Pathol. *15*, 397–400. (29).

VAIL, P. V. and I. M. HALL, 1969, J. Invert. Pathol. *14*, 227–236. (15).

VAN DEN BOSCH, R., 1964, J. Invert. Pathol. *6*, 343. (16).

VAN DER GOOT, P., 1917, Meded. Proefst. Midden-Java. Salatiga *25*, 142. (10).

VAN DER MERWE, 1968, Zool. Anz. *181*, 280–289. (5, 18).

VAN DER PLANK, J. E., 1948, Emp. J. Exptl. Agri. *16*, 134–142. (31).

VAN DER PLANK, J. E. and E. E. ANDERSSEN, 1944, Bull. Dept. Agri. S. Afr. *240*, 1–6. (31).

VANDERVEKEN, J., 1968, Virology *34*, 807–809. (31).

VAN DER WANT, J. P. H., 1954, Thesis, Wageningen Agr. Univ. A. Veenman & Zonen. (25).

VAN DER WANT, J. H. P., 1966, Neth. J. Pl. Pathol. *72*, 111–125. (3).

VANDERZANT, E. S. and R. REISER, 1956, J. Econ. Entomol. *49*, 7–10. (2).

VAN EMDEN, H. F., 1964, Nature, London *201*, 946–948. (31).

VAN EMDEN, H. F., 1966a, Entomol. Expl. Appl. *9*, 444–460. (31).

VAN EMDEN, H. F., 1966b, *Ecology of Aphidophagous Insects*, (Prague, Academia). pp. 227–235. (31).

VAN EMDEN, H. F., 1972, *Handbook of Aphid Technology*. (Academic Press) 344 pp. (7).

VAN EMDEN, H. F., V. F. EASTOP, R. D. HUGHES and M. J. WAY, 1969, Ann. Rev. Entomol. *14*, 197–270. (7).

VAN HOOF, H. A., 1962, Tijdschr. Pl. Ziekten *68*, 391–396. (27).

VAN HOOF, H. A., 1964, Nematologica *10*, 141–144. (27).

VAN HOOF, H. A., 1968, Nematologica *14*, 20–24. (27).

VAN HOOF, H. A., 1970, Neth. J. Pl. Pathol. *76*, 329–330. (27).

VAN HOOF, H. A., 1971, Neth. J. Pl. Pathol. *77*, 30–31. (27).

VAN KAMMEN, A., 1968, Virology *34*, 312–318. (4).

VAN KAMMEN, A., 1971, C.M.I./A.A.B. Descrip. Plant Viruses. No. *47*, (4).

VAN TONGEREN, H. A. E., 1955, Arch. Gesch. Virusforsch. *6*, 158. (1).

VAN TONGEREN, H. A. E., 1965, Trop. Geog. Med. *17*, 339–352. (30).

VARMA, A., 1970, C.M.I./A.A.B., Descrip. Plant Viruses. No. *22*. (4).

VARMA, M. G. R. and M. PUDNEY, 1967, Exptl. Cell Res. *45*, 671. (17).

VARMA, M. G. R. and M. PUDNEY, 1969, J. Med. Entomol. *6*, 432–439. (17).

VARMA, M. G. R. and M. PUDNEY, 1971, Trans. Roy. Soc. Trop. Med. Hyg. *65*, 102–103. (14).

VARMA, M. G. R. and C. E. G. SMITH, 1962, Proc. Symp. Biol. Viruses Tick-borne Enceph. Complex (Smolenice, October 11–14, 1960) pp. 397–400. (18).

VARMA, P. M., 1963, Bull. Natl. Inst. Sci. India, *24*, 11–33. (13).

VASQUEZ, C. and P. TOURNIER, 1962, Virology *17*, 503–510. (26).

VAUGHN, J. L. and P. FAULKNER, 1963, Virology *20*, 484–489. (14).

VERANI, P., M. BALDUCCI and M. C. LOPES, 1970, Folia Parasitol. *17*, 367–374. (18).

VERWOERD, D. W., H. LOUW and R. A. OELTERMAN, 1970, J. Virol. *5*, 1–7. (4).

VIDANO, C., 1970, Virology *41*, 218–232. (26).

VON BONSDORFF, C. H., P. SAIKKU and N. OKER-BLOM, 1969, Virology *39*, 342–344. (4).

VONDRACEK, K., 1957, Fauna C.S.R. (Acad. Sci. Prague) *9*, 1–431. (7).

VONDRACEK, K., 1963, Acta Entomol. Musee. Nat. Prague *35*, 263–290. (7).

VON PROWAZEK, S., 1907, Arch. Protistenk. *10*, 358–364. (2).

VUITTENEZ, A., 1960, C.r. hebd. Séanc. Acad. Agr. Fr. *20*, 89–99. (31).

WAGER, R. and G. BENZ, 1971, J. Invert. Pathol. *17*, 107–115. (14).

WAITE, M. R. F. and E. R. PFEFFERKORN, 1970a, J. Virol. *5*, 60–71. (14).

WAITE, M. R. F. and E. R. PFEFFERKORN, 1970b, J. Virol. *6*, 637–643. (14).

WALKER, D. I., 1964, Prog. Med. Virol. *6*, 111–148. (14).

WALKER, I., 1959, Rev. Suisse Zool. *66*, 569. (16).

WALKINSHAW, C. H., G. D. GRIFFIN and R. H. LARSON, 1961, Phytopathology *51*, 806–808. (27).

WALLACE, H. R., 1963, *The Biology of Plant Parasitic Nematodes*, (Edward Arnold, London). (27).

WALLIS, R. C. and L. WHITMAN, 1967, J. Med. Entomol. *4*, 273–274. (30).

WALLIS, R. L., 1962, J. Econ. Entomol. *55*, 871–874. (8).

WALLIS, R. L., 1967, J. Econ. Entomol. *60*, 904–907. (31).

WALTERS, H. J., 1952, Phytopathology *42*, 355–362. (3, 9).

WALTERS, H. J., 1969, Adv. Virus Res. *15*, 339–363. (9).

WALTON, G. A., 1962, Symp. Zool. Soc. London. 83–156. (5, 18).

WATANABE, H., 1965, J. Invert. Pathol. *7*, 257–258. (21).

WATANABE, H., 1967, J. Invert. Pathol. *9*, 474–479. (21, 29).

WATANABE, H., 1968, J. Invert. Pathol. *12*, 310–320. (15).

WATANABE, H., 1970, J. Invert. Pathol. *15*, 247–252. (15).

WATANABE, H. and M. KOBAYASHI, 1969, J. Invert. Pathol. *14*, 102–103. (14).

WATSON, M. A., 1938, Proc. Roy. Soc. (London). B *125*, 144–170. (25).

WATSON, M. A., 1966, Plant Pathol. *15*, 145–149. (31).

WATSON, M. A. and R. T. PLUMB, 1972, Ann. Rev. Entomol. *17*, 425–452. (25).

WATSON, M. A. and H. L. NIXON, 1953, Ann. Appl. Biol. *40*, 537–545. (25).

WATSON, M. A., R. HULL, J. W. BLENCOWE and B. M. G. HAMLYN, 1951, Ann. Appl. Biol. *38*, 743–764. (31).

WATSON, M. A. and F. M. ROBERTS, 1939, Proc. Roy. Soc. (London). B *127*, 543–576. (3, 25).

WATSON, M. A. and F. M. ROBERTS, 1940, Ann. Appl. Biol. *27*, 227–233. (25).

WATSON, M. A. and R. C. SINHA, 1959, Virology *8*, 139–163. (26).

WATSON, M. A., D. J. WATSON and R. HULL, 1946, J. Agr. Sci. *36*, 151–166. (31).

WAY, M. J., 1954, Bull. Entomol. Res. *45*, 93–112. (10).

WAY, M. J., 1963, Ann. Rev. Entomol. *8*, 307–344. (10).

WAY, M. J. and G. D. HEATHCOTE, 1966, Ann. Appl. Biol. *57*, 409–423. (31).

WEBBE, G., 1961, E. African Med. J. *38*, 239–245. (30).

WECKER, E. and E. SCHONNE, 1961, Proc. Natl. Acad. Sci. U.S. *47*, 278–282. (14).

WEIBENGA, N. H., 1961, Am. J. Hyg. *73*, 350–364. (14).

WEINHEIMER, P. F., R. T. ACTON and E. E. EVANS, 1969, J. Bacteriol. *97*, 462. (16).

WEISER, J., 1965, Bull. World Health Organ. *33*, 586–588. (20).

WEISER, J., 1965, J. Invert. Pathol. *7*, 82–85. (4).

WEISER, J., 1968, J. Invert. Pathol. *12*, 36–39. (20).

WELCH, H. E., 1965, Ann. Rev. Entomol. *10*, 275–302. (12).

WELLINGS, F. M., A. L. LEWIS and L. V. PIERCE, 1972, Am. J. Trop. Med. Hyg. *21*, 201–213. (18).

WELLINGTON, E. F., 1954, Biochem. J. *57*, 334–338. (4).

WELLINGTON, W. G., 1962, J. Insect Pathol. *4*, 285–305. (21).

WENSLER, R. J. D., 1962, Nature, London *195*, 830–831. (25).

WESTAWAY, E. G. and B. M. REEDMAN, 1969, J. Virol. *4*, 688–693. (4, 14).

WETTER, C., 1971, C.M.I./A.A.B. Descrip. Plant Viruses. No. *61*. (4).

WHARTON, R. H., 1962, Bull. Inst. Med. Res. Fed. Malaya. 11. (11).

WHITBY, J. L., J. G. MICHAEL, M. W. WOODS and M. LANDY, 1961, Bacteriol. Rev. *25*, 437. (16).

WHITCOMB, R. F. and L. M. BLACK, 1961, Virology *15*, 136–145. (26).

WHITCOMB, R. F., D. D. JENSEN and JEAN RICHARDSON, 1967, Virology *31*, 539–549. (26).

WHITCOMB, R. F., D. D. JENSEN and JEAN RICHARDSON, 1968, J. Invert. Pathol. *12*, 202–221. (26).

WHITE, J. F., 1971, Current Top. Microbiol. Immunol. *55*, 102–107. (17).

WHITEHEAD, A. G. and D. J. HOOPER, 1970, Ann. Appl. Biol. *65*, 339–350. (12).

WHITFIELD, S. G., F. A. MURPHY and W. D. SUDIA, 1971, Virology *43*, 110–122. (14).

WHITNEY, E. and R. DIEBEL, 1971, Current Top. Microbiol. Immunol. *55*, 138–139. (14).

WIGGLESWORTH, V. B., 1970, J. Reticuloendothelial Soc. *7*, 208. (16).

WILDY, P., 1971, Classification and nomenclature of viruses. Monographs in Virology *5*, (4, 14).

WILLIAMS, A. J., 1954, J. Kans. Entomol. Soc. *27*, 97–99. (16).

WILLIAMS, M. C., J. P. WOODALL and J. D. GILLETT, 1965, Trans. Roy. Soc. Trop. Med. Hyg. *59*, 186–197. (30).

WILLIAMS, P., 1970, Trans. Roy. Soc. Trop. Med. Hyg. *64*, 317–368. (11).

WILLIAMS, R. E., J. CASALS, M. I. MOUSSA and H. HOOGSTRAAL, 1972, Am. J. Trop. Med. Hyg. *21*, 582–587. (18).

WILLIAMS, R. E., H. HOOGSTRAAL, M. N. KAISER, M. I. MOUSSA and J. CASALS, 1973, I. Med 10, (18).

WILLIAMS, R. C. and K. M. SMITH, 1957, Nature *179*, 119–120. (4).

WILLIAMS, R. C. and K. M. SMITH, 1958, Biochim. Biophys. Acta. *28*, 464–469. (2, 4).

WILLMER, E. N., 1965, *In:* Willmer, E. N., ed. Cells and Tissues in Culture. *1*, 143–176. (17).

WILSON, N. S., L. S. JONES and L. C. COCHRANE, 1955, Plant Disease Reptr. *39*, 889–892. (19).

WOLFENBARGER, D. A., 1965, J. Invert. Pathol. *7*, 33–39. (29).

WOOD, J. I., 1953, *In:* Plant diseases, U.S. Dept. Agr. Year book of Agriculture. (3).

WOODARD, D. B. and H. C. CHAPMAN, 1968, J. Invert. Pathol. *11*, 296–301. (20, 29).

WOODRUFF, A. W., 1968, Trans. Roy. Soc. Trop. Med. Hyg. *62*, 446–452. (12, 27).

WOODRUFF, A. W., 1969, *In:* Taylor, A. E. R., ed., *Nippostrongylus* and *Toxoplasma*. Symp. Br. Soc. Parasit. (7th) (London, 1968, Blackwell, Oxford). (12).

WORK, T. H., 1958, Progr. Med. Virol. *1*, 248–279. (1, 18, 30).

WRIGHT, J. W. and R. PAL, 1967, *Genetics of Insect Vectors of Disease*, (Elsevier, Amsterdam, London, New York). (11).

WRIGHT, K. A., 1965, Can. J. Zool. *43*, 689–700. (12).

WRIGLEY, N. G., 1969, J. Gen. Virol. *5*, 123–134. (4).

WRIGLEY, N. G., 1970, J. Gen. Virol. *6*, 169–173. (4).

WYATT, G. R., 1952, J. Gen. Physiol. *36*, 201–205. (4).

WYATT, G. R., T. C. LOUGHHEED and S. S. WYATT, 1956, J. Gen. Physiol. *39*, 853–868. (17).

WYATT, I. J., 1970, Ann. Appl. Biol. *65*, 31–41. (31).

WYSS, U., 1969, Mitt. Biol. Bund Anst. Ld-u Forstw. *136*, 110–126. (12).

WYSS, U., 1970a, Nematologica *16*, 55–62. (12).

WYSS, U., 1970b, Nematologica *16*, 74–84. (12).

WYSS, U., 1971, Nematologica *17*, 505–518. (12, 27).

XEROS, N., 1952, Nature, London *170*, 1073. (2).

XEROS, N., 1954, Nature, London *174*, 562–563. (20).

XEROS, N., 1964, J. Insect Pathol. *6*, 261–283. (2, 14).

YAGITA, H. and Y. KOMURO, 1970, Ann. Phytopathol. Soc. Japan, *36*, 371. (27).

YAMADA, K. and E. SHIKATA, 1969, J. Facult. Agr. Hokkaido. Univ. *56*, 91–102. (26).

YAMAGUTI, N., V. J. TIPTON, H. L. KEEGAN and S. TOSHIOKA, 1971, Brigham Young Univ. Sci. Bull. Biol. *15*, 226. (5).

YANG, Y. J., D. B. STOLTZ and L. PREVEC, 1969, J. Gen. Virol. *5*, 473–483. (14).

YASSIN, A. M., 1968, Nematologica *14*, 419–428. (27).

YASUZUMI, G. and I. TSUBO, 1965, J. Ultrastruct. Res. *12*, 304–316. (14).

YEOMAN, G. H. and J. B. WALKER, 1967, The Ixodid Ticks of Tanzania. A Study of the Zoogeography of the Ixodidae of an East African Country. Commonwealth Institute of Entomology, London. (5, 18).

YIN, F. H. and R. Z. LOCKART, 1968, J. Virol. *2*, 728–737. (4).

YOSHII, H., 1959, Virus *9*, 415–422. (26).

YOSHII, H. and A. KISO, 1959, Virus *9*, 582–589. (26).

YOUNGHUSBAND, H. B. and P. E. LEE, 1969, Virology *38*, 247–254. (14).

YOUNGHUSBAND, H. B. and P. E. LEE, 1970, Virology *40*, 757–760. (14).

YU, F. L., 1956, Mem. Coll. Agr. Natl. Taiwan Univ. *4*, 43–54. (7).

YUNKER, C. E., 1970, Misc. Publ. Entomol. Soc. Amer. *6*, 330–338. (18).

YUNKER, C. E., 1971, Current Top. Microbiol. Immunol. *55*, 113–126. (17).

YUNKER, C. E. and J. CORY, 1967, Exptl. Parasitol. *20*, 267–277. (17).

YUNKER, C. E. and J. CORY, 1968, Am. J. Trop. Med. Hyg. *17*, 889–893. (14).

YUNKER, C. E. and J. CORY, 1969, J. Virol. *3*, 631–632. (14).

YUNKER, C. E., J. L. VAUGHIN and J. CORY, 1967, Science *155*, 1565–1566. (17).

ZAHRADNIK, J., 1963, Die Tierwelt Mitteleuropas (Leipzig) *4*, 1–19. (13).

ZAKORKINA, T. N., 1958, J. Microbiol. Epidemiol. Immunobiol. *29*, 135. (30).

ZEYA, A., 1968, Diploma in Forestry Thesis, University of Oxford, England, 1968. (21).

ZEYEN, R. J. and E. E. BANTTARI, 1969, Phytopathology *59*, 1059. (26).

ZIMMERMAN, E. C., 1948, Insects of Hawaii. vol. 5, Univ. Hawaii Press, Honolulu. (7).

ZUCKERMAN, B. M., W. F. MAI and R. A. ROHDE, 1971, *Plant Parasitic Nematodes*, (Academic Press, New York and London). (12).

General index

Index of virus names